Advances in Intelligent Systems and Computing

Volume 255

Series editor

Janusz Kacprzyk, Warsaw, Poland

For further volumes:
http://www.springer.com/series/11156

About this Series

The series "Advances in Intelligent Systems and Computing" contains publications on theory, applications, and design methods of Intelligent Systems and Intelligent Computing. Virtually all disciplines such as engineering, natural sciences, computer and information science, ICT, economics, business, e-commerce, environment, healthcare, life science are covered. The list of topics spans all the areas of modern intelligent systems and computing.

The publications within "Advances in Intelligent Systems and Computing" are primarily textbooks and proceedings of important conferences, symposia and congresses. They cover significant recent developments in the field, both of a foundational and applicable character. An important characteristic feature of the series is the short publication time and world-wide distribution. This permits a rapid and broad dissemination of research results.

Srikanta Patnaik · Xiaolong Li
Editors

Proceedings of International Conference on Computer Science and Information Technology

CSAIT 2013, September 21–23, 2013, Kunming, China

Editors
Srikanta Patnaik
Computer Science and Engineering
SOA University
Bhubaneswar
India

Xiaolong Li
Electronics and Computer Engineering
Technology
Indiana State University
Terre Haute, IN
USA

ISSN 2194-5357 ISSN 2194-5365 (electronic)
ISBN 978-81-322-1758-9 ISBN 978-81-322-1759-6 (eBook)
DOI 10.1007/978-81-322-1759-6
Springer New Delhi Heidelberg New York Dordrecht London

Library of Congress Control Number: 2013957877

Printed on acid-free paper

Springer is part of Springer Science+Business Media (www.springer.com)

Preface

On behalf of the Program Committee, we welcome you to the International Conference on Computer Science and Information Technology (CSAIT 2013) held during September 21–23, 2013 in Kunming China. The main objective of CSAIT 2013 is to provide a platform for researchers, educators, engineers, and government officials involved in the general areas of CSAIT to disseminate their latest research results and exchange views on the future research directions of these fields. The forum provides an opportunity to exchange computer science-related activities and integrate its practice, application of the academic ideas, and improve the academic depth of computer science and its application.

The mushrooming growth of the IT industry in the twenty-first century determines the pace of research and innovation across the globe. In a similar fashion, Computer Science has acquired a pathbreaking trend by making a swift entry into a number of cross-functional disciplines like Bio-Science, Health Science, Performance Engineering, Applied Behavioral Science, and Intelligence. It seems like the quest of the Homo Sapiens Community to integrate this world with a vision of Exchange of Knowledge and Culture is coming to an end. Apparently the quotation "Shrunken Earth, Shrinking Humanity" holds true as the connectivity and the flux of information remains on a simple command over an Internet protocol address. Still there remains a substantial relativity in both the disciplines which underscores further extension of the existing literature to augment the socioeconomic relevancy of these two fields of study. The IT tycoon Microsoft's address at the annual Worldwide Partner Conference in Los Angeles introduced Cloud Enterprise Resource Planning (ERP), and updated Customer Relationship Management (CRM) software which emphasizes the ongoing research on capacity building of the Internal Business Process. It is worth mentioning here that Hewlett-Packard has been coming up with flying colors with 4G touch pad removing comfort ability barriers with 2G and 3G. If we progress, the discussion will never limit because advancement is seamlessly flowing at the most efficient and state-of-the-art universities and research labs like Laboratory for Advanced Systems Research, University of California. Unquestionably, apex bodies like UNO, WTO and IBRD include these two disciplines in their millennium development agenda, realizing the aftermath of the various application projects like VSAT, POLNET, EDUSAT, and many more. 'IT' has magnified the influence of knowledge management and has congruently responded to the social and industrial revolution.

We have received 325 papers through "Call for Paper," out of which 101 papers were accepted for publication in the conference proceedings through double blind review process. The conference is designed to stimulate the young minds including Research Scholars, Academicians, and Practitioners to contribute their ideas, thoughts, and nobility in these two integrated disciplines. Even a fraction of active participation deeply influences the magnanimity of this international event. I along with Dr. Li must acknowledge your response to this conference. I ought to convey that this conference is only a small step toward knowledge, network, and relationship. The conference is the first of its kind and has been granted with a lot of blessings. We wish all success to the paper presenters. I congratulate the participants for getting selected at this conference. I extend my heartfelt thanks to members of faculty from different institutions, research scholars, delegates, members of the technical, and organizing committee.

Srikanta Patnaik
Xiaolong Li

Organizing Committee

General Chair

Prof. Srikanta Patnaik, Computer Science and Engineering in SOA University, India

General Co-Chair

Dr. Xiaolong Li, Indiana State University, USA
Prof. Xun W. Xu, University of Auckland, New Zealand

Technical Program Committees

Prof. Ivo A. Hümmelgen, Federal University of Paraná, Brazil
Prof. Bo Yang, Shanghai Jiao Tong University, China
Dr. Nicola Fiore, University of Salento, Italy
Dr. Alessandro Bevilacqua, University of Bologna, Italy
Dr. Ahmad Fakharian, Luleå University of Technology, Iran
Dr. Bell Manrique Losada, Universidad de Medellín, Colombia
Dr. Ziad Mohammed Mohammed Ali, South Valley University, Egypt
Dr. M. Joseph, St. Joseph's College of Engineering and Technology, India
Dr. S. M. Riazul Islam, University of Dhaka, Bangladesh
Dr. Daowen Qiu, Sun Yat-sen University, China
Dr. Mehdi Mostofi, East Tehran Branch, Islamic Azad University, Tehran, Iran
Dr. Felix Albu, University of Targoviste, Romania
Dr. Aniruddha Chandra, National Institute of Technology, India
Prof. Waleed H. Abdulla, The University of Auckland, New Zealand
Dr. Mejdi Kaddour, University of Oran, Algeria
Dr. Amjad Alipanah, University of Kurdistan, Iran
Dr. Haider M. AlSabbagh, Basra University, Iraq

Dr. Sangkyun Kim, Kangwon National University, South Korea
Dr. Abdelkrim Haqiq, Hassan 1st University Faculty of Sciences and Techniques, Morocco
Dr. Adib Rastegarnia, Department of Electrical and Computer Engineering, University of Tehran, Iran
Prof. Dr. Muhammad Sarfraz, Department of Information Science, Kuwait University, Kuwait

Contents

Part VIII Computer Applications

Part I
Signal and Image Processing

The PSF Measurement for Motion-Blurred Image Based on Analyzing the Frequency Spectrum

Fang Tang, Ke Zhang and Dian Shimei

Abstract As for restoration of motion-blurred images, whether the Point Spread Function parameter of the blurred image can be accurately estimated directly affects the image restoration results. From analyzing the frequency spectrum of the motion-blurred image, we know that the direction and distance of the dark stripes in the frequency spectrum are relevant with blurred direction and distance of the corresponding blurred image. In this paper, we first detect the boundary of the stripes in the blurred image spectrum using Snake method, combine the Hough transform to compute the direction angle and relative distance of the stripes, and then show the results in polar coordinates. With the computing results, we finally implement the PSF parameter estimation accurately and effectively of motion-blurred images.

Keywords Frequency spectrum · Motion-blurred image · Snake algorithm · Hough transform

1 Introduction

There are many methods to estimate the Point Spread Function parameter of the blurred image. As early as a few years ago, Wang and Zhao [1] proposed that the blur direction estimation from spectrum, but they did not implement. Wan and Lu [2] proposed to estimate the PSF parameter by Hough transform, but is only used in estimating direction.

The Hough transform can connect the interrupt lines, some of the blurred images can be processing, but it will not estimate the fuzzy parameter accurately,

F. Tang (✉) · K. Zhang · D. Shimei
Department of Computer Engineering, Kunming Medical University, Kunming 650500, China
e-mail: tangfang_hj@163.com

S. Patnaik and X. Li (eds.), *Proceedings of International Conference on Computer Science and Information Technology*, Advances in Intelligent Systems and Computing 255, DOI: 10.1007/978-81-322-1759-6_1,

especially for the image with rich boundary information. In 1988, Kass [3] and his partners, Witkin, Terzopoulos, first proposed the Snake model [4] to solve the boundary problem effectively.

In this paper, we discuss the PSF measurement for motion-blurred image based on analyzing the frequency spectrum. First, for the spectrum image binarization, select center pixel as seed for regional growth. Then, use the Snake algorithm to segment the edge of the center bright band that affects the blur parameter, reducing the computation error. Next, we compute the direction angle and the distance of target contour by Hough transform and show the results in polar coordinates. At last, we provide several experiments on motion-blurred parameter estimation.

2 Theoretical Analysis

2.1 The Relationship Between the PSF and Blurred Image Spectrum

Given an image $f(x, y)$, which is in uniform motion [6], set $x_0(t)$ is the movement of x direction, set $y_0(t)$ is the movement of y direction, t is the time, ignoring other factors, and the motion-blurred image is $g(x, y)$:

$$g(x,y) = \int_0^T f[x - x_0(t), y - y_0(t)]\mathrm{d}t \tag{1}$$

And the Fourier transform is

$$\begin{aligned} G(u,v) &= \int_{-\infty}^{+\infty}\int_{-\infty}^{+\infty} g(x,y)\mathrm{e}^{-\mathrm{j}2\pi(\mathrm{ux}+\mathrm{vy})}\mathrm{d}x\,\mathrm{d}y \\ &= \int_{-\infty}^{+\infty}\int_{-\infty}^{+\infty}\int_0^T f[x - x_0(t), y - y_0(t)\mathrm{d}t]\mathrm{e}^{-\mathrm{j}2\pi(\mathrm{ux}+\mathrm{vy})}\mathrm{d}x\,\mathrm{d}y \end{aligned} \tag{2}$$

$$\begin{aligned} G(u,v) &= \int_0^T F(u,v)\mathrm{e}^{-\mathrm{j}2\pi[\mathrm{ux}_0(\mathrm{t})+\mathrm{vy}_0(\mathrm{t})]}\mathrm{d}t \\ &= F(u,v)\int_0^T \mathrm{e}^{-\mathrm{j}2\pi[\mathrm{ux}_0(\mathrm{t})+\mathrm{vy}_0(\mathrm{t})]}\mathrm{d}t \end{aligned} \tag{3}$$

$$\text{Set } H(u,v) = \int_0^T \mathrm{e}^{-\mathrm{j}2\pi[\mathrm{ux}_0(\mathrm{t})+\mathrm{vy}_0(\mathrm{t})]}\mathrm{d}t \tag{4}$$

$$\text{So } G(U, V) = H(u, v)F(u, v) \tag{5}$$

If $x_0(t)$ and $y_0(t)$ were given, the degradation transfer function can get directly by Eq. (4). We assumed that the image did uniform linear motion in the x direction:

$$\begin{cases} x_0(t) = at/T \\ y_0(t) = 0 \end{cases} \tag{6}$$

If $t = T$, $f(x, y)$ moves a, then plug Eq. (6) into the Eq. (4), we get $H(u, v)$:

$$H(u, v) = \int_0^T \mathrm{e}^{-\mathrm{j}2\pi \mathrm{u} \mathrm{x}_0(\mathrm{t})} \mathrm{d}t = \int_0^T \mathrm{e}^{-\mathrm{j}2\pi \mathrm{uat}/\mathrm{T}} \mathrm{d}t = \frac{T}{\pi ua} \sin(\pi ua) \mathrm{e}^{-\mathrm{j}\pi \mathrm{ua}} \tag{7}$$

The Eq. (7) shows that H is zero while $u = n/a$, and n is integer. From the Eq. (5), we know that H and $G(u, v)$ were zero at the same time, H made the spectrum of $G(u, v)$ has some parallel dark strips that is perpendicular to the motion direction; if $u = 0$, the bright band pass is the spectrum center of $G(u, v)$. The bright band width is inversely proportional to the blurred distance [7, 8].

Figure 1 shows that: Fig. 1a is the source image of Lena, Fig. 1b is spectrum image, it has no dark stripe. Figure 1c is the blurred image of Lena with blurred distance is 20 and blurred direction is 40°; Fig. 1d is spectrum image, with parallel dark stripes. Figure 1e also has blurred image with blurred distance 10 and direction 60°; Fig. 1f is its spectrum image, also with parallel dark stripes. And, the distance of dark stripes of Fig. 1f is twice than Fig. 1d; the direction between the parallel dark stripes is related to the blurred direction. Figure 1g is the blurred image of cameraman with blurred distance 10 and blurred direction 60°; Fig. 1h is its spectrum. We found that the distance and direction of dark stripes of Fig. 1h

Fig. 1 Blurred images and their spectrum. **a** Source image. **b** Spectrum image. **c** Blurred with (20, 40°). **d** Spectrum of (**c**). **e** Blurred with (10, 60°). **f** Spectrum of (**e**). **g** Blurred with (20, 40°). **h** Spectrum of (**g**)

is same as Fig. 1f. Thus, we can determine the Point Spread Function of motion-blurred image by the direction and the distance of dark strips of its spectrum image.

2.2 Snake Model

Kass et al. [3] presented the Snake model that is composed of some control points, and those points were connected end-to-end.

As $v(s) = [x(s), y(s)]$, $s \in [0, 1]$, $x(s)$ and $y(s)$ are the coordinates of control point, s is the variable of Fourier transform. We define the energy function is:

$$E_{\text{Snake}} = \int_0^1 E_{\text{Snake}}(v(s))\mathrm{d}s = \int_0^1 [E_{\text{int}}(v(s)) + E_{\text{ext}}(v(s))]\mathrm{d}s \tag{8}$$

E_{int} is the internal energy function:

$$E_{\text{int}}(v(s)) = (\alpha(s)|v_s(s)|^2 + \beta(s)|v_{ss}(s)|^2)/2 \tag{9}$$

E_{ext} is the external energy function:

$$E_{\text{ext}}(v(s)) = E_{\text{img}}(v(s)) + E_{\text{constraint}}(v(s)) \tag{10}$$

In the Eq. (9), $v_s(s)$ is the first derivative of $v(s)$, $v_{ss}(s)$ is the second derivative of $v(s)$, $\alpha(s)$ and $\beta(s)$ are control parameters. E_{img} is the image energy, made the Snake control point no longer left once it nears to boundary, and finish the positioning accurately. $E_{\text{constaint}}$ is the external energy.

From Fig. 1, we found that the bright band of spectrum image shows the boundary information. If the blurred direction is larger, the distance of dark strips is lesser, and the center bright band is narrower. If we want to compute the blurred parameters, the boundary information of bright bands must be known.

2.3 Hough Transform in Polar Coordinate

Hough transform [5] is a kind of point-to-line mapping between image space and the parameter space. If l is a line, ρ is the normal distance, θ is the angle from x axis to normal, and then, the l is:

$$\rho = x\cos(\theta) + y\sin(\theta) \tag{11}$$

Equation (11) shows the Hough transform of (x, y) in polar coordinate, and the (x, y) mapping into a curve in parameter space by Hough transform. After Hough transform, all the curves intersect at (θ', ρ').

As for motion-blurred image, we first get bright bands of spectrum image, then segment the boundaries by Snake algorithm, and next perform the Hough transform. The point in the two dark strips' boundary is the most, it means the curves intersect at (θ, ρ) is the most, and the number of $A(\rho, \theta)$ is also the most. The maximum of $A(\rho, \theta)$ shows in polar coordinate, we can estimate the Point Spread Function.

3 The Point Spread Function Measurement for Motion-Blurred Image

3.1 The Steps for Solving Point Spread Function

Step 1: Compute the Fourier transform of motion-blurred image.
Step 2: Binarize and select seed for regional growth.
Step 3: Determine the initial contour for the result and get boundary of bright band by Snake algorithm.
Step 4: After Hough transform, count points when normal distance is ρ and the angle is θ, return the maximum.
Step 5: Show the result in polar coordinate; the distance of dark stripes and the blurred distance are inverse. Combine the distance of dark stripes and measure the Point Spread Function of motion-blurred image.

3.2 Analysis of Experimental Results

Experiment 1: Fig. 2a is the motion-blurred image, Fig. 2b is spectrum image, Fig. 2c shows the region growth and the image center is seed, and the central bright band is used to determine the initial contour. Figure 2d shows the initial contour accurately by Snake algorithm. After Hough transform, we have known the distance of the two central longer dark stripes in spectrum image which is $\rho_1 - \rho_2 = 8$, the angle is $\theta = 45°$, and Fig. 2e shows the polar plot. Therefore, the PSF is (45°, 30); Fig. 2f shows the result after ten Lucy-Richardson [5].

Experiment 2: Fig. 3a is the motion-blurred image, Fig. 3b is spectrum image, and Fig. 3c shows the region growth. Figure 3d shows the initial contour accurately by Snake algorithm. After Hough transform, we have known the distance of the two central longest dark stripes in spectrum image which is $\rho_1 - \rho_2 = 17$, the angle is $\theta = 60°$, and Fig. 3e shows the polar plot. The PSF is (60°, 15), and Fig. 3f shows the restoration result.

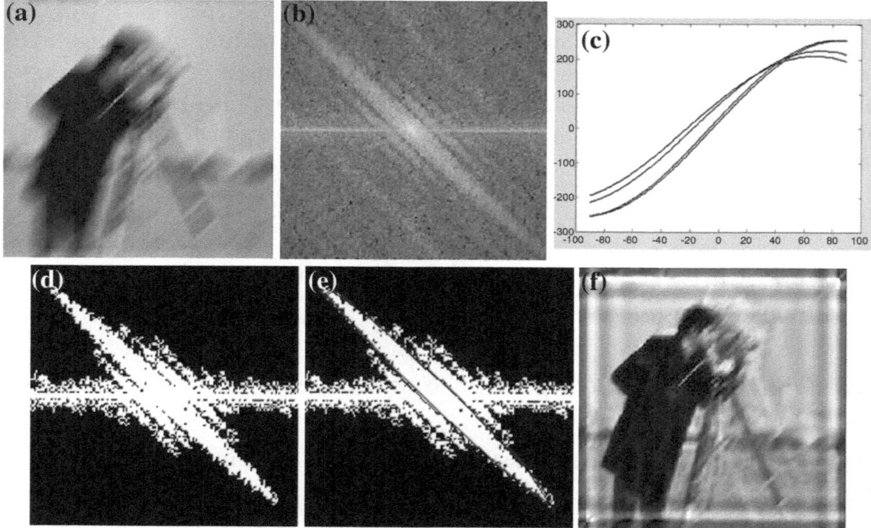

Fig. 2 Motion-blurred image restoration. **a** Blurred image. **b** Spectrum image. **c** The region growth image. **d** Initial contour with Snake. **e** Polar plot. **f** Restoration image

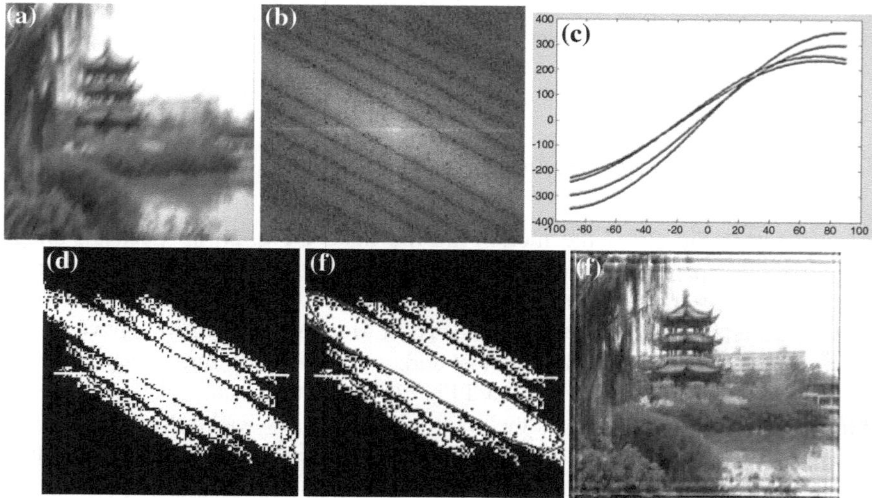

Fig. 3 Motion-blurred scenery restoration. **a** Blurred image. **b** Spectrum image. **c** The region growth image. **d** Initial contour with Snake. **e** Polar plot. **f** Restoration image

The experimental results prove that the method is effective for motion-blurred image restoration and measure Point Spread Function parameter of motion-blurred image accurately.

4 Conclusion

The Point Spread Function is very important for image restoration. In this paper, we discuss the method to estimate the PSF parameter accurately, by selecting the initial contour using Snake algorithm, reducing the influence of high-frequency information and detecting the boundary accurately for bright band to reducing errors. The experimental results prove that the method can compute Point Spread Function parameter of motion-blurred image accurately and have better restoration results.

References

1. Wang, X., Zhao, R.: The PSF estimation for blur image. Comput. Appl. **9**, 40–41 (2001)
2. Wan, L., Lu, L.: The blur image restoration based on spectral parameters and neural nets. Comput. Eng. **5**, 132–134 (2004)
3. Kass, M., Witkin, A., Terzopoulos, D.: Snakes: active contour models. Int. J. Comput. Vision **1**(4), 321–331 (1988)
4. Nie, H., Wu, C.: Snake method based on motion. Comput. Eng. Appl. **44**(28), 166–168 (2008)
5. Zhang, D, et al.: Digital Image Processing. Post & Telecom Press (2009)
6. Gao, M., Chen, S.: The uniform linear blurring image restoration. Comput. Eng. Appl. **6**, 42–45 (2004)
7. Ma, B.: The parameters estimation of motion image. J. Liaoning Univ. **32**(4), 376–378 (2005)
8. Zhu, H., OuYang, G.: Blur parameter identification method based on spectrum correlation with reference image. Chin. J. Sci. Instrum. **8**, 49–51 (2012)

EBR Analysis of Digital Image Watermarking

Fan Zhang and Xinhong Zhang

Abstract An error bit rate (EBR) analysis of digital image watermarking is proposed based on information theory. This work researches how to embed a large number of watermark information in the same time maintaining a low error probability or researches the relationship between watermark payload capacity and EBR. The EBR of watermarking will drop with the decrease in watermark payload capacity. When payload capacity is less than channel capacity, the EBR will keep in a lower level.

Keywords Information theory · Digital watermark · Error bit rate

1 Introduction

In visible watermarking, the information is visible in the picture or video. Typically, the information is text or a logo which identifies the owner of the media. The image on the right has a visible watermark. When a television broadcaster adds its logo to the corner of transmitted video, this is also a visible watermark. In invisible watermarking, information is added as digital data to audio, picture, or video, but it cannot be perceived as such (although it may be possible to detect that some

F. Zhang (✉)
College of Computer and Information Engineering, Henan University, Kaifeng 475001, China
e-mail: zhangfan@henu.edu.cn

X. Zhang
Institute of Image Processing and Pattern Recognition, Henan University, Kaifeng 475001, China
e-mail: zxh@henu.edu.cn

F. Zhang · X. Zhang
Software School, Henan University, Kaifeng 475001, China

S. Patnaik and X. Li (eds.), *Proceedings of International Conference on Computer Science and Information Technology*, Advances in Intelligent Systems and Computing 255, DOI: 10.1007/978-81-322-1759-6_2,

amount of information is hidden). The watermark may be intended for widespread use and is thus made easy to retrieve or it may be a form of Steganography, where a party communicates a secret message embedded in the digital signal. In either cases, as in visible watermarking, the objective is to attach ownership or other descriptive information to the signal in a way that is difficult to remove. It is also possible to use hidden embedded information as a means of coverting communication between individuals [1–3].

The watermarking capacity of digital image is the number of bits that can be embedded in a given host image (original image) [4–6]. The detection performance of watermarking is measured by the error bit rate (EBR) or the error probability. The bit error rate or bit error ratio is the number of received bits that have been altered due to noise, interference, and distortion, divided by the total number of transferred bits during a studied time interval. EBR is a dimensionless performance measure, often expressed as a percentage number. The detection reliability of watermarking closely correlates with two other parameters, which are the watermarking capacity and robustness.

2 Watermarking Communication Model

In the watermarking schemes, the process of watermarking can be considered as a communication process. The image can be considered as the channel in which the watermark messages are transmitted. The watermarking capacity corresponds to the communication capacity of the "watermarking channel." This model can give a rational solution for the analysis of watermarking EBR and capacity. Otherwise, the analysis of watermark EBR and capacity is very difficult without using information theory.

In this paper, we use a simple additive watermark-embedding algorithm,

$$y_i = x_i + w_i + n_i, \tag{1}$$

where x_i denotes the cover image (original image), w_i denotes the watermark information, n_i denotes the noise, and y_i denotes the stego image (watermarked image). In addition, we use w'_I denotes the extracted or recovered watermark.

If P_S denotes the watermark power constraint and P_N denotes the noise power constraint, then according to the well-known Shannon channel capacity formula, the watermarking capacity is:

$$C = W \log_2\left(1 + \frac{P_S}{P_N}\right), \tag{2}$$

where W is the bandwidth of channel. We assume that the size of an image is $N \times N$, the number of pixels is $M = N \times N$. According to Nyquist sampling theory, if we want to correctly express all the pixels, the number of sampling points should be $2W$ at least. So, the bandwidth of this image is $W = M/2$.

Channel capacity is a theoretical limit on the amount of error-free embeddable information, or inversely, on the minimum probability of error attainable for a given message (Shannon capacity). Therefore, capacity is the maximum possible message length for errorless decoding. Greater input message lengths than capacity can be used but a zero EBR will not be attainable.

3 Analysis of Watermarking Channel

In this section, we analyze the relationship between the watermark payload capacity and the EBR. Before the further analysis, we introduce a concept: the watermark payload capacity, C_{PL}.

For a given watermarking algorithm, the payload capacity is the bit length of the embedded watermark, taking no account of the potential redundancy provided by forward error correcting codes for channel coding.

According to the analysis of above section, as the SNR increase, the watermarking capacity will increase, and the watermark detection EBR P_B will decrease. In other words, as the watermark capacity increase, the watermark detection EBR P_B will decrease. But according to Shannon's second theorem, in a discrete memoryless channel, when information transmission rate R is less than capacity C, an error-free transmission is possible, or if we do not send information at a rate greater than the channel capacity, we can transmit information safely in an arbitrary small error rate. Thus, the information capacity defines a fundamental limit on the rate of error-free transmission in the power-limited and band-limited Gaussian channel. When the watermark information transmission rate R is greater than C, an error-free transmission is impossible. Shannon's second theorem is obtained in the discrete channels, but it is also applicable in the continuous channels.

When the information transmission rate R is less than the capacity C, we can use channel coding method to reduce EBR. To enable robust transmission and to against channel loss in digital communication, channel code is used to protect data for storage or retrieval even in the presence of noise (errors). The role of channel coding is to improve the error correction performance in information transmission and to increase transmission reliability. We propose a new watermarking communication model, as shown in Fig. 1.

Here, we have a discussion of watermark EBR according to the model in Fig. 1.

1. In above section, we have distinguished the channel transmission EBR P_E and the detection EBR P_B. P_B only depends on the degree of channels interference but has nothing to do with the probability of watermark itself and the watermark detection program.
2. In the case of channel coding is not used, watermarking channel error bit rate depends on the watermark information EBR.

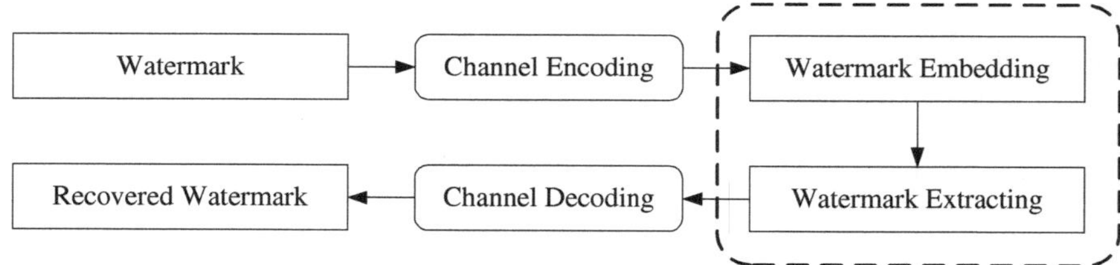

Fig. 1 Watermarking channel model

3. Because the watermark communication SNR is usually small, if channel coding is not used, it is very difficult to achieve an arbitrary small EBR.
4. In the case of channel coding is used, we can view the dashed box.

Figure 1 can be viewed as a black box, which denotes the watermark channel. The EBR of black box channel is P_E. After channel encoding, the coded watermark enters the black box. When the coded watermark outputs from the black box, the channel coding will recover watermark and reduce the final EBR. So, the total error bit rate of this watermark channel depends on both the channel transmission EBR P_E and the channel decoding EBR P_B.

When we use channel coding, the redundant bits are introduced, so we need a higher transmission rates and have to reduce the normalized signal-to-noise ratio E_b/N_0. The performance improvement using channel coding is shown in Fig. 2.

As can be seen from Fig. 2, after channel coding, the system performance is improved. In the condition of same normalized signal-to-noise ratio E_b/N_0, the EBR in the coded system is smaller than it in the codeless system. It is shown as the vertical dot line in Fig. 2. In the condition of same EBR, as shown as the horizontal dot line in Fig. 2, the encoder side requires smaller E_b/N_0 in the coded system than in the codeless system.

The channel coding means that we need redundancy. But, adding redundant bits needs a higher information transmission rate R and requires less energy per bit, which means that we need more bandwidth. However, the bandwidth of watermark channel is limited. The bandwidth of watermark channel only depends on the image itself.

When watermark payload capacity C_{PL} is near to channel capacity limit, enough redundancy bits cannot be provided, and we will unable using channel coding to reduce the EBR. The only way is to reduce the watermark payload capacity C_{PL} and then to reduce EBR. However, if the watermark payload capacity is too small, it is negative because almost no information can be transmitted. Our goal is to embed as much as possible watermark information in images, at the same time, ensure the safety and keep a low EBR. So, what is the suitable number of watermark payload capacity? When the system has a better performance? Next, we will analyze the relationship between the EBR and the payload capacity.

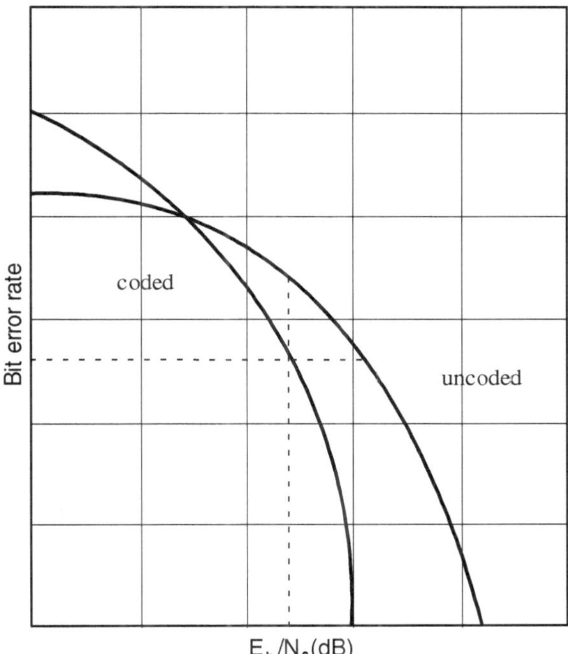

Fig. 2 The performance of channel coding

4 Payload Capacity and EBR of Watermark Detection

In Fig. 1, we establish a simplified watermark communication model. In the case of channel coding not used, we can be view the dashed box in Fig. 1 as a black box. The black box denotes a watermark communication. This model can simplify watermark communication model and the following calculations. We view this black box as a binary symmetric channel (BSC). The cross-transmission probability of this BSC channel is P_E.

In order to enable robust transmission against channel loss, in this watermark communication model, channel coding is introduced to transmit digital watermark sequences over loss channels. There are two different types of channel codes in common use, namely block codes and convolutional codes. The information sequence, message m, is encoded into a sequence c, in which encoded sequence is called code word. c is transmitted over the loss channel, and at the receiver side, sequence c is received and decoded into message m'. Hopefully, we have $m' = m$ for successful decoding.

In the first case, we assume linear block code is used for the error control. Linear code is an important block code in error correction and detection schemes. Linear codes allow more efficient encoding and decoding algorithms than other codes. Linear block code is encoded according to the linear law of the symbol, which keeps a linear relationship between the constraints. The sum of any two code words is also a code word. The encoder of block code divides information sequence into message blocks; each message block contains k information symbols. Each message block is coded independently to a n-tuple code word ($n > k$), and these code words form some binary code word groups. Then $n - k$ redundant bits are inserted in the information bits. This linear block code is called (n, k) block code. The maximum errors bits can be detected in each code word are as follows;

$$t = \left\lfloor \frac{d_{\min} - 1}{2} \right\rfloor, \tag{3}$$

where $d_{\min}$ is the minimum distance between any two different code words.

In the watermarking communication model shown in Fig. 1, the bits' error probability that occur j bit error in an n-bit code word is as follows:

$$p(j, n) = \binom{n}{j} P_B^j (1 - P_B)^{n-j}, \tag{4}$$

$$\binom{n}{j} = \frac{n!}{j!(n-j)!}, \tag{5}$$

where P_E is the EBR of black box watermark channel.

The EBR in the decoder side can be expressed as follow:

$$P_b \approx \frac{1}{n} \sum_{t+1}^{n} j \binom{n}{j} P_B^j (1 - P_B)^{n-j}. \tag{6}$$

In general digital communication channel, we can increase code word length to reduce system EBR, and the increased information can be solved by expanding the bandwidth or increasing transmission time. But in the watermarking channel, the bandwidth is fixed, and there is no transmission time available. So, we cannot increase the code word length indefinitely. This is a characteristic of watermarking channel different from the general communication channel. To get the best EBR, we assume that the maximum possible code word length should be used. This maximum code word length will be decided by the watermarking capacity.

In the second case, we assume that BCH code is used for the error control. Bose, Chaudhuri, Hocquenghem (BCH) code is a class of powerful block codes, and it has the ability to correct multiple errors. Taking into account the capacity of the channel, the available maximum code word length of BCH code is 15. We chose three kinds of BCH codes to calculate EBR of channel coding; all of their code word length is 15. The error correction ability of the three kinds of BCH codes is one, two, and three bits.

According to experimental results, in the case of the Gaussian noise variance is 4, the watermark capacity of Fishing boat image is 89,214 bits, P_E is 0.0918. In these conditions, we calculate the EBR of channel decoder. The experimental results are shown in Table 1.

Table 1 The error bit rate of channel decoder

n	k	t	k/n	EBR
15	11	1	0.733	0.068
15	7	2	0.467	0.0342
15	5	3	0.333	0.012

In Table 1, n is code word length, k is the watermark bits contained in a n-bit code word. t is the maximum number of error bits that can be corrected in each code word. Coding efficiency is defined as k/n, and its reciprocal denotes the code redundancy, and it also determines the payload of watermark.

As can be seen from Table 1, in watermark communication, as the watermark payload capacity C_{PL} decreases, the EBR is improved. When the coding efficiency is 1/3, the EBR dropped from 0.0918 to 0.012. On the contrary, as the watermark payload capacity C_{PL} increases, EBR will increase. We can use Eq. 6 to calculate the EBR of watermarking system. In the condition of using channel coding, we can calculate how many the EBR can be achieved, or in the condition of given a EBR, we can select an optimal encoding method, and achieve the maximum watermark payload capacity.

5 Conclusions

An EBR analysis of digital image watermarking is proposed based on information theory. This work researches how to embed a large number of watermark information in the same time maintaining a low error probability or researches the relationship between watermark payload capacity and EBR. In the case of channel coding is used, the EBR of watermarking depends on the channel decoding EBR and the channel transmission EBR. The EBR of watermarking will drop with the decrease in watermark payload capacity. When payload capacity is less than channel capacity, the EBR will keep in a lower level.

Acknowledgments This research was supported by the Foundation of Education Bureau of Henan Province, China (Grant No. 2010B520003), Key Science and Technology Program of Henan Province, China (Grant Nos. 132102210133 and 132102210034), and the Key Science and Technology Projects of Public Health Department of Henan Province, China (Grant No. 2011020114).

References

1. Cox, I.J., Milller, M.L., Bloom, J.A.: In: Adams, R. (eds.), Digital Watermarking and Steganography, pp. 32–58. Morgan Kaufemann: San Francisco (2001)
2. Moulin, P.: The role of information theory in watermarking and its application to image watermarking. Signal Process. **81**(6), 1121–1139 (2001)

3. Podilchuk, C.I., Delp, E.J.: Digital watermarking: algorithms and applications. IEEE Signal Process. Mag. **4**, 33–46 (2001)
4. Moulin, P., Mihcak, M.: A framework for evaluating the data-hiding capacity of image sources. IEEE Trans. Image Process. **2002**(9), 1029–1042 (2002)
5. Zhang, F., Zhang, X., Zhang, H.: Digital image watermarking capacity and detection error rate. Pattern Recogn. Lett. **1**, 1–10 (2007)
6. Akcakaya, M., Tarokh, V.: Shannon-theoretic limits on noisy compressive sampling. IEEE Trans. Inf. Theory **1**, 492–504 (2010)

Image Descriptors Based on Statistical Thermodynamics and Applications

Kunlun Li, Shangzong Luo, Qi Meng, Yuwei Gao and Hexin Li

Abstract This paper presents a series of new image descriptors based on statistical thermodynamics and discusses their application in content-based image retrieval and image clustering. The paper puts forward image descriptors which represent macro-visual characteristics such as "image energy," "image pressure," "image mass," and "image temperature" according to the analysis-localized sub-system within the statistical thermodynamic theory. We can find a lot of mathematical laws by applying statistical thermodynamic theory in digital image processing. The proposed method has the characteristics of the fast calculation. Experiment verifies the rationality and effectiveness of the proposed method.

Keywords Image descriptor · Image pressure · Image energy · Image mass · Image temperature · CBIR · Statistical thermodynamics

1 Introduction

With the development of network and multimedia technology, we have entered the information age. As a content-rich, intuitive media information, image has caught people's attention gradually [1]. An image has a rich and various content which is simple or complex. Due to the inadequate growth of image understanding and computer at present, there is a semantic gap between users' complex semantics and visual features, and even different people have different understanding to the same image [2]. It can be seen that the image description is one of the problems in the field of computer vision.

K. Li (✉) · S. Luo · Q. Meng · Y. Gao · H. Li
College of Electronic and Information Engineering, Hebei University, Baoding, China
e-mail: likunlun@hbu.edu.cn

S. Luo
e-mail: mrluoshangzong@gmail.com

S. Patnaik and X. Li (eds.), *Proceedings of International Conference on Computer Science and Information Technology*, Advances in Intelligent Systems and Computing 255, DOI: 10.1007/978-81-322-1759-6_3,

In the research of content-based image retrieval (CBIR), it is important to describe the image of machine for image retrieval. The most common method for comparing two images in content-based image retrieval is using an image distance measure [3]. An image distance measure compares the similarity of two images in various dimensions such as color, texture, shape, and others. For example, a distance of 0 signifies an exact match with the query, with respect to the dimensions that were considered. A value greater than 0 indicates various degrees of similarities between the images. Search results then can be sorted based on their distance to the queried image. However, choosing an appropriate representation scheme is only a part of the task [4].

When we are interested in the image texture or color feature [5], we can choose the selected features alternately; sometimes, a variety of means can be used to solve the same problem. In either case, description of methods should be insensitive to translation and rotation for the image [6]. To a large extent, the descriptors proposed in this article can meet one or more of the properties.

Statistical thermodynamics is a branch of physics that applies probability theory, which contains mathematical tools for dealing with large populations, to study the thermodynamic behavior of systems composed of a large number of particles [7]. Their common points are looking for the link between micro-states and macro-states. This discipline has developed very mature. Since the digital image description and statistical thermodynamics are both in solving the problem of micro- and macro-linkages, why do we not apply this discipline method to solve the problem of digital image description? There are many similarities in their analysis method. The Table 1 is an example.

Physics uses energy function to describe macroscopic system level, which is similar to digital image gray histogram analysis method. We can use the moments of gray histogram to describe the texture. $H(r)$ is the gray histogram [8]. And n-order moments can be expressed as follows:

$$Un(r) = \sum_{i=1}^{L} (r_i - m)^n H(r_i) \tag{1}$$

Wherein L represents the number of histogram of the dimensions; m represents the mean value of the histogram. $U_2(r)$ can describe the relative smoothness of the histogram. It is also possible to embody the degree of dispersion of the gray scale.

In this article, a series of image descriptor based on statistical thermodynamics is proposed. Statistical thermodynamics is applied to solving image processing problems. Experiments show that these methods are fast and easy to implement, which can be used to describe the image content.

Table 1 The energy level of the system. ε_i is the independent variable

Energy level	$\varepsilon 1$	……	ε_i	……
Degeneracy	$\omega 1$	……	ω_i	……
The number of particles	n1	……	n_i	……

2 Research Ideas and the Main Content

2.1 Image Descriptor

The low-level feature extraction [9] of digital image is to extract the basic information from pixels. You can find the link between low-level features and image macro-visual characteristics. In statistical thermodynamics, the energy of the system is defined as

If the system is composed of independent particles, the total energy of the system is the sum of energy of each particle. That n_i is the number of particles. It has energy ε_i.

$$E = \sum_{i=1}^{n} n_i \varepsilon_i \tag{2}$$

$$N = \sum_{i=n}^{n} n_i \tag{3}$$

2.2 Image Descriptor: Image Energy

We can conclude that the energy of the digital image is the product of corresponding pixel value and the gray level.

$$e = \sum_{i=1}^{N} n_i \varepsilon_i \tag{4}$$

Wherein n_i is the independent variable of the gray level; ε_i is corresponding to the number of pixels in the gray scale. If the histogram mean is partial right, then the value of the energy is small. If the histogram mean is partial left, then the value of the energy is large.

2.3 Image Descriptor: Image Mass

Similarly, a digital image can correspond to a certain volume of an ideal gas. The digital image pixel values correspond to the speed of the ideal gas molecules. There is another representation of the energy by the law of conservation of e:

$$e = \frac{1}{2} \sum_{i=1}^{n} m v_i^2 \tag{5}$$

Table 2 The energy and mass of image A, B

A image pixel statistics				
Grayscale	1	2	3	4
Pixel number	4	4	4	4
B image pixel statistics				
Grayscale	1	2	3	4
Pixel number	8	0	0	8
The energy and quality of image A, B				
Image energy A	40	Image energy B		40
Image mass A	0.625	Image mass B		0.3125

Where v_i is the gray value of the pixel point. The mass of the ideal gas molecules is m. So combining (4) and (5) m can be obtained:

$$m = \frac{2\sum n_i \varepsilon_i}{\sum v_i^2} \tag{6}$$

In fact, digital image does not have the mass. This parameter is a human-defined. What the visual characteristics have reflected by the m? Suppose there are two digital images, each image with four gray level (Table 2).

2.4 Image Descriptor: Image Pressure

For a digital image, we assume that is a volume of molecular gas; the pixel value is the rate of movement which molecules may have. The image of pressure is

$$p = \frac{nmv_{rms}^2}{3V} \tag{7}$$

In the formula, V is the area of a region of the digital image. V_{rms} is the root mean square of all the pixels in the region, and m is the image mass. The n is the total number of pixels.

2.5 Image Descriptor: Image Temperature

According to the ideal gas equation ($PV = NK_bT$), the expression of the temperature can be obtained as follows:

$$PV = Nk_bT = \frac{nmv_{rms}^2}{3} \Rightarrow T = \frac{mv_{rms}^2}{3k_b} \tag{8}$$

In fact, the image contains a lot of content; the pixel values are not equal everywhere. We just introduced the concept of image temperature to describe the digital image of the "hot and cold" level that is equal to the degree of their "hot and cold" when the temperature characteristics of the two images are equal.

$$\frac{P1}{P2} = \frac{m_1 \cdot v_1^2 \cdot V_2}{m_2 \cdot v_2^2 \cdot V_1} = \frac{\rho_1 \cdot v_1^2}{\rho_2 \cdot v_2^2} = \text{Constant} \tag{9}$$

The contents of the images in the image database have been determined when (*P*, *V*, *n*, *T*) are fixed in the digital image. Similar image has a similar value of (*P*, *V*, *n*, *T*).
In this paper, *NIRFace Image Database* is the data set in the experiment. We selected three "Face Image Database" from the *NIRFace Image Database*. Each "*Face Image Database*" contains about two hundred face images of the same person. There are 470 pictures used in the experiment. The computer calculates the pressure of face images per person. And we calculate the upper and lower parts of the human face that is in order to make the pressure constant.

If the inference is established, the results should meet P1 and P2 linear. The ratio A concentrated in a particular region. The result of the experiment are shown in Figs. 1, 2, 3, 4, 5 and 6; Table 3.

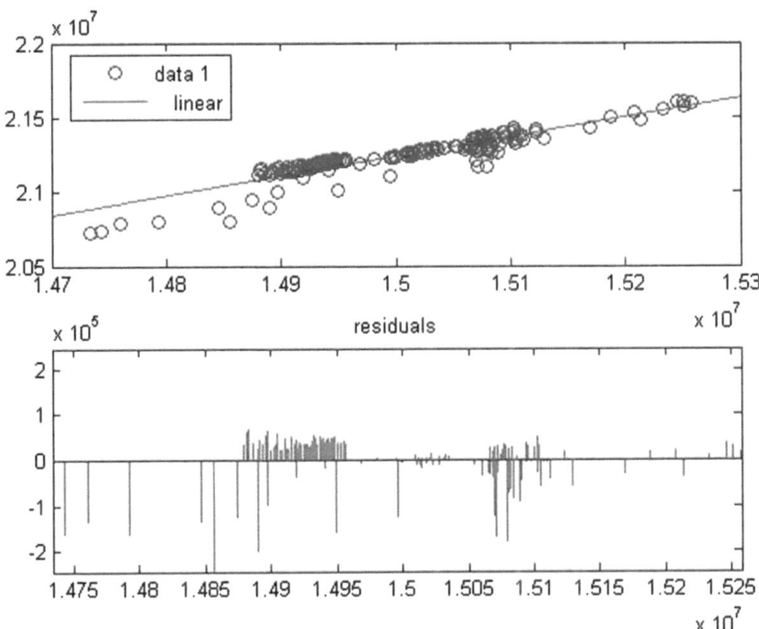

Fig. 1 The linear fitting of P1 and P2/I Image Database

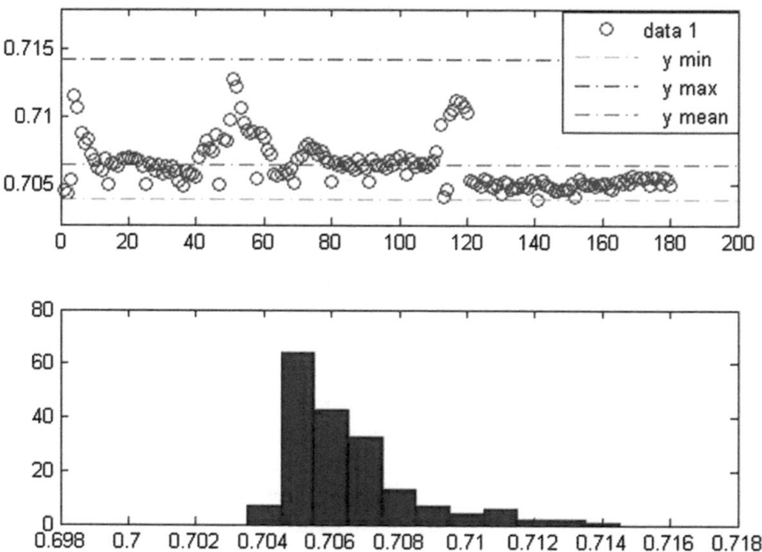

Fig. 2 The ratio of the distribution of the constant A/I Image Database

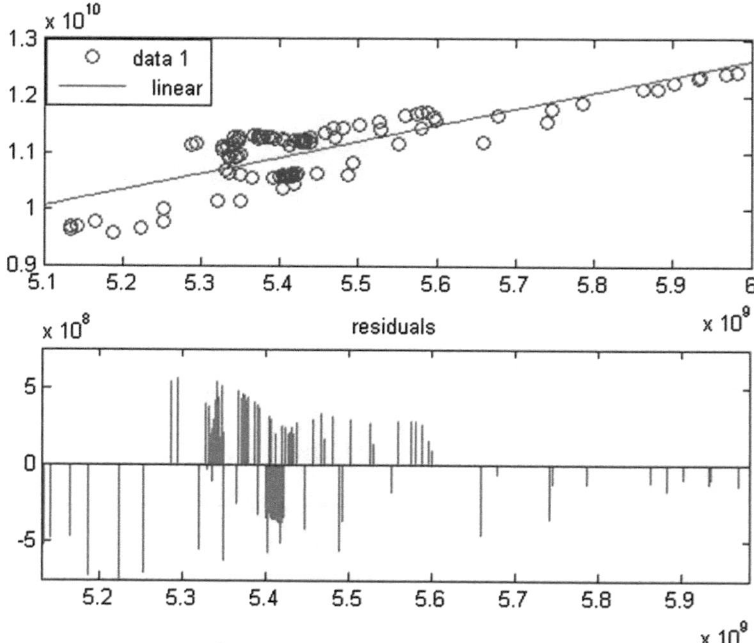

Fig. 3 II Image Database

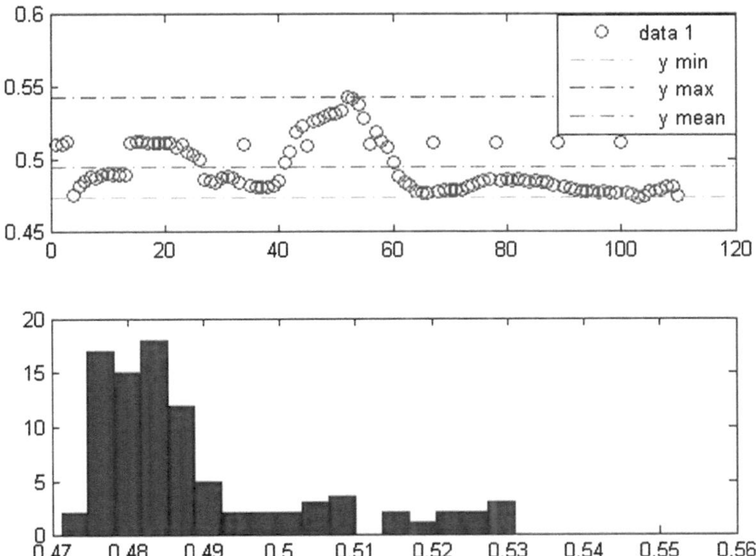

Fig. 4 II Image Database

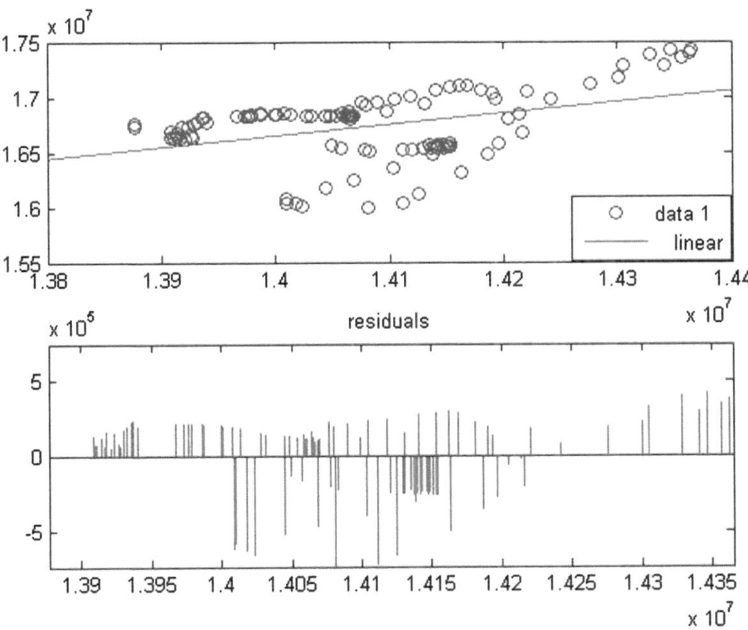

Fig. 5 III Image Database

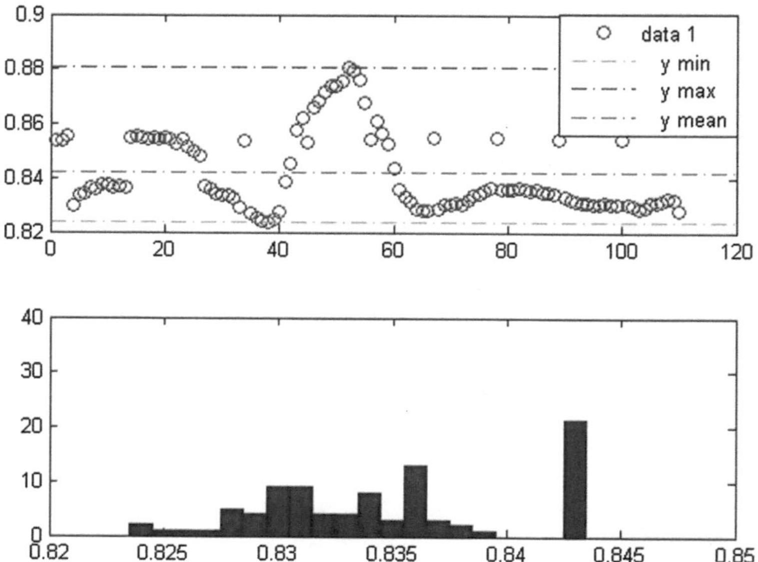

Fig. 6 III Image Database

Table 3 The relationship between R and A

	R	A
I Image Database	0.8102	0.7035
II Image Database	0.7145	0.5425
III Image Database	0.8805	0.5897

Through the above experiment, we can conclude that mathematical laws underlie image pressure characteristics. There was a significant linear correlation between image pressure and image temperature.

3 Experiments and Analysis

The proposed image descriptors are based on statistical thermodynamics. The correctness of theoretical analysis can be validated only by experiments.

3.1 Description of the Digital Image

Figure 7 and Table 4 show that the proposed descriptors can be described the content o the images.

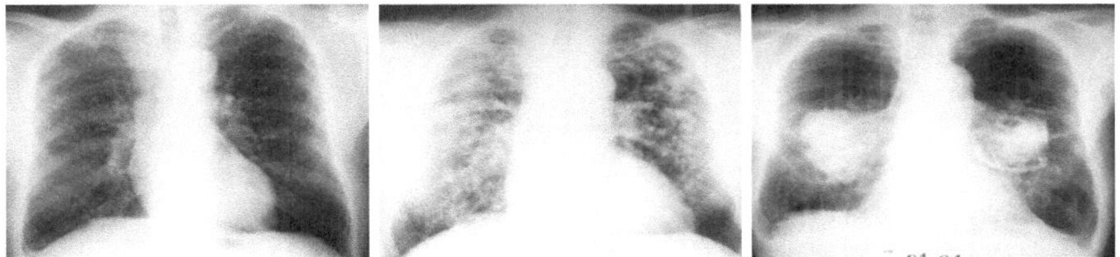

Fig. 7 Different periods of pneumoconiosis X-ray pictures. (Encyclopedia of Respiratory Medicine, 2006, Pages 191–201B. Wallaert)

Table 4 Statistical thermodynamic description of the right lung

Measure	First	Second	Third
E*(e − 07)	5.27412	14.0781	10.5228
N*(e − 03)	112.35	110.25	124.95
m*(e + 02)	3.13	1.22	1.26
P*(e − 05)	1.0487	2.7992	2.0923
T*(e − 05)	2.7596	7.3662	5.5059
S	6.1028	6.7760	7.2858

3.2 Clustering Applications

Cluster analysis or clustering is the task of grouping a set of objects in such a way that objects in the same group are more similar (in some sense or another) to each other than to those in other groups (clusters) [10, 11]. It is a main task of exploratory data mining and a common technique for statistical data analysis used in many fields, including machine learning, pattern recognition, image analysis, information retrieval, and so on (Fig. 8).

3.3 Content-Based Image Retrieval

CBIR, also known as query by image content (QBIC) and content-based visual information retrieval (CBVIR), is the application of computer vision to the image retrieval problem, that is, the problem of searching for digital images in large databases [12, 13]. Content-based image retrieval is opposed to concept-based approaches.

The proposed method is implemented in a CBIR system. In the Windows environment, it has been developed by MATLAB. The image data are selected from the UCI database. The test library has nine types of images, a total of 932 images. Each type of image is not less than 100.

In the experiment, for each test, calculate the similarity with all images in the test library. If the returned images and test images belong to the same semantic class, that is correct (Table 5).

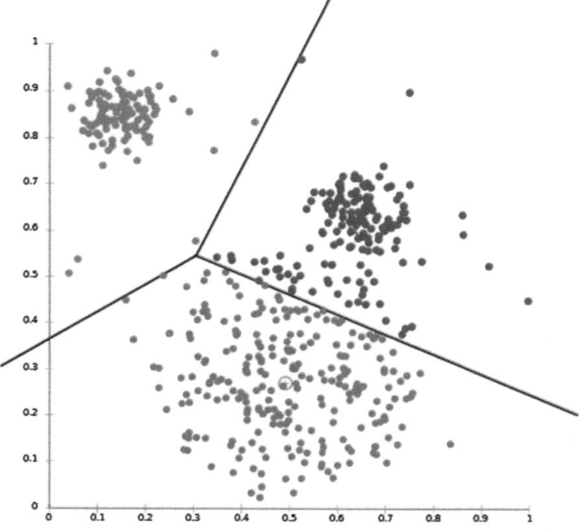

Fig. 8 Clustering examples for image temperature and image pressure

Table 5 The rotation of the image

Class No.	1	2	3	4	5	6	7	8	9
Class	Dinosaur	Building	Stamp	Tools	Women	Bird	Flowers	Seabed	Mountain
Accuracy	0.9	0.6	0.7	0.8	0.6	0.7	0.9	0.7	0.8

Image distance of 0 signifies an exact match with the query, with respect to the dimensions that were considered. The example shows the rationality and validity of methods.

4 Conclusions

In this article, we present a series of new image descriptors based on statistical thermodynamics and discuss them based on content-based image retrieval and clustering applications. The paper puts forward image descriptors which represent macro-visual characteristics such as "image energy," "image pressure," "image mass," and "image temperature" according to the analysis-localized sub-system within the statistical thermodynamic theory. We can find a lot of mathematical laws, by applying statistical thermodynamic theory in digital image processing. The proposed method has the characteristics of the fast calculation. Experiments verify the reasonableness and effectiveness of the proposed method.

Acknowledgments Project supported by the National Nature Science Foundation of China (No.61073121), The National Key Technology R&D Program (No.2013BAK07B04), Natural Science Foundation of Hebei Province of China (No. F2013201170), and Medical Engineering Alternate Research Center Open Foundation of Hebei University (No. BM201102).

References

1. Castells, M.: The power of identity: the information age: economy, society, and culture. Wiley-Blackwell, Hoboken (2011)
2. Joia, P.: Class-specific metrics for multidimensional data projection applied to CBIR. Vis. Comput. **28**(10), 1027–1037 (2012)
3. Hirata, K., Kato, T.: Query by visual example—content based image retrieval. In: Proceedings of the 3rd International Conference on Extending Database Technology: Advances in Database, Vienna, Austria 1992
4. Kekre, H.B.: Sectorization of DCT-DST plane for column wise transformed color images in CBIR. Technol. Sys. Manage. Commun. Comput. Inf. Sci. **145**, 55–60 (2011)
5. Qi, H., Li, K., Shen, Y., et al.: An effective solution for trademark image retrieval by combining shape description and feature matching. Pattern Recogn. **43**(6), 2017–2027 (2010)
6. Tan, X., Triggs, B.: Enhanced local texture feature sets for face recognition under difficult lighting conditions. Image Proc. IEEE Trans. **19**(6), 1635–1650 (2010)
7. Morel, J.M., Yu, G.: ASIFT a new framework for fully affine invariant image comparison. SIAM J. Imaging Sci. **2**(2), 438–469 (2009)
8. Sandler, S.I.: An introduction to applied statistical thermodynamics. Wiley, London (2010)
9. Kapur, J.N., Sahoo, P.K., Wong, A.K.C.: A new method for gray-level picture thresholding using the entropy of the histogram. Comput. Vision Graph. Image Proc. **29**(3), 273–285 (1985)
10. Yang, Y., Xu, D., Nie, F., et al.: Image clustering using local discriminant models and global integration. Image Proc. IEEE Trans. **19**(10), 2761–2773 (2010)
11. Zhao, Z.L., Liu, B., Li, W.: Image clustering based on extreme K-means algorithm. IEIT J. Adapt. Dyn. Comput. **2012**(1), 12–16 (2012)
12. Bian, W., Tao, D.: Biased discriminant Euclidean embedding for content-based image retrieval. Image Proc. IEEE Trans. **19**(2), 545–554 (2010)
13. Gomez R.: Integrating technology in a statistics course for a special program at Florida International University. (2013)

A Novel Image Retrieval Method Based on Fractal Code and Fuzzy Set

Haipeng Li, Feng Li and Yafang Lou

Abstract In this paper, we adopt the advantage of fractal code and classify blocks partitioned from images into four different categories and propose an improved fractal image code as image index label. The method speeds up the course of fractal image encoding without demolishing the perfection of reconstructed image. In applying for image retrieval, with the help of fuzzy classifier, the time consumed for image matching can be reduced obviously. Simulated experiment shows that our proposed method can gain desirable results in image retrieval meantime costing less time.

Keywords Image retrieval · Fractal code · Fuzzy set · Image compress

1 Introduction

Fractal code originates in the fact that our natural environment generally shows self-similarity on a wide scale of some physical parameter. Therefore, the considerable amount of redundancy in image can be reduced by executing fractal encoding.

As per iterated function system (IFS) proposed by Barnsley and Sloan [1], there is a contractive transformation for each image that has the fixed point identical to the image itself. In other words, applying that transform iteratively on an arbitrary

H. Li (✉) · Y. Lou
Computer Science and Technology Department, Zhuhai College,
Jilin University, Zhuhai, China
e-mail: haipengli_jlu@hotmail.com

F. Li
MI Ninth Design and Research Institute of Company Ltd, Changchun, China

S. Patnaik and X. Li (eds.), *Proceedings of International Conference on Computer Science and Information Technology*, Advances in Intelligent Systems and Computing 255, DOI: 10.1007/978-81-322-1759-6_4,

starting image will result converges to the original image. Thus, the image is encoded by the transformation.

With the help of fractal code and IFS, many researchers previously pay more attention on how to improve the performance of compressing ratio of image [2–4], but there exists a problem about higher time consumption brought by complex searching course that cumber this novel method applying in real practice.

In this paper, we proposed a classified method utilizing fuzzy set measure in order to reduce the computational time when executing fractal coding process. Through fuzzy classifier, the blocks in fractal coding process can be classified into four different types; with the help of the classification, fractal coding be speeding up apparently. Experimental results imply that fractal code and classification-based fuzzy set can help improve the accuracy and efficiency of image retrieval.

2 Fractal Coding

Fractal image encoding is based on contractive transformations and a partitioned iterated function system (PIFS) [5] in a two-dimensional metric space. The original input image is first partitioned into nonoverlapping range blocks R with size of $r \times r$ and overlapping domain blocks D with size of $2r \times 2r$; we define R as the denotation of each range block and D as the denotation of each domain block. The gray-level transform can be defined by the below function:

$$W(D) = s_i \tau_i A D_i + o_i I \tag{1}$$

where s_i denotes scaling factor, o_i is an offset, A is the operator that shrinks the domain block via pixel averaging to match the range block size, τ_i is a permutation that shuffles the pixel intensities in the domain block, and I is a block of the same size of spatially contracted domain block D, but with all elements equal to 1.

3 Fuzzy Set Measure

3.1 Fuzzy Classifier

In our proposed method, we cite fuzzy classifier [6] to decrease the search time for matching of range blocks and domain blocks. Assume a block M in image I with $n \times n$ pixels, let $g(i, j)$ denote the pixel value in location (i, j) of the block. Partition it into 9 n/3 $\times$ n/3 subblocks, M_k $(k = 1, 2, \ldots, 9)$, the sum of pixel value of each subblock M_k is defined as:

$$S_i = \sum_{(i,j \in M_k)} g(i, j), \quad k = 1, \ldots, 9. \tag{2}$$

$$S = \frac{\sum_{i=1}^{9} S_i}{9} \tag{3}$$

We define the corresponding pixel value distributed fuzzy set measure stander:
1. Vertical distributed measure:

$$\delta_v(M) = \frac{|(S_1 + S_2 + S_3) - (S_7 + S_8 + S_9)|}{S} \tag{4}$$

2. Horizontal distributed measure:

$$\delta_h(M) = \frac{|(S_1 + S_4 + S_7) - (S_3 + S_6 + S_9)|}{S} \tag{5}$$

3. Diagonal distributed measure:

$$\delta_d(M) = \frac{|(S_1 + S_4 + S_6 + S_9) - (S_2 + S_3 + S_7 + S_8)|}{S} \tag{6}$$

4. Center measure:

$$\delta_c(M) = \frac{S_5}{S} \tag{7}$$

$$\delta_c'(M) = 1 - \frac{S_5}{S} \tag{8}$$

5. Symmetry measure:

$$\delta_s(M) = \frac{|S_1 - S_9| + |S_3 - S_7| + |S_2 - S_8| + |S_4 - S_6|}{S} \tag{9}$$

According to the previous definition of the measure criterion, we can analyze the meaning of each parameter in the fuzzy set measure criterion, epitomize the following results: δ_v denotes the block has apparent edge between upper and lower parts; δ_h denotes the block has apparent edge between left and right parts; δ_d denotes the block has apparent edge existence in diagonal line area; δ_c and δ_c' mean that the block has a more concentrated part in center; δ_s means the block has symmetric feature around central subblock.

With the fuzzy set measuring criterion, we can classify the blocks gained from the partitioned query image into four different classes with following steps:

```
for each block M:
```
$\delta(M) = \min\{\delta_v(M), \delta_h(M), \delta_d(M), \delta_c(M), \delta_c^{'}(M), \delta_s(M)\}$

if ($\delta(M) = \delta_v(M) or \delta_h(M)$) M belong to "central edge blocks";

if ($\delta(M) = \delta_d(M)$) M belong to "diagonal edge blocks";

if ($\delta(M) = \delta_c(M) or \delta_c^{'}(M)$) M belong to "central blocks";

if ($\delta(M) = \delta_s(M)$) M belong to "symmetric blocks";

3.2 Proposed Algorithm

With the help of fuzzy classifier offered in Sect. 3.1, we can improve the fractal coding course and built image database with the following steps:

1. Partition image Q into a set of range blocks with the size of $B \times B$ and into domain blocks with the size of $2B \times 2B$, which compose the domain pool;
2. Classify the range blocks with the measuring criterion offered in Sect. 3.1 into four classifications and repeat the same process in domain pool;
3. Matching process: for each range block, search corresponding domain block to finish fractal coding, but the search district is limited in domain blocks those belonging to the same classification with the current range block;
4. Generate fractal code of each image and store the parameters into database as index label of the image.

4 Similarity Measure

We assume that there are two same-sized image I and image J, defining frac(I) and frac(J) as fractal code of image I and image J, respectively. According to Sects. 2 and 3, we can express frac(I) like this:

$$\text{frac}(I) = \{s_i,\ \tau_i,\ o_i,\ \delta_i;\quad i = 1, \ldots, N\} \tag{10}$$

where N denotes the number of range blocks in image I. In order to measure the similarity between two images, we define the fractal coding distance [7] of blocks in image I and image J:

$$\text{dis}(i, j) = \left\|\text{frac}(I)_i - \text{frac}(J)_j\right\| = \left\|s_i - s_j\right\| + \left\|\tau_i - \tau_j\right\| + \left\|o_i - o_j\right\| \tag{11}$$

where $i = 1, \ldots, N; j = 1, \ldots, M$; M denotes the number of range blocks in image J, here M equates N.

Because we have classified the corresponding image range blocks into four different classifications in previous section, in current process period, we do not have to compute the value of dis(i, j) by searching all range blocks, only do the computational work for the range blocks with the same value of δ that denotes the shape character of range blocks. So, if we define distance between a range block in image I and image J as dis(i, J), we can generate the value of dis(i, J) like this:

```
for(j=0;j<M;j++)
        {  if (δi-δj==0)                        compute
dis(i,j);
             if (dis(i,J)> dis(i,j))              dis(i,J)=
dis(i,j);      }
```

Furthermore, we define the distance between image I and image J like below:

$$\mathrm{dis}(I, J) = \sum_{i=1}^{N} \mathrm{dis}(i, J) \tag{12}$$

So, when we get a minimum value of dis(I, J), we could consider that image I and image J have higher similarity. We utilize the distance between image I and image J as the measuring standard when matching images.

5 Experiment

We firstly simulate experiments to compare the performance of our proposed method with the other researchers in image-compressing ratio. Table 1 shows the compared results from same test environment. From Table 1, we can see that our proposed method can gain not-bad performance in compression ratio (CR) and more desirable performance in reconstructed image measured by PSNR compared with others method [8, 9].

In the course of image retrieval, we assume that there is a query image Q, we have built image database D, where each image with an index label is generated by the method offered in Sect. 3. The index label is presented by fractal image code frac(D_i), $i = 1, 2, \ldots, L$, where L is the number of total images in database D. In our experiment, image database D is composed by 200 images those belong to different types, including buildings, people, animals, vehicle, etc. Figure 1 is the matching result when we specify a query image. The image in up-left of Fig. 1 is the query image. From results, we can see that for the images more similar, the value of distance between two images is more small.

After completing the course of image retrieval, let us consider the time consumption in image-matching course. Assume t_Q as the costing time for calculating frac(Q) of query image Q, and t_{dis} as the time for calculating fractal coding distance between two blocks, n denotes the number of range blocks in image Q,

Table 1 Performance of compression and speed compared with other methods

Image	Tong and Wong's method			No. search method			Our proposed method		
	CR	t(s)	PSNR	CR	t(s)	PSNR	CR	t(s)	PSNR
Lena	0.13	5	34.6	0.12	0.12	34.2	0.12	1.2	35.2
Baboon	0.15	8	25.8	0.13	0.6	24.2	0.13	2.2	29.7

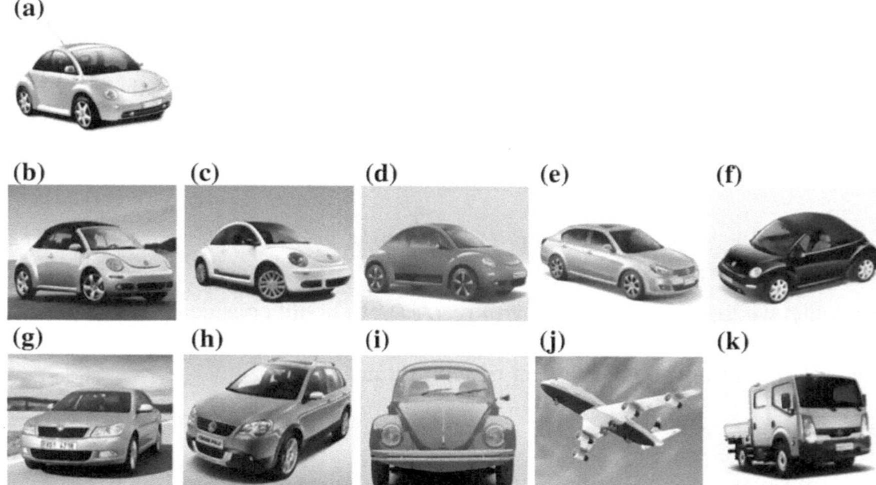

Fig. 1 **a** Query image. **b–k** Matching results sorted by similarity measure

m denotes the number of range blocks in image D_i. So, the computational time T that be cost to complete one query image-matching course in image database is equated by $t_{\text{dis}} \times n \times m \times L$. In our proposed approach, we have classified all the range blocks in an image into four different classifications. Actually in the course of classification, the distribution of the amount of blocks in each class can get balanced, or we can arrive into such amount equilibrium by adding weight parameters when executing classified process. So, in the course of image matching, the value of m can be reduced to $0.25 \times m$, which means just range blocks that belong to same classification need to be computed. Hence, the time consumed during the image retrieval can be saved 75 %.

6 Conclusion

In this paper, we proposed an improved fractal coding method by introducing fuzzy set measurement. By conducting the fuzzy classifier, we classify the range blocks into different categories that characterize the shape feature of range blocks.

Our experimental results indicate that it is an efficient approach to match images in huge multimedia image database by fractal image coding as index label. By classifying the corresponding blocks into different categories with fuzzy classifier, the total time consumed in image matching can be decreased to 75 % than without classification in blocks. Future works include finding more accurate and efficient measuring criteria to test similarities between images.

References

1. Barnsley, M.F., Sloan, A.D.: A better way to compress images. Byte Mag. **13**(1), 215–233 (1988)
2. Vidya, D., Parthasarathy, R., Bina, T.C., Swaroopa, N.G.: Architecture for fractal image compression. J. Syst. Archit. **46**, 1275–1291 (2000)
3. Duh, D.J., Jeng, J.H., Chen, S.Y.: DTC based simple classification scheme for fractal image compression. Image Vis. Comput. **23**, 1115–1121 (2005)
4. Iano, Y., da Silva, F.S., Cruz, A.L.M.: A fast and efficient hybrid wavelet image coder. IEEE Trans. Image Process. **15**, 98–105 (2006)
5. Jacquin, A.E.: Fractal image coding: A review. Proc. IEEE **81**(10), 1451–1465 (1993)
6. Lee, K.F., Gu, W.G., Phua, K.H.: Speed-up fractal image compression with a fuzzy classifier. Signal Proc. Image Commun. **10**, 303–311 (1997)
7. Yan, M., Li, S.: Image retrieval algorithm based on IFS fractal code. Opto-electron. Eng. **33**(2), 81–84 (2006)
8. Furaoa, S., Hasegawa, O.: A fast no search fractal image coding method. Signal Proc. Image Commun. **19**(5), 393–404 (2004)
9. Tong, C.S., Wong, M.: Adaptive approximate nearest neighbor search for fractal image compression. IEEE Trans. Image Process. **11**(6), 605–614 (2002)

Velocity Evaluation in Traffic Accidents Occurred at the Road Intersection Based on Image Processing

Jie Zhang, Hongyun Chen and Hao Wang

Abstract In recent years, video surveillance equipments have been applied widely and installed at all major transport nodes. These equipments record a large volume of day-to-day traffic information, which allows measuring the velocity of accident vehicle based on video. Since the early 1980s, researchers have developed various algorithms to extract speed information from traffic image sequences. In this paper, an algorithm is proposed to evaluate the velocity of turn-drive vehicle. The characteristics of an accident scene including the information of lane marks and accident vehicles are utilized to estimate the state of an accident vehicle. With manual interaction, the state of the accident vehicle, including the vehicle trajectory and its speed when it is turning, can be estimated more robustly and accurately. Experimental results show that the proposed approach can capture the accident vehicle' state accurately and meet the precision demand.

Keywords Traffic accident · Road intersection · Velocity calculation · Image processing

1 Introduction

In recent years, with China's rapid economic and social development, the travel needs of people and the private car quantity are sharply increasing. Construction of basic facilities such as roads gradually speeds up the space. Different classes of

J. Zhang (✉) · H. Chen · H. Wang
Traffic Engineering Department, Research Institute of Highway, Ministry of Transport of China, Beijing, China
e-mail: zhang.jie@rioh.cn

H. Chen
e-mail: hy.chen@rioh.cn

H. Wang
e-mail: hao.wang@rioh.cn

S. Patnaik and X. Li (eds.), *Proceedings of International Conference on Computer Science and Information Technology*, Advances in Intelligent Systems and Computing 255, DOI: 10.1007/978-81-322-1759-6_5,

highways cross each other, and then, varieties of intersections are formed. These road intersections are the only ways of vehicles and pedestrians gathering, steering, and evacuation. These areas are not only the throat of traffic, but also the accident-prone locations. These intersections not only bring the convenience to people but also bring threat to people.

Various types of road intersections and the number of accidents resulting from the loss of lives and property in China 2011 are shown in Table 1 [1].

With the development of video surveillance technology, video surveillance equipments have been applied widely and installed at all major transport nodes, such as important sections and intersections of roads. These equipments record a large volume of day-to-day traffic information, which allows for measuring the velocity of accident vehicle based on video.

Nowadays, there exist many measurement technologies of vehicle speed. Since the early 1980s, researchers have developed numerous algorithms to extract speed information from traffic image sequences. In general, existing algorithms are either tracking-based or virtual-loop-based. Tracking-based algorithms model a vehicle with a close curve or a group of features and track a vehicle within consecutive frames [1–3]. Malik et al. [4] summarized four tracking approaches: 3D-model-based tracking, region-based tracking, active contour-based tracking and feature-based tracking, and they argued that the feature-based tracking approach performs the best in vehicle-tracking applications. Feature-based approach tracks distinguishable points or edges on a vehicle instead of the vehicle itself. Hence, even in presence of partial occlusion, some features of this vehicle remain visible. The average accuracy of speed estimation with this type of approach is above 95 %. A virtual-loop is composed of a few detection lines or a bounding box manually defined [5] or automatically assigned [6]. Through emulating the functionality of an inductive loop detector, the system can detect the optical changes in the loop and generate a signal when a vehicle crosses it. The best performance of speed estimation with such approaches is above 95 %.

Practically, the velocity of vehicles involved in an accident is a crucial evidence for both the traffic police department and the court when dealing with traffic accident cases. It helps identifying the vehicle(s) that causes the accident. However, existing vehicle speed estimation approaches are mainly used for traffic-flow surveillance, which need less precision. Unfortunately, when traffic police

Table 1 Summary of the road intersection accidents (2011)

	Number of accidents	Deaths	Injuries	Direct property damage (Yuan)
T intersection	15,636	3,623	17,793	52,503,174
Four-branch-intersection	20,849	4,881	23,306	75,732,934
Multiple-branch-intersection	2,137	618	2,402	26,502,782
Roundabout	494	94	563	2,312,361
Ramp	1,475	650	1,740	17,110,346
Total	40,591	9,866	45,804	174,161,597

department or the court apportions makes a verdict on an accident, they need to have exact evidence (data) on accident vehicles. So, many new approaches [7, 8] are presented for traffic accident scene investigation, but these approaches all focus on the velocity calculation of straight drive vehicle.

In this paper, a new approach with the help of manual interaction is presented for the velocity calculation of turn-drive vehicle. In this approach, the characteristics of turn-drive vehicle, including the information of accident vehicles, are utilized to estimate the state of accident vehicle.

2 Algorithm Principle

In order to avoid the generation of additional resistance from road surfaces on vehicles as well as excessive tire wear, the steering system of the car must make sure that all wheels of the vehicle are in pure rolling state when the vehicle is turning. Obviously, this can only be achieved when all wheel axis intersect at one point. This intersection point O is called the steering center. The relationship of the deflection angle when the car is turning is shown in Fig. 1.

According to Fig. 1, the turning radius of the four wheels and the centroid can be calculated, which is shown in Table 2.

According to Table 2, the following results can be concluded:

1. When the car is turning left, the least turning radius is left rear wheel and the maximum turning radius is right front wheel. The turning radius of the centroid is between left rear wheel and right front wheel, so the range of the vehicle velocity can be determined.
2. When the car is turning right, the least turning radius is right rear wheel and the maximum turning radius is left front wheel. The turning radius of the centroid is between right rear wheel and left front wheel, so the range of the vehicle velocity can be determined.

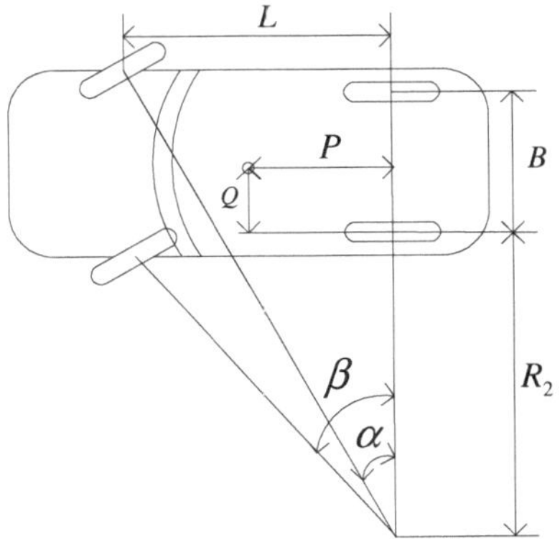

Fig. 1 The relationship of the deflection angle when the car is turning

Table 2 The turning radius of the four wheels and the centroid

	Value of turning radius
Turning radius of the left front wheel	$R_1 = \frac{L}{\sin\beta}$
Turning radius of the left rear wheel	$R_2 = \frac{L}{\tan\beta}$
Turning radius of the right front wheel	$R_3 = \left(\frac{L}{\tan\beta} + B\right)^2 + L^2$
Turning radius of the right rear wheel	$R_4 = \frac{L}{\tan\beta} + B$
Turning radius of the centroid	$R = \left(\frac{L}{\tan\beta} + Q\right)^2 + P^2$

where L vehicle axle base, B wheel center distance, α deflection angle of the left front wheel, and β deflection angle of the right front wheel

3 Main Steps of Algorithm

3.1 Obtain Frame Sequences

Through analyzing the video that contains the information of the whole process of accident, we select parts of the video to process. First, we extract all frame sequences from the video.

Then, the frame rate of this video and the size of all frames are known. In this case, the frame rate is 25 f/s, and the size of frame is 288 × 352 pixel.

3.2 Accurate Velocity of the Vehicle Calculation

The vehicle's state in surveillance video should be transformed to real-world coordinates, which is a 2D–3D mapping and impossible without additional constrictions. Fortunately, in many cases, it can be assumed that the road is plane, and the vehicle moved in this plane surface. The problem of coordinate transformation is essentially a 2D–2D projective mapping formulation. Furthermore, in accident scene, some parameters such as the distances relationship among these lane marks can be measured manually, which help calculating the parameters of projective transformation.

The formulation of projective transformation is shown as:

$$q = H_p p \tag{1}$$

where H_p is the projective mapping matrix. It can also be expressed as:

$$\begin{bmatrix} q_x \\ q_y \\ q_z \end{bmatrix} = \begin{bmatrix} h_{11} & h_{12} & h_{13} \\ h_{21} & h_{22} & h_{23} \\ h_{31} & h_{32} & h_{33} \end{bmatrix} \begin{bmatrix} p_x \\ p_y \\ p_z \end{bmatrix} \tag{2}$$

The matrix H_p contains nine elements, but only has eight independent parameters. It can be solved by four pairs control points.

Through using the above two-dimensional geometric correction technology, movement distance of the left front wheel s_1 and movement distance of the left rear wheel s_2 can be calculated. Then, turning radius of the centroid R and angular velocity of the centroid ω can be obtained:

$$R = \sqrt{(R_2 + Q)^2 + P^2} = \sqrt{\left(\frac{s_2 \times L}{\sqrt{s_1^2 - s_2^2}} + Q\right)^2 + P^2} \tag{3}$$

$$\omega = \frac{\sqrt{s_1^2 - s_2^2}}{L \times T} = \frac{\sqrt{s_1^2 - s_2^2}}{L \times n} \times f \tag{4}$$

Accurate velocity of the vehicle is as formula (5):

$$v = \frac{\sqrt{s_1^2 - s_2^2}}{L \times n} \times f \times \sqrt{\left(\frac{s_2 \times L}{\sqrt{s_1^2 - s_2^2}} + Q\right)^2 + P^2} \tag{5}$$

where, n represents frame count, L represents vehicle axle base, B represents wheel center distance, f represents frame rate.

3.3 Range Velocity of the Vehicle Calculation

The above algorithm is a little complex. In many cases, we only need the range velocity of the vehicle. Through setting reference line, we can make sure the time that the left rear wheel moves through the reference line. Then, we can calculate the range velocity of the vehicle according to the vehicle axle base.

In this case, we extract 75 frame sequences from the video. We set one reference line at the 200th column. Through analyzing the frames which is processed, we capture the moving state of the accident vehicle. At the 32th frame, the left front wheel moves through the reference line. At the 45th frame, the left rear wheel moves through the reference line. The length of vehicle axle base is 2.57 m.

$$v > \frac{L}{n_2 - n_1} \times f \times 3.6 \tag{6}$$

4 Conclusion

This paper presented an approach for the velocity calculation of turn-drive vehicle. The characteristics of an accident scene including the information of land marks and accident vehicles are utilized to estimate the state of an accident vehicle.

With manual interaction, the state of the accident vehicle, including the vehicle trajectory and its speed when it is turning, can be estimated more robustly and accurately. Experimental results show that the proposed approach can capture the accident vehicle state accurately and meet the precision demand.

References

1. Wu, J., Liu, Z., Li, J.: An algorithm for automatic vehicle speed detection using video camera. In: ICCSE, pp. 193–196 (2009)
2. Alefs, B., Schreiber, D.: Accurate speed measurement from vehicle trajectories using AdaBoost detection and robust template tracking. In: Proceedings of ITSC IEEE Conference Intelligent Transport System, pp. 405–412 (2007)
3. Blake, A., Curwen, R., Zisserman, A.: A framework for spatiotemporal control in the tracking of visual contours. Int. J. Comput. Vis. **11**(2), 127–146 (1993)
4. Malik, J., Russell, S., Beymer, D., Coifman, B., Huang, T., Liddy, D., McLauchlan, P.: Traffic surveillance and detection technology development: new traffic sensor technology: final report. University of California—PATH, California (1997)
5. Michalopoulos, P.G.: Vehicle detection video through image processing: the autoscope system. IEEE Trans. Veh. Technol. **40**(1, Part 2), 21–29 (1991)
6. Lai, A.H.S., Yung, N.H.C.: Vehicle-type identification through automated virtual loop assignment and block-based direction-biased motion estimation. IEEE Trans. Intell. Transp. Syst. **1**(2), 86–97 (2000)
7. Tan, H., Zhang, J., Feng, J., Li, F.: Vehicle speed measurement for accident scene investigation. In: IEEE International Conference on E-Business Engineering (2010)
8. Bai, R., Zhang, J., Chen, H.: A velocity measuring algorithm for accident vehicle based on video. In: Proceeding of the 2010 IRAST International Congress on Computer Applications and Computational Science, vol. 12 (2010)

A Hybrid Method for Extracting Liver from 3D CT Image

Xiaolong Song, Qiao Wang and Zhengang Jiang

Abstract As a key technology of computer-aided medical application, medical image segmentation has been a hot topic in image processing field. Computerized tomography (CT) scan of liver is an indispensable technique for clinical diagnosis and treatment of liver diseases. How to extract liver from CT image accurately is a challenging problem to be solved. Because of the difference and complexity of medical image, traditional segmentation methods are not suitable for them. Therefore, this paper presents a method for extracting liver from CT images using three-dimensional region-growing algorithm combined with image morphology. The experimental results show that the proposed method has a high accuracy and lays the foundation for the further accurate segmentation.

Keywords Medical image · Three-dimensional · Segmentation · Region grow · Image morphology

1 Introduction

In the nineteenth century, German physicist Roentgen discovered X-ray and opened the door to modern medical imaging technology. Along with the continuous development of computer technology, physics, electronics, and other disciplines, medical image imaging is becoming more diverse. These images include X-ray imaging techniques, computerized tomography (CT), magnetic resonance imaging (MRI), ultrasonic imaging, infrared thermal imaging (Thermal Imaging), endoscope imaging (Endoscope), micrograph imaging (Micrograph), positron emission tomography (PET), digital subtraction angiography (DSA), and other images from

X. Song · Q. Wang · Z. Jiang (✉)
Changchun University of Science and Technology, School of Computer Science and Technology, Weixing Road, Changchun 7089, China
e-mail: jiaotangmaqiduoqq@gmail.com

S. Patnaik and X. Li (eds.), *Proceedings of International Conference on Computer Science and Information Technology*, Advances in Intelligent Systems and Computing 255, DOI: 10.1007/978-81-322-1759-6_6,

medical imaging equipment. These images have been utilized in clinical applications widely [1].

Because the role of medical imaging in clinical diagnosis and treatment has been more and more significant, medical image segmentation has been a focused research topic in the field of medical image analysis. The task of image segmentation is to extract useful diagnostic information about anatomical structure from CT, MRI, PET, and other modes of medical images with the assistance of computer [2]. Compared with the general image segmentation, it is more difficult for medical image segmentation. As the patients' individual differences, the targets of the segmentation are different in space, spectrum, and intensity. Furthermore, medical image is of complexity and diversity, because the scanning of it is susceptible to a variety of uncertainties. Therefore, accurate segmentation of medical images becomes a challenging problem. In recent years, CT image of liver has been widely applied in clinical applications. It has been an important way for liver function, pathological, and anatomical studies [3]. To extract liver from CT has a significant value for clinical treatment of liver disease.

2 Related Research

In the past years, varieties of image segmentation methods were reported and were applied to the medical image segmentation. These methods can be classified into region-based methods, edge detection methods, hybrid methods, fuzzy theory-based methods, and other segmentation methods [4]. The region-based methods basically utilize the similarity between images to extract object. These methods include threshold segmentation method [5], region-growing segmentation method [6], classifier-segmentation method [7], and clustering algorithms [8, 9]. Edge detection methods use the change of pixel gray value around the edges to segment the target region. These methods distinguish the boundary through edge detection. In this case, parallel differential operators, such as Laplace operator, Prewitt operator, and Gradient operator, are adopted to detect the edge. Region-based segmentation methods often lead to over-segmentation. As the image may be divided into too many regions, it is difficult to complete the segmentation task. The hybrid method takes advantage of the above-two methods. In the specific case, it can get good segmentation results.

Due to the ambiguity and complexity of the medical image, traditional segmentation methods cannot complete the segmentation very well, and they even cause the error segmentation. In particular, the soft tissue in human intra-abdominal organ has the similar density and structure. The liver edges are not clear in some CT-image sequences, which make it more difficult to accurately split. Therefore, this paper presents a method of three-dimensional region-growing algorithm combined with the image morphology. The aim of segmentation is to extract the preliminary three-dimensional segmentation of liver.

3 Proposed Method

In the work, a hybrid method is proposed to extract liver region from CT images. First, an imaging slice that contains liver is chosen from the CT-image sequences, and then a seed point is selected in the corresponding area of liver. According to the growth standards, the three-dimensional region-growing algorithm is used to obtain the initial three-dimensional edge of liver.

Then, opening and closing operations are adopted to optimize the segmentation results. These two operations are combined with expansion and erosion of image morphology.

3.1 Three-Dimensional Region-Growing Algorithm

For a two-dimensional region-growing method, one or more seed points in the target region are selected firstly. Then, according to some growth rules, the seed points grow to the neighborhood directions in the image. But in CT images of liver, target region is a three-dimensional area. There are hundreds of image slices corresponding to target region. For extracting the target region using a two-dimensional region-growing algorithm, not only the seed points are needed by every slice image, but also the spatial information cannot be utilized in the growing process.

In CT-image series, each slice image has a strong correlation, which contains a large amount of structural information. Therefore, if it is supposed to use the traditional 4-neighborhood of two-dimensional region-growing algorithm (Fig. 1), it will cut off the link between the slices and ignore a lot of important information. To solve this problem, this paper adopts 18-neighborhood of three-dimensional region-growing algorithm (Fig. 1). The seed point will grow to 18-neighborhood

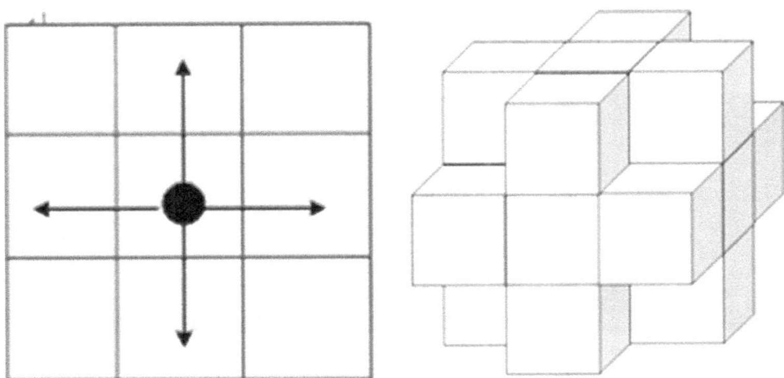

Fig. 1 4-neighborhood and 18-neighborhood

directions at the same time, so that the correlation between the slices can be fully utilized.

After the region-growing procedure finishes, the initial three-dimensional segmentation results of liver will be extracted.

3.2 Image Morphology

Traditional two-dimensional image morphology also cannot deal with the large amounts of information contained in the CT-image slices. For this reason, this work adopts the structure element of three dimension to scan the original image. The third dimension of structure element is defined by the height information of CT-image series of liver.

According to the respective function of opening and closing operation, the work adopts closing operation and takes expansion operation first. The specific operation of expansion and corrosion applies the structure element to scan every pixel of the original image. It will obtain the corresponding pixel value in the output image. Figure 2 shows some experimental results using different structure elements. In Fig. 2a, the structure element adopted in the expansion operation is $7 \times 7 \times 7$ and $7 \times 7 \times 7$ in corrosion operation is. In Fig. 2b, the structure element adopted in the expansion operation is $11 \times 11 \times 11$ and $9 \times 9 \times 9$ in corrosion operation is. In Fig. 2c, the structure element adopted in the expansion operation is $13 \times 13 \times 13$ and $7 \times 7 \times 7$ in corrosion operation is. Experimental results show that the structure selected in Fig. 2c works better.

4 Experimental Results

In order to verify the proposed method, some experiments were performed in this work. The testing data are a group of clinical CT-image series. The purpose of segmentation is to extract the contour of liver.

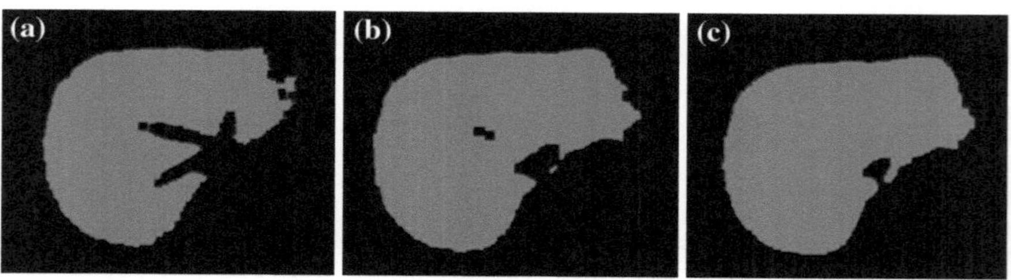

Fig. 2 **a** The structure element adopted in the expansion and corrosion operation is $7 \times 7 \times 7$. **b** The structure element adopted in the expansion operation is $11 \times 11 \times 11$ and $9 \times 9 \times 9$ in corrosion operation is. **c** The structure element adopted in the expansion operation is $13 \times 13 \times 13$ and $7 \times 7 \times 7$ in corrosion operation is

First, the CT images are sharpened using a Laplace transform. In this procedure, the area of pixel gray value changed hugely can be enhanced, and the area of pixel gray value changed smoothly can be reduced. Figure 3 is the original abdominal CT-image slices. Figure 4 shows the results after the Laplace transform. Figure 5a ($n = 136$ slice) is the initial outline of the three-dimensional region-growing algorithm. Because of the cavities in it, the image morphology has been used to deal with the same slice. The result is shown in Fig. 5b. As shown in Fig. 5c, these two images were superimposed. The red part of the picture (gray in binary image) is the differences between the extracted results by the traditional method and the proposed method.

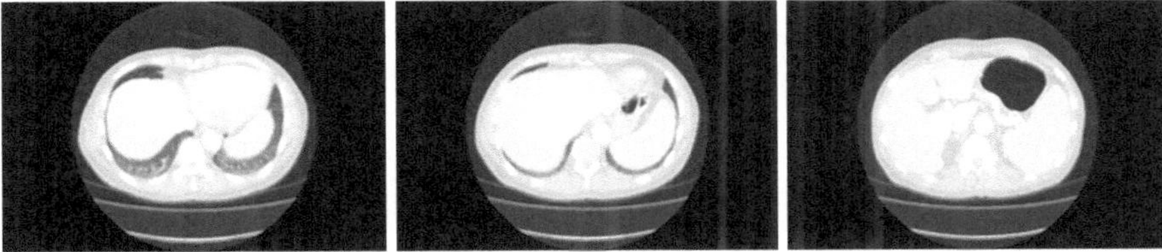

Fig. 3 Original abdominal CT-image series

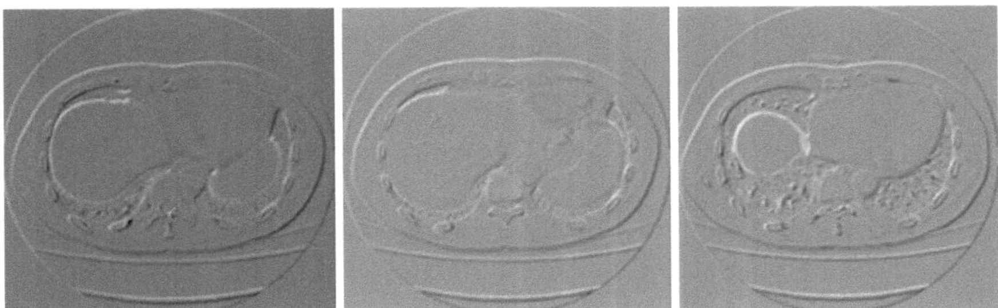

Fig. 4 Images after Laplace transform

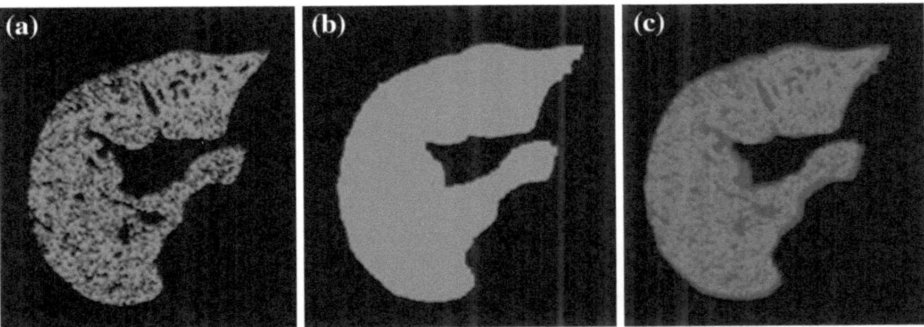

Fig. 5 **a** Result after the three-dimensional region-growing algorithm. **b** Result after the image morphology. **c** Result of the two pictures overlap

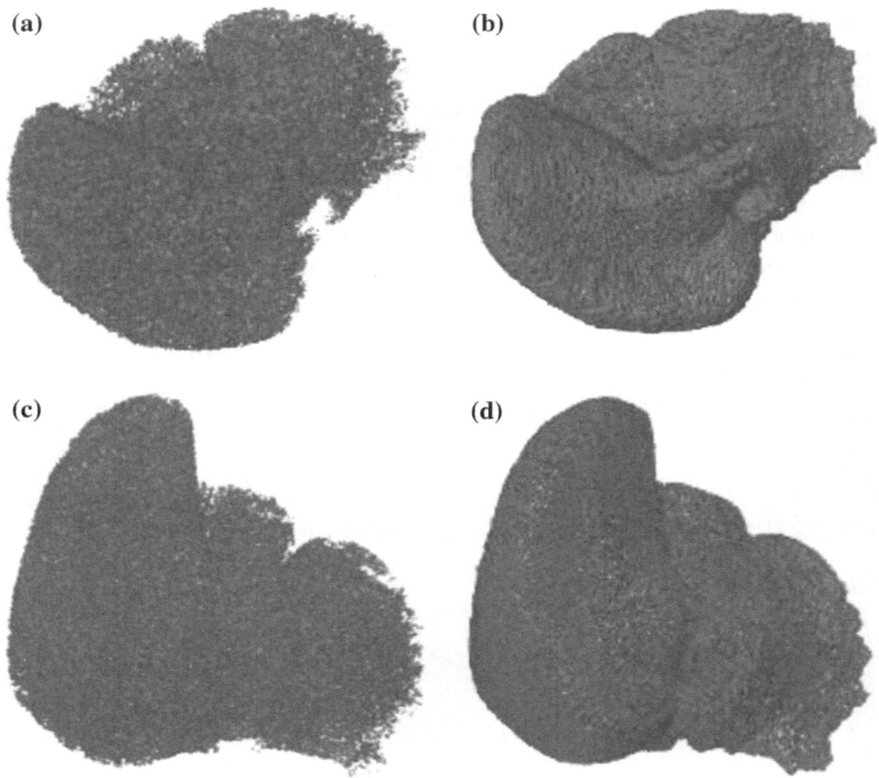

Fig. 6 **a** and **c** are the results of three-dimensional visualization after the region-growing algorithm. **b** and **d** are the results of three-dimensional visualization after the image morphology

As shown in Fig. 5b, the cavities in Fig. 5a have been filled, and the edge of liver is clearer than before.

After all the sequence of images has been cut up, according to the theory of three-dimensional reconstruction, a ray-casting volume rendering software has been used for visualizing three-dimensional data. Figure 6a and c show the results of three-dimensional region-growing algorithm. Figure 6b and d are the results of the proposed method.

5 Summary

This paper presents a hybrid method of three-dimensional region-growth algorithm combined with image morphology. The initial three-dimensional segmentation of liver has been extracted. Experimental results show that the proposed method can extract liver from 3D CT image accurately. It can be utilized to assist doctor to complete the liver segmentation from 3D CT image. How to locate a seed point automatically, to extract liver more accurately, and to reduce the human–computer interaction are our future works.

References

1. Xie Q.: Theory of variational level set method and its application to medical image segmentation. Ph.D. dissertation, Zhejiang University, **10**, 1–111 (2009)
2. Ni, Y.: The technology of medical image segmentation based on snake model. M.S. dissertation, Nanjing University of Aeronautics and Astronautics, **1**, 1–83 (2008)
3. Xu, D., Guo S., Wu X., Cen R., Ou S.: Research progress of liver CT image segmentation techniques. Chinese Med. Equip. J. **30**(3), 34–36 (2009)
4. Lin, Y., Tian, J.: Medical image segmentation review. Pattern Recognit Artif Intell. **15**(2), 192–204 (2002)
5. Sahoo, P.K., Soltani, S., Wang, A.K.C., et al.: A survey of thresholding techniques. Comput. Vis. Graph. Image Process. **41**(1), 233–260 (1988)
6. Pohlman, S., Powell, K.A., Obuchowski, N.A., et al.: Quantitative classification of breast tumors in digitized mammograms. Med. Phys. **23**, 1337–1345 (1996)
7. Pham, D.L., Xu, C., Prince, J.L.: A survey of current methods in medical image segmentation. Annu. Rev. Biomed. Eng. **2**, 315–338 (2000)
8. Lei, T., Sewchand, W.: Statistical approach to X-ray CT imaging and its applications in image analysis—part II: a new stochastic model-based image segmentation technique for X-ray CT image. IEEE Trans. Med. Imaging **11**(1), 62–69 (1992)
9. Liang, Z., MacFall, J.R., Harrington, D.P.: Parameter estimation and tissue segmentation from multispectral MR images. IEEE Trans. Med. Imaging **13**, 441–449 (1994)

A New Image-Fusion Technique Based on Blocked Sparse Representation

Yongping Zhang and Yaojia Chen

Abstract An image-fusion scheme based on blocked sparse representation is presented in the paper. Firstly, the source images are segmented into patches and then the patches are sparsely represented with learned redundant dictionary. Following that, a salient feature of each sparse coefficient vector is calculated by integrating the sparsity and the l^1-norm of the sparse coefficient vector. Next, the sparse coefficient vectors are fused by adopting the weighted average rule in which the weighted factors are proportional to the salient features of the sparse coefficient vectors. Finally, the fusion image is constructed by the fused coefficient vector with the learned redundant dictionary. Experiments show that the fusion algorithm is effective and superior to the method-based wavelet decomposition.

Keywords Sparse representations · Multi-focus images · Sparsity · Fusion rules

1 Introduction

Image fusion is integrating two or more images to create a single image in which all the objects are in focus, such that the new image can get better visibility with high credibility. This has got a significant importance in the image-processing field like segmentation, enhancement, and others.

The existing fusion methods can be divided into three categories, namely, pixel level, feature level, and decision level. So far, most research gives priority to the

Y. Zhang (✉) · Y. Chen
School of Electronic and Information Engineering,
Ningbo University of Technology, Ningbo, Zhejiang, China
e-mail: ypz@nbut.cn

Y. Chen
e-mail: chen_yaojia1115@126.com

S. Patnaik and X. Li (eds.), *Proceedings of International Conference on Computer Science and Information Technology*, Advances in Intelligent Systems and Computing 255, DOI: 10.1007/978-81-322-1759-6_7,

pixel-level fusion, which can be grouped into two aspects, first the special domain-based methods such as coping with image pixels using weighted average scheme, false color-mapping scheme, nonlinear method, Bayesian optimization scheme, artificial neural-network scheme [1] and methods based on transform such as pyramid decomposition, wavelet transform [2], curvelet transform, contourlet transform [3], empirical mode decomposition [4], independent component analysis [5], nonnegative matrix factorization [6], and so on.

Image fusion via multi-resolution transform of the source images has been concentrated as an effective fusion methodology. Firstly in the process, the source images are transformed into another domain. Next the coefficients are fused. At last, the fusion image can be obtained from the inverse transform. The transforms mainly belong to multi-resolution transform in essence, which cannot represent the image signal sparsely enough. On contrary, the sparse representation can accomplish it well because the signal is represented with overcomplete dictionary, namely, which is also deemed to be redundant dictionary. In recent years, the sparse representation have been widely applied to the image processing and analysis tasks including denoising [7], super-resolution [8], restoring and in-painting [9], and face recognition [10], due to its strong ability of sparse expansions.

An image-fusion approach based on sparse representation of each pixel neighborhood has been proposed [11], which really obtained immensely improved results. However, the algorithm brings about heavy computation.

In this paper, we present a new image-fusion scheme based on blocked sparse representation. Firstly, the source images are divided into patches and then are sparsely represented with learned dictionary. Secondly, the coefficient vectors are fused using the weighted average rules in which the weighted factors are calculated by the sparsity in combination with l^1-norm of the coefficient vector. Finally, the fusion image is constructed by the fused coefficient vector with the learned dictionary.

The rest of the paper is organized as follows. Section 2 gives a brief introduction to the sparse representations, dictionary learning, and sparse coding. Section 3 presents the proposed image-fusion framework and the fusion rules adopted in the approach. The experimental results including the performance evaluation are provided in Sect. 4 and the paper is concluded in Sect. 5.

2 Sparse-Coding Model

Recently, the increasing attention has been paid to the research of signal-sparse representation. The aim of sparse representation is to find the sparsest linear combination of the signal atoms in the redundant dictionary to represent the original signal under the condition that it can be compressed. If we view the image signal b as the one-dimensional vector, its sparse representation is also regarded as the low-dimension projection of the signal. The sparse-representation vector could be solved by figuring out the problem:

$$\widehat{x} = \arg\min_{x} ||x||_0 \quad \text{subject to } Ax = b \tag{1}$$

where $||x||_0$ denotes the number of nonzero components in x. $\widehat{x}$ is the sparse representation of the signal b with redundant dictionary A. Generally, the above description is a NP problem, yet solving the problem amounts to finding the solution of the following problem:

$$\widehat{x} = \arg\min_{x} ||x||_0 \quad \text{subject to } ||Ax - b||_2^2 < \varepsilon \tag{2}$$

where ε is a positive constant with small value.

2.1 Selection of the Dictionary

Choosing an appropriate dictionary is a crucial step toward to the sparse representation. There are two possible ways to construct the redundant dictionary. One way is to choose a predefined function as the dictionary, which seems straightforward and undemanding to realize. Another way is to learn a dictionary by using the original signals as the training data. The later can improve the sparseness of signal representation, which is adopted in the present paper.

The most typical design method of overcomplete dictionaries is K-SVD method [12]. It is an iterative method that alternates between sparse coding of the examples based on current dictionary and an update process for the dictionary atoms to better fit the signals. Furthermore, to make the dictionary much better fit the signals and also accelerate the representation process, plenty of improved training algorithms have been proposed [13].

2.2 Sparse Coding of Signals

Beside the choice of redundant dictionary, another key procedure of the sparse-coding model is how to design an effective and fast decomposition algorithm. A number of algorithms have been presented to find the sparse representation over the given dictionary. Those algorithms can be roughly divided into three main categories: greedy pursuit algorithms, l^p-norm regularization-based algorithms, and iterative shrinkage algorithms [14].

The orthogonal matching pursuit (OMP) algorithm is most representative of the greedy class algorithms, in which the corresponding atom is chosen when the inner product between it and the residual signal is the largest at each step. Finally, the loop terminates when the residual is less than the preset or the number of optional atoms exceed the preset number. In the paper, we employ the OMP algorithm and set the residual as the criterion of loop termination. For the sake of accuracy of sparse coding, the parameter ε in the formula (2) should be set as small as possible. Whereas, the smaller the value of ε, the more computation it brings. As a consequence, we should seek to a balance of two aspects.

3 Proposed Fusion Scheme

3.1 Fusion Framework Based on Sparse Coding

The natural images are capable of being represented with redundant dictionary via sparse land model. In the paper, the sparse representation is introduced to the multi-focus image fusion, and a new scheme is developed. Assuming that two source images have been registered, the proposed fusion framework based on sparse land model is shown in the Fig. 1, which includes the following main steps:

Step 1: In order to apply sparse coding with redundant dictionary, the original images are segmented into patches with the block size of 8×8, which is appropriate for the sparse coding [14], maintaining the balance between the sparse coding and the number of the patches. Then, the pixel values of every patch are lexicographic ordered into a column vector (as Fig. 2).

Step 2: The column vectors of two patches corresponding to the source images are represented using OMP over the learned redundant dictionary, resulting in two sparse coefficient vectors.

Step 3: The sparse coefficient vectors are merged into a new sparse coefficient vector as the sparse representation of the expected resultant image according to suitable rules.

Step 4: The final fusion image is reconstructed with the merged coefficient vector and the redundant dictionary.

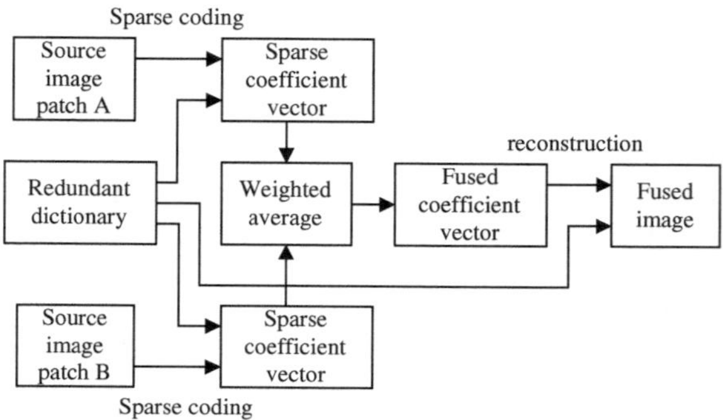

Fig. 1 Schematic diagram of the fusion algorithm proposed

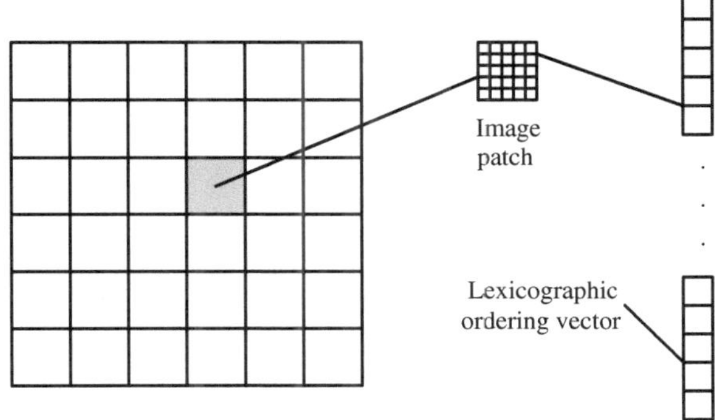

Fig. 2 Preprocess of resource images. Here, the pixel values of each 5 × 5 patch are lexicographic ordered into a 25-dimensional column vector

3.2 Fusion Rule

The fusion rule is of great importance to image fusion. The common rules include average and max-abs rule. In case that we adopt the max-abs rule directly, some blocking artifacts perhaps emerge in the fused image, which can influence image quality. Consequently, we adopt the average rule instead. However, the fusion image may lose part of edge details and the contrast of image may decrease in some degree, while we only use average rules directly.

Taking the above factors into account, here, a weighted average scheme is developed to fuse the two coefficient vectors. It is considered that the l^1-norm of the sparse coefficient vector reflect how much detail information they bring. Meanwhile, the sparsity of the coefficient vector can give expression to its concentration ratio of the detail information. In other words, the larger the l^1-norm and the smaller the sparsity of the coefficient vector, the better the visibility of corresponding image patch. For the above reasons, the weighting factors to be used in coefficients fusion are defined as follows:

$$w^A = \left(\|C^A\|_1/s^A\right)\Big/\left(\|C^A\|_1/s^A + \|C^B\|_1/s^B\right) \tag{3}$$

$$w^B = 1 - w^B, \tag{4}$$

where C^A is the coefficient vector in the sparse representation of image A, C^B is the coefficient vector in the sparse representation of image B, s^A represents the sparsity of the coefficient vector C^A and s^B represent the sparsity of the coefficient vector C^B, respectively. The coefficient vector for reconstructing the expected image F is calculated according to the following equation:

$$C^F = w^A C^A + w^B C^B \tag{5}$$

Fig. 3 Original images. **a** and **c** Image focus on *left*. **b** and **d** Image focus on *right*

4 Experimental Results

Two groups of registered images are tested in the experiment to check the effectiveness of the proposed method. Simultaneously, the results of proposed approach are compared with the wavelet-based fusion method and sparse-coding method using the average fusion scheme (as shown in Fig. 3). The wavelet basis is chosen with bior 2.4 in the wavelet-based method. In addition, we select maximal absolute values while choosing high-frequency coefficients and employ average scheme when choosing the low-frequency coefficients. In the dictionary, learning process of sparse-coding-based approach, we choose 30,000 natural image patches of 8×8 for training and the iterations are set to 80. In the sparse-coding process, the residual is set to 10^{-4} in advance. The results of three methods are shown in Figs. 4 and 5. It can be seen that the quality of the fusion image using proposed method is superior to the other two approaches.

Fig. 4 Fused results of Fig. 3a and b. **a** Wavelet-based method. **b** Sparse-coding-based method using average rules. **c** Proposed method

Fig. 5 Fused results of Fig. 3c and d. **a** Wavelet-based method. **b** Sparse-coding-based method using average rules. **c** Proposed method

Table 1 Values of indicators to evaluate the fused image using different methods

Methods	Figure 3a and b		Figure 3c and d	
	MI	*Q*	*MI*	*Q*
Wavelet	6.4343	0.6900	5.9680	0.6986
SR-average	7.1420	0.6726	6.5751	0.6695
Proposed method	7.2354	0.7125	7.4426	0.7475

At the same time, we employ the objective assessments mutual information *MI* [15] and edge information *Q* [16] to verify the proposed method. *MI* value reflects the information of the fusion image inherits from the source images. The higher the *MI* value, the better results we get. The magnitude of *Q* displays the edge information that the fused image retained from the source images. Similarly, the higher the *Q* value, the better fusion quality the algorithm provides. Table 1 gives the measures indexes *MI* and *Q*, which reveals that the proposed method emerges better performance than other two approaches.

5 Conclusion

A novel multi-focus image-fusion scheme based on blocked sparse representation is presented in the paper. Firstly, the original images are divided into patches and then are sparsely represented with learned dictionaries. Secondly, the coefficients are fused using the weighted average rules in which the weighted factors are calculated with the sparsity in combination with the module of the sparse coefficient vector. Finally, the fusion image is constructed by the fused coefficients with the learned dictionary. The experiments show that the fusion algorithm is effective and superior to the traditional method-based wavelet decomposition.

Acknowledgments This research is supported by NSF of China (No. 61203360) and NSF of Ningbo City (No. 2011B82012).

References

1. Li, S.T., Kwok, J.T., Wang, Y.N.: Multifocus image fusion using artificial neural networks. Pattern Recogn. Lett. **23**(8), 985–997 (2002)
2. Li, H., Manjunath, B.S., Mitra, A.A.: Multisensor image fusion using the wavelet transform. Graph. Models Image Proc. **57**(3), 235–245 (1995)
3. Hang, Q., Guo, B.L.: Mutifocus image fusion using the nonsubsampled contourlet transform. Signal Proc. **89**(7), 1334–1346 (2009)
4. Chen, S.H., Su, H.B., Zhang, R.H.: Improving empirical mode decomposition using support vector machines for multifocus image fusion. Sensors **8**(4), 2500–2508 (2008)
5. Mitianoudis, N., Stathaki, T.: Pixel-based and region-based image fusion schemes using ICA bases. Inf. Fusion **8**(2), 131–142 (2007)
6. Zhang, S., Chen, J., Miao, D.D.: An image fusion method based on WNMF and region segmentation. In: Computational Intelligence and Industrial Application, PACIIA'08, pp. 282–285 (2008)
7. Elad, M., Aharon, M.: Image denoising via sparse and redundant representations over learned dictionaries. IEEE Tran. Image Proc. **15**(12), 3736–3745 (2006)
8. Yang, J.C., Wright, J., Huang, T.: Image super-resolution via sparse representation. IEEE Tran. Image Process **19**(11), 2861–2873 (2010)
9. Marial, J., Elad, M., Sapiro, G.: Sparse representation for color image restoration. IEEE Tran. Image Process **17**(1), 53–69 (2008)
10. Wright, J., Ganesh, A., Yang, A.Y., Ma, Y.: Robust face recognition via sparse representation. IEEE Trans. Pattern Anal. Mach. Intell. **31**(2), 210–227 (2009)
11. Yang, B., Li, S.T.: Multifocus image fusion and restoration with sparse representation. IEEE Trans. Instrum. Meas. **59**(4), 884–892 (2010)
12. Aharon, M., Elad, M., Bruckstein, A.M.: The K-SVD: An algorithm for designing of overcomplete dictionaries for sparse representations. IEEE Tran. Image Process **54**(11), 4311–4322 (2006)
13. Rosenblum, K., Zelnik-Manor, L., Eldar, Y.C.: Dictionary optimization for block-sparse representations. IEEE Tran. Signal Proc. **6**(5) (2012)
14. Yang, J.Y., Peng, Y.G., Xu, W.L., Dai, Q.H.: Ways to sparse representation: An overview. Sci. China Ser. F Inf. Sci. **52**(4), 695–703 (2009)
15. Maes, F., et al.: Multimodality image registration by maximization of mutual information. IEEE Trans. Med. Imaging **16**(2), 187–198 (1997)
16. Xydeas, C.S., Petrovic, V.: Objective image fusion performance measure. Electron. Lett. **36**(4), 308–309 (2000)

The Studying of the Restoration of Defocused Images

Jinglin Wang

Abstract To get a better quality of defocused image restoration, the degradation function for the defocused image is studied first in this paper, second, various degradation functions were compared and analyzed, and last, a model that can represent its degradation accurately is proposed.

Keywords Degradation function · Defocus · Restoration · Spread function

1 Introduction

With the rapid development of science and technology, the application of various imaging sensors is growing rapidly. Digital image processing technology also enters a new era with the application, such as image enhancement, restoration, compression, and recognition and so on. Although many of the imaging system can achieve auto-focusing, the defocus show still exists. For example, the defocus is caused by the relative movement of the camera and the object. Different depths of the objects within the imaging region will result in generating defocused picture. The degradation model is shown in Fig. 1. Image restoration is the inverse process of image degradation [1].

Figure 1, $d(x, y)$ is the degraded image; $f(x, y)$ is the original image; $h(x, y)$ is the extended functions of the imaging system; and $n(x, y)$ is the additive noise. The added noise affect the original image, in some cases, it produces the degradation of the image. If the imaging system is linear shift invariant, the degradation process can be expressed as formula [2] (1):

J. Wang (✉)
Department of Information Engineering, The Second Artillery Engineering University, Xi'an, China
e-mail: Xlfu927@163.com

S. Patnaik and X. Li (eds.), *Proceedings of International Conference on Computer Science and Information Technology*, Advances in Intelligent Systems and Computing 255, DOI: 10.1007/978-81-322-1759-6_8,

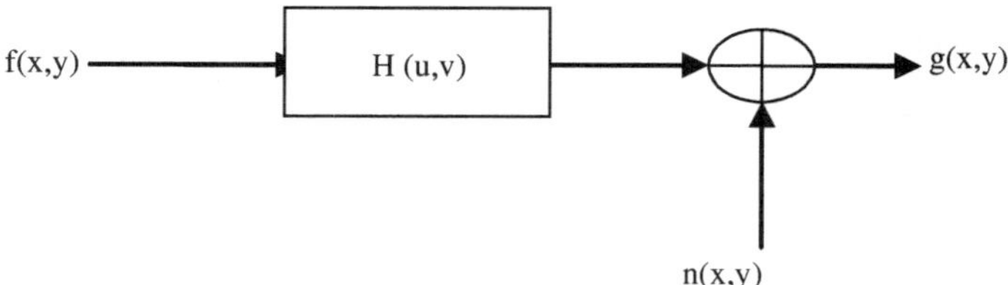

Fig. 1 Degradation model

$$d(x, y) = f(x, y) \times h(x, y) + n(x, y) \tag{1}$$

Expression of the corresponding frequency domain is as formula (1):

$$D(u, v) = F(u, v)H(u, v) + N(u, v) \tag{2}$$

The key problem to make a defocused image restored is how to determinate the function of the image degradation and its parameters. In this paper, some research work of the degradation function is presented.

2 Disk Defocus Model

Disk defocus model was concluded from the research of geometrical optics on the defocus image. And we can conclude from the geometrical optics that the point in the object space can be an image point on the condition that it could be formed in the image space after the ideal-imaging system. However, when the distance made by surface, mirror, or imaging plane in the actual situation does not satisfy Gaussian-imaging formula, the image point turned from the point in the object space will be a small spot. The schematic diagram of Disk defocus model is showed in Fig. 2 [3].

The disk from the point spread function of the naphthalene model can be approximated as a disk function.

$$h_r(x, y) = \begin{cases} 1/(pr^2)x^2 + y^2 = x^2 \\ 0 \end{cases} \tag{3}$$

In the formula (3), the blur radius r that needs to be solved is the only unknown parameter. After radius is determined, the expression of the degradation function can be got in the frequency domain. Then, the frequency-domain expression of the original image can be derived through the corresponding filter, and image restoration can be achieved through inverse Fourier change.

R calculation method will be given below:

From the Fourier transform of degradation model given by the formula [4] (3), we get:

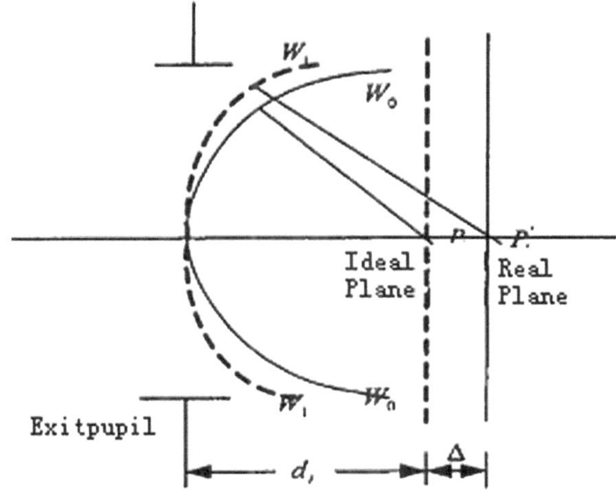

Fig. 2 Schematic model of disk defocus

$$H(u,v) = \frac{2J_1 R\sqrt{\left(\frac{2p}{M}u\right)^2 + \left(\frac{2p}{N}v\right)^2}}{\sqrt{\left(\frac{2p}{M}u\right)^2 + \left(\frac{2p}{N}v\right)^2}} \tag{4}$$

In the formula (4), $J_1[.]$ represents the first class of a first-order Bessel function, $M \times N$ is the size of the two-dimensional Fourier transform. According to the nature of the first class of a first-order Bessel function, we can know that the $H(u, v)$ is located in the frequency domain of the first dark ring, the first zero track is shown as formula (5)

$$2\pi r\sqrt{\left(\frac{u}{M}\right)^2 + \left(\frac{v}{N}\right)^2} = 3.85 \tag{5}$$

In the case of a relatively small noise, we can know by the formula [5] (5) that if found the u, v that is corresponding to the first zero (dark ring) of defocus blurred image Fourier transform, then the r can be obtained by formula (5).

3 Gauss Defocus Model

It is different between Gauss defocus model and Disk defocus model. It is not derived from the knowledge of optics, while an approximation of the model is obtained by consideration of various factors. The formula is expressed as follows [6]:

$$h(x,y) = \frac{1}{2\pi\sigma}\exp\left(-\frac{x^2 + y^2}{2\sigma^2}\right) \tag{6}$$

In the formula (6), the blur radius σ that needs to be solved is the only unknown parameter, which is determined according to the defocus blur image. And then, we can achieve defocus recovery by the filtering. The method that determines the parameters by defocus blurred image's zero-crossing point is not suitable for the Gaussian model. It can be determined by detecting the function of the curve edge in the spatial domain.

First, from the formula (6) of integral formula, we can obtain the line spread function that is still a Gaussian distribution, such as the formula (7) below:

$$l(x) = \frac{1}{2\pi\sigma}\exp\left(-\frac{x^2}{2\sigma^2}\right) \tag{7}$$

Then, from the formula (6) of integral formula, we can obtain the function of the blade edge:

$$e(x) = \int_{-\infty}^{+\infty} l(x)\mathrm{d}x = \frac{1}{\sqrt{2\mathrm{Ps}}}\int_{-\infty}^{+\infty} \exp\left(-\frac{x^2}{\mathrm{Ps}^2}\right)\mathrm{d}x \tag{8}$$

By the characteristics of the Gaussian integral, it can be seen that the $e(\sigma) = 0.84$, $e(-\sigma) = 0.16$. Therefore, if the function of the curve blade ranges from 0 to 1, then the x-axis interval is σ to $(-\sigma) = 2\sigma$, when the value of $E(x)$ is 0.84 or 0.16, the Gaussian parameter is obtained [7].

4 Improved Defocus Model

Improved defocus model has no fixed model, however, we can estimate the defocus point spread function directly according to the circular symmetry of the defocused image. In real life, the naked eye is sensitive to edge response than points and lines, and the straight-edged object response in the imaging system is very easy to be got. Therefore, the line spread function often is to be got by the response of the straight-edge object.

However, in some normal circumstances, a line spread function can not reflect the system characteristics completely, it needs to be measured in different directions. If the system is circularly symmetric, then the direction of the line spread function can clear reflect the system characteristics. After inverse Abel transform in the case of circularly symmetric, the point spread function can be determined by the arbitrary direction of the line spread function.

The point spread function $h(x, y)$ in the case of a circularly symmetric.

$$h(x,y) = h(r) \quad r = \sqrt{x^2 + y^2} \tag{9}$$

Because the line spread function is regardless of the direction in the case of circularly symmetric, so

$$l(x) = \int_{-\infty}^{+\infty} h(r)\mathrm{d}y \tag{10}$$

On the contrary, by the line spread function for the point spread function, we define a new function:

$$q(x) = \int_{-\infty}^{+\infty} l(r)\mathrm{d}y \quad r = \sqrt{x^2 + y^2} \tag{11}$$

The point spread function:

$$h(r) = -\frac{1}{2\pi r}\frac{\mathrm{d}q(x)}{\mathrm{d}r} \quad r = \sqrt{x^2 + y^2} \tag{12}$$

Merging Eqs. (11) and (12), the result is obtained:

$$h(r) = -\frac{1}{\pi}\frac{\mathrm{d}}{\mathrm{d}r}\left(\int_{r}^{\infty} \frac{l(x)}{x(x^2 - r^2)^{\frac{1}{2}}}\mathrm{d}x\right) \tag{13}$$

Formula (13) expresses the inverse Abel transform, and the Abel transform can be obtained from formula (10) where the $l(x)$ is derived by $h(r)$:

$$l(x) = 2\int_{x}^{\infty} h(r)(r^2 - x^2)^{-\frac{1}{2}}\, r\mathrm{d}r \tag{14}$$

In the calculation process, the upper limit of integration can be setted to the blur radius R (because of its circular symmetry, $h(x, y)$ can be nonzero only within a limited circular area) [8].

In the assumptions system, from the point spread function of Gaussian distribution, we can approximately estimate the blur radius R. By the striking of the Gaussian model parameters, we can determine σ, because the value is essentially zero when $|x| > 2\sigma$. Make $R0 = 2\sigma$ for the initial value of the blur radius, and find the real R near $R0$ by Fibonacci optimal search method. Then determine the point spread function and carry on defocus restoration.

5 Conclusion

In fact, most of the light spreads in no line former, which will cause the diffraction of light. Only when the defocus blur is really serious, the dispersion caused by the phase difference is greater than that of diffraction, and the diffusion function of disk model can play a good role. Although the Gaussian model is not derived through strict principles of optics, its zero points in frequency domain are relatively less, the singularity of inverse filtering can be eliminated and to avoid the error of the parameter. Though it is not the most close to actual model, the advantages mentioned are applied widely. And because the improved defocus model is closer to the degradation model, it will restore blurred images better and is more accurate, it is suitable for most of the blurred image restoration.

References

1. Zhenyu, W.: Restoration and identification of defocus blurred image based on in-focus degree. Infrared Laser Eng. **40**, 772–776 (2011)
2. Qin, F.: Blind image restoration based on Wiener filtering and defocus point spread function estimation. In: 5th International Congress on Image and Signal Processing, pp. 360–363, CISP 2012
3. Zhang, Y.J.: Image Processing and Analysis. Tsinghua University Press, Beijing (2005)
4. Cao, M., Sun, Y., Yu, D.: The clarity evaluation function of defocus blurred image. Instrument **22**(3), 259–268 (2001)
5. Sun, H., Zhang, B., Liu, J.H., Li, S.: Defocus blurred image of the Wiener filter restoration. Opt. Technol. **35**(3), 295–298 (2009)
6. He, Z., Zhang, Z., Zhu, G.: Defocused image enhancement technology. Infrared Millimeter Waves **120**(6), 447–450 (2001)
7. Liu, C., Wang, L., He, H.Z., Zhang, X.-F., Zhu, G.-X.: Velet-based decision of defocused image clarity. Comput. Appl. Softw. **7**, 239–240 (2008)
8. Chang, L., Ling, W.: Based on wavelet from focal blurred image clarity judgment. Comput. Appl. Softw. **25**(7), 239–240 (2008)

High-Order Total Variation-Based Image Restoration with Spatially Adapted Parameter Selection

Le Jiang, Jin Huang, Xiao-Guang Lv and Jun Liu

Abstract In this paper, we propose a high-order total variation model to restore blurred and noisy images with spatially adapted regularization parameter selection. The proposed model can substantially reduce the staircase effect, while preserving sharp jump discontinuities (edges) in the restored images. We employ an alternating direction minimization method for the proposed model. Some numerical results are given to illustrate the effectiveness of the proposed method.

Keywords Image restoration · High-order total variation · Augmented Lagrangian method · Staircase · Edge

1 Introduction

Image restoration plays an important part in various areas of applied sciences such as medical imaging, microscopy, astronomy, and film restoration [1, 2]. Typically, the image formation process can be modeled as

$$g = Hf + e, \tag{1}$$

where $f \in \mathbb{R}^{n^2}$ represents the ideal $n \times n$ image, $g \in \mathbb{R}^{n^2}$ represents the observed $n \times n$ image, $e \in \mathbb{R}^{n^2}$ represents the additive noise, and H is an n^2-by-n^2 matrix. The linear model (1) includes two classes of problem: image denoising, where H is

L. Jiang · J. Huang · X.-G. Lv (✉) · J. Liu
School of Mathematical Sciences, University of Electronic Science and Technology of China, Chengdu 610054 Sichuan, People's Republic of China
e-mail: xiaoguanglv@126.com

L. Jiang
School of Science, Huaihai Institute of Technology, Lianyungang 222005 Jiangsu, People's Republic of China

S. Patnaik and X. Li (eds.), *Proceedings of International Conference on Computer Science and Information Technology*, Advances in Intelligent Systems and Computing 255, DOI: 10.1007/978-81-322-1759-6_9,

the identity matrix and image deblurring, where H is typically ill-conditioned matrix representing the blurring phenomena.

It is known that restoring f from g is a typical ill-conditioned inverse problem and a direct solution to the linear system $Hf = g$ often does not yield meaningful solutions. A total variation (TV)-based regularization technique (ROF model), proposed by Rudin et al. [3], is a very popular regularization method for stable and accurate solutions. The main advantage of the ROF model is its capability to recover sharp edges in the restored images. In this approach, we need to solve the following minimization problem:

$$\min_f \| Hf - g \|_2^2 + \lambda \| \nabla f \|_1, \tag{2}$$

where λ is the regularization parameter balancing the trade-off between the data-fitting term $\| Hf - g \|_2^2$ and the regularization term $\| \nabla f \|_1$. The issue of the discrete gradient ∇f can be found in [4, 5]. It is well known that the major difficulty of the ROF model is the high nonlinearity and non-differentiability of the object function. To address the problem, many efficient and robust methods have been proposed; see [6, 7] for more details.

Although the TV regularization is extremely popular in a variety of applications, it has been shown that the TV norm transforms smooth signal into piece-wise constants, the so-called staircase effect. To attenuate the staircase effect, there is a growing interest in the literature for replacing the TV norm by a high-order TV norm. The motivation behind such attempt is to restore potentially a wider of images, which comprise more than merely piece-wise constant regions. Second-order regularization schemes have been considered so far in the literature mainly for dealing with the staircase effect while preserving the edge information in the restored image [8, 9].

2 Problem Formulation

Following a variational approach, we consider the hybrid model:

$$\min_f \frac{1}{2} \| Hf - g \|_2^2 + \lambda(\theta \| \nabla f \|_1 + (1-\theta) \| \nabla^2 f \|_1), \tag{3}$$

where the parameter $\theta \in [0, 1]$ is used to control the balance between the edges and the smooth surface, $\| \nabla^2 f \|_1 = \sum_{i=1}^{n^2} |\nabla^2 f_i|$ with $\nabla^2 f = (f_{xx}, f_{xy}; f_{yx}, f_{yy})$ and $|\nabla^2 f| = \sqrt{f_{xx}^2 + f_{xy}^2 + f_{yx}^2 + f_{yy}^2}$.

Clearly, the regularization λ is an important quantity that controls the properties of the regularized solution and should be chosen with care. Throughout the years, a variety of parameter choice strategies such as the discrepancy principle, the L-curve, and generalized cross-validation (GCV) have been developed [10]. In particular, the TV models with a spatially varying choice of parameters were considered in [11, 12]. Motivated by these works, Dong et al. [13] introduced a spatially dependent regularization parameter selection scheme to enhance image regions containing details while still sufficiently smoothing homogeneous features. The fully automated adjustment strategy of the regularization parameter is based on local variance estimators. In this work, we adopt a similar automated spatially adapted parameter selection technique.

In addition, to emphasize the restoration properties for the second-order TV in smooth regions, we want $0 \leq \theta < 1$. In this work, we adopt the method for updating θ as proposed in [9]. The updating procedure behaves better for our model than the fixed θ. It is because, as the iteration proceeds, the edges and smoothing regions of recovered image are closer to the original image, then the parameter θ computed by the updating scheme can be better suitable for restoration.

Before deriving the algorithm for solving (3), we describe how to choose the spatially adapted regularization parameter. For the sake of simplicity, we consider the continuous case. Similar to [13], we define a normalized filter:

$$\omega(x, y) = \begin{cases} \frac{1}{|\Omega_x^r|}, & \text{if } \|y - x\|_\infty \leq \frac{r}{2} \\ 0, & \text{else} \end{cases}$$

where $x \in \Omega$ is fixed, Ω_x^r is a local window centered at pixel x that is defined by $\Omega_x^r = \{y : \|y - x\|_\infty \leq \frac{r}{2}\}$ and $r > 0$ sufficiently small is the essential width of the filter window. By $F(u)(x)$, we denote the local expected value estimator, which is

$$F(u)(x) = \int_\Omega \omega(x, y)(Hf - g)^2(y)\mathrm{d}y. \tag{4}$$

Hence, we obtain the following TV-based minimization problem with local constraints based on formula (4):

$$\min_{u \in S(\Omega)} \int_\Omega (\theta|\nabla f| + (1 - \theta)|\nabla^2 f|)dx \quad \text{s.t. } F(u) \leq 1 + \varepsilon \quad \text{a.e. in } \Omega \tag{5}$$

where "a.e." stands for "almost everywhere." The proof of the existence of a solution in (5) can be obtained using a similar technique in [13].

In the discrete form, let $\Omega_{i,j}^r$ denote the set of pixel coordinates in a r-by-r window centered at (i, j) with a symmetric extension at the boundary, which is

$$\Omega_{i,j}^r = \{(s + i,\ t + j) : -r - 1/2 \leq s,\ t \leq r - 1/2\}.$$

Hence, we obtain the local expected value estimator:

$$S_{i,j}^r = \frac{1}{r^2} \sum_{(s,t) \in \Omega_{i,j}^r} \left(g_{s,t} - (Hf)_{s,t}\right)^2.$$

Therefore, we can propose the similar update scheme of λ as follows:

$$\frac{1}{\tilde{\lambda}_{i,j}^{k+1}} = \frac{1}{\tilde{\lambda}_{i,j}^{k}} + \delta\max\left((S_k)_{i,j}^{r} - \sigma^2, 0\right), \tag{6}$$

$$\frac{1}{\lambda_{i,j}^{k+1}} = \frac{1}{r^2}\sum_{(s,t)\in\Omega_{i,j}^{r}} \frac{1}{\tilde{\lambda}_{s,t}^{k+1}}, \tag{7}$$

where $\delta > 0$ is a step size.

3 Implementation Details

In this section, we deduce the algorithm for the spatially adapted model (3) based on the augmented Lagrangian (AL) method. Consider the following unconstrained optimization problem:

$$\min_u \Psi(u) + \Phi_1(D_1 u) + \Phi_2(D_2 u), \tag{8}$$

where $\Psi(f) = \frac{1}{2}\|Hf - g\|_2^2$, $\Phi_1(D_1 f) = \lambda\theta\|\nabla f\|_1$, and $\Phi_2(D_2 f) = \lambda(1-\theta)\|\nabla^2 f\|_1$. Introducing new variables $v_i = D_i u$ for $i = 1, 2$, we can rewrite (8) as a constrained problem of the form:

$$\min_{u,v} \Psi(u) + \Phi_1(v_1) + \Phi_2(v_2), \text{ s.t. } D_i u = v_i \quad \text{for } i = 1, 2. \tag{9}$$

The associated AL function for this problem is defined as

$$\mathcal{L}(u, v, b, \eta) = \Psi(u) + \sum_{i=1}^{2}\Phi_i(v_i) + \frac{\eta}{2}\sum_{i=1}^{2}\| D_i u - v_i - b_i \|_2^2, \tag{10}$$

where η is related to the Lagrange multiplier for the constraint in (9). The idea of the AL method is to find a saddle point of $\mathcal{L}$, which is also the solution of the original problem (8). We can use the alternating direction method to iteratively solve the following subproblems:

$$\begin{aligned} u^{k+1} &= \operatorname{argmin}_u \Psi(u) + \tfrac{\eta}{2}\sum_{i=1}^{2}\| D_i u - v_i - b_i \|_2^2, \\ v_i^{k+1} &= \operatorname{argmin}_{v_i} \Phi_i(v_i) + \tfrac{\eta}{2}\| D_i u^{k+1} - v_i - b_i^k \|_2^2, \quad \text{for } i = 1, 2 \\ b^{k+1} &= b^k - \left(Du^{k+1} - v^{k+1}\right) \end{aligned} \tag{11}$$

where $D = \begin{pmatrix} D_1 \\ D_2 \end{pmatrix}$, $b = \begin{pmatrix} b_1 \\ b_2 \end{pmatrix}$ and $v = \begin{pmatrix} v_1 \\ v_2 \end{pmatrix}$.

For the first subproblem, it is required to solve the following normal equation:

$$\left(H^T H + \eta D_1^T D_1 + \eta D_2^T D_2\right)f = H^T g + \eta D_1^T\left(v_1^k + b_1^k\right) + \eta D_2^T\left(v_2^k + b_2^k\right). \tag{12}$$

Under the periodic boundary condition, D_1, D_2 and H have block circulant with circulant blocks (BCCB) structure that can be diagonalized by fast Fourier transforms (FFTs); see [5] for more details. For the second subproblem, we need to solve the following problem:

$$\min_{v_i} \Phi_i(v_i) + \frac{\eta}{2} \| D_i u^{k+1} - v_i - b_i^k \|_2^2, \quad \text{for } i = 1, 2. \tag{13}$$

It is known that the problem can be solved using a shrinkage formula [7].

Above all, we have the following AL method for image restoration with spatially adapted parameter selection.

Algorithm 1. (*AL method for image restoration with spatially adapted parameter selection.*) Input: $f^0 = 0$, λ^0 and $k = 0$.

1. If $k = 0$, employ the AL method with $\lambda = \lambda_0$ to compute:

$$\hat{f}^0 = \text{argmin}_f \frac{1}{2} \|Hf - g\|_2^2 + \lambda(\theta \|\nabla f\|_1 + (1 - \theta)\|\nabla^2 f\|_1),$$

else compute $\hat{f}^k$ by Algorithm 1 with $\lambda = \lambda^k$ and $v^k = g - Hf^k$:

$$\hat{f}^k = \text{argmin}_f \frac{1}{2} \|Hf - v^k\|_2^2 + \lambda^k(\theta \|\nabla f\|_1 + (1 - \theta)\|\nabla^2 f\|_1),$$

2. Update $f^{k+1} = f^k + \hat{f}^k$.
3. Update λ^k based on (6) and (7).
4. Checking the stopping criteria.

4 Numerical Experiments

In this section, we give numerical results to illustrate the performance of the proposed approach for image restoration by comparing the proposed algorithm with the SATV algorithm presented in [13]. All computations were carried out in Matlab 7.10 on a PC with an Intel(R) Core(TM) i3-2130 CPU 3.4 GHz and 4 GB of RAM. The initial guess is chosen to be the zero matrix in all tests. In order to have a fair comparison between SATV and our method, we concentrate on image denoising and use the window size $r = 11$ as reported in [13], i.e., H is the identity matrix. In addition, we choose $\eta = 5$ for our method.

The quality of the restoration results by different methods is compared quantitatively using the peak-signal-to-noise ratio (PSNR) and structural similarity index (SSIM) that is developed by Wang et al. [14] and is a well known quality metric used to measure the similarity between two images.

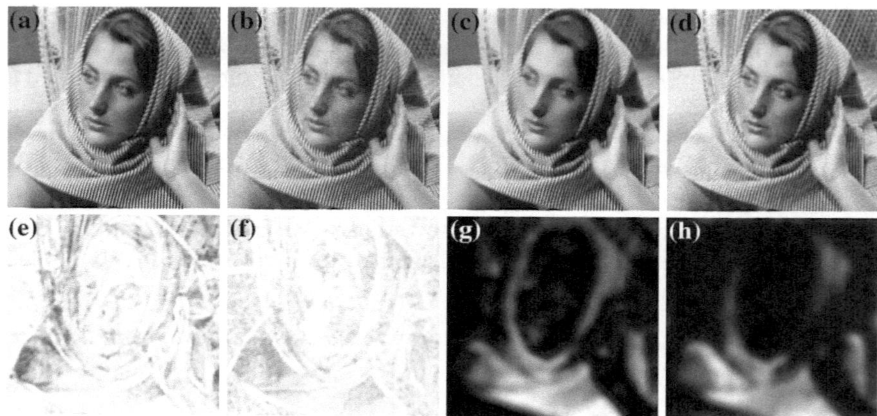

Fig. 1 Results of the Barbara image. **a** Original image. **b** Degraded image. **c** Restored image by SATV (CPU time: 57.09 s, PSNR = 29.33). **d** Restored image by the proposed method (CPU time: 15.38 s, PSNR = 30.00). **e** SSIM map by SATV (SSIM = 0.87). **f** SSIM map by the proposed method (SSIM = 0.91). **g** Final value of λ in SATV. **h** Final value of λ in the proposed method

Suppose f and $\tilde{f}$ are the original image and the restored image, respectively. The PSNR and SSIM are defined as follows:

$$\mathrm{PSNR} = 20\log_{10}\left(\frac{255n^2}{\| \tilde{f} - f \|_2}\right), \quad \mathrm{SSIM} = \frac{\left(2\mu_f\mu_{\tilde{f}} + C_1\right)\left(2\sigma_{f\tilde{f}} + C_2\right)}{\left(\mu_f^2 + \mu_{\tilde{f}}^2 + C_1\right)\left(\sigma_f^2 + \sigma_{\tilde{f}}^2 + C_2\right)},$$

where μ_f and $\mu_{\tilde{f}}$ are averages of f and $\tilde{f}$, respectively. σ_f and $\sigma_{\tilde{f}}$ are the variance of f and $\tilde{f}$, respectively. $\sigma_{f\tilde{f}}$ is the covariance of f and $\tilde{f}$. The positive constants C_1 and C_2 can be thought of as stabilizing constants for near-zero denominator values.

In the first test, we consider the "Barbara" image with size 256 × 256. We add 5 % Gaussian noise to the original image to generate the degraded image. The ideal image and the degraded image are shown in Fig. 1a, b. The restored images by SATV and our method are shown in Figs. 1c and 2d, respectively. It is not difficult to observe that the restored image by our method contains more details. We observe that the CPU time of the proposed method is much less than that of SATV. We show the SSIM maps and the plots of λ in Fig. 1e–h. We see from Fig. 1e, f that the SSIM map of the restored image by our method is whiter that by the SATV algorithm, i.e., our method can get better restoration results.

In the second example, the 256 × 256 "Lena" image shown in Fig. 2a is corrupted with the Gaussian white noise with 5 %. The observed image is displayed in Fig. 2b. The restored images by SATV and the proposed method are shown in Fig. 2c, d, respectively. From the figures, compared with SATV, the proposed method yields better results in the restored image. From Fig. 2e, f, we know that the SSIM value of the restored image by the proposed method is higher

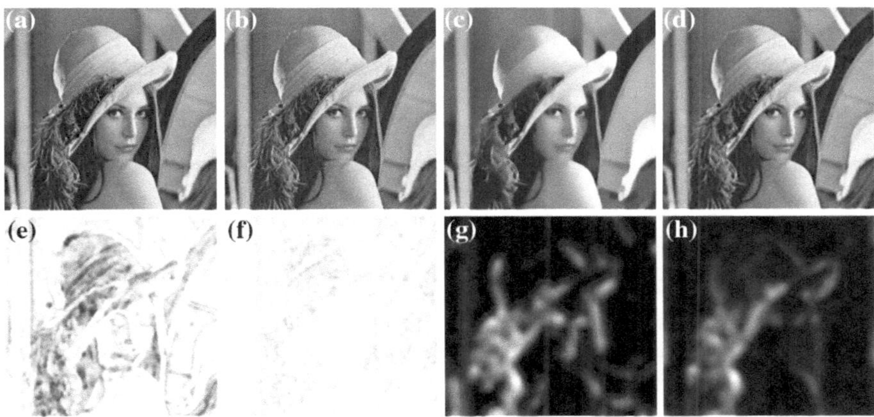

Fig. 2 Results of the Cameraman image. **a** Original image. **b** Degraded image. **c** Restored image by SATV (CPU time: 54.40 s, PSNR = 28.21). **d** Restored image by the proposed method (CPU time: 14.92 s, PSNR = 31.88). **e** SSIM map by SATV (SSIM = 0.82). **f** SSIM map by the proposed method (SSIM = 0.92). **g** Final value of λ in SATV. **h** Final value of λ in the proposed method

than the SATV method. With respect to the SSIM map, CPU time, and PSNR, we obtain that the performance of the proposed method is superior to that of the SATV method.

Acknowledgments This research is supported by NSFC (10871034, 61170311), Sichuan Province Science and Technology. Research Project (12ZC1802) and Research Project of Jiangsu Province (12KJD110001, 1301064B), China Postdoctoral Science Foundation funded project (2013M540454).

References

1. Andrew, H., Hunt, B.: Digital Image Restoration. Prentice-Hall, Englewood Cliffs (1977)
2. Chan, T.F., Shen, J.H.: Image Processing and Analysis: Variational, PDE, Wavelet, and Stochastic Methods. SIAM, Philadelphia (2005)
3. Rudin, L., Osher, S., Fatemi, E.: Nonlinear total variation based noise removal algorithms. Phys. D **60**, 259–268 (1992)
4. Chambolle, A.: An algorithm for total variation minimization and applications. J. Math. Imag. Vis. **20**, 89–97 (2004)
5. Wang, Y.L., Yang, J.F., Yin, W.T., Zhang, Y.: A new alternating minimization algorithm for total variation image reconstruction. SIAM J. Imag. Sci. **1**(3), 248–272 (2008)
6. Vogel, C.R.: Computational Methods for Inverse Problems. SIAM, Philadelphia (2002)
7. Beck, A., Teboulle, M.: Fast gradient-based algorithms for constrained total variation image denoising and deblurring problems. IEEE Trans. Image Process. **18**(11), 2419–2434 (2009)
8. Lefkimmiatis, S., Bourquard, A., Unser, M.: Hessian-based norm regularization for image restoration with biomedical applications. IEEE Trans. Image Process. **21**(3), 983–995 (2012)

9. Lysker, M., Tai, X.C.: Iterative image restoration combining total varition minimization and a second-order functional. Int. J. Comput. Vis. **66**(1), 5–18 (2006)
10. Hansen, P.C., Nagy, J.G. O'Leary, D.P.: Deblurring Images: Matrices, Spectra, and Filtering. SIAM, Philadelphia (2006)
11. Strong, D., Aujol, J.F., Chan, T.F.: Scale Recognition, Regularization Parameter Selection, and Meyer's G Norm in Total Variation Regularization, Technical Report. UCLA, Los Angeles (2005)
12. Almansa, A., Ballester, C., Caselles, V., Haro, G.: A TV based restoration model with local constraints. J. Sci. Comput. **34**(3), 209–236 (2008)
13. Dong, Y.Q., Hintermüller, M., Rincon-Camacho, M.M.: Automated regularization parameter selection in multi-scale total variation models for image restoration. J Math. Imag. Vis. **40**(1), 82–104 (2011)
14. Wang, Z., Bovik, A.C., Sheikh, H.R., Simoncelli, E.P.: Image quality assessment: From error visibility to structural similarity. IEEE Trans. Image Process. **13**(4), 600–612 (2004)

Research of Vehicle Identification Based on Adaptive Corner Detection

Wenju Yuan, Yuankun Jiang and Fang Cai

Abstract The importance of vehicle and type recognition is getting more and more significant in highway toll system and surveillance of city road. Considering the important factors of accuracy and steady performance cannot meet the requirements simultaneously in current vehicle recognition, the article puts forward a kind of method to identify models using adaptive Harris corner detection operator. Compared with the traditional method of corner detection, it can reduce the influence of the noise and angular point recognition error effectively. Through this, we can deal with vehicle recognition problem better.

Keywords Vehicle type recognition · Harris corner · Hausdorff · Image matching

1 Introduction

With the rapid development of modern transportation, the types of models are become more and more, the study of the vehicle quickly identify has become a focus in intelligent transportation system (ITS). The existing vehicle type identification technology mainly uses the vehicle itself, such as the color, length, width, height, contour, and license plate of the car. Due to the diversity of models and influence of noise factors on identification, the vehicle recognition accuracy and

W. Yuan (✉) · Y. Jiang · F. Cai
College of Computer Science and Engineering, Changchun University of Technology, Changchun, China
e-mail: lestatldu@126.com

Y. Jiang
e-mail: lestatldu@163.com

F. Cai
e-mail: foreverqing.tian@163.com

S. Patnaik and X. Li (eds.), *Proceedings of International Conference on Computer Science and Information Technology*, Advances in Intelligent Systems and Computing 255, DOI: 10.1007/978-81-322-1759-6_10,

real-time performance are difficult to meet the actual requirements. The literature [1] used length and height of the car body to simplify the models for "work" type, extracted characteristics of the car body length and higher, then executed neural network analysis to identify. It can solve some simple models for identification, but the identification effect of complex models is not beautiful. The literature [2] extracted the models features using the method of single-scale Harris corner, but it was difficult to accurately extract the full right corners, limiting the use in practice.

The corner contains important local features of recognition target, and it is able to determine the shape of the target image accurately [3]. Though traditional Harris corner detection suppresses the corner metric greatly, the single threshold setting has great impact on the corner detection of the target image and results in a loss of information corner, and thus, it cannot get the car corner profile accurately [4].

In this paper, body image corner is extracted using an adaptive Harris corner method. In order to ensure the uniform and accurate corner extraction, this paper use the method of blocking of target image and eliminating adjacent corner points to avoid threshold setting. Vehicle recognition process roughly: Through corner extraction of standard sample vehicles to get distribution of corner points, and then calculate the Hausdorff distance between the test vehicles and several types of standard sample corner, the same model has the smallest distance. Through experimental analysis, the approach used in this article can be a good solution to the identification of the types of models. The recognition accuracy is also improved.

2 Adaptive Harris Corner Detection

2.1 Harris Corner Detection Principle

Harris C and Stephens MJ proposed Harris corner detection algorithm, it is developed based on Moravec algorithm, and is a operator of point features extraction based on signal. To determine the variations in the image signal based on the local auto-correlation function of the signal is the main idea of the Harris operator. By differential operation and auto-correlation matrix to detect the corner, the operator is easier to calculate. The corner detected becomes more reasonable and accurate [5]. The algorithm R is as follows:

$$R = \det(C) - ktr^2(C) \tag{1}$$

where $Iu_{(x)}, Iv_{(x)}, Iuv_{(x)}$, are partial derivatives and second-order mixed partial derivatives about the gray level of image point x in the direction of u and v; The experience of the k value is usually 0.4 [6].

2.2 Adaptive Harris Corner Detection Algorithm

Traditional corner detection method is defined as follows: When a point Harris operator R is greater than a fixed threshold value T, then this point is corner point. While testing the car body, however, the corner around the threshold would be missed due to the fixed threshold T [7]. Thus, the corner location accuracy is not high, which has a significant impact on the latter part of the identification process. So this article adopted an adaptive Harris corner detection method to apply to the car body corner detection. Specific method is as follows: First, calculate the gradient of the image pixels in the horizontal and vertical directions, the product of the two, and get the value of the four elements in the C.

$$C(x) = \begin{bmatrix} I_u^2(x) & I_{uv}(x) \\ I_{uv}(x) & I_v^2(x) \end{bmatrix} \tag{2}$$

where $I_u^2 = I_u \times I_u$; $I_v^2 = I_v \times I_v$.

Then, partial correlation function obtained by Gaussian filtering is the new C. Calculated the interest values of each pixel corresponding to the original image:

$$R = \left\{ I_u^2 \times I_v^2 - (I_u I_v)^2 \right\} - k\{I_u^2 + I_u^2\}. \tag{3}$$

Thirdly, block the image and detect the present corner in subblock image. The R value of $H(i, j)$ $(C(i, j) = 1,\ i \in M1,\ j \in N1)$, where M_1 and N_1 are the width and height of the image block, respectively) is stored in the array matrix [Sum], where Sum is the size of the array and also represents the number of corner. So there are $k \times$ Sum angular points in Sum, where $k \in (0, 1]$.

This article uses the loop iteration algorithm to solve the k value: Firstly, select a smaller value as the initial value of k in (0, 1), $k = k +$ step for loop iteration, and judge the value of k, and if $k = 1$, the iteration is terminated. Otherwise, it is determined whether there have the retainer corner point on each image block containing the corner points. If it is there, the iteration is terminated and chooses the k value at this time. Finally, use 3×3 operation template to cull adjacent corners on the whole image.

Adaptive corner detection process is shown in Fig. 1; From the comparison chart of initial detection on the body side corner and adaptive Harris corner detector can be seen that the adaptive corner detection is good for eliminating redundant corner points, and the corner point positioning is more accurate, as shown in Fig. 2:

3 Vehicle Classification and Recognition

The vehicle identification methods are divided into two categories: the method based on pattern recognition and image-based matching method. This article adopts the method of image matching. First of all, according to the CCD camera to

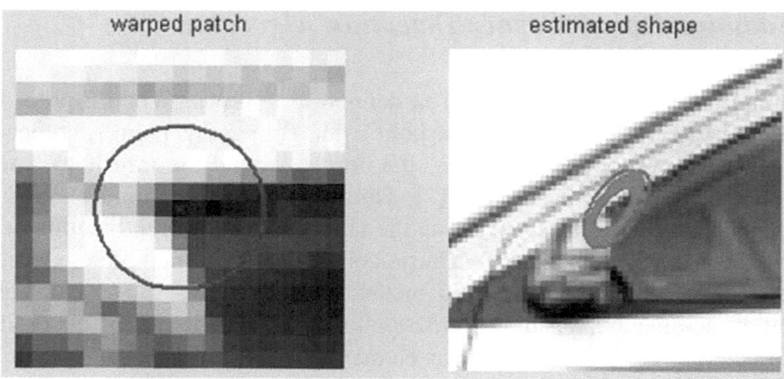

Fig. 1 Adaptive corner detection process

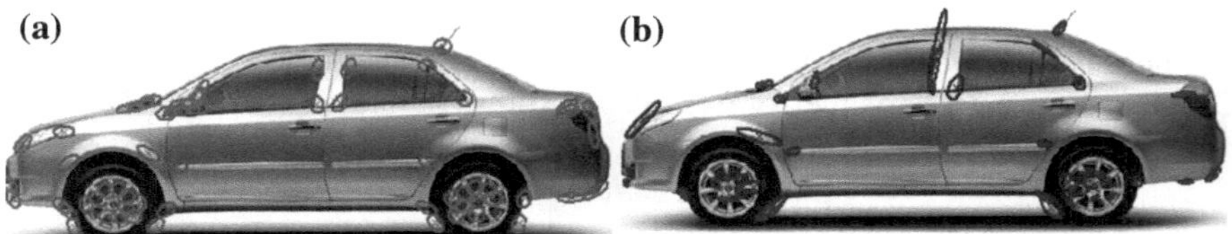

Fig. 2 **a** Fixed threshold corner detection. **b** Adaptive corner detection

obtain the actual situation of image sequences, the vehicle can be divided into three types of cars, buses, trucks, respectively select the standard Harris corner images of the side of the three types of vehicle as the three types of standard sample images. And then obtain target vehicle through background subtraction method. Finally, calculate Harris corner points using Hausdorff distance between target vehicle and standard sample to judge models. Flowchart of vehicle recognition is shown in Fig. 3 [8].

3.1 Vehicle Matching Based on Hausdorff Distance

Hausdorff distance is used to describe a measure of similarity degree between the two sets of point sets. In image recognition, the common Hausdorff distance indicates the similarity degree of the target and template, and the smaller the distance, the more similar the distance between the target and template [9].

Hausdorff distance between A and B is defined as follows:

$$H(A, B) = \max(h(A, B), h(B, A)) \tag{4}$$

$A = \{a_1, a_2, \ldots\}$ and $B = \{b_1, b_2, \ldots\}$ are finite number of two point sets.

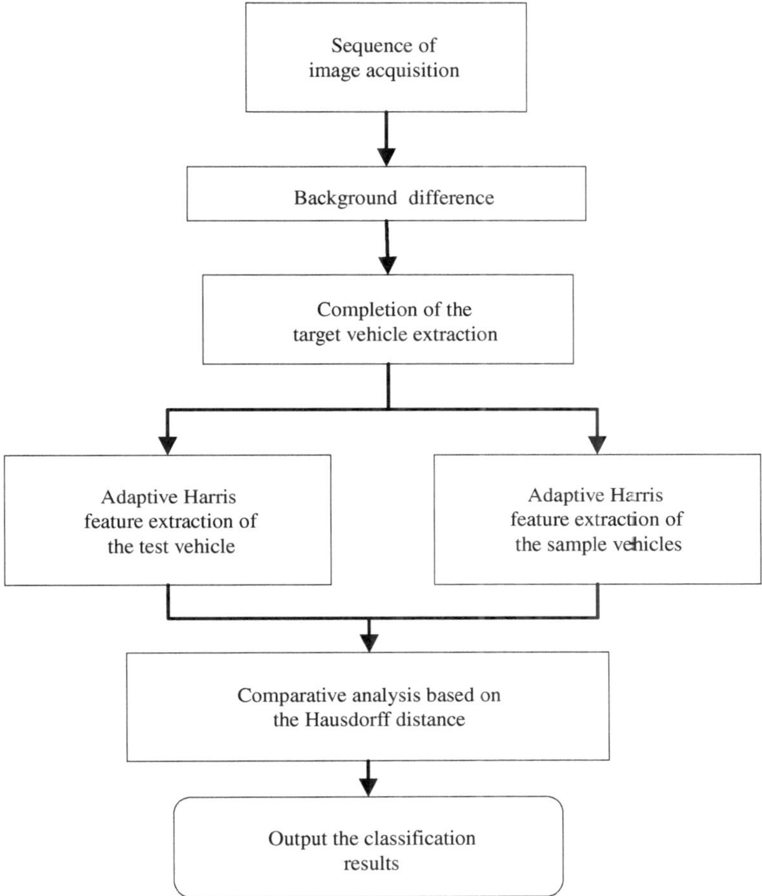

Fig. 3 Model identification flow chart

where, $h(A, B)$ and $h(B, A)$ is referred to the distance between A and B, defined as:

$$\begin{aligned} h(A, B) &= \max_{a \in A}\{\min_{b \in B}\|a - b\|\} \\ h(B, A) &= \max_{b \in B}\{\min_{a \in A}\|a - b\|\} \end{aligned} \tag{5}$$

$H(A, B)$ reflects the degree of not matching between sets A and B.

So we can calculate target vehicle Harris corner points and three kinds of target vehicle Harris corner of Hausdorff distance to determine target models [10]. Comparative analysis of the Hausdorff distance, if the Hausdorff distance of two points is minimum, the vehicles is same model. The distribution of corner points in three kinds of standard sample are shown as follows (Fig. 4).

Fig. 4 Distribution of corner points in three kinds of sample models

3.2 Experimental simulation and analysis

This paper uses the MATLAB simulation platform to simulate the vehicle model identification process based on adaptive corner Hausdorff distance [11]. Experiment 1 analyzes the relationship between the Hausdorff distance and models [12]. Experiment 2 analyzes models of 200 vehicles through the proposed method. Process analysis is as follows:

Table 1 is the matching calculated value of Harris corner between the selected three tested vehicles and three standard samples, we can directly determine recognition models through the Hausdorff distance size. Such as the test vehicle B, the Hausdorff distance of its corners with the standard car corners is 5. Compared with the standard passenger car and truck, its Hausdorff distance is minimum, so it is determined that the vehicle to be tested for cars. Also other two samples of the models are obtained. Table 2 is vehicle recognition results and process time-consuming of the proposed method on 200 samples. Recognition success rate on cars and trucks is higher, and buses recognition rate is relatively low, it has achieved the overall requirements for three kinds of vehicle recognition. The analysis confirmed the validity and accuracy of the method of the proposed corner features extraction on vehicle identification.

Table 1 Hausdorff distance of the vehicle under test and the standard sample

Standard models	Vehicles to be identified	Hausdorff (hi)	Recognition results
Standard cars	A	14	
	B	5	Car
	C	11	
Standard buses	A	22	
	B	19	
	C	6	Bus
Standard truck	A	7	Truck
	B	28	
	C	17	

Table 2 Test results and success rate

Model	Car	Truck	Bus
Sum	78	55	67
Wrong numbers	3	1	5
Recognition rate (%)	96.15	98.18	92.54
Simulation time-consuming	0.258	0.276	0.286

4 Conclusion

In this paper, on the basis of existing problems of models recognition technology, the adaptive Harris corner detection method is applied to vehicle identification based on improving Harris corner algorithm. Compared with traditional methods, the improved method can measure the more evenly distributed corners and more effectively avoid inaccurate identification of vehicle recognition that are caused by the inaccurate identification of corners, with advantages of small amount of calculation and high recognition accuracy. But there were some errors on the Hausdorff distance judgment, and the recognition process of Hausdorff algorithm still needs to be improved.

References

1. Zhao, W., Zhang, Y.: Corner detection technology review. Appl. Res. Comput. **23**(10), 17–19 (2006)
2. Du, Y., Gao, H.: Vehicle type recognition method based on contour matching positioning vehicle. J Yangzhou Univ. Nat. Sci. Ed. **10**(2), 62–65 (2007)
3. Li, J., Zhao, J.: A vehicle type recognition algorithm based on image processing. J. Xi'an Univ. Technol. **29**(3) (2009)
4. Du, H.: Matlab based on the vehicle recognition BP algorithm. Comput. Modernization **20**(5), 03 (2012)

5. Liu, Y., You, Z.: A neural network for image target recognition and vehicle recognition applications. Comput. Eng. **29**(4), 30–32 (2003)
6. Lowe, D.: Distinctive image features from scale-invariant key-points. Int. J. Comput. Vis. **60**(2), 91–110 (2004)
7. Fu, Y., Huang, T.S.: Image classification using correlation tensor analysis. IEEE Trans. Image Process. **17**(2), 226–234 (2008)
8. Murtagh, F., Starck, J.L.: Wavelet and curvelet moments for image classification: application to aggregate mixture grading. Pattern Recogn. Lett. **29**(10), 1557–1564 (2008)
9. Lu, Z.W., Ip, H.H.S.: Image categorization with spatial mismatch kernels. In. IEEE Conference on Computer Vision and Pattern Recognition, pp. 397–404 (2009)
10. Bishop, C.: Pattern Recognition and Machine Learning. Springer, New York (2006)
11. Everingham, M., Gool, L.V., Williams, C., Winn, J., Zisserman, A: The PASCAL Visual Object Classes Challenge 2010 Results (2010)
12. Khvedchenya, E.: Feature descriptor comparison report (2011)

Knowledge-Aided Bayesian Optimum Radar Detector

Hongsen Xie, Jinbo Shi, Huaming Tian, Baokuan Luan and Peng Zhou

Abstract In this paper, we consider the optimum detection in non-Gaussian clutter characterized as spherically invariant random vector (SIRV). SIRV models non-Gaussian vector as a complex Gaussian vector whose variance, the so-called texture, is itself a positive random variable, and the texture describes the non-Gaussianity of clutter. In this paper, using inverse Gamma distribution as the distribution model of texture, the knowledge-aided (KA) Bayesian optimum radar detector (BORD) is proposed. The performance of the KA-BORD is analyzed both on simulated and on real radar data collected by the McMaster University IPIX radar. The result shows that KA-BORD can outperform the other detectors, and its detection performance is close to that of optimum detector.

Keywords Non-Gaussian · Signal detection · Knowledge-aided · Bayesian detection

1 Introduction

In the field of radar signal processing, the signal detection under non-Gaussian clutter is always a focus of research [1, 2]. In many situations, the statistical property of non-Gaussian clutter can be approximated as a compound Gaussian statistical process. The compound Gauss process $c(t)$ can be expressed as the product of Gaussian process $g(t)$ and a non-negative random process $\sqrt{\tau(t)}$. The random process $\sqrt{\tau(t)}$ is called the texture component, and the random process $g(t)$ is called speckle component. The decorrelation time of the texture is much

H. Xie (✉) · J. Shi · H. Tian · B. Luan · P. Zhou
Navigation Engineering Department, Naval Aeronautical Engineering Institute Qingdao Branch, Qingdao, China
e-mail: qdrfeng@sohu.com

S. Patnaik and X. Li (eds.), *Proceedings of International Conference on Computer Science and Information Technology*, Advances in Intelligent Systems and Computing 255, DOI: 10.1007/978-81-322-1759-6_11,

greater than that of the speckle. In radar signal detection, usually the discrete sample of the compound Gaussian process is considered. In a coherent processing interval (CPI), the texture is considered to be completely related and thus indicated with a non-negative random variable $\sqrt{\tau}$, and the speckle can be indicated with a Gaussian distribution vector $\mathbf{g}$; therefore, the clutter vector $\mathbf{c} = \sqrt{\tau} \times \mathbf{g}$ is the so-called spherically invariant random vector (SIRV), that is to say, when this vector is in the generalized spherical coordinates, the distribution of the angle coordinates has nothing to do with the clutter scope distribution.

Knowledge-aided (KA) method is an inevitable choice for further improvement in radar signal processing ability [3–6]. The KA radar can constantly sense the surrounding environment during its operation or constantly readjusts its own parameters in the changing environment, so as to realize the optimum operating performance under a specific environment. The KA radar detector is required to conduct statistical analysis on current clutter environment with certain priori knowledge of the clutter in the signal detection process, so that the structure or parameters of the current detector can be adjusted to improve the detecting performance under specific clutter environment.

2 Bayesian Optimum Radar Detector

Considering detection of signals with unknown scopes in SIRV clutters:

$$\begin{aligned} H_0 &: \mathbf{z} = \mathbf{c} = \mathbf{g}\sqrt{\tau} \\ H_1 &: \mathbf{z} = \mathbf{s} + \mathbf{c} = a\mathbf{p} + \mathbf{g}\sqrt{\tau} \end{aligned} \tag{1}$$

where the bold-faced letters indicate vectors and the vector length is N, indicating the number of pulses in the CPI. The vector $\mathbf{p}$ is the known signal homing vector, and the complex scalar parameter a indicates the attenuation of the signal and its propagation effect, and in this paper, it is supposed to be an unknown definitive parameter. Suppose the Gaussian distribution vector $\mathbf{g} \sim N(\mathbf{0}, \mathbf{M})$; then, the average value for the distribution is 0, and the covariance matrix is the multivariate normal distribution of $\mathbf{M}$. In this paper, the covariance matrix $\mathbf{M}$ is supposed to be known, so the likelihood function can be obtained from the SIRV clutter model:

$$\begin{aligned} p(\mathbf{z}|H_0) &= \int_0^\infty \frac{\tau^{-N}}{(2\pi)^N} \exp\left(-\frac{q_0(\mathbf{z})}{2\tau}\right) p(\tau)\mathrm{d}\tau \\ p(\mathbf{z}|H_1) &= \int_0^\infty \frac{\tau^{-N}}{(2\pi)^N} \exp\left(-\frac{q_1(\mathbf{z})}{2\tau}\right) p(\tau)\mathrm{d}\tau \end{aligned} \tag{2}$$

where $q_0(\mathbf{z}) = \mathbf{z}^H\mathbf{M}^{-1}\mathbf{z}$, $q_1(\mathbf{z}) = q_0(\mathbf{z}-\mathbf{s})$, with the superscript H indicating conjugate transpose of the vector. When the signal scope parameter is an unknown definitive parameter, its maximal likelihood estimation can be used in detection:

$$a_{\mathrm{ML}} = \frac{\mathbf{p}^H\mathbf{M}^{-1}\mathbf{z}}{\mathbf{p}^H\mathbf{M}^{-1}\mathbf{p}} \tag{3}$$

From which we can obtain:

$$q_1(\mathbf{z}) = q_0(\mathbf{z}) - \frac{\left|\mathbf{p}^H\mathbf{M}^{-1}\mathbf{z}\right|^2}{\mathbf{p}^H\mathbf{M}^{-1}\mathbf{p}} \tag{4}$$

With (2)–(4), we can obtain the likelihood ratio detection under SIRV clutters:

$$\lambda(\mathbf{z}) = \frac{p(\mathbf{z}|H_1)}{p(\mathbf{z}|H_0)} = \frac{\int_0^\infty \frac{\tau^{-N}}{(2\pi)^N}\exp\left(-\frac{q_1(\mathbf{z})}{2\tau}\right)p(\tau)\mathrm{d}\tau}{\int_0^\infty \frac{\tau^{-N}}{(2\pi)^N}\exp\left(-\frac{q_0(\mathbf{z})}{2\tau}\right)p(\tau)\mathrm{d}\tau} \tag{5}$$

Considering the Neyman–Pearson rule, the detection threshold is determined by the false alarm rate. In radar signal detection, the distribution function of the texture is usually unknown, so it can be estimated with the data of the K reference units near the detecting distance unit:

$$p_K(\tau) = \frac{1}{K}\sum_{k=1}^{K} p(\tau|\mathbf{z}_k) \tag{6}$$

If $\mathbf{z}_k$ is statistically independent, so the estimated value is agonic, and $\lim_{K\to\infty} p_K(\tau) \to p(\tau)$. With the Bayesian law, we can further obtain:

$$p_K(\tau) = \frac{1}{K}\sum\nolimits_{k=1}^{K} \frac{p(\mathbf{z}_k|\tau)g(\tau)}{\int\limits_0^\infty p(\mathbf{z}_k|\tau)g(\tau)\mathrm{d}\tau} \tag{7}$$

where $p(\mathbf{z}_k|\tau) \sim N(\mathbf{0}, \tau\mathbf{M})$ is known and $g(\tau)$ is the priori distribution of the texture, which is usually unknown. It is quite difficult to choose the appropriate priori distribution, and it is generally required to match the statistical model of the actual distribution as far as possible and to be as simple as possible in mathematical processing. Therefore, in this paper, the inverse gamma distribution (IGD) is taken as the priori distribution of the texture, and in this case, posteriori distribution is also IGD; therefore, the inverse Gamma distribution is also the conjugate prior distribution.

$$g_{\alpha,\beta}(\tau) \triangleq G^{-1}(\tau;\alpha,\beta) = \frac{\beta^\alpha}{\Gamma(\alpha)}\left(\frac{1}{\tau}\right)^{\alpha+1}\exp\left(-\frac{\beta}{\tau}\right) \tag{8}$$

where the IGD distribution parameters α, β, respectively, indicate the scale parameter and the shape parameter, and the selection of parameters represents the amount of priori knowledge for the texture component. When $\alpha, \beta \to 0$, $g(\tau) = 1/\tau$, there is no information priori distribution, or Jeffrey prior. With (8) and the Bayesian law, we can obtain the posteriori distribution of the texture:

$$p(\tau|\mathbf{z}_k) = G^{-1}\left(N + \alpha, \beta + \frac{Q(\mathbf{z}_k)}{2}\right) \tag{9}$$

where $Q(\mathbf{z}) = \mathbf{z}_k^H \mathbf{M}^{-1} \mathbf{z}_k$. Substituting formula (9) into (6), we can get

$$p_K(\tau) = \frac{1}{K}\sum_{k=1}^{K} p(\tau|\mathbf{z}_k) = \frac{1}{K}\sum_{k=1}^{K} \mathrm{G}^{-1}\left(N + \alpha, \beta + \frac{1}{2}Q(\mathbf{z}_k)\right) \tag{10}$$

Substituting (10) into (2), we can get

$$p(\mathbf{z}|\alpha, \beta, H_i) = C \times \sum_{k=1}^{K} \frac{\left[\beta + \frac{1}{2}Q(\mathbf{z}_k)\right]^{N+\alpha}}{\left[q_i(\mathbf{z}) + 2\beta + Q(\mathbf{z}_k)\right]^{2N+\alpha}}; i = 0, 1 \tag{11}$$

where C indicates a constant which has nothing to do with the supposition. From this, we can obtain the likelihood ratio detection:

$$\lambda_{\mathrm{KB-BORD}}(\mathbf{z}; \alpha, \beta) = \frac{p(\mathbf{z}|\alpha, \beta, H_1)}{p(\mathbf{z}|\alpha, \beta, H_0)} = \frac{\sum_{k=1}^{K} \frac{\left[\beta + \frac{1}{2}Q(\mathbf{z}_k)\right]^{N+\alpha}}{\left[q_1(\mathbf{z}) + 2\beta + Q(\mathbf{z}_k)\right]^{2N+\alpha}}}{\sum_{k=1}^{K} \frac{\left[\beta + \frac{1}{2}Q(\mathbf{z}_k)\right]^{N+\alpha}}{\left[q_0(\mathbf{z}) + 2\beta + Q(\mathbf{z}_k)\right]^{2N+\alpha}}} \tag{12}$$

The detection performance is closely related to understanding the parameters α, β. If the texture component of SIRV can be indicated with the inverse Gamma distribution, and the preciseness of the distribution parameter is known, the optimum detection performance can be obtained using formula (12). Therefore, the core for this detector is parameter α, β estimation.

3 Estimation Method for Detector Priori Distribution Parameter

3.1 PDF Fit-Based Method

The most direct estimation for parameters α, β is the statistical analysis on the texture component. Therefore, first the texture component of the clutter should be separated from the speckle component. The extraction process of the texture component is essentially estimation of the coherent length of the texture component. Considering that in the compound Gaussian model, the coherent length of the texture component is much greater than that of the speckle component; therefore, the clutter sequence can be approximated as follows:

$$c(n) \approx \sqrt{\tau(k)} \times g(n); n \in (k - L/2, k + L/2) \tag{13}$$

Thus, the statistics can be constructed as follows:

$$X(k) = \frac{\mathrm{real}\{c(n)\}}{\mathrm{real}\{c(n+k)\}} \approx \frac{\mathrm{real}\{g(n)\}}{\mathrm{real}\{g(n+k)\}} \tag{14}$$

Within the texture decorrelation time, the statistics satisfies the Cauchy distribution, but the scale and position parameters for this distribution are all unknown, so the texture coherent length [7] under a given confidence level with the generalized Kolmogorov–Smirnov test, so as to separate the texture component from the clutter data. With the extracted texture component, we can get the frequency histogram $h(\tau)$ and then construct the following computation:

$$\left(\hat{\alpha}, \hat{\beta}\right) = \min_{\alpha,\beta} \xi = \min_{\alpha,\beta} \int_0^\infty w(\tau)\left|G^{-1}(\tau;\alpha,\beta) - h(\tau)\right.^2 \mathrm{d}\tau \tag{15}$$

Obviously, ξ indicates the mean error of the PDF estimated value, and $w(\tau)$ indicates the weight function. With the random searching algorithm, we can get the parameter ξ, which makes the minimal α, β.

3.2 Method of Fractional Moment

For a random process whose texture meets the inverse Gamma distribution and whose speckle component meets compound Gaussian distribution, its scope distribution can be indicated as follows:

$$f_{|c|}(c) = \frac{2c\beta\Gamma(\alpha+1)}{(\beta c^2+1)^{\alpha+1}\Gamma(\alpha)} \tag{16}$$

where $\Gamma(\cdot)$ indicates the Gamma function. With this distribution, we can obtain the n-order matrix for the clutter scope:

$$m_k(\alpha,\beta) = \left(\frac{1}{\beta}\right)^{k/2} \frac{\Gamma(n/2+1)\Gamma(\alpha - n/2)}{\Gamma(\alpha)} \tag{17}$$

But considering that m_k can be restrained only when $\alpha > k/2$; therefore, in this paper, the parameters are only estimated with 1-order moment and 1/2-order moment (namely, fractional moment), and with statistically independent clutter scope data, the estimated values $\hat{m}_k$ for the 1-order moment and the 1/2-order moment, $k = 1{,}1/2$. Thus, the following computation is constructed:

$$\left(\hat{\alpha}, \hat{\beta}\right) = \min_{\alpha,\beta} \varsigma = \min_{\alpha,\beta}\left[\left(m_1(\alpha,\beta) - \hat{m}_1\right)^2 + \left(\left(m_{1/2}(\alpha,\beta) - \hat{m}_{1/2}\right)^2\right)\right] \tag{18}$$

With the random searching algorithm, we can get the parameters α, β, which makes the minimal ς.

3.3 Method Based on Maximal Likelihood Estimation

For the maximal likelihood estimation method, take the natural logarithm for formula (16) and then take the derivative, respectively, for parameters α, β, and let it be zero. Considering L independently distributed clutter scope data, then we can get

$$\begin{aligned} \frac{\partial \ln\left(f_{|c|}(c)\right)}{\partial \alpha} &= \frac{L}{\alpha} - \sum_{l=1}^{L} \ln\left(\beta c^2(l) + 1\right) = 0 \\ \frac{\partial \ln\left(f_{|c|}(c)\right)}{\partial \beta} &= \frac{L}{\beta} - (\alpha + 1)\sum_{l=1}^{L} \frac{c^2(l)}{\beta c^2(l) + 1} = 0 \end{aligned} \tag{19}$$

Thus, we can get the likelihood estimated value of parameters α, β [8]:

$$\begin{aligned} y\left(\hat{\beta}_{\mathrm{ML}}\right) &= \frac{L\hat{\beta}_{\mathrm{ML}} \sum_{l=1}^{L} \frac{c^2(l)}{\beta c^2(l)+1}}{L - \hat{\beta}_{\mathrm{ML}} \sum_{l=1}^{L} \frac{c^2(l)}{\beta c^2(l)+1}} - \hat{\beta}_{\mathrm{ML}} \sum_{l=1}^{L} \ln\left(\hat{\beta}_{\mathrm{ML}} c^2(l) + 1\right) = 0 \\ \hat{\alpha}_{\mathrm{ML}} &= \frac{L}{\hat{\beta}_{\mathrm{ML}} \sum_{l=1}^{L} \frac{c^2(l)}{\hat{\beta}_{\mathrm{ML}} c^2(l)+1}} - 1 \end{aligned} \tag{20}$$

where the maximal likelihood estimated value for parameter β is the zero point of function $y(\cdot)$.

4 KA-BORD Algorithm Performance Analysis

Before carrying out KA-BORD performance analysis, first several kinds of detectors for SIRV are provided. The asymptotically optimum detector (AOD) is got by substituting the maximal likelihood estimated value of the texture component into the likelihood ratio detection [8, 9]:

$$\lambda_{\mathrm{AOD}}(\mathbf{z}) = \frac{\left|\mathbf{p}^H \mathbf{M}^{-1} \mathbf{z}\right|^2}{\left(\mathbf{z}^H \mathbf{M}^{-1} \mathbf{z}\right)\left(\mathbf{p}^H \mathbf{M}^{-1} \mathbf{p}\right)} \tag{21}$$

The detection performance of this detector is asymptotically optimum, i.e., only when N is great enough, its detection performance can approach the optimum detector. If the KA-BORD parameters $(\alpha, \beta) \to (0, 0)$, we can get BORD [1]:

$$\lambda_{\text{BORD}}(\mathbf{z}) = \frac{\sum_{k=1}^{K} \frac{[Q(\mathbf{z}_k)]^N}{[q_1(\mathbf{z})+Q(\mathbf{z}_k)]^{2N}}}{\sum_{k=1}^{K} \frac{[Q(\mathbf{z}_k)]^N}{[q_0(\mathbf{z})+Q(\mathbf{z}_k)]^{2N}}} \tag{22}$$

That is, BORD is an exception of KA-BORD.

4.1 Algorithm Performance Analysis Based on Computer Simulation Data

In computer simulation, we have considered the K distributes clutter, and the SIRV model for K distribution clutter, whose Texture component is the Gamma distribution, thus we can get the likelihood function:

$$p(\mathbf{z}|H_i) = \frac{2b^{v+N} q_i(\mathbf{z})^{(v-N)/2}}{\pi^N |\mathbf{M}| \Gamma(v) 2^{v+N}} K_{v-N}\left(b\sqrt{q_i(\mathbf{z})}\right); i = 0, 1 \tag{23}$$

where parameters v, b, respectively, indicate the shape parameter and the scale parameter of the K distribution, and $K_n(\cdot)$ is Class II correction Bessel function. From this, we can obtain the K distribution optimum detector (KOD):

$$\lambda_{\text{KOD}}(\mathbf{z}) = \left(\frac{q_1(\mathbf{z})}{q_0(\mathbf{z})}\right)^{(v-N)/2} \frac{K_{v-N}\left(b\sqrt{q_1(\mathbf{z})}\right)}{K_{v-N}\left(b\sqrt{q_0(\mathbf{z})}\right)} \tag{24}$$

Since KOD is the signal detection under the condition that the clutter statistical property is completely known, its detection performance is optimum. In the simulation, different shape parameters for the K distribution are selected, and the scale parameter $b = \sqrt{2v}$, so the mean square value for the texture component of the clutter is equal to 1. The number of pulses in CPI is $N = 4$, and the reference unit number is $K = 8$. we have examined the detection performance of the detector under the false alarm rate $p_{\text{fa}} = 10^{-3}$.

Figure 1 presents the contrast between KA-BORD and KOD detection performance analyses under the K distribution clutter and different shape parameters. For K distribution, the shape parameter v demonstrates its deviation from the Gaussian property; the smaller the v is, the more greatly the K distribution deviates from the Gaussian property. As shown in Fig. 1a, when $v = 0.5$, the non-Gaussian property of the K distribution is remarkable. When the KA-BORD detection performance obtained with the PDF fit method is PD = 0.5, the detection performance is closer to KOD, but the KA-BORD performance obtained with the

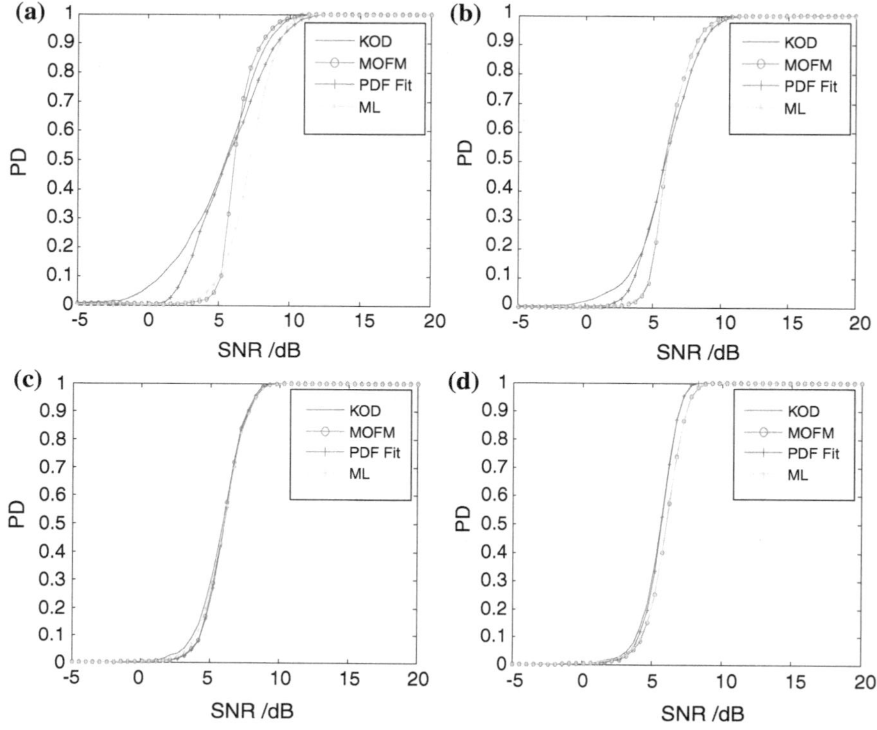

Fig. 1 KA-BORD and KOD detection performances with different parameter estimation methods. **a** v = 0.5, **b** v = 1, **c** v = 2, **d**v = 20

MOFM and maximal likelihood can obtain better performance under larger SNR, and the performance curve is steeper along with the change in the signal-to-noise ratio. Along with the increase in v, the clutter statistical property is closer to Gaussian distribution, so a detection performance more identical to that of KOD can be obtained with the above-mentioned methods, as shown in Fig. 1b, c. However, when v is further increased, the detection performance obtained using the PDF fit method is better, as shown in Fig. 1d, because of the smaller remainder error of the PDF fit. Therefore, with different parameter estimation methods, under different shape parameters, the detection performances obtained by KA-BORD are different.

From computer simulation, we can see that under a condition with more obvious non-Gaussian characteristics, the detection performance of KA-BORD is quite close to that of KOD. This indicates that when N is smaller, its detection performance is better than that of conventional detectors. For AOD, only when N is large enough, its performance can be close to the detection performance of KOD.

Table 1 Measured data parameters

Data number	84	85	86
Central frequency		9.3 GHz	
Pulse width (ns)	200	100	20
Pulse repetition frequency		1,000 Hz	
Glancing angle		0.32°	
Distance (m)	3,000–3,989	3,501–3,996	3,501–3,600
Distance resolution (m)	30	15	3
Radar beam width		0.9°	

4.2 Algorithm Performance Analysis Based on Sea-Clutter-Measured Data

The sea-clutter-measured data come from the IPIX radar data from McMaster University in Canada. The data are collected in 1998 and the survey environment in on the Ontario Lake in Canada. We choose the data of groups 84–86, which are the same as those in literature [2]. These three groups of data, respectively, represent the clutter data measured under low, medium, and high resolutions. Refer to literature [10] and other references for detailed description of the data. We choose the transmitting and receiving data with the same polarization for analysis; for example, HH indicates that transmitting and receiving are all horizontal polarization, and VV indicates that transmitting and receiving are all vertical polarization. Refer to Table 1 for main parameters of the measured data. The algorithm simulation parameters are the same as mentioned above.

Figure 2 provides detection performance contrasts among KA-BORD, BORD, and AOD detectors. Generally speaking, the clutter data under different resolution and different polarization modes, the KA-BORD performance obviously outperforms that of AOD and slightly outperforms that of BORD. But for low-resolution data, for instance, the HH polarization mode, as shown in Fig. 2a, the KA-BORD detection performance obtained with the PDF fit method and the maximal likelihood estimation outperforms that of BORD, but the KA-BORD detection performance obtained with the MOFM is equivalent to that of BORD, but they all outperform that of AOD. With different parameter estimation methods, the detection performances of the KA-BORD are different. But for VV polarization, as shown in Fig. 2b, the KA-BORD detection performance with different kinds of parameter estimation methods is basically the same as that of the BORD, and it outperforms that of AOD. For medium-resolution clutter data, as shown in Fig. 2c, d, the detection performances of KA-BORD and BORD are slightly different with various parameter estimation methods, and under high signal-to-noise ratio conditions, the KA-BORD performance is better. For high-resolution radar clutter data, as shown in Fig. 2e, f, we can find that the advantage of KA-BORD algorithm performance over the BORD detection performance with the PDF fitting method is not obvious, and the KA-BORD detection performances with the MOFM and the maximal likelihood estimation methods are a bit worse. The cause

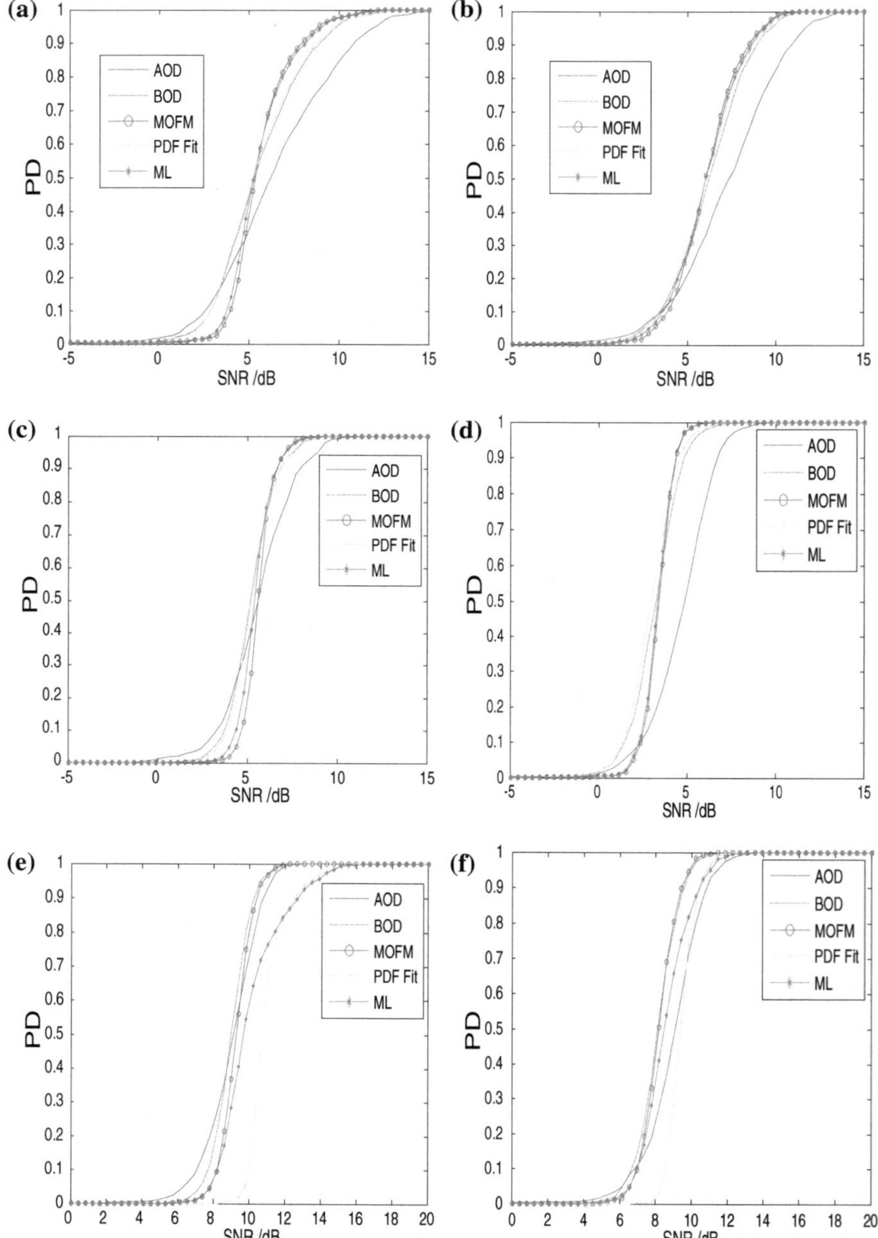

Fig. 2 KA-BORD and KOD detection performance contrast with different parameter estimation methods. **a** data set 85, HH polarizes, **b** data set 84, VV polarization, **c** data set 85, HH polarizes, **d** data set 85, VV polarization, **e** data set 86, HH polarizes, **f** data set 86, VV polarization

of this phenomenon is that for high-resolution radar clutter, it is doubted that the compound Gaussian clutter model should be used, that is to say, the compound Gaussian model should not be used for the statistical property of the clutter.

It can be seen from the above analysis that the performance of KA-BORD is superior to that of AOD, and they are all superior to that of BORD. The effect of different parameter estimation methods on the detection performance of the KA-BORD is different under different clutter environments. This indicates that different parameter estimation algorithms can be selected under different clutter environments, as long as the clutter statistical property satisfies the compound Gaussian model.

5 Conclusion

Radar signal detection in non-Gaussian clutters is an important topic in current field of radar signal processing. This paper, taking the compound Gaussian clutter statistical model as a foundation, presents the knowledge-aided Bayesian optimum radar detector (BORD), provides several parameter estimation methods, and evaluates the effect of different parameter estimation methods on detection performance of the KA-BORD. Based on computer-simulated K distributed clutter data, the analysis result indicates that the KA-BORD detection performance is close to that of KOD. The analysis result of the sea-clutter-measured data analysis indicates that when the clutter statistical property complies with the compound Gaussian clutter model, the KA-BORD detection performance is superior to that of the AOD.

References

1. Jay, E., Ovarlez, J.P., Declercq, D., Duvaut, P.: BORD: Bayesian optimum radar detector. Signal Process. **83**(6), 1151–1162 (2003)
2. Conte, E., Maio, A.D., Galdi, C.: Statistical analysis of real clutter at different range resolutions. IEEE Trans. AES **40**(3), 903–917 (2004)
3. Maio, A.D., Farina, A., Foglia, G.: Knowledge-aided Bayesian radar detectors & their application to live data. IEEE Trans. AES **46**(1), 170–181 (2010)
4. Capraro, G.T., Farina, A., Griffiths, H., Wicks, M.C.: Knowledge-based radar signal and data processing: a tutorial overview. IEEE Signal Process. Mag. **23**(1), 18–29 (2006)
5. Haykin, S.: Cognitive radar: a way of the future. IEEE Signal Process. Mag. **23**(1), 30–40 (2006)
6. Conte, E., Maio, A.D., Farina, A., Foglia, G.: Design and analysis of a knowledge aided radar detector for Doppler processing. IEEE Trans. AES **42**(3), 1058–1078 (2006)
7. Balleri, A., Nehorai, A., Wang, J.: Maximum likelihood estimation for compound Gaussian clutter with inverse Gamma texture. IEEE Trans. AES **43**(2), 775–780 (2007)
8. Conte, E., Lops, M., Ricci, G.: Asymptotically optimum radar detection in compound Gaussian clutter. IEEE transactions on AES **31**(2):617–624, (1995)
9. Sangston, K.J., Gini, F., Greco, M.V., Farina, A.: Structure for radar detection in compound Gaussian clutter. IEEE Trans. AES **35**(2), 445–457 (1999)
10. Conte, E., Bisceglie, M.D., Galdi, C., Ricci, G.: A procedure for measuring the coherent length of the sea texture. IEEE Trans. Instrum. Meas. **46**(4), 836–841 (1997)

Part II
Computer Security

An Improvement of an Identity-Based Key-insulated Signcryption

Guobin Zhu, Hu Xiong, Ruijin Wang and Zhiguang Qin

Abstract As one of the fundamental cryptographic primitives, signcryption can achieve unforgeability and confidentiality simultaneously at the cost significantly lower than the signature-then-encryption approach in terms of computational costs and communication overheads. In view of the damage caused by the secret key leakage, Chen et al. proposed an efficient identity-based key-insulated signcryption (ID-KI-SC) scheme secure in the standard model recently. However, in this paper, we show that their scheme does not achieve the indistinguishability against adaptively chosen ciphertext attacks (IND-CCA2) and existential unforgeability against adaptively chosen message attacks (EUF-CMA). Furthermore, we propose an improved scheme that remedies the weakness of Chen et al.'s scheme.

Keywords Identity-based cryptography · Key-insulated · Signcryption · Standard model

1 Introduction

Signcryption was first initialized by Zheng [1] in 1997 to implement the function of digital signature and public key encryption simultaneously with better efficiency than the signature-then-encryption approach. Since Zheng's pioneering work,

G. Zhu (✉)
School of Computer Science and Engineering, University of Electronic Science and Technology of China, No. 2006, Xiyuan Avenue, West Hi-Tech Zone, Chengdu 611731 Sichuan, People's Republic of China
e-mail: zhugb@uestc.edu.cn

H. Xiong · R. Wang · Z. Qin
School of Computer Science and Engineering, University of Electronic Science and Technology of China, Chengdu 610054, People's Republic of China
e-mail: xionghu@uestc.edu.cn

Z. Qin
e-mail: qinzg@uestc.edu.cn

S. Patnaik and X. Li (eds.), *Proceedings of International Conference on Computer Science and Information Technology*, Advances in Intelligent Systems and Computing 255, DOI: 10.1007/978-81-322-1759-6_12,

dozens of signcryption schemes have been proposed to improve the efficiency and security [2]. However, some issues have to be addressed before this primitive can be applied in practice. Firstly, identity-based public key cryptography (ID-PKC) has been suggested by Shamir [3] to simplify the heavy certificates management in traditional public key cryptography (PKC). Different from conventional PKC, the public key of the user can be obtained easily from its identity information, such as email address or social insurance number, in ID-PKC. Therefore, the certificate binding user's identity and its public key in traditional PKC is no longer needed in the ID-PKC. It is natural to integrate the notion of signcryption and ID-PKC to enjoy their merits at the same time. Inspired by the idea of bilinear pairings, the first identity-based signcryption (ID-SC) scheme has been proposed by Malone-Lee in [4]. ID-SC has then received a lot of attention from the cryptography community [5, 6]. Secondly, most ID-SC schemes are only proven secure in the random oracle model [7]. Taking into account of the limitation and criticism on the random oracle models [8], ID-SC scheme provably secure in the standard model draws a great interest. Therefore, Jin et al. [9] suggested a provably secure ID-SC scheme without random oracles recently. Unfortunately, this works has been shown to offer neither IND-CCA2 property nor EUF-CMA property by Li et al. [10]. Thirdly, key exposure seems to be inevitable when the cryptographic primitive is deployed in the insecure environment. To solve all of these problems, Chen et al. [11] proposed the first identity-based key-insulated signcryption (ID-KI-SC) by integrating the notion of key-insulated mechanism [12] and ID-SC [9] recently. The security of their scheme is proved in the standard model. However, in this paper, two attacks are proposed to show that Chen et al.'s scheme cannot achieve the IND-CCA2 and EUF-CMA resilience. Furthermore, we propose an improved scheme that remedies the weakness of Chen et al.'s scheme, and the security of our improved scheme has been proved in the standard model.

The rest of this paper is organized as follows: In Sect. 2, we analyze the Chen et al.'s scheme. The improved scheme is given in Sect. 3. Finally, the conclusions are given in Sect. 4.

2 Analysis of the Chen et al.'s Scheme

2.1 Review of the Chen et al.'s Scheme

We describe Chen et al.'s ID-KI-SC scheme [11] as follows:

In order to create identities and messages of the desired length, two collision-resistant hash functions $H_u : \{0,1\}^* \rightarrow \{0,1\}^{n_u}$ and $H_v : \{0,1\}^{n_m} \rightarrow \{0,1\}^{n_v}$ are chosen, where $n_u, n_m, n_v \in_R \mathbb{Z}$. Let F be a pseudorandom function (PRF) which outputs a κ-bit string $F_s(x)$ with the input of a κ-bit seed s and a κ-bit argument x.

Setup: Select a pairing $\hat{e} : G \times G \rightarrow G_T$ where the order of G and G_T is q of size κ. Let g be a generator of G. Randomly select $\alpha \leftarrow_R Z_q$, compute $g_1 = g^{\alpha}$, and pick $g_2 \leftarrow_R G$. Also select randomly the following elements: $u', m' \leftarrow_R G$,

$u_i \leftarrow_R G$, and $m_j \leftarrow_R G$ for $i = 1, \ldots, n_u$ and $j = 1, \ldots, n_v$, respectively. Then, define $\mathbf{U} = (u_i)$, $\mathbf{M} = (m_i)$. Define $V : \Gamma \rightarrow G_T$ to be a bijective function, where Γ is a subset of $\{0,1\}^{n_u+n_m+n_v}$ with p elements. Next, define V^{-1} as the inverse mapping of V. The public parameters are $(G, G_T, \hat{e}, g, g_1, g_2, Y = \hat{e}(g_1, g_2), u', \mathbf{U}, m', \mathrm{M}, H_u, H_v, V)$ and master secret is $\mathrm{msk} = g_2^{\alpha}$.

Extract: Define $\mathbf{u}$ to be an identity in the form of a bit string of length n_u, and $\mathbf{u}[i]$ to be the i-th bit of $\mathbf{u}$. Let $\mathcal{U}_{\mathbf{u}} \subseteq \{1, \ldots, n_u\}$ be the set of indices i such that $\mathbf{u}[i] = 1$. Define $\mathbf{w}_{\mathbf{u},0}$ to be the output of $H_u(\mathbf{u}\|0)$ and $\mathbf{w}_{\mathbf{u},0}[i]$ to be the i-th bit of $\mathcal{W}_{\mathbf{u},0}$. Let $\mathcal{W}_{\mathbf{u},0} \subseteq \{1, \ldots, n_u\}$ be the set of indices i such that $\mathbf{w}_{\mathbf{u},0}[i] = 1$. Choose a helper key $\mathrm{HK}_{\mathbf{u}} \leftarrow \{0,1\}^{\kappa}$ and compute $k_{u,0} = F_{\mathrm{HK}_{\mathbf{u}}}(0)$. To generate the initial secret key regarding to the user with identity $\mathbf{u}$, the PKG randomly picks $r_u \leftarrow_R Z_q$ and computes:

$$\begin{aligned} d_{\mathbf{u},0} &= \left(d_{\mathbf{u},0}^{\langle 1\rangle}, d_{\mathbf{u},0}^{\langle 2\rangle}, d_{\mathbf{u},0}^{\langle 3\rangle}\right) \\ &= \left(g_2^{\alpha}\left(u' \prod_{i \in \mathcal{U}_u} u_i\right)^{r_u}\left(u' \prod_{i \in \mathcal{W}_{u,0}} u_i\right)^{k_{u,0}}, g^{k_{u,0}}, g^{r_u}\right) \end{aligned}$$

HelperUpt: Similar to the **Extract** algorithm, define $\mathbf{w}_{\mathbf{u},t}$ to be the output of $H_u(\mathbf{u}\|t)$ and $\mathbf{w}_{\mathbf{u},t}[i]$ to be the i-th bit of $\mathbf{w}_{\mathbf{u},t}$, and let $\mathcal{W}_{\mathbf{u},t} \subseteq \{1, \ldots, n_u\}$ be the set of indices i such that $\mathbf{w}_{\mathbf{u},t}[i] = 1$. Besides, define $\mathbf{w}_{u,t'}$ to be the output of $H_u(\mathbf{u}\|t')$ and $\mathbf{w}_{\mathbf{u},t'}[i]$ to be the ith bit of $\mathbf{w}_{\mathbf{u},t'}$, and let $\mathcal{W}_{\mathbf{u},t'} \subseteq \{1, \ldots, n_u\}$ be the set of indices i such that $\mathbf{w}_{u,t'}[i] = 1$. Compute $k_{u,t} = F_{\mathrm{HK}_{\mathbf{u}}}(t)$ and $k_{u,t'} = F_{\mathrm{HK}_{\mathbf{u}}}(t')$. The helper generates the key-update information $\mathrm{UI}_{\mathbf{u},t',t}$ regarding the identity $\mathbf{u}$ from period t' to t as follows:

$$\begin{aligned} \mathrm{UI}_{\mathbf{u},t',t} &= \left(\mathrm{UI}_{\mathbf{u},t',t}^{\langle 1\rangle}, \mathrm{UI}_{\mathbf{u},t',t}^{\langle 2\rangle}\right) \\ &= \left(\left(u' \prod_{i \in \mathcal{W}_{u,t}} u_i\right)^{k_{u,t}} \Big/ \left(u' \prod_{i \in \mathcal{W}_{u,t'}} u_i\right)^{k_{u,t'}}, g^{k_{u,t}}\right). \end{aligned}$$

UserUpt: After receiving the temporary private key, $d_{\mathbf{u},t'} = (d_{\mathbf{u},t'}^{\langle 1\rangle}, d_{\mathbf{u},t'}^{\langle 2\rangle}, d_{\mathbf{u},t'}^{\langle 3\rangle})$ regarding the identity $\mathbf{u}$ and period t' generated by the **Extract** algorithm, and the key-update information $\mathrm{UI}_{\mathbf{u},t',t} = \left(\mathrm{UI}_{\mathbf{u},t',t}^{\langle 1\rangle}, \mathrm{UI}_{\mathbf{u},t',t}^{\langle 2\rangle}\right)$ regarding the identity $\mathbf{u}$ from period t' to period t generated by the algorithm **HelperUpt**, the user $\mathbf{u}$ computes the temporary private key regarding to identity $\mathbf{u}$ and period t as follows:

$$\begin{aligned} d_{\mathbf{u},t} &= \left(d_{\mathbf{u},t'}^{\langle 1\rangle} \cdot \mathrm{UI}_{\mathbf{u},t',t}^{\langle 1\rangle}, \mathrm{UI}_{\mathbf{u},t',t}^{\langle 2\rangle}, d_{\mathbf{u},t'}^{\langle 3\rangle}\right) \\ &= \left(g_2^{\alpha}\left(u' \prod_{i \in \mathcal{U}_u} u_i\right)^{r_u}\left(u' \prod_{i \in \mathcal{W}_{u,t}} u_i\right)^{k_{u,t}}, g^{k_{u,t}}, g^{r_u}\right) \end{aligned}$$

Signcrypt: Define $\mathbf{m} \in \{0,1\}^{n_m}$ to be a bitstring representing a message. Equipped with the temporary private key as $d_{\mathbf{a},t} = (d_{\mathbf{a},t}^{\langle 1 \rangle}, d_{\mathbf{a},t}^{\langle 2 \rangle}, d_{\mathbf{a},t}^{\langle 3 \rangle})$ in period t, Alice signcrypt a message $\mathbf{m}$ to Bob as follows. First, Alice picks $r_m, r_t' \leftarrow_R Z_q$ randomly, lets $r_t = r_t' + k_{a,t}$, and chooses $\mathbf{r} \leftarrow_R \{0,1\}^{n_v}$ satisfying $\mathbf{a}\|\mathbf{m}\|\mathbf{r} \in \Gamma$. Define $\mathcal{M}_{\mathbf{m}} \subset \{1,\ldots,n_v\}$ to be the set of indices j for which the j-th bit of $H_v(\mathbf{m})$ is different from that of $\mathbf{r}$, i.e., $\mathcal{M}_{\mathbf{m}} = \{j \in Z : H_v(\mathbf{m})[j] \oplus \mathbf{r}[j] = 1\}$. After that, Alice computes:

$$\sigma^{\langle 1 \rangle} = Y^{r_m} \cdot V(\mathbf{a}\|\mathbf{m}\|\mathbf{r}), \quad \sigma^{\langle 2 \rangle} = g^{r_m},$$

$$\sigma^{\langle 3 \rangle} = \left(u' \prod_{i \in \mathcal{U}_b} u_i\right)^{r_m}, \quad \sigma^{\langle 4 \rangle} = \left(u' \prod_{i \in \mathcal{W}_{b,t}} u_i\right)^{r_m},$$

$$\sigma^{\langle 5 \rangle} = d_{\mathbf{a},t}^{\langle 1 \rangle} \cdot \left(u' \prod_{i \in \mathcal{W}_{a,t}} u_i\right)^{r_t'} \cdot \left(m' \prod_{i \in \mathcal{M}_m} m_i\right)^{r_m}$$

$$= g_2^{\alpha} \left(u' \prod_{i \in \mathcal{U}_a} u_i\right)^{r_a} \cdot \left(u' \prod_{i \in \mathcal{W}_{a,t}} u_i\right)^{r_t} \cdot \left(m' \prod_{i \in \mathcal{M}_m} m_i\right)^{r_m}$$

$$\sigma^{\langle 6 \rangle} = d_{\mathbf{a},t}^{\langle 2 \rangle} \cdot g^{r_t'} = g^{k_{a,t} + r_t'} = g^{r_t}, \quad \sigma^{\langle 7 \rangle} = d_{\mathbf{a},t}^{\langle 3 \rangle} = g^{r_a}$$

Finally, Alice sends the ciphertext $(t, \sigma) = (t, (\sigma^{\langle 1 \rangle}, \sigma^{\langle 2 \rangle}, \sigma^{\langle 3 \rangle}, \sigma^{\langle 4 \rangle}, \sigma^{\langle 5 \rangle}, \sigma^{\langle 6 \rangle}, \sigma^{\langle 7 \rangle}))$ to Bob.

Unsigncrypt: After receiving a ciphertext $(t, \sigma) = (t, (\sigma^{\langle 1 \rangle}, \sigma^{\langle 2 \rangle}, \sigma^{\langle 3 \rangle}, \sigma^{\langle 4 \rangle}, \sigma^{\langle 5 \rangle}, \sigma^{\langle 6 \rangle}, \sigma^{\langle 7 \rangle}))$, Bob decrypts it as follows.

1. Compute $V^{-1}\left(\sigma^{\langle 1 \rangle} \cdot \hat{e}\left(d_{\mathbf{b},t}^{\langle 2 \rangle}, \sigma^{\langle 4 \rangle}\right)\hat{e}\left(d_{\mathbf{b},t}^{\langle 3 \rangle}, \sigma^{\langle 3 \rangle}\right) / \hat{e}\left(d_{\mathbf{b},t}^{\langle 1 \rangle}, \sigma^{\langle 2 \rangle}\right)\right) \rightarrow \mathbf{a}\|\mathbf{m}\|\mathbf{r}$.
2. Generate $\{j \in Z : H_v\ (\mathbf{m})\ [j] \oplus \mathbf{r}[j] = 1\} \rightarrow \mathcal{M}_{\mathbf{m}}$.
3. Accept the message if the following equation holds:

$$\hat{e}\left(\sigma^{\langle 5 \rangle}, g\right) = Y \cdot \hat{e}\left(\sigma^{\langle 7 \rangle}, u' \prod_{i \in \mathcal{U}_a} u_i\right) \hat{e}\left(\sigma^{\langle 6 \rangle}, u' \prod_{i \in \mathcal{W}_{a,t}} u_i\right) \hat{e}\left(\sigma^{\langle 2 \rangle}, m' \prod_{i \in \mathcal{M}_m} m_i\right).$$

Note: The original scheme in [11] had typos in the **Signcrypt** and **Unsigncrypt** algorithms. Instead of writing $\sigma^{\langle 4 \rangle} = (u' \prod_{i \in \mathcal{W}_{b,t}} u_i)^{r_m}$ in the **Signcrypt** algorithm, it was written as $\sigma^{\langle 4 \rangle} = (w' \prod_{i \in \mathcal{W}_{b,t}} w_i)^{r_m}$. Instead of writing $V^{-1}\left(\sigma^{\langle 1 \rangle} \cdot \hat{e}(d_{\mathbf{b},t}^{\langle 2 \rangle}, \sigma^{\langle 4 \rangle})\hat{e}(d_{\mathbf{b},t}^{\langle 3 \rangle}, \sigma^{\langle 3 \rangle}) / \hat{e}\left(d_{\mathbf{b},t}^{\langle 1 \rangle}, \sigma^{\langle 2 \rangle}\right)\right)$ in the step 1 of **Unsigncrypt** algorithm, it was written as $V^{-1}(\sigma^{\langle 1 \rangle} \cdot \hat{e}(d_{\mathbf{b},t}^{\langle 3 \rangle}, \sigma^{\langle 4 \rangle})\hat{e}(d_{\mathbf{b},t}^{\langle 2 \rangle}, \sigma^{\langle 3 \rangle}) / \hat{e}(d_{\mathbf{b},t}^{\langle 1 \rangle}, \sigma^{\langle 2 \rangle}))$. These typos have been corrected in our review to maintain the consistency.

2.2 Analysis

Attack the IND-ID-KI-SC-CCA2 property. According to the IND-ID-KI-SC-CCA2 property in [11], the adversary $\mathcal{A}$ generates two equal length plaintexts $\mathbf{m}_0$, $\mathbf{m}_1$, time period t^* and two identities **a** and **b** on which it wants to be challenged. The challenger $\mathcal{B}$ chooses a bit $\gamma \in_R \{0,1\}$ and generates the ciphertext $(t^*, \sigma^*) = \left(t^*, \left(\sigma^{\langle 1\rangle *}, \sigma^{\langle 2\rangle *}, \sigma^{\langle 3\rangle *}, \sigma^{\langle 4\rangle *}, \sigma^{\langle 5\rangle *}, \sigma^{\langle 6\rangle *}, \sigma^{\langle 7\rangle *}\right)\right)$. To guess which message is signcrypted in the ciphertext (t^*, σ^*), $\mathcal{A}$ chooses $r' \leftarrow_R \mathbb{Z}_p^*$ randomly and computes $\sigma^{\langle 1\rangle'} = \sigma^{\langle 1\rangle *}, \sigma^{\langle 2\rangle'} = \sigma^{\langle 2\rangle *}, \sigma^{\langle 3\rangle'} = \sigma^{\langle 3\rangle *}, \sigma^{\langle 4\rangle'} = \sigma^{\langle 4\rangle *}, \sigma^{\langle 5\rangle'} = \sigma^{\langle 5\rangle *}\left(u' \prod_{i \in \mathcal{U}_a} u_i\right)^{r'}$, $\sigma^{\langle 6\rangle'} = \sigma^{\langle 6\rangle *}$ and $\sigma^{\langle 7\rangle'} = \sigma^{\langle 7\rangle *} g^{r'}$. $\mathcal{A}$ then makes an unsigncryption query on the ciphertext $\left(\sigma^{\langle 1\rangle'}, \sigma^{\langle 2\rangle'}, \sigma^{\langle 3\rangle'}, \sigma^{\langle 4\rangle'}, \sigma^{\langle 5\rangle'}, \sigma^{\langle 6\rangle'}, \sigma^{\langle 7\rangle'}\right)$ regarding the period t^* and identities **a** and **b**. If $\mathcal{B}$ responds $\mathbf{m}_0$ as the response to this query, $\mathcal{A}$ wins the game by returning $\gamma' = 0$. Otherwise, $\mathcal{A}$ wins the game by returning $\gamma' = 1$. Therefore, the Chen et al's ID-KI-SC scheme can not achieve IND-ID-KI-SC-CCA2.

Attack the EUF-ID-KI-SC-CMA property. According to the EUF-ID-KI-SC-CMA property in [11], the adversary $\mathcal{A}$ chooses a message $\mathbf{m} \in \{0,1\}^{n_m}$ and two identities **a** and **b** in time period t. At first, $\mathcal{A}$ makes a signcryption query on (**m**, **a**, **b**, t) and a temporary private key on **b**. Upon receiving these queries, the challenger $\mathcal{B}$ returns $(t, \sigma) = \left(t, \left(\sigma^{\langle 1\rangle}, \sigma^{\langle 2\rangle}, \sigma^{\langle 3\rangle}, \sigma^{\langle 4\rangle}, \sigma^{\langle 5\rangle}, \sigma^{\langle 6\rangle}, \sigma^{\langle 7\rangle}\right)\right)$ and $d_{\mathbf{b},t} = \left(d_{\mathbf{b},t}^{\langle 1\rangle}, d_{\mathbf{b},t}^{\langle 2\rangle}, d_{\mathbf{b},t}^{\langle 3\rangle}\right)$ to $\mathcal{A}$. Secondly, $\mathcal{A}$ can be obtained **r** by computing $\mathbf{a}\|\mathbf{m}\|\mathbf{r} = V^{-1}\left(\sigma^{\langle 1\rangle} \cdot \hat{e}\left(d_{\mathbf{b},t}^{\langle 3\rangle}, \sigma^{\langle 4\rangle}\right) \hat{e}\left(d_{\mathbf{b},t}^{\langle 2\rangle}, \sigma^{\langle 3\rangle}\right) / \hat{e}\left(d_{\mathbf{b},t}^{\langle 1\rangle}, \sigma^{\langle 2\rangle}\right)\right)$. After decrypting **m** and **r**, $\mathcal{A}$ then generates $\mathcal{M}_{\mathbf{m}} = \{j \in Z : H_v(\mathbf{m})[j] \oplus \mathbf{r}[j] = 1\}$. Thirdly, $\mathcal{A}$ chooses a new message $\mathbf{m}'$ and computes $\mathbf{r}'$ satisfying $\mathcal{M}_{\mathbf{m}'} = \mathcal{M}_m$ where $\mathcal{M}_{\mathbf{m}'} = \{j \in Z : H_v(\mathbf{m}')[j] \oplus \mathbf{r}'[j] = 1\}$. Especially, from $i = 1$ to $i = n_v$, if $i \notin \mathcal{M}_{\mathbf{m}}$, then $\mathbf{r}'[i] = H_v(\mathbf{m}')[i]$; otherwise, $\mathbf{r}'[i] = \overline{H_v(\mathbf{m}')[i]}$. At last, $\mathcal{A}$ computes $\sigma^{\langle 1\rangle'} = \sigma^{\langle 1\rangle} V^{-1}(\mathbf{a}\|\mathbf{m}\|\ \mathbf{r}) V(\mathbf{a}\|\mathbf{m}'\|\mathbf{r}'), \sigma^{\langle 2\rangle'} = \sigma^{\langle 2\rangle}, \sigma^{\langle 3\rangle'} = \sigma^{\langle 3\rangle}, \sigma^{\langle 4\rangle'} = \sigma^{\langle 4\rangle}, \sigma^{\langle 5\rangle'} = \sigma^{\langle 5\rangle}, \sigma^{\langle 6\rangle'} = \sigma^{\langle 6\rangle}$ and $\sigma^{\langle 7\rangle'} = \sigma^{\langle 7\rangle}$. In this case, $\mathcal{A}$ forges a valid signature $\left(\sigma^{\langle 1\rangle'}, \sigma^{\langle 2\rangle'}, \sigma^{\langle 3\rangle'}, \sigma^{\langle 4\rangle'}, \sigma^{\langle 5\rangle'}, \sigma^{\langle 6\rangle'}, \sigma^{\langle 7\rangle'}\right)$ on message $\mathbf{m}'$ regarding the identities **a** and **b** in time period t.

3 An Improved Scheme

To remedy the weakness in [11], we describe an improved ID-KI-SC scheme based on [13] in this paper.

Setup, **Extract**, **HelperUpt**, **UserUpt**: These algorithms are the same as those in Chen et al.'s scheme.

Signcrypt: Define $\mathbf{m} \in G_T$ to be a bitstring representing a message. Equipped with the temporary private key as $d_{\mathbf{a},t} = \left(d_{\mathbf{a},t}^{\langle 1 \rangle}, d_{\mathbf{a},t}^{\langle 2 \rangle}, d_{\mathbf{a},t}^{\langle 3 \rangle}\right)$ in period t, Alice signcrypts a message $\mathbf{m}$ to Bob as follows. First, Alice picks $r_m, r_t' \leftarrow_R Z_q$ randomly, and lets $r_t = r_t' + k_{a,t}$. After that, Alice performs the following steps:

1. Compute $\sigma_1 = \mathbf{m} \cdot Y^{r_m}, \sigma_2 = g^{r_m}$, $\sigma_3 = \left(u' \prod_{i \in \mathcal{U}_b} u_i\right)^{r_m}, \sigma_4 = \left(u' \prod_{i \in \mathcal{W}_{b,t}} u_i\right)^{r_m}$, $\sigma_5 = d_{\mathbf{a},t}^{\langle 2 \rangle} \cdot g^{r'_t} = g^{k_{a,t} + r_t'} = g^{r_t}$, $\sigma_6 = d_{\mathbf{a},t}^{\langle 3 \rangle} = g^{r_a}$
2. Compute $m = H(\sigma_1, \sigma_2, \sigma_3, \sigma_4, \sigma_5, \sigma_6, \mathbf{a}, \mathbf{b})$ where $H : \{0,1\}^* \rightarrow \{0,1\}^{n_v}$ is a collision-resistant hash function, and let $\mathcal{M}_m \subset \{1, \cdots, n_v\}$ be the set of indices i such that $m[i] = 1$;
3. Compute

$$\begin{aligned}
\sigma_7 &= d_{\mathbf{a},t}^{\langle 1 \rangle} \cdot \left(u' \prod_{i \in \mathcal{W}_{a,t}} u_i\right)^{r_t'} \cdot \left(m' \prod_{i \in \mathcal{M}_m} m_i\right)^{r_m} \\
&= g_2^{\alpha} \left(u' \prod_{i \in \mathcal{U}_a} u_i\right)^{r_a} \cdot \left(u' \prod_{i \in \mathcal{W}_{a,t}} u_i\right)^{r_t' + k_{a,t}} \cdot \left(m' \prod_{i \in \mathcal{M}_m} m_i\right)^{r_m} \\
&= g_2^{\alpha} \left(u' \prod_{i \in \mathcal{U}_a} u_i\right)^{r_a} \cdot \left(u' \prod_{i \in \mathcal{W}_{a,t}} u_i\right)^{r_t} \cdot \left(m' \prod_{i \in \mathcal{M}_m} m_i\right)^{r_m}
\end{aligned}$$

4. Output $(t, \sigma) = (t, (\sigma_1, \sigma_2, \sigma_3, \sigma_4, \sigma_5, \sigma_6, \sigma_7))$.

Unsigncrypt: On input (t,σ), this algorithm outputs $\mathbf{m}$, or $\perp$ (in case the signcryptext is not valid) as follows:

1. Compute $m = H(\sigma_1, \sigma_2, \sigma_3, \sigma_4, \sigma_5, \sigma_6, \mathbf{a}, \mathbf{b})$, and let $\mathcal{M}_m \subset \{1, \cdots, n_m\}$ be the set of indices j such that $m[j] = 1$, where $m[j]$ is the jth bit of m;
2. Output

$$\mathbf{m} = \sigma_1 \cdot \frac{\hat{e}\left(d_{\mathbf{b},t}^{\langle 2 \rangle}, \sigma_4\right) \hat{e}\left(d_{\mathbf{b},t}^{\langle 3 \rangle}, \sigma_3\right)}{\hat{e}\left(d_{\mathbf{b},t}^{\langle 1 \rangle}, \sigma_2\right)}$$

if

$$\hat{e}(\sigma_7, g) = Y \cdot \hat{e}\left(\sigma_6, u' \prod_{i \in \mathcal{U}_a} u_i\right) \hat{e}\left(\sigma_5, u' \prod_{i \in \mathcal{W}_{a,t}} u_i\right) \hat{e}\left(\sigma_2, m' \prod_{i \in \mathcal{M}_m} m_i\right)$$

and $\perp$ otherwise.

It is obvious to observe that our improved scheme preserves the same computation efficiency and signcryptext size of Chen et al.'s scheme [11].

3.1 Security Proof

Theorem 1 *Our improved ID-KI-SC scheme achieves IND-ID-SC-KI-CCA2 under the DBDH assumption in the standard model. Specifically, if there is an adversary* $\mathcal{A}$ *that is able to distinguish two valid ciphertexts during the game defined in* [11] *with advantage at least* ε *when running in time at most t and asking at most* q_e *extract queries,* q_t *temporary private key queries,* q_s *signcryption queries and* q_u *unsigncryption queries, then there is a distinguisher* $\mathcal{B}$ *that solves the DBDH problem in time* $t' \leq t + O((q_e + q_s + q_t + q_u)n_u t_m + (q_e + q_s + q_t)t_e + q_u t_p)$ with advantate $\varepsilon' \geq \frac{\varepsilon}{54q_s(q_e+q_t+q_s)^2(n_u+1)^2(n_v+1)}$, where t_m, t_e, and t_p denote the time for a multiplication, an exponentiation in G and a pairing computation, respectively.

Theorem 2 *Our ID-KI-SC scheme can achieve EUF-ID-KI-SC-CMA property in the standard model, assuming that the computational Diffie-Hellman assumption holds in groups G. Concretely, if there exists an EUF-ID-KI-SC-CMA adversary* $\mathcal{A}$ *that is able to produce a forgery during the game defined in* [11] *with advantage at least* ε *when running in time at most t and asking at most* q_e *extract queries,* q_t *temporary private key queries,* q_s *signcryption queries and* q_u *unsigncryption queries, there exists a challenger that can solve an instance of the CDH problem in time* $t' < t + O((q_e + q_s + q_t + q_u)n_u t_m + (q_e + q_t + q_s)t_e + q_u t_p)$ *with advantage* $\varepsilon' > \frac{\varepsilon}{27q_s(q_e+q_t+q_s)2(n_u+1)2(n_v+1)}$, *where* t_m, t_e, *and* t_p *denote the same quantities as in Theorem 1.*

The security proof is omitted due to the page limit.

4 Conclusion

We have showed that Chen et al.'s ID-KI-SC scheme does not offer IND-CCA2 and EUF-CMA properties. Then, we proposed an improved scheme which preserves the same computation efficiency and signcryptext size. We proved that the improved scheme satisfies confidentiality and unforgeability without random oracles.

Acknowledgments The authors would like to acknowledge National Natural Science Foundation of China under Grant Nos. 61003230, 61370026, Chongqing Key Lab of Computer Network and Communication Technology under Grant No. CY-CNCL-2012-02, the national key scientific and technological special project of China under Grant No. 2011ZX03002-002-03.

References

1. Zheng, Y.: Digital signcryption or how to achieve cost (signature & encryption) $\ll$ cost (signature) + cost(encryption). In: Proceeding Advances in Cryptology-CRYPTO'97, LNCS, vol. 1294, pp. 165–179 Springer, Heidelberg (1997)
2. Dent, A.W., Zheng, Y.: Practical Signcryption, pp. 1–269. Springer, New York (2010)
3. Shamir, A.: Identity-based cryptosystems and signature schemes, In: Proc. Advances in Cryptology-CRYPTO'84, LNCS, vol. 196, pp. 47–53, Springer, Heidelberg (1984)
4. Malone-Lee, J.: Identity based signcryption, Cryptology ePrint Archive, Report 2002/098, 2002. Available from: http://eprint.iacr.org/2002/098 (2002)
5. Chow, S. S. M., Yiu, S. M., Hui, L. C. K et al.: Efficient forward and provably secure ID-based signcryption scheme with public verifiability and public ciphertext authenticity, In: ICISC 2003, LNCS, vol. 2971, pp. 352–369 Springer, Heidelberg (2004)
6. Libert, B,. Quisquater, J.J.: A new identity based signcryption schemes from pairings, In: Proceeding 2003 IEEE Information Theory Workshop, pp. 155–158, Paris, France (2003)
7. Bellare, M., P. Rogaway.: Random oracles are practical: A paradigm for designing efficient protocols, In: Proc. 1st ACM CCS, pp. 62–73. ACM Press. (1993)
8. Bellare, M., A. Boldyreva, A. Palacio.: An uninstantiable random-oracle-model scheme for a hybrid-encryption problem, In: EUROCRYPT 2004, LNCS, vol. 3027, pp. 171–188, Springer, Heidelberg (2004)
9. Jin, Z., Wen, Q., Du, H.: An improved semantically-secure identity-based signcryption scheme in the standard model. Comput. Electr. Eng. **36**, 545–552 (2010)
10. Li, F., Liao, Y., Qin, Z.: Analysis of an identity-based signcryption scheme in the standard model. IEICE Trans. Fundam. Electron. Commun. Comput. Sci. **94-A**, 268–269 (2011)
11. Chen, J., Chen, K., Wang, Y., et al.: Identity-based key-insulated signcryption. Informatica **23**, 27–45 (2012)
12. Dodis, Y., Katz, J., Xu, S., Yung, M.: Strong key-insulated public key cryptosystems, In: Proceeding Advances in Cryptology-Eurocrypt' 02, LNCS 2332, pp. 65–82 Springer, Heidelberg (2002)
13. Li, X., Qian, H., Weng, J., Yu, Y.: Fully secure identity-based signcryption scheme with shorter signcryptext in the standard model. Math. Comput. Model. **57**, 503–511 (2013)

Stability Analysis of a Rapid Scanning Worm Propagation Model with Quarantine Strategy

Yong Yang, Yinling Niu, Fangwei Wang and Honggang Guo

Abstract Rapid scanning worms are a great threat to Internet infrastructure. To effectively defend against them, this paper proposed an epidemic SEIQV model with quarantine and vaccination strategies. Through analysis of this model, its stability condition is obtained: When the basic reproduction number is less than or equal to one, our model is stable at its worm-free equilibrium where worms finally get eliminated. Simulation results show that quarantine strategy is efficient, in terms of the number of infected hosts and reducing worm propagation speed.

Keywords Network security · Worm propagation model · Basic reproduction number · Stability analysis · Endemic equilibrium

1 Introduction

Rapid scanning worms can replicate themselves and actively infect other hosts with certain vulnerability via Internet. Numerous worms have appeared on the Internet whose goal is to compromise the confidentiality, integrity, and availability

Y. Yang
Network Information Center, Yunnan University, Kunming, China
e-mail: yy@ynu.edu.cn

Y. Niu
Hebei Branch, CERNET Company Limited, Shijiazhuang, China
e-mail: niuyl@cernet.com

F. Wang (✉) · H. Guo
Computer Network Center, Hebei Normal University, Shijiazhuang, China
e-mail: fw_wang@hebtu.edu.cn

H. Guo
e-mail: ghg@hebtu.edu.cn

S. Patnaik and X. Li (eds.), *Proceedings of International Conference on Computer Science and Information Technology*, Advances in Intelligent Systems and Computing 255, DOI: 10.1007/978-81-322-1759-6_13,

of infected computing systems. With the emergence of the Internet of things, worms have become a great threat to our work and daily life, caused tremendous economic losses. How to combat worms effectively is an urgent issue confronted with defenders. Therefore, it is necessary to comprehend the long-term behavior of worms and to propose effective strategies to defend against Internet worms.

Based on the infectivity between a worm and a biological virus, some epidemic models [1–6] were presented to depict the propagation of worms. All of them assume that infected hosts in which the worm resides are in an exposed state which cannot infect other hosts. Actually, an infected host that is in latency can infect other hosts by means of some methods. Recently, more research attention has been paid to the combination of worm propagation models and defense strategies to study the prevalence of worms, e.g., worm immunization and quarantine strategy [7]. However, it does not take exposed state into account.

This paper proposes an extended SEIQV worm attack model, which is appropriate for measuring the effects of security countermeasures on worm propagation. Contrary to existing models, our model takes the exposed state and defense strategy into consideration. Using the reproduction number, we derive global stabilities of the worm-free equilibrium and endemic equilibrium. Furthermore, simulation results show the effectiveness of our model. Finally, equilibrium points are confirmed by plots.

2 Mathematical Model Formulation

The total host population N is partitioned into five groups, and any host can potentially be in any of these groups at any time t: the susceptible (S), exposed (E), infectious (I), quarantined (Q), and vaccinated (V). The total number of population N at time t is given by $N(t) = S(t) + E(t) + I(t) + Q(t) + V(t)$. The state transition diagram of the quarantine mode is shown in Fig. 1.

Based on the compartment model presented in Fig. 1, our model is described by the following system of differential equations:

$$\begin{cases} S'(t) = \mu N - \beta SI - \rho S - \mu S, \\ E'(t) = \beta SI - (\mu + \omega + \alpha_1)E, \\ I'(t) = \omega E - (\mu + \alpha_2 + \gamma)I, \\ Q'(t) = \alpha_1 E + \alpha_2 I - (\mu + \theta)Q, \\ V'(t) = \rho S + \gamma I + \theta Q - \mu V, \end{cases} \tag{1}$$

where β is the worm infection rate, ρ is the transition rate in which some susceptible hosts can be directly patched into the vaccinated state, ω is the transfer rate between the exposed and the infectious. Some exposed and infectious ones can be detected by a misuse detection system and then constantly quarantined at rates α_1, α_2, respectively. Some hosts in the quarantined state become vaccinated ones by repairing and then patching at rate θ. Some infectious hosts can be

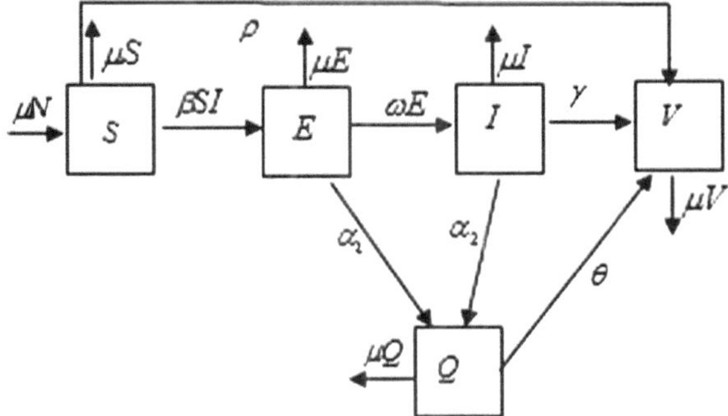

Fig. 1 State transition diagram of the quarantine model

detected and then manually patched at rate γ. The hosts in each of the five groups encounter death at rate μ. Meanwhile, some hosts enter the network at the same rate μ. Since the birth rate is equal to the death one, the population can obviously remain unchanged.

The first four Eqs. in (1) have no dependence on the fifth one, and therefore, the fifth Equation can be omitted. Thus, the system (1) can be rewritten as the following four-dimensional system:

$$\begin{cases} S'(t) = \mu N - \beta \mathrm{SI} - \rho S - \mu S, \\ E'(t) = \beta \mathrm{SI} - (\mu + \omega + \alpha_1)E, \\ I'(t) = \omega E - (\mu + \alpha_2 + \gamma)I, \\ Q'(t) = \alpha_1 E + \alpha_2 I - (\mu + \theta)Q. \end{cases} \tag{2}$$

Summing equations in (2) yields $(S+E+I+Q)' = \mu[N-(S+E+I+Q)] - \rho S - \theta Q \leq \mu[N-(S+E+I+Q)]$, then it follows that $\limsup_{t\to\infty}[S(t)+E(t)+I(t)+Q(t)] \leq N$, thus the set $\Omega = \{(S,E,I,Q) \in R_+^4 : S+E+I+Q \leq N\}$ is positively invariant for (2). Therefore, we will study the global stability of (2) on the set Ω.

For $I = 0$, we can easily obtain the worm-free equilibrium $P_0 = (\mu N/(\rho+\mu), 0, 0, 0)$.

For $I^* > 0$, the unique endemic equilibrium is: $P^* = (S^*, E^*, I^*, Q^*)$, where

$$S^* = \frac{(\omega+\alpha_1+\mu)(\alpha_2+\mu+\gamma)}{\omega\beta}, \quad I^* = \frac{\omega\beta pN - (\rho+\mu)(\omega+\alpha_1+\mu)(\alpha_2+\mu+\gamma)}{\beta(\omega+\alpha_1+\mu)(\alpha_2+\mu+\gamma)}$$
$$E^* = \frac{(\alpha_2+\mu+\gamma)I^*}{\omega}, \quad Q^* = \frac{[\alpha_1(\alpha_2+\mu+\gamma)+\omega\alpha_2]I^*}{[\omega(\theta+\mu)]}.$$

By applying the method of the next generation matrix in [8], we obtain the basic reproduction number of system (2)

$$R_0 = \beta\mu N/[(\rho+\mu)(\omega+\alpha_1+\mu)]. \tag{3}$$

3 The Stability Analysis for Equilibriums

3.1 Worm-Free Equilibrium and Its Stability

Lemma 1 The worm-free equilibrium P_0 is locally asymptotically stable in Ω if $R_0 < 1$ and unstable if $R_0 > 1$.

Proof The Jacobian matrices of the system (2) at P_0 is

$$J(P_0) = \begin{Bmatrix} -(\rho+\mu) & -\beta\mu N/(\rho+\mu) & 0 & 0 \\ 0 & \beta\mu N/(\rho+\mu) - (\omega+\alpha_1+\mu) & 0 & 0 \\ 0 & \omega & -(\alpha_2+\mu+\gamma) & 0 \\ 0 & \alpha_1 & \alpha_2 & -(\theta+\mu) \end{Bmatrix}. \tag{4}$$

The characteristic equation of Eq. (4) is:

$$(\lambda+\rho+\mu)(\lambda-\beta\mu N/(\rho+\mu)+\omega+\alpha_1+\mu)(\lambda+\alpha_2+\mu+\gamma)(\lambda+\theta+\mu)=0. \tag{5}$$

From Eq. (5), the characteristic values are obtained by: $\lambda_1 = -(\rho+\mu)$, $\lambda_2 = \beta\mu N/(\rho+\mu) - \omega - \alpha_1 - \mu$, $\lambda_3 = -(\alpha_2+\mu+\gamma)$, $\lambda_3 = -(\theta+\mu)$.

According to Routh–Hurwits criteria [9], the worm-free equilibrium P_0 is stable if and only if all of the characteristic values are less than 0. Obviously, λ_1, λ_3, λ_4 are negative. Therefore, the stability of (2) relies on whether $\lambda_2 < 0$. By a direct computation of λ_2, the stability condition is obtained

$$R_0 = \beta\mu N/[(\rho+\mu)(\omega+\alpha_1+\mu)] < 1.$$

That is, $R_0 < 1$ is the necessary and sufficient condition that the system (2) is locally asymptotically stable at the worm-free equilibrium P_0. Otherwise, the system (2) is not stable when $R_0 > 1$. The proof is completed.

Lemma 2 When $R_0 \le 1$, the worm-free equilibrium P_0 is globally asymptotically stable in Ω. When $R_0 > 1$, all solutions starting in Ω and sufficiently close to P_0 move away from $\{P_0\}$.

Proof From the first equation of system (2), we can obtain $S'(t) \le \mu N - (\rho+\mu)S(t)$. When $t \to \infty$, we can obtain $S(t) \le \mu N/(\rho+\mu)$. Consider the following Lyapunov function: $L = \omega E + (\omega+\alpha_1+\mu)I$.

Its derivative along the solutions in (2) is:

$$\begin{aligned}
L' &= \omega E' + (\omega + \alpha_1 + \mu)I' \\
&= \omega(\beta SI - (\omega + \alpha_1 + \mu)E) + (\omega + \alpha_1 + \mu)(\omega E - (\alpha_2 + \mu + \gamma)I) \\
&\le \omega\beta SI - (\omega + \alpha_1 + \mu)I = \omega I(\beta S - (\omega + \alpha_1 + \mu)) \\
&\le \omega I\left(\frac{\beta\mu N}{\rho + \mu} - (\omega + \alpha_1 + \mu)\right) = (\omega + \alpha_1 + \mu)\omega I(R_0 - 1) \le 0.
\end{aligned}$$

Furthermore, $L' = 0$ if and only if $I = 0$ or $R_0 = 1$. Thus, the largest compact invariant set in $\{(S, E, I, Q)|L' = 0\}$ is the singleton $\{P_0\}$. When $R_0 \le 1$, the global stability of P_0 follows from LaSalle's invariance principle [10]. LaSalle's invariance principle [10] implies that P_0 is globally asymptotically stable in Ω. When $R_0 > 1$, it follows from the fact $L' > 0$ if $I > 0$. This completes the proof.

3.2 Endemic Equilibrium and Its Stability

Now, we investigate the local stability of the endemic equilibrium P^* . The Jacobian matrix of system (2) at the endemic equilibrium P^* is:

$$J(P^*) = \begin{pmatrix} -\beta I - \mu - \rho & 0 & -\beta S & 0 \\ \beta I & -\mu - \omega - \alpha_1 & \beta S & 0 \\ 0 & \omega & -\mu - \alpha_2 - \gamma & 0 \\ 0 & \alpha_1 & \alpha_2 & -\mu - \theta \end{pmatrix}.$$

Thus, the eigenfuction of $J(P^*)$can be denoted as

$$(\lambda_1 + (\mu + \theta))(\lambda^3 + C_1\lambda^2 + C_2\lambda + C_3) = 0, \tag{6}$$

where

$$\begin{aligned}
C_1 &= \alpha_1 + \alpha_2 + 3\mu + \rho + \gamma + \omega + \beta I, \\
C_2 &= (\alpha_1 + \alpha_2 + 2\mu + \gamma + \omega)(\mu + \rho + \beta I),. \\
C_3 &= \omega\beta^2 SI.
\end{aligned}$$

By a direct calculation, we obtain that $AB - C > 0$. According to the theorem of Routh–Hurwitz, the endemic equilibrium P^* is locally asymptotically stable. This completes the proof.

Lemma 4 If $R_0 > 1$, the unique positive equilibrium P^* of model (2) is globally asymptotically stable in Ω.

Proof It is easy to see that the model (2) has unique positive equilibrium P^* if $R_0 > 1$ holds. Then, we consider the following Lyapunov function [11] defined as:

$$L(t) = \int_{S_1^*}^{S} \frac{x - S_1^*}{x} \mathrm{d}x + \int_{E_1^*}^{E} \frac{x - E_1^*}{x} \mathrm{d}x. \tag{8}$$

The time derivative of *L*(*t*) along the solution of Eq. (3) is given by

$$\begin{aligned} L'(t) &= \left(\frac{S - S^*}{S}\right) S' + \left(\frac{E - E^*}{E}\right) E \\ &= \left(1 - \frac{S^*}{S}\right) [\mu N - \beta \mathrm{SI} - \mu S - \rho S] + \left(1 - \frac{E^*}{E}\right) [\beta \mathrm{SI} - (\mu + \alpha_1 + \omega) E]. \\ &\le -\mu N \left(\frac{S}{S^*}\right) \left(\frac{S^*}{S} - 1\right)^2 \le 0. \end{aligned}$$

Thus, we prove that the endemic equilibrium P^* is globally stable. This completes the proof.

4 Numerical Simulations

Let $N = 75{,}000$, $S_0 = 50{,}000$, $I_0 = 25{,}000$, $\beta = 4{,}000/2^{32}$, $\mu = 0.00001$, $\omega = 0.005$, $\alpha_1 = 0.002$, $\alpha_2 = 0.007$, $\gamma = 0.006$, $\theta = 0.09$, and $\rho = 0.002$, where $R_0 = 0.499$. We can see the result in Fig. 2. From Fig. 2, we can clearly see that the tendency of the worm propagation is depressive, which is consistent with Lemmas 1 and 2. Finally, all infected hosts vanish and the population, in the long term, is in a vaccinated state.

Let $N = 75{,}000$, $S_0 = 74{,}990$, $I_0 = 10$, $\beta = 4{,}000/2^{32}$, $\mu = 0.001$, $\omega = 0.008$, $\alpha_1 = 0.002$, $\alpha_2 = 0.002$, $\gamma = 0.00006$, $\theta = 0.09$, and $\rho = 0.002$, where $R_0 = 2.117$. We can see the result in Fig. 3. From Fig. 3, the number of susceptible,

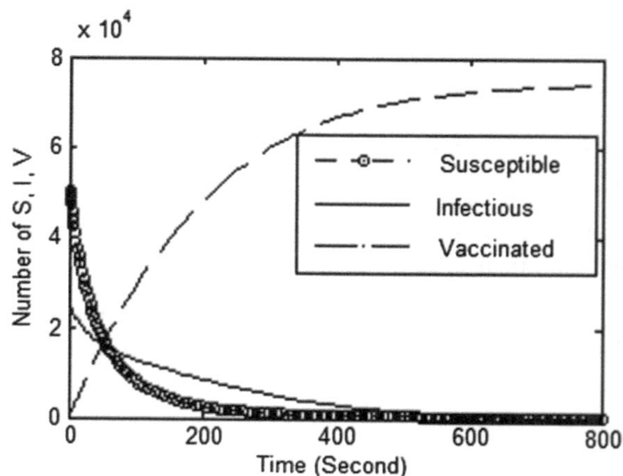

Fig. 2 Stability of worm-free equilibrium

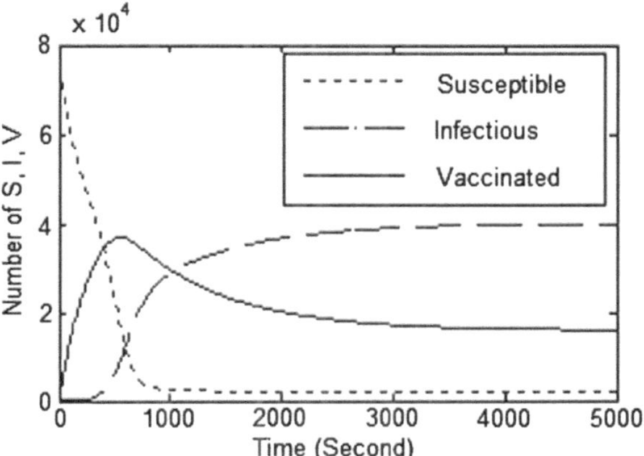

Fig. 3 Stability of endemic equilibrium

infected, and vaccinated hosts eventually become positive values between 0 and N. The worm does not disappear. Finally, all states reach their equilibriums. This is fully consistent with the conclusions of Lemmas 3 and 4.

Let $N = 75{,}000$, $S_O = 74{,}990$, $I_O = 10$, $\beta = 4{,}000/2^{32}$, $\mu = 0.001$, $\omega = 0.5$, $\alpha_2 = 0.002$, $\gamma = 0.0006$, $\theta = 0.009$, and $\rho = 0.002$. When α_1 is equal to 0.002, 0.004, 0.006, 0.008, respectively, We can see the result in Fig. 4. From Fig. 4, we can see that the quarantined rate α_1 does not play an important role in decreasing the number of infected hosts infected by worms.

Let $N = 75{,}000$, $S_O = 74{,}990$, $I_O = 10$, $\beta = 4{,}000/2^{32}$, $\mu = 0.001$, $\omega = 0.008$, $\alpha_1 = 0.002$, $\alpha_2 = 0.002$, $\gamma = 0.00006$, $\theta = 0.09$, and $\rho = 0.002$, where $R_O = 2.117$. We can see the result in Fig. 5. We can obtain that the quarantined rate α_2 plays an important role in containing the infected hosts. The quarantined rate relies mainly on the accuracy of intrusion detection systems. We can improve the efficiency and false positive of intrusion detection systems to obtain a larger quarantined rate.

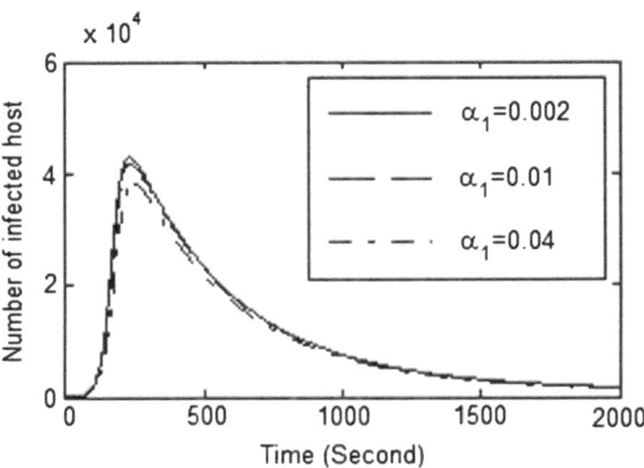

Fig. 4 Effect of quarantined rate α_1

Fig. 5 Effect of quarantined rate α_2

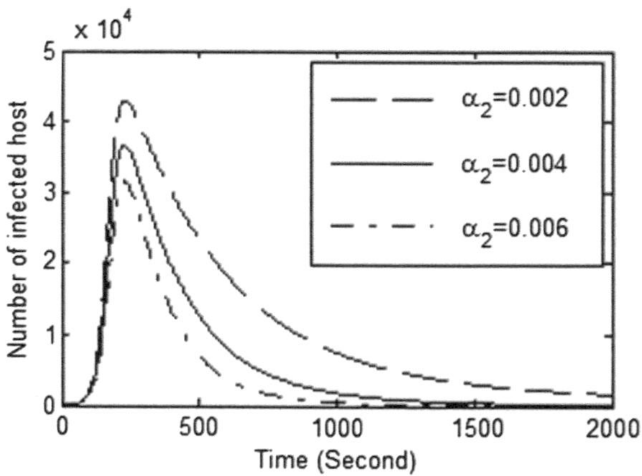

5 Conclusions

This paper proposed an extended epidemic model to model rapid scanning worms, which takes quarantine and vaccination strategies into consideration. Firstly, we obtain the basic reproduction number determining whether the worm extinguishes. Secondly, the global asymptotic stabilities of our model have proved by the use of the Lyapunov function. When the reproduction number is less than or equal to one, the proposed model has only a worm-free equilibrium which is globally stable, it implies the worm dies out eventually; when the reproduction number is larger than one; our model has a unique endemic equilibrium which is globally stable, and it implies that the worm persists in the whole host population and tends to a steady state. Finally, some numerical examples are given to verify our conclusions.

Acknowledgments This research was supported by the Natural Science Foundation of China under No. 61272541, China Postdoctoral Science Foundation of China under No. 2013M532018, and Foundation of Hebei Normal University under No. L2010B21.

References

1. Ren, J., Yang, X., Zhu, Q., Yang, L., Zhang, C.: A novel computer virus model and its dynamics. Nonlinear Anal. Real World Appl. **13**(1), 376–384 (2012)
2. Ren, J., Yang, X., Yang, L., Xu, Y., Yang, F.: A delayed computer virus propagation model and its dynamics. Chaos, Solitons Fractals **45**(1), 74–79 (2012)
3. Toutonji, S.M., Yoo, Park, M.: Stability analysis of VEISV propagation modeling for network worm attack. Appl. Math. Model. **36**(6), 2751–2761 (2012)
4. Fan, W., Yeung, K.H., Wong, K.Y.: Assembly effect of groups in online social networks. Physica A **392**(5), 1090–1099 (2013)

5. Yang, L., Yang, X., Zhu, Q., Wen, L.: A computer virus model with graded cure rates. Nonlinear Anal. Real World Appl. **14**(1), 414–422 (2013)
6. Mishra, B.K., Pandey, S.K.: Dynamic model of worms with vertical transmission in computer network. Appl. Math. Comput. **217**(21), 8438–8445 (2011)
7. Yao, Y., Guo, L., Guo, H., Yu, G., Gao, F., Tong, X.: Pulse quarantine strategy of internet worm propagation: modeling and analysis. Comput. Electr. Eng. **38**(9), 1047–1061 (2012)
8. van den Driessche, P., Watmough, J.: Reproduction numbers and sub-threshold endemic equilibrium for compartmental models of disease transmission. Math. Biosci. **180**(1), 29–48 (2002)
9. Robinson, R.C.: An Introduction to Dynamical System: Continuous and Discrete. Pearson Education Inc. Press, New York (2004)
10. LaSalle, J.P.: The Stability of Dynamical Systems, Regional Conference Series in Applied Mathematics. SIAM, Philadelphia (1976)
11. Xu, W., Zhang, Z.: Global stability of SIR epidemiological model with vaccinal immunity and bilinear incidence rate. Coll. Math. **19**(6), 76–80 (2003)

A Fuzzy Bayesian Approach to Enhance SCADA Network Security

Shu Jin, Tangjun Dan, Li Zhang and Liu Liu

Abstract To enhance the intrusion detection system with more accuracy and less false-positive rate while still providing acceptable performance and adaptivity, a Bayesian anomaly intrusion detection system using fuzzy probability assignment is presented. After categorizing the security-related system events and properties into four models represented by their corresponding fuzzy membership functions, the real-time probabilities of specific security-breaching events are calculated and the decision of whether the system supervised is under attack is made from the synthesis of the probabilities generated. A Bayesian belief network algorithm is presented to synthesize the real-time fuzzy probabilities at runtime and proved to be effective by simulations. Compared with previous works that employs the threshold methods in identifying attacks, the algorithm describes the probabilities of security events more accurately through utilizing the continuous fuzzy probability model and scales better for modeling various kinds of system security properties in normal system behavior profiling.

Keywords SCADA security · Anomaly detection · Fuzzy probability assignment · Bayesian belief network

S. Jin (✉) · T. Dan · L. Liu
Nanjing SAC-Metso Control System Co., Ltd, XingHuo Road #8,
Nanjing 210032 JiangSu, People's Republic of China
e-mail: jinshu@acm.org

T. Dan
e-mail: tangjun-dan@sac-china.com

L. Liu
e-mail: liu-liu@sac-china.com

L. Zhang
Nanjing SAC Rail Transit Engineering Co., Ltd, XingHuo Road #8,
Nanjing 210032 Jiangsu, People's Republic of China
e-mail: li-zhang@sac-china.com

S. Patnaik and X. Li (eds.), *Proceedings of International Conference on Computer Science and Information Technology*, Advances in Intelligent Systems and Computing 255, DOI: 10.1007/978-81-322-1759-6_14,

1 Introduction

Invented by scientists with good will to exchange information freely with each other despite distances, Internet (based on TCP/IP protocol suites) is fundamentally vulnerable to malicious exploitations. With the wide spread use of the World Wide Web, instant messaging, and many other communication applications, Internet plays an increasingly important role in practically every aspect of human activities. On the other side of the coin, all sorts of security breaches and network attacks cooked by people with not that good intensions flooded the Internet and caused great loss, the distributed denial-of-service (DDoS) attacks launched against Yahoo.com and Amazon.com in year 2000 annoyed a major number of Web surfers and filled the cyberspace with horror. As network security issues attract more concern every single minute, the data communication within/between computerized industrial automation systems is left victim to the network attacks without even the least care necessary.

Most network security systems fall into the following two categories: (1) static security enhancements such as data encryption, message digesting, public key certificate authentication, and firewalls, which protect the system effectively against exterior hostilities; (2) dynamic attack countermeasures of real-time intrusion detection and mitigation, which may shield the system extensively even from the attacks launched inside the proprietary local network. Intrusion detection systems (IDS) recently in use are architected according to two models, Misuse and Anomaly. A misuse IDS profiles every bit of known malicious actions and keeps trying to recognize one at run time, while an anomaly IDS figures out and records the characteristic variables when systems work properly in perfect health and strive to identify some abnormities in real time.

As a mechanism used for probabilistic machine reasoning, Bayesian belief network may generate pretty good results after fed with priori knowledge requested and is proved to be effective [1]. In the training stage, a variety of variables relevant to system security are recorded and filtered to profile the normal working status of host system [2–5], when put to use, the IDS is collecting characteristic variables continuously to calculate probabilities of specific anomaly aspects as according to different fuzzy membership functions designated, with all those probabilities synthesized, the IDS will decide whether the system is in good health or under some kind of attacks, whatever their type.

2 Architecture

As illustrated in Fig. 1, the working model of the IDS contains two stages: system profiling and the real-world anomaly intrusion detection. A proper set of system properties are chosen as according to their relevance to specific system security characteristics and monitored continuously to collect a data set large enough for

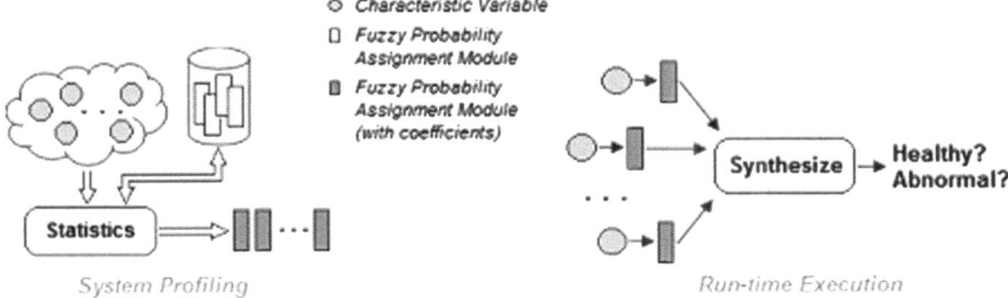

Fig. 1 Working model of the IDS

analyze and extract coefficients applied to the corresponding fuzzy membership functions, which are created to profile the normal system behavior; after put into operation, the system properties are scanned continuously as the input of the fuzzy probability assignment functions, the result will be passed to the Bayesian belief network synthesizer, which is responsible to make a real-time decision of whether the host system is under attack or in good health.

3 Normal System Behavior Profiling

To profile the normal behavior of a system in operation, four models are suggested to map out system security-related properties. For each property A_i to be monitored, if it is considered abnormal (represented as A_i), the system may be out of order either (noted by $\neg I$); let us simplify the problem with $P(I/A_i) = P(A_i)$, then $P(\neg I/A_i) = P(\neg A_i) = 1 - P(A_i)$. Likewise, in the fuzzy Bayesian algorithm, $P(A_i/\neg I)$ and $P(X/E_{pi})$ can be calculated similarly, with the support of statistic knowledge such as $P(A_i)$, $P(E_{ci}/X)$ and $P(P_{mn})$ acquired during the normal system behavior profiling procedure.

1. Trapezium Distribution

As defined in formula (1), the model is applied for the system properties with a relatively clear cut and continuous region of legal values, such as the number of log entries recorded in a specific time period. Once a real-time value falls outside, it is highly suspected that the system undergoes some abnormal situation

$$\neg A_i(x) = \begin{cases} e^{-\left(\frac{x-a}{\sigma}\right)^2}, & a < x \\ 1, & a \le x \le b \\ e^{-\left(\frac{x-a}{\sigma}\right)^2}, & x > b \end{cases} \tag{1}$$

2. Right-Side Gaussian Distribution

The model describes the system properties with their legal values limited within a upper bound, which indicates that the more a value exceeds the upper limit, the more likely the host system is under some kinds of attacks. For example, the uptime of a host, the login/out counts and RTDB access rate in a specific period of time

$$\neg A_i(x) = \begin{cases} 1, & x \leq a \\ e^{-\left(\frac{x-a}{\sigma}\right)^2}, & x > a \end{cases}. \tag{2}$$

3. Left-Side Gaussian Distribution

Like its right-side counterpart, left-side distribution models the system properties with their legal values limited within a lower bound, which indicates that the more a value exceeds the lower limit, the more likely a system is in an abnormal situation. The time interval recorded between logons is a good example for this model

$$\neg A_i(x) = \begin{cases} 0, & x \leq a \\ 1 - e^{-\left(\frac{x-a}{\sigma}\right)^2}, & x > a \end{cases}. \tag{3}$$

4. Gaussian Distribution

Gaussian distribution model is introduced to fit the system properties with the majority of their legal values limited within a Gaussian bell curve, which indicates that any value drops outside the curve will be treated as a signal of abnormity as according to its distance against the center. RTDB access interval, CPU utilization percentages, data points I/O load, network flow, and many other system properties agree with the Gaussian model

$$\neg A_i(x) = e^{-\left(\frac{x-u}{\sigma}\right)^2}. \tag{4}$$

Figure 2 is the diagrams of the four distribution models. Compared with the simple probabilistic induction models that derive their decision against thresholds, continuous membership functions generate more smooth results, which significantly improve the accuracy and adaptivity of system behavior profiling.

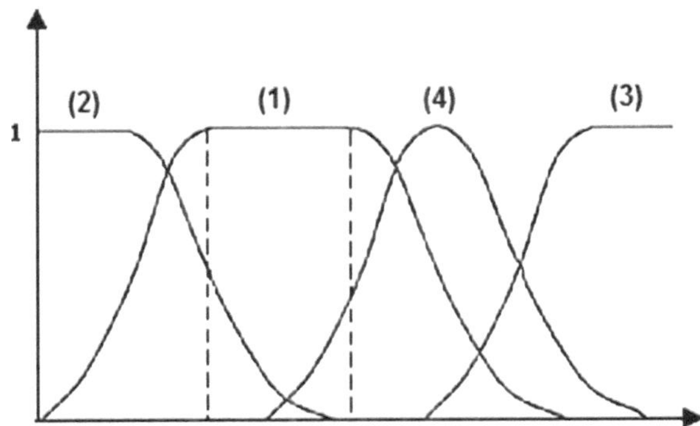

Fig. 2 Membership functions for normal system behavior profiling

4 Anomaly Detection Algorithm

As a decision making model based on conditional probabilities, Bayesian belief network derived its ability from a directed acyclic graph (DAG) connected with interdependent nodes representing entities in the problem space that are linked through edges, which direct from the casuals to the resultants. With each edge assigned the numerical value of the conditional probability between the two end nodes, a Bayesian belief network integrates all the knowledge of interrelations between the facts concerned. With priori and posterity information combined, Bayesian belief network is perfect in modeling cooperating entities interconnected with each other with individual degrees of relevancy in a smooth way.

Illustrated in Fig. 3 is the Bayesian belief network architected for the anomaly IDS for enhancing SCADA network security. Target variable of X represents the system security status, whether a host system is normal ($X = I$) or under some kinds of attacks ($X = \neg I$). $E_p = \cup E_{pi}$ covers all the security-related system properties concerned, which constitutes the priori knowledge of the Bayesian belief network. Once polled with a real-time value E_{pi} = val, a fuzzy probability $P(X = I/E_{pi}$ = val) is assigned to the corresponding edge. Similarly, for each

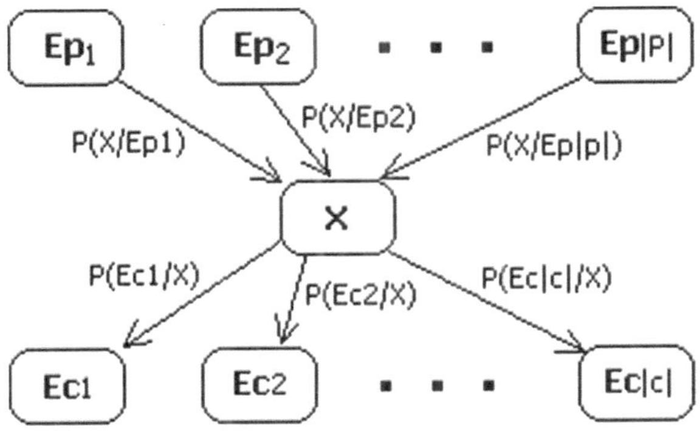

Fig. 3 Bayesian network for anomaly detection

posterity property determined by X, which values will be impacted during security breaches and described as $P(E_{ci} = \text{val}/X)$, $E_c = \cup E_{ci}$. Once all the data relevant is collected as $E = E_c \cup E_p$, a decision whether the system runs properly or not ($X = I$ or $X = \neg I$) can be made as according to the principle introduced by formula (5)

$$P(X/E) \propto P(E_c/X)P(X/E_p). \tag{5}$$

To calculate $P(E_c/X)$, take no account of the interrelations between priori properties, each E_{ci} is thus considered independent as $P(E_{ci}/E_{cj1}, E_{cj2} \ldots E_{cjm}, X) = P(E_{ci}/X)$, $j_k \neq i(k \in 1 \ldots m)$, $r_c = P(E_c/X)$ results (6), in which formula $P(E_{ci}/X)$ indicates the influences X impacted on each posterity property E_{ci}. Specific values of $P(E_{ci}/X)$ can be tuned by system administrators as parameters through their knowledge acquired regarding their relevance, sensitivity, and priority in view of system security

$$\begin{aligned} P(E_c/X) &= P(E_{c1}, E_{c2}, \ldots E_{c|c|}/X) \\ &= P(E_{c1}/X)P(E_{c2}/X)\ldots P(E_{c|c|}/X) \\ &= \prod_{I=1}^{|c|} P(E_{ci}/X) \end{aligned} . \tag{6}$$

Instrumented by [6], $P(X/E_p)$ is calculated and summed up on the complete set of different input priori properties, with P_{mn} for a specific value of P_m in its legitimate span. Have the independence assumption in mind, $P(P_{ij}/E_{p1}, E_{p2}, \ldots E_{pi}, \ldots E_{p|p|}) = P(P_{ij}/E_{pi})$, $P(P_1, P_2 \ldots P_{|p|}/E) = \prod P(P_i/E)$, we have

$$\begin{aligned} r_p &= P(X/E_p) = P(X/E_{p1}, E_{p2}, \ldots, E_{p|p|}) \\ &= \sum_{\text{all i,j,...,k}} P(X/P_{1i}, P_{2j}, \ldots, P_{|p|k})P(P_{1i}, P_{2j}, \ldots, P_{|p|k}/E_{p1}, E_{p2}, \ldots, E_{p|p|}) \\ &= \sum_{\text{all i,j,...,k}} P(X/P_{1i}, P_{2j}, \ldots, P_{|p|k})P(P_{1i}, E_{p1})P(P_{2j}/E_{p2}), \ldots, P(P_{|p|k}/E_{p|p|}) \\ &= \sum_{\text{all i,j,...,k}} P(X/P_{1i}, P_{2j}, \ldots, P_{|p|k})P(P_{1i})P(P_{2j}), \ldots, P(P_{|p|k}) \end{aligned} \tag{7}$$

In formula (7), each $P(P_{mn}/E_{pm})$ represents a possible value of the priori node P_m, which can be figured out through performing statistics calculations on the data collected in the training stage. As indicated by formula (5), we have

$$P(I/E) \propto r_{pI}, P(I/E) \propto r_{cI}, P(\neg I/E) \propto r_{p\neg 1}, P(\neg I/E) \propto r_{c\neg I}$$

substitute $X = I$, $X = \neg I$ in formula (6) and (7) and then normalize it as follows:

$$P(I/E) = \frac{r_{c1}r_{p1}}{r_{c1}r_{p1} + r_{c2}r_{p2}}, \; P(\neg I/E) = \frac{r_{c2}r_{p2}}{r_{c1}r_{p1} + r_{c2}r_{p2}}$$

To make the final decision of whether the host system is healthy or in an abnormal state, $P(I/E)$ is divided by $P(\neg I/E)$ as suggested by formula (8). If the quotient R is greater than δ (typically set to 1), the computer monitored may well be suffering from some malicious manipulations; while if $R < \delta$, the host system is more likely to work properly. Generated with all the priori and posterity knowledge synthesized, the decision made is much more expressive and will better fit computing systems with diversified hardware and software environments as compared to the simple threshold approaches.

$$R = \frac{P(I/E)}{P(\neg I/E)} = \frac{r_{c1} r_{p1}}{r_{c2} r_{p2}}. \tag{8}$$

5 Simulations and Results

Prototyped with Microsoft Visual C++ 2003 and put to run as a background daemon service on the test bed machine (IBM NetVista 3305HC3, Pentium 4 2.4G CPU, 256M RAM, 10/100M Ethernet) recording a set of system properties for a continuous 30 days, four properties are chosen to profile the normal behavior of the host system: (1) Logon Frequency, right-side Gaussian, 0–3 times a day, legitimate range (0–25); (2) Operation Time, trapezium distribution, 3–11 h/day, legitimate range (0–24); (3) CPU Utilization Percentage, Gaussian, around 7 % typically, legitimate range (0, 100 %); (4) RTDB Data Points Accessing Rate, Gaussian, around 60 ops/s for continuous historical data recording and HMI updates, legitimate range (0, 10^5 ops/s). With the first two set as preconditions and the latter two consequents, a Bayesian belief network is constructed with the parameters retrieved.

A simulator program that traverses the legitimate ranges of the four properties to generate all the possible scenarios is introduced in Table 1, while the entire anomaly detection mechanism raises more alarms than the human supervisor, the Bayesian belief network approach outperforms its counterparts through providing the lowest false-positive rate, which is supported by the results suggested in [7] as well.

Table 1 Performance comparisons

	Abnormal (%)	Positive (%)	False positive (%)
Human	75.3	24.7	N/A
Simple probabilistic	80.9	19.1	7.4
Weighted simple probabilistic	84.6	15.4	12.4
Bayesian belief network	79.2	20.8	5.2
Threshold approach	83.5	16.5	10.9

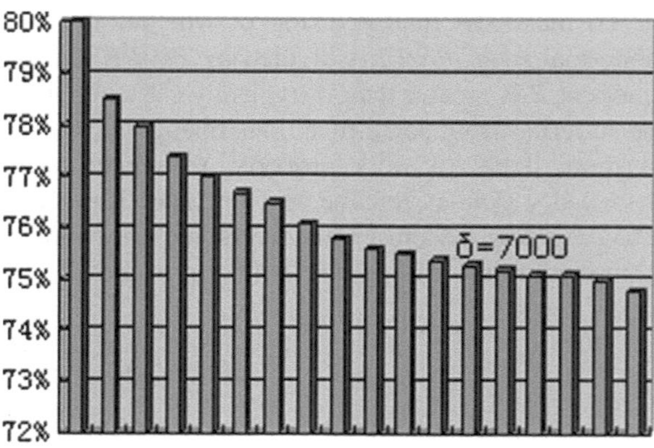

Fig. 4 Positive rate with δ

As the empirical parameter to filter out false positives, δ may be determined both with user knowledge and real-world benchmarking results. In the prototype implementation, 18 integers from 5 through 15,000 are tested to figure out the optimized value for the host system, which is around 7,000 as illustrated in Fig. 4.

Acknowledgments Many thanks go to Mr. LE Ling-zhi for his invaluable assistance. Thank you Mr. LIU Xiao-meng, Mr. XU Bin, and Mr. ZHU Neng-fei; without your encouragement and support, the work cannot be accomplished as smoothly.

References

1. Mitchell, T.M.: Machine Learning. McGrall-Hill Press, New York (1997)
2. Puttini, R.S., Marrakchi, Z., Mé, L.: A Bayesian classification model for real-time intrusion detection. AIP Conf. Proc. **659**, 150–162 (2003)
3. Hastings, J.R.: Incremental Bayesian segmentation for intrusion detection. MIT Master's Thesis (2003)
4. Kruegel, C., Mutz, D., Robertson, W., et al.: Bayesian event classification for intrusion detection. In: Proceedings of 19th Annual Computer Security Applications Conference (2003)
5. Gowadia, V., Farkas, C., Valtorta, M.: PAID: a probabilistic agent-based intrusion detection system. Comput. Sec. **24**(7), 529–545 (2005)
6. Duda, R.O., Hart, P.E., Stork, D.G.: Pattern Classification. Wiley Press, NJ (2000)
7. Burroughs, D.J., Wilson, L.F., Cybenko, G.V.: Analysis of distributed intrusion detection systems using Bayesian methods. In: Proceedings of IPCCC (2002)

Trusted Network Access Authentication Scheme Based on the Label

Yu Wang, Yu Duan and Fei Wang

Abstract Trusted network is a new direction of Internet research. On the basis of trusted network, this paper puts forward a trusted terminal that takes the USBKey as a trusted root and designed a trusted label that used to access authentication. Then, against the background of the trusted local area network (LAN), this paper describes the initial access authentication process and the sustainable authentication process when the trusted terminal access the trusted LAN, meanwhile, analyzes its safety performance.

Keywords Trusted network · USBKey · Label · Access authentication

1 Introduction

At present, with the rapid expansion of the computer, the information network presented an explosive development situation. The Internet not only plays a key role in the entire world's economic and social development, but also becomes a fundamental part of people's Daily life. At the same time, the lack of effective means of identity authentication and behavior regulation, and difficult to online tracking, forensics, positioning attack source, all cause the network security problems emerge in endlessly, that seriously threat the security, stable operation, and sustainable development of network. Therefore, how to build a secure and a controlled trusted network has become a focus now [1, 2]. All the truthfulness of the information source [3] is an important research content of network security architecture. The access authentication mechanism is an effective and safe method

Y. Wang (✉) · Y. Duan · F. Wang
Department of Information Equipment, Academy of Equipment, Beijing, China
e-mail: duanyu_gogo@163.com

Y. Duan
e-mail: dyaiydb@163.com

S. Patnaik and X. Li (eds.), *Proceedings of International Conference on Computer Science and Information Technology*, Advances in Intelligent Systems and Computing 255, DOI: 10.1007/978-81-322-1759-6_15,

that guarantees the truthfulness of the information source. The network access authentication can not only check the real identity of the users who access to the network, but also audit some unsafe activities of the access user and limit its malicious behavior. Therefore, only from the perspective of the source, and solved the access authentication, the management and control problems of user and device can effectively solve the problems of network security and build a trusted computing network environment [4].

Based on the trusted network, this paper proposes an access authentication scheme that is on the basis of a trusted label. This scheme tries to control the access of terminals that are the source of unsafe factors in the network and ensures the security and integrity, so as to realize trusted and controlled of the network. The scheme adopts the trusted labels produced by trusted terminal to bind the access user information and terminal ID, realized the user's identity information and real corresponding relationship of user terminals, and the user's terminal sustainable certification process ensures the credibility of the terminal.

2 Summary of the Trusted Networks

Literature [1] pointed out that "trusted network should be that the behavior of the system and the results are predictable should do that the behavior and state can be monitored, the behavior results can be evaluated, and in addition, the abnormal behavior can be controlled. In particular, the credibility of the network should include a set of properties: from the user's point of view, it needs to guarantee the security and survivability of the service; from the designer's point of view, it needs to provide the control of the network. Different from security, survivability, and controllability in the traditional concept that is dispersed and isolated, trusted network will be under the target that the network is credibility, integrate the three basic attributes (security, survivability, controllability), and around the trust maintenance and behavior controlled between the network components become an organic whole." Trusted terminal will restart access the trusted network when it re-engages; then, gateway will control terminal to access. By the means of access authentication of trusted terminal, it can realize effective extension of trusted network and reduce the potential security risks that the untrusted terminal access to the network.

The trusted terminals in this paper use UKey as the trusted root of security platform, build up a trusted chain, and coordinate system software to ensure the safety of the terminal. The descriptions are shown in Fig. 1. The trusted chain takes the UKey as a trusted root. After the machine power on, the BIOS boot and identify with UKey mutually, and further to measure the integrity of the MBR, OSLOADER, OS kernel. Only succeeded check all component, then it can start run and guide the next component boot. Thus the terminal uses the UKey as a trusted root, builds a integrated trusted chain, completely pass the trust one by one, and provides a safe running environment for application. If the integrity of the checking component was broken, system will stop run.

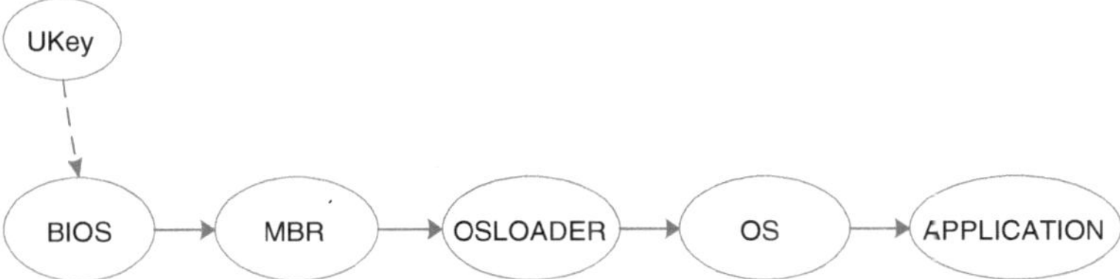

Fig. 1 Trusted terminal trust chain

3 Trusted Network Access Authentication Scheme Based on the Label

3.1 Scene Setting

Here, existing the following scenario shown in Fig. 2, the local area network (LAN) consists of trusted terminal, the exceptional devices, server, and gateway. Depending on the type of business, the trusted terminal in LAN can be divided into different kinds of security domain; The exceptional device is terminal which have not be configured the USBKey, but the management center allows it exists in the LAN and authorized to it; in order to facilitate unified management, all the servers in the LAN will be into a business server domain; The gateway outside the LAN holds the firewall functions, not only can stop the illegal terminals connected to the LAN, and using the label technology can control abnormal traffic.

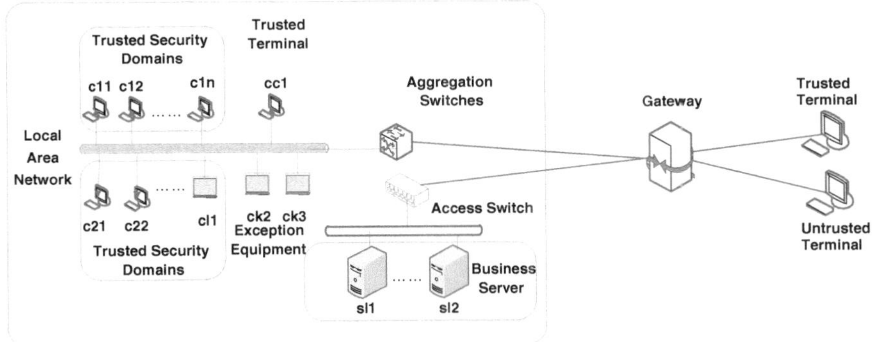

Fig. 2 Trusted LAN

3.2 Trusted Label

Trusted label is a trusted tag which is produced by the network filter driver of the trusted terminal and loaded to packet; the tag includes some information such as terminal ID, user's information. In order to prevent IP packets to be tampered with attacker, it need to add the source IP address and destination IP address of packets, when use the Ks that is produced by the terminals and gateway to work out the message verification code. That is to say, use the session key Ks and HMAC algorithm to calculate the information of the label, the source IP address and destination IP address of packets and to work out the message verification code which is also one of the contents of the trusted. The structure of the trusted label is shown in Fig. 3. The terminal ID and user information are the unique information and saved in the UKey. Trusted label representing the terminal is a reliable terminal.

3.3 Access Authentication

The scheme of trusted network access authentication based on the label includes two parts: the initial access authentication process and the sustainable certification process of label. The specific process is shown in the Fig. 4.

(a) The initial access authentication process

As shown in the Fig. 4, if the trusted terminal which has been inserted into the UKey will access to the trusted LAN, the gateway and terminal must determine the identities with each other, then the terminal can be accessed. In the process of initial authentication, it mainly carries on the authentication and produces a session key. Identity authentication is initiated by a trusted terminal and adopt the peer-to-peer identity authentication protocol based on public key, which shown in the Fig. 5, the session key that negotiation generates is saved in each memory. The identity authentication uses the method of "challenge/response."

Step 1: The terminal a starts to send its identity tag and the public key digital certificate to the gateway B;

Step 2: After the gateway B receives the public key digital certificate of terminal A, it will get the public key of A, then the gateway B uses A's public key to encrypt the B's identity and the random data which produced by B, and finally B sends the encrypted information and the B's public key digital certificate to the terminal A;

Terminal ID	User Information	Reserved Field	Check Value

Fig. 3 The structure of the trusted label

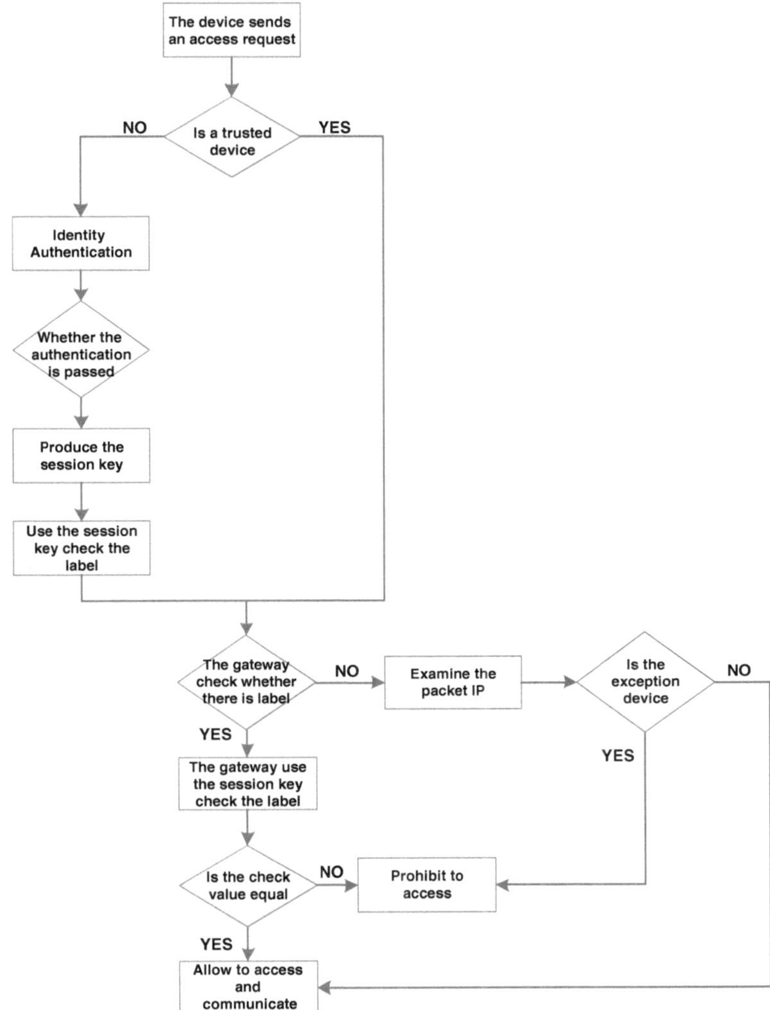

Fig. 4 Access authentication flow chart

Step 3: After the terminal A receives the information, it will use B's public key digital certificates to get the gateway B's public key and then uses B's public key to encrypt the information includes identification and the random number of A, the random number B sends to A, and the session key and sent to the gateway B;

Step 4: B used his private key to decrypt the encrypted information and get the session key, and finally, B will use the session key encrypt the random number that produced by A and sent to A, then both sides end of the peer authentication.

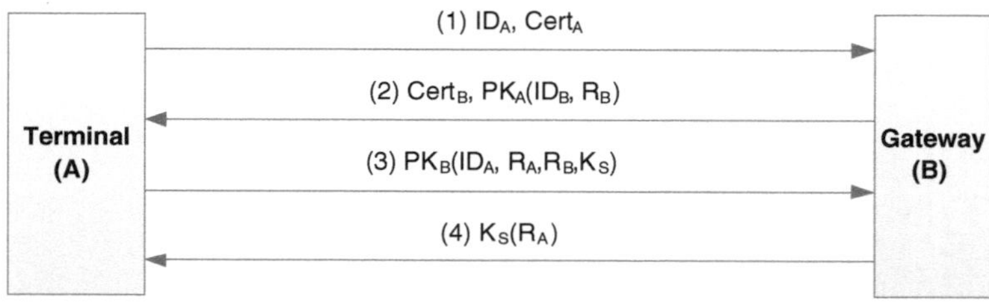

Directions:
1. ID is the identity mark of the both authentication and use to prevent the parallel session attack;
2. R is the random number and use to prevent replay attack;
3. Cert is the public key digital certificate;
4. PK is the public key;
5. Ks is the symmetric encryption session key;

Fig. 5 Peer-to-peer identity authentication based on the public key system

After the initial access authentication finished successfully, the session key generated, the terminal will use the session key and the hash algorithm to calculate the source IP address, destination IP address of packets, UKey ID, user information, the reserved field and generate a calibration value, then load it on the label and become the message authentication code of label.

Additionally, because an untrusted terminal cannot produce the trusted labels, the gateway will do the exceptional authentication, as shown in Fig. 4. For untrusted terminal, according to the IP address of packet, the gateway can determine whether it belongs to the exceptional equipment which was allowed to access, if allows, then it can access to communicate; otherwise, access is prohibited.

(b) Label the sustainable authentication process

In order to ensure the lasting credibility after the terminal access, Fig. 4 shows the sustainable certification process after the initial certification was successfully finished. Specifically, this process is the gateway to use the session key to verify the label of the trusted terminal that passed the initial access authentication. Recalculate the label information, source IP address, and destination IP address of the packet, if the recalculated calibration value is equal to the value that originally exists in the trusted label, then the verification label was successfully finished and allowed to access, or not, prohibit the communication.

4 Security Analysis

The trusted network access certification based on labels is a scheme that was proposed on the basis of the trusted network to ensure the truthfulness of the information source. This scheme uses "trusted label" of the trusted terminal to bind the terminal ID and user information and, because of the sustainable authentication made up the shortage of security after the trusted terminal accessed the network, improves the controllability and management of the trusted network. The security properties are as follows:

(a) Double-way authentication
In the process of initial access authentication, the peer-to-peer authentication based on the public key system uses the public key digital certificates of the terminal and the gateway to do the double-way authentication. At the same time, each other adopts the way that uses the public key of opposite side to encrypt a random number to ensure the truthfulness of information, because only each other's private key can decrypt the random number. This way ensured the double-way authentication of the terminals and gateway.

(b) The truthfulness of information sources
The public key digital certificate Cert of the trusted terminal when it access to the network and the sustainable authentication after accessed to the network can ensure the truthfulness of the information source effectively.

(c) Lasting credibility
After complete the initial access authentication process, the gateway uses the label on the packet of the terminal to do sustainable certification, it can ensure the credibility of the terminal.

5 Conclusion

In order to better solve the problem of the existing network security, we must do the protection from the source. On the basis of trusted computing, this paper proposes a trusted network access authentication scheme, including the initial access authentication process and the label's sustainable authentication process. In the two parts, the identity authentication of initial access process mainly adopt the method of "challenge/response" and use the public key digital certificate to achieve a reliable double-way authentication of the terminals and networks; In addition, the way that gateways do the label's sustainable certification fully ensure the lasting credibility and the truthfulness of the terminal's identity. But, if the gateway must do all sustainable authentication that the trusted terminal complete the initial access authentication, it will reduce the efficiency of the gateway; therefore, in order to ensure the efficiency of gateway to verify the label, and to assure the normal operation of the network, the next step of work will take place on the verification of label, namely according to different situation to adjust the verification of label method of the gateway.

References

1. Lin, C., Peng, X.H.: The research of trusted network. Chin. J. Comput. **28**(5), 751–758 (2005)
2. Lin, C., Ren, F.Y.: Controllable, trustworthy and scalable new generation internet. J. Softw. **15**(2), 1815–1821 (2004)
3. Wu, J., Bi, J., Li, X.: A source address validation architecture (SAVA) Tested and Deployment Experience. In: IETF Internet Standard, RFC 5210, (2008)
4. Liu, W., Yang, L., Dai, H., Hou, B.: A new network access control method and performance analysis of authentication session. Chin. J. Comput. **30**(10), 1806–1812 (2007)

A Mobile Terminal Authentication Scheme Based on Symmetric Cryptographic Mechanisms

Ying Li, Guifen Zhao, Liping Du and Jianwei Guo

Abstract In order to achieve secure access to mobile applications, in this paper, authentication scheme of application layer based on symmetric cryptography is proposed. In the scheme, seed key are pre-stored on both sides of the authentication. In the authentication process, one-time authentication code is generated by authentication protocol in SD key and certificated in authentication center to confirm the mobile user's identity. Hardware encryption devices are utilized to store and protect seed key and authentication protocol by security mechanism of encryption devices. Analyses show that the scheme is efficient and safe.

Keywords Mobile terminal · Authentication · Symmetric cryptographic mechanisms

1 Introduction

Mobile networks enable users free to access to the network anytime, anywhere that makes it be used in a wide range of fields. With the large number of intelligent mobile terminals access to mobile networks, mobile commerce services and applications have been developed rapidly. Such services need to prevent unauthorized users access to service resources while preventing legitimate users from maliciously consuming service resources. To authenticate the user's identity before providing services can effectively prevent illegal user's access to services resources.

Y. Li (✉) · G. Zhao · L. Du · J. Guo
Beijing Key Laboratory of Network Cryptography Authentication, Beijing Municipal Institute of Science & Technology Information, No. 138 Xizhimenwai Street, Xicheng District, Beijing, China
e-mail: shai_wang@hotmail.com

S. Patnaik and X. Li (eds.), *Proceedings of International Conference on Computer Science and Information Technology*, Advances in Intelligent Systems and Computing 255, DOI: 10.1007/978-81-322-1759-6_16,

Compared with the PC, mobile terminals are with the features of weak-capacity CPU, small memory, and low access communication speed. Recently, although the mobile terminals' performance has been improved quickly, complicated authentication protocol cannot still run well on the mobile terminals for its poor performance in application experience. So, an authentication scheme of low traffic and low complexity of computation is demanded.

At present, almost all security solutions on mobile terminals are based on public-key cryptosystem. Classic SET protocol and WTLS apply public-key cryptography to complete authentication and digital signature functions, and the protocol is too complex to achieve for mobile terminals in spite of high security [1, 2]. ECC is a relatively high-speed public-key algorithm, so in recent years, it began to be applied in many schemes based on public-key cryptosystem as an alternative to RSA. In article [3], an efficient authentication mechanism designed for wireless mobile terminals by exploiting ECC-based trust delegation mechanism is put forward. But for the disadvantages of asymmetric cryptography, such as high complexity, long packet length, and low speed, the above schemes cannot work well in mobile networks.

Symmetric ciphers have begun to be applied in some authentication schemes in mobile networks for its small overhead and high speed. L-WTLS achieves authentication by replacing ECC with symmetric ciphers [2]. But in L-WTLS, the public-key institutional framework is still used, and numerous consultations and data exchange are processed because the symmetric key is hard to be distributed and managed.

Hardware encryption devices are generally used to store secret keys and encrypt data. Password-authenticated key exchange protocols based on smart card have been widespread concerned. In articles [4–7], secure sim card are used to enhance security of authentication protocol, while in [8, 9] SD key is used for the same purpose. But the above schemes are still based on asymmetric cryptography and public-key certificates. The schemes using SIM card is more secure than the schemes using SD key, but the former must cooperate with operators and cannot be employed in an independently application system. Just for this reason, SD key is exploited in this paper to structure the authentication scheme.

2 System Architecture

The architecture of authentication system consists of three parts: mobile terminal, data center of application system, and authentication center, shown in Fig. 1. Mobile terminal and data center exist in the original business systems.

In mobile terminal, A SD key, secure digital memory card, is embedded. Authentication protocol is stored in SD key and called by application client to generate a one-time authentication code. The data center is the business center, in which a middleware is deployed to exchange information with data center to provide authentication function. The middleware transponds authentication code

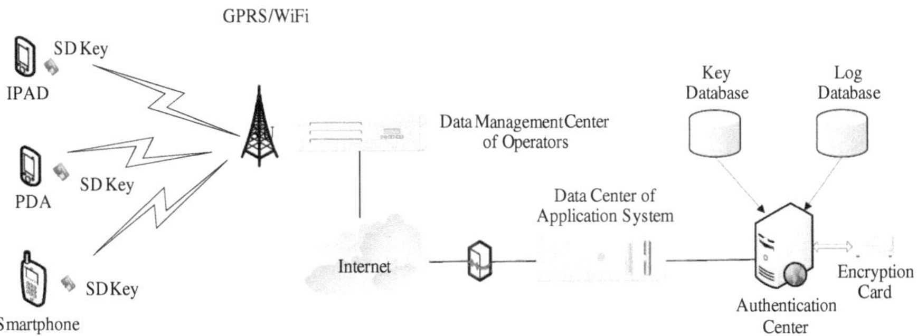

Fig. 1 System architecture

got from data center to authentication center, In other words, the middleware is the data bridge between the data center and the authentication center. A PCI encryption card is deployed in authentication center, by which authentication center certificates the authentication code. In authentication center, key database and log database are installed to save secret key and authentication log information.

Authentication center is designed to a trusted third party as a distributer of SD keys in mobile terminals and an arbiter of authentication. For most of present mobile application services, it is necessary to establish a centralized platform to receive, handle and response service request, so it is feasible to set up a manage center connected directly with data center as a trusted third party to confirm the legitimacy of the mobile user.

3 Authentication Process

How to store and update the symmetric key is a problem in symmetric cryptography mechanism. In the scheme, the problem is solved using hardware encryption devices and a specific-authentication protocol. The detail of the scheme is shown in Fig. 2.

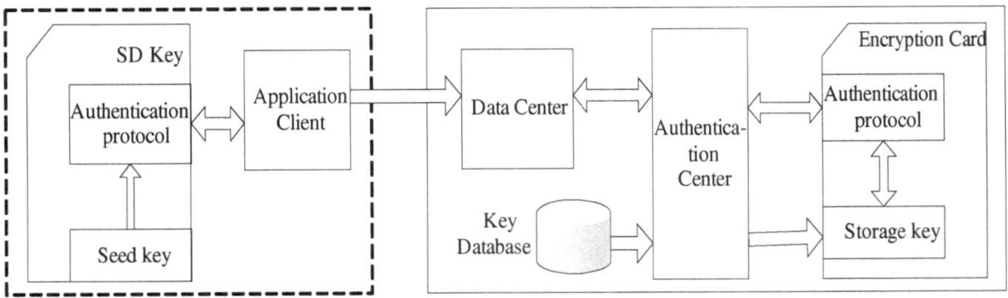

Fig. 2 Authentication Scheme

In SD key of the mobile terminal, the authentication protocol and seed key are stored. When authenticating, a one-time authentication code is generated as terminal authentication credentials by authentication protocol, in which seed key is selected, combined, and symmetric encrypted.

In the authentication center side, seed key is stored in the form of cipher text in key database. When authentication center receives authentication request, secret seed key and authentication code are input into encryption card. In the encryption card, secret seed key is decrypted by storage key and authentication code is regenerated by authentication protocol.

The basic principle of the authentication scheme is that: at both ends of authentication, the same shared seed key is pre-stored. When authenticating, two authentication codes are produced on the both sides by applying the same authentication protocol with the same input of operational factor and seed key. Then, the compared result of the two authentication codes is the certification result.

3.1 Mobile User Registration

In authentication center, random sequence of 256 bytes $A = [a_0 \quad a_1 \quad \ldots \quad a_{255}], a_i \in (0, 255)$ is produced as shared seed key by encryption card. Shared seed key for each mobile user are different from each other. The shared seed key is written in a SD key with mobile user's identity while it is saved in key database of authentication center after encrypted by storage key stored in encryption card. After SD key is distributed and embedded in corresponding mobile user' terminal, the registration is finished. Only registered mobile user can be certificated successfully.

3.2 Authentication

When a mobile user requests some resources and services from application system, it must be authenticated to obtain permission. The authentication process is described and shown in Fig. 3

- Random numbers $R = [r_0 \quad r_1 \quad \ldots \quad r_{15}], \mathrm{r_i} \in (0, 15)$ and $T = [t_0 \quad t_1 \quad \ldots \quad t_{15}], \ldots t_i \in (0, 15)$ are produced in SD key.
- Mobile user identity ID_m and pre-shared seed key A are read to memory from storage area of SD key.
- A secret key $K_e = [k_{e0} \quad k_{e1} \quad \ldots \quad k_{e15}], k_{ei} \in (0, 255)$ is computed by selecting 16 bytes from A with the input of R and T and combining them together.
- K_e is encrypted by itself to generate authentication token $\mathrm{AT}_m = [\mathrm{at}_0 \quad \mathrm{at}_1 \quad \ldots \quad \mathrm{at}_{15}], \mathrm{at}_i \in (0, 255)$.

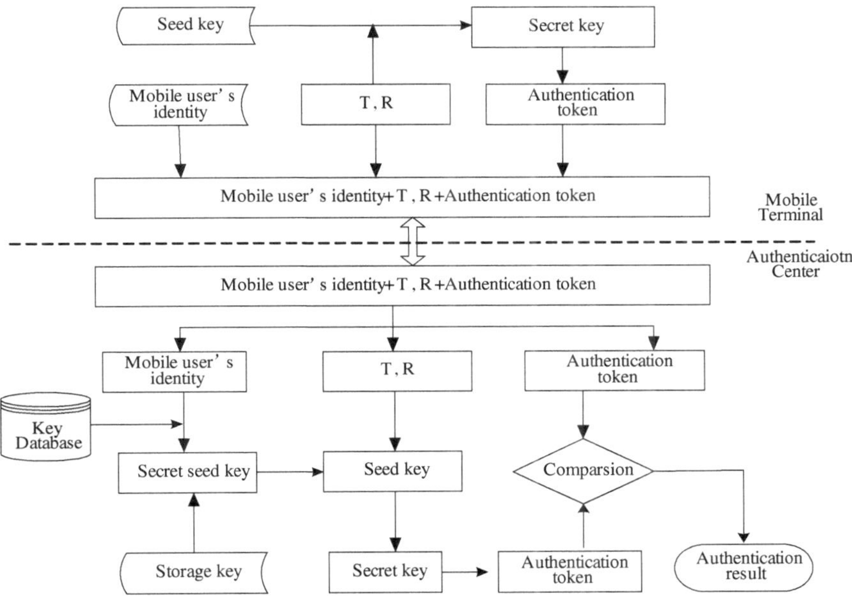

Fig. 3 Authentication Process

- $\mathrm{AC}_m = \{\mathrm{ID}_m||T||R||\mathrm{AT}_m\}$ is produced as authentication code of mobile user and sent to data center.
- The authentication center receives AC_m and splits it into ID_m, T, R, and AT_m.
- Secret seed key A_s was drawn from key database by ID_m and input into encryption card with T, R, and AT_m.
- In encryption card, secret seed key A_s is decrypted by storage key to seed key A'.
- A secret key $K_e' = [k_{e0} \quad k_{e1} \quad \ldots \quad k_{e15}], k_{ei} \in (0, 255)$ is computed by selecting 16 bytes from A' with the input of T and R and combining them together.
- $K'e$ is encrypted by itself to generate authentication token $\mathrm{AT}_a = [\mathrm{at}_0 \quad \mathrm{at}_1 \quad \ldots \quad \mathrm{at}_{15}], a_{ti} \in (0, 255)$.
- Compare AT_m with AT_a, and the compared result is the result of authentication.

3.3 Mobile User Unregistration

If the mobile terminal of one mobile user is reported the loss, the identity of mobile user can be unregistered by deleting mobile user information and shared seed key saved in authentication center. When a mobile user is unregistered, it no longer can be successfully authenticated until a new SD Key for the mobile user is distributed and embedded in a mobile terminal.

4 Scheme Analyze

4.1 Security Analyze

In the scheme, some important security arrangements are taken as below:

- Encryption devices of high security are used to ensure the safe of seed key and authentication process. Seed key and authentication protocol cannot be read out of encryption devices, and authentication is processed in encryption device, which makes hackers disable to analyze and trace the procedure of generating authentication code.
- One-time authentication token is dynamic generated by seed key cooperating with parameters T and R. T and R are randomized every time. By this means, authentication token possesses a property of one time and some attacks such as interception and fabrication are avoided.

Therefore, this scheme possesses both of forward secrecy and backward secrecy. Since seed key and authentication parameters are randomly generated, authentication token computed according to seed key and parameters are random. Both of the previous authentication token and subsequent authentications token cannot be deduced from the authentication token this time.

4.2 Performance Analysis

Authentication scheme performance is mainly reflected in the computation complexity and communication overhead. The scheme put forward in this article is compared with the other scheme L-WTLS based on symmetric cipher in the above aspects. They are tagged, respectively, to A-Sk and L-WTLS.

Because the computational complexity of dot product is much greater than symmetric cryptographic algorithm and other operations in Table 1, computational expenses of L-WTLS is much greater than that of A-Sk.

In the scheme L-WTLS, server's certificate is transmitted to client one time and the length of a certificate is about 1024 bytes. So, the total communication bandwidths of transmitted messages of L-WTLS are at least 1136 bytes in the period of three times communication as shown in Table 2.

Table 1 Comparisons of the computational expenses

Scheme name	Symmetric encryption times	Symmetric decryption times	Hash times	Dot-product times	Mod times
L-WTLS	2	2	6	8	1
A-Sk	2	2	0	0	0

Table 2 Comparisons of the communication expenses

Scheme name	Communication times	Communication bandwidth
L-WTLS	3	>1136 bytes
A-Sk	2	80 bytes

5 Conclusion

In this paper, a mobile terminal authentication scheme based on symmetric cryptographic mechanisms is proposed. In the scheme, mobile user's identity is confirmed in a safe and efficient way using hardware cryptographic devices and a specific-security authentication protocol. Hence, the scheme is suitable for mobile secure payment service, such as M-commerce and mobile banking service, and other mobile business fields with high-security requirement.

Acknowledgments The authors wish to thank the helpful comments and suggestions from my director and colleagues in Beijing Key Laboratory of Network Cryptography Authentication. This work is supported by the Program of Large-scale Network Authentication Center affiliated to Beijing Municipal Institute of Science & Technology Information (No. PXM2012_178214_000005).

References

1. Zhang, R.:Research and Implementation on Mobile Payment System Based on SET Protocol. Northwest University. (2008)
2. Xiang, W., Tao, L., Wang, T.: Secure and efficient WTLS handshake protocol. J Comput. Appl. **28**(11), 2798–2800 (2008)
3. Zhou, T., Li, Q., Zheng, D.: Wireless network mobile authentication protocol analysis and improvement. Comput. Appl. Software **29**(3), 19–22 (2012)
4. Yu, D.: The Research of Sim Card Dignity Attestation System Based on Oms Mobile Terminal. Beijing University of Posts and Telecommunications. (2010)
5. Liu, N., Deng, Z.: Study on security problems and measures of mobile e-commerce based on short message service system. Inf. Secur. Technol. **3**(9), 18–21 (2012)
6. Liang, R., Lin, P.: Research intelligent terminal security based on sim cards. China New Telecommun. **14**(12), 33–35 (2012)
7. Zheng, Y., He, D.K., He, M.X.: Trusted computing based user authentication for mobile equipment. Chinese J. Comput. **29**(8), 1255–1263 (2006)
8. Yan, S.: The Research and Implementation on Secure Access Solutions of Mobile Terminal Based on Smart Cards. Beijing Jiaotong University. (2008)
9. Zhao, Z.: The Research on Secure Access Solutions of Mobile Terminal Based on Encryption SD Cards. Department of Electric Engineering Southeast University. (2010)

Wormhole Detection Algorithm Based on RTT and Neighborhood Information

Jun Liu, Xiuping Liu, Xianghong Jiang and Mingbo Sha

Abstract We propose a detection algorithm based on time delay and topology. Node uses RTT to detect its suspicious neighbors and establish its neighbor list. Then, node finds the abnormal network topology with its neighborhood information. If three or more nodes occur, which are mutually non-1-hop neighbors, in the intersection of the two nodes which are 2-hop neighbors, there is wormhole attack. Then, the false neighbors are isolated. The simulation results show that the algorithm has high detection rate.

Keywords Wormhole attack · RTT · Lookup algorithm · False neighbor · Isolation

1 Introduction

Wormhole attack [1] is one kind of severe malicious attack mainly for routing protocols. Two or more colluding malicious nodes first create a private channel called "tunnel". A malicious node records packets overheard and tunnels them to a

J. Liu (✉) · X. Liu (✉)
College of Information Science and Engineering, Northeastern University, Shenyang, China
e-mail: liujun@ise.neu.edu.cn

X. Liu
e-mail: liuxiuping5081425@126.com

X. Jiang
College of the PLA Air Force Early Warning, Wuhan, China
e-mail: 13808625516@139.com

M. Sha
ALLWIN Telecommunication Co., Ltd., Shenyang, China
e-mail: shamingbo@126.com

S. Patnaik and X. Li (eds.), *Proceedings of International Conference on Computer Science and Information Technology*, Advances in Intelligent Systems and Computing 255, DOI: 10.1007/978-81-322-1759-6_17,

remotely located colluding node. The colluding node replays the information received. Wormhole attack can be categorized into two types: exposed and hidden wormhole attacks [2]. In this paper, we focus on the hidden wormhole attack. In the hidden wormhole attack, the malicious nodes are not visible on the route and legal nodes are also not aware of malicious nodes involved. So encryption and authentication have no effect on such a wormhole attack, and it is more difficult to be detected. Several approaches have been proposed to detect hidden wormhole attack.

Wang et al. [3] proposed an approach using the minimum hop estimated. This approach needs positioning device such as GPS, and the detection rate is not high when the distance of source and destination node is far. Thaier Hayajneh et al. [4] proposed a protocol using the routing difference between neighbor nodes. It is very important to choose sensitivity parameters and path in the process of detection. Farid Nait—Abdesselam et al. [5] proposed an approach using delay of per-hop transmission. But link congestion, queuing delays, etc. also can cause delays which make the wormhole difficult to detect. Dang et al. [6] proposed a kind of reference broadcasting approach. It also did not consider the delay by link congestion. Maheshwari et al. [7] proposed an approach by looking for a forbidden network. It is suitable for network of larger density. Wang et al. [8] proposed an algorithm using neighborhood information called WDI. It is also insufficient. It ignores the case that the first random selected node and the detection node are on the same side of the "tunnel," and they are attacked by wormhole in the process of lookup. This case cannot detect the wormhole. It deletes all nodes attacked by wormhole from the network which makes the network not able to communicate normally.

Different from the previous approaches using one feature separately, this paper using delay and topology proposes a wormhole detection algorithm based on RTT and neighborhood information.

2 Wormhole Detection Algorithm Based on RTT and Neighborhood Information

The proposed algorithm mainly uses round trip time (RTT) [9] and the demonstration proposed by Wang et al. [8] that if there are three or more nodes, which are mutually non-1-hop neighbors, in the intersection of the neighbor sets of i and j where i and j are 2-hop neighbors, then i and j must be attacked by a wormhole as shown in Fig. 1. The algorithm needs to solve three problems: The first is how to detect suspicious neighbor nodes, the second is how to select nodes in the process of looking for abnormal topology, and the third is how to isolate its false neighbors after detecting wormhole attack. For the first two problems, the algorithm introduces RTT. It uses RTT to detect its suspicious neighbors and select node in the process of looking for abnormal topology. For the third problem, the detection node will delete its false neighbors from its neighbor list and notice its real neighbors within 2 hops.

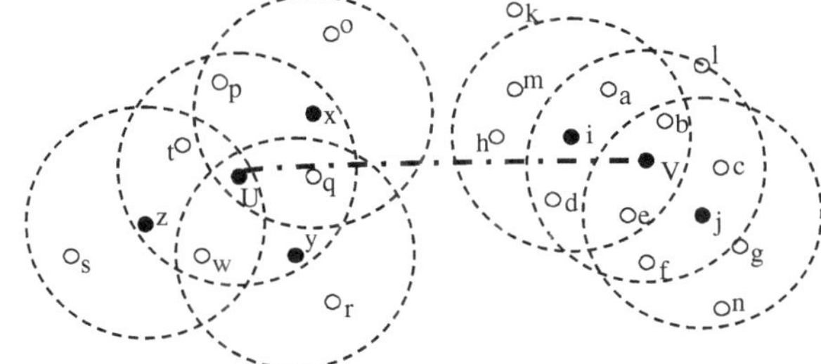

Fig. 1 The abnormal network topology

Assume that each node applies the UDG model, each node has the same communication radius R, the replaying radii of wormhole nodes are the same as R, the links are bidirectional, nodes are randomly and uniformly deployed, private/public keys have been deployed in the network, every node is able to sign REQ, REP, and its wormhole notification message packets with its private key, and each legal node can verify its signature with the public key, to ensure the integrity of the information.

Since each node has the same communication radius, the maximum RTT of each node is also the same. The following formula is used to calculate the maximum RTT:

$$\mathrm{RTT}_{\max} = 2R/V \tag{1}$$

R is the maximum transmission range of each node. V is the propagation speed of the wireless signal.

It can be found that messages between two false neighbors must transmit through the two communication radius of malicious nodes and "tunnel" according to the principle of wormhole attack. If the RTT between two nodes satisfies the formula (2), the nodes are suspicious.

$$\mathrm{RTT} \geq 2 \times \mathrm{RTT}_{\max} \tag{2}$$

2.1 Detecting Suspicious Neighbors

Each node in the network maintains its 1-hop list and 2-hop list. Nodes are exchanging HELLO and ACK packets for populating their neighbor lists. The HELLO packets contain the following field: source ID to specify who initiates this broadcast; sequence number to distinguish this HELLO packet from others initiated from the same source node; hop_limit to limit the flooding range of this HELLO packet; and path to record the path this HELLO packet has traversed so far. The ACK packet contains the following fields: source ID specifies who sends this ACK packet; destination ID specifies who should receive this ACK packet; sequence number refers to which HELLO packet this ACK is replying to; and path specifying the path for forwarding this ACK packet from sender to receiver. The details of the HELLO and ACK exchange are described in the following steps:

Step 1: The source node broadcasts a HELLO packet to its 1-hop neighbors and records the local time of the broadcast t_s. The source ID field of this HELLO packet is referring to the node itself; the sequence number field is assigned with a random number; hop_limit field is set 2; the path field includes the source node itself initially.

Step 2: A node receiving a HELLO packet first checks whether it has processed the same HELLO packet before. If so, the HELLO packet will simply be dropped. Otherwise, it replies immediately with a ACK packet. The sequence number field of this ACK packet is the same as the HELLO packet; path is the reverse of the path specified from the HELLO packet's path field. At the same time, the node modifies the HELLO packet's path field by appending itself to the end of the path field and decreases the hop_limit field in the HELLO packet by 1. If the hop_limit field is not 0, it rebroadcasts this modified HELLO packet.

Step 3: After receiving the ACK packet, the source node can easily judge the ACK from 1-hop neighbor or 2-hop neighbor by checking the path field. If it is from a 2-hop neighbor, simply put it into the 2-hop list; if it is from a 1-hop neighbor, it records the received time t_d and calculates the RTT between two nodes using formula (3).

$$\mathrm{RTT}_{s \to d} = t_s - t_d \tag{3}$$

Table 1 shows the 1-hop list of node s. Compare value represents the ratio of RTT and $\mathrm{RTT}_{\max}$. If it satisfies the formula (2), Compare value is True; otherwise, Compare value is False. True indicates that the corresponding 1-hop neighbor is a suspicious neighbor. False indicates that the corresponding node is a trusted neighbor.

2.2 Looking for Abnormal Topology

If there is suspicious neighbor in a node's 1-hop list, the node as a detection node finds whether there is abnormal topology. The detection node first sends REQ packets to all its neighbors within two hops. The recipients reply with REP back to the detection node. In a REP packet, the 1-hop list of the recipient is attached. After receiving REP, the detection node records the information. Then, it visits each node in its 2-hop list to find whether there are three or more nodes, which are

Table 1 The 1-hop list of node s

Node ID	Node 1	Node 2	…	Node n
RTT	RTT	RTT	…	$RTT_{s \to n}$
Compare	True/False	True/False	…	True/False

mutually non-1-hop neighbors among $C_1(i, j)$. $C_1(i, j)$ represents the intersection of detection node i and its 2-hop neighbor j. The details of the process are as follows:

Step 1: Node i first computes $C_1(i, j)$, then chooses the node with maximum RTT from $C_1(i, j)$. By removing the selection node and its 1-hop neighbors from $C_1(i, j)$, a new set $C_1'(i, j)$ is obtained.

Step 2: If $C_1'(i, \mathrm{j})$ is empty, there are no three or more nodes, which are mutually non-1-hop neighbors among $C_1(i, j)$, returning the detection node a FALSE. The detection node continues to find the intersections with other 2-hop neighbors. If $C_1'(i, j)$ is not empty, node i again chooses the node with maximum RTT from $C_1'(i, j)$. By removing the selection node and its 1-hop neighbors from $C_1'(i, j)$, a new set $C_1''(i, j)$ is obtained.

Step 3: If $C_1''(i, j)$ is not empty, 3 nodes are found, which are mutually non-1-hop neighbors, returning the detection node a TRUE and finishing the process of lookup. If $C_1''(i, j)$ is empty, node i chooses another node from $C_1'(i, j)$ and repeats the process of removing till $C_1''(i, j)$ is empty, returning the detection node a FALSE. The detection node continues to find the intersections with other 2-hop neighbors. If every time the return value is FALSE, it finishes the process of lookup.

If the return value is TRUE, the detection node and its 2-hop neighbor are both attacked by wormhole and there are the same false neighbors in their 1-hop lists. So the detection node puts the nodes in which Compare values are both True in the 1-hop list of the detection node and its 2-hop neighbor into a DET list.

2.3 Isolating False Neighbors

The detection node deletes nodes in the DET list from its 1-hop list, doesn't use the links and messages through the links are dropped. At the same time, sends a wormhole notification message packet to its neighbors in the updated 1-hop list. The message contains source ID, destination ID, sequence number, and false neighbors ID.

A node receiving the message first checks whether it has processed the same message before. If so, the message will simply be dropped. Otherwise, it checks whether its 1-hop neighbor list has nodes in false node ID field. If not, the message will be dropped. If nodes in the 1-hop neighbor list are the same with nodes in false node ID field, it deletes the same nodes from its 1-hop neighbor list. At the same time, it modifies the source ID and destination ID of the message and sends the modified message to its neighbors in the updated 1-hop neighbor list.

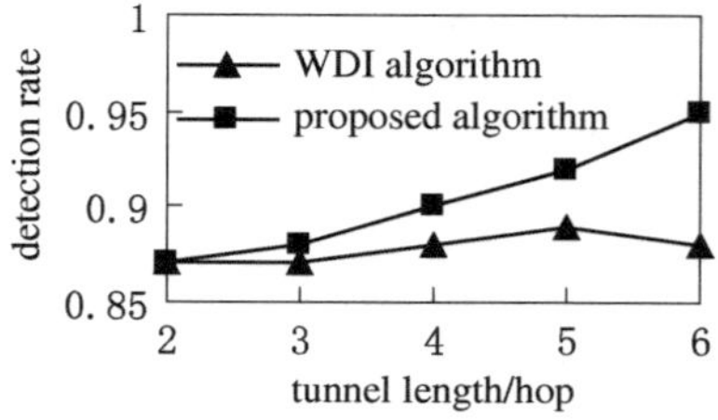

Fig. 2 Detection rate

3 Simulation and Performance Analysis

We use ns-2 simulator to evaluate the performance of our proposed detection algorithm. The simulated area is limited in a square field of size 1500 × 1500 m in ad hoc network for randomly generating network topology with 100 nodes. Transmission range of each node is set to 250 m. The two attack nodes are selected randomly among the nodes in the formed network. For prevention of statistical biases, we performed 100 times experiments with each variable setting and averaged the results. For simplicity, we choose AODV to simulate.

Figure 2 shows the curve of detection rate changing with the increasing tunnel length. Detection rate is the probability which can detect the wormhole attack. As can be seen, the detection rate of WDI algorithm almost does not change with the increasing tunnel length. Then, the detection rate of proposed algorithm increased with the increasing tunnel length, and it was higher than WDI algorithm. This is because of the introduction of RTT. The longer the tunnel, the RTT between false neighbors is larger. At the same time, using RTT to select node also avoids the situation that the selected node and the detection node are at the same side of the "tunnel" which can not detect the wormhole. Therefore, the detection rate of proposed algorithm is high when the tunnel is long.

Figure 3 shows the curve of false alarm rate changing with the increasing tunnel length. False node alarm means that the proposed algorithm can correctly detect wormhole attacks but may recognize some clean nodes as false neighbors and falsely isolate them. As can be seen, the false alarm rate of WDI algorithm almost does not change with the increasing tunnel length. Then, the false alarm rate of proposed algorithm decreased gradually with the increasing tunnel length and below WDI algorithm. This is because of the introduction of RTT. The RTT between real neighbors is smaller than between false neighbors. Therefore, the false alarm rate of proposed algorithm can be reduced when the tunnel is long.

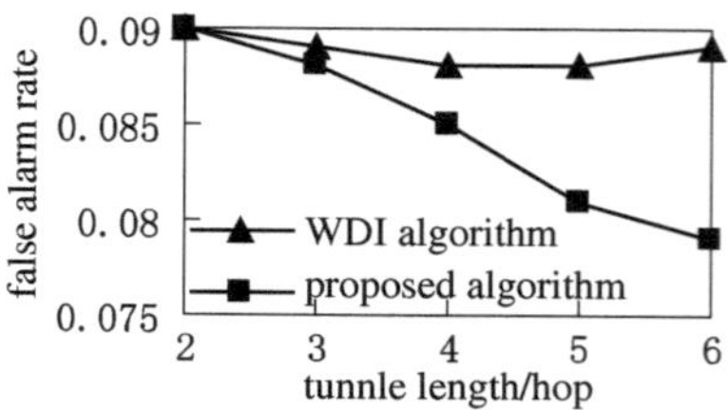

Fig. 3 False alarm rate

4 Conclusion

The proposed detection algorithm uses the delay mechanism and topology, compared with WDI algorithm proposed by Yun Wang et al. With the introduction of RTT, in the process of finding abnormal topology using the RTT value to select nodes, which reduces time and number of search, saves network resources, reduces time of detection, and reduces the false alarm rate; At the same time, it also avoids the situation that the selected node and the detection node are at the same side of the "tunnel" which not detect the wormhole. So it improves the detection rate. Finally, the algorithm isolates the false neighbors to avoid nodes within 2-hop exchange information with false neighbors and ensures the integrity of information. The algorithm does not require special hardware, and it is very easy to implement.

Acknowledgments The National Natural Science Foundation of China (61151002,60939002); The Fundamental Research Funds for the Central Universities (N110404033).

References

1. Hu, Y.C., Perring, A., Johnson, D.B.: Wormhole attacks in wireless networks. IEEE J. Sel. Areas Commun. **24**(2): 370–380 (2006). doi:10.1109/JSAC.2005.861394
2. Alshamrani, A.S.: PTT: Packet travel time algorithm in mobile ad hoc networks. 2011. 25th IEEE International Conference on Advanced Information Networking and Applications Workshops (WAINA 2011), Singapore, Singapore, 22–25 March 2011
3. Wang, X., Wong, J.: An end-to-end detection of wormhole attack in wireless ad-hoc networks. 31st Annual International Computer Software and Applications Conference, Beijing, China, 23–27 July 2007
4. Hayajneh, T., Krishnamurthy, P., Tipper, D.: DeWorm: A simple protocol to detect wormhole attacks in wireless ad hoc networks. 3rd International Conference on Network and System Security, pp. 73–80, 19–21 October 2009
5. Nait-Abdesselam, F., Bensaou, B., Yoo, J.: Detecting and avoiding wormhole attacks in optimized link state routing protocol. 8th IEEE Wireless Communications and Networking Conference, Kowloon, China, 11–15 March 2007
6. Nguyen, D., Lamont, L.: A simple and efficient detection of wormhole attacks. New Technologies, Mobility and Security, Tangier, Morocco, 5–7 November 2008
7. Maheshwari, R., Gao, J., Das, S.: Detecting wormhole attacks in wireless networks using connectivity information. Proceedings of 26th Annual Joint Conference on IEEE Computer and Communication Societies (INFOCOM'07), Anchorage, Alaska, USA, 12 May 2007
8. Wang, Y., Zhang, Z., Wu, J.: A distributed approach for hidden wormhole detection with neighborhood information. IEEE International Conference on Networking, Architecture, and Storage (NAS 2010), Macau, China, 15–17 July 2010
9. Shin, S.Y., Halim, E.H.: Wormhole attacks detection in MANETs using routes redundancy and time-based hop calculation. International Conference on ICT Convergence (ICTC), Jeju Island, South Korea, 15–17 October 2012

Intrusion Detection Algorithm for the Wormhole Attack in Ad Hoc Network

Jun Liu, Huiting Chen, Zhong Zhen and Mingbo Sha

Abstract The wormhole attack is a kind of attack that is focusing on routing protocols and having big destructiveness. We propose a path-tracking method based on detecting nodes. It deploys a number of detecting nodes on the edge of the network to find all disjoint paths between them, then to identify a suspicious path that has the smallest hops according to the characteristic of wormhole attack. The detecting node sends a tracing packet to intermediate nodes through the suspicious path, to find and isolate infected nodes. Simulation results show that this detection algorithm resists the wormhole attack effectively and ensures the network security.

Keywords Wormhole · Attack detection · The shortest path · Path tracking

1 Introduction

Wormhole attack which is against the network route is a kind of attack that two malicious nodes collude to establish a high-quality and bandwidth private tunnel to transmit information and perform only one hop in the routing path. Therefore, the

J. Liu (✉) · H. Chen
College of Information Science and Engineering, Northeastern University, Shenyang, China
e-mail: liujun@ise.neu.edu.cn

H. Chen
e-mail: 1559343403@qq.com

Z. Zhen
Communication Institute, Equipment Academy of Air Force, Beijing, China

M. Sha
ALLWIN Telecommunication Co, Ltd., Shenyang, China
e-mail: shamingbo@126.com

S. Patnaik and X. Li (eds.), *Proceedings of International Conference on Computer Science and Information Technology*, Advances in Intelligent Systems and Computing 255, DOI: 10.1007/978-81-322-1759-6_18,

path through the wormhole nodes can attract more and more traffic; on this basis, it leads to more attack, such as packet loss, tapping, and other attacks. And it also causes neighbor information list confusion of nodes which is near the wormhole tunnel end, thus disrupting the network topology. Since malicious node which at wormhole tunnel's end is reproducing real data packages, wormhole attack can bypass the key mechanism detection. Therefore, this attack is not only devastating to the network but also very difficult to detect. The wormhole attack is classified as display and implicit attack based on whether nodes (at both ends of tunnel) add their ID addresses to the IP packet. The paper is aimed at the implicit attack that nodes near the wormhole tunnel will be treated as direct neighbor to each other.

Many detection researches have been proposed for the wormhole attack. Hu et al. introduced a method of adding location information to the data packet and then calculating distance between nodes by use of GPS to detect the wormhole attack in [1]. It also derived to a new way called end-to-end wormhole detection algorithm [2]. A "packet leashes" mechanism demands all nodes to have the tight synchronization clock, and it adopts an effective authentication protocol TIK to detect in [3]. Some researches use round trip time (RTT) to estimate the distance between two nodes to detect wormhole attacks such as [4, 5]. Reference [6] proposed a trust neighbor evaluation method by monitoring the time to forward the packet of neighbor nodes, thus preventing the "pseudo neighbors," a neighbor-based detection way by the analysis of the irrational network topology in [7]. Paper [8] presented that statistical analysis on the number of neighbor, and this approach is based on infected nodes that will increase the number of their neighbors. Reference [9] used directional antennas to prevent wormhole attack by identifying correct neighbors.

2 Wormhole Path-Tracking Algorithm Based on Detecting Nodes

This paper presents a detection algorithm, decorates some source and destination nodes (detecting nodes) in the network, and uses routing discovery process between them to find the suspicious route with abnormally less hops. Then, the detecting node sends a tracing packet to intermediate nodes of the suspicious path to identify infected nodes. Detection principle is shown in Fig. 1.

2.1 The Specific Detection Algorithm Structuring

1. We deploy some pairs of source and destination nodes (detecting nodes) in the network (S/D), distributing detecting nodes on the edge of the network to conduct the entire network's detection, and each source node S knows its matching destination node D's ID number.

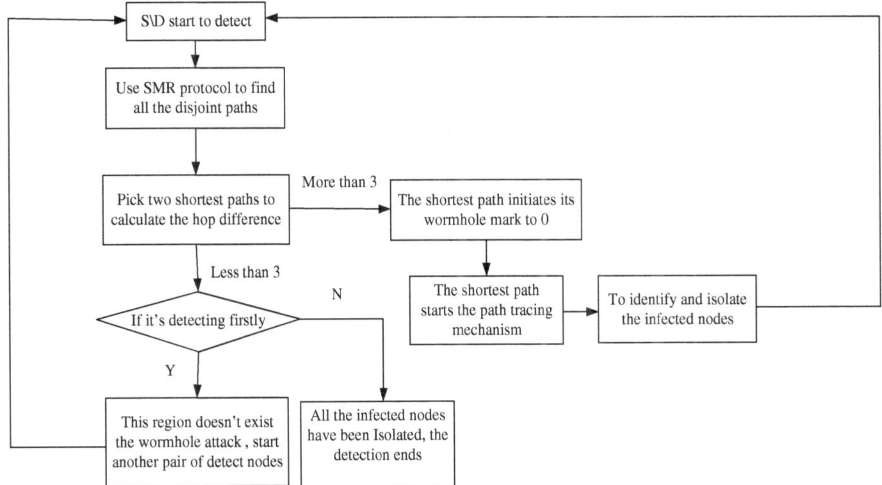

Fig. 1 The detection principle

As shown in Fig. 2, black for detecting nodes (source and destination node pairs), red for wormhole malicious nodes, between the malicious nodes connecting a private tunnel, and rest are normal nodes. The detection is conducted for every once in a while, and each pair of detecting nodes' testing time is not conflicting, in order to save network resource, we conduct just one pair detection at a time.

2. When the detection starts, we choose one pair of detecting nodes ($S_1 \backslash D_1$) as an example. The source node S_1 broadcasts RREQ; in the routing discovery process, we use split multipath routing (SMR) protocol to find out all disjoint paths to the destination node D_1, as shown in Fig. 3, and gather all the routing information in node $D_{1.}$
3. The destination node D_1 chooses suspicious path with the least hops from all the summary routings. As wormhole's "tunnel" performs only one jump in the route, hop counts of the wormhole path are very small compared with other paths. Noticing that if each pair of detecting nodes starts path-tracking mechanism as soon as finding out the path with least hops, this will lead to a waste of resource. Therefore, in order to avoid over costing, node D_1 still needs to pick out the penultimate less hop path and calculate the difference of these two paths' hop counts, then compare the hop difference with the threshold value. If hop difference is less than the threshold, we consider that this shortest path is normal. So, this region does not exist the wormhole attack and then immediately starts the next test of another pair of detect nodes. If hope difference is greater than or equal to the threshold, this shortest path initiates its wormhole mark to 0 and starts the path-tracking mechanism.

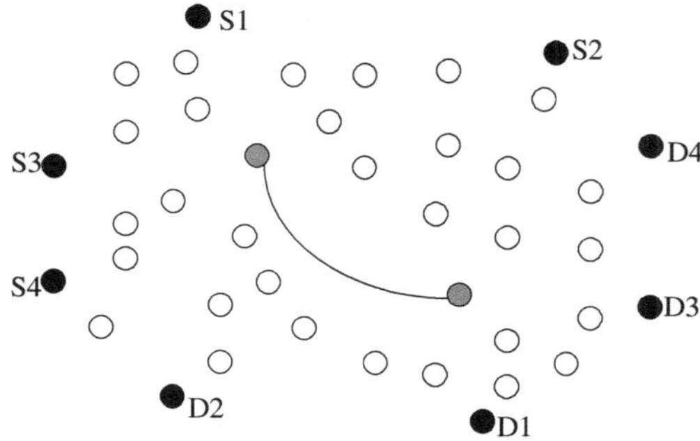

Fig. 2 Nodes' distribution map

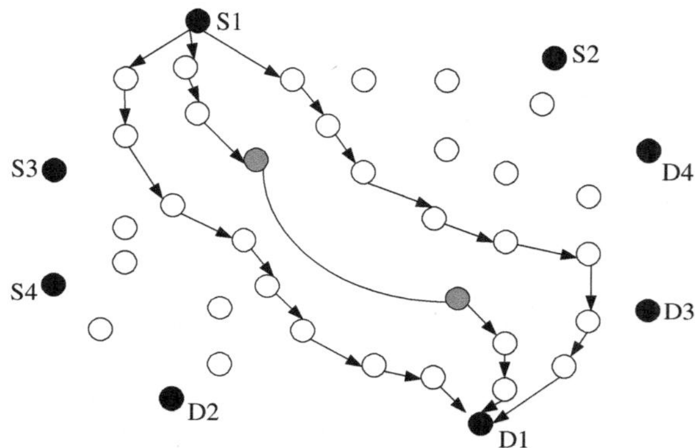

Fig. 3 Routing discovery of detection nodes

The threshold value is 3, as shown in Fig. 4. Wormhole tunnel (B to C) distance is larger than transmission range, so the distance between A and B is greater than three hops. Because of the implicit wormhole attack, A and D look like only one hop in the routing performance, so the number of hop of normal path subtracts the number of hop of wormhole path which must be larger than three hops.

4. When the shortest path initiates its wormhole mark to 0, it starts the path-tracking mechanism. The destination node D_1 sends a tracing packet to intermediate nodes along this path and starts the timer at the same time. Intermediate node reverts a tracing reply immediately after receiving, while passing the tracing packet to the next hop along the route. When the destination node D_1 receives the tracing reply

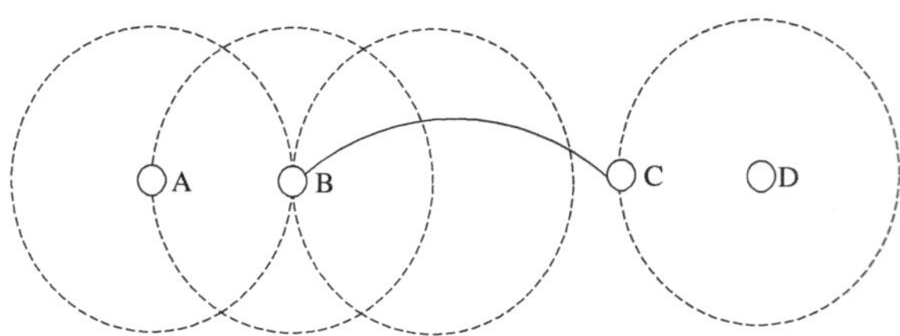

Fig. 4 Shortest hops of wormhole tunnel

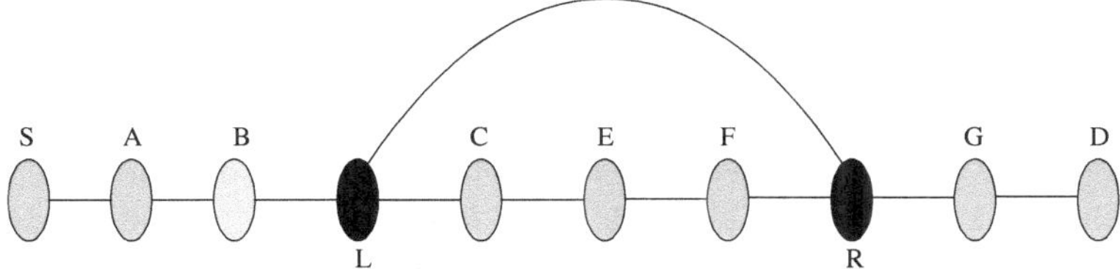

Fig. 5 Wormhole path

Table 1 Intermediate nodes' RTT of the wormhole path

D	G	B	A	S
The actual number of hops	1	2	3	4
RTT (ms)	10	50	60	70

packet of intermediate nodes, it records RTT of the packet. While the tracing packet arrives to the source node S_1, it will be discarded. Finally, the destination node D_1 sequentially records packets' RTT of intermediate nodes on the path, as shown in Fig. 5.

In normal circumstances, the tracing packet's RTT from D1 to intermediate nodes will increase linearly with the number of hops; if we find that just increasing one jump but with suddenly a larger add of RTT, then this node is more likely to be infected node of the wormhole attack. As shown in Table 1, intermediate nodes G and B differ only one hop but with a longer RTT, directly from 10 to 50 ms, we can determine that there is wormhole attack between B and G. Now, the destination node D1 issues a statement that B and G are infected nodes, then all neighbors of these two nodes mark them as infection and no longer send them messages about routing, thereby isolating B and G, as shown in Fig. 6.

5. When B and G have been isolated, S_1 again starts wormhole detection to the node D_1 and finds out other infected nodes by the same wormhole attack until no longer infected nodes are found.

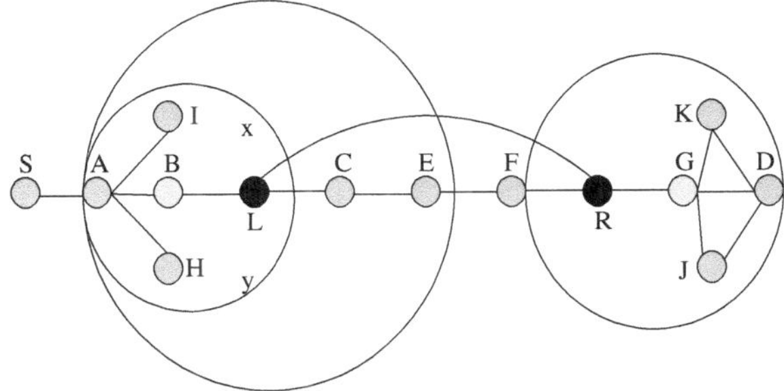

Fig. 6 Influence sphere of wormhole malicious nodes

3 The Simulation

We use NS2 to simulate the detection algorithm. There are about 50 nodes at the 1,000 × 1,000 m^2 ad hoc network, including three pairs of wormhole tunnel distributed, and we arrange eight pairs of detecting nodes at the edge of the network which run under AODV protocol.

Figure 7 shows the change of total packet dropped in wormhole attack and after detection under the AODV protocol. When the wormhole exists in the network, due to its characteristic of less hops, nodes always choose wormhole path in the routing discovery process, thus the packet loss will increase dramatically. As shown in Fig. 7, lost packets increase almost linearly with the network load. After the detection, all infected nodes are isolated, so that wormhole paths can be bypassed in the routing discovery process. So, data packets can be transmitted smoothly, and the number of lost packets reduces substantially contributing to the network optimization.

As can be seen in the Fig. 8, the wormhole path that has short hops can always be found in the routing discovery process. But, packets through the wormhole path are always dropped or changed making the communication failed. So, the network has to maintain its routing continually, and its overhead increases greatly as follows. After detection, all wormhole paths are avoided, although there is some cost in the detection algorithm itself, but as shown in Fig. 8, the overhead of the detection algorithm is very small .

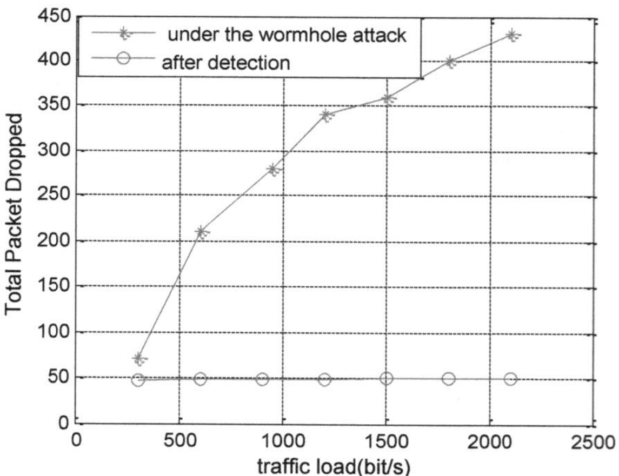

Fig. 7 Simulation result of total packet dropped of the network

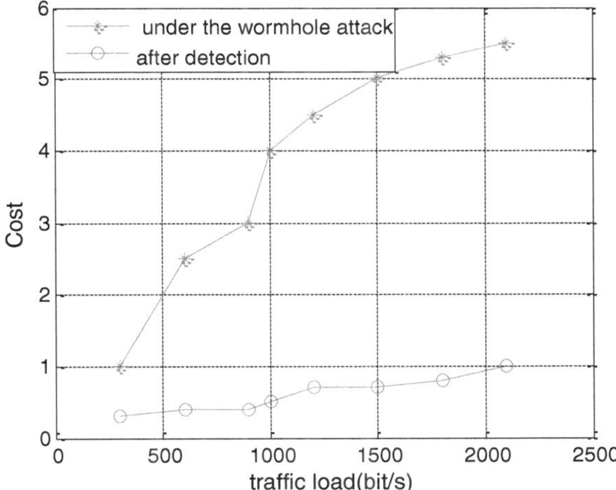

Fig. 8 Simulation of network overhead

4 Conclusion

According to characteristics of the wormhole attack, we propose a wormhole path-tracking algorithm based on detecting nodes. This method arranges some detecting nodes on the edge of the network and uses SMR protocol to find all the disjoint paths between detect nodes. Then, pick two short paths to compare, whether the difference of two paths' hops are less than threshold, the shortest path is the normal, and the area does not exist wormhole; if it exceeds the threshold, the shortest path starts wormhole-tracking mechanism to identify and isolate infected nodes. Simulation results show that this algorithm can detect the wormhole attack effectively.

Acknowledgments The National Natural Science Foundation of China (61151002, 60939002); the Fundamental Research Funds for the Central Universities (N110404033).

References

1. Yu, Y., Govindan, R., Estrin, D.: Geographical and Energy Aware Routing: A Recursive Data Dissemination Protocol for Wireless Sensor Networks. University of California at Los Angeles Computer Science Department, Technical Report UCLA/CSD- TR-,01-,0023, May 2001
2. Wang, X., Wong, J.: An end-to-end detection of wormhole attack in wireless ad-hoc networks. In: 31st Annual International Computer Software and Applications Conference (COMPSAC 2007). doi:10.1109/COMPSAC.2007.63
3. Hu, Y.C., Perrig, A., Johnson, D.B.: Packet leashes: a defense against wormhole attacks in wireless networks. In: The 22nd Annual Joint Conference of the IEEE Computer and Communications Societies. Kluwer Academic Publishers, San Francisco, pp. 1976–1986. doi:10.1109/INFCOM.2003.1209219 (2003)
4. Zhen, J., Srinivas, S.: Preventing replay attacks for secure routing in ad hoc networks. In: Proceedings of 2nd Ad Hoc Networks and Wireless (ADHOCNOW'03), pp. 140–150 (2003)

5. Tun, Z., Maw, A.H.: Wormhole attack detection in wireless sensor networks. World Acad. Sci, Eng. Technol. **46**, 2008 (2008)
6. Pirzada, A.A., McDonald, C.: Circumventing sinkholes and wormholes in wireless sensor networks. In: International Conference on Wireless Ad Hoc Networks (IWWAN) (2005)
7. Wang, Y., Zhang, Z., Wu, J.: A distributed approach for hidden wormhole detection with neighborhood information. In: Proceedings of the 2010 IEEE International Conference on Networking, Architecture, and Storage (NAS 2010), pp. 63–72. doi:10.1109/NAS.2010.22 (2010)
8. Buttyán, L., Dora, L., Vajda, I.: Statistical wormhole detection in sensor networks. Second European Workshop on Security and Privacy in Ad Hoc and Sensor Networks (ESAS 2005), Visegrad, Hungary, 13–14 July 2005, pp. 128–141
9. Hu, L., Evans, D.: Using directional antennas to prevent wormhole attacks. In: Proceedings of the 2004 Symposium on Network and Distributed System Security Symposium, San Diego, USA (2004)

Ad hoc Eavesdropping Algorithm-Based Network Coding

Jun Liu, Fei Fei Wang, Shao-hua Li and Sheng-li Li

Abstract In view of ad hoc open wireless channel as well as the characteristics of easy to eavesdrop, the secure network coding algorithm about non-eavesdropping is presented. The algorithm based on random network coding, using the character of one-way trapdoor function, achieves the function of non-eavesdropping by passing the value of a one-way trapdoor function, which attackers do not have and cannot get the original information. In view of pollution attack of network coding in use, a kind of pollution prevention algorithm is put forward, which makes use of the fact that the linear combination of message is still in the message space, in the source node encrypts orthogonal vectors, and in destination checks whether dot product between coding vector and orthogonal vector received is zero to check pollution. The algorithm is simple and does not limit the number of hacking channel and use the secret channel.

Keywords Eavesdropping prevention · Pollution prevention · Secure network coding

J. Liu (✉) · F. F. Wang
College of Information Science and Engineering, Northeastern University, Shenyang, China
e-mail: liujun@ise.neu.edu.cn

F. F. Wang
e-mail: wffifkui@sina.com

S. Li
Communication Institute, Equipment Academy of Air Force, Beijing 100096, China
e-mail: lishaohua@sina.com

S. Li
Shenyang Selection Center of Air Force Pilot Bureau, Shenyang 110015, China

S. Patnaik and X. Li (eds.), *Proceedings of International Conference on Computer Science and Information Technology*, Advances in Intelligent Systems and Computing 255, DOI: 10.1007/978-81-322-1759-6_19,

1 Introduction and Related Research

Network coding was first proposed in 2000 by R. Alshwede, the purpose of which is to achieve a theoretical maximum transmission capacity. But with the deepening of the research, the network coding has also showed the advantages of other aspects, such as the use in wireless networks: saving energy consumption, increasing throughput and transmission reliability, in the same time, poses a significant security problems, mainly including two aspects: hacking and pollution.

Cai and Yueng [1] present an interception model of secure network coding and presents a method to construct a secure network coding. Hareda and Yamamoto [2] proposed a better secure network coding algorithm, which makes the eavesdropper only eavesdrop the last several components when the number of edges is more than *k*. Zhang [3] proposed a simplified secure network coding scheme on the base of Hareda and Yamamoto. Ho [4] proposed to detect Byzantine with random network coding, which transform the original packet with hash and adds the results into the packet in the source node, and the destination node comparing the decoded packet with hash value to judge whether the data have been modified. Levente [5] put forward a kind of algorithm to detect pollution attack without redundant information, which is not actual in the network, because of its premise that the amount of coding vector is more than the number of minimum encoding vectors. Vilela [6] presented a weak secure coding method with encrypting part of coding vector. Jaggi et al. [7] has also presented a secure network coding method in the network with both eavesdropping and pollution attack existing at the same time, but their encoding system aimed mainly at the pollution problem, did not achieve the goal of prevent hacking. Ngai and Yang [8] constructed a secure error-correcting coding which can prevent both pollution attack and the hacking attack at the same time. The disadvantage is destination node needs to know all of the source information collection in their model, which is not practical in many cases, and the error correction capability of the model is limited. Zhou Yajun presented a secure correction network coding using the metric of message space, which not only needs a secret channel, but also has poor security aiming at universal attacker. Oggier F and Fathi H designed effective homomorphic signature system according to pollution attacks problem; however, the algorithm needs to know the entire network topology, without considering the resistance to pollution attacks, which is not suitable for large-scale network and wireless network.

This paper puts forward a scheme of resistance to eavesdropping attack and pollution attack, according to the characteristics of ad hoc limited resource, which achieve functions of both the hacking and authentication through encrypting trapdoor function values and orthogonal vectors one way.

2 Ad hoc Hacking Attack Resistance of Secure Network Coding

2.1 Network Model

For an acyclic multicast network G = (V, E), V is the set of points, E is the set of channel and source produce the following information per unit time, in which A_i is the packet:

$$X = \begin{pmatrix} x_{11} & x_{12} & \cdots & x_{1n} \\ x_{21} & x_{22} & \cdots & x_{2n} \\ \cdots & \cdots & \cdots & \cdots \\ x_{m1} & x_{m2} & \cdots & x_{mn} \end{pmatrix} = \begin{pmatrix} X_1 \\ X_2 \\ \cdots \\ X_m \end{pmatrix}$$

2.2 Attack Model

Considering the simple model of single source and destination, to which multiple sources and destination is similar, only need a record in packet headers to write the source node ID. Define the source node to Alice, and the destination node to Bob, the attacker to Calvin. Calvin can eavesdrop any channel to obtain information, and A_i stands for the global coding vector of the eavesdropper, A_iX for the stealing information.

2.3 Algorithm Ideas

Considering the radio characteristics of wireless ad hoc making secret channel established and the number of hacking channel limited difficult, one-way trapdoor function is proposed to solve the problem of hacking, which uses the proof that network coding package is still the linear combination of original packets to transfer orthogonal vector-valued to detect pollution, while Bob checks whether dot product between coding vector and orthogonal vector received is zero to decide whether receiving or not.

2.4 Algorithm Ideas

The coding algorithm of Alice: Alice encodes X through the following steps:

1. Choose a random number w and calculate the value of one-way trapdoor function h(w).
2. Calculate $i = 1, 2, \ldots, m$ through the following function:

$$x_{i1}^{'} = x_{i1} + w$$
$$x_{i2}^{'} = x_{i2} + w$$
$$x_{i3}^{'} = x_{i3} + w$$
$$\ldots$$
$$x_{in}^{'} = x_{in} + w$$

3. Use the schmidt orthogonalization method to calculate a orthogonal vector of the original information $Z = (z_1, z_2, \ldots, z_{m+n})$, that is, $Z \times X_i = 0$, computation formula is as follows:

$$\varepsilon_1 = X_1$$
$$\varepsilon_2 = X_2 - \frac{(\varepsilon_1, X_2)}{(\varepsilon_1, \varepsilon_1)}\varepsilon_1$$
$$\ldots\ldots$$
$$\varepsilon_{m+1} = X_{m+1} - \frac{(\varepsilon_1, X_{m+1})}{(\varepsilon_1, \varepsilon_1)}\varepsilon_1 - \frac{(\varepsilon_2, X_{m+1})}{(\varepsilon_2, \varepsilon_2)}\varepsilon_2 - \ldots\ldots - \frac{(\varepsilon_m, X_{m+1})}{(\varepsilon_m, \varepsilon_m)}\varepsilon_m$$

After that $Z = \varepsilon_{m+1}$.

4. Obtaining-information nodes public key and ellipse encryption algorithm will be Z and h(w) encryption for, specific steps are as follows:

$$C_1 = r \times e_1$$
$$C_2 = P + r \times e_2$$

Sender by specific algorithm will definitely Z and h(w) mapped to a point P on the elliptic curve and calculate the cipher text press type:That is $Z' = (C_1, C_2)$ where Z' is encrypted for transmission.

5. The first packet sent the packet of (E, X', Z'), while the rest send (E, X'), in which E is a matrix code vectors, $X' = \left(X_1', X_2', \ldots, X_m'\right)$ is anti-bugging vectors, Z' is the encrypted data.

Bob's decoding algorithm: Bob decodes through the following steps

Suppose that the message Bob received is (W, Y, Z''), then Bob decoded as follows:

1. If Bob received enough packets, then it can decide to get X' using the following formula: $X' = W^{-1}Y$.
2. Bob gets Z' using elliptic curve encryption algorithm, namely $P = C_2 - d \times C_1$, then a point on the elliptic curve is mapped to Z', from which we can get Z and $h(w)$.
3. Bob calculates w using the trapdoor and $h(w)$, then get $X_i(i = 1, 2, \ldots, m)$ by the following formula:

$$\begin{aligned} x_{i1} &= x'_{i1} - w \\ x_{i2} &= x'_{i2} - w \\ &\vdots \\ x_{in} &= x'_{in} - w \end{aligned}.$$

4. Bob verifies whether the formula is right or not, if right, then the information received is not contaminated; otherwise, Bob received or description contaminated packet and discard it.

3 Secure Proof

Here, we prove the safety of the pollution attacks first.

Theorem 1 For any message $w = (c_1, c_2, \ldots, c_n, w_1, w_2, \ldots, w_m)$ received, the necessary and sufficient condition that it is judged as pollution messages is $(w_1, w_2, \ldots, w_m) \cdot (z_1, z_2, \ldots, z_m) \neq 0$ where $(z_1, z_2, \ldots, z_m)$ is the orthogonal vectors in the information space.

Proof If the information $(w_1, w_2, \ldots, w_m)$ is not contamination, then $w' = (w_1, w_2, \ldots, w_m)$ should be the linear combination of the original information, that is, $w\prime = \sum_{i=1}^{l} c_i x_i$. The orthogonal vector in the information space should be orthogonal with every packet, namely $Z \times X_i = 0$, that is to say, $(w_1, w_2, \ldots, w_m) \cdot (z_1, z_2, \ldots, z_m) = 0$. We make use of reductio ad absurdum to talk about the probability of the general appearance that an attacker can find a contaminated message w_0, which is fit for the formula $Z \times w_0 = 0$.

(1) First, we can be bold said that the question of searching the polluted information $w_0 = (w_{01}, w_{02}, \ldots, w_{0n}, w_{0n+1}, \ldots, w_{0m+n})$, which makes the formula $Z \times w_0 = 0$ right is equal to solving the discrete logarithm problem of elliptic, because it requires to break the elliptic curve encryption algorithm.
(2) Next, we discuss the success probability of guessing orthogonal vectors Z, assuming that the attacker has cracked the elliptic curve encryption algorithm, then he guess orthogonal vectors, because the orthogonal vector transmitted is only a random orthogonal vectors which the source node choose in countless

orthogonal vectors, so the probability that the attacker guess orthogonal vectors is negligible.

(3) Then, we analyze the security of the algorithm for the anti-eavesdropping attacks.

The algorithm uses trapdoor function to transmit the random number instead of the traditional secret channel, which makes the eavesdropper listen to the value $h(w)$, unknown the random number used, furthermore, he cannot recover the original information while in the established protocol, if the eavesdropper has tapped into the random number, he can use a hash function to calculate the function easily and recover original information moreover, thus proving the algorithm is safer than the original scheme based on a secret channel.

4 Communication Overhead

Assuming that the source can send one packet each time, the communication overhead consists of two parts, one is the coded vector, the other is the orthogonal vector carried, of which the first is universal in all the algorithms based on network coding, and of which the rest transmit only once, the bandwidth is $\frac{1}{m}$ compared with the traditional algorithm. Its safety performance can be shown in Tables 1 and 2.

Table 1 Anti-bugging performance

	Need secret channel whether or not	Limit the eavesdropping ability whether or not	Randomization method
Document 1, etc.	Yes	Yes	Random matrix is multiplied with the original information
Algorithm in this article	No	No	Random matrix is added with the original information

Table 2 Pollution prevention performance

	Detect the pollution whether or not	Computing category	Signature whether or not
Document 10, etc.	Detect almost all the pollution	Modulo, multiplication, exponentiation	Yes
Algorithm in this article	Detect almost all the pollution	Adder, multiplication	No

5 Conclusion and Outlook

Ad hoc network is a non-central, self-organization heterogeneous network with dynamic topology and limited resource, which makes traditional security methods-based cryptography unfit; at the same time, the broadcasting feature of ad hoc makes the establishment of secret channel quite difficult. This article will combine cryptography with information theory to propose a new algorithm based on network coding. The algorithm is improved based on random network coding, which use the characteristic of one-way trapdoor function and pass the values of trapdoor function for anti-eavesdropping; at the same time, using the fact that orthogonal vector received should be orthogonal with vectors in message space to detect pollution. Comparison shows that algorithm in this article does not limit the number of tapping channels, do not use the secret channel, and is suitable for ad hoc network with a small amount of computation.

Acknowledgments The National Natural Science Foundation of China (61151002, 60939002); The Fundamental Research Funds for the Central Universities (N110404033).

References

1. Cai, N., Yueng, W.R.: On the optimality of a construction of a secure network codes. In: proceedings of 2008 IEEE International SymPosium on Information Theory, Toronto, Canada, (PP. 116–170), July 2008
2. Harada, K., Yamamoto, H.: Strongly secure linear network coding. IEICE Trans. Fundam. Electron. Commun. Comput. Sci. **91**(10), 2720–2728 (2008)
3. Zhang, Y. The Research of An Advanced Secure Network Coding, China Academic Journal Electronic Publishing House. All rights reserved. http://www.cnki.net
4. Ho, T., Leong, B., Koetter, R., et al. Byzantine modification detection in multicast networks using randomized network coding. In: proceedings of 2004 IEEE International Symposium on Information Theory, Chicago, Illinois, June 2004
5. Czap, L., Vajda, I.: Detection and recovery from pollution attacks in coding based distributed storage schemes. IEEE Trans Dependable and Secure Comput. **6**, 824–838 (2011)
6. Vileia, J.P., Lima, L., Barros, J. Light weight security for network coding. Proceedings of the IEEE International Conference on Communications (ICC), pp. 1750–1754, IEEE Press, Beijing (2008)
7. Jaggi, S., Langberg, M., Ho, T., et al. Correction of adversarial errors in networks. IEEE International Symposium on Information Theory, pp. 1455–1459. Adelaide, Australia (2005)
8. Ngai, C.K., Yang, S. Deterministic secure error-correcting (SEC) network codes. IEEE Information Theory Workshop, pp. 96–101. (2007)
9. Jain, K.: Security based on network topology against the wire tapping attack. IEEE Wireless Commun. **11**(1), 68–71 (2004)
10. Siavoshani, M.J., Fragouli, C.:Onlocating byzantine attackers. In: 4th Workshop on Network Coding, Theory and Applications, pp. 1–6. (2008)

Improved Halftone Visual Cryptography by Random Grids

Zhuoqian Liang and Meihua Yang

Abstract Halftone visual cryptography (HVC) is a technique to share secret image into a group of halftone shadows with pleasing visual quality. However, pixel expansion problem is serious in existing HVC schemes. In this paper, a random grid-based HVC scheme is proposed to solve this problem. The proposed method adopts complementary shares to prevent non-relevant information from showing on the decoded image. The size of halftone share generated by the proposed scheme is the same as the original secret image. Extensive experimental results are provided, demonstrating the effectiveness and advantages of the proposed schemes.

Keywords Visual cryptography · Visual secret sharing · Random grid · Halftone

1 Introduction

Visual cryptography (VC), which is also called visual secret sharing (VSS), is a novel type of secret sharing for images proposed by Naor and Shamir [1]. VC is a paradigm for cryptographic schemes to encrypt a secret image into noise-like images (called shares or shadows) and to decrypt the secret image visually by stacking sufficient shares together via human visual system without the aid of any computational devices. Particularly, a secret image is encoded into n random-looking shares in a k out of n VC scheme. The secret image is visually recovered by superimposing any k or more shares, whereas the stacked results of any $k-1$ or less give no clue about the secret.

Z. Liang (✉)
College of Information Science and Technology, Jinan University, Guangzhou 516032, People's Republic of China
e-mail: tliangzq@jnu.edu.cn

M. Yang
91630 PLA Troops, Guangzhou 510320, People's Republic of China

S. Patnaik and X. Li (eds.), *Proceedings of International Conference on Computer Science and Information Technology*, Advances in Intelligent Systems and Computing 255, DOI: 10.1007/978-81-322-1759-6_20,

Extensive investigations on VC have been conducted since the pioneer work of Naor and Shamir was presented. Construction of VC for general access structures was proposed in [2]. Optimal contrast and minimum pixel expansion were discussed in [3, 4]. Moreover, constructions for sharing gray-level/color images were proposed in [5, 6].

However, shares generated by the above-mentioned VC schemes are random-looking. It is difficult for participants to identify these noise-like shares. On the other hand, this type of shares may lead to suspicion of secret communication. To deal with this issue, extended VC (EVC) scheme [7] and halftone VC schemes [8, 9] were introduced. Meaningful shares are constructed by these methods [7–9]. Further, random grid-based VC methods [10–16] were proposed to generate image size invariant VC.

However, the mentioned halftone visual cryptography (HVC) schemes encode each secret pixel into an array of subpixels in each halftone share. For example, a 4×4 block in each halftone share is utilized to share one secret pixel. Pixel expansion problem in HVC is more serious than that in conventional VC. In this paper, novel constructions for HVC using random grids are proposed. The size of halftone share is the same as the original secret image. Pixel expansion problem is solved by the proposed methods.

The remaining sections of this paper are organized as follows: The proposed construction for HVC by random grids is presented in Sect. 2. Section 3 demonstrates extensive experimental results to illustrate the effectiveness and advantages of the proposed method. Concluding remarks are given in Sect. 4.

2 The Proposed Schemes

In this section, a construction for RG-based HVC without pixel expansion is proposed. The proposed method conceals a secret image into a group of halftone shares which contain a pair of complementary shadows. To reveal the secret image, the pair of complementary halftone shares is required.

A binary secret image and multiple gray-level cover images are considered as the inputs. In the first method, two types of pixels are produced in the halftone shares: (1) secret information pixels (SIPs) that carry secretly shared information and (2) cover information pixels (CIPs) that are generated by error diffusion algorithm. To visually reveal the secret image, it is required that all qualified shares that contain a pair of complementary shares are stacked together. The aim of employing a pair of complementary shares is to prevent the visual information on the share from showing on there covered image. In such a complementary pair, one share is called the primary share which is a natural image while the other one, called the complementary share, is generated by reversing all the CIPs in the primary share. Halftone share construction of the first method consists of three steps: (1) random position selection, (2) distribution of SIPs, and (3) generation of halftone shares.

2.1 Random Position Selection

The first step of constructing halftone shares is to determine the initial positions of SIPs. Prior to selecting the positions, the number of SIPs should be fixed. Factor $\gamma\,(0 \leq \gamma \leq 1)$ is introduced to denote the number of SIPs on each halftone share. Assumed that the size of secret image is $M \times N$, the number of SIPs on each halftone share is given by $\gamma \times M \times N$. Determine the number of SIPs that is of great importance to HVC since a large number of SIPs would lead to decreasing visual quality of the halftone shares while a fraction of SIPs would lower the contrast of the revealed secret image. When the number of SIPs is determined, positions of SIPs are randomly selected. For the ease of description, $S_{\text{RPS}} = \{(i,j)|(i,j)$ is the randomly selected position$\}$ is used to denote the set of all randomly selected positions.

2.2 Homogeneous Distribution of SIPs

Since the positions of SIPs are randomly selected, some of them may cluster together. To achieve pleasing visual quality of the halftone shares, SIPs should be distributed as homogeneously as possible and be maximally separated from each other. In this case, a void-and-cluster algorithm [17] is adopted to manipulate the randomly selected positions. The terms minority pixel and majority pixel are used. When less than half of the pixel is black, there are minority pixels, and the majority pixels are white. The reverse is true when more than half of the pixels are white. Moreover, the terms cluster and void refer to the arrangement of minority pixels on the background of majority pixels. A cluster is tight grouping of minority pixels, and a void is a large space between minority pixels. In this paper, pixels at the randomly selected positions are minority pixels, and the pixels at nonselected positions are majority pixels. To produce homogeneous distribution of SIPs, minority pixels are added in the center of the largest voids, and majority pixels are added in the center of the tightest clusters.

To find the largest void and the tightest cluster, a void-and-cluster-finding filter is used, as given by:

$$M(i,j) = \sum_{p=-\frac{W}{2}}^{\frac{W}{2}} \sum_{q=-\frac{W}{2}}^{\frac{W}{2}} G(p,q)\, D(i+p, j+q) \tag{1}$$

where $M(i, j)$ is the minority pixel density (m.p.d.) at position (i, j), $G(p, q)$ is a two dimensional Gaussian filter, W is the window width of the filter, and $D(i + p, j + q)$ is the distribution of SIPs, as defined by:

$$D(i+p, j+q) = \begin{cases} 1, & \text{if } D(i+p, j+q) \in S_{\text{RPS}} \\ 0, & \text{otherwise.} \end{cases} \tag{2}$$

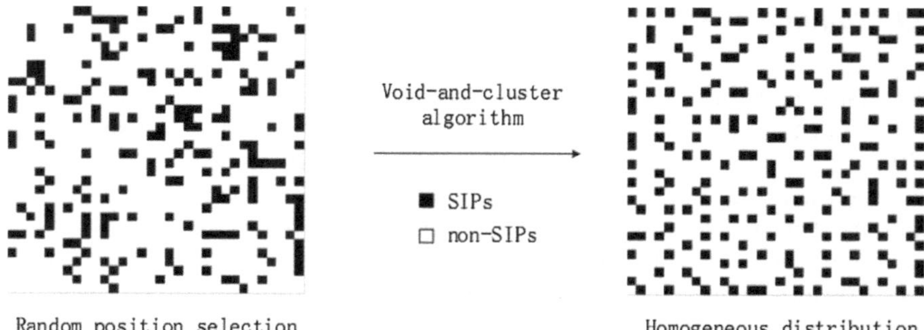

Fig. 1 An example of generating homogeneous distribution of SIPs using void-and-cluster algorithm where the image size is 32 × 32 and $\gamma = 0.2$

The two dimensional Gaussian filter is given by $G(p, q) = e^{-\frac{p^2+q^2}{2\sigma^2}}$, where σ is a scalar constant that provided best results at $\sigma = 1.5$ in [17]. The center of the largest void is the majority pixel with the lowest m.p.d, and the center of the tightest cluster is the minority pixel with the highest m.p.d., as denoted by $M^v(i, j)$ and $M^c(m, n)$, respectively. For each iteration, position (i, j) is added into the set S_{RPS} and the position of (m, n) is deleted from S_{RPS}. Once again, new positions of the largest void and tightest cluster are identified, and the set S_{RPS} is modified. The processing terminates when the position (m, n) deleted from S_{RPS} leads to creation of largest void. Finally, homogeneous distribution of SIPs is obtained. The newly formed position set is represented by S_{HDS}. An example of constructing homogeneous distribution of SIPs is shown in Fig. 1, where the image size is 32 × 32 and γ is set to 0.2.

2.3 Generation of Halftone Shares

The last step is to assign the values to SIPs and CIPs for generating the halftone shares. The assignment of SIPs is determined by the threshold RG-based VC scheme. For each position $(i, j) \in S_{\text{HDS}}$, n-shared pixels are constructed by:

$$[r_1, \ldots, r_n] = \text{Random\_Grid}(S(i, j)) \tag{3}$$

where procedure random grid is implemented by Chen and Tsai's method and $S(i, j)$ denotes the pixel value at position (i, j) of the secret image. Then, the n-shared pixels $r_1, \ldots, r_n$ are assigned to the associated n halftone shares $H_1(i, j), \ldots, H_n(i, j)$.

When the assignment of SIPs is complete, error diffusion algorithm is conducted to produce halftone shares from the gray-level cover images. For position $(i, j) \notin S_{\text{HDS}}$, the halftone shares are computed by error diffusion. Let $g_x(i, j)$ be the pixel of the xth gray-level image, the input to the thresholding block is calculated by:

$$d_x(i,j) = g_x(i,j) - \sum_{k,l} f(k,l)e_x(i-k, j-l) \tag{4}$$

where $f(k, l) \in F$, F is the error filter, and $e_x(i - k, j - l)$ is the quantization error at position $(i - k, j - l)$. The output halftone pixel $H_x(i, j)$ is 1 when $d_x(i, j) > 0.5$ and is 0 when $d_x(i, j) \leq 0.5$. The quantization error given by

$$e_x(i,j) = H_x(i,j) - d_x(i,j) \tag{5}$$

is further diffused to the neighboring pixels that are not processed.

To be noticed, only the CIPs are processed by error diffusion. The SIPs are preset by RG-based threshold VC. They are not processed by the thresholding block. But, the quantization error between the SIP value and the input to the thresholding block is calculated and is diffused to the neighboring unprocessed pixels.

The complementary halftone share is constructed by two steps: (1) the SIPs in the complementary share are determined by RG-based threshold VC and (2) the CIPs in the complementary share are generated by reversing the CIPs in the primary halftone share.

3 Simulation Results and Discussions

In this section, experimental results are shown in the illustrations to demonstrate the effectiveness and advantages of the three proposed methods. Secret image used in these experiments is illustrated in Fig. 2a. Three grayscale cover images are shown in Fig. 2b–d. All the images consist of 512 × 512 pixels.

One example is a (3,3)-threshold case of the proposed method with $\gamma = 0.15$, as shown in Fig. 3. Three halftone shares are illustrated in Fig. 3a–c. PSNRs of halftone shares “Earth” and “Lena” are 23.75 and 28.61 dB, respectively. The stacked results of any two of the three halftone shares are shown in Fig. 3d–f. Nothing about the secret image is disclosed. The stacked result of the three shares is demonstrated in Fig. 3g. The secret image is visually reconstructed. Contrast of the revealed secret image is $6/41$.

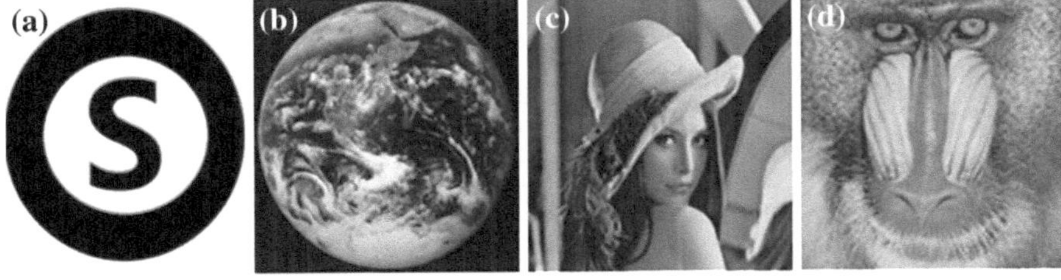

Fig. 2 A secret image and three grayscale cover images. **a** Secret image of size 512 × 512, **b**–**d** three grayscale cover images “Earth”, “Lena”, “Baboon” of size 512 × 512

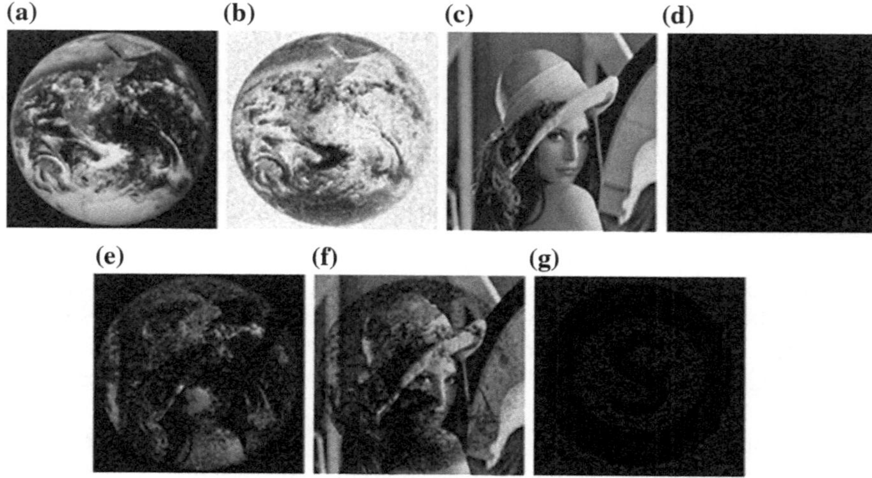

Fig. 3 A (3,3)-threshold HVC of the first method with = 0.15. **a** Halftone share "Earth", PSNR = 23.75 dB, **b** complementary halftone share of "Earth", **c** halftone share "Lena", PSNR = 28.61 dB, **d** stacked result of (**a**) and (**b**), **e** stacked result of (**a**) and (**c**), **f** stacked result of (**b**) and (**c**), **g** stacked result of (**a**), (**b**) and (**c**), contrast is $6/41$

4 Discussions

Detailed comparisons of PSNRs between normal halftone images and halftone shares generated by the proposed method are provided in Table 1. All the halftone shares are with pleasing visual quality. On the other hand, visual quality of halftone shares generated by the second method is slightly lower than the other two methods since RBPs are randomly inserted into the halftone share.

Factor γ determines the visual quality of the halftone share and the contrast of the reconstructed secret image. Smaller γ leads to better visual quality but introduces lower contrast. To provide pleasing visual quality, γ should be smaller than 0.2. In order to obtain relatively hight contrast of the decoded secret image for different thresholds, γ can be increased appropriately.

Features' comparisons between the proposed methods and relative methods are demonstrated in Table 2. Pixel expansion problem is solved in the proposed methods. Meanwhile, code book is not required in the share construction phase. And, halftone share with pleasing visual quality can be generated by the proposed schemes as well.

Table 1 Comparisons of PSNRs between normal halftone images and halftone shares generated by the proposed method

Halftone images	Error diffusion	The proposed method
Earth	24.14	23.75
Lena	31.15	28.61
Baboon	26.54	–

Table 2 Features' comparisons between the proposed method and relative methods

Schemes	Features				
	Pixel expansion	Encryption method	Code book needed	Type of share	Visual quality of shares
[7]	Yes	VC-based	Yes	Meaningful	Medium
[8]	Yes	VC-based	Yes	Halftone	High
[9	Yes	VC-based	Yes	Halftone	High
11]	No	RG-based	No	Meaningless	–
Ours	No	RG-based	No	Halftone	High

5 Conclusions

This paper proposes an approach for constructing a (*k, n*)-threshold HVC scheme by random grids. In the proposed method, SIPs are randomly selected and manipulated by void-and-cluster algorithm to form homogeneous distribution. For the proposed method, a pair of complementary halftone shares is needed to block non-relevant information showing on the decoded image. The proposed method is effective, and advanced merits such as no pixel expansion and no code book required are achieved.

References

1. Naor, M., Shamir, A.: Visual cryptography. Lect. Notes Comput. Sci. **950**(1), 1–12 (1995)
2. Ateniese, G., Blundo, C., De Santis, A., Stinson, D.: Visual cryptography for general access structures. Inf. Comput. **129**(2), 86–106 (1996)
3. Hofmeister, T., Krause, M., Simon, H.: Contrast-optimal k out of n secret sharing schemes in visual cryptography. Comput. Comb. 176–185 (1997)
4. Blundo, C., De Santis, A., Stinson, D.: On the contrast in visual cryptography schemes. J. Cryptology **12**(4), 261–289 (1999)
5. Blundo, C., De Santis, A., Naor, M.: Visual cryptography for grey level images. Inf. Process. Lett. **75**(6), 255–259 (2000)
6. Shyu, S.: Efficient visual secret sharing scheme for color images. Pattern Recogn. **39**(5), 866–880 (2006)
7. Ateniese, G., Blundo, C., Santis, A., Stinson, D.: Extended capabilities for visual cryptography. Theoret. Comput. Sci. **250**(1–2), 143–161 (2001)
8. Zhou, Z., Arce, G., Di Crescenzo, G.: Halftone visual cryptography. IEEE Trans. Image Process. **15**(8), 2441–2453 (2006)
9. Wang, Z., Arce, G., Di Crescenzo, G.: Halftone visual cryptography via error diffusion. IEEE Trans. Inf. Forensics Secur. **4**(3), 383 (2009)
10. Kafri, O., Keren, E.: Encryption of pictures and shapes by random grids. Opt. Lett. **12**(6), 377–379 (1987)
11. Chen, T., Tsao, K.: Threshold visual secret sharing by random grids. J. Syst. Softw. **84**, 1197–1208 (2011)
12. Wu, X., Sun, W.: Random grid-based visual secret sharing for general access structures with cheat-preventing ability. J. Syst. Softw. **85**(5), 1119–1134 (2011)

13. Wu, X., Sun, W.: Visual secret sharing for general access structures by random grids. IET Inf. Secur. **6**(4), 299–309 (2012)
14. Wu, X., Sun, W.: Improving the visual quality of random grid-based visual secret sharing. Signal Process. **93**(5), 977–995 (2013)
15. Wu, X., Liu, T., Sun, W.: Improving the visual quality of random grid-based visual secret sharing via error diffusion. J. Vis. Commun. Image Represent. **24**(5), 552–566 (2013)
16. Wu, X., Sun, W.: Generalized random grid and its applications in visual cryptography. IEEE Trans. Inf. Forensics Secur. **8**(9), 1541–1553 (2013)
17. Ulichney, R., et al.: The void-and-cluster method for dither array generation. In: Proceedings of SPIE, vol. 1913, pp. 332–343 (1993)

Statistical Tests for Combined Secret Key Sequence

Guifen Zhao, Ying Li and Liping Du

Abstract The randomness and amount of secret key is important to the security of key management. A combined secret key method is presented and statistical tests are performed. Generate a set of combined secret keys according to the combined secret key generation algorithm using a smart card which performs compression mapping to compose a binary sequence of a given length for statistical test. The Frequency Test, Runs Test, and Serial Test are presented in detailed. The statistical conclusion is that the combined secret key sequence is accepted as random. Consequently, combined secret key is proved secure enough for secret key generation.

Keywords Statistical test · Combined secret key · Randomness · Key generation · Smart card

1 Introduction

The key management for any encryption algorithm is quite important, and the security of encryption key and decryption key depend on the total amount and randomness. Consequently, different random number generators with excellent randomness performance are adopted to generate encryption key and decryption key.

G. Zhao (✉) · Y. Li · L. Du
Beijing Key Laboratory of Network Cryptography Authentication, Beijing Municipal Institute of Science and Technology Information, Beijing, China
e-mail: gfzh@hotmail.com

Y. Li
e-mail: shai_wang@hotmail.com

L. Du
e-mail: duliping_419@163.com

S. Patnaik and X. Li (eds.), *Proceedings of International Conference on Computer Science and Information Technology*, Advances in Intelligent Systems and Computing 255, DOI: 10.1007/978-81-322-1759-6_21,

For asymmetric encryption algorithm, for example RSA, random numbers are also used while calculating public key and private key according to two large prime numbers selected randomly. For symmetric encryption algorithm, random numbers or pseudo-random numbers are introduced to generate seed key, which can establish a relatively random key space [1].

When random numbers are used in the key generation process, all values should be generated randomly or pseudo-randomly so that all possible combinations of bits and all possible values are equally likely to be generated. If a sequence of numbers is not random at all, the regular bits can be deduced easily, which means that the cryptographic application using the sequence is under the environment of threats. The generated encryption keys based on random numbers need enough randomness and amount. So the generated key sequence should perform statistical tests to verify it is random or not.

The statistical test methods include National Institute of Standards and Technology (NIST) Special Publication 800-22, FIPS 140, the Diehard test suite, CryptX statistical software package. [2–4]. The NIST Test Suite provides 15 statistical methods to test the randomness of binary sequences: Frequency (Monobit) Test, Frequency Test within a Block, Runs Test, Tests for the Longest-Run-of-Ones in a Block, Binary Matrix Rank Test, Discrete Fourier Transform (Spectral) Test, Nonoverlapping Template Matching Test, Overlapping Template Matching Test, Maurer's "Universal Statistical" Test, Linear Complexity Test, Serial Test, Approximate Entropy Test, Cumulative Sums (Cusums) Test, Random Excursions Test, and Random Excursions Variant Test [5].

The test sequence is composed with large amounts of combined secret key which is generated by a smart card with USB interface according to combined secret key algorithm. Several statistical tests mentioned in NIST SP800-22 and corresponding assessments are performed.

2 Binary Sequence Generation for Test

2.1 Combined Secret Key Generation

Because it is hard to guarantee the security of secret key while distributing and using, cryptographic applications need design secure key exchange protocol. Moreover, key update is also a key point to be solved. A combined secret key method is proposed to achieve secure secret key management [6].

Combined secret key is generated by mapping of key seeds according to control parameters. The element of key seeds is a 1-byte nonnegative integer generated by random number generator equipped on the smart card with USB interface. Consequently, all elements of key seeds are part of Z_{256}, i.e., key seeds matrix S shown as follows is a residual matrix of Z_{256}.

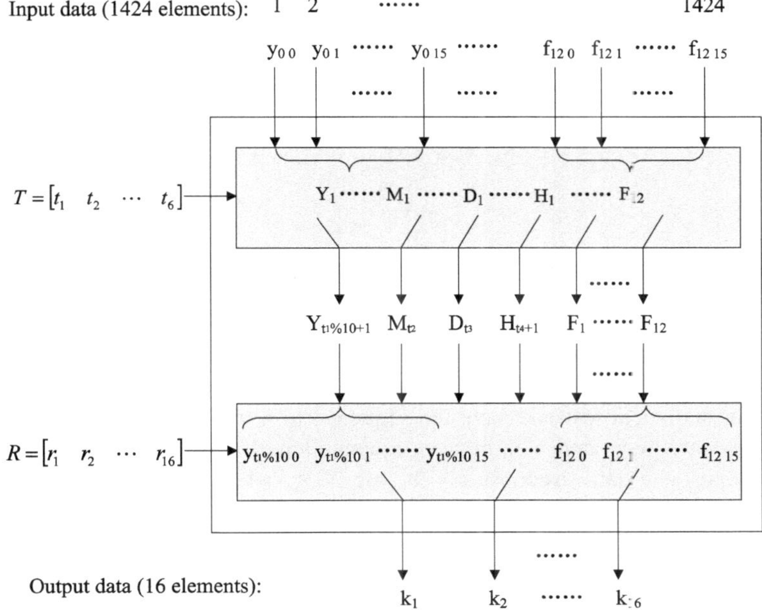

Fig. 1 Compression mapping

$$S = \begin{bmatrix} s_{11} & s_{12} & \dots & s_{1n} \\ s_{21} & s_{22} & \dots & s_{2n} \\ \vdots & \vdots & & \vdots \\ s_{m1} & s_{m2} & \dots & s_{mn} \end{bmatrix}$$

where $m = 89$ and $n = 16$, namely the key seeds of each user total 1,424 bytes.

The secret key for 3DES, SM1, etc. algorithm amounts to 16 bytes. The combined secret key algorithm performs compression mapping from 1,424 input elements mentioned above, i.e., elements of key seeds matrix, to obtain 16 output elements, i.e., combined secret key, shown in Fig. 1.

While part elements are discarded to get output key data after compression mapping, all the input data cannot be deduced on the basis of output data, i.e., the compression mapping is irreversible. The control parameters of compression mapping are time stamp and random numbers. The time stamp is generated by system clock of servers, including year, month, day, hour, minute, and second. The time stamp matrix T is defined as follows:

$$T = [\, t_1 \quad t_2 \quad \dots \quad t_6 \,]$$

where $t_1 \quad t_2 \quad \dots \quad t_6$ separately correspond with year, month, day, hour, minute, and second element of time stamp. Each element is a nonnegative integer.

Random numbers, totally 16 hexadecimal random numbers, are generated by random number generator equipped on the smart card with USB interface. Each random number is a nonnegative integer between 0 and 16, i.e., the random number matrix R is a residual matrix of Z_{16}.

$$R = \begin{bmatrix} r_1 & r_2 & \ldots & r_{16} \end{bmatrix}.$$

2.2 Generate and Save the Binary Sequence

According to the combined secret key generation algorithm, generate a set of combined secret key to compose a binary sequence of zeros and ones of a given length for statistical test and save the sequence to a file. Totally 8,500 combined secret keys are generated, i.e., binary sequence length is 1,088,000. The sequence of bits exists as a data structure: $\varepsilon = \varepsilon_1, \varepsilon_2, \ldots, \varepsilon_n$, where ε_i is 1 or 0, and n is the length of the sequence.

3 Statistical Test and Assessment for CSK Sequences

Several statistical tests according to NIST Test Suite are presented in detailed in this section. For each statistical test, a set of P-values and relevant intermediate values are produced. For a fixed significance level, a certain percentage of P-values are expected to indicate failure. For example, if the significance level α is chosen to be 0.01, then about 1 % of the sequences are expected to fail. A sequence passes a statistical test if the P-value $\geq \alpha$ and fails otherwise. Based on these P-values, a conclusion regarding the quality of the sequences can be made.

3.1 Frequency (Monobit) Test

The most basic test focuses on the proportion of zeroes and ones for the entire sequence. The Frequency Test is that of the null hypothesis: in a sequence of independent identically distributed Bernoulli random variables, i.e., ε, the probability of ones is 1/2, that is, the number of ones and zeros in a sequence should be approximately the same. All subsequent tests are conditioned on having passed this first basic test.

While being tested, the zeros and ones of the input sequence (ε) are converted to values of -1 and +1 and are added together to produce $S_n = X_1 + X_2 + \cdots + X_n$ where $X_i = 2\varepsilon_{i-1} = \pm 1$. If the sequence is random, then the plus and minus ones will tend to cancel one another out so that the test statistic will be about 0. The test

Table 1 Computational information of Frequency (Monobit) Test

Computational item	Value	Conclusion
S_n	−976	
S_n/n	−0.000897	
P_{value}	0.349429	Success

is derived from the well-known Central Limit Theorem for the random walk S_n. According to the Central Limit Theorem, this classical result serves as the basis of the simplest test for randomness. According to the test based on the statistic s, evaluate the observed value s_{obs}:

$$s_{\text{obs}} = \frac{|S_n|}{\sqrt{n}}. \tag{1}$$

If $z = \frac{s_{\text{obs}}}{\sqrt{2}}$ is distributed as normal, then $|z|$ is distributed as half normal. And then, calculate the corresponding P-value, which is

$$P_{\text{value}} = \text{erfc}\left(\frac{s_{\text{obs}}}{\sqrt{2}}\right) = 2[1 - \Phi(s_{\text{obs}})]. \tag{2}$$

Here, erfc is the complementary error function:

$$\text{erfc}(z) = \frac{2}{\sqrt{\pi}} \int_z^{\infty} e^{-u^2} du. \tag{3}$$

The Frequency Test result is shown in Table 1.

Since the $P_{\text{value}} \geq 0.01$, the conclusion is that the sequence is random.

3.2 Runs Test

The focus of Runs Test is the total number of runs in the sequence, where a run is an uninterrupted sequence of identical bits, i.e., consecutive 1 s or consecutive 0 s. A run of length k consists of exactly k identical bits and is bounded before and after with a bit of the opposite value. The purpose of the Runs Test is to determine whether the number of runs of ones and zeros of various lengths is as expected for a random sequence. In particular, this test determines whether the oscillation between such zeros and ones is too fast or too slow.

The Runs Test carries out the Frequency Test as a prerequisite. Firstly, compute the pretest proportion π of ones in the input sequence:

$$\pi = \frac{\sum_j \varepsilon_j}{n} \tag{4}$$

Table 2 Computational information of Runs Test

Computational item	Value	Conclusion
π	0.499551	
V_n(obs)	544,768	
P_{value}	0.140640	Success

If it can be shown that $\pi - 1/2 < \tau$ where $\tau = 2/\sqrt{n}$, then the prerequisite Frequency Test is passed. And then, compute the total number of runs across all n bits, i.e., the total number of zero runs plus the total number of one runs:

$$V_n(\text{obs}) = \sum_{k=1}^{n-1} r(k) + 1 \tag{5}$$

where $r(k) = 0$ if $\varepsilon_k = \varepsilon_{k+1}$, and $r(k) = 1$ otherwise.

Finally, compute P-value:

$$P_{\text{value}} = \text{erfc}\left(\frac{|V_n(\text{obs}) - 2n\pi(1-\pi)|}{2\sqrt{2n}\pi(1-\pi)}\right) \tag{6}$$

The test result is shown in Table 2.

Since $P_{\text{value}} \geq 0.01$, accept the sequence as random

3.3 Serial Test

The Serial Test focuses on the frequency of all possible overlapping m-bit patterns of given length across the entire sequence. The purpose of this test is to determine whether the number of occurrences of the 2^m m-bit overlapping patterns is approximately the same as would be expected for a random sequence.

Firstly, extend the sequence by appending the first $m-1$ bits to the end of the sequence for length of n. Here, the block length $m = 16$. Determine the frequency of all possible overlapping m-bit blocks, all possible overlapping $(m - 1)$–bit blocks and all possible overlapping $(m - 2)$-bit blocks. Let $v_{i_1 \ldots i_m}$ denote the frequency of the m-bit pattern $i_1 \ldots i_m$; let $v_{i_1 \ldots i_{m-1}}$ denote the frequency of the $(m - 1)$-bit pattern $i_1 \ldots i_{m-1}$; and let $v_{i_1 \ldots i_{m-2}}$ denote the frequency of the $(m - 2)$-bit pattern $i_1 \ldots i_{m-2}$. Compute:

$$\begin{aligned} \psi_m^2 &= \tfrac{2^m}{n} \sum_{i_1 \ldots i_m} \left(v_{i_1 \ldots i_m} - \tfrac{n}{2^m}\right)^2 \\ \psi_{m-1}^2 &= \tfrac{2^{m-1}}{n} \sum_{i_1 \ldots i_{m-1}} \left(v_{i_1 \ldots i_{m-1}} - \tfrac{n}{2^{m-1}}\right)^2 \\ \psi_{m-2}^2 &= \tfrac{2^{m-2}}{n} \sum_{i_1 \ldots i_{m-2}} \left(v_{i_1 \ldots i_{m-2}} - \tfrac{n}{2^{m-2}}\right)^2 \end{aligned} \tag{7}$$

Table 3 Computational information of Serial Test

Computational item	Value	Conclusion
m	16	
ψ_m^2	65870.185412	
ψ_{m-1}^2	32862.750118	
ψ_{m-2}^2	16443.271529	
$\nabla\Psi_m^2$	33007.435294	
$\nabla^2\Psi_m^2$	16587.956706	
P_{value1}	0.174732	Success
P_{value2}	0.130136	Success

Compute $\nabla\Psi_m^2$ and $\nabla^2\Psi_m^2$ to measure how well the observed frequencies of m-bit patterns match the expected frequencies of the m-bit patterns. The obtained values are:

$$\begin{aligned}\nabla\Psi_m^2 &= \psi_m^2 - \psi_{m-1}^2 \\ \nabla^2\Psi_m^2 &= \psi_m^2 - 2\psi_{m-1}^2 + \psi_{m-2}^2\end{aligned} \tag{8}$$

Compute P-value 1 and P-value 2:

$$\begin{aligned}P_{\text{value1}} &= \text{igamc}\left(2^{m-2}, \frac{\nabla\Psi_m^2}{2}\right) \\ P_{\text{value2}} &= \text{igamc}\left(2^{m-3}, \frac{\nabla^2\Psi_m^2}{2}\right)\end{aligned} \tag{9}$$

The test result is shown in Table 3.

Since both P-value 1 and P-value 2 were ≥ 0.01, the conclusion is that the sequence is random

4 Conclusions

The binary sequence is composed with a set of combined secret key generated according to corresponding algorithm on the basis of a 1,424-bytes key seeds. The key seeds and random numbers are generated by random number generator provided by a smart card with USB interface. The combined secret key sequence has passed the Frequency Test, Runs Test, Serial Test and other tests, i.e., the sequence is accepted as random. The randomness of combined secret key is proved by these statistical tests. Therefore, combined secret key is proved secure enough for secret key generation and update.

Acknowledgments Authors would like to appreciate the support from the Program of Large-scale Network Authentication Center affiliated to Beijing Municipal Institute of Science & Technology Information (No. PXM2012_178214_000005) and the Innovation Group for Internet Real-name System (No. IG201003C2). And also thank the helpful suggestions from the director and colleagues in Beijing Key Laboratory of Network Cryptography Authentication.

References

1. Zhang, Y.: The Random Number Generators and the Methods for Randomness Test. Univ. Electron. Sci. Technol. China. (2006)
2. Fips Pub 140-1.: Security Requirements for Cryptographic Modules. Federal Information Processing Standards Publication, Jan 1994
3. Li-Min, F., Deng-Guo, F., Chen, H.: On the parameter selection of randomness test. J. Commun. **30**(1), 1–6 (2009)
4. Chen, X., Shu-Wang, L.: One way to read and random test of random sequence. Comput. Appl. **22**(9), 7–9 (2002)
5. Rukhin, A., Soto, J., Nechvatal, J., Smid, M., Barker, E., Leigh, S., Levenson, M., Vangel, M., Banks, D., Heckert, A., Dray, J., Vo, S.: Special Publication 800-22 Revision 1a: A Statistical Test Suite for Random and Pseudorandom Number Generators for Cryptographic Applications, April 2010
6. Hu, X., Zhao, G.: A CSK-based solution for person authentication. In: The Seventh Wuhan International Conference on E-business: Unlocking the Full Potential of Global Technology, pp. 244–249 (2008)

Part III
Communication and Networking

Interactive Satellite Remote Education System Based on Bluetooth and GPRS

Xiangyu Bai and Shufang Wang

Abstract Because of the advantage of broadcast and the lower-cost satellite, tele-education is an important distance education method. However, the current satellite-based distance education systems use one-way reception mode, so it is unable to meet the needs of the user interactive learning, which limits the application of satellite-based distance education. Against the limitations of the existing satellite remote education system, this paper proposes a communication method based on Bluetooth and general patio radio service (GPRS). The end-user sends requests to the mobile phone via Bluetooth; mobile phones carrying the requests move to the place where there is mobile network coverage, then mobile phone sends the requests to the gateway by GPRS/GMS, and finally, the requests are forwarded by the gateway to the broadcast server. This approach is compatible with existing satellite-based distance education system. Since the scheme is based on the widely deployed wireless communication network, which made it easily to deploy and cost less. .

Keywords Distance education · Bluetooth · GPRS · Return communication method

1 Introduction

China has a vast territory, and great gaps exist between development of regional economic and the distribution of educational resources. Therefore, modern distance education is a great important long-term development strategy in our country

X. Bai (✉) · S. Wang
College of Computer Science, Inner Mongolia University, Hohhot 010021, China
e-mail: baixiangyu@imu.edu.cn

S. Wang
e-mail: 13848170847@126.com

S. Patnaik and X. Li (eds.), *Proceedings of International Conference on Computer Science and Information Technology*, Advances in Intelligent Systems and Computing 255, DOI: 10.1007/978-81-322-1759-6_22,

[1]. The satellite remote education has the advantage of the broadcast and the distribution of large amounts of data in wide area in lower costs, reducing the cost of implementation of distance education, particularly suited to sparsely populated, economically less developed areas to achieve educational information [2]. At present, the remote education system usually consists of resource storage network, satellite radio network, and terminal node group.

However, the one-way characteristic of satellite communications has limitations. The satellite remote education application has been limited. Satellite radio one-way networks use "infusion teaching" way to restrain students' study enthusiasm. In addition to different schools, different user demand for resource is different. For instance, primary and middle schools in different regional schools and the national school demand for resource are not the same, and there are great changes with time. So the user hopes to be able to choose resources according to their own situation. In such cases, the one-way broadcast communication of satellite-based distance education system must be changed to bidirectional communication. Thus, reverse communication mechanism is necessary (Fig. 1).

In 1999, the Ministry of Education put forward a general outline of modern distance education. It is pointed out that modern distance education should have a variety of return links, so the user can implement interactive functions. At present the interactive systems can be divided into two categories. (1) the use of other means or devices to achieve external interactive; (2) use of two-way satellite terminal system to achieve internal interactive communication. Through the outside, the interactive communication technology has two modes to realize interaction: (1) through the satellite radio and the Internet (telecommunication network) outside the interactive communication way; (2) through the satellites plus ground cable network together with Internet outer interaction. The above two modes are dependent on the support of network infrastructure, often achieved by means of Internet access. However, in many mountainous areas, to carry out

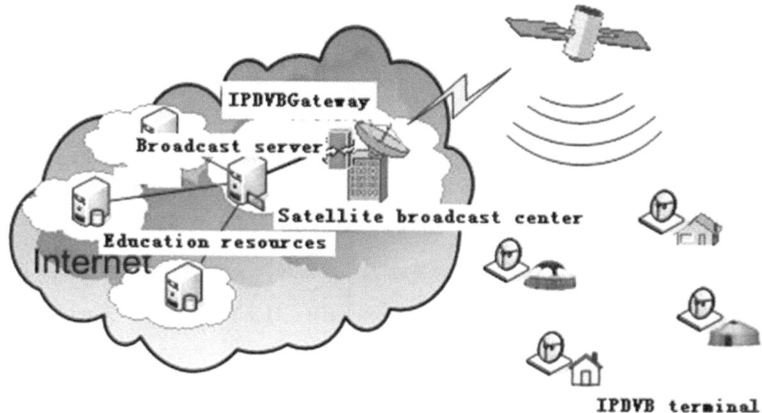

Fig. 1 Satellite remote education system schematic diagram

Internet access services will require significant investment. VSAT system at the same time is able to point to multipoint of unidirectional radio and point to multipoint two-way communication; in the Tsinghai University in Yunnan, two remote education sites are using this way. But the VSAT system is suitable for the unit to group for deployment and does not apply for individual users. Therefore, according to the delay-tolerant network (DTN) storage, carry forward thought [3] put forward a kind of suitable system for individual users of satellite remote education interactive method; this method mainly uses Bluetooth and general patio radio service (GPRS) technology.

Bluetooth is a short-range communications technology. It can be used to exchange information between the mobile phone, PDA, wireless headphones, notebook computer, peripherals and other equipment. The technology established a general Radio Air Interface (Radio-Air-Interface) and makes an open standard for the control software [4]. Bluetooth enables wireless communication technology closely, so that the portable devices from different manufacturers can connect each other and operate same function. According to the statistics, Bluetooth technology in the intelligent mobile phone is supported permeability by over 95 %. At present, the intelligent mobile phone in the market share has more than half. We use it as a Bluetooth communication process of a path in the road without the deployment of other infrastructure; the implementation is simple, and it is easy to operate.

GPRS (General Packet Radio Service) is available to provide GSM mobile phones users a mobile data services. GPRS can be said that the continuation of GSM, by adding some hardware modules and software upgrades. GPRS forms a new network logical entity, which providing end-to-end, wide area wireless IP connection. GPRS can support TCP/IP protocol and x.25 protocol. According to the statistics, at present, as a representative of wireless Internet access service, GPRS has relatively high coverage rate, such as Inner Mongolia mobile wireless network signal has been realized to 90 % of the natural village coverage [4]. Therefore, on the link we adopt GPRS as based network. It can make full use of the existing resources and increases the chances of successfully passing back with less investment.

By the end of 2011, China's mobile phone penetration is per one hundred people having 73.6 telephones [5]. Therefore, using the widespread mobile Internet way to realize satellite remote education, user request information is submitted, meaning to realize the two-way remote education based on mobile communication will be the most convenient method in the agricultural and pastoral areas. Direct use of mobile phones to transfer the user request information is restricted by the percentage of network coverage, because terminal node location may not have network signal coverage. At the same time, the computer system and mobile phone system are not also mutually compatible and unable to directly communicate, resulting in the user use disorder. This paper proposes a new communication method for reverse backhaul communications in satellite distance education systems, and the method combines the Bluetooth communications, data storage -and -carrier technology and mobile GPRS technology.

2 The Return Communication System

2.1 The Basic Frame of Method

The way it works as follows: 1.A terminal user t receives resources from satellite broadcast. 2. User creates some problems when learning the resources. The problem to be solved are organized and stored in the local computer. 3. When a mobile phone moves into the communication range of Bluetooth signal, the computer use Bluetooth mode to transfer the information stored itself. 4. Then, the mobile phone carries the information, and continues to move. 5. If the mobile phone comes into the coverage of GSM network, it will automatically submit information to the satellite broadcast center using GPRS. This is a new mechanism to achieve external interactive for satellite-based distant education. It can overcome the weakness of insufficient network coverage, and it's a very simple way to provide the data transmission services for terminal computers. Due to low cost of hardware and daily use, the mechanism is suitable for users to carry out interactive distant education in agricultural and pastoral areas.

The process of each node in the interaction method is shown in Fig. 2.

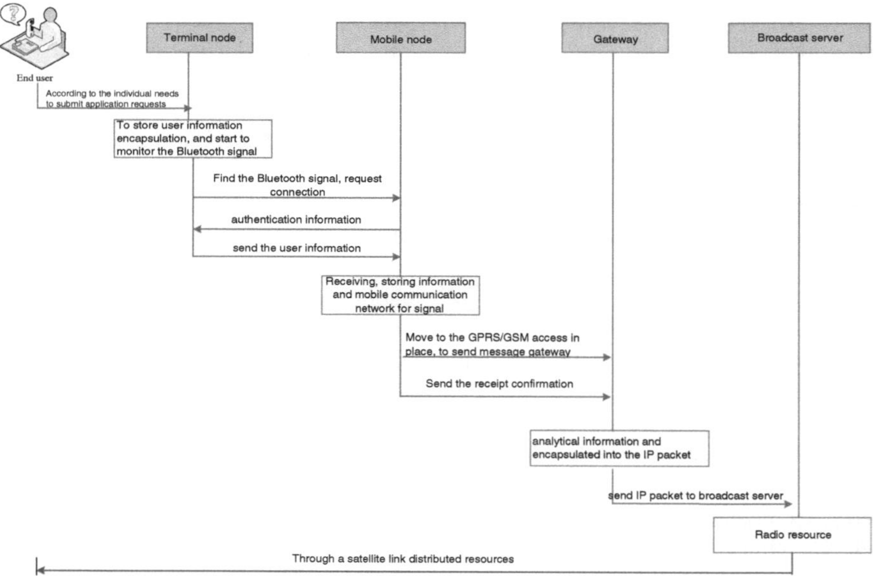

Fig. 2 The interaction process between nodes

2.2 Interactive Satellite Remote Education System Based on Bluetooth and GPRS

Based on Bluetooth and GPRS satellite remote education, user-interactive communication system can support both broadcast and multicast business and foreign mutual two-way communication business; it consists of distributed resource server group, administration server, broadcast server, IPDVB gateway, terminal node, mobile phones, and gateway composition; the structure is as shown in Fig. 3.

2.3 Data Interaction Process of System

Interaction patterns based on Bluetooth and GPRS/GSM satellite remote education system of the interaction of the data processing is as follows:

1. Broadcast server list of education resources and the commonly used teaching resources in the form of IP packets; periodically send IPDVB gateway.
2. After encapsulating IP packets to TS stream, IPDVB gateway uses DVB-S satellite channel to transmit them.

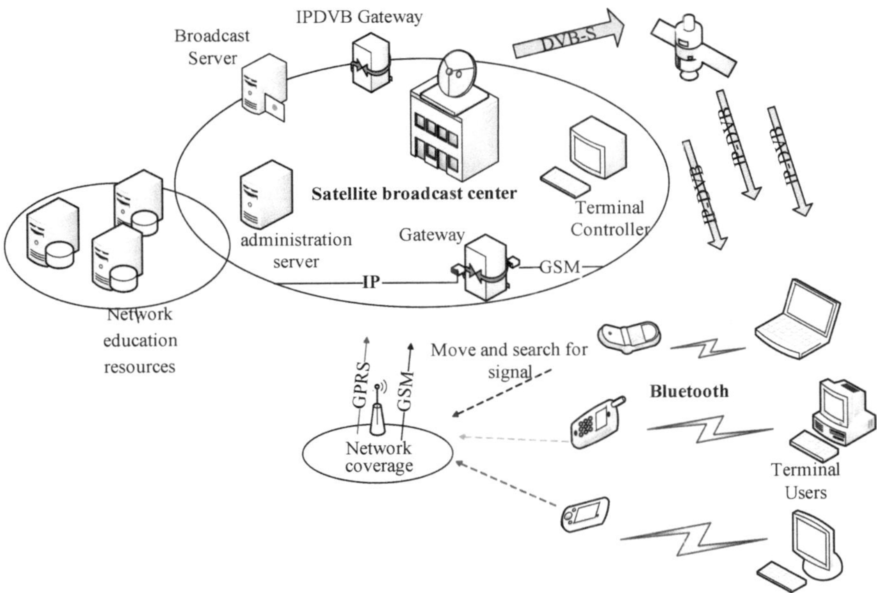

Fig. 3 Interaction patterns based on Bluetooth and GPRS/GSM satellite remote education system

3. Terminal node gets the TS stream through DVB-S receiving system. Then IP packet is formed after de-encapsulation, and organized into education resources which stored in the local terminal teaching software system of local computer as a web page, video, voice, images and other forms of teaching.
4. Remote user education in the process of learning, it needs to communicate information or resources through teaching software system oriented to submit requests.
5. Terminal node classifies user's requests and encapsulates them to proper format. It will establish a connection with a nearby cell phone via Bluetooth, and then a secure authentication process between the node and mobile phone happen. If the authentication fails, the phone is deemed untrustworthy terminal node and the connection fails. If the authentication is successful, the terminal nodes will transfer the information requested by the user to the mobile phone via Bluetooth.
6. Mobile phone stores user request information and mobile search network signal.
7. Mobile having Internet through GPRS/GSM sends the information to the gateway.
8. Gateway receives the information requested by users and de-encapsulates them. The information are treated according to the type and sent to the appropriate broadcast server respectively.

3 The Design and Implementation of Core Modules

We proposed a new method which on the basis of the original satellite remote education system puts forward a way to back. In the method, involved units are the terminal node, gateway, and mobile nodes (mobile phone).

3.1 Terminal Node

The terminal node information processing capabilities include the following:

1. Processing resources. The process of the information includes classification, coding and compression of education resources. The computer encodes and compresses user's request according to the type of information. So, the data during Bluetooth communication and GPRS transmission can be minimized to reduce communication costs and power consumption of mobile phone users.
2. Transmitting information. The node establishes communication with mobile phone based on Bluetooth and chooses different way of information processing and transmission according to the characteristics of different users.
3. Authenticating with mobile phone.

4. Message, that is, depending on the different user's phone, prompt the user to choose suitable mode of operation, according to different operating modes corresponding operation hints.

The terminal node uses the Windows system and is equipped with a Bluetooth adapter. It communicates with mobile nodes nearby and forward messages based on Bluetooth. Terminal node software mainly includes the following features: (1) the packaging information, compressed package on the user's request, (2) the information stored, packaged data stored in local disk around waiting for the mobile node, (3) authentication, identification of mobile phone verification code, and (4) forwarding data, to forward the packets through Bluetooth to the phone.

3.2 Gateway

The function of gateway is to receive SMS or MMS from GSM/GPRS through the mobile node and send confirmation messages to mobile node through the GSM, then classify the received information, packed information which was handled into IP packets, and then send it to the corresponding broadcast server by the enteric socket.

The gateway node uses the Windows system and is equipped USB port wireless modem which supporting GPRS / GSM. Gateway receives the SMS and MMS messages sent by mobile nodes via modem. Gateway software mainly contains the following functions: (1) analyzes data packets, check the integrity of the received packet, de-encapsulation the checked packet, and read the request type of the data; (2) package information, packets the pack to IP format, (3) send confirmation, after receiving the data and completed checking send confirmation to the mobile node, and (4) forward packets, depending on the type of request information to forward the packets through Ethernet to the corresponding broadcast server.

3.3 Mobile Node (Mobile)

Mobile node receives the information sent by terminal node via Bluetooth, stores the information in local database, and sent confirmation message to terminal node. The mobile node starts to search around mobile communication network signal after receiving the information, sends it to the gateway through the GPRS/GSM, and then deletes the message after receiving the confirmation sent back by the gateway.

Mobile node uses smartphone which supports Bluetooth, and the function of Store-and-Forward information is achieved on android2.3. The software system of mobile node mainly includes the following features: (1) Storing information. The received information is stored locally and inserted into the transmit queue in accordance with the priority of information. (2) Sending acknowledgment. When the integrity of the information received is confirmed, mobile sends a confirmation message to the terminal node via Bluetooth. (3) Forwarding information. When the node detected the signal of surrounding network, it selects the transition mode according to the type of information and sends the message to the gateway.

4 Conclusions

The interactive program of remote education proposed in this paper has many characteristics such as strong applicability, low cost, etc. And the new method is not only easy to promote but is also compatible with existing systems. We developed related software on android mobile phone platform to meet the requirements of mobile phone users. For those mobile phones that with no Bluetooth device or does not support the software we developed, users can send relevant demands by SMS or MMS. Manually sending text messages and multimedia messages using GPRS or transmission via Bluetooth have a significant price advantage compared to leasing satellite channel and laying the Internet. The users using Android mobile phones just need to install the related software package, and the existing terminal users simply need to add a Bluetooth adapter. Now, the system has been carried out in more than 20 primary schools used and tested. In the future, we will focus on their application needs to improve the system performance.

References

1. Yan, J., Wang, T.: Satellite remote education in the development of China's satellite application. Satell. Appl. May 2007
2. Zhang, X., Li, J., Fang, Y.: Satellite communications become the main means of remote education in less developed 225 areas. Digit. Commun. World Sept 2007
3. Fall, K.: Delay-tolerant network architecture for challenged internets. In: Proceedings of ACM SIGCOMM'03, pp. 27–34. ACM, New York (2003)
4. Inner Mongolia news: mobile company to strengthen the construction of the enhancement of improving customer satisfaction. http://inews.nmgnews.com.cn/system/2011/07/12/010621866.shtml, 12 July 2011
5. Bluetooth SIG: Bluetooth profile specification (Version 1.1), 22 Feb 2001

A Study of Smart Grid Communication Architecture

Xiaoxue Liu and Huayang Cao

Abstract Efficient communication is critical to smart grid. In this paper, smart grid (SG) communication requirements have been analyzed and a communication architecture for SG called smart grid information communication architecture (SGICA) that has finer granularity of communication network has been put forward. Furthermore, the information communication technologies for constructing networks of SGICA are proposed. Finally, the communications challenges that SG may face are discussed, and IP-based communications are suggested for SGICA.

Keywords Smart grid · Communication architecture · SGICA · IP-based communications

1 Introduction

Built on integrated and high-speed two-way communication networks, smart grid (SG) highly combines the newest information and computer-control technologies with the current infrastructure of transmission and distribution, to form a new intelligent power grid. In 2001, America proposed the concept of "Intelligrid."

Xiaoxue Liu, female, born in 1990, graduate student, research on smart grid communication, School of Computer, National University of Defense Technology.

X. Liu (✉) · H. Cao
School of Computer, National University of Defense Technology, Changsha 410073, China
e-mail: xiaoxue@nuaa.edu.cn

H. Cao
e-mail: huayang.cao@gmail.com

S. Patnaik and X. Li (eds.), *Proceedings of International Conference on Computer Science and Information Technology*, Advances in Intelligent Systems and Computing 255, DOI: 10.1007/978-81-322-1759-6_23,

In 2005, Europe set up SG Technology Forum pointing out that information and communication technology are the critical areas needed to be studied firstly. In 2006, IBM puts forward the SG solutions regarding the data collecting and processing as the key factor that highly relies on efficient communication architecture. Jiandong Wu, a Chinese energy expert, gave the concept of "Interactive Grid" based on both open and interconnected information mode. So, efficient and reliable communications on which various functions of SG rely are important for SG.

With the SG developing and scale of the network becoming larger, new intelligent applications have been appearing constantly that bring more complicated and strict demands to communication network. So, the current power system communication architecture will not be able to support the SG communications well, and new communication architecture needs putting forward. On this, some researches have been done in the view of Things architecture and transport model. Paper [1] proposed a typical future Things Architecture Model (U2IoT) with a reflecting concept of human central nervous system and social structure. It also gave a systematic security architecture that combines the network world, physical world, and awareness of the human society into the U2IoT asynchronously. Paper [2] puts forward an all-IP wireless sensor network architecture including an all-IP adaptation method and four network protocols. However, all those architectures and methods cannot be applied into SG directly because they did not take the special features of SG into account. Cisco has studied the structure of power grid and proposed the reference model Grid Blocks for the future SG communication architecture [3]. But to apply Grid Blocks into Chinese power industry, much more work needs to be done that is also what we have been focusing on.

In this paper, we firstly made a comprehensive analysis of the communication requirements of SG. Then, we put forward the smart grid information communication architecture (SGICA) by referring to Cisco Grid Blocks and combining our own research on Chinese power industry. Finally, we discussed the challenges in SG communications and suggested IP-based communications for SGICA.

2 SG Communications Requirements

Functions of the SG, such as advanced metering infrastructure (AMI), demand response (DR), supervisory control and data acquisition (SCADA), and so on, produced new communication requirements. These functions with a large amount of data need real-time, high-speed, and two-way communication, which therefore requires bandwidth, delay, and reliability strictly.

Bandwidth—The bandwidth of the communication network not only determines the data transfer rate, but also largely affects the real-time communication. Applications in SG are numerous and different tasks demand different network bandwidths: those tasks transmitting larger amount of data and have higher requirement for real time, often need a higher bandwidth, such as the SCADA. To determine the bandwidth requirements for the SG communication, network is extremely important, because it directly affects the choice of transmission medium (optical fiber, radio waves, coaxial cable, etc.) and communication technologies (3G, LTE, WIMAX, etc.).

Delay—Transmission delay, one of the most important communication requirements in SG, affects the real-time communication directly. The scale of SG is so large that different tasks demand different transmission delays. It can be higher if the data are used for the coordinated control of the entire system while data for local analysis or emergency response require a lower delay [4].

Reliability—Communication reliability refers to the ability of communication network to meet communication needs of the users continuously under the condition that various destructive factors coexist [5]. It concerns the successful completion of different tasks in SG, thus determining the stability of the system. For example, power outages may lead to interruption of grid communication, thereby further affect the recovery of the fault. So, it is very important to ensure the extremely high reliability of the SG communication network.

Mobility—The emerging function in SG requires large-scale coverage of the communication network in order to support the mobile communication. For example, electric vehicles (EVs) will charge at different locations, including home, office, public, or private sites in the long-distance travel. In addition, remote meter reading and line inspection, emergent command, etc., all demanding mobile communication, requires the communication network to support wide area seamless coverage and roaming mobility.

Security—Secure information storage and transportation are critical for power utilities, especially for billing purposes and grid control. Some data transmitting in the SG, concerning users' privacy and business secrets, therefore belonging to sensitive information, demands high confidentiality; serious consequences will come out once data of some application are tampered by an attacker. For example, in the process of SCADA, an attacker intercepts the collected abnormal data and tampers them as normal, then sends them back to the control center, which will be not able to find out the abnormal condition in the grid, thus leading to inappropriate decision. In addition, in the automatic distribution process, if the communication security could not be guaranteed and distribution network suffers from malicious attacks, it would be likely to bring out large-scale blackouts.

In summary, the large-scale SG has different functions that need distinct communication conditions. So while planning and allocating the communication resources, we should think about the different requirements of all kinds of functions in the SG so as to make full use of the communication resources. In addition, a SG should be scalable enough to facilitate the operation of the power grid [6]. Many smart meters, smart data collectors, and renewable energy resources are joining the communications network. Hence, SG should handle the scalability with the integration of advanced Web services, reliable protocols with advanced performance, such as self-configuration and security aspects.

3 SGICA

To support the planning, designing, and running of the power company toward the power system, Cisco has provided a set of technology architecture Grid Blocks, including series of products from reference model to guiding for design and achievement. However, the Grid Blocks are short in complicated structure and low operability. We have streamlined and perfected the Grid Blocks architecture combining the current communication condition of power industry in China, so as to put forward the SGICA. It divides the power communication infrastructure into ten logical layers to support the networking of the whole power transmitting link. It also defines cooperations across regions. Figure 1 shows each communication layer in SGICA.

The layers, from the bottom to the top of Fig. 1, include:

Prosumer (producer and consumer)—This layer covers all the devices and systems that is not part of the utility infrastructures that can impact the power system and interact with the utility. This layer includes networks managing distributed generation and storage, responsive loads in residences or commercial/industrial facilities, on board electric vehicle networks, and so on [3].

Distribution—This layer is located between primary distribution substations and end users, broken into two levels:

- *Distribution level* 2—This lower level is composed of dedicated networks such as the "Last Mile" network, neighborhood area network (NAN), and so on. They can service the infrastructures including smart metering, distribution automation, and electric vehicle charging.
- *Distribution level* 1—This level can support multiple services including aggregating all kinds of networks in level 2 and connecting to the primary distribution substations directly so as to achieve the distributed smart distribution. This level can also provide peer connection to the field area networks (FANs).

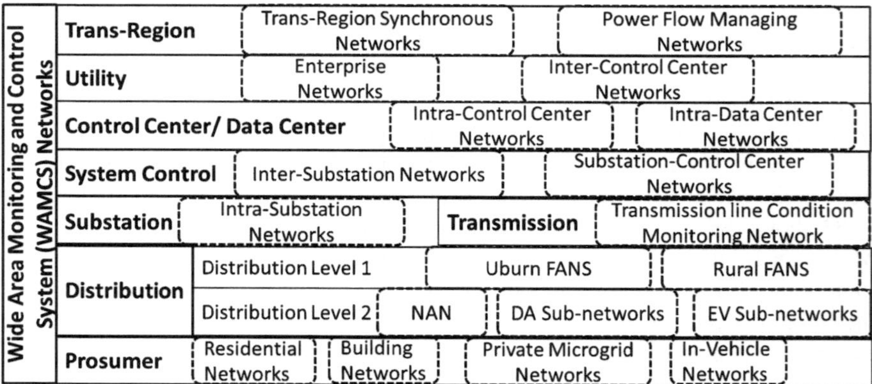

Fig. 1 SGICA

Transmission—This layer covers the local area networks (LANs) or wide area networks (WANs) consisting of the transmission line condition monitoring systems that are interconnected. These layer networks are connected to the system control networks directly to realize the condition monitoring and decision control for the transmission lines.

Substations—This layer comprises all the intra-substation networks from the relatively simple secondary substations to the complicated primary substations that can provide critical low latency functions such as remote protection [3].

System Control—This layer is composed of all the WANs that can provide interconnections between substations or between substation and control center. To achieve high-performance control, this layer demands strict latency and emergency response. System control networks mainly provide networking communication services for related systems such as SCADA, remote protection, peer connections between distributed smart substations, and so on.

Intra-Control Center/Intra-Data Center—This layer consists of all the intra-control center networks and intra-data center networks of the utility. They have the same logical level while the control center requires the real-time system higher than security and connectivity. They both can provide connectivity for the neighboring system control networks and utility networks.

Utility—This layer comprises enterprise networks and inter-control center networks of the utility. Because the utility usually has several control centers and its enterprise networks cross a large geographical area, this layer networks consist of metropolitan area networks (MANs) and FANs.

Trans-Region—This layer is composed of the trans-region synchronous networks used for power exchanging between different regions and the power flow managing networks.

Wide Area Monitoring and Control System (WAMCS) Networks—This layer contains the power management units (PMUs) networks. These networks must be connected to the other layer entities and these connections usually be realized by special network planning.

SGICA can provide finer granularity of communication network than other SG communication architecture so as to support special communication requirements. In addition, SGICA can also support the interaction and interoperation of different regional networks, which will contribute to determine the scope of the network layer to be upgraded to not impacting the other layers.

4 Information and Communication Technologies for SGICA

The current information and communication technologies divided into wired communication technology and wireless communication technology all can be applied into SG communications.

Classification according to the distance	Wireless personal Area Network (WPAN)	Wireless Local Area Network (WLAN)	Wireless Metropolitan Area Network (WMAN)	Wireless Wide Area Network (WWAN)
Standard based on	IEEE802.15	IEEE802.11	IEEE802.16	IEEE802.20
Access Technologies	Bluetooth, RFID, UWB, Zigbee, Zigbee, Z-Wave, NFC etc.	Wi-Fi, WMN etc.	WiMAX、3G、LMDS、MMDS、Cluster communications etc.	Satellite communication, Point to point macrowave communication etc.

Fig. 2 Wireless communication technologies used to SGICA

The wired communications include optical fiber communication, power line communication (PLC), and so on. Having high-transmission rate, bandwidth, and reliability, the wired communication technologies can be used for constructing the backbone network of the SGICA. Especially, when it comes to system control and control center/data center layers of which the communication networks need low latency and high reliability, the wired communication technologies will be a good choice.

Certainly, the wired communication also has many shortcomings such as poor mobility, small coverage, and vulnerability to natural disasters. Fortunately, the wireless communication technologies can make up these shortcomings well at the same time bringing high bandwidth and long-distance transmission as well as fast deployment.

Figure 2 shows that the wireless communication technologies are the important parts of SGICA. All kinds of wireless communication technologies with different coverage ranges and data access speeds have strong complementarity for each other [7] when applied into SGICA. Bluetooth and UWB can achieve both close and ultrahigh-speed wireless access so that they can be used in private micro-grid networks and residential networks of the prosumer layer in SGICA. WLAN can achieve relatively long distance and high-speed data access and it can be applied to distribution automation networks (distribution level 2), intra-substation networks (substation layer), and intra-control/data center networks (control center/data center layer) etc. 3G can provide wide area seamless coverage and strong roaming mobility, so it can be used for in-vehicle networks (prosumer layer) and EV sub-networks (distribution layer) that demand strong mobility; In addition, 3G can also be used to FANs (distribution level 1), inter-substation networks (system control layer), and enterprise networks (utility layer) etc. WiMAX can be applied to distribution level 2 of SGICA because it is able to offer excellent access service for "Last Mile" networks. Besides, satellite communication can be used as emergency communications.

Furthermore, wireless mesh networks, completely different from traditional wireless networks, have brought many benefits including fast and easy deployment, high bandwidth, flexible structure etc. It has redundancy mechanism and rerouting function, so the whole network will not be influenced when some devices in the SG get any fault.

All in all, the wired communication technologies make SGICA satisfy the SG communication with high speed and good reliability; while all kinds of wireless communication technologies let SGICA be able to support mobile communication which thus brings SG many new functions, such as emergent responses to disasters and intelligent office. Additionally, by applying information security technologies such as encryption, access control, and data backup etc., SGICA will achieve more secure information communications.

5 IP-Based Communications for SG

Connecting numerous devices and providing kinds of services, SG consists of multiple power subsystems, which usually has its own dedicated communication protocols such as SCADA [8] that stops free and secure communication between different power subsystems and further impact the overall scheduling and running of the grid.

So we propose IP-based communications for SGICA. IP can provide the only one address or mark for each device in the SG in order to interconnect different kinds of subsystems or networks. Users can visit each device in every step of the SG including generation, transmission, substation, distribution, and dispatching, which will achieve the two-way communications of SG [9].

Nowadays, IPv4 addresses are about to be used up while IPv6 will supply an infinite number of addresses and further brings easy addressing and routing to SG communications. In addition, the working groups routing over lossy and low-power networks (RoLL) and 6LoWPan (IPv6 over Low power WPAN) of Internet Engineering Task Force (IETF) are studying the low-power IPv6 network that will be applied into home automation and building automation, that has great prospects in the SG communications.

6 Summary

In this paper, we firstly made a comprehensive analysis of the communication requirements of SG. Then, based on Cisco Grid Blocks, we propose the new SG information communication architecture SGICA, which can support special communication requirements. Then information communication technologies are discussed and proposed for constructing communication networks of SGICA. Finally, through discussing the challenges, SG communications may face and superiorities IP-based communications will bring, we suggest the IP-based communications for SGICA. We hope this paper will help the development of SG in China.

Acknowledgments This project is supported by National Natural Science Foundation of China (61170285).

References

1. Ning, H., Liu, H.: Cyber-physical-social based security architecture for future internet of things. Sci. Res. **2**, 1–7 (2012)
2. Hong, S., Kim, D., Ha, M., et al.: SNAIL: an IP-based wireless sensor network approach to the internet of things. IEEE Wirel. Commun. **17**, 34–42 (2010)
3. Grid Blocks Architecture. http://www.cisco.com/web/strategy/energy/gridblocks_architecture.html
4. Moslehi, K., Kumar, A.B.R., Shurtleff, D., Laufenberg, M., Bose, A., Hirsch, P.: Framework for a self-healing power grid. In: IEEE on Power Engineering Society General Meeting, vol. 3, p. 3027, 12–16 June 2005. doi:10.1109/PES.2005.1489709
5. Zhang, XY., Liang, XJ.: Discussion on communication reliability definition. J. Beijing Univ. Posts Telecommun. **20**(2), 30–34 (1997)
6. Gungor, V.C., Hancke, G.: Industrial wireless sensor networks: challenges, design principles, and technical approaches. IEEE Trans. Ind. Electron. **56**(10), 4258–4265 (2009)
7. The Analysis of Wireless Communication Technologies in Smart Grid. http://www.hqew.com/tech/sheji/664545.html
8. Internet Protocol Architecture for Smart Grid. http://www.docin.com/p-489391842.html
9. Research on Application of IPv6 in the Smart Grid. http://wenku.baidu.com/view/61085c166c175f0e7cd13788.html

Green Router: Power-Efficient Router Design

Yi Kai, Bin Liu and Jianyuan Lu

Abstract High speed routers in Internet are becoming more powerful, as well as more energy hungry. In this paper, we present a power-efficient router named Green Router. Packet processing capacities in different line-cards, mainly including IP lookup and forwarding engines, are shared and modulated according to traffic loads. Unoccupied capacities are powered off in order to save power. Green Router separates a line-card into two parts: network interface card (DB) and packet processing card (MB), connected by a two-stage switch fabric. Traffic from all the DBs shares all the MBs in Green Router, thus can be aggregated to a few active MBs on demand and other inactive MBs can be shutdown to save power. We give Green Router's architectural design and propose a flow-slice-based dynamic allocation policy. Real-trace-driven experiments show that Green Router can save significant power and the impact of QoS is small.

Keywords Router · Power-efficient

1 Introduction

Recent studies reveal that the power consumption of information and communication technologies (ICT) varies from 2 to 10 % of the worldwide power consumption [1]. Moreover, there is a trend of a notable increase in the ICT power

Y. Kai (✉) · B. Liu · J. Lu
Department of Computer Science and Technology, Tsinghua University, Beijing, China
e-mail: kaiyi02@gmail.com

B. Liu
e-mail: lmyujie@gmail.com

J. Lu
e-mail: lujy@foxmail.com

S. Patnaik and X. Li (eds.), *Proceedings of International Conference on Computer Science and Information Technology*, Advances in Intelligent Systems and Computing 255, DOI: 10.1007/978-81-322-1759-6_24,

consumption in the future. To this extent, network devices consume a considerable amount of power. But Internet transmission links exhibit relatively low average utilization. The average utilization of today's backbone links is reported to be 40 % or even less [2]. And network traffic normally has the 24-hour effect, higher in the daytime and lower at night [3, 4]. But routers are conventionally designed to meet the worst-case traffic demand which results in low-power efficiency. Meanwhile, according to Bolla et al. [5], packet processing engines and forward engines are the dominant power consumers in modern core routers.

In this paper, we propose to redesign a power-efficient router at the architectural level named Green Router. Our vision is to let all the interfaces share all the processing engines/capacities in different line-cards dynamically and adaptively to create room for power saving. A line-card in router is separated into two parts:

- Motherboard (MB) for packet processing, including IP lookup engines and forwarding engines;
- Daughterboard (DB) consisting of network interfaces.

Traffic from all DBs is aggregated to a few MBs on demand, and other MBs can be shutdown to save power. Figure 1 illustrates an example of Green Router's power-saving gain. We assume a four-port traditional router, as shown in Fig. 1a; all the line-cards should keep working although the line-cards' utilizations are low. While in Green Router, as illustrated in Fig. 1b, only two active MBs are demanded to process the traffic from four ports, the other two MBs can be shutdown.

2 Architectural Design

The key idea of Green Router's power saving is sharing MBs' processing capacities when traffic load is light. The implementation of interconnection between MBs and DBs is important. It is not practical to aggregate traffic from all

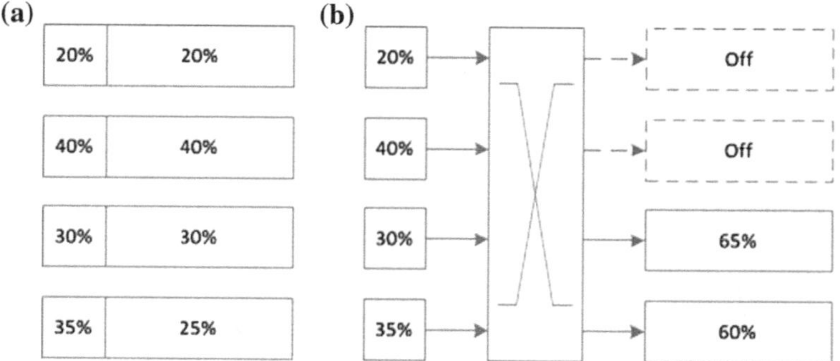

Fig. 1 An illustration of Green Router. **a** Traditional router. **b** Green Router

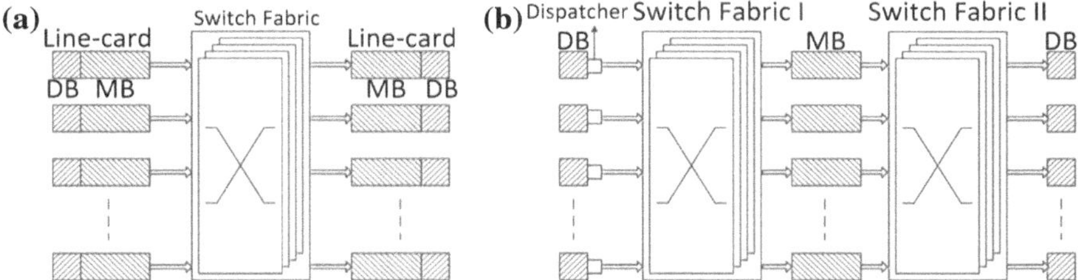

Fig. 2 Architecture comparison of routers. **a** Traditional router. **b** Green Router

DBs which can feed up to 40 Gbps traffic per interface, or even higher. So an extra switch fabric is used.

In traditional routers, as shown in Fig. 2a, each line-card, as an indiscerptible binding entity functionally, is connected to switch fabric. When a line-card receives a packet from its network interface, it looks up the routing table and forwards the packet to the destination line-card via the switch fabric.

Differently, in Green Router, as shown in Fig. 2b, an extra switch fabric is used. The two switch fabrics are named Switch Fabric I and Switch Fabric II, respectively. When a packet arrives at a DB, the dispatcher determines which MB will process the packet, and then, the packet is sent to the selected MB via Switch Fabric I. The MB processes the packet and forwards it to the destination DB via Switch Fabric II according to routing information.

The dispatchers and the extra switch fabric may incur additional power consumption. So they are powered on only when the traffic load is light. They are turned off when the traffic load is heavy, and Green Router will work as a traditional router.

3 Operation Mechanism

3.1 Dynamic Sleep and Wakeup

When traffic volume is increasing, Green Router should wake up some sleeping MBs immediately to avoid packet loss. On the other hand, Green Router should shutdown some idle MBs when traffic drops down to conserve power. We assume that powering on or waking up a MB costs no extra power and time currently to simplify the analysis.

Green Router makes decisions when to sleep or wake up a MB according to the average load. *THAL_H* and *THAL_L* are two thresholds of the average load of all active MBs in Green Router.

- When the average load is lower than *THAL_L*, Green Router will try to sleep a MB. The MB whose load is lightest is chosen, and dispatchers will try not to send packets to the chosen MB.
- When the average load of active MBs is higher than *THAL_H*, Green Router will wake up a MB immediately.

3.2 Dispatching Policy

Many applications work on per-flow basis. But Green Router's sharing idea may lead to out-of-order if packets in the same flow are processed by different MBs. To improve the QoS performance, "flow-slice" is defined. If the interval between two consecutive packets in a flow is larger than the maximal processing time of a packet in MBs, this flow can be split at the gap point and the two sub-flows can be processed by different MBs without causing packet out-of-order. We name these new-cut sub-flows as "flow-slices." Flow-slice-based dispatching is far more accurate than flow-based dispatching, because the number of flow-slices is much larger than the number of flows [6].

A flow-slice-based dynamic allocation approach to dispatch traffic from DBs to MBs is proposed. The packets in a same flow-slice will be processed by a fixed MB, so packet out-of-order can be avoided. Dispatcher is a key component in Green Router to implement dispatching policy. A flow table is used in dispatcher to record flows' states. Every table entry records the MB which is processing the flow, and the arrival time of the latest packet in the flow.

After a packet arrived, the dispatcher looks up the flow table:

- If the packet belongs to a new flow, the flow will be assigned to the active MB whose load is lightest, and a new entry will be inserted into the flow table.
- If the packet belongs to an existing flow, the dispatcher will judge if the flow can be cut according to the arrival time of the latest packet in the flow:
  - If the flow can be cut at this gap point, a new flow-slice will be created, and the new flow-slice will be assigned to the active MB whose load is lightest.
  - If the flow cannot be cut, which means the packet belongs to the last flow-slice, then the packet will be dispatched to the MB which is processing the flow-slice.

Because the traffic volume of flow-slices is unpredictable, the aggregated traffic volume of flow-slices which are assigned to the same MB may exceed the capacity of the MB. So a FIFO is used in front of every MB. While a packet arrives and the selected MB is full, the packet will be buffered in the FIFO.

4 Modeling and Evaluation

4.1 Modeling

Green Router's power saving comes from the reduced power expense of the turned-off MBs (but MBs are powered on/off dynamically), while paying power cost to the extra switch.

Power consumption proportions in a core router are summarized in Table 1, according to the authors' estimation in [5]. MBs which are designed for packet processing including IP lookup, forwarding engines, and buffers are dominant power consumers. Power consumption of switch fabric cannot be ignored either.

Traditional router's and Green Router's power consumptions can be expressed approximately as Eq. 1. N represents the number of line-cards in a traditional router which also represents the number of MBs and DBs in Green Router; M represents the number of active MBs in Green Router; P_{TR} and P_{GR} represent the power consumption of traditional router and Green Router, respectively; and P_{Ctrl}, P_{MB}, P_{DB}, and P_{Switch} represent the power consumption of control plane, a single MB, a single DB, and a switch fabric, respectively. $P_{P\text{ and }H}$ and $P'_{P\text{ and }H}$ represent the power consumptions of power and heat management. We can have the power consumption in two kinds of routers as below:

$$\begin{aligned} P_{TR} &= P_{P\,\text{and}\,H} + P_{Ctrl} + N \cdot (P_{MB} + P_{DB}) + P_{Switch} \\ P_{GR} &= P'_{P\,\text{and}\,H} + P_{Ctrl} + N \cdot P_{DB} + M \cdot P_{MB} + 2P_{Switch}. \end{aligned} \tag{1}$$

We assume the power consumption of power and heat management is directly proportional to the power consumption of other modules in a router. And according to the proportion of power consumption in Table 1, we can get the relationship between the number of active MBs and the power consumption of Green Router compared to traditional routers:

$$P_{GR} = \frac{\left({}^{M}\!/_{N} \cdot 0.37 + 0.38 \right)}{0.65} P_{TR} \tag{2}$$

The problem of power saving in Green Router becomes how many MBs can be shutdown. *M/N* in Eq. 2 means the number of active MBs out of all MBs. Apparently, M, the number of active MBs, is determined by the real-time traffic

Table 1 Power consumption proportions in core router

Module	%	Component	%
Power and heat management	35	–	–
Control plane	11	–	–
Data plane	54	MB: IP lookup, forwarding engines	37
		DB: I/O	7
		Switch fabric	10

load and the two thresholds *THAL_H* and *THAL_L* which are mentioned in Sect. 3.1. Equation 2 also shows that if the percentage of active MBs is larger than 72.97 %, Green Router with an extra switch fabric will cost more power than traditional routers. So if the traffic volume is heavy enough and over 72.97 % of MBs are active, Green Router should shutdown the extra switch fabric, stop sharing, and then work as a traditional router.

4.2 Experiment and Evaluation

In this section, we evaluate Green Router's power-saving gain, as well as the packet delay performance by applying real traffic traces. The results show that it is able to achieve considerable power savings with little impact on QoS performance.

Two different real traces for the evaluations are used as shown in Table 2. Traces are all monitored from OC-192 links and the durations are all 60 min. Every trace is cut into 10 pieces and a 10-port router is constructed. These 10 pieces are assumed to arrive at Green Router's 10 different ports at the same time. On the other hand, we modify the throughputs of these two original traces to imitate the 24-hour effect. Sine-wave filter for throughput is used to modify traces. The modified traces' information is also shown in Table 2.

Power Savings Fig. 3 illustrates the results of experiments using two original traces, respectively, where the number of active MBs varies while traffic load changes although the fluctuation is not obvious. Compared with traditional routers,

Table 2 Summarization of simulations' results

Trace	Average load (%)	Power saving (%)
Chicago	31.61	16.82
San Jose	80.50	0.32
Modified Chicago	19.27	27.00
Modified San Jose	53.02	9.69

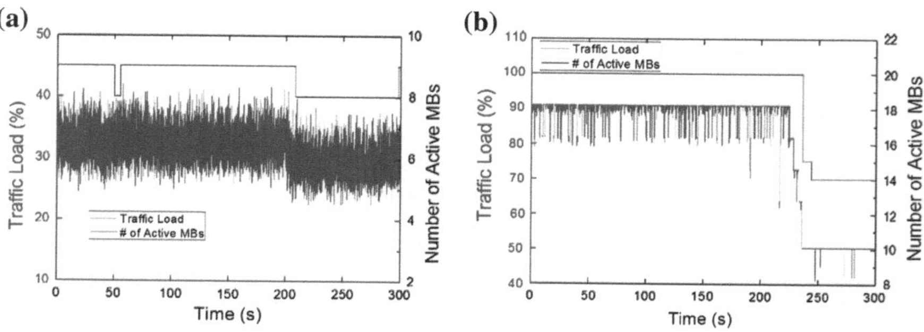

Fig. 3 Simulation results: dynamic number of active MBs with original traces. **a** Chicago. **b** San Jose

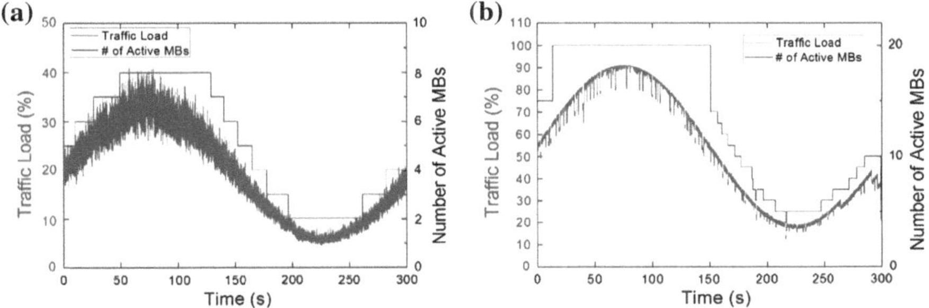

Fig. 4 Simulation results: dynamic number of active MBs with modified traces. **a** Chicago. **b** San Jose

Green Router expenses less power, due to parts of the processing capacity are powered off. The number of active MBs and Eq. 2 are used to estimate Green Router's power consumption. Table 2 summarizes power saving using the two original traces.

Figure 4 illustrates the results of experiments using two modified traces. And Table 2 also summarizes power-saving gain using these two modified traces. Adaptation of active MBs is more clearly in Fig. 4.

We can see from Figs. 3 and 4, when the average utilization is low, as the trace from Chicago, Green Router can save significant power. But when the average utilization is high, as the trace from San Jose, the power-saving gain is accordingly decreased because of the extra switch fabric. When the traffic load is heavy, as we discussed in Sect. 3, the power-saving gain from sleeping MBs is less than the power consumption of extra switch fabric. So Green Router will wake up all MBs, shutdown the extra switch fabric and work like a traditional router.

Extra Delay While applying flow-slice-based dispatching, FIFOs are used to avoid overflow. So extra delay is introduced. Figure 5 illustrates the cumulative distribution function (CDFs) of the extra packet delay over the two original traces.

Fig. 5 CDF of extra delays

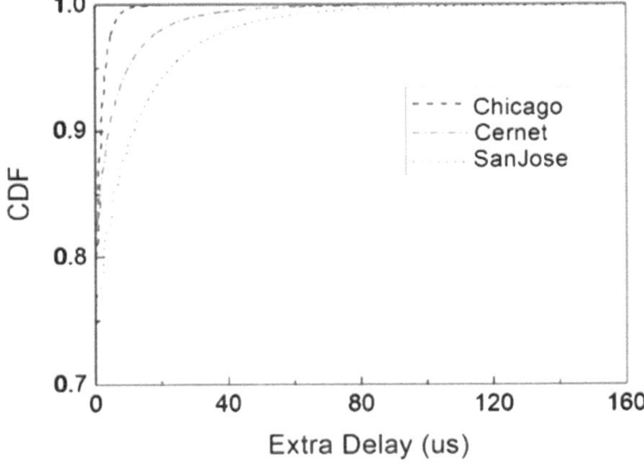

Most of the packets experience very low extra delays, and over 95 % packets' extra delay is less than 50 us. We can see the extra packet delay by adding the FIFO is minor.

5 Conclusion

Low average utilization and redundant processing capacity in today's core routers provide opportunities for power saving. By unbinding network interfaces and packet processing engines in line-cards, traffic is aggregated to on-demand active MBs, so we can deactivate other MBs for saving power. In this paper, we design a two-stage switch fabric to interconnect DBs and MBs in a fully sharing manner. Then, we design a flow-slice-based dynamic allocating algorithm to dispatch the traffic among active MBs. Evaluations based on real network traffic show that Green Router is able to achieve considerable power saving while guaranteeing the network QoS performance. Maybe Green Router's power-saving gain looks not so incredible. But the base of comparison is the power consumption of an entire router which is very considerable.

References

1. Webb, M.: Smart 2020: Enabling the low carbon economy in the information age. Clim. Group London (2008)
2. Guichard, J., Faucheur, F.L., Vasseur, J.P.: Definitive MPLS Network Designs. Cisco Press, USA (2005)
3. Thompson, K., Miller, G.J., Wilder, R.: Wide-aera internet traffic patterns and characteristics. IEEE Netw. **11**(6), 10–23 (1997)
4. Labovitz, C.: What europeans do at night. http://ddos.arbornetworks.com/2009/08/what-europeans-do-at-night/ (2009)
5. Bolla, R., Bruschi, R., Davoli, F., Cucchietti, F.: Energy efficiency in the future internet: a survey of existing approaches and trends in energy-aware fixed network infrastructures. Commun. Surv. Tutorials **13**, 223–244 (2011)
6. Kandula, S., Katabi, D., Sinha, S., Berger, A.: Dynamic load balancing without packet reordering. ACM SIGCOMM Comput. Commun. Rev. **37**, 51–62 (2007)
7. Equinix-chicago. CAIDA. http://www.caida.org/data/monitors/passive-equinix-chicago.xml (2012)
8. Equinix-sanjose. CAIDA. http://www.caida.org/data/monitors/passive-equinix-sanjose.xml (2012)

Design and Implementation of Web Service for Scenario Data Service System in War Gaming

Zhihua Cao, Xiaofeng Hu, Guochun Zhang and Xueya Wang

Abstract Scenario data is the basis for much military training, war gaming, planning, and scenario data service system is also an important component of war-gaming system. It provides an easy interface to store, compile, read and write, transfer scenario data needed in war gaming, especially, trainees in war gaming need to query and search data which could assist them to make combat plan, etc. This led us to design and implement Scenario Data Service System. In this paper, we introduce the architecture of system and approach of technology. Then, we focus on e-design and implementation of Web service, as it is the key technique and the core of system. In the second half of paper, we introduce the Web service process and then expatiated on the modules of Web layer.

Keywords Web service · Scenario data · War gaming · Lucene · Design and implementation

---

Z. Cao (✉) · X. Wang
Institute of Postgraduate, National Defence University, Beijing, China
e-mail: caozhihua509@gmail.com

X. Wang
e-mail: xfhu@vip.sina.com

X. Hu · G. Zhang
Department of Information Combat and Command Training,
National Defence University, Beijing, China
e-mail: XiaofengHu@sina.com.cn

G. Zhang
e-mail: Zhanggch1@163.com

S. Patnaik and X. Li (eds.), *Proceedings of International Conference on Computer Science and Information Technology*, Advances in Intelligent Systems and Computing 255, DOI: 10.1007/978-81-322-1759-6_25,

1 Introduction

Scenario data service system in war gaming [1] is mainly to provide the trainees with the data related to scenario and military theory which they want to query and browse. The scenario data includes the quality and quantity of forces, performance data of weapons and equipment, target data, damage data, and a variety of prototype data, etc. Military theory contains operational theory, military geographic information, examples of battles, and exercise information, and so on.

Normally, data and information needed in war gaming are huge, and trainees from different sides in different places require differently. So, to solve this problem, the data service system should be a network and an integrated system. It contains various data and supplies multifunctions (such as, data management, browsing and querying, full-text searching, and scenario editing, etc.). It has five specific objectives: (a) Develop a variety of ways for the trainees to browse and query data of scenario, troops and weapons, and documentation retrieval service; (b) Meet the requirements of scenario data querying which trainees in remote and off-site needed; (c) Because data displayed for different sides and users is also different, classification must be reasonable; (d) The interface must be friendly and easy to operate; and (e) It must be convenient for data and information maintenance. If the data and information are modified, the user's interface updates synchronously. Those objectives are also our innovations.

2 Approach of Technology

The system provides scenario data querying and browsing for trainees of different sides based on Web. We designed a framework used for joint operations and multisides, multiparties' war gaming, which meets the issues of integrate organizations, management and knowledge services derived from mass data. Due to the huge amount of data of the system, which involves troop's data, equipment data and logistical data, it is difficult to design Web pages for troops and equipment separately. The Web pages cannot update synchronously when data (such as, the quantity of one troop) changed in the database. Therefore, in order to reduce the Web design work at the same time, the system based on structure database can quickly transform and generate performance framework and user interface automatically. It meets the requirements of querying the latest data in the process or stage of war gaming.

Scenario data service system in war gaming generates the entire system automatically. We use the functions of input and output in .NET to achieve the purpose. First of all, in connection with the forms of data related to system, we store the design of text as template, then read and control the content displayed in the template through programming. The system outputs the data into .html documents to build the entire system. The whole process is illustrated in Fig. 1.

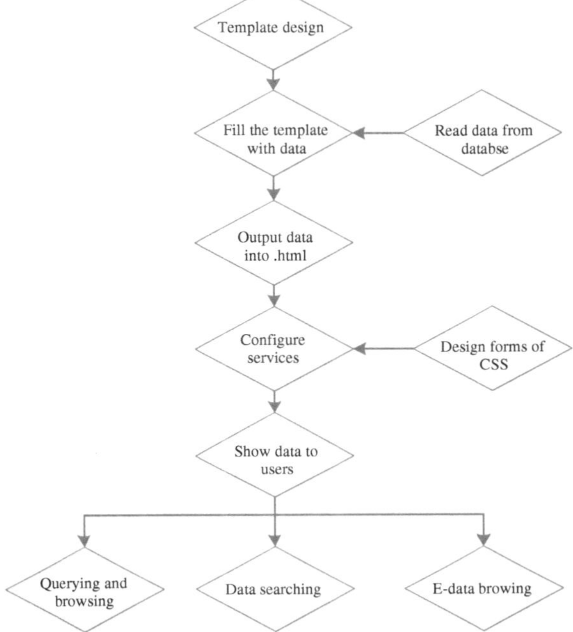

Fig. 1 System generation process

Because we adopt Web service interface in service-oriented architecture (SOA) [2], and we supply interfaces for other system, our system has strong flexibility. We adopt standard and unified naming rule. For safety reasons, we also adopt encryption techniques that insure data transfer.

3 Architecture of the System

Scenario Data Service System is mainly composed of the client application layer, the Web layer, and the business logic and data service layer.

Scenario Data Service System in war gaming is oriented to the management and the secure accessing of scenario data and information. It is also oriented to applying of those data and information, providing multidimensional, integrated, dynamic, and network synthetic data and information services. The system is generally composed of the client application layer, Web layer, and business logic and data service layer. The Fig. 2 shows the architecture of the entire system.

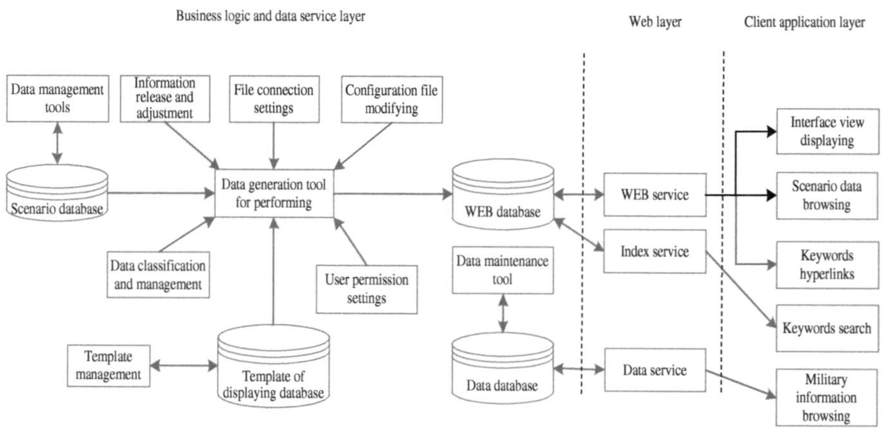

Fig. 2 Scenario Data Service System architecture

4 Design and Implementation of Web Layer

The Web layer includes Web service, index service, and data service. Among them, the Web service provides services for interface views, scenario data browsing, and the keywords hyperlinks. Index service serves the keyword-searching module. Information service serves the module of military data browsing.

The Web layer generates representation logic and receives feedbacks from the user' clients. These clients in presentation layer are HTML client and Web client. For client request, the presentation layer generates the corresponding response.

4.1 Web Service Design

In this section, we explain the Web service process firstly. Then, we introduce the main function of subsystem and the workflow of monitor module.

Web service process: Web service process is shown in Fig. 3. When client enter an URL or click on a hyperlink, a TCP/IP connection will be established between Web service and client. And then, Web service goes into a request–response cycle stage, waiting for a client request. After that, Web service will examines the request and map the URL requests into particular data and then find data file from the database. If data file exists, then send it to the client; if not found, returns an error message.

Monitor module design: Monitor module is one of the most important modules of HTTP server, which includes several parts, such as the server-side operation socket, port binding, client request, and monitoring. It is also a key component for server running.

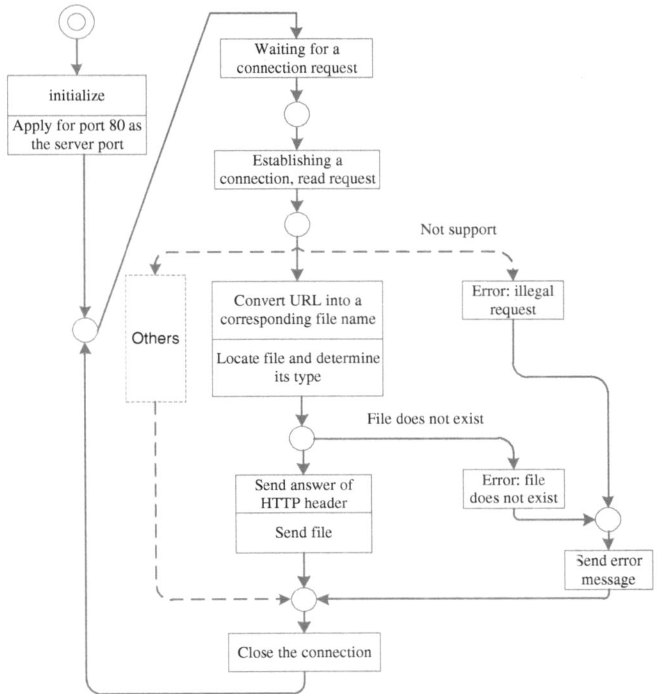

Fig. 3 Web service process

First, function WSAStartup () is called in the initialization process. This function initializes the Socket DLL in application. Only this function is called successfully, application can then call other API functions in Windows Socket DLL.

Second is socket creation. After initializing WinSock DLL, it is necessary to create a listening Socket, so we call function Socket () to create this Socket and define the communication protocol used by the Socket. If called successfully, returns 0, else INVALID_SOCKET.

Third is bind port. Next, we should develop an address and port for Socket created by server-side, so the client knows which port will be connected to one address. For that, we have to call function bind (), the function returns 0 if successful, otherwise return SOCKET_ERROR.

Fourth is monitor. After the server-side Socket object is bidden, a queue of monitors must be established in server-side to receive connection requests from client. Function listen () sets server-side Socket into monitoring state and set the maximum number of connections that can be established.

Fifth, server-side accepts connection requests from client. When client requests for connection, the server-side hwnd will receive a message sent by Winsock Stack that is defined by us. Here, we can analyze the lParam and then call related functions to handle this event. In order to accept other client request, we use the function accept () that creates a new Socket to connect the client. The original monitor Socket continues to listen, waiting for another request.

Sixth is closing the Socket connection. Either ends of connection (server or client) can stop connection, as long as calling function Closesocket (). In addition, we should call function WSACleanup () to notify the Winsock Stack release of resources occupied by the Socket corresponding with function WSAStartup ().

Design and implementation of response module: Users can access the server by input URL in a browser (IE, Firefox, etc.); the server will analyze the requests, including client's address, port, and requesting data files.

4.2 Index Service Design

Index service includes two functions: One is to generate index database according to the performance database; the other is to search and locate data containing index content quickly which is based on user input. The indexing service framework is shown in Fig. 4. It is composed of four parts: document-parsing module, automatic Chinese word segmentation module, full-text indexing module and full-text retrieval module.

Document-parsing module: First of all, document-parsing module reads and analyzes data from performance database, and then, automatic Chinese word segmentation module segments the word, finally index database is generated by full-text-indexing module. When users search, full-text retrieval module finds and returns the information to the users.

Users can access the server by input URL in a browser, and then, the server will analyze the requests, including client's address, port, and data files requested.

Document-parsing module design: This module can parse performance data file effectively, it facilitates full-text indexing. The final form of performance data file is HTML, so its main function is to analyze, identify, and serialize various makeup of HTML.

Automatic Chinese word segmentation module design: It chooses words from performance data file after being parsed by document-paring module and prepares

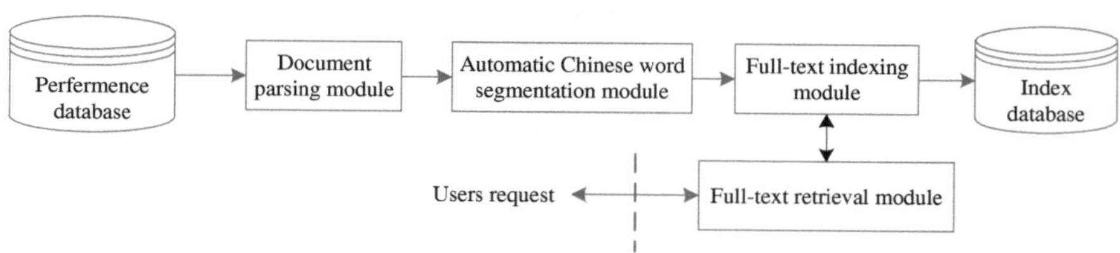

Fig. 4 Index service framework

for full-text indexing. After analysis of various types of data modules, the next step is to extract the appropriate data indicated by makeup as the phrase.

For example: the data template fragment is

<td class= “propertytd”>side </td>
<td class= “valueofpro” >ToBeReplaced_side</td>
The formation of performance data document:
<td class= “propertytd” >side </td>
<td class = “valueofpro” > Red </td>

As described above, it can extract phrase ‘red’ by makeup: valueofpro, td, /td, >, and <.

This module can parse performance data file effectively; it facilitates full-text indexing. The final form of performance data file is HTML.

Full-text-indexing module design: After two front steps, full-text-indexing module establishes an XML documents index. It adopts Lucene [3] to finish the work. It categorizes XML by the meaning of makeup and creates index document; that is convenient for user to search document by different makeup.

The generation process of index database is shown in Fig. 5.

Full-text retrieval module design: This module deals with the user’s query effectively and executes a search function to help users to find information they need.

Full-text retrieval module uses Lucene as the underlying technology. It deals with user’s query string, then searches index files according to the content, ranks, and returns the results to client. The processing steps as follows:

Step 1: Retrieve keywords into the Lucene query object. Keywords are passed to Query Parsel class, and it will be converted into Query object.

Step 2: Retrieval: The retrieval procedure calls index reader Index Reader and reads index file, then matches search words in index file according to form of the Query object. Index Reader accesses index files according to the path settled. If there are multiple paths, retrieval system creates a query thread for each index file. Each thread completes the matching process in the corresponding index file independently. Searching results are collected at last. Code of constructing query multithread is as follows:

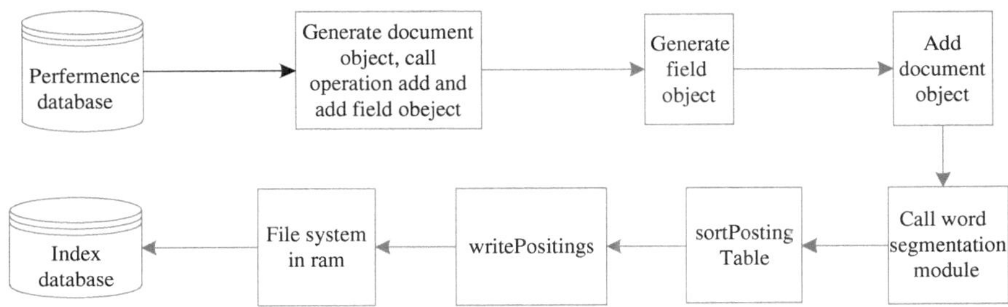

Fig. 5 Generation process of index database

Index Searcher [] searchers=new Index Searcher [2];
Searchers [0]=new Index Searcher(indexpath1);
Searchers [1]=new Index Searcher(indexpath2);
Paralle Multi Searcher Searcher=new Paralle Multi Searcher (Paralle Multi Searcher).

Step 3: Output retrieval results: Retrieval results for outputting include hit records and the total number of records. The retrieval system is optimized after first searched; it does not read out all records (Document), but sorts the first 100 results of docID in cache by correlation and returns them. Therefore, even if the total number is big, results of Lucene does not occupy much memory. When result is returned to the client, the pages that the users query will be shown in a browser window.

5 Conclusion

In this paper, we present a full design of Scenario Data Service System and the implementation method of its core component Web service module-based SOA. The SOA approach improves interoperability and flexibility. The system demonstrates and leaves behind a reuseable data service toolset that assists trainees in war-gaming exercises as well as developers. We have applied this system to war gaming many times, and it has been endured real testifying.

Acknowledgments This paper is sponsored by National Natural Science Foundation of China (Grant NO. 61174156 and 61203140).

References

1. Xiofeng, H., Jiabin, F.: Introduction of War Gaming Exercises. National Defence University Press, Beijing (2012)
2. Mayott, G., Self, M., Miller, G.J., McDonnell, J.: SOA Approach to Battle Command: Simulation Interoperability. PSISDG, Orlando (2010)
3. Michael, M.C., Erik, H., Otis, G.: Lucene in Action, 2nd edn. Manning Publication Co., Stanford (2010)

Gateway of Internet of Things for Intelligent Warehouse Management System Based on Embedded Web Server

Senbin Yang, Rong Tao, Wei Tan and Wenhua Zhang

Abstract Aiming to access and control perception devices through Web page remotely and conveniently, an Internet of Things Gateway (IOTGW) and its prototype system based on embedded Web server are presented. According to the typical use scenario and application requirements of warehouse management, embedded Web server is adopted as a lightweight approach for accessing perception devices and interacting with heterogeneous networks, and the hardware and software architectures of IOTGW based on ARM chip and embedded Web server are proposed. Experiments show that the IOTGW can address the heterogeneity of different sensor networks and diversity of protocols in the perception and network layer and realize the functionality of IOT management and control.

Keywords Gateway · Internet of things · Embedded Web server · Intelligent warehouse management system

1 Introduction

Internet of Things (IOT) is an integrated part of future Internet and could be defined as a dynamic global network infrastructure with self-configuring capabilities based on standard and interoperable communication protocols where

S. Yang (✉)
Special Operations College, Xi'an, China
e-mail: ysb-007@163.com

R. Tao · W. Tan · W. Zhang
Xi'an Communications Institute, Xi'an, China
e-mail: biaojie-007@163.com

W. Tan
e-mail: weitanl@163.com

W. Zhang
e-mail: zwh198628@163.com

S. Patnaik and X. Li (eds.), *Proceedings of International Conference on Computer Science and Information Technology*, Advances in Intelligent Systems and Computing 255, DOI: 10.1007/978-81-322-1759-6_26,

physical and virtual "things" have identities, physical attributes, virtual personalities, and use intelligent interfaces and are seamlessly integrated into the information network. With the development of IOT technologies, the most important IOT application areas cover modern agriculture, infrastructure construction, public security, environment protection, intelligent industry, business service and other fields [1].

In the architecture of IOT, the traditional mobile communication network and Internet are mainly used in the transmission of information among computers or people, while the wireless sensor network (WSN) can realize the short distance communication among the objects by constructing wireless networks in ad hoc manners. However, it is difficult to connect the WSN and mobile communication networks or the Internet with each other because it lacks uniform standardization in communication protocols and sensing technologies, and the data from WSN cannot be transmitted in long distance with the limitation of WSN's transmission protocols [2]. Therefore, a new type of network equipment called the Internet of Things Gateway (IOTGW) is invented, whose goal is to carry out data communication between the two kinds of networks by coping both the protocol transformation and the device heterogeneity throughout IOT network.

There are already some researches on the design and implementation of IOT-GW system. Zhu et al. [2] proposed an IOT-Gateway system based on Zigbee and GPRS protocols according to the typical IOT application scenarios and requirements from telecom operators, and the implementation of prototyping system based on ARM9 and Python was given. In order to reduce or weaken the effects of IOT on backbone networks from a traffic perspective and attain no discount on functions in a CobraNet-based digital broadcast system (CDBS), an enhanced IOTGW for CDBS is presented in [3]. The IOTGW was emulated on ARM9 and Linux OS, and it can provide functions that were conventionally implemented by a PC platform or a server. In order to integrate real-world things into the existing Web, Guinard made devices an integral part of the Web by using HTTP as application layer, instead of using the Web protocols merely as a transport protocol [4]. Two alternative methods to enable representational state transfer (REST)-based interaction with embedded devices were given: Devices are directly made part of the Web by implementing a web server on them directly, or devices are connected through a Smart Gateway that translates requests across protocols. Riedel et al. [5] used Web Service-based interface descriptions paired with a model driven approach to achieve a high flexibility at a low runtime overhead when designing message-based communication within an IOT. The experiments with an industrial servicing use case showed that Web Services and standard HTTP communication-based gateways for sensor nodes can integrate multiple concurrent IOT systems. In a word, IOTGW plays a leading role in the IOT systems, and the technologies of ARM and WSN are usually used when it is designed and implemented. However, the embedded Web server and other protocols in the perception layer such as radio frequency identification (RFID) and global position system (GPS) are less considered.

The rest of the paper is organized as follows: Sect. 2 introduces the use scenario and functional requirements of IOT Gateway in Intelligent Warehouse Management System. Section 3 proposes the hardware and software architecture of multiprotocol IOTGW based on embedded Web server and finally some concluding remarks are made in Sect. 4.

2 Use Scenario and Functional Requirements of IOTGW

2.1 IOT Architecture of Intelligent Warehouse Management System

Among the applications of IOT, intelligent warehouse management system (IWMS) is an important agricultural servicing use case. The IOT-based architecture of IWMS can be divided into three layers: perception layer, network layer, and application layer, which is shown in Fig. 1.

Perception Layer. The perception layer aims to acquire, collect, and process the data from the physical things, which consists of two parts: the sensor device and WSN.

The former one includes RFID label, GPS, camera, and sensor nodes and so on. The latter one is a self-organizing wireless network with protocol of Lonworks,

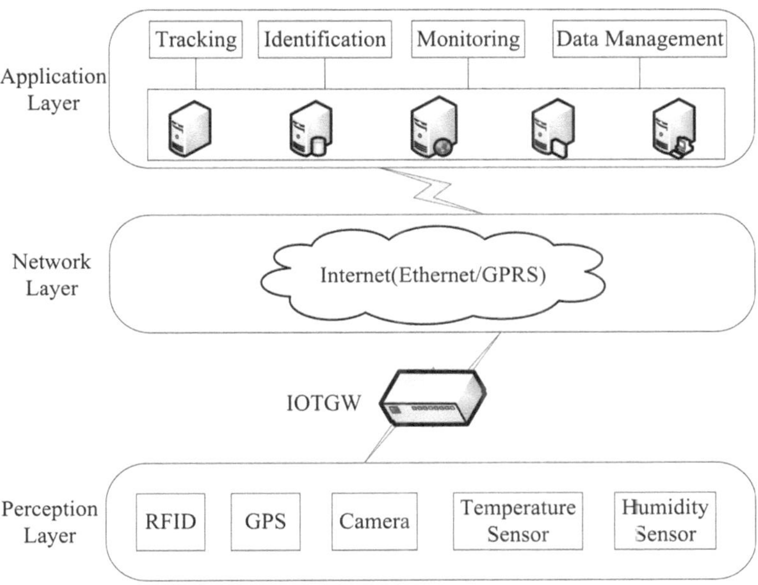

Fig. 1 Internet of things-based architecture of IWMS

ZigBee, 6LowPAN, Rubee, or Bluetooth, which links many sensor nodes distributed in a large area.

Network Layer. The network layer aims to transfer data in a large area or long distance, which is constructed based on the Internet (via Ethernet and GPRS) to realize the integration of the perception and communication network. Thus, the data collected from perception layer can be transferred successfully to remote destination. Long-range wired and wireless communication technologies, network techniques are important in this layer.

Application Layer. Data processing and services providing are two major purposes of the application layer. The data from network layer is handled by corresponding management systems and then various services including tracking, identification, monitoring, and data management will be provided to all kinds of users [6].

2.2 Functional Requirements of IOT Gateway

As shown in Fig. 1, IOT Gateway is the bridge to connect the perception layer with network layer, which has the following functional requirements:

Broad Access. Currently short-range communication technology standards are diverse and incompatible, such as Zigbee, Rubee, Bluetooth, etc. In order to enable "things" to interact and communicate among themselves and with existed or evolving communication infrastructures, the sensor devices should be integrated through IOTGW. In the use case of IWMS, the IOTGW can access devices with different protocols including RFID, GPS, USB, and Zigbee.

Protocol Conversion. As mentioned above, Internet based on Ethernet and GPRS is used to transfer the data collected from perception layer to remote destination, and its core protocol is HTTP/IP. However, the protocols in the perception layer and network layer are different in many aspects such as data formats, data rate, and data meaning. In order to exchange information between these two layers, the IOT Gateway should fuse heterogeneous networks and support protocol interworking seamlessly including conversion of data messages, events, commands, and time synchronization.

Powerful Management. Effective management is vital for keeping the network up and running smoothly. Meanwhile, the management of IOT Gateway not only means sensor node management in the subnet, but also means the gateway device management. The former one aims to acquire the node's identification, status, and properties, and to realize remote start-up, shutdown, control, and diagnosis. The latter one aims to realize the gateway device's configuration, supervision, upgrade, and maintenance.

3 Design of Multiprotocol IOT-Gateway Based on Embedded Web Server

3.1 Design Elements and Principles

The IOT Gateway is designed for settling with the heterogeneity between different perception layer protocols (e.g., between Zigbee and RFID) and between a perception layer protocol and a network layer protocol (e.g., between Zigbee and GPRS), and managing the IOTGW itself and perception devices. The IOTGW is designed with three main goals in mind: simplicity, extensibility, and modularity [7]. Simplicity and extensibility refer to users can extend and customize the I-OTGW to their needs. Modularity means that internal components of the IOTGW can interact only through small interfaces, thus allowing the evolution and exchange of individual parts of the system.

An embedded Web server is a component of a software system that implements the HTTP protocol. The embedded Web server technology is the combination of embedded device and Internet technology, which provides a flexible remote device monitoring and management function based on Internet browser and it has become an advanced development trend of embedded technology. Through this embedded Web server, user can access their equipments remotely [8]. There are a few advantages to using embedded Web server to design IOTGW: (1) HTTP is a well studied cross-platform protocol, and there are mature implementations freely available; (2) Web browsers are readily available with all computers and mobile phones, and other utility softwares are needless in the application layer; (3) with the usage of Web service, interacting with a sensor node becomes as easy as typing a URI in a Web browser. Consequently, embedded Web server can be adopted as a lightweight approach for accessing perception devices and interacting with heterogeneous networks. Moreover, the general problem of using embedded Web services is that even lightweight implementations are too resource heavy for many IOT systems.

3.2 Architecture of Hardware and Software

According to above elements and principles, hardware architecture of IOT-gateway system is shown in Fig. 2.

The hardware architecture is composed of five major modules: minimum ARM system, interface module, transmission module, LCD touch screen, and power source, each responsible for a well defined set of tasks. Minimum ARM system is the kernel of the IOTGW, which provides the utilities of processing, control, and storage. Embedded Web server with TCP/IP protocol suite is built in the ARM processor, and some web application programs, take protocol conversion, message parsing, and device management for example, are written in ARM. The interface

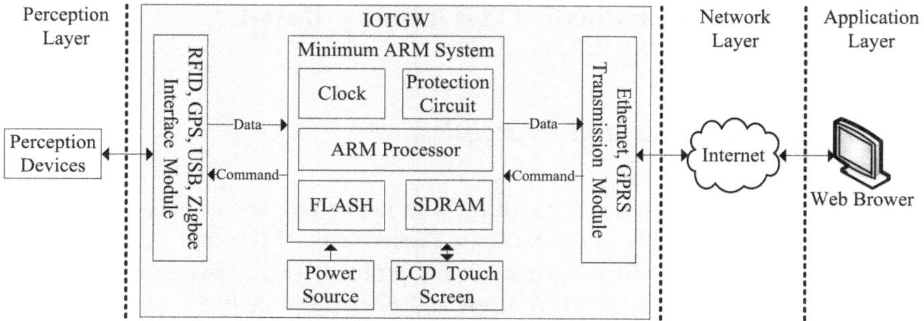

Fig. 2 Hardware architecture of IOT-gateway system

module makes the IOTGW accessible to different perception devices, which employ the protocol or interface of RFID, GPS, USB, and Zigbee. With the help of software, these devices can be controlled and managed by appointing unique IP address, and their heterogeneity is shielded. The Ethernet and GPRS transmission module realizes the data transmission based on Internet by the means of wired and wireless, respectively. The heterogeneity between these two protocols and perception devices is shielded too, so data and command can be exchanged and understood between perception layer and network layer. The LCD touch screen is used for output and input of data, command, or status, which are useful during equipment debugging and testing. After the IOTGW is in proper working order, the LCD touch screen can be removed. The power source supplies direct current (DC) for all IOTGW components, such as 3.3, ±5 V, and the source type contains electric supply and battery.

In order to implement the IOTGW's functionalities, the software is designed as Fig. 3.

Embedded operating system (EOS) is installed on the hardware platform firstly, and then embedded Web server and embedded database are constructed under the EOS, three functional blocks including data exchange, protocol conversion, and equipment management are programing finally.

Web Application Programs
Data Exchange
Command Mapping
Command Agent
Data Transceive
Protocol Conversion
Message Conversion
Message Parsing
Protocol Parsing
Equipment Management
Equipment Manipulation
Configuration Management
Device Driver
Embedded Web Server
Embedded Database
Embedded Operating System
ARM Hardware Platform

Fig. 3 Software architecture of IOT-gateway system

Fig. 4 Interface of video surveillance in the browser

4 Conclusions and Future Work

In the implementation of IOTGW, AM3352 + Linux + AppWeb + SQLite + C is selected as the system platform and tools on the basis of comprehensive comparison of the performances [9, 10]. In the IWMS system, the communication between the server and the client contains the device status, user data, control commands, and other sensitive information, so it is necessary to configure the secure transmission system in the embedded Web server. Therefore, the secure sockets layer (SSL) protocol is added in AppWeb to establish an encrypted data connection.

The AppWeb server can be visited in PC browser by “https://192.168.1.15:4443,” as shown in Fig. 4.

IOT is a huge global information system composed of hundreds of millions of objects that can be identified, sensed, and processed based on standardized and interoperable communication protocols. The IOT system can intelligently process the objects' state, provide management and control for decision-making, and even make them cooperate with each other autonomously without human's intervention. IOT Gateway plays an important role in IOT applications, which facilitates the seamless integration of objects and Internet, and the management and control with perception devices. This paper presents a prototyping implementation of IOT Gateway for IWMS based on embedded Web server, which realizes data forwarding, protocol transformation, device management, and control. According to the structure and requirements of IOTGW for IWMS, the hardware and software architecture of IOTGW are designed using the technology of ARM and embedded Web server. The hardware design is flexible, making it usable for many applications, such as smart home, industrial monitoring, smart grid, environment monitoring, etc. In future works, in order to meet the application requirements, advanced functions of IOT Gateway including fault handling and security management will be considered.

Acknowledgments This work has been funded by Advance Research Project of Xi'an Communications Institute under Grant no. XATYB013. The authors would like to thank Prof. Yanpu Chen for his help and valuable comments.

References

1. Atzori, L., Iera, A., Morabito, G.: The internet of things: a survey. Comput. Netw. **15**, 2787–2805 (2010). doi:10.1016/j.comnet.2010.05.010. (Accessed 2 February 2013)
2. Zhu, Q., Wang, R.C., Chen, Q., Liu, Y., Qin, W.J.: IOT gateway: bridging wireless sensor networks into internet of things. In: Proceedings of IEEE/IFIP International Conference on Embedded and Ubiquitous Computing, Hong Kong, December 2010
3. Jiang, X.Y., Li, D.S., Nie, S.B., Luo J., Lu Z.H.: An enhanced IOT gateway in a broadcast system. In: Proceedings of 9th International Conference on Ubiquitous Intelligence and Computing and 9th International Conference on Autonomic and Trusted Computing, Fukuoka, Japan, September 2012
4. Guinard, D., Trifa, V.: Towards the web of things: web mashups for embedded devices. In: Proceedings 2nd Workshop on Mashups, Enterprise Mashups and Lightweight Composition on the Web, Madrid, Spain, April 2009
5. Riedel, T., Fantana, N., Genaid, A., Yordanov, D., Schmidtke H.R., Beigl M.: Using web service gateways and code generation for sustainable IOT system development. In: Proceedings of Internet of Things, pp. 1–8, doi:10.1109/IOT.2010.5678449. (Accessed 12 February 2013)
6. Chen H., Jia X.Q., Li H.: A brief introduction to IOT gateway. In: Proceedings IET International Conference on Communication Technology and Application, Beijing, China, October 2011
7. Trifa V., Wieland S., Guinard D., Bohnert T.: Design and implementation of a gateway for web-based interaction and management of embedded devices. In: Proceedings of 2nd International Workshop on Sensor Network Engineering, Marina del Rey, CA, USA, June 2009
8. Chhatwani, S., Khanchandani, K.B.: Embedded web server. Int. J. Eng. Sci. Technol. **2**, 1233–1238 (2011). (Accessed 21 June 2012)
9. Texas Instruments Incorporated.: AM335x ARM Cortex-A8 Microprocessors. 11 January 2013. Accessed January 15, 2013. https://www.ti.com/
10. Embedthis Software.: Appweb for Dynamic Applications. 24 October 2012. Accessed 13 April 2013. https://appwebserver.org/

Research of Wireless Transmission Strategy for High-Speed Railway

Daquan Wu and Ning Zhang

Abstract Due to the wireless base station coverage is not high along the high-speed railway lines, the wireless signal around is not stable. It is difficult to guarantee reliable wireless data transmission in high-speed train just depending on the hardware environment. For common solutions, there is no control over data transmission on application layer, and data transmission depends on hardware completely. So, the reliability of data transmission is poor, and efficiency is low in common solutions. To solve these problems, we propose a novel wireless data transmission method, predicting the next state of signal in software layer. The new transmission strategy in this paper adopts the control strategy of wireless data communication based on prediction. With predicting wireless signal strength along the railway reasonably and adopting data transmission control accordingly, it makes data transmission more secure. This is significant to wireless data transmission of high-speed railway.

Keywords Wireless data transmission · High-speed railway

1 Introduction

With the rapid development of high-speed railway information technology, real-time information of the train in motion is requested to send to the ground as soon as possible. So, the data communication based on public wireless network in high-speed-moving train becomes particularly important. Railway informationization is the main symbol of railway modernization. For all communication platforms of

D. Wu (✉) · N. Zhang
Bejing JiaoTong University, Beijing 100044, China
e-mail: 11120485@bjtu.edu.cn

N. Zhang
e-mail: nzhang1@bjtu.edu.cn

S. Patnaik and X. Li (eds.), *Proceedings of International Conference on Computer Science and Information Technology*, Advances in Intelligent Systems and Computing 255, DOI: 10.1007/978-81-322-1759-6_27,

integrated information on train, public mobile communication network is the foundation of railway informationization [1].

Now, most of the existing wireless base stations in our country have shortcomings, such as low coverage, susceptible to interference, and so on. There is no wireless signal in some mountainous areas, tunnels, and sparsely populated areas. Because of the high speed of the high-speed train, the change of the network signal is accelerated; some big data packets cannot be sent normally [2–5]. To address this problem, we propose to provide a mechanism in software layer in order to increase the reliability and the transfer efficiency of data transmission. The carrier of the research in this paper is the wireless transmission device on train developed by our research center, which controls the receiving and sending of run-time information of the train in motion and communicates with the server on the ground to deliver information by wireless network.

2 Research Environment and Content

Network operators provide wireless base stations for data transmission of high-speed train, but the coverage is low. And, wireless signal strength is inversely proportional to the square of the distance to wireless base station, so the larger the distance, the weaker the signal it gets. At the same time, reasons below will lead to the signal attenuation:

- Closed environment of the high-speed train;
- The high speed of the train;
- When the train is going through mountainous areas and canyons.

In conclusion, the high-speed train network environment is unstable, and the packet loss rate of data transmission is high. So, we adopt the method of adding transmission control strategy in communication software layer to guarantee reliable data transmission.

As seen in Fig. 1, when the train is close to the data transmission base station, the signal is strong. Conversely, the longer the distance is, the weaker the signal is. In Fig. 1, the area with the largest loop number of coil has the largest signal strength.

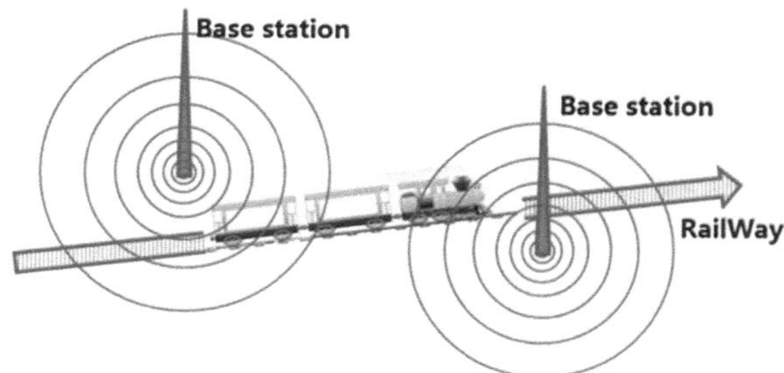

Fig. 1 Signal along railway

Based on the transmission service from the wireless base stations provided by the public network operators, we study how to achieve reliable wireless data communication between train and ground in this paper. We improve the existing wireless data transmission methods to adapt to the environment of the high-speed rail better. According to the experiments results, our method can increase the success rate of data transmission and the size of data packet being sent in per unit time.

3 The Existing Wireless Data Transmission Ways

The existing wireless data transmission ways [6] are usually thought to be of two kinds:

- Configure wireless unit as a network, then build socket connection in the application to transfer data.
- Depend on serial port completely, control wireless unit using a special command set to finish dial-up access and socket connection.

Both of the two ways totally rely on the wireless hardware modules when sending data. In application layer, data packets are put in a queue and will be resent if they are sent unsuccessfully in the last time.

4 The Drawbacks of the Existing Wireless Data Transmission Ways in High-Speed Railway

In the urban areas or in the good wireless coverage areas, the existing ways are feasible to transmit data in both mobile devices and static devices. The existing wireless modules provide good support for TCP protocol of socket.

Fast-moving train often goes through the sparsely populated areas where signal coverage is low and instable. Along with the Doppler effect generated by fast moving, the signal strength in high-speed train will be reduced.

To send data relying on the wireless module entirely, the phenomena in Fig. 2 may occur.

As shown in the Fig. 2, it would waste a lot of time to dial and establish TCP connection in the condition of weak signal for the existing wireless data transmission. Consequently, data packets could not be sent even though wireless signal gets much stronger during this period, which is obviously unreasonable.

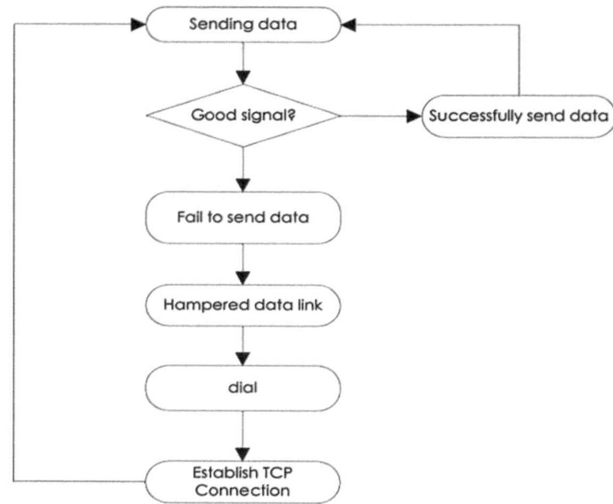

Fig. 2 Process of sending data

5 Propose the New Way of Wireless Data Transmission

Though the wireless signal in the high-speed train has a rapid change and the relative signal strength is weak, the strength value of the wireless signal is still a continuous curve on the time axis. With this in mind, we can make essential and reasonable predictions about the strength of the wireless signal in next time. If our prediction shows that the wireless signal is weak in next time, we would wait until it gets better instead of send data immediately.

Since network connection will be kept at the moment of network interruption, data packets could be sent in the present time without dialing and establishing network connection again if we predict the signal strength is good in next time. It is time saving, but this approach requires setting the network connection hold time to a maximum to avoid network interruption.

So, the wireless data transmission in this paper can save the time of dialing and establishing network connections. It can also improve the actual data sending time and the efficiency of data transmission [7]. Wireless unit needs to get IP address only once when the train is in an area with wireless network signal coverage. The approach in this paper is effective only if the train is not in the dead zone of tunnel signal light.

6 The Signal Strength Prediction Algorithm

First of all, the condition of the algorithm is that the value of signal strength is smooth. In other words, the train is not going through the tunnel and gorge, which can cause great change of the signal value (Fig. 3).

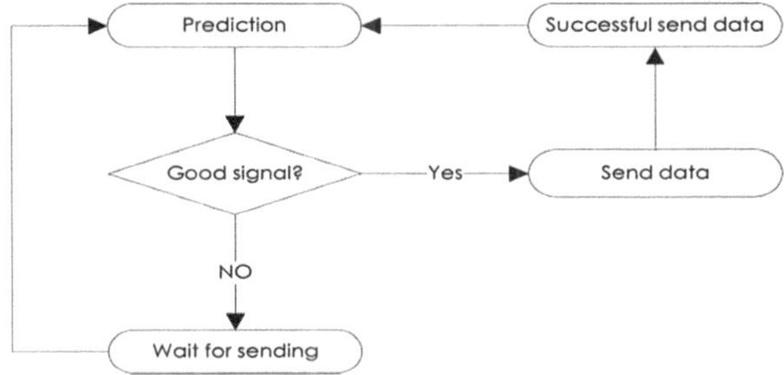

Fig. 3 Process of sending data with prediction

Divide the time line into several points with regular intervals, use X_n to express the nth clock tick of signal value, and use X_{n-1} to express signal value at the last clock tick of X_n, X_{n+1} could be expressed as the signal value at the next clock tick of X_n, shown in Fig. 4. Define M as the lower limit of signal strength value to send data successfully.

The signal strength in the actual situation is a continuous curve, so it would not change a lot in a short span. The increment of signal value for every span during every sending period can make contribution to our prediction. The increment helps us to control the prediction of next time.

The signal strength prediction algorithm is executed as follows:

- Calculate the average value $T^{'}$ of continuous decline times of signal strength in total time. Analyzing the real data of wireless transmission of high-speed train, we can find that the continuous decline times (test per 1 s) of signal strength is stable. In the Fig. 5 (ignore the unit temporarily), $T^{'}$ is calculated as follows:

$$T^{'} = \frac{1}{6}(3 + 1 + 2 + 3 + 2 + 2) = 2.1 \tag{1}$$

- The increment of signal value in next time is calculated:

$$V_n = X_n - X_{n-1} \tag{2}$$

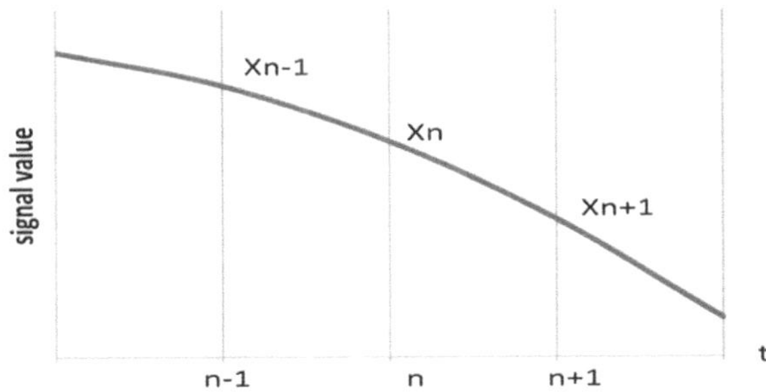

Fig. 4 Define of X

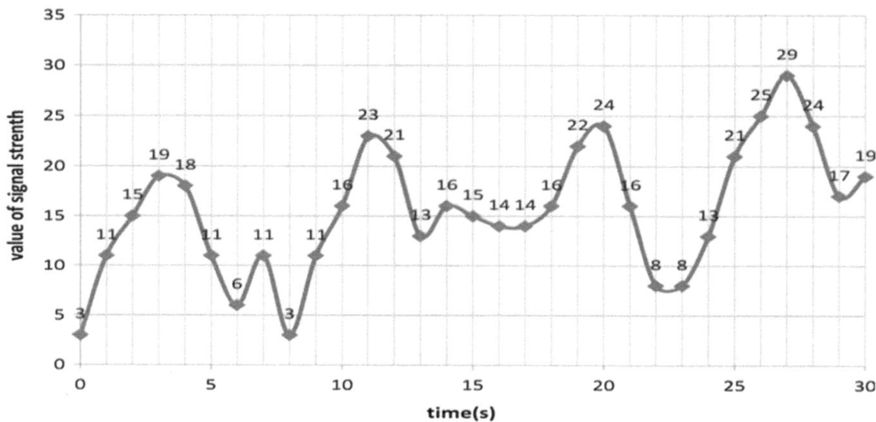

Fig. 5 The value of signal strength(example)

$$V_{n+1} = (1 - \alpha) \times V_n + \alpha \times V_{n-1} \tag{3}$$

α is the balance parameter between V_n and V_{n-1}. It can be given by experience.

- Calculate the original value of X_{n+1} of prediction:

$$X_{n+1} = X_n + V_{n+1} \tag{4}$$

- If X_{n+1} is smaller than M and V_{n+1} is a negative number, the prediction of signal is inaccurate. Valley value may appear in the situation. The prediction of signal value should be added but decreased in practice. So, we need to consider two cases as below:
- If the current value T (subtract current signal value from last peak value) is greater than T', negative the value of V_{n+1} and calculate the X_{n+1} again. $X_{n+1} = X_n - V_{n+1}$. It happens when the valley value appears at current time; we should increase the value of our prediction.

- If the current value T (subtract current signal value from last peak value) is smaller than T', program should send data in the next time. We can set X_{n+1} to be a constant value that is smaller than M.
- In other cases, which means X_{n+1} is greater than M or V_{n+1} is a positive number, the prediction of signal is accurate.
- Send data when the signal value is greater than M; otherwise, wait for sending data in next time.

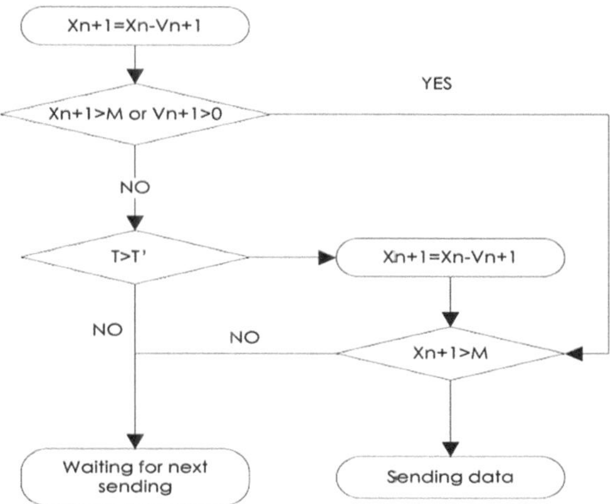

Fig. 6 Prediction algorithm

Controlling the sending program with our prediction algorithm, we can control the sending time by the predicted value of signal strength. Time of building network is saved. The prediction algorithm makes great efforts for predicted value when time goes near the valley value. If we do not have any control in this case, we would lose some time for sending data. For example, in sixth clock tick in Fig. 5, the current signal value X_6 is 6(ignore the unit temporarily), without step 4, 5, 6, the predicted value for X_7 will be approximate 1. In this case, data cannot be sent, but it is 11 in reality, which is good enough for sending data (Fig. 6).

7 Experimental Measurement

The results are obtained from SIM548C module, which can return the value of signal strength. The unit of signal strength value of SIM548C module is defined as follows:

- 0 represents −113 dBm or less
- 1 represents −111 dBm
- 2…30 represent −109 dBm…−53 dBm
- 31 represents −51 dBm or greater

We get the average time of dialing and establishing TCP connection per time for 200 tests (signal strength value is between 15 and 18, the maximum is 31, and the minimum is 0), and results are shown below (Table 1):

In order to make the signal strength value in our experiments close to the actual data of high-speed train, the following approaches are adopted to obtain original data of signal strength for experiment.

Table 1 Time to establish network connection

Content time	Content time (s)
Dialing	1.8
Establishing TCP connection	2.0
Total	3.8

- We installed the SIM548C chip and application devices in a car, and the car travelled at a speed of 70 km/h in the suburbs of Beijing. We obtain signal strength value per second.
- The signal strength value is obtained every 3 s based on the original data, which conformed to situation in the high-speed train travelled at a speed of 300 km/h. Due to the attenuation of wireless signal inside the high-speed train, the value 10 is subtracted from all the signal strength value. We collected the data for 1,000 s.

The signal strength value of a small part (Fig. 7).

The interval for detection of signal value is 1 s and attempt to send data per second at the same time (For the existing ways of wireless data transmission, data are sent per second ignoring the signal strength value. For the way based on prediction, make prediction per second to decide whether to send or not). The maximum speed of high-speed train must not exceed 300 km/h, and the total test time is 1,000 s, namely 16.7 min. The experimental data designed satisfies the above two requirements.

When the wireless signal strength value goes down to 10, data could not be sent. So, the lower limit of signal strength value to send data successfully is 10. There are 150 times of undeliverable case for data. Data with 92 bytes are sent per time, and the sending time interval is assumed to be 1 s. In the experiment, set

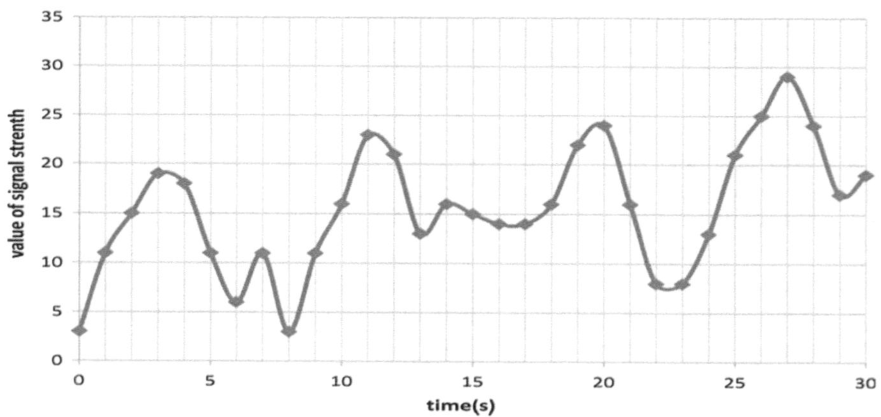

Fig. 7 The signal strength value in partial time

Table 2 Result of experiment

	Existing method	New method based on signal prediction
Total test time (s)	1,000	1,000
Total number of packets sent	1,000	850
Number of packets sent successfully	720	780
Success rate (%)	72	91.76
Number of packets sent per second	0.72	0.78

0.125 as the value of α, this is an experience value in previous statistics experiment from our laboratory. The count of continuous decline times is 267 in our experiment data. The calculated T_0 is 2.3.

We design simulation program based on the prediction algorithm and reasonable assumptions mentioned above. By running the simulation program, the following measured results are obtained (Table 2).

These results show that wireless data transmission strategy based on prediction can improve the success rate of sending data and increase the number of packets sent in per unit time. The success rate of sending data is increased by 19.76 %, and the number of packets sent in per unit time is increased by 0.06 in test, which indicates that the data transmission strategy based on prediction is effective.

8 Conclusion

The research of our paper is only under the condition of normal sections of railway. When the signal changes smoothly on normal path, our strategy can improve the success rate of data transmission and increase the size of the data packets sent per unit time. Because of the abrupt change of signal strength when the train goes in and out of the tunnel or gorge, the prediction algorithm will be disabled. Some addition controlling can make the algorithm perfect. This may be effective by referencing the GPS information.

Acknowledgments This work has been supported by Project NO. 2012AA040912 of the National High-Technology Research & Development Program of China.

References

1. Fang, W., Liu, Z., Liang, X., Liu, F., Huang, S., Li, L: Research and implementation of an AAA system for railway WiMAX communication networks. Int. J. Digital Content Technol. Its Appl. **5**, 238–246 (2011)
2. Aggelis, K., Louvros, S.: GPRS performance optimization with pre-empted packet queue analysis. Design and Technology of Integrated Systemsin NanoscaleEra, 2011 6th International Conference on (2012)

3. Ming, F., Zhu, X., Torres, M., Anaya, L., Patanapong-pibul, L.: GSM/GPRS bearer's efficiency analysis for machine type communications. IEEE 75th, Vehicular Technology Conference (VTC Spring) (2012)
4. Lu, G., Xiang, R., He, D.: Study on high-speed railway station operation simulation system. Information and Computing (ICIC), 2010 Third (2010)
5. Aguado, M., Onandi, O., Augstin, P.S., Higuero, M., Taquet, E.J.: WiMAX on rails. IEEE Veh. Technol. Mag. **3**, 47–56 (2008)
6. Lu, Y., Hu, J.: Design of tracker for electric vehicle based on GPRS and GPS. Int. J. Digital Content Technol. Its Appl. **6**, 85–92 (2012)
7. Jung, Y.C., Song, H.G., Moon, S.T.: Efficient indexing methods of continuously moving object in wireless networks. MobiWac '11 Proceedings of the 9th ACM international symposium on Mobility management and wireless access (2011)

Power Consumption Analysis and Modeling of Mobile Communication Architecture

Andong Zhang, Shuowen Zhang, Pengcheng Zhu and Xiaohu You

Abstract The reduction of the power consumption of mobile communication system is an important issue nowadays. Traditional base station structure is considered to consume too much energy and at the same time, distributed antenna system (DAS) has been recognized as an energy-efficient base station architecture. In this paper, we investigate power consumption model of both structures. We compare the power consumption of serving a random user at the cell edge when varying the cell radius and also investigate the power consumption of both systems when the number of antenna increased. Results show that (1) DAS costs less energy in every scenario and has a better cell-edge performance. (2) Improving the efficiency of power amplifier (PA) or reducing the average distance between the antenna and the user is the key factor of reducing the power consumption of DAS architecture.

Keywords DAS · Base station structure · Power consumption · Cell size

A. Zhang · S. Zhang · P. Zhu · X. You (✉)
National Mobile Communications Research Laboratory, Southeast University, Nanjing 210096, People's Republic of China
e-mail: xhyu@seu.edu.cn

A. Zhang
e-mail: adzhangseu@gmail.com

S. Zhang
e-mail: shwzhang8@gmail.com

P. Zhu
e-mail: p.zhu@seu.edu.cn

S. Patnaik and X. Li (eds.), *Proceedings of International Conference on Computer Science and Information Technology*, Advances in Intelligent Systems and Computing 255, DOI: 10.1007/978-81-322-1759-6_28,

1 Introduction

Nowadays, the global mobile communication industry is growing rapidly and the global number of mobile phone subscribers approaches 6 billion. However, the phenomenal growth of mobile communication industry leads to an expense of considerable energy consumption. The growing concern over the power consumption aspect of wireless and cellular networks has triggered a new research initiative in academia and industry that can collectively be referred to as "green communication" techniques [1]. Recent surveys on the energy consumption of cellular networks reveal that around 80 % of the energy required for the operation of a cellular network is consumed at base station (BS) sites [2]. There are currently more than four million BSs serving mobile users, each consuming an average of 25 MWh per year. The large amount of energy costs of mobile communication systems has drawn an emerging trend of attention among the network operators and regulatory bodies such as 3 GPP and ITU [3, 4]. The authors in [1] estimate that of the total power consumption of a typical cell site, 43 % is consumed by the cooling system followed by 41 % by the BS itself. Radio-over-fiber (RoF)-distributed antenna system (DAS) technique has been considered to be a promising solution due to its cost-effective remote antenna unit (RAU) and functionally simple architecture compared with the traditional wireless systems [5]. In this paper, we demonstrate the power model of both systems and also provide extensive simulation results to compare the power consumption. Besides, efforts have been made to compare the power consumption when changing the cell size and antenna numbers of both systems. Finally, we conclude this article.

2 Power Model

2.1 Traditional Base Station

Traditional base station system utilizes centralized structure, namely, every cell has a base station located in the center and several antennas are in charge of signal transmission and receiving. Base station controller (BSC) and base transceiver station (BTS) are the two main parts of a traditional base station. The power consumption of the BSC is neglectable compared with that of BTS. A BTS contains a few transceivers (TRXs) that serve several antenna elements. A base station transceiver consists of a small-signal radio frequency (RF) transceiver module, a power amplifier (PA), a baseband (BB) unit, a DC–DC power supply, an AC–DC unit acting as mains supply and an active cooling system [6].

Power Amplifier and Antenna Interface: Power amplifier is a vital component of the transceiver acting at the last stage of a TRX. PA mainly amplifies high-frequency modulated signals in order to meet transmission power requirements. In general, base station of traditional architecture is often situated at different physical locations than the antennas, a feeder loss of about ($\sigma_{\text{feeder}} = -3$ dB needs to be added. Power consumption of the power amplifier is therefore

$$P_{PA} = \frac{P_{out}}{\eta_{PA} \cdot (1 - \sigma_{feed})} \tag{1}$$

Rectifier: The rectifier conducts signal conversion from alternating currency to direct currency and has a latest rectifiers with a newly developed structure could reach up to 97 %. The power consumption of a rectifier is related not only to the output power but also to its efficiency η_{rec} and is thus given by

$$P_{total} = \frac{P_{in}}{\eta_{rec}} \tag{2}$$

Baseband Unit (BB): The BB takes charge of processing of baseband signals. The power consumption of signal processing including channel coding, duplexing, modulation, and frequency expanding is related to N_c, the number of antenna elements (AE). In the LTE-Advanced testbed implementation [7], about 10 % of the overall analog and digital processing power is due to uplink channel estimation and roughly 3 % are due to uplink- and downlink-MIMO processing. The former scales linearly with N_c, the quadratic term of N_c represents the cost of MIMO-signal processing, we assume P_{SPB} is 58 (*Watt*). The power consumption of BB is thus

$$P_{SP} = P_{SPB}\left(0.87 + 0.1\, N_c + 0.03^2\, N_c\right) \tag{3}$$

RF Transceiver: Parameters with highest impact on the RF energy consumption, P_{RF}, are the required bandwidth, the allowable signal-to-noise-and-distortion ratio (SiNAD), and the resolution of the analog-to-digital conversion.

Power Supply and Cooling: The power consumption of the cooling system is positively related to total output energy. Cooling factor $\sigma_{cooling}$ is used to represent this relationship. Losses are caused by DC–DC power supply, mains supply is also linearly with the power consumption of the other components and can be described by the loss factors $\sigma_{power\ supply}$, we can have

$$P_{cooling} = \left(\sigma_{cooling} + \sigma_{power\ supply}\right) P_{total} \tag{4}$$

and we assume $\sigma_{power\ supply} + \sigma_{cooling}$ is 0.35.

Total Power Consumption: Consider a traditional base station with N_{TRX} transceivers, its total power consumption is

$$\begin{aligned} P_{total} = {} & N_{TRX}\left(1 + \sigma_{cooling} + \sigma_{power\ supply}\right) \frac{P_{out} + P_{RF}}{(1 - \sigma_{feed})\eta_{rec}} \\ & + \frac{P_{SPB}(0.87 + 0.1\, N_c + 0.03^2\, N_c)}{\eta_{rec}} \end{aligned} \tag{5}$$

2.2 Distributed Antenna System

A RoF DAS system can be divided into three parts as: central processing entity, converter, and remote antenna units. In the following each part is analyzed.

Central Processing Entity: In the DAS structure, it is mostly the same as the signal processing part of traditional base station. A CoMP cluster in DAS performs the same MIMO-signal processing as a micro-BS, but we do not deploy CoMP in this paper so we can directly take Eq. 3 to get the power consumption of central processing entity in the following [6]:

$$P_{\mathrm{SP}} = 0.87\, P_{\mathrm{SPB}} + P_{\mathrm{SPB}}\left(0.1\, N_{\mathrm{c}} + 0.03^2\, N_{\mathrm{c}}\right) N_{\mathrm{c}} \tag{6}$$

Converter: In our analysis, the laser-drive amplifier, laser diode, and photodiode can be considered to have fixed power consumptions and laser cooler is not necessary. We can calculate the power consumption of the laser diode by estimating the bias current that is around 40 mA and voltage (around 2 V) resulting in 80 mW electrical power [8]. So the power consumption of downlink is

$$P_{\mathrm{converter}} = P_{\mathrm{laser\ diode}} + P_{\mathrm{laser\ drive\ amplifier}} \tag{7}$$

Antenna Units: The structure of remote antenna units can be simplified to amplifiers that have much smaller efficiency and energy consumption. The power consumption is modeled by the required output power, we take the number 2.2 % here [9].

$$P_{\mathrm{rau}} = \frac{P_{\mathrm{out}}}{\eta_{\mathrm{rau}}} \tag{8}$$

Total Power Consumption: Consider a DAS with N_{AE} antenna working, its total power consumption is

$$P_{\mathrm{total}} = N_{\mathrm{AE}}(P_{\mathrm{Rau}} + P_{\mathrm{converter}}) + P_{\mathrm{SP}} \tag{9}$$

3 Channel Model

Consider a cellular system that consists of a circle cells with radius 1 km, the cell is located at the origin of the *x*–*y* plane. The following assumptions (Ass) are made as follows:

AS-1. In the circle cell, the traditional BS and DAS are loaded with only one MT. The traditional BS consists of multiple transceivers and the MT chooses the transceiver that costs the least energy. For the DAS, the MT also chooses the most energy-saving RAU.

AS-2. The DAS system consists of *N* RAUs, and each RAU is equipped with L antennas. The MT uses *M* antennas. Thus, $M \leq L$ is assumed in this paper, to simplify the model, we set $M = 1$ and $L = 1$.

AS-3. The locations of the RAUs are fixed. For the traditional BS, all the transceivers are located at the center of the cell. The RAUs of the DAS system and the location of the MT are arranged uniformly distributed in the corresponding cell.

In mobile cellular scenarios, the radio propagation can be characterized by three independent phenomena: path loss variation with distance, large-scale shadowing, and small-scale fading. All of them will be incorporated in this paper. Moreover, it is assumed that the wireless channel experiences frequency-flat fading. In a strict sense, the wireless channel is frequency selective. However, in the case of OFDM, each sub-channel can be assumed to be non-frequency selective. The fading coefficients remain quasi-static within some time interval (called a block) and changes independently between blocks.

Therefore, the wireless channel from MT to the *n*-th RAU in Cell-1 can be modeled as an $L \times M$ random matrix

$$H_{n,k} = \sqrt{cd_{n,k}^{-\alpha} s_{n,k} W_{n,k}}$$

where $cd_{n,k}^{-a}$ denotes the path loss. $d_{n,k}$ is the distance(in km) between MT-*k* and the *n*-th RAU in the Cell; α is the path loss exponent, typically taking a value between 3.0 and 5.0; and *c* is the median of the mean path loss at the reference distance of 1 km; $s_{n,k}$ is usually a log-normal distributed shadowing variable and we do not consider this impact in our analysis. $W_{n,k} \in \mathbb{C}^{L \times M}$ represents the small-scale fading. $W_{n,k}$ are independently and identically distributed(i.i.d.) circularly symmetric complex Gaussian random variables with zero mean and unit variance. The random variables $d_{n,k}$, $s_{n,k,}$ and matrices $W_{n,k}$ are assumed to be independent of each other and independent for all *n, k*.

Using the typical receiver sensitivity (−120 dBm), the transmitting signal power of the base station must be above the receiver sensitivity after taking into account the propagation loss over a distance of *d* we can obtain the maximum cell side lengths at different transmitting powers.

4 Analysis and Comparison

In Sect. 2, we have studied the components of power consumption for both systems, and we set up the simulation and propagation model. In this section, we will investigate the power consumption when varying the cell size (BS coverage) and the number of antennas.

As we can see from Fig. 1, when the number of antennas is the same (nine antennas), it is obvious that the traditional base station consumes more energy and the power consumption of traditional base station tends to increase quicker than the DAS system. It is also obvious that the transmit power of DAS is much lower

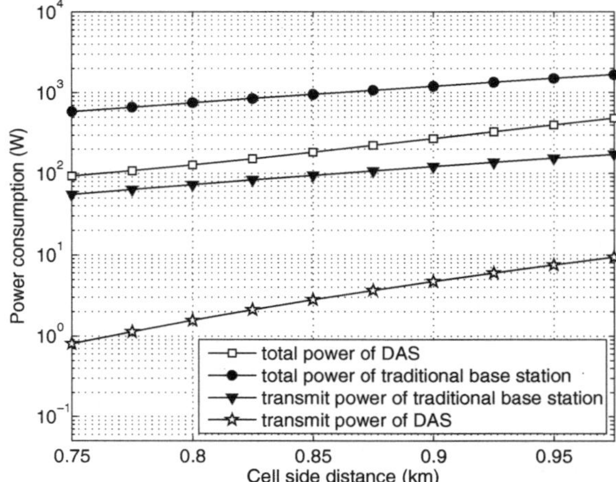

Fig. 1 Power consumption(W) versus cell side length(km)

because DAS decrease the average distance between user and antennas dramatically. For both mobile system structures, transmit power only accounts for ten percent of total power consumption, the efficiency of PA is a key effect of total power consumption (Fig. 2).

When the position of the users are random, the number of the antennas affects the power consumption, it decreases for the both system when the number of antennas grows. For the traditional base station, as the number of antennas increases, base station can choose the best one according to the propagation model. For the DAS, more antennas not only give more choices but also spread over the cell, so the average distance between user and antennas is much smaller and the edge performance is much better.

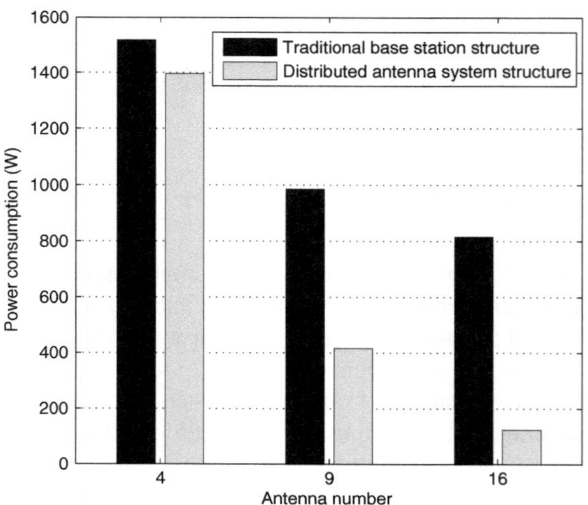

Fig. 2 Power consumption of two types of system (W) versus antenna number

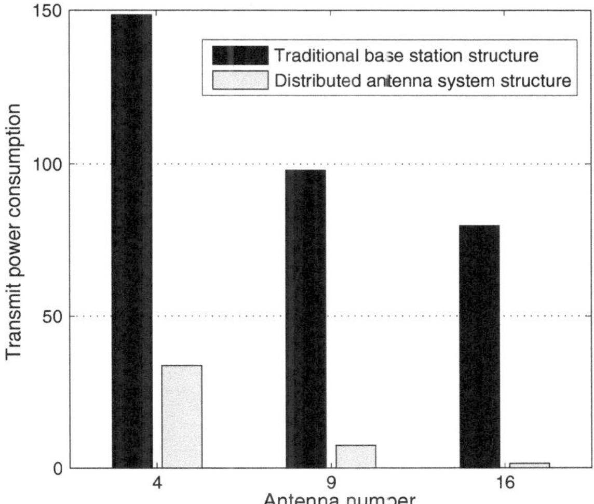

Fig. 3 Transmit power of two types of system (*W*) versus antenna number

5 Conclusion

In this paper, we analyze the system model of traditional base station structure and DAS structure. We proposed power consumption model of both systems and compared the performances when varying the cell radius and antenna numbers and found that the DAS always saves energy in every scenario. For example, when the cell radius is the same, the power consumption of DAS structure is about 10 % of traditional base station. There are two main reasons, for the most part, why DAS structure saves energy: (1) DAS structure makes the average distance between user and antenna much smaller; (2) the transmit power only make up 10 % of total power consumption. The feeder loss, low efficiency PA, and cooling system cost a lot energy. Respectively, simulations show that if DAS structure can improve the PA efficiency, the structure can save more energy and have better cell-edge performance (Fig. 3).

Acknowledgment This work was supported by the Natural Science Foundation of China under grant 61101086 and the National Key Special Program under grant 2012ZX03001036-004.

References

1. Hasan, Z., Boostanimehr, H., Bhargava, V.K.: Green cellular networks: A survey, some research issues and challenges. J. Commun. **13**, 524–540 (2011)
2. Fehske, A., Fettweis, G., Malmodin, J., et al.: The global footprint of mobile communications: The ecological and economic perspective. J. Commun. Mag. **49**(8), 55–62 (2011)
3. GPP TR 32.826, Telecommunication management; Study on energy savings management (ESM), (Release 10), Mar 2010. http://www.3gpp.org/ftp/Specs/html-info/32826.htm

4. ITU-T Focus Group on Future Networks (FG FN), FG-FN OD-66, Draft Deliverable on "Overview of Energy Saving of Networks", Oct 2010. http://www.itu.int/dmspub/itut/oth/3A/05/T3A050000660001MSWE.doc
5. Wake, D., Nkansah, A., Gomes, N.J.: Radio over fiber link design for next generation wireless systems. J. Lightwave Technol. **28**(16), 2456–2464 (2010)
6. Auer, G., Giannini, V., Desset, C., et al.: How much energy is needed to run a wireless network? J. Wireless Commun. **18**(5), 40–49 (2011)
7. Enablers for Ambient Services and Systems Part C -Wide Area Coverage (EASYC), project website. http://www.easy-c.com
8. Crisp, M., Penty, R.V., White, I.H., Bell, A.: Wideband radio over fiber distributed antenna systems for energy efficient in-building wireless communications. In: Vehicular Technology Conference (VTC 2010-Spring), pp. 1–5, IEEE 71st (2010)
9. Ezzeddine, A.K., Huang, H.C.: 10 W ultra-broadband power amplifier. In: Microwave Symposium Digest IEEE MTT-S International, pp. 643–646, IEEE (2008)

A New Ants Routing Algorithm in Ad Hoc Networks with GPS

Wang Anbao and Zhu Bin

Abstract We propose a new routing algorithm based on ant colony optimization for MANETs with global positioning system (GPS) and heuristic methods. In mobile ad hoc networks, every node in different positions has different probabilities to forward the ant to the next hop, so as to greatly reduce the overhead of the packets used for maintaining the routing information. When a node forwards an ant to launch a process to construct a path from the source to the destination, it may get more different paths. Our algorithm selects one or two of them recorded in the node's local routing table for the robustness reason. When a link is disconnected, a mechanism is taken to repair the path so as to create an alternative path to promote the robustness of the routing algorithm. Simulation results show that our algorithm achieves good packet delivery ratio with low communication delay.

Keywords Routing algorithm · Mobile ad hoc networks · ACO · Robustness

1 Introduction

It was reported that OLSR do not scale well in heterogeneous networks, many authors propose optimizations to OLSR in order to limit the amount of control traffic in wireless networks [1–3]. Our work tries to modify the ant colony algorithm using heuristic strategies and methods, such as the location information of the nodes from the protocol of global positioning system (GPS), and let the nodes in deferent areas should have deferent probabilities to forward ant to the next hop. In order to improve the pertinence of the routing algorithm and find the position of

W. Anbao (✉) · Z. Bin
School of Computer & Information, Shanghai Second Polytechnic University, Shanghai 201209, China
e-mail: abwang@it.sspu.cn

S. Patnaik and X. Li (eds.), *Proceedings of International Conference on Computer Science and Information Technology*, Advances in Intelligent Systems and Computing 255, DOI: 10.1007/978-81-322-1759-6_29,

destination nodes more quickly, the heuristic factor to the basic ant colony algorithm is added as one of the parameters for the selection of next hop [3–7].

2 The Analyses of Routing Algorithms

2.1 Mathematical Analyses of the Routing in Mobile Ad Hoc Network

Let l be the Ad Hoc network radius, and d be the node's propagation radius. The hop counts are as follows: $(0, 1, \ldots, l/d)$, this is used only for simple to model. n be the number of nodes in the whole ad hoc networks, $\rho = n/\pi l^2$ is the density of the nodes in ad hoc networks; The number of nodes within 1-hop is: $\text{count}_{\text{hop}_1} = n\left(\frac{d}{l}\right)^2$. The number of nodes within 2-hops is: $\text{count}_{\text{hop}_2} = n\left(\frac{2d}{l}\right)^2 - \text{count}_{\text{hop}_1} = 3n\left(\frac{d}{l}\right)^2$. And so on, the number of nodes within m-hops is: $\text{count}_{\text{hop}_m} = n\left(\frac{\text{md}}{l}\right)^2 - \text{count}_{\text{hop}_{m-1}} - \cdots - \text{count}_{\text{hop}_1} = (2m-1)n\left(\frac{d}{l}\right)^2$

Assume that each node has been equipped with the device of GPS, so the ad hoc network radius (l) and the number (n) of nodes in the whole Ad Hoc network can easily be worked out if we know the position of each node according to the GPS information. For easy analyze to the problems, we assume that all the nodes are on the circle. A node generates an ant to construct the route for the other nodes in the whole networks, and it broadcasts the information to all its neighbors, which is just $1-$ hop range from the node itself.

In Fig. 1, in order to unicast or multicast packets to other nodes, the node n_s wants to generate the ants to propagate in the whole ad hoc network and workout the routing for itself. Let us name the propagation range in the circle as track_1, $\text{track}_2, \ldots, \text{track}_n$, and the node on the track_1 (circle) should generate an ant with the probability of p, so as to cover the area between the track_1 and track_2. From the mathematical computation, we can easily get the result that we just select three nodes (in Fig. 1, n_0, n_1, and n_2) from all of the nodes in the track_1, and it can make the nodes' propagation range intersecting at the track_1 circle, and the most area is covered in track_2.

There are still some areas that cannot be covered by the nodes selected in track_1, while those areas can be covered by the nodes selected from the track_2 which is display in the Fig. 1. In the track_2, we select six nodes (n_3, n_4, n_5, n_6, n_7, and n_8) which broadcast the routing information for the source node n_s, the areas which cannot be covered by the nodes selected from the track_1 mostly can be covered by the nodes selected from the track_2, if we select more nodes in track_2, the whole area will be perfectly covered.

From the knowledge above, in track_1, there are $n(d/l)^2$ nodes, every node in the track_1 generates the ant with the probability of $3/(n(d/l)^2)$ or more, and nodes in the

Fig. 1 The selected nodes in $track_2$

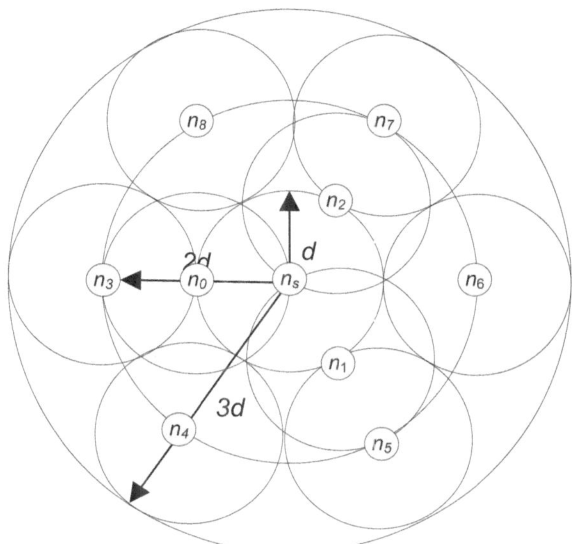

$track_2$ with the probability of $6/(3n(d/l)^2)$. The behaviors of the ants were modified, so as to make the ants have the broadcast abilities and the ant at the nodes that does not generate new ant will die at once.

2.2 Our Algorithm about the Behaviors of the Ant

The main idea of our protocol is as follows: nodes in the mobile ad hoc network know the location of their neighbors within one- or two-hop distance, so we have to create a table to maintain the coordination of each node for future use. In order to use the ant colony optimization in our algorithm, for every node u, it has to maintain a table to store the amount of pheromone on a link in the view of itself, and a two dimension array is used to realize this kind of function.

To find a path from a source node s to a destination node d, it is the responsibility of s to create a forward ant and broadcast it to all its neighbor nodes; the ant goes through the network to search a route from s to the d, when the ant reaches the next neighbor node u, whether node u will forward the ant to its neighbor n is decided on three steps:

1. The number of pheromone P_{nd} in its local table;
2. The node broadcasts the ant to all its neighbor nodes with some kind of probability p according to whether the node n is on the $track_1$ or $track_2$;

$$p = \left\{ \begin{array}{ll} 3/n\left(\frac{d}{l}\right)^2 & \text{if } n \text{ is on track}_1 \\ 6/3n\left(\frac{d}{l}\right)^2 & \text{if } n \text{ is on track}_2 \end{array} \right\}$$

If the node u knows the number of the neighbor nodes which are on track_1, it should choose the bigger one from the value of $n(d/l)^2$ and the number of the neighbor nodes, similarly, If the node u knows the number of the neighbor nodes which are on track_2, it should choose the bigger one from the value of $3n(d/l)^2$ and the number of the $2-$ hop neighbor nodes

3. If the node u does not locate on the track_1 or track_2 area, it should select a node according to the pheromone in u's pheromone array: the ant moves to the next neighbor node w selected from u's neighbors with the probability P_{wn}. Where neighbor_u is a set of neighbors of u, and α is a parameter of randomness represented by a constant real value. If u has no pheromone for the destination d (i.e., $\forall\ w \in \text{neighbor}_u$, $P_{wd} = 0$), then the ant broadcasts to all neighbors of u.

$$P_{wd} = \frac{(P_{wd})^{\alpha}}{\sum_{x \in \text{neighbour}_u} (P_{wd})^{\alpha}}, \quad a \geq 0$$

3 Constructing the Routing Information

In order to minimize the delay of finding an optimal routing from the source to the destination and reduce the number of the routing messages in mobile ad hoc network, in our algorithm, the life cycle of the ant is constrained to be two or three hops; after the ant has jumped two or three hops, it exchanges the information with the node which it resided in. There are two kinds of conditions, respectively, one is that if the routing information getting from the node is enough to construct a full routing path from the source to the destination, so the path can be created which represents a path from the source to the destination, then the routing information is recorded in the PREQ message, the ant sends a Path REPly (PREP) message to the source node in response to a PREQ message. The PREP message also just acts like in AODV(RFC3561), which contains a routing path got from the PREQ message. The PREP message toward the source node is forwarded by the intermediate node along the reverse path obtained from the PREP message.

The ant should modify the number of pheromone data in the node. Because we limited the hop count of the ants, the number of pheromone data in the nodes from the source to destination which is greater than two or three hops should be modified by the method of relay strategies [4].

When the information got from the intermediate nodes is not enough to construct a valid routing path, the node which the ant is residing in creates a next generation ant; this kind of ant acts just like the ant generated before, and it takes

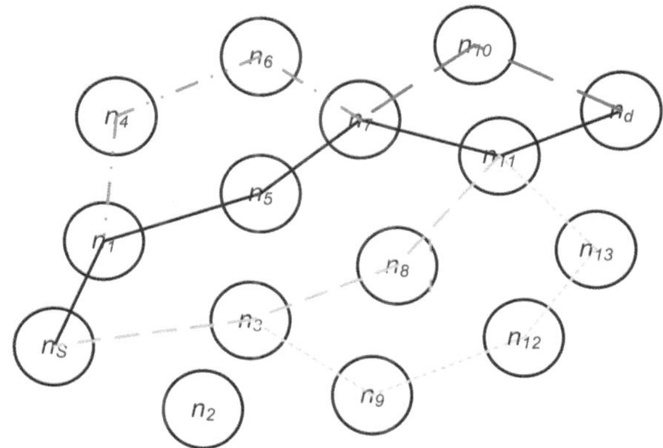

Fig. 2 Different path created by different ants

the current node as the source node and the original destination node as its destination node. Current node should records the routing information, which the old ant has constructed. If the next generation ant has got the full path information, the path is created which represents a route from the current to the destination node, the path information is also recorded in the PREQ message, and then the ant sends a Path REPly (PREP) message to the upstream nodes in response to a PREQ message. The PREP message also just acts like in AODV, which contains a routing path, got from the PREQ message. The PREP message toward the node which the current generation ant has been created there is forwarded by the intermediate node along the reverse path obtained from the PREP message. At the same time, the ants should modify the number of pheromone data in the node. The current node can use the routing information from the first generation and the second generation ants to construct path information from the original source to the destination and send back the PREP message to the original source node. If current generation ant still cannot get the full path from the current node to the destination within two or three hops, it repeats the steps just like the previous generation ant has done.

A lot of ants flood out from the node which it resides in, so may be there are more than one ant who get the full path from the source to destination, in Fig. 2, the ants have created five paths which are marked with different colors and line styles. The ant firstly arrives at the source, and the full path information from the source to the destination is carried with it. In general, the path the ant carried with may be the best path from the source to destination that is because the time used for constructing the path is the shortest. The source node firstly modifies its routing table, sends a broadcasting message to all its neighbors to notify that the path has been created, and then other Path REPly (PREP) message to the source node in response to a PREQ message is not needed now, this can reduce the number of routing packets in the network. In some cases, the source node may receive more than one PREP message, and it represents more different paths to the destination; in this circumstance, we chose two of them recorded in the routing table for robust

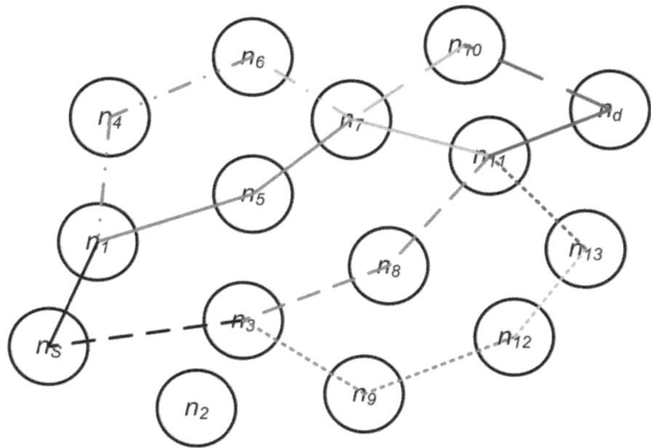

Fig. 3 The different hops for ants to find different paths

reason. Different ants need different hops to get the total path information. In Fig. 3, different hops are marked, respectively, with different colors, 1 − hop is marked as blank, and 2 − hop is marked as purple, and so on.

In mobile ad hoc networks, nodes move randomly, or its power may be closed in anytime, so the routing path which has been created may get invalid in anytime, and it may break sometimes especially in sparser networks.

To alleviate path breaks in mobile ad hoc networks, in the process of constructing routing path, when the ant arrives at the destination, the destination also records the path from the source to the destination in its routing table, when the source sends a packet; the packet traces the path it goes through from the source to the destination. Note that this path may be different from the path found by the source during the PREQ/PREP phase. This is because forwarding packet along the path may break by some reasons; the packet may select a path that is nearly the same as the path in the former routing table. When the packet arrives at the destination, it compares the path information resided in its routing table form the source to destination with the path traced by the packet. If it is different, this is because the path may break by some reasons; the packet selects a path that is different from the path found by the source during the PREQ/PREP phase. The destination then creates an ant taking with a new path message to the source; the source uses the new path message to modify the routing information. In Fig. 3, the source node n_s wants to send packets to n_d; the path it selected from its routing table is $(n_s \rightarrow n_1 \rightarrow n_5 \rightarrow n_7 \rightarrow n_{11} \rightarrow n_d)$, when the packets go through the network, if the node n_{11} is closed and the node n_7 selects the node n_{10} to forward the packets to the destination n_d, as the packets arrived at the destination, the destination checks the packet header; if it finds that the actual path it goes through has been changed, the destination will launch the process to help the source to recover the routing path from the source to the destination using the algorithm described above. Using upper method can greatly improve routing performance by reducing path re-establishment as much as possible [8].

4 Simulation Results

We compare the performance of our algorithm with the well-known protocol, OLSR, which is supported in NS. Different numbers of nodes are placed in a rectangular area of 1,000 × 1,000, they move according to the random-direction-2d-mobility model. In this model, each node selects a random destination within the simulation area and moves. Simulations were run for a total of 1,000 s each time. The data traffic was generated by 20 constant bit rate (CBR) sources. We use the 802.11 protocol at the MAC layer. The radio propagation distance is set to 250 m, and the data rate is 5.5 Mbit/s.

It shows the end-to-end delay of our algorithm in comparison with OLSR routing protocols. Our algorithm gets better results than OLSR. Because the heuristic methods are used during the ant forward and backward period, and moreover, the path information is also stored in the intermediate node, therefore, the ant can use this kind of routing information resided in the intermediate node during the constructing path procedure; this can reduce the overhead of the ant to find a path between the source and the destination. Since the paths are readily accessible in the whole process, it also benefits from the fast end-to-end packets transmission. In our algorithm, when an ant is created by the source node, all the nodes in the 1-hop propagation area can sense the ant, but only three of them can receive and make the ant forward to the next hop, at the same time the ants lifecycle is limited to two or three hops; this can greatly reduce the number of packets used for finding a path. Another reason is that the ants can traverse on the links or ignore the links according to the heuristic information by choosing the next hop according to the pheromone in pheromone table in the intermediate node. These reasons allow our algorithm to produce better end-to-end delay results than other routing protocols.

In our algorithm, the upstream node of the broken link may have an alternative path to the destination, or it can firstly buffer all the packets it receives, and using the routing information in its local routing table to construct a new path, it also can create a new ant to build a new path from itself to the destination. If the node successfully finds an alternative path to the destination, it will send all the packets buffered before to the destination via the newly created path within the tolerable time, meanwhile, a new notification ant is created and sent to the source in order to let the source node know the difference between the old path information and the newly created one, while In OLSR the ants cannot select any links to travel and the data packet is dropped at that node.

5 Conclusion

A routing algorithm was presented which adapts ant colony optimization methods for MANETs. In order to make ant colony optimization routing as valid, some heuristic information is used in such a dynamic network environment, such as the location information of each node in the MANETs which are got from the GPS protocol, the number of one or two hops of nodes that is got from the mathematic method and other analyzed methods, etc., and the intermediate node gives the fullest information or other methods to construct the routing path.

In the algorithm, a recovery mechanism is used to reduce the overhead in path reconstructing and alleviate the packets loss rate. Our work in progress is to elaborate the algorithm and make the algorithm even more suitable for MANETs. It is also necessary to further reduce the overhead used in routing path creating period by improving the efficiency of our algorithm.

References

1. Ge, Y., Lamont, L., Villasenor, L.: Hierarchical OLSR—a scalable proactive routing protocol for heterogeneous ad hoc networks. In: Wireless and Mobile Computing, Networking and Communications, vol. 3, pp. 17–23 Aug 2005
2. Sachin, S.: P-OLSR: position-based optimized link state routing for mobile ad hoc networks. In: Local Computer Networks, pp. 237–240 Oct 2009
3. Villanueva-Pena, P., Kunz, T., Dhakal, P.: Extended topology knowledge for localization and quality of service with OLSR. In: Wireless and Mobile Computing, Networking and Communications, pp. 449–456 (2006)
4. Quintero, A., Li, D., Castro, H.: A location routing protocol based on smart antennas for Ad Hoc networks. J. Comput. Netw. Appl. **30**, 614–636 (2007)
5. Giruka, V., Singhal, M.: Two scalable location service protocols for wireless Ad Hoc networks. Pervasive Mob. Comput. **2**, 262–285 (2006)
6. Wang, H., Shi, Z., Ge, A., Yu, C.: An optimized ant colony algorithm based on the gradual changing orientation factor for multi-constraint QoS routing. Comput. Commun. **32**(4), 449–456 (2009)
7. Kadono, D., Izumi, T., Ooshita, F., Kakugawa, H., Masuzawa, T.: An ant colony optimization routing based on robustness for Ad Hoc networks with GPSs. Ad Hoc Netw. **8**, 63–76 (2010)
8. Giruka, V., Singhal, M.: A self-healing on-demand geographic path routing protocol for mobile ad-hoc networks. Ad Hoc Netw. **5**, 1113–1128 (2007)

Design and Implementation of Online Monitoring System for Large-Scale Mechanical and Electrical Equipment Based on Embedded Techniques

Weimin Bi and Xuefeng Ruan

Abstract Monitoring of working status for large-scale instruments or equipment is of great concern (Zhang et al. 2007; Chen 2009), and establishment of such system is helpful to reduce manual intervention and ensure the proper functioning of the instruments. We have introduced the embedded technique for monitoring method and realized a new online monitoring system for large equipment.

Keywords Large-scale instruments and equipment · Working status monitoring · Embedded techniques

1 Introduction

Monitoring system concerned with large-scale instruments is not common [1, 2]. Since there are a lot of different kinds of large equipment, only one kind or simple function monitoring system cannot work. In order to solve the problem, we have to apply various methods or hybrid techniques [3, 4]. Therefore, here we will expound a new framework designed for the online monitoring system specially for large equipment.

W. Bi (✉)
Office of Laboratory and Equipment Management, Wuhan University, Wuhan 430072, China
e-mail: awmbi@whu.edu.cn

X. Ruan
School of Power and Mechanical Engineering, Wuhan University, 430072 Wuhan, China
e-mail: xf-ruan@whu.edu.cn

S. Patnaik and X. Li (eds.), *Proceedings of International Conference on Computer Science and Information Technology*, Advances in Intelligent Systems and Computing 255, DOI: 10.1007/978-81-322-1759-6_30,

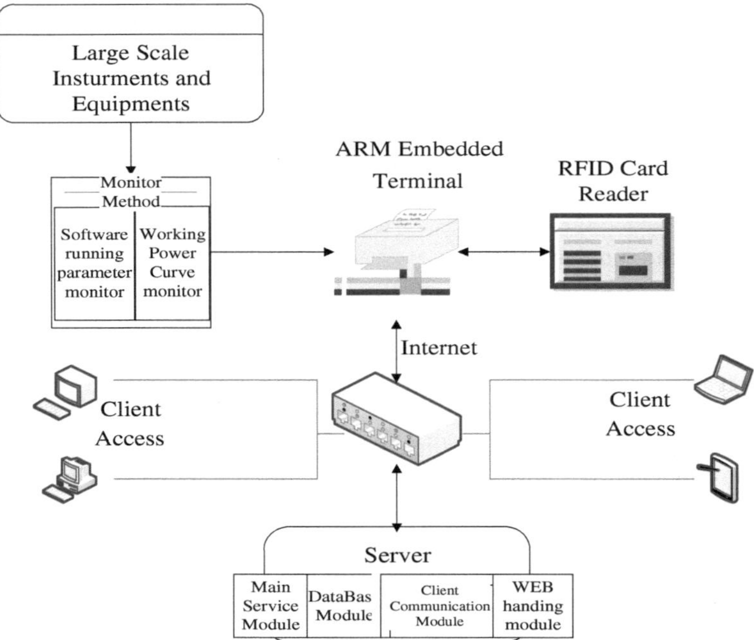

Fig. 1 System-block diagram

2 Characteristics of System

The basic principle of the system is shown in the Fig. 1. Large equipment work-status monitoring system consists of a monitoring module, Advanced RISC machines (ARM)-embedded terminals, RFID card identification modules, central data processing module, as well as real-time display, and query analysis system. The monitoring module records the instrument operating parameters that includes the information collected by software or power current value obtained from current monitor device. ARM-embedded terminal is the most important core processing unit, on the one hand, it communicates with the monitoring module to get the instruments status data, on the other hand, it is also in charge of controlling RF-card-reader device. ARM terminal can get the card ID through the card-reader device, then the ID information and the data of instruments status are all uploaded to the server. The central data processing module on the server parses the uploaded data packet and receives the real-time information of the user and instrument, and updates the data corresponding to database as well. Display and query analysis module display the real-time status information for large equipment, or develop a user's query within a certain date, statistical functions, and form the result to understand and reference the large equipment work status and performance statistics.

3 Design and Technology Solution

Original intention of this system is to ease the workload of the crew that use the large equipment and reduce manual intervention. Therefore, the system should be small, exquisite, fast, and convenient to use with its installation configuration. Of course, the system must ensure the functional richness and reliability as well. We adopt the embedded technology for the purpose. The Embedded technology is the non-PC system developed rapidly in recent years. It is characterized by the size, function, and power. The system with strict targeted software and hardware is available for cutting configuration. The final system with strong integration, functionality, and scalability has a rich communication interface. Combined with the actual needs, the system, ARM9 with S3C2440 central processor is applied [5, 6].

S3C2440-embedded system based on the ARM920T CPU core with 300, 400, and 533 MHz-operating frequency, its interior has a 32 KB buffer, which has a 16 KB instruction command buffer and 16 KB of data buffer. S3C2440 chip itself has 130 general-purpose input and output I/O interface, 24 external interrupt sources, external Interrupt 6 hardware timer (one watchdog timer), 4-channel DMA (Direct Memory Access) controller, all the way to the DMA-controller exclusive 4–256 K color LCD controller. Two-way USB host, USB-slave device. Three-way UART (Universal Asynchronous Receiver/Transmitter Universal Asynchronous transmission). SPI, I2C, and I2S bus interface, as well as the 8-channel 10-bit accuracy A/D converter, etc.

Under this hardware architecture, we use the LCD controller an external control 4.3-inch 256-color touch LCD display, USB host identification reader module communication (also can be extended to the camera, GSM module), the A/D module is used in current detection threshold of the large-scale equipment, UART transmission is used in data communication on the host–computer monitoring of large equipment and ARM-terminal SPI bus, I/O interfaces, and external terminal is used in additional auxiliary detector monitoring such as ultrasound, infrared detection. Screen interface with the camera is connected to the actual situation to observe large instrument.

Based on the embedded operating system software features, and in order to achieve the operating system of cross-platform applications, we select the embedded Linux-operating system, well-known open-source operating system. Linux in recent years have made great development of the graphical interface and hardware-driver support. Linux system itself supports multiple CPU and the hardware platform with different kernels. The system design for the development of embedded Linux system is involved with the following aspects:

In the ARM-terminal application, the remoted data transmission related to the linux-socket technology, ARM-terminal socket session automatically connected to the server listening port after the start of the terminal.

ARM-terminal application is required to coordinate the collection of data to send, receive, and display a variety of functions in the server feedback data, therefore, the program must be multi-threaded. Here, we established new linux thread to receive and send network data for each socket session.

The interprocess variables shared with the data communication has been one of the difficulties in the Linux programing, in this case, memory-sharing technique is used for the communication between background program and the display program. We develop the embedded display application base on QT interface, and QTimer class is used to check sharing memory for display data at a certain frequency. The embedded user interface is shown in Fig. 2.

For some of the equipment, especially of mechanical instrument, in order to achieve the purpose of monitoring instruments work state, we use the principle of measuring instruments inductor current. The ARM motherboard itself contains two A/D converter channels and could be called under linux environment as follows:

adc_fd = open ("/ dev / adc", 0);/ / initialize ARM development board comes with A/D conversion module and obtain the corresponding file descriptor.

For different large equipments, the standby current is not the same, so the pre-configuration is needed, when the current value of the recording apparatus standby adc_value0, the measuring current code into a thread, and continuously detecting the change in current of the instrument, when the value is greater than of the adc_value * 105 %, the recording apparatus would turn into working condition, and otherwise status recorded as the standby state.

RF card is also called non-contact IC card, encryption, and decryption-recognition technology combines RF-communication technology and IC contactless passive way to achieve both. Compared with the traditional IC card, it is more friendly to the environment, more secure and reliable.

An RF-card factory has a unique identifier, which is usually referred to the hardware ID (HID), in the system, the terminal reads the user card's HID number, then through the network, then by means of the network, this HID number is sent to the server, server service program soon finds a match in the database, and the

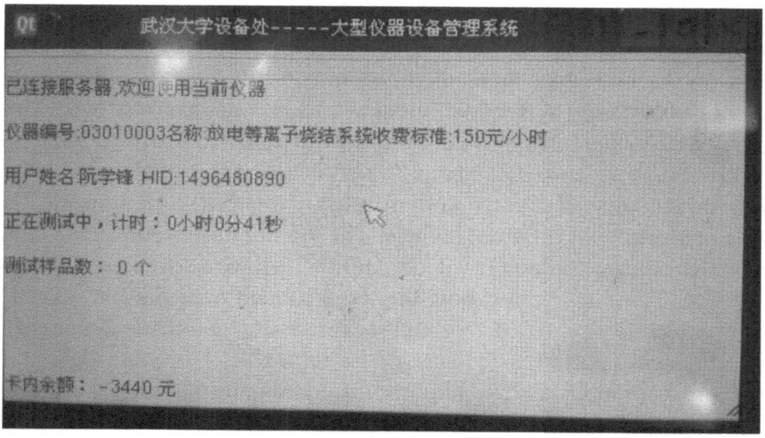

Fig. 2 User interface on touch screen of embedded terminal

corresponding HID number of other user information will be fed back to the ARM terminal. This design is to avoid the user's key information or private information being stored on the card, all user information is centrally stored on the server.

Socket communications, also known as TCP/IP communication, is one of the most common network communications, which has a very good universal standards can be completed from the communication between the different operating systems such as linux and windows [7, 8]. The server side can listen to the same port to connect multiple clients by creating a multi-threaded, or thread pool, and data transmission. Both blocking or non-blocking connection between ARM-terminal and the server is a new session thread, monitoring the thread, to ensure that large equipment ARM terminal transit after real-time data exchange with server. Server-side program using VC++ programing, socket class inherits the default class and adds a record session information in its structure. As follows:

```
typedef struct tagHeadBody
{  SOCKET sock; //sockt main body
  char chIP[20]; //Client ip
  int port; //Client port number
  int live; //live identification of session
  int clienttype; //ARM terminal type
  int armid; //ARM terminal serial number
  char clientid[20]; //user id
}MYHEADBODY;
```

With a reliable communication protocol, working status data of large equipment could be handled and showed at the same time without being interrupted, as is shown in Fig. 3.

Fig. 3 Real-time working status of large equipment

With the increasingly high degree of precision and degree of automation of large-scale equipment, the absolute majority of the instruments are equipped with a specialized computer control measurement software. Therefore, as well as hardware current detection methods, monitoring these respective equipment operating software is another important way of statistics and digging equipment and the crew-working condition.

Here, we detect each operating software's key API function call, to trace the running of the instrument. Using hook technology, hooking the operating thread, and then modifying the import symbol table (the PE head) in the correlation function address replacing the key API address by the custom-defined function address. Such as:

Replace Api Address (hModule, "Create File A", "kernel32.dll", (DWORD *) & oldCreateFileA, (DWORD) MyCreateFileA, type);

4 Summary

During the processing, we researched and established a reliable and efficient working status monitor system of large equipment. The system can handle multiple mechanical or electrical equipment of itself and show the statistic result or the real-time working status online.

Acknowledgement The authors would like to acknowledge the financial support from the Fundamental Research "Benefit evaluation system for large instruments in university. NO. SY2012-01".

References

1. Gu, Z.Y., Zheng, J.J., He, Y., Liu, J.: In: Advances in Electrical Engineering & Electrical Machines, vol. 133, pp. 85–90 (2011)
2. Zhang, J.Y., Cai, W., Huang, X.X., Xie, J.: In: 2009 Ninth International Conference on Hybrid Intelligent Systems, 978-0-7695-3745-0/09, p. 373 (2009)
3. Zhang, Z.L., Zeng, X.L., Cai, W.: In: The Eighth International Conference on Electronic Measurement and Instruments, 1-4244-1135-1/07, pp. 4–92 (2007)
4. Chen, H.Y.: In: 2009 International Conference on New Trends in Information and Service Science, 978-0-7695-3687-3/09, p. 822 (2009)
5. Hong, Y.: In: Microcomputer Information, 32-014 (2007)
6. Li, Z., Lin, Y., Yang, Y.: Chin. J. Sci. Instrum. S1–134 (2005)
7. Ming, X.: In: Circuits, Communications and Systems, 2009. PACCS '09, 978-0-7695-3614-9, p. 775 (2007)
8. Zhou, Z.B., Bian, F.L.: Sci. Surv. Mapp. 04–028 (2006)

Application of Wireless Sensor Networks Based on ZigBee in Taxi Dispatching System

Weibin Wang and Yadong Yu

Abstract Intelligent taxi dispatching system (ITDS) is a transportation service system using information technology, communication technology, and network technology. It will reduce the cost of communication and have low power to apply wireless sensor network based on ZigBee technology. The scheme was introduced to design vehicle terminal using ZigBee module.

Keywords Intelligent taxi dispatching system · ZigBee · Wireless sensor network

1 Introduction

There were four technical innovations in development of taxi dispatching system (TDS). The first generation of TDS based on the radio communications complete the dispatching task by real-time voice. It has been eliminated step by step because of its low degree of automation. The second generation of TDS was built on the base of computer and LAN technology. It processed information by computer and called each other by radio. But because of rare frequency resources, it has been eliminated too. With the development of GPS, the third generation of TDS appeared. Dispatching center arranged for passenger with the latest taxi according to the position of passenger and taxi. The fourth generation of TDS integrated GPRS, GPS, GIS, and intelligent dispatching algorithm. It raised working efficiency, but communication cost needs to be paid to operating agencies.

W. Wang (✉) · Y. Yu
Jiaxin University, Jiaxin, Zhejiang, China
e-mail: 1613761080@qq.com

Y. Yu
e-mail: yuydfriend@sohu.com

S. Patnaik and X. Li (eds.), *Proceedings of International Conference on Computer Science and Information Technology*, Advances in Intelligent Systems and Computing 255, DOI: 10.1007/978-81-322-1759-6_31,

In recent years, wireless sensor network based on ZigBee technology attracted increasing attention [1, 2]. It will be wildly used in national defense notion, military, traffic management, and medical field [3, 4]. Wireless sensor network based on ZigBee have some merits such as low power consumption and cost. Communication cost would be reduced and maintenance of system would be simplified. The scheme applying ZigBee to TDS was described in this paper. The rest of the paper is organized as follows. Features and functions of REX3D ZigBee were introduced in the Sect. 2. In Sect. 3, design of the hardware and software of vehicle terminal was presented followed by a summary and conclusion.

2 Introduction of REX3D ZigBee Module

2.1 Features of REX3D ZigBee Module

REX3D is a small high-sensitivity ZigBee Module which complies with specification of IEEE802.15.4 [5, 6]. It can be used in some fields such as wireless sensor network and data collection system [7, 8]. Supply voltage of the module is from 2.1 to 3.6 V. Working frequency is divided into sixteen frequency bands in the range of 2,400–30 MHz. In open area away from obstacles, the signals were transmitted over a distance of 2,000 m under the support of power amplifier. Internal architecture of the module is shown in Fig. 1.

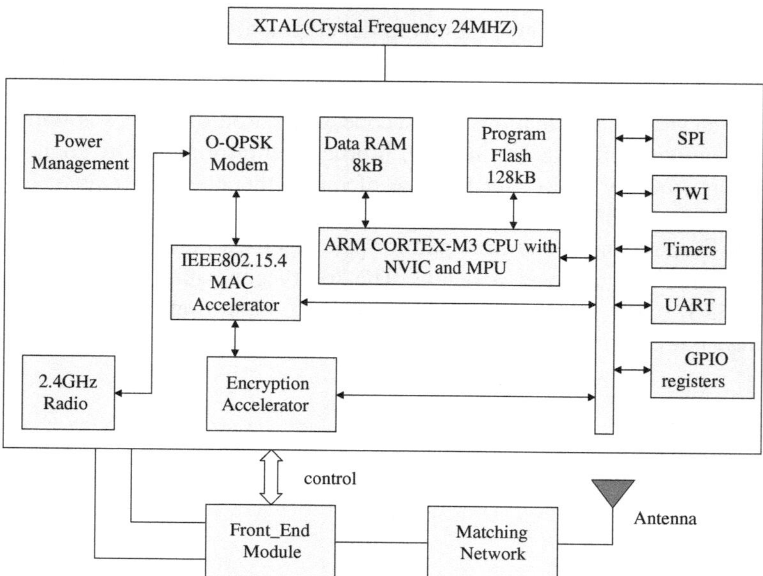

Fig. 1 Internal architecture of REX3D ZigBee

2.2 Interface of REX3D ZigBee Module

REX3D module provides enough interfaces such as six high-resolution analogue input channels and a standard USART interface for further development. Only the USART interface needs to be used if the module is used to transfer data. If hardware flow control is not used, the TX port of the USART interface is connected to the RX port of external MCU, while the RX port of the USART interface is connected to the RX port of external MCU. External MCU may configure the work parameters of REX3D module through AT instruction by USART.

2.3 Function of REX3D ZigBee Module

The work mode of REX3D module may be one of the three modes which include COO Router and ZED. COO node is coordinator which builds and manages the ZigBee network. Only one COO node is in one network. COO node can provide protocol routing and manage 32 child nodes like ZED. Router node is full-function node of ZigBee network. It can relay the data sent by other node. ZED node is terminal node of ZigBee network. Active sleep is its most obvious feature, but it cannot relay data. ZED node is awaked periodically. It sends request to parent node. If there are data need to be sent, parent node will send data to ZED node while it receives the request.

3 The Design of Vehicle Terminal Based on REX3D ZigBee

3.1 Design of Hardware

There are six parts in the vehicle terminal which include power REX3D ZigBee module, GPS module, MCU module, and human–computer interface. Structure of the hardware is shown in Fig. 2.

Power module converts car battery pack of 24 to 3.3 V DC which is supplied for other parts of the circuit. Driver can input the choice which shows whether drivers are willing to obey scheme or not through keyboard. Passenger information which is sent by the dispatch center is displayed in LCD.

STM32F103RB was selected as the core processor, and STM32 family of microcontrollers uses ARM Cortex-M3 kernel with powerful performance. It is a cost-effective microprocessor produced by company of ST specifically for embedded development. In its interior, there are two USART interfaces which can communicate with the GPS module and ZigBee module. STM32F103RB receives location information from GPS module through the USART 1. STM32F103RB

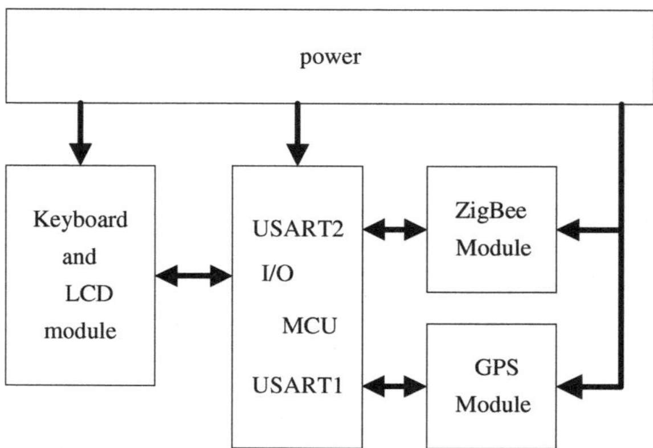

Fig. 2 Structure of the hardware

sends location information of itself using ZigBee module by the USART 2 when the dispatch center needs it.

EB-3631 which is the third-generation high-sensitivity module of SIRF was used to receive GPS signal receiver module. It has 20 channels and high receiver sensitivity (−159 dBm). It outputs standard NMEA0183 signal through USART interface. The module sends positioning data once per second including sentences such as $GPGGA, $GPGSA, $GPGSV, and $GPRMC. Sentence $GPRMC was used in the design which was included the information of target such as longitude, latitude, speed, the direction of movement angle, year, month, hours, minutes, seconds, and milliseconds.

REX3D ZigBee module works in routing mode. It possesses the capability of transparent transmission. The ZigBee modules have the functions of data sending and data receiving. MCU only needs to send data to the ZigBee module or receive data from the module through the USART; then, data are transported to destination node by the module according to the address.

3.2 Design of Software

In order to improve the clarity and reusability of code, the embedded operating system uC/OS-II was applied. uC/OS-II is a priority-based preemptive hard real-time kernel and it has the characteristic of easy to transplant and retrench. It contains basic functions of task scheduling, task management, time management, memory management, and inter-task communication and synchronization. Besides main task, the design consisted of three subtasks: the subtask of location update, the subtask of communications between terminal and dispatch center, and the subtask of instruction processing by dispatch center. After initialization of the operating system, each task was created and the operating system was started.

Fig. 3 Flowchart of the main task

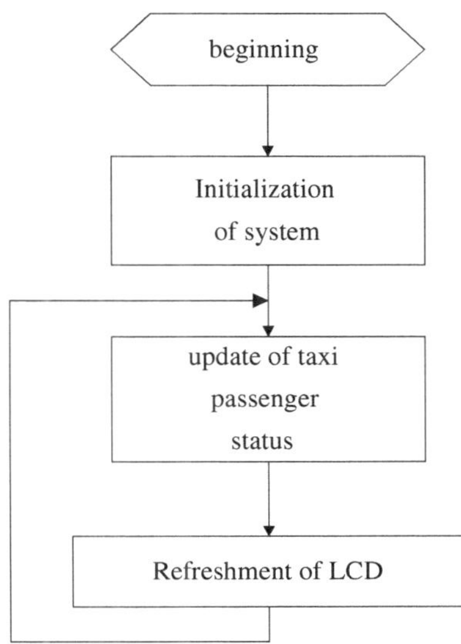

1. Main task

The main task executed initialization of the system, including the setting of the clock, the setting of I/O pin, initialization of the peripheral, initialization of interrupt vector, and starting the children tasks. In addition, the main task also executed the update of taxi passenger status and controlling of LCD. Flowchart of the main task was shown in Fig. 3.

2. Subtask of location updating

Dispatch center needs to obtain the real-time data of position and direction of taxi for high effective management. The subtask of location updating got the position and the traveling direction of the taxi according to the serial data received from the GPS module. The USART1 of MCU worked in interrupt mode. The subtask of location update flowchart is shown in Fig. 4.

3. Subtask of communication between terminal and dispatch center

Terminal receives the data from the ZigBee module through the USART2 in MCU, obtaining the instructions issued by the dispatch center.

The operation mode of ZigBee module is set by MUC through the AT commands. There are the three AT commands which can be used to set baud rate, parity bit, the network identifier, communication channel, and so on.

Application program accesses module through the serial port. Each frame in link layer consists of five fields shown in Table 1.

All frames start with the flag sequence 0x2A and end with the flag sequence 0x23. The length field indicates the length of payload field. The payload field contains the valid bytes. The check field is the lowest byte of sum of the payload data. Content of payload was shown in Table 2.

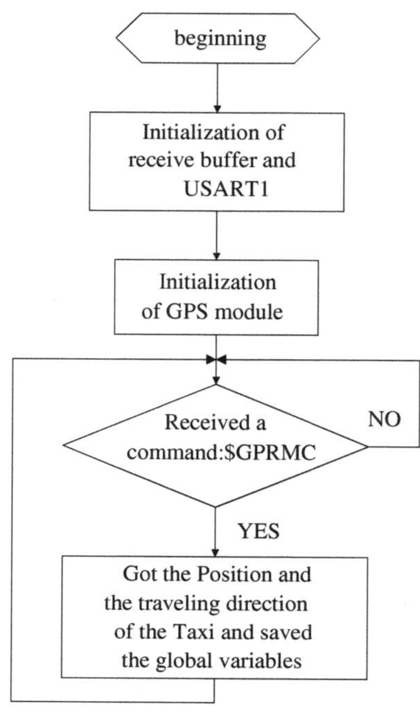

Fig. 4 Flowchart of the subtask of location updating

Table 1 Format of frame in link layer

1 byte	1 byte	n bytes	1 byte	1 byte
0x2A	Length	Payload	Check	0x23

Table 2 Content of payload

2 bytes	6 bytes	4 bytes	4 bytes	4 bytes	6 bytes	2 bytes	2 bytes	n bytes
Frame control	Reserved	Source address	Reserved	Target address	Reserved	Cluster ID	Reserved	ADF

The frame control field is filled with 0x8841. Source address is the lower 4 bytes of MAC address which is the address of source node. Target address is the lower 4 bytes of MAC address which is the address of destination node. Cluster ID is the command identifier used to distinguish between different commands. ADF is the content of the application layer.

Communication protocol between terminal and the dispatch center was defined in the application layer. It consists of six fields shown in Table 3.

The beginning marker is defined with the character "$", and the data length is the number of bytes of the command code and ALD. The command code denotes the scheduling order. The ALD field is filled with the content of command while it is sent by the dispatch center, but if it is sent by the terminal, the ALD field is

Table 3 Content of payload

1 byte	2 bytes	1 byte	n bytes	1 byte	2 bytes
Beginning flag	length	Command code	ALD	Check num	Ending flag

Table 4 The meaning of the command code

Command code	Meaning
0x01	To report the state of the taxi
0x02	To report current position and direction of movement
0x03	To report position of passenger who want to take a taxi

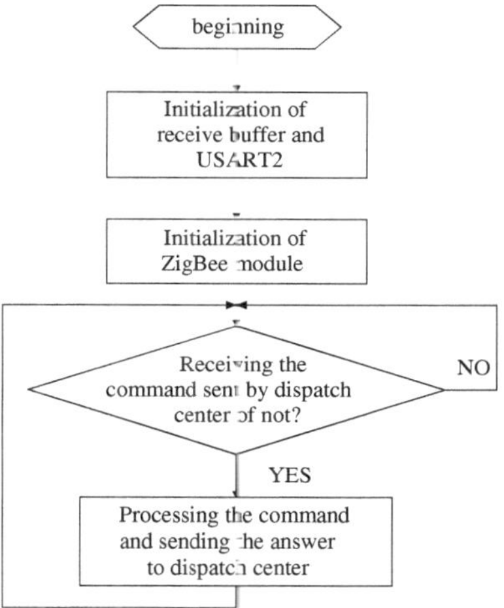

Fig. 5 Flowchart of the communications subtask between the terminal and dispatch center

meaning that the response for the command. Checksum byte is the lowest byte of sum of the length of the ALD, the command code, and bytes of data. The ending flag is CRLF. There are three main commands. The meaning of the command code was shown in Table 4.

The flowchart of the communications subtask between the terminal and dispatch center is shown in Fig. 5.

4 Conclusions

The system cooperating with the software of server side of the dispatch center can effectively reduce passenger's waiting time, reduce the rate of free driving of the taxi, and improve the operating efficiency of the taxi system.

Acknowledgments This work was financially supported by the Science Technology Department of Zhejiang Province, China, under Grant Number 2012C31003 in 2012.

References

1. Mraz, L., Cervenka, V., Komosny, D., Simek, M.: Comprehensive performance analysis of ZigBee technology based on real measurements. Wireless Pers. Commun. Vol. **71**, (2013)
2. Jiang, F.-C., Wu, H.-W., Yang, C.-T.: Traffic load analysis and its application to enhancing longevity on IEEE 802.15.4/ZigBee sensor network. J. Supercomputing **62**(2), 895–915 (2012)
3. Chen, L.-J., Sun, T., Liang, N.-C.: An evaluation study of mobility support in ZigBee networks. J. Sig. Process. Syst. **59**(1), 111–122 (2010)
4. Kim, D.-H., Song, J.-Y., Lee, S.-H., Cha, S.-K.: Development and evaluation of Zigbee node module for USN. Int. J. Precis. Eng. Manuf. **10**(5), 53–57 (2009)
5. ZigBee Alliance: ZigBee specification, V1.0, Dec 2004
6. ZigBee Alliance: http://www.ZigBee.org (2010). Accessed 6 May 2010
7. Lin, S., Liu, J., Fang, Y.: ZigBee based wireless sensor networks and its applications in industrial. In: Proceedings of the 2007 IEEE International Conference on Automation and Logistics, pp. 1979–1983, NJ USA (2007)
8. Wheeler, A.: Commercial applications of wireless sensor networks using Zigbee. IEEE Commun. Mag. **45**, 70–77 (2007) (Topics in Ad Hoc and sensor networks)

Pollution Resistance Network Coding Research for Ad hoc Network

Jun Liu, Chang Liu, Hui Liu and Xiang-jun Liu

Abstract Network coding technique was applied to ad hoc network transmission mechanism, which has the advantages of enhancing network throughput and balancing load. In view of the inherent vulnerability of network coding against pollution attack, combined with node identity authentication system, a detecting pollution attack scheme based on message authentication code (MAC) was proposed. The simulation shows that the scheme has high detection efficiency and low computational complexity of authentication information generation and verification. It has saved the computational overhead; therefore, this paper is more suitable for the resource-constrained ad hoc network.

Keywords Ad hoc network · Network coding · Pollution attack · Authentication codes

J. Liu (✉) · C. Liu
College of Information Science and Engineering, Northeastern University, Shenyang, China
e-mail: liujun@ise.neu.edu.cn

C. Liu
e-mail: liuchang5620@163.com

H. Liu
Chinese People's Liberation Army in the 764 Factory Military Agent's Room, Jinan, China
e-mail: lh_186@126.com

X. Liu
China National Software & Service Co., Ltd., Beijing 100081, China
e-mail: liu_xiangjun@sina.com

S. Patnaik and X. Li (eds.), *Proceedings of International Conference on Computer Science and Information Technology*, Advances in Intelligent Systems and Computing 255, DOI: 10.1007/978-81-322-1759-6_32,

1 Introduction

Network coding theory [1] is a hot topic in the field of wireless communications in recent years. As it allows the intermediate node to encode the received information and transmit it, the destination node receives a certain number of linearly independent data, and then it can decode and recover the original message. Compared with the traditional "Store–Forward" transfer mode, its advantage is that it can not only improve network throughput and load balancing, it can also reduce the frequency of the transmission of information so that the energy costs can be reduced [2]. Therefore, combining the network coding technology with ad hoc network transport mechanism can improve much network performance effectively.

However, with the openness for wireless communication network, the malicious nodes may arbitrarily altered or falsified messages. Random noise often damages the message integrity. These may result in a packet being polluted. When polluted packet is encoded and transmitted to another node, the other node codes message in the same way and it makes pollution diffused [3]. When ad hoc network uses network coding technology for information transmission, the pollution attacks harm the transmission of messages greatly. Therefore, designing secure network coding which can combat pollution attacks with ad hoc network is very urgent and meaningful.

2 Related Research

For traditional network, it uses digital signature to ensure data integrity. However, the source node cannot calculate in advance all possible linear combinations of the signature. Ho et al. [4] proposed a scheme that can detect the presence of pollution attacks. The message sent by the source node is belong to finite field. All the message vectors consist of three parts: the global encoding vector, source message, and hash value. The preimage of hash value is the source message. After receiving the data packets, destination node decodes and calculates its hash value. Comparing with the received hash value, destination node checks whether the message is polluted. The disadvantage of the scheme is that it only verifies the integrity of the message in the destination node. If the authentication fails, source node will need to retransmit the message which will result in a larger network overhead.

Krohn et al. [5] uses homomorphic hash function to prevent pollution attacks. The main idea is that source node calculates the hash value of the original message with homomorphic hash function and then sends it to intermediate nodes and destination node by additional secure channel. When receiving coded vectors, a node calculates its homomorphic hash value and compares it with the previously received hash value. If certain conditions are met, the node further encodes and sends them to the downstream nodes, or discards it and asks the uplink to resend. The scheme's main drawback is the need for additional secure channel transmission of the original message's homomorphic hash value. The homomorphic hash function itself has a higher computational complexity. The intermediate nodes need to calculate homomorphic hash for a large number of packets, which is undoubtedly a huge challenge to this resource-constrained ad hoc network. Zhou et al. [6] ensures the integrity of transmitted messages by using homomorphic hash functions based on Diffie Hellman. While avoiding the use of additional secure channel, as long as the attacker gathers enough information, it will be able to get the secret number of hash function and distort the message.

Aiming at the shortcomings of the existing technologies, this article has combined with the idea of node identity authentication and used message authentication code (MAC) to authenticate the network coding vectors hop by hop. So the spread of pollution message can be restrained effectively. While ensuring low computational complexity to reduce communication overhead, it is more suitable for the resource-constrained ad hoc network. And it does not require additional secure channel to distribute authentication information.

3 Pollution Resistance Network Coding Scheme Based on MAC

In order to detect malicious nodes, the identity authentication function is required in the system. As the ad hoc network is without a trusted third party which is able to show corresponding certificates for the node identify, the scheme has adopted the identity authentication system [7]. Assumption of this article is that the identity authentication system is secure. Meanwhile, a node with valid identity will not become internal malicious nodes in the communication process. The security of the system parameters depends on the security of the authentication system.

3.1 System Initialization Process

The communication link between a source node and a sink node has been completed. The identity authentication scheme is that entity identity of each node is regarded as its public key. Therefore, each node does not need to ask or save other node's public key before communication. It has saved the storage space for all the nodes to a certain extent.

3.2 Source Node

Source node in network chooses a public pseudo-random sequences generator f and randomly generated a pair of safety parameters u, r. u is the selected seeds of pseudo-random sequence generator. r is an element of the finite field q. The original message M sent by source node is divided into vectors $\overrightarrow{l_1}, \overrightarrow{l_2}, \ldots, \overrightarrow{l_n}$,

$\overrightarrow{l_i} = (v_{i,1}, \ldots v_{i,n})c$. Each element $v_{i,k}, k \in [1, m]$ is in finite field q. The message M can be expressed as

$$M = \begin{bmatrix} \overrightarrow{l_1} \\ \overrightarrow{l_2} \\ \vdots \\ \overrightarrow{l_n} \end{bmatrix} = \begin{bmatrix} w_{1,1} & w_{1,2} & \cdots & w_{1,m} \\ w_{2,1} & w_{2,2} & \cdots & w_{2,m} \\ \vdots & \vdots & \ddots & \vdots \\ w_{n,1} & w_{n,2} & \cdots & w_{n,m} \end{bmatrix} \quad (1)$$

In order to transmit all the later coding coefficient vector with the message, the source node will extend $\overrightarrow{l_i}, i \in [1, n]$ to be as below:

$$\overrightarrow{w_i} = (w_{i,1}, \ldots w_{i,m}, \overbrace{0, \ldots 0}^{i-1}, 1, \overbrace{0, \ldots 0}^{n-1}) = (w_{i,1}, \ldots w_{i,m}, w_{i,m+1}, \ldots w_{i,m+n}) \quad 1 \le i \le n \quad (2)$$

The source node calculates message authentication code MAC($\overrightarrow{w_i}$) for each original vector $\overrightarrow{w_i}$ using safety parameters u and r as well as the selected pseudo-random sequence generator *f*. As described below:

The source node generates $\overrightarrow{u_i} = (u_1, u_2, \ldots, u_{m+n}) \in F_q^{m+n}$ bit by bit using pseudo-random sequence generator f and secret seeds u.

The source node calculates authentication value for $\overrightarrow{w_i} = (w_{i,1}, \ldots, w_{i,m+n}), 1 \le i \le n$ using safety parameters u and r:

$$c(\overrightarrow{w_i}) = \overrightarrow{w_i} \cdot \overrightarrow{u}^T + \sum_{k=m+1}^{m+n} w_{i,k} \cdot r = \overrightarrow{w_i} \cdot \overrightarrow{u}^T + r \quad (3)$$

The encrypted secret numbers u and r by the descending nodes' public key and the message authentication value $c(\overrightarrow{w_i})$ form a MACs together.

$$\mathrm{MAC}(\overrightarrow{w_i}) = \{(r, u)_{\mathrm{pk_j}}, c(\overrightarrow{w_i})\} \quad (4)$$

PK_i is the public key for downside node i of the source node. $(\ldots)_{\mathrm{pk}_i}$ is the encrypted content by PK_i.

In order to prevent the source message from being eavesdropped, the source node codes original vector $\overrightarrow{w_i}, i \in [1, n]$ by random linear coding to get the new coding vector:

$$\overrightarrow{e} = \sum_{i=1}^{n} \eta_i \cdot \overrightarrow{w} \quad (5)$$

$\overrightarrow{e} = (e_1, \ldots e_m, \eta_1, \ldots, \eta_n) = (e_1, \ldots, e_m, e_{m+1}, \ldots, e_{m+n})$, $\eta_i \in F_q$ is the coding coefficient. Therefore, authentication value of the new coding vector is obtained as follows:

$$c(\overrightarrow{e}) = \overrightarrow{e} \cdot \overrightarrow{u}^T + \sum_{k=m+1}^{m+n} e_k \cdot r = \left(\sum_{i=1}^{n} \eta_i \cdot \overrightarrow{w_i} \cdot \overrightarrow{u}^T\right) + \left(\sum_{i=1}^{n} \eta_i \cdot r\right)$$
$$= \sum_{i=1}^{n} \eta_i \cdot c(\overrightarrow{w_i}) \quad (6)$$

$c(\overrightarrow{e})$ and $(r, u)_{\mathrm{pk}_j}$ together form authentication codes for coding vector $\overrightarrow{e}$:

$$\mathrm{MAC}(\overrightarrow{e}) = \{(r, u)_{\mathrm{pk}_j}, c(\overrightarrow{e})\} \quad (7)$$

The generated MAC will be attached to the corresponding coding vector and sent to the downside nodes together.

3.3 Intermediate Node

After receiving the source messages, the intermediate nodes obtain **r** and **u** according to their own private key. Then, they use the public pseudo-random sequences generator f and the seed u to generate $\overrightarrow{u} = (u_1, u_2, \ldots, u_{m+n}) \in F_q^{m+n}$ bit by bit. At last, they calculate the corresponding message authentication value:

$$c(\overrightarrow{e})' = \overrightarrow{e} \cdot \overrightarrow{u}^T + \sum_{k=m+1}^{m+n} e_k \cdot r \quad (8)$$

The intermediate nodes compare $c(\overrightarrow{e})'$ with $c(\overrightarrow{e})$. If verification is successful, they continue to code and generate new authentication codes, whereas they will discard the information and send the retransmission request.

3.4 Destination Node

When the destination node receives n linearly independent and integrity verified coding vector, it can restore the original message.

Decoding method for original message is as follows:

$$E_{n\times(m+n)} = \begin{bmatrix} e_{1,1} & e_{1,2} & \cdots & e_{1,m+n} \\ e_{2,1} & e_{2,2} & \cdots & e_{2,m+m} \\ \vdots & \vdots & \ddots & \vdots \\ e_{n,1} & e_{n,2} & \cdots & e_{n,m+n} \end{bmatrix} = (P_{n\times m}, Q_{n\times n}) \quad (9)$$

The matrix $P_{m\times n}$ is encoded vector matrix. The matrix $Q_{n\times n}$ is the coefficient matrix. Original message is $M = Q_{n\times n}^{-1} \cdot P_{n\times m}$.

4 Simulation Analyses

In order to illustrate the effectiveness of the algorithm, simulation experiment is done based on the network model. Simulation software NS2 is used. The simulation scenario is as follows: network layer uses the AODV protocol. 40 nodes are randomly and uniformly distributed in the area of the square. The node movement is in the range of 1,500 × 1,500 m. Data flow type is Constant Bit Rate (CBR). The packet size is 512 byte. The finite field size is 256.

In view of the efficiency of this scheme for detecting pollution coding vector, the simulation parameter is as follows: malicious nodes are, respectively, set to 0–5. The number of source node is 5. Data rates of legitimate nodes and malicious nodes are 10 per second. The simulation results are as shown in Fig. 1: with the increasing malicious nodes, network throughput decreases obviously in the network coding scheme without using any pollution resistance. When malicious nodes are to five, network throughput is only around 23 %. This greatly weakens the original intention of network coding. However, in the network coding scheme based on MACs, network throughput keeps above 97 % with the increasing malicious nodes. Therefore, the latter can better resist pollution attacks.

Computation complexity for verifying code vector can be embodied in the end-to-end delay of message. The simulation parameter is as follows: data rate is, respectively, 10, 20, 30, 40, 50, 60, and 70 (number/s). The simulation results are as shown in Fig. 2: the end-to-end delay increases significantly in both literature [5] and the proposed scheme. The validation process of the proposed scheme mainly depends on the vector inner product operation, while the computational complexity of literature [5] mainly depends on modular exponentiation computation. Therefore, under the same data rate, the end-to-end delay performance in the proposed solution is better.

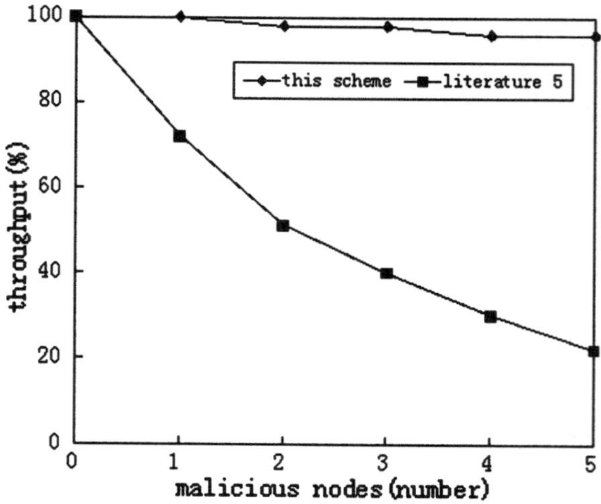

Fig. 1 Comparing throughput

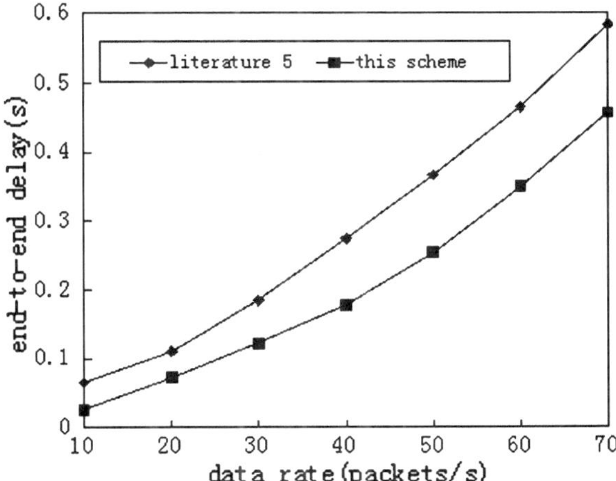

Fig. 2 Comparing end-to-end delay

5 Conclusion

Network coding has broken traditional information transmission method. For ad hoc network, designing secure network coding to meet its characteristic requirements is very meaningful. Focused on pollution resistance, this paper proposed a detecting pollution attack scheme based on MACs, which has high detection efficiency and low computational complexity. The future direction of the research is aiming at how to design a scheme which can defense proactive attack and not reduce the performance of the system. This will present a bigger challenge for the network coding in the safety aspects of the research.

Acknowledgments The National Natural Science Foundation of China (61151002, 60939002); The Fundamental Research Funds for the Central Universities (N110404033).

References

1. Ahlswede, R., Cai, N., Li, S.Y., Yeung, R.W.: Network information flow. IEEE Trans. Inf. Theory **46**(4), 1204–1216 (2000)
2. Deb, S., Effors, M., Ho, T.: Network coding for wireless applications: a brief tutorial. In Proceeding of International Workshop on Wireless Ad-hoc Networks. IWWAN, London (2005)
3. Siavoshani, M.J., Fragouli, C., Diggavi, S.: On locating byzantine attackers. In Network Coding Workshop: Theory and Applications (2008)
4. Ho, T., Leong, B., Koetter, R., et al.: Byzantine modification detection in multicast networks with random network coding. IEEE Trans. Inf. Theory **54**(6), 2798–2803 (2008)
5. Krohn, M.N., Freedman, M.J., Mazieres, D.: On-the-fly verification of rateless erasure codes for efficient content distribution. IEEE, pp. 226–240 (2004)

6. Zhou, Y., Li, H., Ma, J.: Secure network coding against the contamination and eavesdropping adversaries. http://arxiv.org/pdf/0805. 2286 (2008)
7. Wang, X., Hao, Z.: Identity-based without a trusted center key management scheme in ad hoc network. Comput. Secur. **6**, 13–15 (2010)

A Hybrid MAC Protocol with QOS Guarantee in Ad hoc Network

Jun Liu, Zhen Wang, Yan Huo and Yu Wang

Abstract A hybrid MAC protocol that has QOS guarantee in Ad hoc network is proposed. First, this protocol takes volume and the delay requirement of service for the parameter and forms business priorities of nodes. After that, the nodes dynamically form the priority table of time slots according to business priorities of nodes. Nodes compete for time slots according to the dynamical priority table of time slots. The simulation results show that this hybrid MAC protocol can assure the QOS in Ad hoc network.

Keywords QOS · Hybrid MAC protocol · Priority of time slots · Dynamical

1 Introduction

In the traditional Ad hoc network, MAC protocol is divided into competitive type of MAC protocols and distributive type of MAC protocols and hybrid MAC protocols. Among them, competitive MAC protocol has a good performance in the network with light load. However, with the increase in network load, more and more serious conflicts happen, which leads to high packet loss rate and lower

J. Liu (✉) · Z. Wang
College of Information Science and Engineering, Northeastern University, Shenyang, China
e-mail: liujun@ise.neu.edu.cn

Z. Wang
e-mail: 990956700@qq.com

Y. Huo
People's Liberation Army 95923 Unit, Beijing 100101, China
e-mail: Huoyan2012@126.com

Y. Wang
Air Force Engineering University of PLA, Xi'an, China

S. Patnaik and X. Li (eds.), *Proceedings of International Conference on Computer Science and Information Technology*, Advances in Intelligent Systems and Computing 255, DOI: 10.1007/978-81-322-1759-6_33,

network throughput; distributive type of MAC protocol has good performance when network load is high, but when the network load is light, this type of MAC protocol will lead to the waste of resources [1]. In order to apply to the dynamic change in network load, many hybrid MAC protocols which are based on the distributive type of MAC protocol have been put forward. A hybrid MAC protocol uses fixed allocation of time slot when network load is high. However, competitive mechanism is added when network load is low. For example, ADAPT protocol [2] and FPRP protocol [3] are firstly put forward. Then, a hybrid MAC protocol gradually becomes worldwide research focus. In the aspect of improving network throughput, a lot of hybrid MAC protocols have been put forward. For example, literature [4] uses synchronous RTS/CTS method to solve the hidden terminal and exposed terminal problems, which improves the network throughput and literature [5–7] use the method of dividing a frame into time slots to avoid conflicts, which improve the network throughput. But they are not involved with QOS guarantee and with the improvement of multimedia services; the requirements of the network performance change is more and more important. Therefore, a hybrid MAC protocol with QOS guarantee in Ad hoc network is put forward.

2 Hybrid MAC Protocol with QOS Guarantee in Ad Hoc Network

2.1 Structure of Frame

Frame of the protocol includes the statement stage, packing and forwarding stage, the broadcast stage, and the data transfer phase. The frame structure is shown in Fig 1. Each node respectively has a fixed slot in the three stages. In the statement stage, nodes in their respective slot send a statement grouping. Through this statement, the nodes will know their one hop and active neighbors. In the package forwarding stage, nodes will pack the data received in the broadcast stage and broadcast the data. After packing forwarding stage, nodes will know their all active nodes within two hops and know their competitive time slot and delay requirement of service. After the exchange of information, management node will determine the priority table upon the receipt information of volume and the delay requirement of service. In the hierarchical structure of the network, management node can be served by the cluster head nodes. In the broadcast phase, the priority table of

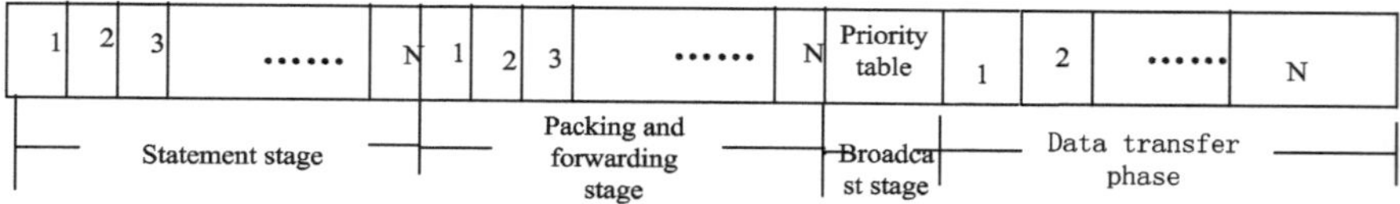

Fig. 1 Structure of Frame

time slots will be broadcast. In the data transfer phase, after active nodes compete for the time slots according to the priority table, data will be transferred in the corresponding time slots.

We analyze the protocol from the statement stage, packing forwarding stage, broadcast stage, and data transfer phase.

2.2 *Process of Protocol*

2.2.1 Statement Stage

The statement stage is divided into *N* small time slots (*N* is the number of nodes); active nodes will send corresponding statement grouping in their own time slots. Nodes that are not active are in the listening state. Statement grouping is shown in Fig 2.

Each packet includes a guard time before and after the statement, the source node's ID, the service type flag, and traffic information. Among them, the guard time is used to solve the problem that is not synchronized caused by time drifts, service type and traffic information are in the form of the quantitative data.

2.2.2 Packing Forwarding Stage

After statement stage, the nodes know one-hop node's information. For data transmission, the data will conflict only within the two-hop nodes. So, we need to know the two-hop node's information, and then, nodes will compete for the data slots. The role of packing forwarding phase is to pack the information gathered in the statement stage and then broadcast them in their corresponding slot. After this stage, each node will receive the active nodes' information within two hops.

2.2.3 Broadcast Stage

During the broadcast stage, the management nodes first aggregate the information collected in the statement stage and in the packing forwarding stage. Then, service type and traffic information will be weighted to obtain the business priorities of nodes. After that, management nodes form the priority table of time slots according to the business priorities of nodes and broadcast this table to each node.

Guard time	Source node ID	Service type flage	Traffic information	Guard time

F

Fig. 2 Statement grouping

2.2.4 Data Transfer Phase

During the data phase, the active nodes transfer the data in their corresponding time slots.

2.3 Formation of Priority Table

2.3.1 Formation of Business Priorities Table of Node

Here, we make the size of the business's volume and the type of business for data integration, assuming that the volume of business is "bi" and the business type of the node is "u_i". If the delay requirement is higher, the "u_i" will be smaller, and the priority of the node will be higher. If the volume of business is larger, the "bi" will be smaller. The principles of the prioritization are as follows:

- If delay requirement of business is higher, the priority is higher;
- If nodes' business has the same delay requirements, the node that has greater volume has a higher priority than other nodes.

Table 1 shows priority table based on the business types/the amount of business volume. The arrows in the figure indicate the task priority order, and along the direction of the arrow, the priority gets lower. For the business type, we use the descending method—if the business type is smaller, the priority is higher. For the volume of business, we use the ascending method—if the volume of business is greater, the priority is higher.

As is shown in Fig 1, each task has a priority class in the priority table:

$$p = i + j \tag{1}$$

Here, "i" and "j" denote the position of the type and the volume of business in the sequence. In Table 1, the tasks on each diagonal line have the same priority level P. Although they have the same priority level of the task, the order must be in accordance with the direction of the arrow. That is to say, for business with the

Table 1 Priority table of time slots

Nodes	Time slot				
	1	2	3	4	5
1	5	2	3	1	4
2	1	5	2	4	3
3	3	4	5	2	1
4	4	3	1	5	2
5	2	1	4	3	5

same priority level, the priority inclines to business with higher delay requirement. The task priority "p" can be calculated according to the following formula.

$$p = (i + j - 1) \times (i + j - 2) + i \tag{2}$$

The "p" is smaller, the priority of the task is higher. In Table 1, digital superscript says node number. In this way, each node business priority has been determined.

2.3.2 Creation of the Priority of Time Slots

After establishing the priority table of nodes, the management nodes generate the priority of time slots according to the priority of nodes. The process of the creation should follow certain rules:

- The master node with the highest priority slot;
- The nodes that have higher priority of nodes have the higher priority of time slots;
- The cross-column of the priority table of time slots cannot have the same number;
- In the process of forming the priority table of time slots, if there is a conflict, the node order of priority will exchange with the previous node that will not conflict with each other.

Table 1 shows the formed priority table of time slots according to Fig. 3.

2.4 Competitive Process of Slots

After the broadcast stage, the node will depend on the information gathered in forwarding phase and the node priority table from the management node to allocate the time slots. First of all, the node depends on the collected information to determine the slot which it can compete with, including the non-active nodes

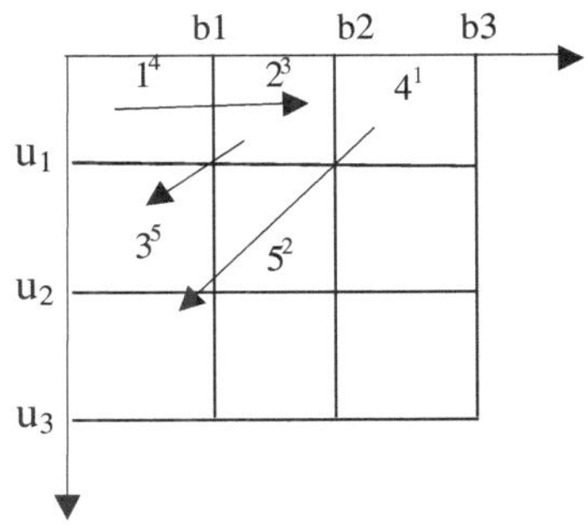

Fig. 3 Business priority table of nodes

Fig. 4 Network topology

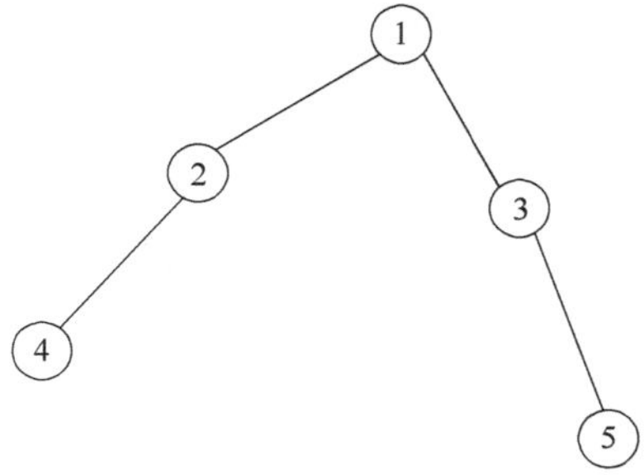

within two-hop and the main slot beyond two-hop nodes. Then, it will base on the slot priority table to allocate the time slots. In this process, the node will depend on the business volume to determine whether the node needs to compete for the extra slots. There, we set a threshold "L" of the business volume. If the business volume exceeds this threshold, it means that the node needs to compete for the extra slots. On the contrary, there is no need to compete for extra slots. If a node needs to compete for additional time slots, it can achieve additional time slots by competing with other nodes after searching the slot priority table. If the node does not need additional slots and it can compete for extra time slots, the node will give up competing for those extra time slots, and those time slots will be assigned to other nodes.

The network topology is shown in Fig 4, in which active nodes are 3, 4, and 5, and the three nodes' volume is very big. Business delay requirement of nodes 3 and 4 is higher than node 5.

Its current business priorities table of node and priority of time slots are as shown in Tables 1 and 2, respectively. The assignment of time slots is as shown in Table 2.

In Table 2, the time slot 2 and time slot 1 is, respectively, assigned to node 3 and node 4. Node 5 that has low delay requirement does not compete for time slot 1. Time slot 2 is assigned to node 5 because node 2 is not within two-hop of node 5. In the table, those front time slots are assigned to those nodes that have high latency requirement. In this case, it can meet the QOS requirement of a network in a certain extent.

Table 2 The allocation of time slot

Active nodes	Nodes that can be competed	Volume	Compete for the extra time slots	Results
3	1, 2, 4	Bigger than L	yes	2, 3, 4
4	1, 2, 5	Bigger than L	Yes	1, 3, 4, 5
5	1, 2, 4	Bigger than L	Yes	2, 4, 5

3 Simulation and Performance Analysis

Using OPNET simulation software, the simulation time is 600 s. The produce probability of each node in the simulation data follows the Poisson distribution. The topology has 16 nodes in the form of random distribution. In this protocol, the control slot duration is set to 2 ms. The length of information slot is 8 ms. The throughput rate, packet overhead, and the number of data packet transmission are analyzed.

In Fig 5, when the contract rate is small, the throughput rate of all protocols is high; for 802.11, the throughput rate decreases with the increasing contract rate due to the more and more conflict. For TDMA, because there is no conflict, the throughput cannot decrease with the increase in the contract rate. For hybrid MAC, when the contract rate is small, it can realize the slots reuse, so the throughput is higher than TDMA, but with the increase in the contract rate, the performance will be close to the TDMA.

In Fig 6, for 802.11, with the increase in the contract rate, each node will participate in the competition, which leads to more and more conflict. So, the cost

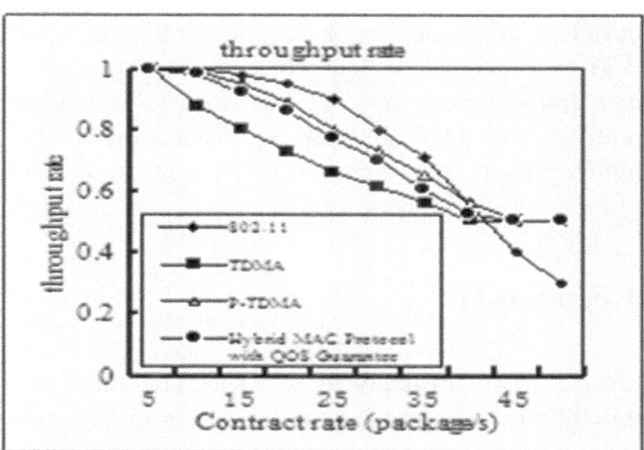

Fig. 5 Throughput rate

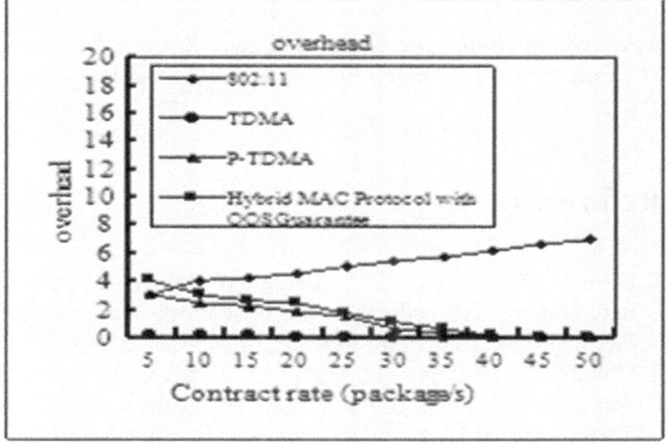

Fig. 6 Overhead

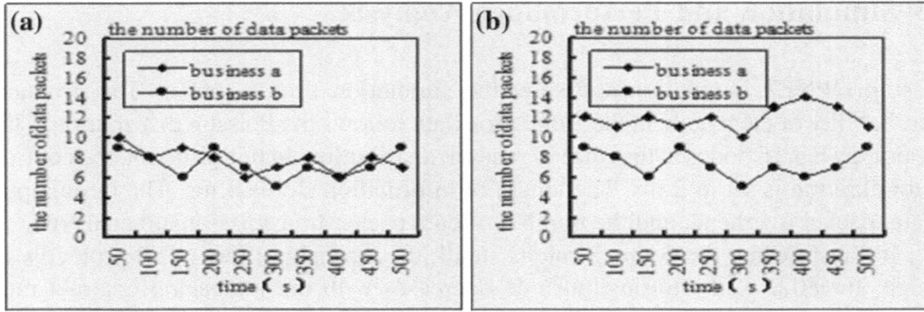

Fig. 7 The number of data packets. **a** P-TDMA. **b** Hybrid MAC protocol with QOS guarantee

will increase with the increasing contract rate. For TDMA, nodes do not need to compete with each other, so the cost is nearly zero. For hybrid MAC protocol, when the contract rate is small, nodes compete for time slots, and with the increasing contract rate, the performance will be close to the TDMA, so the overhead of hybrid MAC will decrease with the increase in the contract rate.

In Fig 7, business "a" stands for the business that has the delay requirement; business "b" stands for the business that has no delay requirement. In Fig 6a, for P-TDMA, because it does not consider the delay request, the average number of data transmission for different business is similar. In Fig 6b, for hybrid MAC protocol with QOS guarantees, business with higher latency requirement has the higher priority to be sent, so the average number of business "a" is more than "b".

4 Summary

Hybrid MAC protocol with QOS guarantee not only improves the network throughput by improving the utilization rate of slots and spatial reuse, but also can satisfy the multimedia business requirements which need time delay by dynamically adjusting the competing priorities of the node slot.

Acknowledgments The National Natural Science Foundation of China (61151002,60939002); The Fundamental Research Funds for the Central Universities (N110404033).

References

1. Herman, T., Tixeuil, S.: A distributed TDMA slot assignment algorithm for wireless sensor networks. In: Proceedings of the First Workshop on Algorithmic Aspects of Wireless Sensor Networks, vol. 6, Turku, Finland, pp. 45–58 (2004)
2. Chlamtac, I., Farago, A., Myers, A.D., Syrotiuk, V.R., Zaruba, G.: ADAPT: a dynamically self-adjusting media access control protocol for ad hoc networks. Paper Published in Global Telecommunications Conference, Rio de Janeireo, pp. 11–15, Dec (1999)

3. Zhu, C.X., Corson, M.S.: A Five Phase Reservation Protocol (FPRP) for Mobile Ad Hoc Networks. Paper Published in INFCOM'98. Seventeenth Annual Joint Conference of the IEEE Computer and Communications Societies. Proceedings IEEE, San Francisco, CA, pp. 322–331 (1998)
4. Ni, J., Tan, B.R., Srikant, R.: Q-CSMA: Queue-length based CSMA/CA algorithms for achieving maximum throughput and low delay in wireless networks. Paper Published in Networking, IEEE Transactions on, vol. 20. pp. 825–836, Aug (2012)
5. Gexin, P., Shengli, X.: A conflict-free fixed TDMA algorithm based on dynamic slot allocation. In China: Paper Published in Information Security and Communications Privacy, pp. 115–120, Nov (2005)
6. Jianyao, N., Yong, X.: An improved TDMA algorithm based on dynamic slot allocation used in Ad Hoc networks. In China: Paper Published in the Mobile Communications, vol. 10. pp. 83–86 (2008)
7. Bin, J., Shang, S.: Dynamic TDMA time slot allocation research based on hash algorithm. In China: Paper Published in Communication Technology, vol. 45. pp. 248–251 (2012)

Enhancing Source Location Privacy in Energy-Constrained Wireless Sensor Networks

Guangbao Tan, Wei Li and Jie Song

Abstract Research community has done much work to tackle the privacy preservation problem of wireless sensor networks in recent years. Many privacy-related issues have been addressed by security mechanism; however, location privacy problem has not been addressed effectively. The location privacy problem, which allows an adversary to determine the location of nodes of interest to him, is quite different from content privacy which could be protected by encryption and authentication. In this paper, an enhanced directed random walk (EDROW) technique is proposed. This approach introduces a ring, which makes an adversary to backtrack to the origin of the sensor communication more difficult, to enhance the directed random walk route. The results of simulation demonstrate that the enhanced directed random walk technique is a powerful approach for preserving source location privacy.

Keywords Source location · Wireless sensor network · Directed random walk · Ring · Ring nodes · Privacy

1 Introduction

A wireless sensor network consists of a large number of spatially distributed, energy-constrained sensor nodes and one or more powerful sink nodes. Wireless sensor networks have a great potential to be widely used in both military and civilian applications, such as battlefield reconnaissance, environmental monitoring, and smart home monitoring.

However, wireless communication media is a broadcast media; anybody with a proper wireless receiver can intercept wireless sensor network communications.

G. Tan (✉) · W. Li · J. Song
Department of Computer Science and Technology, Anhui University, Hefei, China
e-mail: itgb1989@gmail.com

S. Patnaik and X. Li (eds.), *Proceedings of International Conference on Computer Science and Information Technology*, Advances in Intelligent Systems and Computing 255, DOI: 10.1007/978-81-322-1759-6_34,

How to preserve privacy in wireless sensor networks has become one of the major issues which may jeopardize the successful deployment of wireless sensor networks. The privacy threats that exist for wireless sensor networks may be classified into two broad classes: content-oriented privacy threats and contextual privacy threats [1]. Content-oriented privacy threats are any means by which an adversary can determine the meaning of the exact content of packets being sent over the wireless sensor network. Contextual privacy threats are any methods by which an adversary can determine the location of a communicating entity which is a critical entity of the wireless sensor network.

Source location privacy is an important security issue. In many scenarios, the location of the message initiator is sensitive and needs to be protected.

To preserve source location privacy in energy-constrained wireless sensor networks, we propose a three-phase routing process. In the first routing phase, the message is sent from the source node and forwarded by normal sensor nodes in directed random fashion until it reaches a ring. Then, the message is forwarded on the ring for random hops until it reaches a specified ring node. Lastly, the ring node forwards the message to the sink with directed random walk (DROW).

2 Related Work

The location privacy preservation problem in wireless sensor networks has drawn the attention of researchers in recent years. Many approaches have previously been designed to prevent an adversary from tracking the location of source.

Ozturk et al. [1] first introduced the Panda-Hunter model to formalize the source location privacy problem in wireless sensor networks. They also proposed the phantom routing technique which is based on both single-path routing and flooding [1–3]. The phantom routing scheme is comprised of two phases. The first phase is a walking phase which may be a sector-based or a hop-based directed random walk [2]. The second phase, which meant to deliver the message to the sink, is a flooding or single-path routing stage [2]. The phantom routing scheme has received a lot of attention in research community. Some researchers consider it works well while against local adversaries [4, 5], while others find weakness in it [6–8]. Much work has been done to review and improve the scheme [5, 9]. Recently, some opportunistic routing schemes have been proposed, such as in [10, 11].

The idea of delivering packets by directed random walk has been proposed in [2, 3, 12]. Yao and Wen [12] proposed a scheme called DROW. It addressed a problem that random walk is inefficient in making a fake phantom source far away from the actual source and improved the safety period of source by trying to create enough paths from source to sink. However, some sensor nodes in the particular regions have no enough paths to sink; they could be traced by an adversary easily for the starkness of pronounced paths. To address this problem, we introduce a ring, called a mule ring, which will not only create enough paths for the vulnerable sensor nodes, but also mislead adversaries to the incorrect direction. The rest of this paper is organized as follows. In Sect. 3, the system and the adversary models are presented. Routing scheme is described in Sect. 4. Subsequently in Sect. 5, performance analysis and experiment results are presented. Lastly, conclusion in Sect. 6.

3 Models and Design Goals

3.1 The System Model

Our system is similar to the explanatory Panda-Hunter Game that was introduced in [1]. We make the following assumptions about the wireless sensor network:

- A sink and a number of sensor nodes are in the network. The network is evenly divided into small grids. Sensor nodes are uniformly distributed on grids. The whole network is fully connected through multi-hop communications.
- There is only one sink node. The information of the sink node is in the open, and the sink node is the destination location that all data packets will be delivered to.
- The content of each packet will be encrypted. However, the encryption operation is beyond the scope of this paper. The interested readers can get detailed information of encryption operation from such treatise as [13, 14].

3.2 The Adversaries Model

For the capabilities of an adversary, we assume that an adversary has the following characteristics [12]:

- The adversary only executes passive attacks which only involve eavesdropping work. It will not interfere with the proper functioning of the network.
- The adversary has unlimited energy resource. It can determine the location of immediate sender and then move to the sender without much delay.
- The adversary has a hearing radius which is equal to the sensor node transmission radius generally.

3.3 Overview of the Proposed Scheme

In our scheme, every data packet will experience three phases: a random walk, which is a directed random walk route, was proposed in [12] and a subsequent ring route to deliver the packet to a specified ring node after the packet is forwarded by ring nodes for random times and lastly the directed random walk route to deliver the packet to the sink.

After the routing setup scheme, every node setting up its multiple parent nodes and multiple child nodes and a ring is generated. The ring, called a mule ring, is comprised of a number of ring nodes. The ring is sink-centric. The ring nodes are further divided into relay ring nodes and normal ring nodes. The relay ring nodes and normal ring nodes are cross-arranged on the ring. Data packets will be transmitted on the ring in a given direction, called ring direction, which may be clockwise or anticlockwise direction. Only relay ring nodes can deliver data packets to the sink directly; normal ring nodes just can forward the data packets to their successors which are the relay ring nodes. Data communication between adjacent relay ring nodes is done with the help of the normal ring node, between the two relay ring nodes, which relays data for the two relay ring nodes, and vice versa. When a normal node, which is out of the ring, has a data packet to transmit, it just sends the data packet to the sink by DROW route scheme. Before the data packet reaches the sink, it will reach a ring node. The ring node firstly chooses a direction as the ring direction, and then it selects a random number H_{relay} which specifies the data packet that will be transmitted by H_{relay} relay ring nodes on the ring before the data packet leaves the ring for the sink; lastly, it forwards the data packet to its successor. The data packet will be transmitted on the ring until it reaches the specified relay ring node. The specified relay ring node will forward the data packet to the sink by DROW route scheme. For a normal node in the ring, the only difference from a node out of the ring is that it will select one child node from multiple child nodes instead of multiple parent nodes to forward a data packet. The data packet will firstly be delivered to the ring before to the sink likewise.

An example is given in Fig. 1, where S indicates a source node in the network, R_1, R_2, R_{k-1}, R_k and R_n are five relay ring nodes, and SINK is the sink node. Normal ring nodes between relay ring nodes are not presented for simplicity. S sends a data packet to R_1 in directed random fashion, and then, R_1 selects an anticlockwise direction as the ring direction, a random number $H_{\mathrm{relay}} = k$, and forwards the data packet on the ring to R_2 in ring route fashion; lastly, R_k delivers the data packet to SINK in directed random way.

The detailed prescription of the proposed three-phased routing will be described in the subsequent sections.

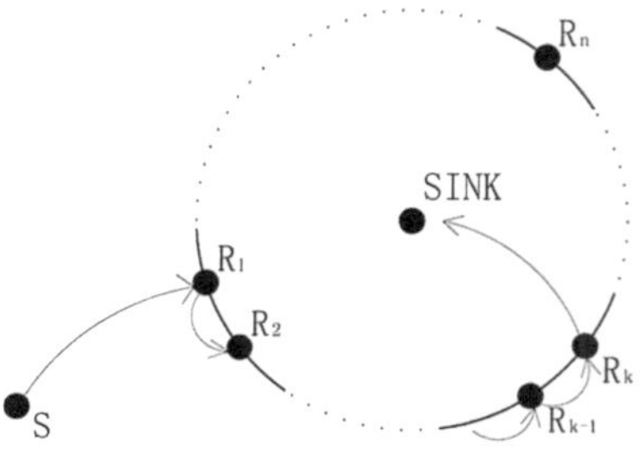

Fig. 1 Illustration of three-phase routing

4 Proposed Source Location Privacy Scheme

For the purpose of simplifying the discussion and analysis of source location privacy in wireless sensor networks, we select the DROW. It got a decent source location privacy protection to some extent. However, it has been observed that the performance of DROW routing varies in relatively different locations between source and sink. Some relatively vulnerable locations exist in the sensing filed. In order to solve the problem, we proposed a novel scheme which combines DROW with a mule ring.

4.1 Directed Random Walk

If the source-sink route is fixed, an adversary can easily reach the source by tracing back the route [9]. In order to introduce non-determinacy to the source-sink route, the DROW routing technique was proposed in [12] to enhance source location privacy.

In DROW, when a source node sends a packet, the packet is forwarded, based on unicast, to one of its parent nodes which is randomly selected. The intermediate node forwards the received packet to one of its parent nodes randomly likewise. The core idea behind DROW routing is creating a larger number of redundant paths from source to sink and attracting the adversary toward a path, which may be no packet transmitted on it in some future time, to retard the progress of tracing back the route.

How to set up multiple parents for each node is not detailed in this paper; the interested readers can refer Ref. [12]. As for multiple children, the only difference from setting up multiple parents is that a sensor node selects all neighbor nodes whose level value is greater than its level value as its child nodes.

4.2 DROW Combines with the Mule Ring

If the number of hops between source and sink is $h = n$ hops, the number of paths from the source to the sink is between 1 and $C_n^{\lceil n/2 \rceil}$ in DROW. $C_n^{\lceil n/2 \rceil}$ is defined as follows:

$$C_n^{\lceil n/2 \rceil} = \binom{n}{\lceil n/2 \rceil} = \frac{n(n-1)\cdots(n-\lceil n/2 \rceil+1)}{(\lceil n/2 \rceil)!} \quad (1)$$

Obviously, some source nodes have no enough paths, from themselves to sink, to deter the adversary to backtrack them. Besides, data packets that originate from source always be transmitted in some relatively stationary directions, which can leak direction information of source and make the adversary backtrack source more easily.

To avoid the weakness of DROW, we propose a novel scheme which modifies DROW by introducing a mule ring. We call this scheme enhanced directed random walk (EDROW). We construct a ring whose center is the sink and radius is r hops. The ring consists of N_{ring} sensor nodes. The ring enhances the privacy greatly and introduces limited overhead and time delay. On average, the ring will increase the hop count per message about $2 \times r$. In order to confuse the adversaries effectively, it should be selected carefully.

When a ring node receives a data packet that is not transmitted from a ring node, in other words, the data packet does not come from the precursor of the ring node but from a normal node, the ring node will do two things before forwarding the data packet: firstly, selects a direction which specifies the forward direction of the data packet on the ring; secondly, generates a random number H_{relay} which points out the data packet to be forwarded by how many relay ring nodes before leaving the ring for the sink. H_{relay} When a normal ring node receives a data packet from its precursor which is a relay ring node, it will deliver the data packet to its successor which is a relay ring node. When a relay ring node receives a data packet from its precursor which is a normal ring node, it will check the value of H_{relay}; if the value is zero, it will deliver the data packet to the sink; otherwise, it will set the value of H_{relay} to H_{relay} minus one and then deliver the data packet to its successor which is a normal ring node.

For maximizing paths randomly, we adopt a strategy that a ring node selects the ring direction which is opposite to the direction which it selected last time. In this way, the adversary always cannot easily reach the ring, for the strategy always makes its previous progress invalid by attracting it to a new location which may be far away from the ring compared to the current location to the ring.

To avoid leaking direction information of source, we will select a proper H_{relay} to make the paths, from source to sink, more dispersive and random. We select H_{relay} from a set in which upper and lower bounds are one and a quarter of N_{ring}, respectively.

In our scheme, the network would be evenly divided into small grids as shown in Fig. 2a. In order to simplify the analysis of the path diversity from source to sink and the ring routing in the wireless sensor network, we make the mule ring to be a standard rectangle as shown in Fig. 2a. The intersection of two diagonals of the rectangle is the location of the sink. The hop count between the sink and every relay ring node is r, and that of every normal ring node is $r + 1$, such as $r = 5$ as shown in Fig. 2a. Since only relay ring nodes can forward data packets to sink directly; we can further simplify the model, which is presented in Fig. 2a, by omitting normal ring nodes; and then get an easier analytical model of path diversity which is presented in Fig. 2b.

So, in our scheme, if the number of hops between source and sink is $h = n$ hops and the radius size of the ring is r hops, there are $r + 1$ relative locations between source and sink. Figure 3a–f provides the possible source locations for $h = 10$ hops and $r = 5$ hops. The average number of paths from the source to the sink for every relatively location is given by

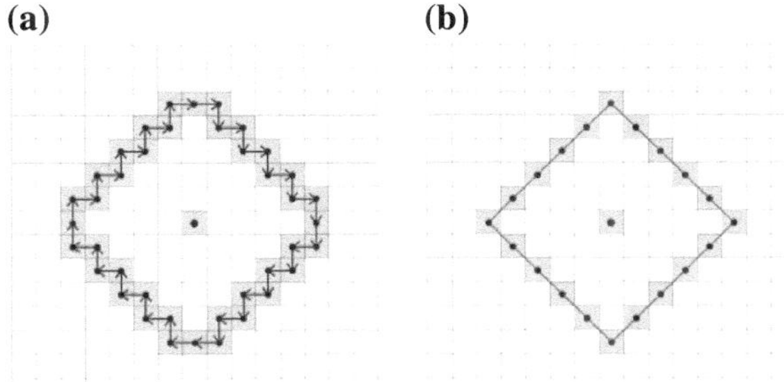

Fig. 2 Illustration of the mule ring model. **a** Illustration of ring routing. **b** Simplification of ring model

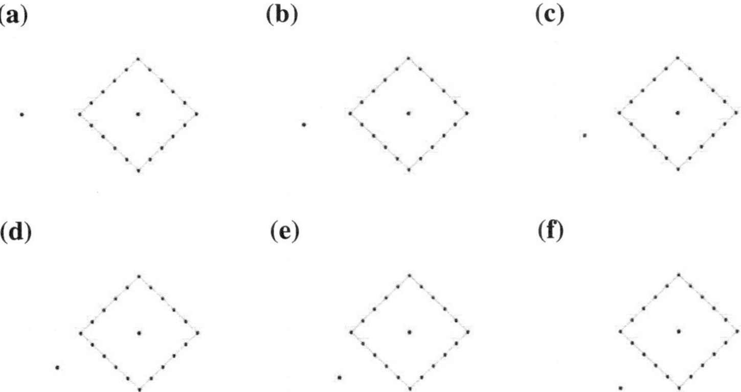

Fig. 3 Relative location between source and sink ($h = 10$ hops, $r = 5$ hops)

$$N_i = \begin{cases} \left(2\sum_{p=0}^{\frac{r}{2}-1} C_r^p + C_r^{\frac{r}{2}} - 1\right) \Big/ r + \sum_{q=0}^{i} C_{n-r}^q & 0 \le i \le r,\ r\ is\ \text{even.} \\ \left(2\sum_{p=0}^{\lceil\frac{r}{2}\rceil-1} C_r^p - 1\right) \Big/ r + \sum_{q=0}^{i} C_{n-r}^q & 0 \le i \le r,\ r\ is\ \text{odd.} \end{cases} \tag{2}$$

In Eq. (2), N_i is the average number of paths, from source to sink, of the ith relative location. It consists of two parts: paths from source to the ring and paths from the ring to the sink. Figure 4 provides the average number of paths about six relative locations between source and sink when the distance between source and sink is $h = 10$ hops and ring radius is $r = 5$ hops. We improve the number of paths, from source to sink, in the vulnerable case in DROW effectively.

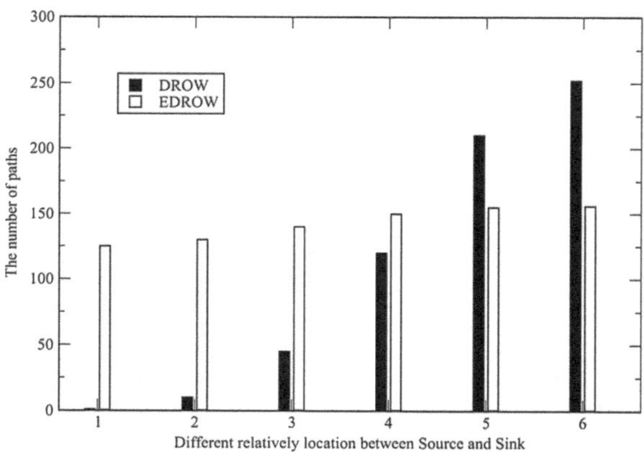

Fig. 4 The paths for different relative location

5 Performance Analysis and Experiment Results

The EDROW protocol was evaluated through a simulation of a wireless sensor network target tracking application which uses a discrete event simulation tool called OMNET++. Each simulation of the application consisted of 10,000 sensor nodes which were uniformly distributed on a 100×100 grid. Originally, the adversary stays near the sink. It will move to the new location when it detects a new packet transmitted from the location.

5.1 Performance of EDROW

The safety period is the time that an adversary needs cost to reach the source, which is measured by the number of new messages that the source has sent before the adversary reaches the source.

The performance of EDROW and DROW is compared in Fig. 5 when $h = 10$. To avoid introducing overmuch energy consumption, we select, on the premise that the approach can provide sufficient protection for the source, a proper radius size of the ring which is $r = 5$ hops. The detailed influences caused by radius size of the ring will be described in the next section. In Fig. 5a, average number of hops of per message in EDROW is greater than that of DROW, which is caused by the ring routing. Figure 5b shows a performance improvement which is achieved by employing the ring routing. Obviously, we get a great improvement, more than twice safety period compared to that of DROW, by introducing limited energy overload.

The scrupulous readers may find an interesting phenomenon, the safety period increased with the number of paths, from source to sink, in DROW, however, inversely in EDROW. The difficulty of the adversary tracking the source is not only influenced by the number of paths, but also influenced by the dispersion

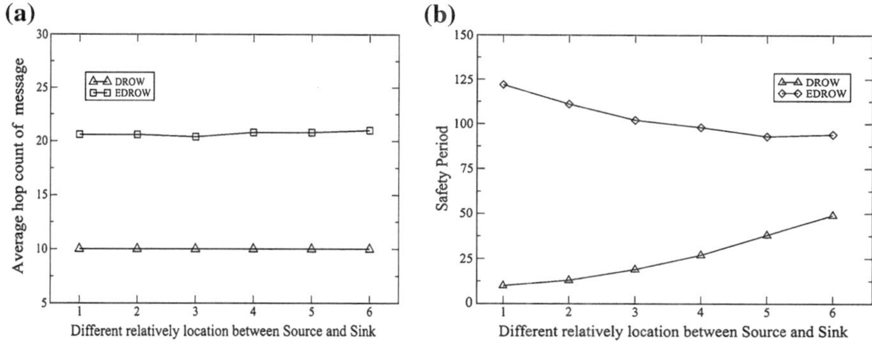

Fig. 5 Performance evaluation ($h = 10$, $r = 5$). **a** Average hop count. **b** Safety period

degree of the paths. Paths joint to each other, which makes the adversary backtrack the source easily. For the size of the set of disjoint paths in EDROW more than that of DROW, the former always gets a better performance than the latter. In Fig. 5b, the more paths produced, the worse EDROW performs, because more and more paths joint to each other. We conclude that the greater the dispersion degree of the paths, the greater the safety period.

5.2 The Ring Radius

In this section, we will analyze the safety in EDROW for different radius size of the ring. The distance from source to sink is $h = 20$ hops, and radius size of the ring changes from 5 to 9 hops in the simulation. For EDROW can provide sufficient protection for the source when ring radius is $r = 9$ hops, there is no need to enlarge the ring radius any more.

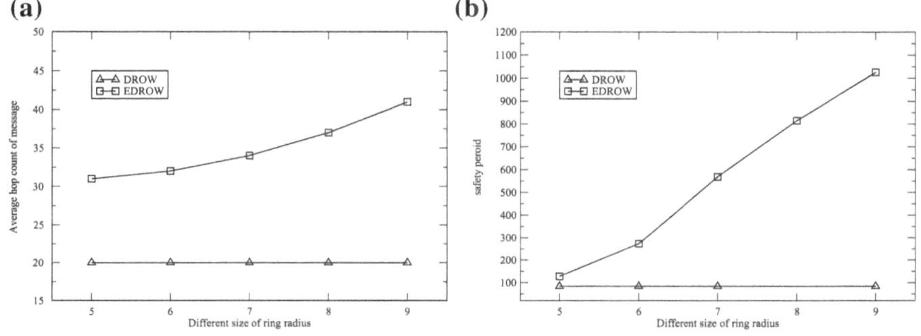

Fig. 6 Influence of ring radius size ($h = 20$). **a** Average hop count. **b** Safety period

In Fig. 6a, the average number of hops per message enlarged with the ring radius. Figure 6b shows that the safety period increased with the ring radius. From Fig. 6, we can see that the safety period grew faster than the ring radius.

Although EDROW introduced a little more time delay and overhead, just related with ring radius r, compared with DROW in which each message is forwarded to the sink along the shortest path, it got a great improvement of the safety period. The average hop count per message in EDROW is about $2 \times r$ greater than that of DROW. However, with expansion of scale of sensor network, the delay and overhead caused by the ring routing have less and less effect on performance of the routing scheme. Besides, we can tune the value of the ring radius r to balance the trade-off between energy consumption and privacy protection. Generally, we should select a ring radius r which is less than 10 hops.

6 Conclusion

In wireless sensor networks, source location privacy is important for target tracking application. An adversary may be traced back to the source location hop by hop with the help of some devices. The performance of DROW routing is not satisfying in some relative locations between source and sink. In this paper, an enhanced directed random walk method, called EDROW, was proposed to solve this problem. From the results of the simulation evaluation, this approach can get a sufficient source location privacy protection effectively.

In future, we will make further efforts to improve the dispersion degree of the paths and reduce overload in EDROW.

References

1. Ozturk, C., Zhang, Y., Trappe, W.: Source-location privacy in energy-constrained sensor network routing. In: Proceedings of the 2nd ACM Workshop on Security of ad hoc and Sensor Networks, pp. 88–93 (2004)
2. Kamat, P., Zhang, Y., Trappe, W., Ozturk, C.: Enhancing source-location privacy in sensor network routing. In: Proceedings of 25th IEEE International Conference on Distributed Computing Systems (ICDCS 2005), pp. 599–608 (2005)
3. Ozturk, C., Zhang, Y., Trappe, W., Ott, M.: Source-location privacy for networks of energy-constrained sensors. In: Proceedings of Second IEEE Workshop on Software Technologies for Future Embedded and Ubiquitous Systems, pp. 68–72 (2004)
4. Mehta, K., Liu, D., Wright, M.: Location privacy in sensor networks against a global eavesdropper. In: IEEE International Conference on Network Protocols (ICNP 2007), pp. 314–323 (2007)
5. Wei-ping, W., Liang, C., Jian-xin, W.: A source-location privacy protocol in WSN based on locational angle. In: IEEE International Conference on Communications (ICC'08), pp. 1630–1634 (2008)

6. Suarez-Tangil, G., Palomar, E., Ramos, B., Ribagorda, A.: An experimental comparison of source location privacy methods for power optimization in WSNs. In: Proceedings of the 3rd WSEAS International Conference on Advances in Sensors, Signals and Materials, pp. 79–84 (2010)
7. Lightfoot, L., Li, Y., Ren, J.: Preserving source-location privacy in wireless sensor network using STaR routing. In: IEEE Global Telecommunications Conference (GLOBECOM 2010), pp. 1–5 (2010)
8. Wang, H., Sheng, B., Li, Q.: Privacy-aware routing in sensor networks. Comput. Netw. **53**, 1512–1529 (2009)
9. Zhang, L.: A self-adjusting directed random walk approach for enhancing source-location privacy in sensor network routing. In: Proceedings of the 2006 International Conference on Wireless Communications and Mobile Computing, pp. 33–38 (2006)
10. Spachos, P., Song, L., Bui, F.M., Hatzinakos, D.: Improving source-location privacy through opportunistic routing in wireless sensor networks. In: 2011 IEEE Symposium on Computers and Communications (ISCC), pp. 815–820 (2011)
11. Li, N., Raj, M., Liu, D., Wright, M., Das, S.K.: Using data mules to preserve source location privacy in wireless sensor networks. In: Distributed Computing and Networking, pp. 309–324. Springer (2012)
12. Yao, J., Wen, G.: Preserving source-location privacy in energy-constrained wireless sensor networks. In: 28th International Conference on Distributed Computing Systems Workshops (ICDCS'08), pp. 412–416 (2008)
13. Traynor, P., Kumar, R., Choi, H., Cao, G., Zhu, S., La Porta, T.: Efficient hybrid security mechanisms for heterogeneous sensor networks. IEEE Trans. Mobile Comput. **6**, 663–677 (2007)
14. Chan, H., Perrig, A.: PIKE: Peer intermediaries for key establishment in sensor networks. In: INFOCOM 2005. Proceedings of 24th Annual Joint Conference of the IEEE Computer and Communications Societies, pp. 524–535 (2005)

Research on Resource Discovery Method for Networks with Unstructured Peer Model

Bingbing Xue, Weihua Yu and Hua Huo

Abstract Unstructured peer model is widely used in many fields, but the common blind search method, such as flooding, generates lots of redundant messages. The limited bandwidth can easily cause network congestion. For the existing problems in traditional discovery method, a resource discovery method based on peer interest similarity is put forward. Based on grouping, similar peers are divided into domain on the basis of peer interest similarity. Query message is forwarded in the same group of the present domain at first, and network topology is adjusted dynamically during searching. Then, route hops are reduced and the amount of redundant messages is decreased effectively. Simulation experiment analyzes and confirms the searching efficiency of the method.

Keywords Unstructured peer model · Resource discovery method of network · Interest similarity · Index

1 Introduction

Peer-to-peer (P2P) network mainly uses decentralized topology structure, directly share resources among various nodes and no longer distinguish server and client. So it can make full use of the processing power and resources contained in the network node. In P2P networks, the dissemination of information is more quick

B. Xue (✉) · W. Yu · H. Huo
Electronic Information Engineering School, Henan University of Science and Technology, Luoyang, China
e-mail: xhyxbb@163.com

W. Yu
e-mail: haustywh@163.com

H. Huo
e-mail: hhuo@mail.xjtu.edu.cn

S. Patnaik and X. Li (eds.), *Proceedings of International Conference on Computer Science and Information Technology*, Advances in Intelligent Systems and Computing 255, DOI: 10.1007/978-81-322-1759-6_35,

and the network bandwidth utilization is optimized, and the positioning of the resources is a precondition to network information sharing among users. Therefore, the resource discovery mechanism of overlay network is the key point of P2P technology research. It becomes pivotal to improve the network scalability and to solve the network bandwidth.

According to the resource discovery mechanism, P2P overlay network can be roughly divided into hybrid, structured and unstructured P2P network. Hybrid network system still cannot get rid of the characteristics of centrality, Napster is typically representative, resources are maintained through a few service nodes, and limitations exist such as network bottlenecks and single point of failure. Structured P2P mainly use distributed discovery strategy based on distributed hash table (DHT) [1], and typical systems are Chord, CAN, Tapestry, Pastry, etc. This discovery mechanism has high discovery efficiency to specific resource, but needs dealing with dynamic characteristics that network nodes change frequently by paying greater price. It suits smaller network applications. Because distributed and unstructured P2P network, such as Gnutella [2], has no clear control in the topological structure and the dynamic changes of the node will not affect the network structure, it suits large P2P network applications. However how to make points' physical location consist with the upper overlay network [3], and reduce search redundancy [4] become the hot spots.

2 Problem Analysis

In distributed unstructured P2P network, the nodes and resources are randomly distributed; the resource requester is difficult to position nodes with resources accurately and can only forward and spread news through the neighbor nodes, until to the node with resources. The most common unstructured P2P resource discovery method is flooding [5], random walks [6], etc. These methods who using blind news forwarding mechanism can locate resources quickly and have better tolerant ability. But the node may receive the same resource request repeatedly and massive query data traffic is produced with the increase of network size.

Gnutella is the typical unstructured P2P network. It uses flooding search method. When node requests resources, it sends the request to neighbor nodes of the request nodes. If the neighbor nodes have nodes with resources, they response source node. Otherwise, all the neighbor nodes will continue to forward news request to their respective neighbor nodes, until they find the goal node or news reached the maximum set range by jogging. So the flooding is also called broadcast search, as shown in Fig. 1. Node F and node G both have the resources requested of node A. This search system is simple to realize, and inquired news can spread to the whole network. With the increase in network size, data increase exponentially and caused a lot of redundancy and influence bandwidth seriously. To avoid consuming network resources unlimitedly by this broadcast way, it generally increase jogging limitation of news by setting Time-to-Live (TTL) value. If every jump increases in

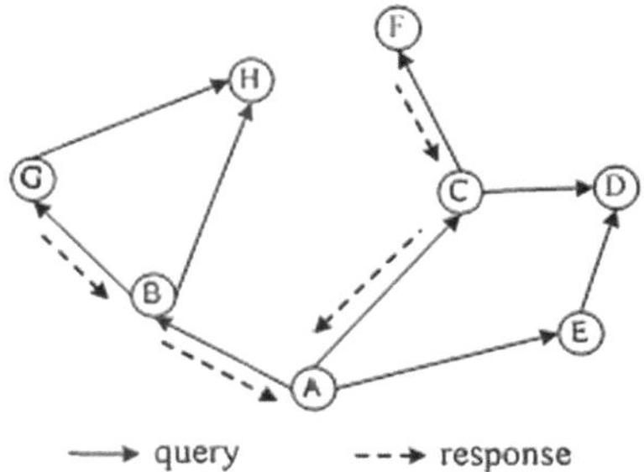

Fig. 1 Search pattern

overlay network, the TTL value decreases 1. When the TTL value reduced to 0, it does not continue to forward the message.

In random walk methods, the node will no longer deliver request news to all the neighbors, but will select k neighbor nodes to forward this message randomly; if neighbor nodes did not have request resource, they will transmit received one to a neighbor node, and it has less news flow by this method, but actually increased the time delay of network searching. For resource discovery problems in unstructured P2P network, modified search algorithms are put forward successively, such as k-random walker [7], modified BFS. Based on studying relevant optimization strategy, this paper proposed a resource discovery method based mechanism.

At first, similar interest nodes are formed a domain. Then according to the interests' similarity of the nodes, the formation of nodes in the domain is changed dynamically. And the efficiency of routing is improved by the friendly degrees among domains.

3 Mechanism of Interested Similar

Firstly, it defines the relevant network model and the interest similarity between the nodes and provides resource discovery method based on this interest similarity. In the process of transferring information, node maintains the corresponding interest degree index and adjusts the members of domain dynamically according to their respective interest degree index. In the situation of resource discovery of domain failure, it reduces redundancy in news forwarding process through the friendly degree index of domain orientating route.

3.1 Covering Network Model

There is no strict limitation in cover network topology for unstructured P2P system, considering the matching with physical network, firstly groups the similar physical nodes and then divides nodes in close proximity to interest similarity in

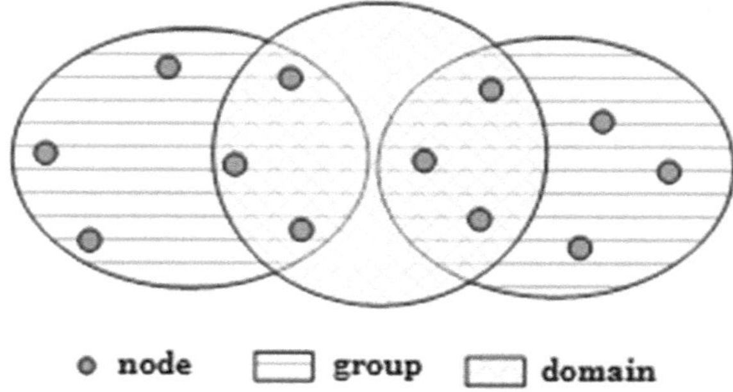

Fig. 2 The division of group and domain

the same domain according to interest similarity between the nodes, forming overlay network. It chooses the nodes with superior performance in the group and domain as super-nodes, keeps the related information such as number and resource of ordinary nodes in this group or domain, maintains the friendly degree index table of the domain, and records the ordering on the domain, which contacts with the table frequently. In addition to keeping its resource information, the ordinary node of domain needs to maintain interest similarity index, recording some node information that is close to interest similarity degree. Division of the group and the domain is shown in Fig. 2. Resource discovery is firstly made in the nodes of the same group in domain. If it is failure in the same group in domain, it is made in the nodes of other groups in the domain and finally considers cross-realm search.

The interest similarity between the nodes reflected the degree of correlation in resources between them. The degree of the nodes' correlation is higher, and the possibility of their contacting and visiting is bigger. It gives a reference for routing. Below is the definition of interest similarity of nodes.

3.2 Interest Similarity of Nodes

For the resources which every node in the network has, its feature can be marked through the corresponding key word, give key word for resources the corresponding weights referring to the $T_F * ID_F$ weighted technology [8]. The importance degree in the resources S_j of key words K_i is shown as W_{ij}; $S_j = (W_{1j}, W_{2j}, \ldots, W_{nj})$ and $q = (W_{1q}, W_{2q}, \ldots, W_{nq})$ are shown as resource S_j and vector of search requests Q, and the calculation formula of similarity between S_j and Q is shown in Eq. (1):

$$\mathrm{sim}(s_j, q) = \frac{s_j \cdot q}{|s_j| \times |q|} = \frac{\sum_{i=1}^{n} W_{ij} \times W_{iq}}{\sqrt{\sum_{i=1}^{n} W_{ij}^2} \times \sqrt{\sum_{i=1}^{n} W_{iq}^2}} \tag{1}$$

In real network environment, if the node D_i launched the request of resources S_k many times in a certain time, the possibility that it launched another request of resources S_k in the later time is very large and the possibility that node D_j that has the resources S_k responds D_i again increases. If the time that node D_i gains response of the request from D_j in unit time is N_{ij}, then take smaller value of resource number m between nodes D_i and D_j; the resource vectors D_i and D_j are expressed as $d_{ki} = (W_{1i}, W_{2i}, \ldots, W_{ni})$ and $d_{kj} = (W_{1j}, W_{2j}, \ldots, W_{nj})$, on the basis of formula (1), giving nodes D_i and D_j interest similarity $h(i, j)$, as shown in Eq. (2):

$$h(i,j) = N_{ij} \times m^{-1} \times \sum_{k=1}^{m} \mathrm{sim}(d_{ki}, d_{kj}) \tag{2}$$

Because N_{ij} is dynamically changed in the process of resource discovery, the interest similarity it maintains is also changed. On the basis of fixed scale in domain, nodes of domain adjust according to the change in the index table, optimize system topology dynamically, and thus cluster nodes with high interest similarity of system in the same field.

3.3 Resource Discovery Strategy

The node scale of domain is obviously smaller than the group's in the network model analyzed previously. The node in the same domain can be divided into the same group and also can be divided into different groups. While discovering the resources, the request message will be delivered among the nodes of the same group of the domain firstly. When the discovery in the same domain failed, the delivery of the message will happen in other groups. If the discovery between the groups of the domain does not work still, it will consider to search among friendly adjacent domains, which visit frequently. The specific methods are as follows:

(1) If search sponsors D_q first visit the super-node S_1 in the domain, it inquires the node information index table of S_1 that records the information of the node number of the domain, ID of common node in the domain and ID of the group of the node. Identify the node of the group D_q in S_1 through the information index table and search request resources by adopting the method of *K*-order random walk discovery method in these nodes. A series of dynamic regulations will be taken if resource discovery is success, such as updating the value of interest similarity of target node and D_q, to offer the basis for the adjustment of neighbor nodes.

(2) When search for the same group in the domain is failed, it inquires index table of super-nodes of group in S_1 that contains all super-nodes of groups in the domain. These super-nodes of groups arrange increasingly according to the physical distance with S_1, by selecting the groups closer to S_1 in the domain to carry out the resource discovery process, the same with the same group search.

(3) If the discovery of all groups in domain S_1 is failed, it inquires the friendly degree index table of domain of S_1 (adjacent domain tables), and adjacent

domain table records all the domains which exchange with S_1 frequently, which are recorded in index table of the nodes of group of S_1. Namely, the domains are friendly with domain S_1 in a high degree. These domains arrange diminishingly according to visit frequency. The search turns to the most friendly domain, through the adjacent domain table. The search will be continued by the chosen method, which is one of the two methods mentioned above, according to the specific circumstance, until resource is fail to obtain or the discovery is fail.

In the process of message routing, it detects whether the resource of D_q is obtained or not whenever leapfrogging every three steps. In order to avoid to producing too much information redundancy, news forward is stopped when the resource has been obtained. Update the index information that is maintained by the corresponding nodes and regulate the network topology dynamically whenever the target is found or failed.

4 Experimental Analysis

The performance of resource discovery method in peer-to-peer network can be assessed by the search time, use ratio of bandwidth and the news redundancy. To test query performance of discovery methods based on interest similarity degree in this paper, we use degree of peer interest similarity (DPIS) to say the method and use Java language in P2P simulated environment to produce the topology of network with different node scales, average level of connectivity of node is 10, according to the two indexes that average queried time delay and redundant news, simulate and contrast queried performance of three methods that Flooding, Random Walk and DPIS. Figure 3 shows the average queried delay time comparison of three methods with the same resource request under different network scales. Along with the increase in the number of nodes, blind search method of flooding

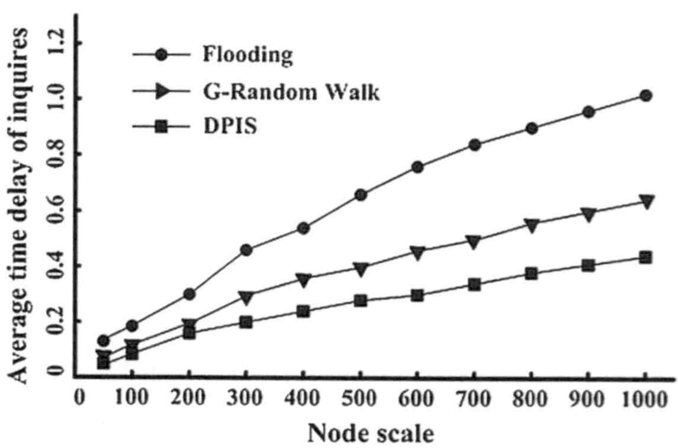

Fig. 3 Average time delay of inquires contrast

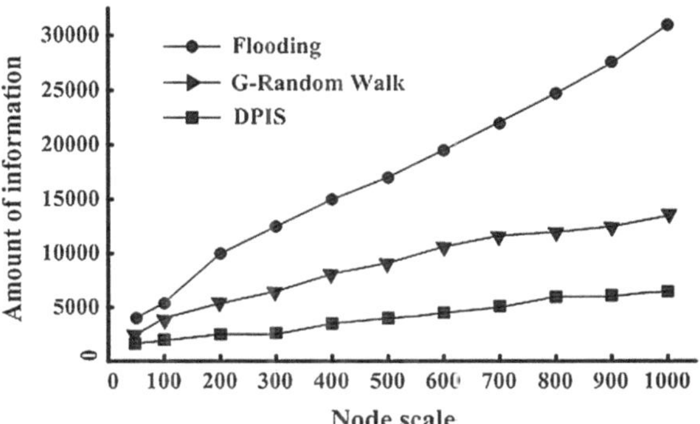

Fig. 4 Amount of information contrast

produces the large amount of redundant news, significantly increased the queried time delay. Local routing of DPIS walks randomly by the k order. Its tendency of average time delay is similar to Random Walk. But it registers the nodes in the groups and the domains through the interest similarity. Because route hop number reduces, the average query time delay reduces significantly.

News forwarding number manifests r network bandwidth utilization ratio from the other hand. The less the queried news was transpond, the less information content of network was produced, the higher of queried performance of network is. Check the news forwarding number in three discovery method by changing the scale of network, compared to the resources consumption of network. Figure 4 shows the comparison of news forwarding number of three methods in different nodes scale in the network, because node forward news to all the neighborhood in the process of flooding search, news number increase quickly with the node scale. Random Walk selects neighbor to forward message randomly, the network scale has small effect on the news number. DPIS method considers topological match while its node is registered in different group and domain according to their interest similarity. So the amount of news forward is still maintained in relatively stable range although node scale increases.

5 Conclusions

Through transmitting queried information in the network, blind searching of unstructured P2P networks can constantly spread information to each node, the search strategy has the characteristic of high rate of coverage and easy to achieve, but it also has a lot of redundancy news, limiting the bandwidth utilization. Aim at the problems of resource discovery in unstructured P2P networks, we put forward a resource discovery method based on the interest similarity of nodes, and it groups nodes according to their physical location, clusters nodes with similar interest to form domain, adjusts network topology dynamically in process of resource discovery, and optimizes discovery performance. We compared the method with the performance of two blind search methods through simulation and analyzed the simulation results. But for searching information in unstructured overlay network, the problems still need further study in the future work, such as positioning resource accurately, improving the recall ratio, and so on.

Acknowledgments This work was supported by grants (No. 60743008) from the National Natural Science Foundation of PRC and the Henan science research project (104300510063).

References

1. Li, Y., Feng, Y.: Study of data search in DHT P2P networks. Appl. Res. Comput. **10**, 226–228 (2006). (in Chinese)
2. Ripeanu, M.: Peer-to-peer architecture case study: Gnutella network. In: Proceedings of 1st International Conference on Peer-to-Peer Computing, pp. 99–100 (2001)
3. Hao, X., Song, O.: Overlay topology optimization in unstructured P2P systems. Comput. Eng. Appl. **46**(22), 144–146 (2010). (in Chinese)
4. Wang, X., Xue, L., Jia, D.: Unstructured P2P resources search mechanism based on ant colony optimization. Comput. Eng. **35**(7), 189–190 (2009). (in Chinese)
5. Song, J., Lei, G., Zhang, X.: Light flooding: Minimizing redundant message and maximizing the scope of peer-to-peer search. IEEE Trans. Parallel Distrib. Syst. **19**(5), 601–614 (2008)
6. Liu, Z., Zheng, T., Wu, W.: Trust path search algorithm based on random walk. Comput. Eng. **35**(18), 156–158 (2009). (in Chinese)
7. Lv, Q., Cao, P., Cohen, E., Li, K., Shenker, S.: Search and replication in unstructured peer-to-peer networks. In: Proceedings of the International Conference on Supercomputing, pp. 84–95 (2002)
8. Chen, X., Chen, D., Le, J.: Government information resource retrieval algorithm based on metadata semantic relevance oriented ranking. Comput. Eng. Appl. **47**(25), 121–125 (2011). (in Chinese)

VLSI Design of Reconfigurable Cipher Coprocessor Supporting both Symmetric and Asymmetric Cryptographic Algorithms

Chaoxuan Tian, Jialiang Zhu, Weiwei Shan and Xingyuan Fu

Abstract A novel reconfigurable cipher coprocessor (RCP) is designed with supporting both symmetric and asymmetric algorithms. First, a memory-sharing S-box is proposed to provide a reconfigurable S-box with reduced hardware resources. Then, arbitrary permutation unit, reconfigurable arithmetic operation unit, and shift unit are designed. All the operation units are combined with control module, configuration registers, data interconnect bus, and other parts to form a RCP coprocessor, which can implement different cryptographic algorithms by changing the configuration to adapt different application scenarios. The reconfigurable cipher coprocessor that can realize DES, 3DES, AES, IDEA, RC6, and RSA algorithms is integrated with a 32-bit CPU, 32K SRAM, and other peripherals. The simulation results show that the RCP has advantages in resource usage and flexibility with a relevant performance.

Keywords Reconfigurable cipher coprocessor · Integrated circuit · Symmetric cryptographic algorithms · Asymmetric cryptographic algorithms

C. Tian · J. Zhu · W. Shan (✉) · X. Fu
National ASIC System Engineering Research Center, Southeast University, Nanjing 210096, China
e-mail: wwshan@seu.edu.cn

C. Tian
e-mail: tcxuan21@gmail.com

J. Zhu
e-mail: zjl.raul@gmail.com

X. Fu
e-mail: dafufu@yeah.net

S. Patnaik and X. Li (eds.), *Proceedings of International Conference on Computer Science and Information Technology*, Advances in Intelligent Systems and Computing 255, DOI: 10.1007/978-81-322-1759-6_36,

1 Introduction

Nowadays, information, digitization, and networking has become more and more important. Thus, information security has become an important issue. Cryptographic algorithms-based hardware is the foundation of the information security systems. Especially, the symmetric and asymmetric cryptographic algorithms are widely used in various fields.

Currently, cryptographic chips supporting some single algorithms are widely used [1–5]. These chips can only support fixed algorithms. Reconfigurable cipher coprocessor (RCP) is a kind of cryptography chip that uses reusable hardware resources to realize different cryptographic algorithms, thus greatly improves the flexibility and expandability of cryptographic chips [6, 7].

In this paper, firstly, reconfigurable process units are designed by extracting the common arithmetic operation units of symmetric and asymmetric cipher algorithms. A memory-sharing S-box is proposed to provide a reconfigurable S-box with reduced hardware resources. Then, an arbitrary permutation unit, reconfigurable arithmetic operation unit, and shift unit are designed. Based on these operation units, a reconfigurable cipher coprocessor that can support DES, 3DES, AES, IDEA, RC6, and RSA algorithms is designed. Simulation results show that it has much more flexibility and expandability than ordinary cryptographic chips.

2 Implementation Analysis

Common symmetric and asymmetric cipher algorithms such as DES, AES, IDEA, and RSA have many similarities in the structure and operational characteristics. Although the specific realizations of each cipher algorithm are different, they are composed of similar basic logical and arithmetic operations.

After in-depth analyzing of a variety of symmetric and asymmetric cipher algorithms as shown in Table 1, the related computational characteristics are summarized as follows:

1. Cryptographic arithmetic operations are mostly unsigned integer operations and have a number of logical operations. The algorithm does not use floating point or fixed point type and commonly includes S-box, permutation, modular multiplication, modular addition, modular subtraction, shift, and exclusive-or and other operations;
2. A large number of logical operations (XOR, AND, OR, NOT) are used in cryptographic algorithms, logic operation bit width is usually 32-bit and 64-bit, 32-bit is mainly used;
3. Cryptographic algorithms often use logical shift; the shift patterns have both fixed and variable mode; the bit width of shift generally varies from 1 to 32; the shift width is mostly 16-, 28-, 32-, 64-, and 128-bit;

Table 1 Basic arithmetic operations of symmetric and asymmetric ciphers

Algorithms		Operations						
		Width	S-box	Shift	Permutation	Logical operation	Mod-add	Mod-multi
Symmetric cipher algorithm	DES/3DES	64	6×4	28	64-64/32-48 48-32/56-48	32	–	–
	AES	128	8×8	32	–	32	–	–
	IDEA	64	–	128	–	16	2^{16}	$2^{16} + 1$
	RC6	128	–	32	–	32	2^{32}	2^{32}
	SMS4	128	8×8	32	–	32	–	–
	Serpent	128	4×4	32	128-128	32	–	–
	Two fish	128	8×8	32	–	32	2^{32}	–
Public-key algorithm	RSA	512	–	–	–	–	–	2^{512}
	RSA (Improved)	512	–	32	–	32	2^{32}	–

4. S-box substitution operation exists in vast majority of the symmetric cipher algorithms, including the following patterns: 4×4, 6×4, 8×8, 8×32; currently 8×8 S-box substitution mode is most widely used;
5. Block ciphers have a lot of big-bit-wide permutation operations, the permutation bit width is from 32 to 128 in which 32 and 64 are mainly used.
6. Modular add and modular multiplication are also commonly used in cryptographic algorithms, especially for the public-key algorithms.

3 Design of Reconfigurable Cipher Coprocessor

3.1 Design of Reconfigurable Arithmetic Units

According to the computing features of symmetric and asymmetric ciphers, seven reconfigurable arithmetic units are designed, which are reconfigurable permutation unit, reconfigurable S-box unit, reconfigurable modular add unit, reconfigurable shift unit, reconfigurable modular multiplication unit, reconfigurable logic arithmetic unit, and reconfigurable mix-column unit. Some important reconfigurable arithmetic units are designed as follows.

3.1.1 Reconfigurable S-box Unit

S-box is the only nonlinear operation module that provides the necessary block cipher confusion effect. In different cipher algorithms, S-box is built in different way, either logic-based implementation or lookup table (LUT)-based implementation, which uses RAM or ROM to store the substitution data. In this paper, a memory-sharing S-box is proposed based on lookup table with support of 4×4, 6×4 and 8×8 substitution modes. Eight 64×4-bit register arrays make up the sharing memory.

3.1.2 Reconfigurable Permutation Unit

Permutation operations in cryptographic algorithms provide confusion spread function. Currently used permutations in a variety of cryptographic algorithms are symmetrical permutation, compression permutation, and extended permutation. The operation bit width ranges from the 32-bit to 128-bit. In order to meet the different requirements of a variety of algorithms, the design of reconfigurable permutation unit must be able to complete 128-bit permutation.

This paper uses BENES network structure to achieve replacement function [6]. BENES is widely used as a nonblocking interconnection interactive network in the field of communication networks, it can achieve any bit width and any permutation operation bit in theory.

3.1.3 Reconfigurable Shift Unit

The above analysis shows that shift operation is widely used in the various types of cryptographic algorithms. Shift operation includes four operating modes: ring shift left, ring shift right, logical shift left, and logical shift right. The 32-bit shift operation is mostly used, and then is the 16-bit and 8-bit. The reconfigurable shift unit of this design supports 8-bit, 16-bit, and 32-bit input data shift to achieve 1–32 arbitrary bits shift operation. When the input data is 32-bit, reconfigurable shift unit processes normally; while the input data is 16-bit or 8-bit, it will first automatically do filling operation, then do the corresponding operations. For instance, when input data is the 8-bit data [7:0], it will use the fill operation to change the input data into 32-bit {data0 [7:0], data0 [7:0], data0 [7:0], data0 [7:0]}, then intercept the low 8-bit of the output data as the shift result.

The structure of reconfigurable shift unit is shown in Fig. 1:

3.2 RCP Coprocessor Architecture

The architecture of reconfigurable cipher coprocessor is shown in Fig. 2. It contains seven modules, which are reconfigurable units, key registers (KEY MEM), control module (FSM), buffer module (Buffer MEM), config registers (Config_MEM), data interconnect (ICN) bus, and output module.

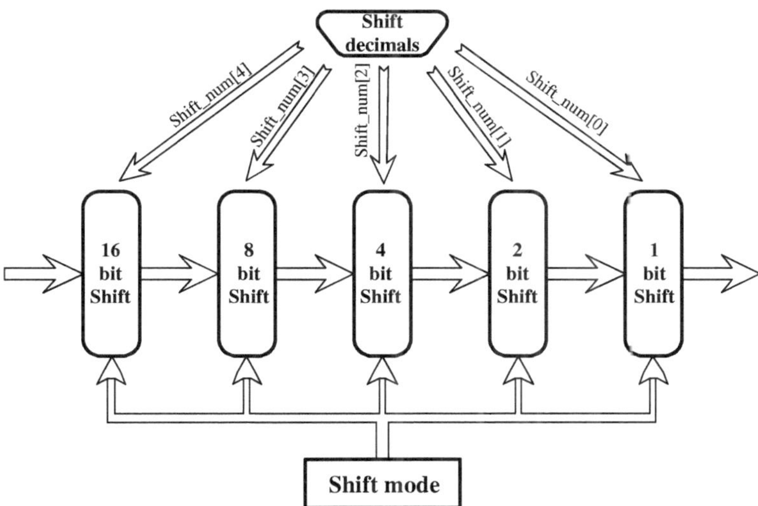

Fig. 1 Reconfigurable shift unit structure

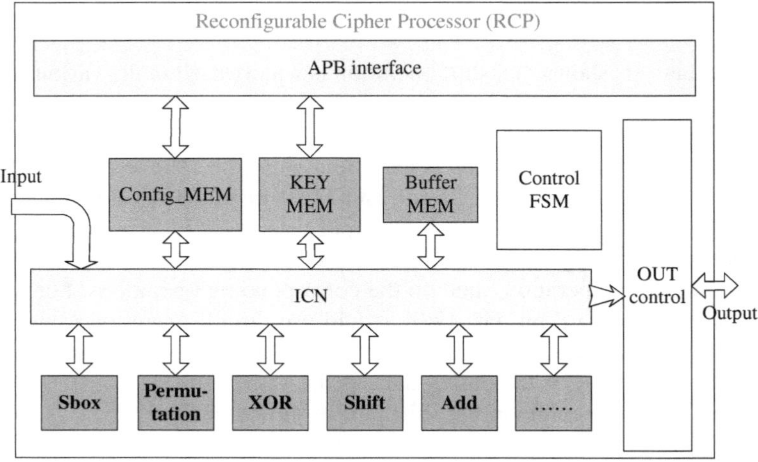

Fig. 2 The architecture of RCP

3.3 Configuration Flow

Different symmetric cipher algorithms can be realized on it by different configurations. Following are the steps to achieve a given cipher algorithm using RCP as shown in Fig. 3.

Firstly, the given algorithm will be analyzed to prepare the config information for the RCP.

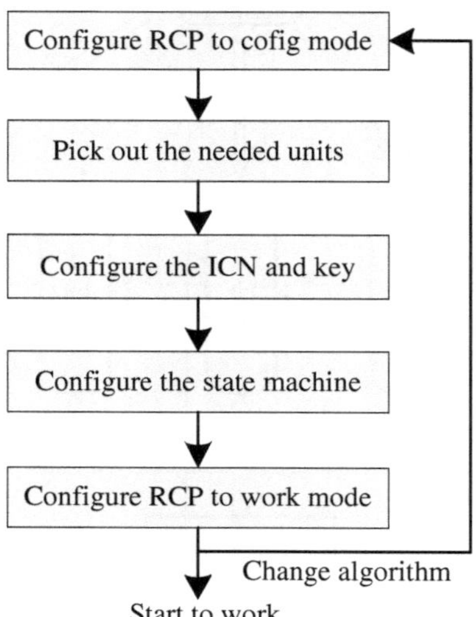

Fig. 3 Configuration flow of RCP

Secondly, the needed reconfigurable units will be picked out and configuration will be written to them. These units will complete the operation for the given algorithm.

Thirdly, the config data of ICN will be inputted to make the reconfigurable units connected in the same way of given algorithm. In the same time, the key data should be written to RCP and different algorithms need different key data.

Fourthly, the config data that state machine needed will be written. The data decide how many clock cycles the RCP need to finish in one encryption, how to handle the input and output data and control the procedure of encryption. Then, the state machine will make the whole RCP run orderly and efficiently.

After configuration, the RCP will be changed to work mode. Now plaintext and enable signal can be inputted. The RCP will run orderly and output cipher text in fixed clocks. If the algorithm wanted to be changed, the steps above will be done again with different config data.

4 Simulation and Results

4.1 SoC Architecture

The proposed RCP can be configured to realize five symmetric cipher algorithms (DES, 3DES, AES, IDEA, and RC6) as well as one asymmetric algorithm (RSA). Reconfigurable cipher coprocessor is integrated into a system-on-chip (SoC) based on a 32-bit ARM CPU and mounted on the APB bus, as shown in Fig. 4. The SoC is designed and synthesized under 0.18 μm CMOS process. Its working frequency is 100 MHz with a supply voltage of 1.8 V. Synthesis using EDA tool of Synopsys Design Compiler shows that the total logic gate of RCP is 28,390, which is a small number compared with the 32-bit CPU.

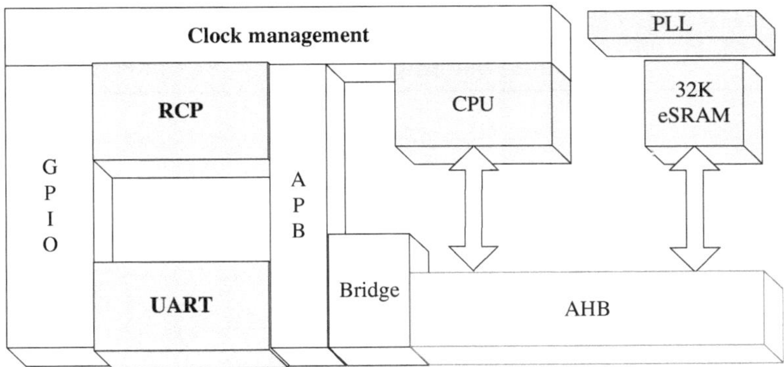

Fig. 4 SoC chip layout graph

The module of SOC structure is listed as follows:

1. RCP: our reconfigurable cryptographic processor.
2. CPU: a 16/32-bit CPU, supporting JTAG debugging program.
3. MEMORY: 32K SRAM on the chip.
4. PMC: Controlling the system clock, 20M outside crystal oscillator, supporting 1/2/4/8/16/32-bit frequency divider.
5. PLL: Supporting frequency multiplication, maximum 100 MHz.
6. GPIO: supporting data input and output.

4.2 Realization of Algorithms and Simulation Result

In this paper, DES, AES, RC6, IDEA, and RSA algorithms are realized on RCP. The functions of the above realizations are first verified by simulating a complete encryption process, and then compare the encryption results with the correct results. A large number of data are used in verification; and a few data are listed in Table 2 for illustration.

The simulation waveforms of DES and RSA on RCP are shown in Figs. 5 and 6 for further illustration. RCP is executing a full encryption process with 100 MHz frequency. The simulation results show the function correctness.

Table 2 Function verification of RCP

Algorithms	Plaintext	Key	Encryption result	Results
DES	4E6F7720	AAAAAAAA	BD3292FA	✔
	69732074	AAAAAAAA	56DCA3D0	
AES	FF	AAAAAAAAAAAAAAAA	4E76CE7873D82660	✔
		AAAAAAAAAAAAAAAA	87FD10AB4FE0BD1D	
RC6	0	0	36a5c38f78f7b156	✔
			4edf29c11ea44898	
IDEA	FF	BBBBBBBBBBBBBBBB	93F1973122172F2F	✔
		BBBBBBBBBBBBBBBB		

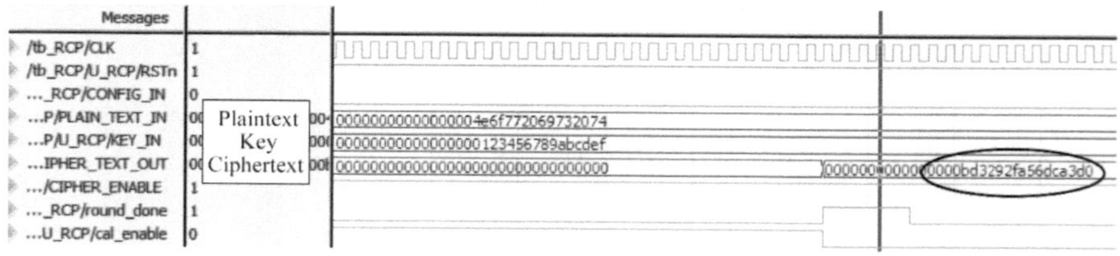

Fig. 5 Simulation result of DES on RCP

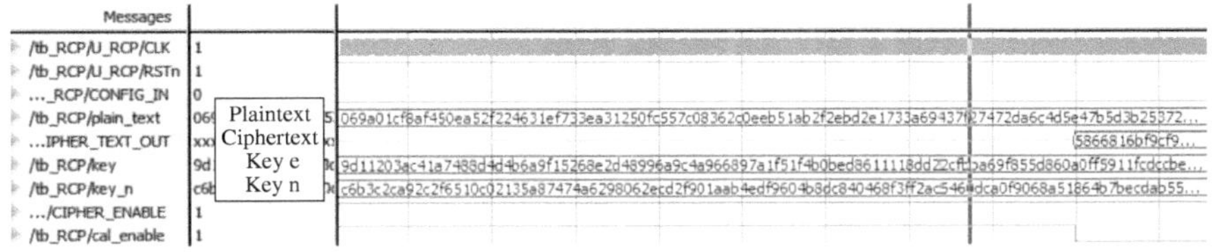

Fig. 6 Simulation result of RSA on RCP

As seen from above, the proposed reconfigurable cryptographic coprocessor has advantages in supporting multiple cipher algorithms with a relevant performance. And it has the potential to be extended to support more symmetric cipher algorithms, with high flexibility and expandability.

5 Conclusion

In this paper, a reconfigurable cipher coprocessor is designed and implemented, which uses reusable hardware resources to support multiple cryptographic algorithms. Different cryptographic algorithms can be completed by changing its hardware structure to match different application requirements. This cryptography chip has a big advantage of flexibility and expandability.

Acknowledgments This work was sponsored by the National Natural Scientific Foundation of China (Grant No. 61006029) and Qing Lan Project.

References

1. Patel, D., Muresan, R.: Triple-DESASIC module for a power smart system-on-chip architecture. Canadian Conference on Electrical and Computer Engineering, pp. 1069–1072 (2006)
2. Wang, C., Heys, H.M.: Using a pipelined S-box in compact AES hardware implementations. 8th IEEE International Conference (NEWCAS), pp. 101–104 (2010)
3. Chen, Y.L., et al.: Design and implementation of reconfigurable RSA cryptosystem. International Symposium on VLSI Design, Automation and Test, pp. 1–4, April 2007
4. Muthukumar, B., et al.: High speed hardware implementation of an elliptic curve cryptography (ECC) co-processor. Trendz Inf Sci Comput, 176–180 (2010)
5. Jithendra, K.B., et al.: FPGA implementation of secure time shared hash stream cipher. International Conference Computational Intelligence and Communication Networks (CICN), pp. 381–385 (2011)
6. Yan, W., You, K., Han, J., Zeng, X.: Low-cost reconfigurable VLSI implementation of the SMS4 and AES algorithms. IEEE 8th International Conference on ASIC, pp. 135–138 (2009)
7. Smyth, N., McLoone, M., McCanny, J.V.: Reconfigurable processor for public-key cryptography. IEEE Signal Processing Systems Design and Implementation, pp. 7803–9333 (2005)

Intelligent Monitoring System of Special Vehicle Based on the Internet of Things

Guohou Cao, Xiaoqiang Yang and Huanliang Li

Abstract The main objective of this work is the development of an intelligent monitoring system based on the Internet of Things (IoT). The core of the system is embedded in ARM microcontroller of LPC2478 together with technology of Internet of Things, which has so far applied in fields of emergency response, intelligent shopping, smart product management, and military field sensing and so on. This system can implement fault detection, processing, memory storage, and alarming of the operating status parameters, such as temperature of cooling system, oil temperature, oil pressure, engine rotation speed, oil temperature of torque converter as well as oil level of sump by means of acquiring the operating status data from various testing sensors and the like. Specifically, it can accomplish functions of special vehicle's orientation and track using GPS/GPRS, directing the transport routes, output the special vehicle accumulative weights, transmitting operating status to monitor center, querying history data, remote maintenance support, and querying history data. The system can record various types of operating parameters and send them to monitoring center by network. It is therefore applicable to many different types of engineering vehicle.

Keywords Internet of things · Network of sensors · Monitoring system · Embedded equipment · Special vehicle

G. Cao
Engineering Scientific Research and Design Institute of Chengdu Military Region, Kunming, China
e-mail: yangsecond@126.com

X. Yang (✉) · H. Li
PLA University of Science and Technology, Nanjing, China
e-mail: yangthird@126.com

S. Patnaik and X. Li (eds.), *Proceedings of International Conference on Computer Science and Information Technology*, Advances in Intelligent Systems and Computing 255, DOI: 10.1007/978-81-322-1759-6_37,

1 Introduction

This paper focuses on the design of intelligent special vehicle and traffic-control system based on the Internet of Things (IoT), aiming at introducing the IoT into road vehicles monitoring management and realizing the effective supervision and control of the vehicles [1].

The Internet of Things refers to uniquely identifiable objects and their virtual representations in an Internet-like structure. It integrates various information transmission equipment, for example, radio-frequency Identification (RFID) apparatus, infrared sensor, GPS, laser scanner, home appliance, security equipment, and the like, with internet so as to create a huge network. Thereby all the things are linked together with the network so that the identification, management, and monitoring are implemented conveniently. This application can provide users with ubiquitous full-service in terms of IoT-integrated application [2].

In a general way, IoT includes three layers at least: perception layer, which transforms information of things to readable digital signals with RFID, sensors, etc.; network layer, which sends digital signals to corresponding platforms via network; application layer, which unscrambles and applies digital signals through corresponding software. Various objects, user contexts, and scenarios demand different networks, different information processes, and different application methods [3, 4].

2 Hardware Architecture

2.1 General Structure

IOT is formed by three layers. The bottom is perception layer, whose function is cognizing and collecting information of objects. The middle is transportation layer. It consists of mobile phone networks, fixed telephone networks, broadcasting networks, and closed IP data networks for each carrier. The top is application layer, where abundant applications run. Typical applications includes smart traffic, precise agriculture, intelligent logistics, smart industry, environment protection, mining monitor, remote nursing, safety defense, smart government, etc. The monitoring system based on IoT in this work has adopted advanced embedded operating system, technology of GPS and GPRS. It can resolve the difficulty in operating status monitoring and management such that the working efficiency of special vehicle is greatly improved. The working status monitoring system is composed of terminal equipment, monitoring center and its management client as well as GIS client. The architecture is described in Fig. 1.

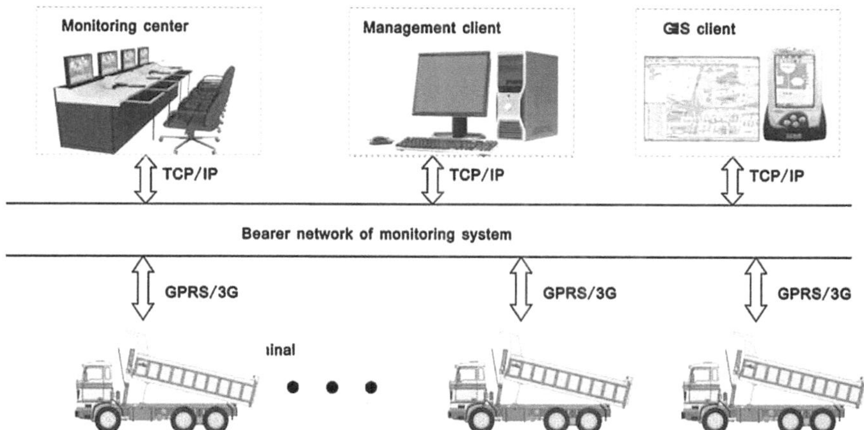

Fig. 1 General architecture of monitoring system

2.2 Hardware Structure

The hardware platform of monitoring system mainly includes core circuit board of LPC2473, current excitation module, multiplexer, LCD interface circuit, USB interface, signal conditioning circuit, SD card, storage and Flash, GPS module, GPRS module, Ethernet interface as well as video interface. The hardware framework is illustrated in Fig. 2.

Operation status parameters of every subsystem (such as diesel engine, transmission box, torque converter, oil circuit, air circuit, etc.) of the special vehicle including temperature, pressure, voltage, oil level of sump, engine rotation speed, vehicle speed, etc. can be converted to voltage or current signals through the sensors mounted on every test point of the special vehicle. These voltage or current signals are then converted to voltage values through the front end amplifier and then be transmitted to the input port of the device. The ADC converter of the device then acquires these parameters and sends it to the LPC 2478 via internal bus. In the meantime, it processes these data in real time and outputs corresponding diagrams and digital signals to the LCD, gives out alarms if any and save the fault status data in the SD card [5].

3 General Design of Software Framework

The software framework consists of vehicle terminal module, monitoring center (data processing module), management client of monitoring center, and GIS client. The software block diagram is described in Fig. 3. The vehicle terminal module,

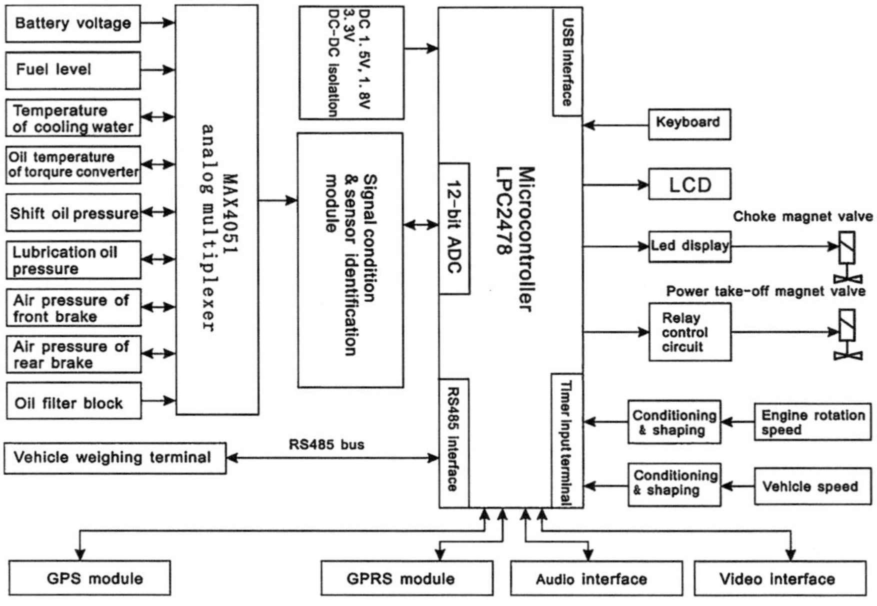

Fig. 2 Hardware framework of the monitoring system

which runs in the hardware platform of vehicle terminal, implements corresponding function of terminal service. Its progress involves database module, GPS module, message agent module, NMS Agent (Network Management Station) module, power management module, update module, photograph module as well as parameter configuration module that are all operated in terms of the core service function module. The message agent module of graphical user interface (GUI) progress communicates with the message agent module of main progress via message queue.

The function of management client includes:

- Offer system entrance to user login server.
- Take charge of the information present, including object position, image (video information) and the like.
- Implement query and display of history data generated by real-time information, including object's history tracking, video graph, and replay output of data.
- Carry out the setting of GIS data for terminal controller.
- Employ the simple control of device.

In view of the main usage of monitoring system, the GIS interface panel is developed to be the principle daily human–machine interactive interface. It implements the function of geographical information system as well as the real

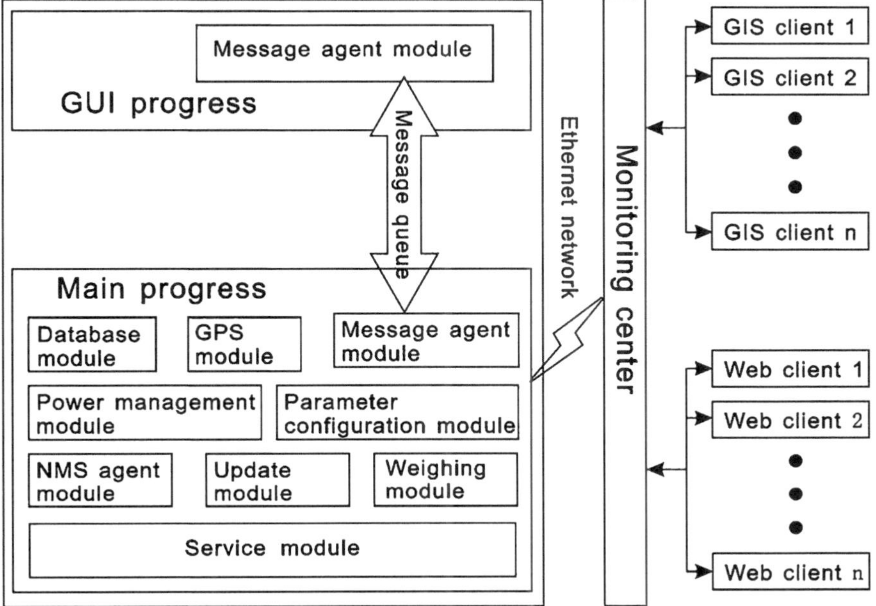

Fig. 3 Software framework of monitoring system

time and dynamic display of operating status of special vehicle. In addition, it also performs the usual operating parameters configuration of vehicle terminal.

The terminal controller is mounted in the special vehicle, and it implements a variety of functions, such as GPS, GIS and real-time data acquisition of operating status. It communicates with the remote center server via wireless technique. The center server is a set of integrated management system.

4 Key Technologies

4.1 Main Software Control Module

The main control module includes two sections: one is the principle progress program for the implementation of various service logic processing; the other is the GUI progress program for user interface. The function of principle progress involves receiving external command, dispatching received message to corresponding service module for handling, dealing with various messages, and employing service logic.

The message dispatching is used to be the basic framework of principle progress program, which includes communication module, message agent module, and service module.

4.2 Vehicle Weighing Module

The vehicle weighing module usually communicates with the vehicle weighing system via RS485 serial interface and Modbus protocol. The module is master with the vehicle weighing terminal acting as slave. The input of weighing module includes the corresponding weighing data imported from weighing terminal via RS485 interface and external weighing operating command from message agent module. The output of weighing module includes three disciplines: the weighing data are transmitted to monitoring center via network; the weighing result is recorded to database; the relative command is dispatched to weighing terminal via RS485 interface. The function of weighing module includes that it establishes data link with weighing terminal via RS485 interface and reads data regularly from weighing terminal via RS485 interface, the result is delivered to the module itself after data parsing; the module receives weighing instruction coming from monitoring system via message agent module and carry out it; the module sends the weighing result to monitoring center via network and saves them to database for system query.

4.3 GPS Module

GPS module is mainly used to acquire geographical information parameters, for example, latitude, altitude, and elevation. It also acquires time for time calibration of the system. The input of GPS module is the GPS information data transmitted from the GPS device via network. The output of GPS module includes: the GPS information data are sent to control center via network; the GSP information data are sent to GUI module via message agent module; the relative warning message are saved to database; the relative warning message are sent to control center. The handling process of GPS module is carried out by means of receiving data information from GPS device via a dedicated network connection, and the message is parsed by GPS data parsing module for further processing. The relative warning information is recorded into database. So, the GPS information is sent to GUI module via that of message agent, as shown in Fig. 4.

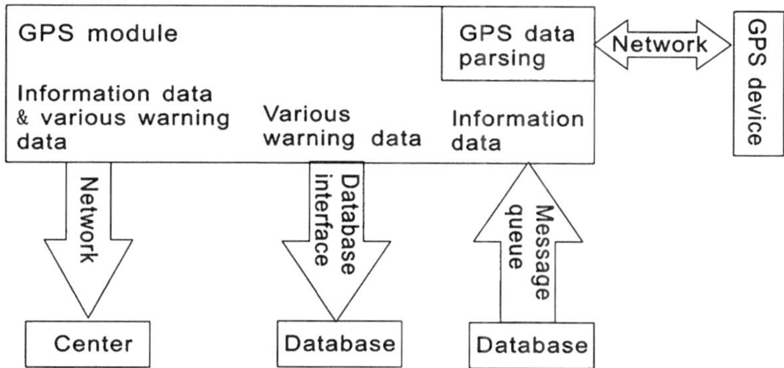

Fig. 4 Block diagram of GPS module

4.4 Communication Module

The communication module is used to accomplish communication with monitoring center. In this monitoring system, the communication function is developed in accordance with GPRS.

Because the GPRS device is regarded as a network one, the monitoring system can communicate with it via socket operation just like the common connecting operation of network. A TCP connection must be established properly when system is initializing. The terminal client is connected to center server in terms of the assigned center IP and monitoring port number. After connection is established, it can communicate with monitoring center.

4.5 Main GUI Module

The main GUI module is the human–machine interface of the monitoring system. This module acquires various operating status information from data acquisition system and displays relative parameters in LCD screen. During the working process of the special vehicle, the display panel can remind the user of special vehicle's operating status in real time and tell the user how to perform the troubleshooting operation. The user can select different functional modules through pressing corresponding keys on the film keyboard and perform the detection operation according to the instructions displayed in the "operation instructions" box. The main interface panel of the device is illustrated in Fig. 5.

Fig. 5 Main GUI panel

5 Conclusion

The intelligent monitoring system of special vehicle is designed according to IoT technology. It can implement real time monitoring and management to large amount of special vehicle so as to improve the working efficiency. Compared with traditional electronic monitoring system, this system has many advantages such as large storage capacity, high work stability, high reliability, and friendly operation interface. It can keep operating parameters of the special vehicle comprehensively and accurately also help the maintenance personnel to carry out fault diagnosis as well as trend analysis and improve the quality of special vehicle. Furthermore, it can also monitor the special vehicles effectively and instantaneously as well as provide necessary services to the vehicles. It may give many helps in the analysis of traffic accidents and mastery for conditions about road traffic.

Acknowledgments This work was financially supported by the Jiangsu Provincial Youth Science Research Foundation of China (BK2012061) and the Science Research Foundation of College of Field Engineering, PLA University of Science and Technology (KYGYZLYY1305).

References

1. ITU-T Y.2002 (Y.NGN-UbiNet): Overview of ubiquitous networking and of its support in NGN. Oct 2009
2. Baoyun, W.: Research review of IOT technologies. Periodical Electron. Meas. Instumentation **23**(12), 112–112 (in chinese)
3. Amardeo, C., Sarma, J.G.: Identities in the future: Internet of things. Wireless Pers. Commun. **49**(3), 353–363 (2009)
4. Zhen-xia, W., lai-shen, X.: Modern logistics monitoring platform based on the internet of things. International Conference on Intelligent Computation Technology and Automation. Changsha. pp. 726–731 (2010)
5. Guo, S., Zhang, Q.: Design of multi-channel data acquisition system of special vehiclery based on MSP420F419 microcomputer. Mining Process. Equip. **34**(1), 94–96 (2006) (in Chinese)

Data Flow Incompatible Vulnerability Analysis and Detection for Networked Collaborative Design Business Process

Huaizhi Yan, Wenwen Ye and Jihu Zhang

Abstract This paper presents a data flow incompatible vulnerability detection method for networked collaborative design (NCD) business process. First, we analyze the collaborative design network composition and its business process characteristics, while the business process and data flow is discussed. Then, combined with the characteristics of cooperative design workflow, the causes of the data flow incompatible vulnerabilities are analyzed and description method of the activity diagram data flow incompatible vulnerability is given. According to the business process model of active incompatible vulnerability, business process models were constructed for its detection, and vulnerability detection algorithm is proposed based on the description logic, which is used to detect collaborative design business process data flow incompatible vulnerability. Finally, the experiment and conclusion are given here.

Keywords Networked collaborative design · Business process · Incompatible vulnerability detection

H. Yan (✉) · W. Ye · J. Zhang
Beijing Key Laboratory of Software Security Engineering Technology,
School of Software, Beijing Institute of Technology, South Zhongguancun Street,
Haidian District, Beijing, China
e-mail: yhzhi@bit.edu.cn

W. Ye
e-mail: two_ye@hotmail.com

J. Zhang
e-mail: Zhangjihu@hotmail.com

S. Patnaik and X. Li (eds.), *Proceedings of International Conference on Computer Science and Information Technology*, Advances in Intelligent Systems and Computing 255, DOI: 10.1007/978-81-322-1759-6_38,

1 Introduction

Along with the rapid development of the manufacturing, industry division continues to be refined, the application of information technology is widely used in manufacturing industry, more and more manufacturing companies have chosen networked collaborative method for product design and production [1]. This collaborative work method and the application of appropriate information system make the manufacturing enterprises complete a large workload and complexity of the high-production tasks [2, 3]. Enterprise use information network systems for collaborative design, collaborative manufacturing work, in order to save the cost of manufacturing, strengthen the interaction along different regions and departments, and finally improve efficiency of the manufacturing [4].

However, the traditional enterprise business process is not standardized at all, which brings the information technology facing a severe challenge particularly for security issues [5]. In fact, the current business process model is mainly based on the sequence of operations on the basis of arrangements and coordination. Successful business process management depends on effective data process design, modeling, and analysis [6]. From the control flow perspective, because those models do not regulate the flow of data, resulting in the error still occurs, they can generate some errors named data flow anomalies [7].

Therefore, the goal of this paper is to propose a method for the business process data flow anomaly detection. Using this method, we can analyze the data flow anomalies in the business process of networked collaborative design (NCD).

2 Business Process, Data Flow and Incompatible Vulnerability and Their Relationship in NCD

2.1 Networked Collaborative Design

In order to accomplish the corporate design using the enterprise information system platform, production and management of specific activities, different subsystems should be created according to the different activities and the specific needs of different businesses within the enterprise. Among them, the collaborative network is designed for different designers to complete the design-related tasks together for the common design goal.

The NCD system can be summarized as: "to work as the center of the project management, product design, production and management-integration platform," including the two core concepts of collaborative and platform.

The collaborate includes the coordination among projects, collaborative product design, collaborative production process, as well as between project management, product design, production process of three collaboration. Within the enterprise, headquarters and branch offices may be located in geographically different locations, so each subinformation system participating in collaborative design is heterogeneous hardware and software. At the same time, platform includes the realization of the platform and effective management of the project management of the collaborative project, centralized collaborative product-design platform for product-data management, production-management platform for collaborative production process life cycle management. As to data issues, NCD includes the collaborative design and interactive, collaborative design-process management and collaborative design-related data management three main aspects.

2.2 Business Process, Activities, and Data Flow in NCD

In short, the business process is a set of input into output of inter-related or interacting activities, which has a different computer definition called workflow model. The workflow model can transform the actual business process into the formal definition handled by computer. The business process is intended to facilitate the control of workflow-management system.

Workflow is a computational model of business process, which brings the logic and rules of the work before and after the organization together into the right model to express and calculate in the computer. Therefore, the proper process model is the basic guarantee to achieve business objectives. Effective business process management depends on the flow of data corresponding to the accurate and detailed investigation and analysis. The main problem of workflow to solve is: to achieve a business goal, automatically transfer documents, information or tasks according to some predetermined rules between many participants using computer. Simply, the workflow is a series of interlocking, automatic task. The purpose of verifying the correctness of the workflow model is to establish the correctness of the workflow model according to some principles for validation.

Activity is the basic unit of the business process, each business process will include a number of activities that follows a certain logic execution. Business processes and data flow in an accurate analysis of normalization depends on the business process involved a clear definition of the respective data items.

Data flow is used to indicate the value of an intermediate data flow, which is a tool for simulating system data-transmission process in the system. The data flow model reflects the data dependencies of activities between the process models, it can ensure the correctness of data modeling, solving the mapping relationship between some heterogeneous data and the data flow scheduling; data-transmission mode is data interactive strategy between the process engine operation and service, as well as the protocol for sending and receiving data. The data flow model can clearly express the process model of each activity that conferred by the input, output parameters between the conversion, transfer, reflect the activities of the information–exchange relationship.

2.3 *Incompatible Vulnerability*

In fact, the different participants in NCD may mutually do the exclusive operations at the same time, which causes the incompatible vulnerability, while the execution of a process may require all previous activities completed, so that there will be uncoordination and bring it into conflict affecting the system's business process executions.

There are such security problems as incompatible vulnerability to be solved from the point of view of applications and enterprise layer in NCD. So, on the task-oriented point of view, from the task (activities) perspective to build security model and security mechanisms to provide dynamic and real-time data security management in the process of task handling.

3 The Description of Data Flow Incompatible Vulnerability

3.1 *Causes of Incompatible Vulnerability in the Data Flow*

The definition of incompatible vulnerability gives an incompatible vulnerability semantic description and definition. If the data flow can not be specified correctly in the business process system, the error and incompatible vulnerability will be occurred, this is called as data flow anomalies. In NCD, data access control is not static, but is along with the mission change context. So, we must take the incompatible vulnerability priority in workflow-environment protection to the information problem. In workflow environment, processing the data is associated with the last processing as well as the corresponding access control. So, data access control is a context-dependent access-control model. At the same time, data access control not only can carry out the different access-control strategies to different workflows, but also for the same different task instances workflow implement different access control strategy.

Data loss and data redundancy are the main causes of incompatible vulnerability. Data Loss: When a data item is accessed before initialization, data loss occurs with exception. Data redundancy refers to data duplication, it can be said to be the same data storage phenomena in different data in the file.

3.2 The Activity Diagram-Based Data Flow Incompatible Vulnerability Description

This paper use UML-activity diagrams to describe the operation behavior. Workflow activity diagram is used to.schedule for each task or activity of achieving business goals. Activities can be performed manually and tasks can also be automated. It can perform a unit of work. Activity diagram is a special form of state diagrams.

Now, we can analyze the data flow incompatible vulnerabilities: in a business process, if there are different versions of the same data item, it creates a data incompatible vulnerability. When the data incompatible vulnerability occurs, it is very hard to judge which version of the data items. In a business process system, there is more than one event to initialize the same data item, it causes data flow incompatible vulnerability abnormalities.

4 Data Flow Incompatible Vulnerability Detection Algorithm and its Steps

There is no effective business process specification to detect and prevent data flow anomalies before. Because of lacking data flow analysis, the business process data flow is difficult to avoid errors indeed. A process that contains the data flow error will lead to unexpected interruptions in the process, leading to high-running costs in debugging and modifying.

4.1 Building Detection Model

In order to establish the incompatible vulnerability detection method, we built TBox and ABox firstly that were based on description logic. During this process, we decompose the network business process model so as to get the number of activities. Because the business process model design is very complex, such causes as invalid input or state changes are likely to form active incompatible vulnerabilities, resulting in business process deadlock. Through that decomposition, our goal is to obtain incompatible vulnerability in data issues disguising in the specific activities.

Description logic is a knowledge representation method, it not only can be expressed in the field of knowledge, also has the reasoning mechanism, which can derive implicit knowledge. In this paper, the description logic to express the knowledge base is composed of two major parts of TBox and ABox. TBox defines the structure of the specific-domain knowledge (business process) and containing a series of axioms. ABox contains an instance of the TBox concept.

The main attributes of TBox for business process flow are trigger event, process results, and execution rule sets. In this paper, we can set up the business process TBox.

4.2 Data Flow Incompatible Vulnerability Detection

Before detecting the data flow incompatible vulnerability, we can decompose the orient network into a base one and several subnetworks so as to achieve the complicated business process network model reduces to a simple model purposes. Then, data flow incompatible vulnerability detection algorithm can be given as follows:

(a) Take the ABox activities relationship set and the decomposed networks as initialization of the detection;
(b) Access the whole units in the ABox activities relationship set. When a unit is an "and" node, while the node is invalid, then obtained an Incompatible Vulnerability. When a unit is a "or" node, while the node is invalid, then obtained a potential risk.
(c) If an analyzed node is a subnet, then analyze the subnet following the above steps until the activities relationship set was fully iterated.

Then, we can obtain the data flow incompatible vulnerability node and the potential risk node. Furthermore, the particular type of the incompatible vulnerability can be analyzed combining with the control flow and the detailed data resources.

5 Conclusions

In the NCD business processes, data flow incompatible vulnerabilities often lead to serious security risks. Lacking of data flow analysis can cause unavoidable errors in the business process flow. This paper analyzes the characteristics of network business process flows collaborative design based on the business process activities and explored the relationship between data flows, analyzes the data flow and workflow root causes of incompatible vulnerability, using the activity diagram method achieves a description of the data flow incompatible vulnerability. On this basis, business process models constructed by TBox and ABox are given based on description logic to construct the incompatible vulnerability detection algorithms. The method can effectively detect the NCD business processes data flows incompatible vulnerability. However, this method has not been tested in the business process optimization and business process reengineering, which need to be studied further.

Acknowledgments This work was supported by the Project of China National Ministry under the grant no. A2120110006, the Project of China "863" Plan under the grant no. 2009AA01Z433, and BIT Basic Research Fund.

References

1. Serigio, N., Fabio, N.: A concurrent engineering decision model: management of the project activities information flows. Int. J. Prod. Econ. **54**(2), 115–127 (1988)
2. Bettig, B., Shah, J.: Derivation of a standard set of geometric constraints for parametric modeling and data exchange. CAD **33**(1), 17–33 (2001)
3. Carballo, J.A., Director, D.W.: Application of constraint-based heuristics in collaborative design. In: Design Automation Conference, 18–22 June 2001
4. Sputhwest Research Institue. SwRI: Cyber security, In-formation Assurance. http://www.swri.edu/4org/d10/compsys/cybersec/default.htm. Feb 2002
5. Sun, B., Zhang, R., Wang, Y., Wang, R.: Fuzzy collaborative design conflict resolution method. Comput. Eng. Appl. **45**(3), 84–86 (2009-3)
6. Li, M., Yang, Y., Li, J., et al.: A preliminary study on synchronized collaborative design based on heterogeneous CAD system. In: Computer Supported Cooperative Work in Design, the 8th International Conference, vol. 2, pp. 22–27, 26–28 May 2004
7. Cheng, J.: Workflow conflict detection algorithm based on description logic. Mech. Electr. Eng. Mag. **25**(3), 40–42 (2008-3)

RFID Tag Anticollision Scheme Using Modular Arithmetic

Zhongyuan Qin, Yongxin Zheng, Yuying Wang and Jie Huang

Abstract In order to solve the problem of multiple tag collision in RFID system, a RFID tag anticollision scheme using modular arithmetic is proposed in this paper. Firstly, the number of tag is estimated, and then modular arithmetic is taken on the tag's ID bit-by-bit. The tags with the same remainder are classified into a group. After K times' modular arithmetic, all tags are divided into $2K$ groups. The tags in each group have few collision bits, and the group ID is the slot that tags will be read. Then, tags sent their ID according to the corresponding slot. Collision tags can be further identified with binary tree algorithm. The efficiency of tag identification is greatly improved using this scheme. It is suitable for identifying massive tags in RFID system.

Keywords RFID · Tag anticollision · Modular arithmetic

1 Introduction

RFID is a noncontact automatic identification technology. The basic principle is to realize the automatic recognition of objects or commodity using the transmission characteristics of the radio frequency signal or space coupling (inductance or electromagnetic coupling). Compared with other automatic identification technologies (Bar code technology, Optical recognition, Biological recognition technology,

Z. Qin (✉) · Y. Zheng · Y. Wang · J. Huang
Information Science and Engineering School, Southeast University, Nanjing, China
e-mail: zyqin@seu.edu.cn

Y. Zheng
e-mail: demonshir@sina.com

Y. Wang
e-mail: guyuexuan437@163.com

J. Huang
e-mail: jhuang@seu.edu.cn

S. Patnaik and X. Li (eds.), *Proceedings of International Conference on Computer Science and Information Technology*, Advances in Intelligent Systems and Computing 255, DOI: 10.1007/978-81-322-1759-6_39,

including iris, face, voice, and fingerprint), RFID has many advantages. For example, its anti-interference ability is stronger, it can provide large amount of information, it can read or write out of the visual range, it can be used for a long time, etc. It is widely used in logistics, supply chain, animals and vehicle identification, entrance guard system, library management, automatic charging and manufacture, etc., [1]. Till now, the main problem of RFID system is that the tag recognition rate is too low, when tag collision happens. Tag collision is that when multiple tags exist in the same radio frequency channel at the same time, the reader cannot read the tags.

At present, the two most widely used algorithm to solve the RFID tag collision are frame slot ALOHA algorithm and the binary search algorithm. Due to its simple and practical applicability, the frame slot ALOHA algorithm is adopted in more aspects. For example, ISO/IEC18000-6 Type A protocol and EPC Class 1 both use ALOHA algorithm.

2 Related Works

2.1 *Probabilistic Anticollision Algorithm*

ALOHA algorithm is a kind of information communication transmit–receive algorithm, which is random time division multiple access. It uses information frame to express the channel. Frame is a time quantum containing a number of slots, which is required by the reader. Information frame can be divided into more than one slot. Slot is the length of time caused by sending their own ID. When a slot is occupied by only one tag, the tag can be identified correctly. Collision happens when a slot is occupied by two or more tags; therefore, the reader cannot identify them properly. The reader will skip if the slot is empty.

The throughput of original ALOHA algorithm is rather low, only 18.4 %. Frame slot ALOHA algorithm (FSA) is expansion of the original ALOHA algorithm. For FSA algorithm, when frame length is approximately equal to the number of unidentified tag, it will achieve its maximum throughput, about 36.8 % [2]. There are further improvements based on ALOHA algorithm, such as Dynamic Slotted ALOHA algorithm [3], which dynamically adjust the length of frame according to the tag number to ensure the maximum throughput efficiency. In one word, when the quantity of tag is small, the identification efficiency of probabilistic anticollision algorithm is not high, and a large amount of slots are occupied.

2.2 *Binary Tree Algorithm*

Binary tree algorithm requires confirming the precise location of the data collision bit. Therefore, we must have the appropriate coding method. Manchester code signifies 0 with rising edge and 1 with the falling edge and do not allow any change state of 0 in the process of data transmission. If we use ASK modulation mode, when two (or more) readers send different value at the same time, the corresponding Manchester code of rising and falling edge offset each other, the carrier received is uninterrupted, this causes an error state. Thus, the collision position can be determined.

For anticollision algorithm based on binary tree algorithm, many schemes are proposed in recent years, e.g., bit-by-bit tree (BBT) [4] algorithm. Its basic idea is that the reader sends the request command to ask the tag sending back its ID. The responsive tag sends only one bit every time. If there is no collision, the reader will store the received bit in its memory and then request the next bit; If there is collision, the collision tags can be divided into two branches, namely branches 0 and 1. The reader chooses one branch and requests the next bit. The above process is repeated until every bit of the tag's ID is identified. Gao et al. proposed a binary search algorithm based on backward type [5]. When a tag is identified, it gets the next command according to the last request command parameters, which is to shorten the identification process greatly. It will improve the recognition efficiency mainly by reducing reader command in the basic algorithm and tag redundant data in response information. Although this kind of certainty anticollision algorithm has high recognition rate, it needs to send all or part of the tag's ID to search, and there will be heavy traffic between tag and reader if the tag's ID is too long [6].

3 Design

In order to improve the identification efficiency when a large number of tags exist, RFID tag anticollision scheme using modular arithmetic is proposed. Grouping mechanism is used step-by-step until each group has a small amount of tags. Generally, the tag number in each group is less than four. Using this method, a collision environment will be formed that is conducive to tag identification. It makes the tags of each group have more identical bits and less collision bits. It can efficiently identify tags, combining the characteristics of Manchester coding and the binary search algorithm. The steps of tag identification are shown in Fig. 1.

3.1 Estimating the Number of Tags

Firstly, we need to estimate the number of tags. Since this is a well-studied problem, existing method [7] is adopted in our scheme. It is a compound estimation algorithm combining the maximum likelihood estimation and the moment-based estimation to enhance the estimation accuracy. Interested readers can refer to [7] for more details.

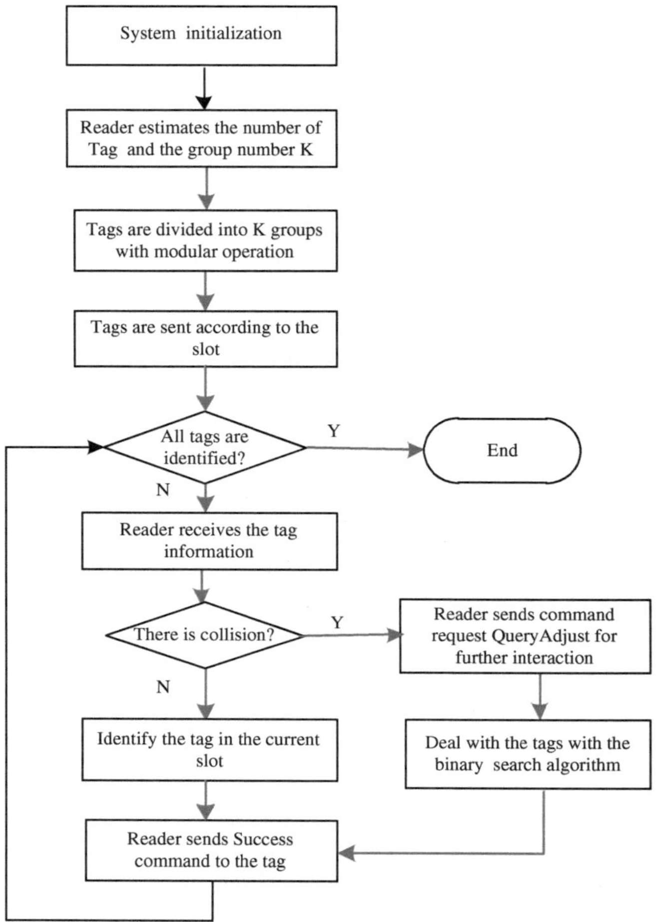

Fig. 1 The steps of tag identification

3.2 Grouping Tags

1. Reader sends the number of group

Reader calculates the number of tag groups $k = \text{floor}(\log_2 N)$, where N is the number of tags estimated in the first step. At the beginning of a frame, reader sends query command with k to all tags. Assuming that the number of tags is $N = 100$, and the number of tag group is $k = 6$, reader sets up the number of slot $2^k = 64$, namely frame length $L = 64$.

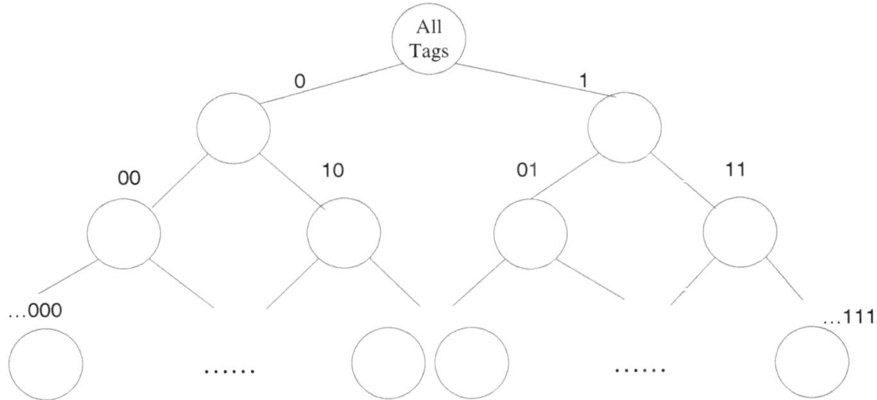

Fig. 2 Dividing tags into groups

2. Dividing tags into groups

According to the group number, tags are divided into groups using modular 2 operation. It is operated as follows: After the tag receives k from reader, it will set the counter to k and begin k time's module operation. Because modular 2 operation is equivalent to moving one bit to the right, tag ID is enough to explain in the following. As shown in Fig. 2, in the first group, we divide tags with 0 remainder into one group, which will be sent in the first 32 slot, and divide tags with 1 remainder into another group, which will be sent in the latter 32 slot. We divide tags into groups according to the lowest bit of their ID. After this operation, tags will store 0 or 1 in the highest bit of temporary register, then the counter minus 1. In the second group, tags are divided into four groups "00", "10", "01", and "11". The tag will store 0 or 1 in the second highest bit of temporary registers, and so on. After six times, tags are divided into 64 groups and sent in 64 slots. It can be found that there is a reverse relation between the last k bits of tag ID and their sending slots. For instance, the last k bits of tag ID is 101100, and then, it will be sent at slot 001101. Tag group sequence is stored in a temporary register, and it will be used in tag matching.

When all tags are divided into groups, the tag will test whether the current slot is matching with the group number. If it does, tag sends its ID, else the tag sets its state to silence.

4 Simulation and Analysis

In this section, we simulate the communication between the reader and the tags in terms of the number of queries, assuming the scenarios where tag IDs in RFID system. Next, we will show an example. Assume that the number of tags is 100, the length of tag ID is 10 bits, and the tag IDs are generated uniformly, such as:

$$\text{TagID} = \begin{bmatrix} 916 & 219 & 4 & 899 & 240 & 250 & 652 & 310 & 839 \\ 897 & 958 & 396 & 811 & 159 & 631 & 705 & 87 & 535 \\ 894 & 265 & 236 & 842 & 497 & 153 & 231 & 657 & 562 \\ 291 & 620 & 712 & 279 & 409 & 359 & 774 & 134 & 892 \\ 286 & 493 & 772 & 667 & 148 & 685 & 127 & 928 & 868 \\ 504 & 664 & 954 & 123 & 733 & 805 & 760 & 186 & 416 \\ 15 & 315 & 131 & 436 & 553 & 764 & 405 & 513 & 890 \\ 864 & 523 & 960 & 50 & 910 & 320 & 1{,}022 & 784 & 332 \\ 158 & 27 & 907 & 645 & 815 & 889 & 643 & 867 & 243 \\ 835 & 866 & 283 & 69 & 720 & 80 & 683 & 419 & 09 \\ 1{,}010 & 496 & 591 & 14 & 437 & 823 & 106 & 410 & 611 \\ 273 \end{bmatrix}$$

After calculation, we get the group number $k = 6$. All tags are divided into 64 groups and sent in 64 slots. The first ten slots are shown in Table 1. First of the table is the slot number, and each row of the rest columns stores tags ID. As there is a corresponding relationship between slot number and the tag ID, tags only need to send part of their ID. Taking tags in the second slot, for example, the last 6 bits of ID is 100000 (reverse), so the tag with ID code 928 only needs to send the first four bits 1110 (Assume the bit of ID code is 10).

According to the character of the Manchester code combined with binary search algorithm. Taking the second slot, for example, after the reader receives information from tags, getting the signal of X1XX, the reader sends Query Adjust command with an additional parameter 0111. The three tags in the current slot check whether the sending code is less than 0111. If it is, then the tag sets the counter to zero and responds with its ID, else the tag sets the counter to 1 and turns into wait. After reader identifies the tag successfully, the reader sends Query Adjust command with an additional parameter 1111 again, the remaining two tags set counter to 0 and responds. The following process is as same as the binary search algorithm. In the second slot, identification cost five rounds of interaction between the reader and tag. If there are two tags in the same slot, we only need three rounds of interaction. After 64 slots in one frame, the reader identifies all tags.

Table 1 The first ten slots after grouping

1	960	320	0
2	928	416	864
3	784	720	80
4	240	496	0
5	712	0	0
6	0	0	0
7	664	0	0
8	504	760	0
9	4	772	0
10	868	0	0

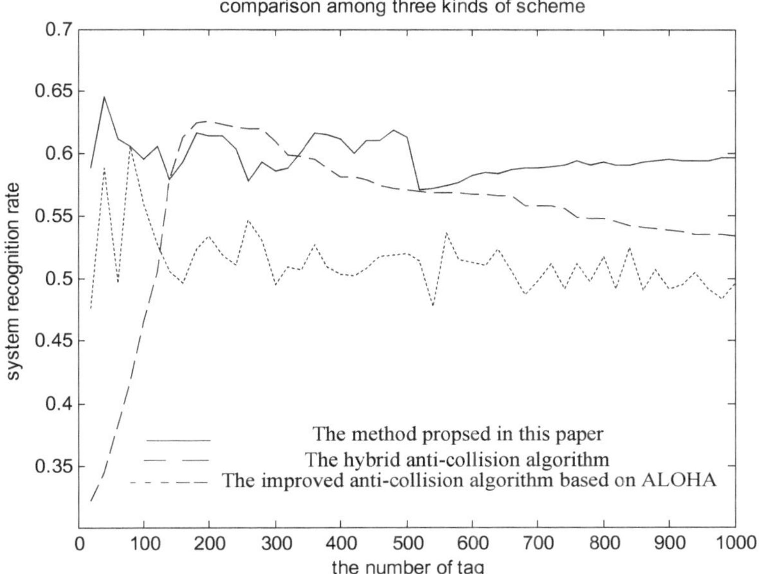

Fig. 3 Comparison among three kinds of schemes

In the comparison among an improved anticollision algorithm based on ALOHA [8], a new hybrid anticollision algorithm [9] and our scheme is taken using Matlab. The number of tag is 100–1,000. The simulation results is shown in Fig. 3. It shows the system recognition rate proposed in this paper is higher than that in the other two schemes. Specially, when the number of tag increases, the effect is more obvious.

5 Conclusion

In this paper, we propose a novel scheme to identify the massive tags efficiently. In this, scheme tags are divided into groups combined with the properties of the Manchester code. The tags in each group have few collision bits, and the group ID is the slot that tags will be read. As there is a corresponding relationship between slot number and tag ID, this reduces the ID length sent by tags. Experiments show that the efficiency of tag identification is improved.

Acknowledgments This work is supported by Information Security Special fund of National Development and Reform Commission (project name: Development of Security Test Service Capabilities in Wireless Intelligent Terminals), the National High Technology Research and Development Program of China (863 program), under Grant 2013AA014001.

References

1. Dind, Z.: Research and Realization on Technologies of RFID. University of Science and Technology of China, Anhui (2009)
2. Li, M., Qian, Z., Xu, Z.: Slot-predicting based ALOHA algorithm for RFID anti-collision. J. Commun. **32**(12), 43–50 (2011)
3. Jiang, L., Lu, G.: Research on anti-collision algorithm for RFID systems. Comput. Eng. Appl. **43**(5), 29–32 (2007)
4. Mateu, L., Moll, F.: Review of energy harvesting techniques and application for microelectronics. In: Proceedings of SPIE, pp. 359–373 (2009)
5. Yi, G., Guiling, S., Weixiang, L., et al.: Wireless sensor node design based on solar energy supply. In: Proceedings of the 2nd International Conference on Power Electronics and Intelligent Transportation System, pp. 203–207 (2009)
6. Liu, M., Xu, Z.: Analysis of RFID multi-tag identification defect and anti-collision algorithm. Microcontroller Embed. Syst. Appl. **1**, 18–20 (2010)
7. Li, Z., Zhang, C., Huang, M., Shi, H.: A compound tag number estimation scheme for aloha based anti-collision algorithm in RFID networks. In: IET International Conference on Communication Technology and Application, pp. 742–746, ICCTA (2011)
8. Wensheng, S., Rongqing, C.: An improved anti-collision algorithm based on ALOHA in RFID system. Commun. Netw. **37**(9), 107–110 (2011)
9. Jinhui, G., Xiaoyan, Z.: A new hybrid anti-collision algorithm in RFID system. Comput. Technol. Appl. **37**(12), 130–136 (2011)

Dynamic Vegas: A Competitive Congestion Control Strategy

Keren Zhou, Qian Yu, Zhenwei Zhu and Wenjia Liu

Abstract TCP Vegas is regarded as balanced congestion control strategy, however, shortcomings such as low bandwidth utilization and not gain fairness when sharing link with TCP Reno. We reinvestigate these issues and propose a modification strategy called Dynamic Vegas, which changes initial TCP Vegas' in slow start and congestion avoidance phase. It dynamically chooses slow start algorithm and adjust decrease/increase rate in congestion avoidance phase according to specific network environment. Experiments show that within a single link network, it performs as well as TCP Vegas. Additionally, in Multi-link environment, it achieves strong competitiveness against TCP Reno and gains fairness in throughput of all senders in the end of transmission.

Keywords Slow start · Congestion avoidance · Competitiveness · Fairness

1 Introduction

With the development of network communication technology, there are mainly 4 versions of TCP protocol: TCP Tahoe, TCP Reno, TCP NewReno and TCP SACK.

TCP Tahoe which is the earliest version includes three basic algorithms: "slow start", "congestion avoidance" and "fast retransmit". TCP Reno is added the "fast recovery" on the basis of TCP Tahoe. TCP NewReno modifies "fast recovery" process by taking a situation that a large quantity of data packages was lost in a

K. Zhou (✉) · Q. Yu · Z. Zhu · W. Liu
School of Software, Yunnan University, Kunming 650091, China
e-mail: robinho364@gmail.com

Q. Yu
Key Laboratory in Software Engineering of Yunnan Province, Kunming, China

S. Patnaik and X. Li (eds.), *Proceedings of International Conference on Computer Science and Information Technology*, Advances in Intelligent Systems and Computing 255, DOI: 10.1007/978-81-322-1759-6_40,

send window. TCP SACK which modifies the acknowledgement mechanism only retransmits the data packages lost.

A new congestion control strategy-TCP Vegas is invented by [1]. TCP Vegas detects whether network is congested or not by testing RTT to change the size of congestion window. For example, Vegas decreases the size of congestion window when detects that network is congested when RTT is becoming larger. Conversely, if RTT is becoming smaller, Vegas increases the size of congestion window.

The biggest advantage of TCP Vegas is the congestion mechanism which relies on RTT. It provides a more accurate way to predict available bandwidth and let TCP Vegas to be a fair and efficient protocol.

However, TCP Vegas cannot be used in large scale network, because it could not compete with other strategies such as TCP Reno [2]. The reason is that routers in network buffer data packages will lead the increment of RTT, whereas, it does not always mean that the network is congested. Thus, transmission delay will increase, RTT cannot provide an accurate measurement of the transmission rate. Back to the original TCP Vegas mechanism, to avoid congestion, the congestion window decreases. In the end, it can share fairness with TCP Reno. This situation is mostly happened in the wireless network.

Our study which can be divided into slow start phase and congestion avoid phaseis aiming at improving the competitiveness of TCP Vegas. In slow start phase, our strategy dynamically choosing slow start algorithm. In congestion avoid phase, we adjust the congestion window by the ratio of differs about predictive size and real size to improve TCP Vegas's efficiency.

2 Issues With TCP Vegas

TCP Vegas maintains two values: *ActualRTT* and *BaseRTT*, and values are changed each time receiving a valid ACK.

$$Expected = \frac{Cwnd}{BaseRTT} \tag{1}$$

$$Actual = \frac{Cwnd}{ActualRTT} \tag{2}$$

$$Diff = Expected - Actual \tag{3}$$

Expected is the expected transmission speed, *Actual* is the actual transmission speed. *Diff* is the diff of expected speed and actual speed.

In the congestion phase, Vegas defines two constant α and β. The size of the congestion window is increased by 1 when $Diff < \alpha$; when $Diff > \beta$, it decreases by 1. Otherwise, it remains unchanged.

$$Cwnd = \begin{cases} Cwnd + 1 & (Diff < \alpha) \\ Cwnd - 1 & (Diff > \beta) \\ Cwnd & (\alpha > Diff > \beta) \end{cases} \tag{4}$$

This algorithm will adjust the value of *Diff* to be $\alpha > Diff > \beta$, which indicates that resource of the network is balanced. And for the reason that α and β are set manually, it could flexibly fit specific network environment.

Although TCP Vegas can provide better performance than other protocols, and it has less jitter and retransmission, there are still many drawbacks about TCP Vegas, such as fairness, rerouting, and unfair treatment of connections [3].

Fairness is concerned when TCP Vegas and TCP Reno are both in the network, TCP Reno's *cwnd* is increasing consistently until a packet is lost. However, it leads to the increment of actual RTT measured by TCP Vegas, which results in the decrement of *cwnd* of TCP Vegas.

Rerouting is about traditional TCP Vegas performs badly when transmission route is changed. A method [4] called "active spurring" is provided to solve the problem, which has a basic idea that break down the balance when *cwnd* is stabilized.

3 Dynamic TCP Vegas

We present our improvement of TCP Vegas in this section. The core of our modification is dynamically choosing a strategy to fit the bandwidth.

In slow start phase, our algorithm detects the bandwidth of the route to decide which strategy be used.

In congestion control phase, the original TCP Vegas algorithm adjusts the value of *cwnd* by a constant number, no matter how much the ratio of expected transmit speed to current transmit speed.

In our modified algorithm, *cwnd* is dynamically changed according to current congestion degree. The recovery algorithm of dynamic Vegas is the same as that of Vegas. We discuss details of our adjustment in the following part.

3.1 Slow Start

As mentioned before, the slow start strategy is to dynamically choose algorithm. Additionally, a new algorithm called "mid speed start" is introduced to promote the performance of dynamic TCP Vegas. To begin with, we detect the bandwidth of the transmission route. Two variable delta and alpha in TCP Vegas is used to separate the situation into three categories as follow:

If delta is less than alpha, it indicates that utilization of network bandwidth is low. Thus, we apply "mid speed start" to quickly utilize the bandwidth.

The "mid speed start" algorithm is divided into following three stages:

The first stage. Set the value of *ssthresh*, and set the *cwnd* with half of *ssthresh*. The initial increase value of *cwnd* is *cwnd/2*.

The second stage. The value of *cwnd* increases with half of its increasing value in previous RTT.

The third stage. *cwnd* increases by one each RTT.

Intuitively, it is a reverse of the traditional slow start algorithm in TCP Reno which begins at a low speed, and goes exponentially. Thus, when both dynamic Vegas and TCP Reno are applied in the network, our algorithm has strong competitiveness.

If *delta* is greater than *alpha* and less than *beta*, we use traditionally start strategy in TCP Reno to utilize bandwidth.

If *delta* is greater than *beta*. Obviously, the transmission route is congested. The slow start strategy in Dynamic Vegas increases*cwnd* every two RTT.

Algorithm SlowStart()

```
If receive first ack in slowstart then //we determine slow-start algorithm when first ack arrives
    if delta<alpha then//indicate that bandwidth is sufficient
        cwnd=ssthresh/2
        flag=1
    else if alpha<delta<beta then//indicate that bandwidth is as we suppose
        flag=2
    else//bandwidth is to some extent congested
        flag=3
else if cwnd<ssthresh then//applying dynamic vegas
    if flag=1 then
        if cwnd/2>1 then//mid-speed start
            incr=cwnd/2
        else
            incr=1
    else if flag=2 then//common start
        incr=1
    else//vegas start
        incr add 1 for every other RTT
```

3.2 Congestion Control

Recall TCP Vegas's strategy which adjusts the sending rate to keep network transmission steady. Weakness in that strategy is modifying the value of *cwnd* by a constant number so that the speed is changed slowly. Thus, a more adaptive idea is to modify *cwnd* dynamically by *delta* and *alpha*.

$$ratio = \begin{cases} \frac{delta-beta}{beta}, & delta > beta \\ \frac{alpha-delta}{alpha}, & delta < alpha \end{cases} \tag{5}$$

If *delta* is greater than *beta*, which means network bandwidth is congested, the greater value of *delta*, the quicker ratio is decreasing. If *delta* is less than *alpha*, which indicates that network bandwidth is not utilized, sending rate should be increased. The closer *delta* to zero, the higher the *ratio*. However, since we are in the congestion control phase, the increasing speed could not be too high. If *delta* is greater than *alpha* and less than *beta*, which reflects that the actual sending rate is close to current sending rate, thus, *cwnd* remains unchanged.

Algorithm Congestion()

```
If delta>beta then
   ratio=(delta-beta)/beta
   incr=-ratio//decrease cwnd by ratio
   if cwnd<2 then
      cwnd=2
else if delta<alpha then
   ratio=(alpha-delta)/alpha
   incr=ratio//increase cwnd by ratio
else
   in
```

4 Simulation

In this section, we set up two simulations by using NS-2 software.

The first one is a single link TCP simulation. Throughput between traditional TCP Vegas and Dynamic Vegas are compared.

The second one is a multiple TCP connection simulation which consists of several receivers and senders. We have compared attributes such as *throughput* and *cwnd* of TCP Vegas and Dynamic Vegas in both heterogeneous and homogeneous network.

Table 1 The comparison between vegas and dynamic vegas's throughput capacity

Properties	Vegas	Dynamic Vegas
Throughput	3948.0 kbps	3933.6 kbps

4.1 Single TCP Connection Simulation

First we test the performance of Vegas and Dynamic Vegas in single high bandwidth-delay connection environment, comparing their time interval from slow start phase to congestion avoidance phase and their throughput capacity.

Link sender to router1 and router2 to receiver has a bandwidth of 100 Mbps with initial RTT of 1 ms, and router1 to router2 has a bandwidth has a bandwidth of 1 Mbps with initial RTT of 40 ms. Assuming that window size is 1,024, routers' cache size are 50. The simulation runs for 40 s so as to make the connection time believable and stable. Table 1 shows the comparison of the average performance between Vegas and Dynamic Vegas.

In the condition of single link, the performance of the Vegas and Dynamic Vegas are basically identical.

4.2 Multiple TCP Connection Simulation

This simulation will compare and analyze the bottleneck link's throughput when Vegas and Dynamic are in the network simultaneously. Link bandwidth for sender to router1 and router2 to receiver are 10 Mbps, the propagation delay is 1 ms. Set router1 to router2 link bandwidth 1 Mbps, the delay as 48 ms. Assuming window size to be 1,024, router cache to be 50. Running time for this simulation is 200 s, specific topology is shown in Fig. 1 (Table 2).

Vegas tends to finish the slow start phase which has an exponential growth of congestion window as soon as possible, and then enters the congestion avoidance phase until the congestion window into balance. Because of the congestion

Fig. 1 Multiple TCP connections topology

Table 2 The compassion between Vegas and Dynamic Vegas when reach steady state in different networks

Properties	Vegas	Dynamic Vegas
Cwnd (Homo)	10 packets	20 packets
Cwnd (Hete)	5 packets	25 packets
Throughput (Homo)	500 kbps	500 kbps
Throughput (Hete)	200 kbps	500 kbps

window has a linear growth in congestion avoidance phase, so it takes a lot of time of Vegas to make congestion window tends to the balance. Vegas has a lot of problems such as reroute, continue to congestion and so on. And a heterogeneous network composed of Vegas and TCP Reno exists compatibility issues and causes the decrease in network's throughput which composed of Vegas and Reno.

Table 2 shows performance of Dynamic Vegas compare with Vegas in both heterogeneous and homogenous network. We find that TCP Vegas and TCP Dynamic Vegas have almost consistent performance in the single TCP connection simulation. In multiple TCP connection simulation, both of them have almost same throughput in homogeneous network. But Dynamic Vegas has a lager congestion window than Vegas, so that it will have advantages of throughput in later period. The performance of TCP Dynamic Vegas are significantly better than TCP Vegas in heterogeneous network, which reflected in the size of the congestion window and the competitiveness of the throughput. And we can realize Dynamic Vegas' great stability of congestion window phase in heterogeneous network. Dynamic Vegas' congestion window does not make too big change when Reno joining this network, which improves TCP Vegas in competitiveness and stability.

We also find some problems in our algorithm. For example, it will lead to some packets lossbecause "mid-speed-start" has high bandwidth in start phase. But we do not consider that serious defects. Because Reno also has the same problem in its start phase. We will solve this problem for further improvement in our future research.

5 Conclusion

In this paper, we come up with a modified version of TCP Vegas, TCP Dynamic Vegas, which is better than TCP Vegas in some conditions. For example, it have advantages of achieving higher throughput in multiple connection simulation. Moreover, TCP Dynamic Vegas has gained greater competitiveness than TCP Vegas in heterogeneous network. And Dynamic Vegas' congestion window does not make significant change when Reno joins in the network, which improves TCP Vegas' stability and fairness. But, there are still some problems in this algorithm, and we will solve these problems for further improvement in our future research.

Acknowledgments This work is supported by Scientific Research Fund of Yunnan Provincial Department of Education under Grant No. 2012C108; Education Innovation Fund of Software School of Yunnan University under Grant No. 2010EI13 and No. 2010EI14.

References

1. Brakmo, L.S., Peterson, L.L.: TCP Vegas: End to end congestion avoidance on a global Internet. IEEE J. Sel. Areas Commun. **13**(8), 1465–1480 (1995)
2. Mo, J., La, R.J., Anantharam, V., Walrand, J: Analysis and comparison of TCP Reno and Vegas. IEEE proceedings of eighteenth Annual Joint Conference of the IEEE Computer and Communications Societies, INFOCOM'99, vol. 3 (1999)
3. Srijith, K.N., Lillykutty, J., Ananda, A.L.: TCP Vegas-A: solving the fairness and rerouting issues of TCP Vegas. Conference Proceedings of the IEEE International Performance, Computing, and Communications Conference 2003
4. Zeng, Y., Wei-Jun, JZ: A rerouting issue with TCP Vegas and its solution. Comput. Sci. 8(2006), 009 (2006)

The Research of Web Parallel Information Extraction Based on Hadoop

Songyu Ma, Quan Shi and Lu Xu

Abstract Big data that are driven by three major trends such as cloud computing, social computing, and mobile computing are reshaping the business process, IT infrastructure and our capture of the enterprise, customer and Internet information and use. To extract the big data in the Internet, the enterprise needs a scalable, flexible, and manageable data infrastructure. Therefore, this paper is based on the Hadoop framework, to analyze and design the large data information extraction system. Measurement shows that the huge amounts of data extraction on the basis of cluster have great improvement in performance compared with single extraction, with high reliability and scalability. What is more? The research of this paper will provide better technical solutions to Web information extraction and sensitive information.

Keywords Hadoop · Web information extraction · Crawler · Parallel indexing

1 Introduction

Large-scale Web information extraction is the extraction of users' interested data from a mass of Web. According to the certain strategy, specific computer program is used to collect information from the Internet and make the organization and processing of information.

With the rapid development of network technology, the large data on the Web is in the form of exponential growth, which makes the Web as the world's largest collection of data, and information extraction based on large-scale Web has been the focus of research scholars both at home and abroad. In the era of big data, the

S. Ma · Q. Shi (✉) · L. Xu
School of Computer Science and Technology, Nantong University, Nantong 226019
Jiangsu, China
e-mail: sq@ntu.edu.cn

S. Patnaik and X. Li (eds.), *Proceedings of International Conference on Computer Science and Information Technology*, Advances in Intelligent Systems and Computing 255, DOI: 10.1007/978-81-322-1759-6_41,

normal method of data extraction already cannot adapt to the existing data scale and Web page form. Therefore, it needs a scalable, high stability and reliability design that can adapt to large-scale Web data extraction system, so as to meet the massive Web application and the environmental requirements. Therefore, this article will use cluster parallel computing and distributed storage ability based on Hadoop [1], which will consume lots of computational resources through network distribution to multiple nodes. It is currently an effective solution.

2 System Architecture and Function

Web information extraction based on Hadoop cluster system architecture is aimed to realize the goal with high scalability and stability and make various functions of the system realize modular in the form of cluster. Finally, using CDH (Cloudera's distribution including Apache Hadoop, visualization cluster management tools) [2], we build a visual management tool to manage the cluster. Overall, system architecture is divided into four subcomponents: capture, organization, analysis, and decision. The system's overall architecture design is shown in Fig. 1.

As shown in Fig. 1, there are two kinds of monitoring modules that are designed in the whole system, which are internal feedback and external feedback. Internal feedback is responsible for monitoring the workflow task of running status and the failure to restart switch or server downtime. On the other hand, external feedback is responsible for the report of the cluster running status and task execution status to the system administrator.

2.1 *Capture Workflow Design*

The behavior of the capture layer is similar to Web crawler, through the design of program on a parallel crawler. Capture controller layer includes the crawler controller and workflow. Capture workflow can be divided into three states:

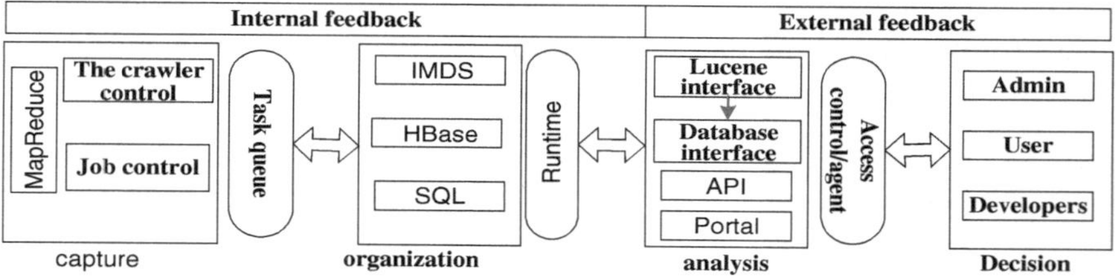

Fig. 1 The overall system architecture diagram

1. Seeder is to be responsible for the management of the URL, that is, the directional seed site URL, and ever catch all the URLs, scalable storage page fingerprints, history fetching information, etc;
2. Harvester are multiple, grasping task by competition, and paying attention to download, and then passing the content to the Collector;
3. The Collector has two tasks, which are parsing HTML, collecting new link, and giving the parsed content according to the need to pass to the index system and storage system. The collection of new connections is passed to the Seeder module.

2.2 The Design of Layer Based on the Organization of the Memory Database

A data operation queue is achieved by memory database. When the application requests database, it will interact with the main memory database firstly and make request to enquire in operation. Data operation queue query operations are prior to the update and insert operations. In addition, set the status of protection process monitoring task queue. When an exception occurs to the task queue, it will feedback to the application to take corresponding measures. When the database can be a normal response, it executes the team operation and returns the result. The design of data operation queue is made to adapt to the high concurrent data access, which is still able to maintain the normal operation of service.

2.3 Analysis Layer and the Design of Policy Makers

Analysis and policy makers are to provide regular data operation interface and analyze the data processing. Through the use of visual interface, or API of the system, mass information is extracted by the data analysis and statistics. Mainly based on artificial intelligence, machine learning, pattern recognition, statistics, database, and the application of visualization technology, etc., it can be highly automated analysis of known data, make the inductive reasoning, and reveal previously unknown and potentially valuable information hidden from massive data, tap potential model, so as to help the decision maker to make the right decisions. Implement the data query using open source Lucene full-text search engine combined with Hadoop, in a distributed manner to achieve the function of query and index. By partitioning the Lucene command to execute relevant nodes, we improve the query speed.

3 The Key Technology of System Implementation

3.1 Hadoop

Hadoop is mainly composed of distributed file system (HDFS) [3], MapReduce [4], and distributed storage system (Hbase). By infrastructure, it is a distributed system developed by the software foundation of open source parallel computing programming tools, supporting data-intensive distributed applications. Hadoop implements a calculation model, which is known as Map/Reduce. The calculation model of the application is divided into many small pieces, and each piece can be executed on any node in the cluster or to perform. Hadoop provides the HDFS for the use in each of the distributed cluster nodes to store data. It provides very high aggregate bandwidth for cluster.

3.2 IMDS

In-memory database system (IMDS) [5] refers to as opposed to the disk database for all data access control in memory. In disk database, although there is certain cache mechanism, it cannot avoid switching from a peripheral to the memory, and this kind of exchange loss is fatal. Because of memory read/write speed (dual-channel DDR3-1333 up to 9,300 MB/s, generally about 150 MB/s) disk, random access time can be in nanoseconds. Taking advantage of the high read/write database can greatly improve the performance of the system.

3.3 Lucene

Lucene is an open source framework of high performance, scalable full-text search engine [6], which provides complete query engine and index engine. Unlike search engines, Lucene does not have the capture function. It can only be for local existing index data indexing. According to the content index, it can quickly retrieve index information. Using the open source framework, full-text search engine can quickly build a private search engine.

4 The Design of Web Information Extraction Algorithm Based on Hadoop Cluster

4.1 Extraction Algorithm Design

In view of the mass Web information extraction, this paper makes use of the technology of MapReduce in large-scale network forum to directional extraction

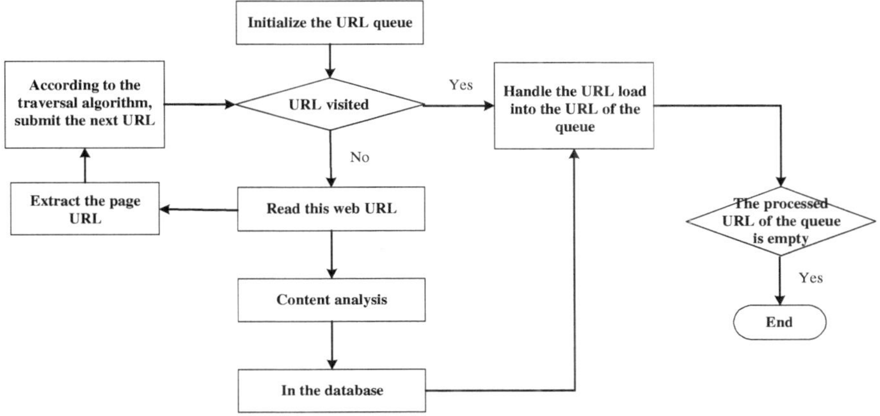

Fig. 2 Design flowchart of algorithm

of information based on the Hadoop framework. The main process is to access the target Web site, Web access code, and structured Web page code. Then, the extracted information is stored in a text file. Design algorithm process is shown in Fig. 2.

4.2 Distributed Indexing Algorithm Design

The test cases are mainly to test for the same amount of data in a cluster and single environment index time comparison, and the design algorithm process is as follows:

Step 1: The Map contents of the input file for each Map task need to request data in database;

Step 2: Read the value of the input file and instantiate the database objects and the Lucene index object;

Step 3: According to the document value into the database query, data column uses Lucene interface for the index operation;

Step 4: Statistical time-consuming and remove the previously created index.

5 System Test and Analysis

5.1 Experimental Environment

The framework of cluster based on Hadoop was measured. Here is built a Hadoop cluster with nine nodes, with Linux operating system kernel of Centos operating system and CDH visual cluster management tools. The whole cluster are in the same network segment, and the nodes use the blade server virtualization 2 × 8 core processor 96 G memory node from the data center each node has a dual core processor and memory 4 G, using gigabit network between nodes.

5.2 Cluster Extraction Test

In order to illustrate the efficiency better, here adopts the comparison of single and cluster extractions. The number of nodes in different is tested and analyzed, divided into 1, 3, 5, 9 nodes, composed of four groups of test examples. The extraction effect of the number of running time and extraction on Web pages is shown in Fig. 3. The abscissa is to extract the number of pages, ordinate is to extract the time (in second). It is obviously shown that it can effectively save the extracted time through cluster extraction.

5.3 Distributed Index Test

In the distributed index tests, it still takes the single node compared with multiple nodes, using the design of 1, 3, 5, and 9 nodes test cases. On the amount of indexing the 40, 80, 140, 200 (in ten thousands) of the data, the format of the data for the title with the body of the HTML code, test results are shown in Fig. 4, the abscissa is the number of nodes, the ordinate is indexed by the time consumption (in second).

It can be seen from the Fig. 4 that using cluster testing than single speed a lot, and the cluster has a better fault tolerance, stability, and the characteristics of easy to management.

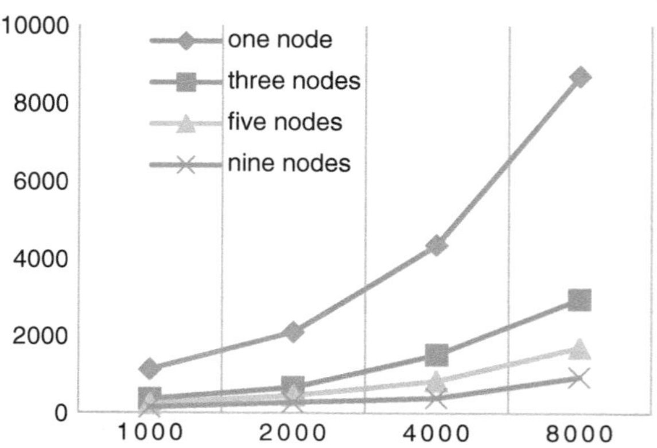

Fig. 3 Single extraction and cluster test results

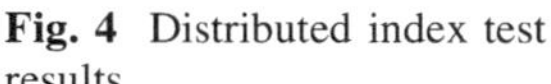

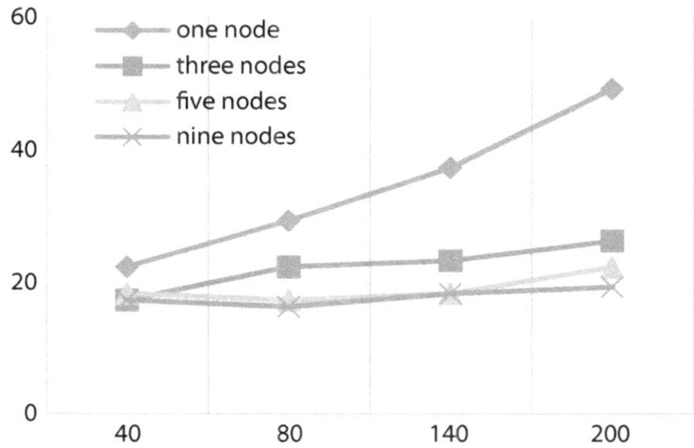

Fig. 4 Distributed index test results

6 Conclusion

This paper introduces the related technology of large-scale Web information extraction based on Hadoop cluster. The system architecture is designed to extract Web information, which is based on detailed analysis and design of trapping layer, organization layer, IMDS memory data analysis of four process layer, and decision layer. At the same time, the mass of Web information extraction flowchart of the algorithm is given. Finally, based on the detailed design, data extraction and distributed index are tested and analyzed. The results show that, with the cluster mode, data extraction compared with stand-alone extraction has great improvement and has high reliability and high scalability. This study provides better technical solutions to public opinion monitoring for a mass of Web information extraction and sensitive information.

In this paper, the follow-up will continue to study the distributed crawler working mode and the definition of workflow. The Hadoop crawler and distributed crawler routine will be compared and analyzed. On this basis, the detailed design of distributed index system architecture and the related data query interface is made for upper application, which has high coupling between the platform and the application.

Acknowledgments This work was supported by the National Natural Science Foundation of China (No. 61171132), Natural Science Foundation of Jiangsu Province (No. BK2010280), Graduate Students Research and Innovation Plan of Jiangsu (No. CXZZ13_0868), Students in Jiangsu Province Innovation Training Program (201310304064Y), Applying Study Foundation of Nantong (BK2011003, BK2011069, BK2012001, BK2012034), and Nantong technology platform projects (CP2013001).

References

1. PoweredBy-Hadoop Wiki (EB/OL). http://wiki.apache.org/hadoop/PoweredBy. Accessed on 17 Aug 2013
2. Chen, Y., Alspaugh, S., Katz, R.H.: Interactive analytical processing in big data systems: A cross industry study of MapReduce workloads. In Proceedings of VLDB, pp. 1802–1813 (2012)
3. Chang, F., Dean, J., Ghemawat, S., et al.: Bigtable: A distributed storage system for structured data. ACM Trans. Comput. Syst. **26**(2), 205–218 (2008)
4. Dean, J., Ghemawat, S.: MapReduce: Simplified data processing on large clusters. Commun. ACM **51**(1), 107–113 (2005)
5. In-Memory Database Systems—Questions and Answers (EB/OL). http://www.mcobject.com/in_memory_database. Accessed on 17 Aug 2013
6. Chen, L., Eugenio, D.B.: A Lucene and maximum entropy model based hedge detection system. Fourteenth Conference on Computational Natural Language Learning, Proceedings of the Shared Task, pp. 114–119 (2010)

A Novel Class of Periodic Complementary Sequence Sets over 8-QAM+ Constellation

Fanxin Zeng, Xiaoping Zeng, Zhenyu Zhang and Guixin Xuan

Abstract Based on a ternary complementary sequence set $\mathrm{PCS}_3^{N,2M}$, the number and period of whose sub-sequences are $2M$ and N, respectively, and its reversal, we propose a method so as to convert it into an 8-QAM+ periodic complementary sequence set (CSS) $\mathrm{PCS}_{8-\mathrm{QAM}+}^{N,2M}$. In contrast to the existing methods, new method's advantage lies in that the resultant 8-QAM+ periodic CSSs have same the number and period of sub-sequences as the ternary CSSs employed. On the contrary, the known methods cannot do so. As a consequence, all the methods can provide a large number of candidates of 8-QAM+ periodic CSSs so as to arrive at the requirement of applications in communications, such as channel estimation and synchronization.

Keywords Complementary sequence sets · 8-QAM+ constellation · Ternary · Period · Autocorrelation

1 Introduction

A QAM+ constellation, which was first investigated by Boztaş and Parampalli [1] in 2010, means an expanded quadrature amplitude modulation (QAM) constellation with the number "0". Although the signals over an expanded constellation

F. Zeng (✉) · Z. Zhang · G. Xuan
Chongqing Key Laboratory of Emergency Communication,
Chongqing Communication Institute, Chongqing 400035, China
e-mail: cqfzengx@gmail.com

Z. Zhang
e-mail: emzhenyuzhang@hotmail.com

F. Zeng · X. Zeng
College of Communication Engineering, Chongqing University, Chongqing 400044, China
e-mail: zxp@cqu.edu.cn

S. Patnaik and X. Li (eds.), *Proceedings of International Conference on Computer Science and Information Technology*, Advances in Intelligent Systems and Computing 255, DOI: 10.1007/978-81-322-1759-6_42,

have a drawback when they are transmitted, that is, the transmission of zero is set to a pause for a lasting symbol period [1], which implies the loss of communication energy, and their advantages are apparent. For example, it is well known that perfect QPSK sequences exist only at the lengths 4, 8, and 16 [2]; nevertheless, there are fairly rich perfect QPSK+ sequences [1], [3]. An 8-QAM+ constellation Ω_8^+ is defined as

$$\Omega_8^+ = \{0, 1, 1+j, j, -1+j, -1, -1-j, -j, 1-j\}. \tag{1}$$

To the best of authors' knowledge, up to now, lots of sequences over 8-QAM+ constellation have been known. More clearly, those sequences include almost perfect 8-QAM+ sequences [1], 8-QAM+ periodic complementary sequence sets (CSSs) [4], perfect 8-QAM+ sequences [5], perfect 8-QAM+ arrays [6], 8-QAM+ ZCZ sequence [7], and so forth.

In fact, 8-QAM+ symbols can be expressed by the direct product $\{-1, 0, 1\} \times \{-1, 0, 1\}$, which were discovered by the present authors in 2012 [4]. More clearly, a mapping ϕ from the direct product $\{-1, 0, 1\} \times \{-1, 0, 1\}$ to 8-QAM+ constellation Ω_8^+ can reach this goal.

$$\phi: \begin{array}{ll} \{-1,0,1\} \times \{-1,0,1\} & \rightarrow \Omega_8^+ \\ (a,b) & \rightarrow \phi(a,b), \end{array} \tag{2}$$

where

$$\begin{array}{ll} \phi_1 : \phi_1(a,b) = aj + b, & \phi_2 : \phi_2(a,b) = aj - b, \\ \phi_3 : \phi_3(a,b) = -aj + b, & \phi_4 : \phi_4(a,b) = -aj - b. \end{array} \tag{3}$$

On the other hand, in order to implement the goal of this paper, the ternary CSSs must be available. Fortunately, there are a large number of the existing ternary CSSs [8–16].

The rest of this paper is organized as follows. In Sect. 2, some necessary concepts will be given. And the existing methods for yielding 8-QAM+ periodic CSSs will be simply recalled in Sect. 3. In the following section, the major results will be stated. Finally, the concluding remark will appear in Sect. 5.

2 Preliminaries

Definition 1 Let $\underline{u}_r = \{u_r(t)\}_{t=0}^{N-1}$ and $\underline{v}_s = \{v_s(t)\}_{t=0}^{N-1}$ be two complex sequences with each of period N, more clearly, $\underline{u}_r = \{u_r(t)\} = (u_r(0), u_r(1), u_r(2), \cdots, u_r(N-1))$ and $\underline{v}_s = \{v_s(t)\} = (v_s(0), v_s(1), v_s(2), \cdots, v_s(N-1))$. For time shift τ, we refer to

$$R_{u_r,v_s}(\tau) = \sum_{t=0}^{N-1} u_r(t)\overline{v_s(t+\tau)} \tag{4}$$

as a periodic correlation function between the sequences $\underline{u}_r$ and $\underline{v}_s$. When $u = v$ and $r = s$, $R_{u_r,u_r}(\tau)$ is referred to as a periodic autocorrelation (PAC) function, otherwise, a periodic cross-correlation (PCC) function, where $\bar{x}$ denotes the complex-conjugate of x.

Definition 2 Let a complex sequence $\underline{u} = (\underline{u}_0, \underline{u}_1, \ldots, \underline{u}_{M-1})$ consist of M sub-sequences with each of period N. If sum of PACs of all sub-sequences in the sequence $\underline{u}$ satisfies

$$\sum_{r=0}^{M-1} R_{u_r,u_r}(\tau) = \begin{cases} \sum_{r=0}^{M-1} E_{u_r} & \tau \equiv 0 \pmod N \\ 0 & \tau \not\equiv 0 \pmod N, \end{cases} \tag{5}$$

we refer to the sequence $\underline{u}$ as a periodic CSS, which is denoted by $\mathrm{PCS}_{\mathrm{H}}^{N,M}$, that is, $\mathrm{PCS}_3^{N,M}$ for ternary ones and $\mathrm{PCS}_{8-\mathrm{QAM+}}^{N,M}$ for 8-QAM+ ones, where E_{u_r} is energy of the sub-sequence $\underline{u}_r$, namely, $E_{u_r} = \sum_{t=0}^{N-1} |u_r(t)|^2$. In particular, if $M = 2$, we call the sequence $\underline{u}$ a complementary sequence pair.

Definition 3 Let $\underline{u} = (\underline{u}_0, \underline{u}_1, \ldots, \underline{u}_{M-1})$ and $\underline{v} = (\underline{v}_0, \underline{v}_1, \ldots, \underline{v}_{M-1})$ be two $\mathrm{PCS}_{\mathrm{H}}^{N,M}$. For $\forall\ \tau$ ($|\tau| \le N-1$), if the following condition

$$\sum_{l=0}^{M-1} R_{u_l,v_l}(\tau) = 0 \tag{6}$$

holds, we say that these two CSSs are the mate to each other.

Definition 4 Let $\underline{u}_r = (u_r(0), u_r(1), u_r(2), \ldots, u_r(N-1))$ be a complex sequence. We refer to the sequence

$$(u_r(N-1), u_r(N-2), \ldots, u_r(1), u_r(0)) \tag{7}$$

as the reversal of the sequence $\underline{u}_r$, which is denoted by $\tilde{\underline{u}}_r$. Apparently, we have that $\tilde{\underline{u}}_r(t) = u_r(N-1-t)$ $(0 \le t \le N-1)$.

Lemma 1 [2]. *Let $\underline{u}_r$ and $\underline{v}_k$ be two ternary sequences. Hence, we have*

$$R_{\tilde{u}_r,\tilde{v}_k}(\tau) = R_{v_k,u_r}(\tau). \tag{8}$$

The properties of periodic complementary sequences referred to in this letter are given as follows.

Lemma 2 [2]. *Let $\underline{u} = (\underline{u}_0, \underline{u}_1, \ldots, \underline{u}_{2M-1})$ be a ternary CSS with $2M$ sub-sequences. Then we have*

1. *Interchanging any number of sub-sequences in $\underline{u}$ also produces a ternary CSS.*
2. *Reversing any number of sub-sequences in $\underline{u}$ also produces a ternary CSS.*
3. *The sequence*

$$(\underline{\tilde{u}}_1, -\underline{\tilde{u}}_0, \ldots, \underline{\tilde{u}}_{2M-1}, -\underline{\tilde{u}}_{2M-2}) \tag{9}$$

is the mate of the sequence $\underline{u}$*, where the symbol "* $-\underline{u}_1$*" means to negate the sequence "* $\underline{u}_1$*." In other words, for* $\forall\,\tau$*, we have*

$$\sum_{l=0}^{M-1}[R_{u_{2l},\tilde{u}_{2l+1}}(\tau) - R_{u_{2l+1},\tilde{u}_{2l}}(\tau)] = 0. \tag{10}$$

Definition 5 For the sequences $\{u(t)\}$ and $\{v(t)\}$ with each of period N, we construct a new sequence, called an interleaved sequence and denoted by$I[\,u(k), v(k)]$ or $I[\underline{u}, \underline{v}]$, as follows.

$$I[\underline{u}, \underline{v}] = (u(0), v(0), u(1), v(1), \ldots, u(N-1), v(N-1)), \tag{11}$$

which implies that an interleaved sequence has a period $2N$.

3 Existing Constructions

Construction 1 [4].

Based on the known ternary CSSs and the mappings $\phi_1 - \phi_4$, 8-QAM+ CSSs can be constructed by the following steps:

Step 1: Choose a ternary $\mathrm{PCS}_3^{N,M} = \underline{u} = (\underline{u}_1, \underline{u}_2, \ldots, \underline{u}_M)$, where the period N of its sub-sequences is even.

Step 2: Choose an integer l $(1<l<4)$ and construct 8-QAM+ sequence set $\underline{q}^l = (\underline{q}_1^l, \underline{q}_2^l, \ldots, \underline{q}_M^l)$, where $q_r^l(t) = \phi_l(u_r(t+\delta_{l,r,1}), u_r(t+\delta_{l,r,2}))(1 \le r \le M)$.

Step 3: Choose integers $\delta_{l,r,1}$ and $\delta_{l,r,2}(1 \le r \le M)$ in Step 2 so as to satisfy the following condition:

$$\delta_{l,r,1} \equiv \delta_{l,r,2} (\mathrm{mod}\, N/2). \tag{12}$$

Then, the resultant 8-QAM+ sequence set $\underline{q}^l$ is $\mathrm{PCS}^{N,M}_{8-\mathrm{QAM}+}$.

Construction 2 [4].

Based on the known ternary CSSs and the mappings $\phi_1 - \phi_4$, 8-QAM+ CSSs can be constructed by the following steps.

Step 1: Choose a ternary CSS $\mathrm{PCS}_3^{N,M} = \underline{u} = (\underline{u}_1, \underline{u}_2, \ldots, \underline{u}_M)$.

Step 2: By making use of the mappings ϕ_1 and ϕ_2, construct two 8-QAM+ sequence sets $\underline{q}^l = \left(\underline{q}_1^l, \underline{q}_2^l, \ldots, \underline{q}_M^l\right)$, where $q_r^l(t) = \phi_l\big(u_r\big(t+\delta_{l,r,1}\big), u_r\big(t+\delta_{l,r,2}\big)\big)$ $(1 \le r \le M \,\mathrm{and}\, l = 1, 2)$.

Step 3: Choose integers $\delta_{l,r,1}$ and $\delta_{l,r,2}(1 \le r \le M \,\mathrm{and}\, l = 1, 2)$ in Step 2 so as to satisfy the following condition.

$$\delta_{1,r,2} - \delta_{1,r,1} \equiv \delta_{2,r,2} - \delta_{2,r,1} (\mathrm{mod}\, N). \tag{13}$$

Hence, the resultant 8-QAM+ sequence set $\underline{q} = (\underline{q}^1, \underline{q}^2)$ is $\mathrm{PCS}^{N,2M}_{8-\mathrm{QAM}+}$.

Construction 3 [4].

Based on the known ternary CSSs, the interleaving technique, and the mappings $\phi_1 - \phi_4$, 8-QAM+ CSSs can be constructed by the following steps:

Step 1: Choose a ternary CSS $\mathrm{PCS}^{N,M}_3 = \underline{u} = (\underline{u}_1, \underline{u}_2, \ldots, \underline{u}_M)$ with odd period N.

Step 2: By making use of the mappings ϕ_1 and ϕ_2, construct an 8-QAM+ sequence set $\underline{q} = \left(I[\underline{q}^1_1, \underline{q}^2_1], I[\underline{q}^1_2, \underline{q}^2_2], \ldots, I[\underline{q}^1_M, \underline{q}^2_M]\right)$, where $q^l_r(t) = \phi_l(u_r(t+\delta_{l,r,1}), u_r(t+\delta_{l,r,2}))(1 \le r \le M \text{ and } l = 1,2)$ and $I[\underline{v}_k, \underline{v}_r]$ denote an interleaved sequence constructed by the sequences $\underline{v}_k$ and $\underline{v}_r$.

Step 3: Choose integers $\delta_{l,r,1}$ and $\delta_{l,r,2}(1 \le r \le M \text{ and } l = 1,2)$ in Step 2 so as to satisfy the following condition:

$$\begin{cases} \delta_{1,r,2} - \delta_{1,r,1} \equiv \delta_{2,r,2} - \delta_{2,r,1} (\mathrm{mod}\ N) \\ \delta_{2,r,1} - \delta_{1,r,2} \equiv \delta_{1,r,1} - \delta_{2,r,2} + 1 (\mathrm{mod}\ N) \end{cases} \tag{14}$$

As a consequence, the resultant 8-QAM+ sequence set $\underline{q}$ is $\mathrm{PCS}^{2N,M}_{8-\mathrm{QAM}+}$.

Obviously, if period N is even, no solution exists in Eq. (4). On the contrary, its general solution is

$$\begin{cases} \delta_{2,r,1} \equiv (N-1)/2 + \delta_{1,r,1} + 1 (\mathrm{mod}\ N) \forall \delta_{1,r,1} \\ \delta_{2,r,2} \equiv (N-1)/2 + \delta_{1,r,2} + 1 (\mathrm{mod}\ N) \forall \delta_{1,r,2}. \end{cases} \tag{15}$$

4 New Construction

First of all, we need to transform a ternary CSS into an 8-QAM+ periodic CSS. Let $\underline{u} = (\underline{u}_0, \underline{u}_1, \ldots, \underline{u}_{2M-1})$ be a $\mathrm{PCS}^{N,2M}_3$, whose entries take on values in the set $\{-1, 0, 1\}$. By making use of the mappings ϕ_1 and ϕ_2, we can construct the 8-QAM+ sequences $\underline{p}_k = \{p_k(t)\}(0 \le k \le 2M-1)$, which are given by $k = 2r$ and $k = 2r+1$, respectively, as follows.

$$\begin{cases} p_{2r}(t) = \phi_1(u_{2r}(t+\delta_{1,r,1}), u_{2r+1}(t+\delta_{1,r,2})) \\ p_{2r+1}(t) = \phi_2(\tilde{u}_{2r+1}(t+\delta_{2,r,1}), \tilde{u}_{2r}(t+\delta_{2,r,2})), \end{cases} \tag{16}$$

where the integers $\delta_{l,r,1}$ and $\delta_{l,r,2}(l = 1,2)$ satisfy $-N+1 \le \delta_{l,r,1}, \delta_{l,r,2} \le N-1$, and $0 \le r \le M-1$.

In comparison with the known constructions in Ref. [4], the new construction utilizes both the ternary $\underline{u}$ and the reversal of its sub-sequences, whereas only the ternary CSS $\underline{u}$ for the former. Just so, the new construction overcomes the former's shortcomings; in other words, the new construction transforms a ternary CSS into an 8-QAM+ periodic CSS with unaltered both period and number of sub-sequences.

Construction 4

Based on the known ternary CSSs, their reversals, and the mappings $\phi_1 - \phi_2$, 8-QAM+ CSSs can be constructed by the following steps.

Step 1: Choose a ternary CSS $\mathrm{PCS}_3^{N,2M} = \underline{u} = (\underline{u}_0, \underline{u}_1, \ldots, \underline{u}_{2M-1})$ with even number 2M of sub-sequences.

Step 2: By making use of the mappings ϕ_1 and ϕ_2, construct an 8-QAM+ sequence set $\underline{p} = \left(\underline{p}_0, \underline{p}_1, \ldots, \underline{p}_{2M-1}\right)$ by Eq. (16).

Step 3: Choose integers $\delta_{l,r,1}$ and $\delta_{l,r,2}(1 \le r \le M \text{ and } l = 1, 2)$ in Step 2 so as to satisfy the following condition.

$$\delta_{1,r,2} - \delta_{1,r,1} \equiv \delta_{2,r,2} - \delta_{2,r,1} (\mathrm{mod}\ N), \tag{17}$$

where $0 \le r \le M - 1$.

As a consequence, the resultant 8-QAM+ sequence set $\underline{p}$ is $\mathrm{PCS}_{8-\mathrm{QAM}+}^{N,2M}$.

Proof Choose an integer r in the range 0 to $M - 1$ and consider two PAC functions $R_{p_{2r},p_{2r}}(\tau)$ and $R_{p_{2r+1},p_{2r+1}}(\tau)$ as follows. By the definition of the 8-QAM+ sequences in Eq. (16), the autocorrelation of sub-sequence $\underline{p}_{2r}$ can be calculated by

$$\begin{aligned} R_{p_{2r},p_{2r}}(\tau) &= \sum_{t=0}^{N-1} p_{2r}(t)\overline{p_{2r}(t+\tau)} \\ &= \sum_{t=0}^{N-1} [ju_{2r}(t+\delta_{1,r,1}) + u_{2r+1}(t+\delta_{1,r,2})] \cdot [-ju_{2r}(t+\delta_{1,r,1}) + u_{2r+1}(t+\delta_{1,r,2})] \\ &= R_{u_{2r},u_{2r}}(\tau) + R_{u_{2r+1},u_{2r+1}}(\tau) + j[R_{u_{2r},u_{2r+1}}(\tau + \delta_{1,r,2} - \delta_{1,r,1}) - R_{u_{2r+1},u_{2r}}(\tau + \delta_{1,r,1} - \delta_{1,r,2})]. \end{aligned} \tag{18}$$

With the same argumentation as in Eq. (18), we have

$$\begin{aligned} R_{p_{2r+1},p_{2r+1}}(\tau) = R_{\tilde{u}_{2r},\tilde{u}_{2r}}(\tau) + R_{\tilde{u}_{2r+1},\tilde{u}_{2r+1}}(\tau) + j[R_{\tilde{u}_{2r},\tilde{u}_{2r+1}}(\tau + \delta_{2,r,1} - \delta_{2,r,2}) \\ - R_{\tilde{u}_{2r+1},\tilde{u}_{2r}}(\tau + \delta_{2,r,2} - \delta_{2,r,1})]. \end{aligned} \tag{19}$$

In accordance with Lemma 1 and the condition in Eq. (17), apparently, for $\forall\ \tau$ and $r = 0, 1, \ldots, M - 1$, we have

$$\begin{cases} R_{u_{2r},u_{2r+1}}\left(\tau + \delta_{1,r,2} - \delta_{1,r,1}\right) = R_{\tilde{u}_{2r+1},\tilde{u}_{2r}}\left(\tau + \delta_{2,r,2} - \delta_{2,r,1}\right) \\ R_{u_{2r+1},u_{2r}}\left(\tau + \delta_{1,r,1} - \delta_{1,r,2}\right) = R_{\tilde{u}_{2r},\tilde{u}_{2r+1}}\left(\tau + \delta_{2,r,1} - \delta_{2,r,2}\right), \end{cases} \tag{20}$$

Due to the fact that the sequence $\underline{u}$ is a ternary CSS and Lemma 1, we have

$$\begin{aligned}&\sum_{r=0}^{M-1}\left[R_{u_{2r},u_{2r}}(\tau)+R_{u_{2r+1},u_{2r+1}(\tau)}\right]\\&=\sum_{r=0}^{M-1}\left[R_{\tilde{u}_{2r},\tilde{u}_{2r}}(\tau)+R_{\tilde{u}_{2r+1},\tilde{u}_{2r+1}(\tau)}\right]\\&=\begin{cases}2\sum\limits_{k=0}^{2M-1}E_{u_k} & \tau\equiv 0 \pmod N\\ 0 & \tau\not\equiv 0 \pmod N).\end{cases}\end{aligned}\tag{21}$$

By summarizing the above, we have

$$\sum_{k=0}^{2M-1}R_{p_k,p_k}(\tau)=\begin{cases}\sum\limits_{k=0}^{2M-1}E_{u_k} & \tau\equiv 0 \pmod N\\ 0 & \tau\not\equiv 0 \pmod N).\end{cases}\tag{22}$$

which suggests that the 8-QAM+ sequence $\underline{p}$ is $\mathrm{PCS}_{8-\mathrm{QAM}+}^{N,2M}$.

According to the known optimal ternary CSSs, listed in Tables 1 and 2 in Refs. [8] and [11], respectively, the proposed 8-QAM+ periodic CSSs up to period 15 are given in Table 1, where the 8-QAM+ periodic CSSs with periods 13 and 15 result from the ternary CSSs in [11], the other from the ternary CSSs in [8], and here, we take on $(\delta_{1,r,1},\delta_{1,r,2},\delta_{2,r,1},\delta_{2,r,2})=((N-1)/2,0,0,(N+1)/2)$ for N odd or (N/2,0, 0, N/2) for N even $(0\le r\le M-1)$. Apparently, all the resultant 8-QAM+ periodic CSSs are of unaltered both periods and numbers of sub-sequences. And in Table 1, the symbol $\begin{pmatrix}a\\b\end{pmatrix}$ denotes the element $a+b_j$ in the 8-QAM+ constellation. Finally, Table 2 summarizes the parameters N's and M's of four constructions.

Note: Let $\underline{u}$ be a ternary perfect sequence with period N. Then, $(\underline{u},\underline{\tilde{u}})$ is a $\mathrm{PCS}_3^{N,2}$, which is an input of Construction 4.

Table 1 The 8-QAM+ periodic CSSs from construction 4 up to period 15

N	8-QAM+ periodic CSSs
2	$\begin{bmatrix} \binom{1}{1} & \binom{-1}{1} \\ \binom{1}{1} & \binom{1}{-1} \end{bmatrix}$
3	$\begin{bmatrix} \binom{1}{1} & -j & \binom{1}{1} \\ \binom{1}{-1} & -1 & \binom{1}{-1} \end{bmatrix}$
4	$\begin{bmatrix} \binom{1}{1} & \binom{1}{-1} & \binom{-1}{1} & \binom{1}{1} \\ \binom{1}{-1} & \binom{1}{1} & \binom{-1}{-1} & \binom{1}{-1} \end{bmatrix}$
5	$\begin{bmatrix} 1 & \binom{1}{1} & -j & \binom{-1}{1} & \binom{1}{1} \\ \binom{1}{-1} & \binom{1}{1} & -1 & \binom{1}{-1} & -j \end{bmatrix}$
6	$\begin{bmatrix} \binom{1}{1} & 1 & \binom{-1}{1} & \binom{-1}{1} & j & \binom{-1}{-1} \\ \binom{-1}{1} & 1 & \binom{1}{1} & \binom{1}{1} & -j & \binom{1}{-1} \end{bmatrix}$
7	$\begin{bmatrix} \binom{-1}{1} & j & -j & j & \binom{-1}{-1} & -j & \binom{-1}{-1} \\ \binom{-1}{1} & -1 & \binom{-1}{1} & 1 & -1 & 1 & \binom{1}{1} \end{bmatrix}$
8	$\begin{bmatrix} \binom{-1}{-1} & \binom{-1}{1} & \binom{1}{1} & \binom{1}{-1} & \binom{-1}{-1} & \binom{1}{-1} & \binom{-1}{-1} & \binom{1}{-1} \\ \binom{-1}{-1} & \binom{-1}{1} & \binom{-1}{-1} & \binom{-1}{1} & \binom{-1}{-1} & \binom{1}{-1} & \binom{1}{1} & \binom{-1}{1} \end{bmatrix}$
9	$\begin{bmatrix} 1 & \binom{1}{1} & \binom{1}{1} & \binom{-1}{-1} & j & \binom{-1}{1} & \binom{-1}{1} & \binom{1}{1} & \binom{-1}{-1} \\ \binom{-1}{1} & \binom{1}{-1} & \binom{1}{1} & \binom{1}{1} & 1 & \binom{-1}{1} & \binom{1}{-1} & \binom{1}{-1} & -j \end{bmatrix}$
10	$\begin{bmatrix} \binom{1}{1} & \binom{-1}{-1} & \binom{-1}{-1} & \binom{-1}{-1} & \binom{-1}{1} & \binom{-1}{1} & \binom{-1}{-1} & \binom{1}{-1} & \binom{1}{1} & \binom{-1}{-1} \\ \binom{-1}{1} & \binom{1}{-1} & \binom{-1}{-1} & \binom{-1}{1} & \binom{1}{1} & \binom{1}{1} & \binom{-1}{1} & \binom{-1}{1} & \binom{-1}{1} & \binom{1}{-1} \end{bmatrix}$
11	$\begin{bmatrix} 1 & 1 & \binom{1}{1} & \binom{-1}{1} & -j & j & j & \binom{-1}{1} & \binom{-1}{1} & \binom{1}{-1} & -1 \\ j & \binom{-1}{-1} & \binom{1}{1} & \binom{1}{1} & 1 & 1 & -1 & \binom{1}{1} & \binom{1}{-1} & j & -j \end{bmatrix}$
12	$\begin{bmatrix} 1 & 1 & \binom{1}{-1} & \binom{-1}{-1} & j & -j & \binom{-1}{1} & \binom{1}{1} & \binom{1}{1} & \binom{1}{-1} & \binom{-1}{1} & \binom{1}{1} \\ \binom{1}{-1} & \binom{1}{1} & \binom{-1}{-1} & \binom{1}{-1} & \binom{1}{-1} & \binom{1}{1} & -1 & 1 & \binom{-1}{1} & \binom{-1}{-1} & -j & -j \end{bmatrix}$
13	$\begin{bmatrix} -1 & \binom{1}{1} & \binom{1}{-1} & \binom{1}{-1} & j & 1 & j & \binom{-1}{-1} & \binom{1}{1} & \binom{1}{1} & \binom{-1}{1} & 0 & \binom{-1}{1} \\ \binom{1}{1} & 0 & \binom{1}{1} & \binom{1}{-1} & \binom{1}{-1} & \binom{-1}{1} & 1 & -j & 1 & \binom{-1}{-1} & \binom{-1}{-1} & \binom{1}{-1} & j \end{bmatrix}$
14	$\begin{bmatrix} \binom{1}{-1} & \binom{1}{-1} & \binom{1}{1} & \binom{-1}{-1} & \binom{-1}{1} & 1 & \binom{1}{1} & \binom{1}{1} & \binom{-1}{1} & \binom{1}{1} & \binom{-1}{1} & \binom{-1}{-1} & j & \binom{-1}{1} \\ \binom{1}{1} & 1 & \binom{-1}{1} & \binom{1}{1} & \binom{1}{-1} & \binom{1}{1} & \binom{1}{-1} & \binom{1}{-1} & -j & \binom{1}{1} & \binom{-1}{1} & \binom{1}{-1} & \binom{-1}{-1} & \binom{-1}{-1} \end{bmatrix}$
15	$\begin{bmatrix} \binom{1}{1} & \binom{1}{-1} & -1 & \binom{1}{-1} & \binom{1}{1} & \binom{1}{-1} & -1 & \binom{-1}{1} & \binom{-1}{1} & j & \binom{1}{-1} & \binom{-1}{1} & \binom{1}{1} & j & \binom{-1}{1} \\ \binom{1}{1} & 1 & \binom{1}{-1} & \binom{1}{1} & \binom{-1}{-1} & 1 & \binom{1}{1} & \binom{1}{1} & j & \binom{-1}{-1} & \binom{1}{-1} & \binom{-1}{-1} & j & \binom{-1}{-1} & \binom{1}{-1} \end{bmatrix}$

Table 2 Parameters N's and M's of four constructions

Types	Construction 1	Construction 2	Construction 3	Construction 4
PCS_3	N (even)	N	N (odd)	N
$PCS_{8\text{-QAM+}}$	N	N	$2N$	N
PCS_3	M	M	M	$2M$
$PCS_{8\text{-QAM+}}$	M	$2M$	M	$2M$

5 Conclusion

This paper presents a new construction for yielding 8-QAM+ periodic sequences, whose advantage is to unalter the original number and length of sub-sequences. But, the ternary CSSs employed must have even number of sub-sequences, which is a drawback in the proposed method. However, the union of all the constructions can provide rich candidates for various applications.

Acknowledgements This work was supported by the National Natural Science Foundation of China (NSFC) under Grants 60872164, 61002034, 61171089, and 61271003, and the Ministry of Industry and Information Technology of China (No. Equipment [2010]307).

References

1. Boztaş, S., Parampalli, U.: Nonbinary sequences with perfect and nearly perfect autocorrelations. In: Proceedings of ISIT 2010, pp.1300–1304. IEEE, Austin, Texas (2010)
2. Fan, P.Z., Darnell, M.: Sequence Design for Communications Applications, Wiley (1996)
3. Zeng, F.X., Zeng, X.P., Zeng, X.Y., Zhang, Z.Y., Xuan, G.X.: Several types of sequences with optimal autocorrelation properties. IEICE Trans. Fundam. **E96-A**(1), 367–372 (2013)
4. Zeng, F.X., Zeng, X.P., Zhang, Z.Y., Xuan, G.X.: 8-QAM+ periodic complementary sequence sets. IEEE Commun. Lett. **16**(1), 83–85 (2012)
5. Zeng, F.X., Zeng, X.P., Zeng, X.Y., Zhang, Z.Y., Xuan, G.X.: Perfect 8-QAM+ sequences. IEEE Wireless Commun. Lett. **1**(4), 388–391 (2012)
6. Zeng, F.X., Zeng, X.P., Zhang, Z.Y., Xuan, G.X.: Perfect quaternary arrays based on perfect binary arrays. In: Proceedings of ICISE2011, pp. 4818–4822. Yangzhou, China (2011)
7. Li, Y.B., Xu, C.Q.: Zero correlation zone sequence sets over the 8-QAM+ constellation. IEEE Commun. Lett. **16**(11), 1844–1847 (2012)
8. Gavish, A., Lempel, A.: On ternary complementary sequences. IEEE Trans. Inf. Theory **40**(2), 522–526 (1994)
9. Gysin, M., Seberry, J.: Multiplications of complementary pairs. Australas. J. Combin. **14**, 165–188 (1996)
10. Koukouvinos, C.: On ternary complementary sequences. Bull. Inst. Comb. Appl. **22**, 99–101 (1998)
11. Gysin, M.: Seberry, J: On ternary complementary pairs. Australas. J. Combin. **23**, 153–170 (2001)
12. Doković, D.Z.: Base sequences, complementary ternary sequences, and orthogonal designs. J. Combin. Designs. **4**(5), 339–351 (1996)
13. Craigen, R., Koukouvinos, C.: A theory of ternary complementary pairs. J. Combin. Theory Ser. A **96**, 358–375 (2001)
14. Craigena, R., Georgioub, S., Gibson, W., Koukouvinos, C.: Further explorations into ternary complementary pairs. J. Combin. Theory Ser. A **113**, 953–965 (2006)
15. Craigen, R., Gibson, W., Koukouvinos, C.: An update on primitive ternary complementary pairs. J. Combin. Theory Ser. A **114**, 957–963 (2007)
16. Craigen, R.: Boolean and ternary complementary pairs. J. Combin. Theory Ser. A **104**, 1–16 (2003)

Conditional Diagnosability of Twisted-Cube Connected Networks

Xiaoyan Li, Lishan Lu and Shuming Zhou

Abstract The growing size of a multiprocessor system increases its vulnerability to component failures. It is crucial to locate and replace the faulty processors to maintain a system's high reliability. The fault diagnosis is the process of identifying faulty processors in a system through testing. Through fault tolerance analysis of the multiprocessor system based on twisted-cube connected network TN_n, we derive the conditional diagnosability of the system, which is about three times of its classical diagnosability under the comparison model.

Keywords Conditional diagnosability · Comparison diagnosis model · Twisted-cube connected network

1 Introduction

In recent years, large-scale distributed systems and loosely coupled multiprocessor systems have been developed for many critical applications in military, commerce, and scientific computing. The growing size of the systems increases their vulnerability to component failures. To maintain one system's high availability, it is crucial to locate the faulty processors therein efficiently and then replace them with spare ones. The process of identifying faulty processors in a system by analyzing the outcomes of available interprocessor tests is the so-called system-level diagnosis. The foundation of system diagnosis and an original diagnostic model,

X. Li · L. Lu · S. Zhou (✉)
School of Mathematics and Computer Science, Fujian Normal University, Fuzhou, 350007 Fujian, China
e-mail: zhoushuming@fjnu.edu.cn

S. Zhou
Key Laboratory of Network Security and Cryptology, Fujian Normal University, Fuzhou 350007 Fujian, China

S. Patnaik and X. Li (eds.), *Proceedings of International Conference on Computer Science and Information Technology*, Advances in Intelligent Systems and Computing 255, DOI: 10.1007/978-81-322-1759-6_43,

namely the PMC model, were established in a classic paper by Preparata et al. [1]. Along with their diagnostic model, Preparata et al. proposed the one-step diagnosis whose target is to identify the exact state of all nodes. The comparison-based diagnosis models, first proposed by Malek [2] and Chwa and Hakimi [3], have been considered to be a practical approach for fault diagnosis in multiprocessor systems. In these comparison-based models, the same job is assigned to a pair of processors in the system and their outputs are compared by a central observer. This central observer performs diagnosis using the outcomes of these comparisons. Maeng and Malek [4] extended Malek's comparison approach to allow the comparisons carried out by the processors themselves. Sengupta and Dahbura [5] developed this comparison approach such that the comparisons have no central unit involved. The diagnosabilities of hypercube enhanced hypercube [6, 7], crossed hypercube [8], and the locally twisted cubes [9] under the comparison diagnosis model are upper bounded by their regular vertex degree. Zheng et al. [10] investigated the diagnosability of Star graph S_n under the comparison diagnostic model. Lin et al. [11] introduced the conditional diagnosis and obtained that the conditional diagnosability of Star graphs S_n is $3n - 7$ which is about three times larger that the classical diagnosability of star graphs. This idea was attributed to Lai et al. [12] who is the first to use a restricted diagnosis strategy. They obtained that the conditional diagnosability of the hypercubes Q_n is $4n - 7$ under the PMC model. Recently, Zhou et al. [13] have determined the conditional diagnosability of multiprocessor system based on dual-cube DC_n under the comparison model. In this paper, we establish the conditional diagnosability of twisted-cube connected networks TN_n $(n \geq 7)$ under the comparison diagnosis model.

2 Terminologies and Preliminaries

2.1 Definitions and Notations

Throughout this paper, we use a graph $G = G(V, E)$ to present a self-diagnosable system, where each node $u \in V$ denotes a processor and each edge $(u, v) \in E$ denotes a link between nodes u and v. Let S be a subset of $V(G) \cup E(G)$. The subgraph of G induced by S, denoted by $G[S]$, is the graph with the vertex set $S \cap V(G)$ and the edge set $\{(u, v) | (u, v) \in E(G), u, v \in S\}$. For any subset $F \subset V$, the notation $G - F$ represents the graph obtained by removing the vertices in F from G and deleting those edges with at least one end vertex in F simultaneously. If $G - F$ is disconnected, then F is called a vertex-cut or a separating set. The components of G are its maximal connected subgraphs. The connectivity $\kappa(G)$ of $G = G(V, E)$ is the minimum number of nodes whose removal results in a disconnected or a trivial graph. For any node u of G, denote by $N(u)$ the set of all its neighboring nodes, i.e., $N(u) = \{v | (u, v) \in E\}$. For any subset $F \subset V$. Let $N(F) = \bigcup_{u \in F} N(u) - F$, $N[F] = N(F) \cup F$. For brevity, $N[u] = N(u) \cup \{u\}$, $N(\{u, v\})$ and $N[\{u, v\}]$ are written as

$N(u, v)$ and $N[u, v]$. The symmetric difference of the two sets F_1, F_2 is defined as the set $F_1 \Delta F_2 = (F_1 - F_2) \cup (F_2 - F_1)$.

The connectivity $\kappa(G)$ is an important parameter to measure the fault tolerance of the network, while it has an obvious deficiency in that it tacitly assume that all elements in any subset of G can potentially fail at the same time. To compensate for this shortcoming, it would seem natural to generalize the classical connectivity by introducing some conditions or restrictions on the set separating set S and/or the components of $G - S$. Consequently, Esfahanian introduced the concept of the restricted cut and studied the restricted connectivity of hypercube [14]. Furthermore, Latifi [15] explored the diameter of hypercube under this restricted fault model. A restricted vertex set S is a restricted vertex-cut if $G - S$ is disconnected, and no component is an isolated vertex. The restricted vertex connectivity of a graph G, denoted by $\kappa'(G)$, is the minimum cardinality of a restricted vertex-cut.

2.2 The Conditional Comparison Diagnosis Model

Fault diagnosis is an important step in the design of multiprocessor systems. The comparison diagnosis strategy can be modeled as a multigraph $M = (V, C)$ where V is the same node set defined as in G, C is the labeled edge set. A labeled edge $(u, v)_w$ is said to belong to C if (u, v) is an edge labeled by w, which implies that the processors u, v are compared by processor w. Since different comparators can compare the same pair of processors, M is a multigraph. Denote the comparison result as $r((u, v)_w)$ such that $r((u, v)_w) = 0$ if the outputs of u, v agree; and $r((u, v)_w) = 1$ if the outputs disagree. If the comparator w is fault-free and $r((u, v)_w) = 0$, the processors u and v are fault-free; while $r((u, v)_w) = 1$, at least one of the three processors u, v, w is faulty. If the comparator w is faulty, the comparison result is unreliable. The collection of the comparison results defined as a function s: $C \rightarrow \{0, 1\}$, is called the syndrome of the diagnosis. A subset $F \subset V$ is said to be compatible with a syndrome s if s can arise from the circumstance that all vertices in F are faulty and all vertices in $V - F$ are fault-free. A system is diagnosable if for every syndrome s, there is unique subset $F \subset V$ that is consistent with it. The maximum number of faulty vertices that the system G can guarantee to identify is called the diagnosability of G, written as $t(G)$. A faulty comparator can lead to unreliable results, so a set of faulty vertices may produce different syndromes. Let $\sigma_F = \{\sigma | \sigma \text{ is compatible with } F\}$. Two distinct subsets F_1, F_2 of $V(G)$ are said to be indistinguishable if and only if $\sigma_{F_1} \cap \sigma_{F_2} \neq \emptyset$; otherwise, F_1, F_2 are said to be distinguishable. There are several different ways to verify whether a system is t-diagnosable under the comparison approach. The following lemma given by Sengupta and Dahbura [5] is necessary and sufficient condition ensuring distinguishability.

Lemma 1 [5] *Let G be a graph. For any two distinct subsets* F_1, F_2 *of V(G),* (F_1, F_2) *is a distinguishable pair if and only if at least one of the following conditions is satisfied:*

1. *there are two distinct vertices* $u, w \in V(G) - (F_1 \cup F_2)$ *and there is a vertex* $v \in F_1 \Delta F_2$ *such that* $(u, v)_w \in C$;
2. *there are two distinct vertices* $u, v \in F_1 - F_2$ *and there is a vertex* $w \in V(G) - F_1 \cup F_2$ *such that* $(u, v)_w \in C$; *or*
3. *there are two distinct vertices* $u, v \in F_2 - F_1$ *and there is a vertex* $w \in V(G) - F_1 \cup F_2$ *such that* $(u, v)_w \in C$.

Lin et al. [11] introduced a new restricted diagnosability called conditional diagnosability. They considered the situation that no faulty set can contain all the neighbors of any vertex in a system. A faulty set $F \subset V(G)$ is called a conditional faulty set if $N_G(v) \not\subset F$ for every vertex $v \in V(G)$. A system $G(V, E)$ is said to be conditionally t-diagnosable if F_1 and F_2 are distinguishable for each pair of distinct conditional faulty set F_1 and F_2 of $V(G)$ with $|F_1| \le t$, $|F_2| \le t$. The maximum value of t such that G is conditionally t-diagnosable is called the conditional diagnosability of G, denoted by $t_c(G)$. It is trivial that $t_c(G) \ge t(G)$.

3 Fault Tolerance of Twisted-Cube Connected Networks

Recently, advances in VLSI circuit technology made it possible to build a large parallel and distributed system involving thousands or even tens of thousands of processors. Topological design is very important in a parallel and distributed system. Among the deigned structures, the hypercube Q_n is a popular one with many attractive properties. To reduce the communication diameter without adding extra edges, a family of variants of the hypercubes have been proposed, such as the crossed cube CQ_n [8], the Mobius cube MQ_n [16], and the locally twisted-cube LTQ_n [17] and the twisted-cube TQ_n [18]. Wang and Zhao [19] presented a new variant of the hypercube—the twisted-cube connected network. This (n-dimensional) variant, denoted by TN_n, is an n-regular graph and has the same number of vertices and edges as that of the hypercube. TN_n is constructed with the twisted 3-cube as basic module, and the diameter of TN_n is just $\lceil (n+1)/2 \rceil$ which is about half that of Q_n.

If $u = u_n u_{n-1} \ldots u_2 u_1$ is a binary string of length n, we call u_i the ith bit of u, and the complement of u_i will be denoted by $\overline{u_i}$. e_i denotes an n-bit binary string where the ith bit is 1 and all other bits are $0 (i = 1, 2, \ldots, n)$. For any binary string $u = u_n u_{n-1} \ldots u_2 u_1$, $v = v_n v_{n-1} \ldots v_2 v_1$ and integer k, define

1. $u + v = a_n a_{n-1} \ldots a_2 a_1$, where $a_i = u_i + v_i (\bmod 2)$;
2. $ku = b_n b_{n-1} \ldots b_2 b_1$, where $b_i = ku_i (\bmod 2)$.

Definition 1 [19] The n-dimensional twisted-cube connected network TN_n is recursively defined as follows (see Fig. 1):

1. TN_1 is a K_2, TN_2 is a 4-cycle, and TN_3 is the twisted 3-cube;
2. TN_n $(n \geq 4)$ consists of two $(n-1)$-dimensional twisted-cube connected networks TN_{n-1}^0, TN_{n-1}^1: for $i = 0, 1$, $V(\mathrm{TN}_{n-1}^i) = \{u_n u_{n-1} \ldots u_2 u_1 | u_n = i\}$, and two vertices $u = u_n u_{n-1} \ldots u_2 u_1 \in V(\mathrm{TN}_{n-1}^0)$, $v = v_n v_{n-1} \ldots v_2 v_1 \in V(\mathrm{TN}_{n-1}^1)$ are adjacent in TN_n if and only if $v = u + e_n + \sum_{t=1}^{q_n} u_{2t-1} e_{2t} + \sum_{t=3}^{n-1} u_t e_2$, where $q_n = \lceil n/2 - 1 \rceil$.

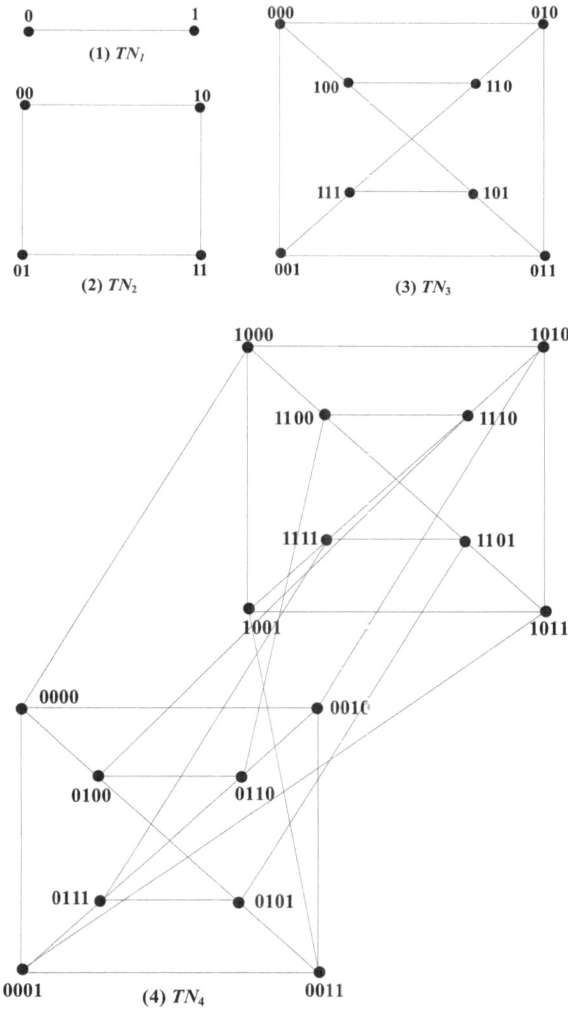

Fig. 1 The twisted-cube connected networks TN_1–TN_4

It is easy to see that $\mathrm{TN}_n = \mathrm{TN}_{n-1}^0 \oplus \mathrm{TN}_{n-1}^1 = L \oplus R$. For any vertex $u \in L$, we use $\overline{e}(u) = (u, \overline{u})$ to denote the crossed edge which is incident to $u \in L$. $\overline{u} \in R$ and $u \in L$ are called each other's pair vertex.

The following lemmas are helpful throughout the paper.

Lemma 2 [19] $\kappa(\mathrm{TN}_n) = n$.

Lemma 3 *The restricted vertex connectivity of* TN_n *is* $\kappa'(\mathrm{TN}_n) = 2n - 2$.

Proof Consider any pair adjacent vertices $u, v \in \mathrm{TN}_n$. Because TN_n do not has a cycle of length less than four, $|N(u, v)| = 2n - 2$. If these $2n - 2$ nodes are faulty, the edge (u, v) will be disconnected from the remaining graph and the faulty set satisfies the condition that each node has at least one fault-free neighbor, so $\kappa'(\mathrm{TN}_n) \leq 2n - 2$. Now, we show that $\kappa'(\mathrm{TN}_n) \geq 2n - 2$. It suffices to prove that $\mathrm{TN}_n - F$ with $|F| \leq 2n - 3$ is connected under the restricted condition that $N(v) \not\subset F$ for $\forall v \in \mathrm{TN}_n$. Let $F_0 = F \cap \mathrm{TN}_{n-1}^0$, $F_1 = F \cap \mathrm{TN}_{n-1}^1$. Because $F_0 \cap F_1 = \phi$ and $|F| \leq 2n - 3$, it implies that either $|F_0| \leq n - 2$ or $|F_1| \leq n - 2$. Without loss of generality, assume that $|F_1| \leq n - 2$. By applying Lemma 2, $\mathrm{TN}_{n-1}^1 - F_1$ is connected. It suffices to show that each fault-free node in $\mathrm{TN}_{n-1}^0 - F_0$ is connected by a path to the connected subgraph $\mathrm{TN}_{n-1}^1 - F_1$. Observe that each vertex $x \in \mathrm{TN}_{n-1}^0 - F_0$ has a pair vertex $\overline{x} \in \mathrm{TN}_{n-1}^1$. If $\overline{x} \notin F_1$, x may connect to $\mathrm{TN}_{n-1}^1 - F_1$ through the crossed edge $(x, \overline{x})$, we are done. So we suppose that $\overline{x} \in F_1$. Under the conditional faulty set assumption, $N(x) \cap V(\mathrm{TN}_{n-1}^0 - F_0)$ is not empty because $N(u) \not\subset F$ for any $u \in V(\mathrm{TN}_n)$. Thus, there exists $y \in N(x) \cap V(\mathrm{TN}_{n-1}^0 - F_0)$. If $\overline{y} \notin F_1$, we are done. Now assume $\overline{y} \in F_1$, and let $S = N(x, y) \cap \mathrm{TN}_{n-1}^0$, $F' = F - \{x, y\}$. Obviously, $|S| = 2n - 4$, $|F'| \leq 2n - 5$, and each vertex $s_i \in S = N(x, y) \cap \mathrm{TN}_{n-1}^0$ has a pair vertex $\overline{s_i} \in \mathrm{TN}_{n-1}^1$, so there exists some $s_0 \in S = N(x, y) \cap \mathrm{TN}_{n-1}^0$ such that $\overline{s_0} \notin F_1$. Thus, the node $x \in \mathrm{TN}_{n-1}^0 - F_0$ is connected to the subgraph $\mathrm{TN}_{n-1}^1 - F_1$.

Lemma 4 *Let u be any vertex of* TN_n *(* $n \geq 3$*). Then,* $\kappa(\mathrm{TN}_n - N[u]) = n - 2$.

Proof This is clearly true for the case $n = 3$. Now, we assume that it is true for $k = n - 1 \geq 3$ and show it is true for $k = n$. Without loss of generality, we set $u \in V(\mathrm{TN}_{n-1}^0)$. Let $F_0 = F \cap \mathrm{TN}_{n-1}^0$, $F_1 = F \cap \mathrm{TN}_{n-1}^1$. Clearly, $\kappa(\mathrm{TN}_n - N[u]) \leq \delta(\mathrm{TN}_n - N[u]) = n - 2$ for $\forall u \in V(\mathrm{TN}_{n-1}^1)$. So it suffices to prove $\kappa(\mathrm{TN}_n - N[u]) \geq n - 2$, and for any set F of $n - 3$ vertices in the graph $\mathrm{TN}_n - N[u]$, the subgraph $\mathrm{TN}_n - N[u] - F$ is connected.

Case 1 $|F_0| = n - 3$ and $|F_1| = 0$.

By the fact of $|\{\overline{u}\} \cup F_1| \leq n - 2$, and TN_{n-1}^1 is $(n - 1)$-connected, $\mathrm{TN}_{n-1}^1 - \{\overline{u}\}$ is still connected. Observe that each vertex $x \in \mathrm{TN}_{n-1}^0 - N[u] - F_0$ has a pair vertex $\overline{x} \in \mathrm{TN}_{n-1}^1 - \overline{N[u]}$ (where $\overline{N[u]} = \{\overline{v} \in \mathrm{TN}_{n-1}^1 | \overline{v}$ has exactly one paired vertex $v \in N[u] \subset \mathrm{TN}_{n-1}^0$ such that $(v, \overline{v})$ is a crossed edge), and $\overline{x} \notin F$, so x may connect to $\mathrm{TN}_{n-1}^1 - \overline{N[u]}$ through the crossed edge $(x, \overline{x})$. Hence, $\mathrm{TN}_n - N[u] - F$ is connected.

Case 2 $|F_0| \le n - 4$ and $|F_1| \ge 1$.

By the fact that $|\{\overline{u}\} \cup F_1| \le n - 2$, $\mathrm{TN}_{n-1}^1 - \{\overline{u}\} - F_1$ is still connected. $\mathrm{TN}_{n-1}^0 - N[u] - F_0$ is also connected by induction. Observe that there are $2^{n-1} - n$ crossed edges between $\mathrm{TN}_{n-1}^0 - N[u]$ and $\mathrm{TN}_{n-1}^1 - \overline{N[u]}$. Since $2^{n-1} - n > n - 3 = |F|$ (where $n \ge 4$), there must exist a crossed edge $(x, \overline{x})$ between $\mathrm{TN}_{n-1}^0 - N[u] - F_0$, and $\mathrm{TN}_{n-1}^1 - \overline{N[u]} - F_1$. So $\mathrm{TN}_n - N[u] - F$ is connected.

Lemma 5 *Let* $\{u, v\}$ *be a pair of adjacent vertices of* TN_n *(* $n \ge 3$*). Then the connectivity of* $\mathrm{TN}_n - N[u, v]$ *is not less than* $n - 3$.

Proof Let $F_0 = F \cap \mathrm{TN}_{n-1}^0$, $F_1 = F \cap \mathrm{TN}_{n-1}^1$ with the restriction that $|F| \le n - 4$. The claim is clearly true for the case $n = 3$. This is clearly true for the case $n = 3$. For example, if $(u, v) \in \{(000, 010), (001, 011), (100, 110), (111, 101)\}$, the connectivity of $\mathrm{TN}_3 - N[u, v]$ is 1; otherwise, the connectivity of $\mathrm{TN}_3 - N[u, v]$ is 0. Now, assume it is true for $k = n - 1 \ge 4$ and it suffices to prove that it is true for $k = n$. Let $\{u, v\}$ be arbitrary pair of adjacent vertices in TN_n. In view of the location of $\{u, v\}$, we examine as follows.

Case 1 $u \in \mathrm{TN}_{n-1}^0$, $v \in \mathrm{TN}_{n-1}^1$ ($u \in \mathrm{TN}_{n-1}^1$, $v \in \mathrm{TN}_{n-1}^0$, similarly).

Applying Lemma 4 to $\mathrm{TN}_{n-1}^0 - N[u]$, and $\mathrm{TN}_{n-1}^1 - N[v]$, respectively, we derive that they are both $n - 3$ connected. So both of $\mathrm{TN}_{n-1}^0 - N[u] - F_0$ and $\mathrm{TN}_{n-1}^1 - N[v] - F_1$ are connected. Observe that there are $2^{n-1} - 2n + 1$ crossed edges between $\mathrm{TN}_{n-1}^0 - N[u]$ and $\mathrm{TN}_{n-1}^1 - N[v]$. Since $2^{n-1} - 2n + 1 > n - 4 = |F|$ $(n \ge 4)$, there must exist a crossed edge (x, x') between $\mathrm{TN}_{n-1}^0 - N[u] - F_0$ and $\mathrm{TN}_{n-1}^1 - N[v] - F_1$. So $\mathrm{TN}_n - N([u, v]) - F$ is connected.

Case 2 $u, v \in \mathrm{TN}_{n-1}^0$ ($u, v \in \mathrm{TN}_{n-1}^1$, similarly).

Case 2.1 $|F_0| \le n - 5$

By induction, $\mathrm{TN}_{n-1}^0 - N[u, v] - F_0$ is connected. $\mathrm{TN}_{n-1}^1 - \{\overline{u}, \overline{v}\} - F_1$ is connected because of $|\{\overline{u}, \overline{v}\} \cup F_1| \le n - 2$. Observe that there are $2^{n-1} - 2n + 2$ crossed edges between $\mathrm{TN}_{n-1}^0 - N[u, v]$ and $\mathrm{TN}_{n-1}^1 - \{\overline{u}, \overline{v}\}$. Since $2^{n-1} - 2n + 2 > n - 4 = |F|\,(n \ge 4)$, there must exist a crossed edge $(x, \overline{x})$ between $\mathrm{TN}_{n-1}^0 - N[u, v] - F_0$ and $\mathrm{TN}_{n-1}^1 - \{\overline{u}, \overline{v}\} - F_1$. So $\mathrm{TN}_n - N([u, v]) - F$ is connected.

Case 2.2 $|F_0| = n - 4$. Then, $|F_1| = 0$.

By applying Lemma 2, $TN_{n-1}^1 - \{\overline{u}, \overline{v}\}$ is connected. Observe that each vertex $x \in \mathrm{TN}_{n-1}^0 - N[u] - F_0$ has a pair vertex $\overline{x} \in \mathrm{TN}_{n-1}^1 - N[v]$, and $\overline{x}$ is fault-free. As a result, $\mathrm{TN}_n - N([u, v]) - F$ is connected.

Lemma 6 *Let F be a set of at most $3n - 6$ vertices of TN_n ($n \geq 5$). Under the conditional fault model, i.e., $N(u) \not\subset F$ for $\forall u \in V(\mathrm{TN}_n)$, $\mathrm{TN}_n - F$ satisfies one of the following conditions:*

1. *$\mathrm{TN}_n - F$ is connected; or*
2. *$\mathrm{TN}_n - F$ has two connected components, one of which is K_2, and the other one has $2^n - |F| - 2$ vertices.*

Proof Let $F_0 = F \cap \mathrm{TN}_{n-1}^0$, $F_1 = F \cap \mathrm{TN}_{n-1}^1$ with the restriction that $|F| \leq 3n - 6$. Since $N(u) \not\subset F$ for $\forall u \in V(\mathrm{TN}_n)$, two possibilities need to be investigated:

Case 1 For any pair adjacent vertices $\{u, v\}$ of TN_n such that $N(u, v) \not\subset F$.

Case 1.1 Either $|F_0| \geq 2n - 4$ or $|F_1| \geq 2n - 4$.

Without loss of generality, we may assume that $|F_1| \geq 2n - 4$. Then, $|F_0| \leq 3n - 6 - |F_1| \leq n - 2$. Since $\kappa(\mathrm{TN}_{n-1}^0) = n - 1$, $\mathrm{TN}_{n-1}^0 - F_0$ is connected. The key is to show that there exist a path connecting u to $\mathrm{TN}_{n-1}^0 - F_0$ for any vertex $u \in \mathrm{TN}_{n-1}^1 - F_1$:

If $\overline{u} \notin F$, then u can connect to $\mathrm{TN}_{n-1}^0 - F_0$ through the crossed edge $(u, \overline{u})$; otherwise, since $N(u) \not\subset F$ for any vertex $u \in \mathrm{TN}_n$, there exists one vertex $v \in N(u) \cap (\mathrm{TN}_{n-1}^1 - F_1)$. If $\overline{v} \notin F$, then it succeeds as above; otherwise, since $N(u, v) \not\subset F$ for any pair adjacent vertices $\{u, v\} \subset \mathrm{TN}_{n-1}^1$, there exists one vertex $w \in N(u, v) \cap (\mathrm{TN}_{n-1}^1 - F_1)$ such that $\overline{w} \notin F$ (otherwise, $|F| \geq 3n - 5$, a contradiction), so u may connect to $\mathrm{TN}_{n-1}^0 - F_0$ through the crossed edge $(w, \overline{w})$ (see Fig. 2).

Case 1.2 $|F_0| \leq 2n - 5$ and $|F_1| \leq 2n - 5$.

If one of the two subgraphs $\mathrm{TN}_{n-1}^0 - F_0$, $\mathrm{TN}_{n-1}^1 - F_1$, such as $\mathrm{TN}_{n-1}^1 - F_1$, is connected. The discussion is similar to that of Case 1.1. Now, we may assume that $\mathrm{TN}_{n-1}^0 - F_0$, $\mathrm{TN}_{n-1}^1 - F_1$ both are disconnected. By Lemma 3, we know that there exists one vertex $u \in \mathrm{TN}_{n-1}^0 - F_0$ such that $N(u) \cap \mathrm{TN}_{n-1}^0 \subset F_0$; similarly, there

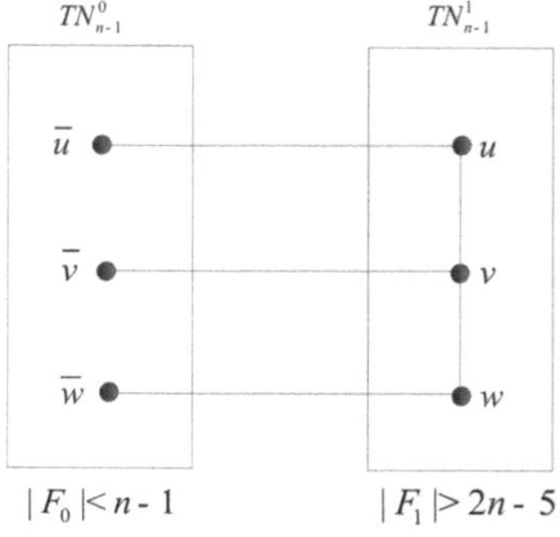

Fig. 2 Illustration of Case 1.1

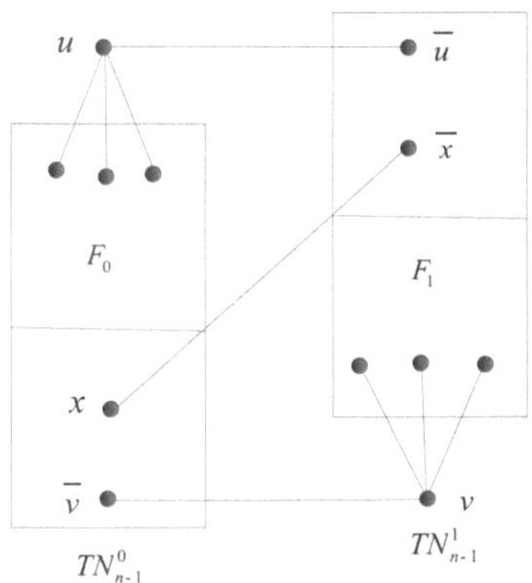

Fig. 3 Illustration of Case 1.2

exists some $v \in \mathrm{TN}^1_{n-1} - F_1$ such that $N(v) \cap \mathrm{TN}^1_{n-1} \subset F_1$. Since $|F_0 - N[u]| \le 2n - 5 - n = n - 5$, $\mathrm{TN}^0_{n-1} - F_0 - \{u\}$ is connected by Lemma 4, similarly $\mathrm{TN}^1_{n-1} - F_1 - \{v\}$ is connected.

We now need to show that these four components $\{u\}$, $\{v\}$, $\mathrm{TN}^0_{n-1} - F_0 - \{u\}$, and $\mathrm{TN}^1_{n-1} - F_1 - \{v\}$ constitute exactly one connected component: First, $\bar{u} \notin F$, otherwise, $N(u) \subset F$, which contradicts the assumption that $N(u) \not\subset F$ for any vertex $u \in \mathrm{TN}_n$; Second, $v \ne \bar{u}$; otherwise, $N(u, v) \subset F$, which contradicts the assumption that $N(u, v) \not\subset F$ for any pair adjacent vertices $\{u, v\} \subset \mathrm{TN}^1_{n-1}$. We have $\bar{u} \in \mathrm{TN}^1_{n-1} - F_1 - \{v\}$ by the connectness of $\mathrm{TN}^1_{n-1} - F_1 - \{v\}$; similarly, $\bar{v} \in \mathrm{TN}^0_{n-1} - F_0 - \{u\}$.

Now, we show that $\mathrm{TN}^0_{n-1} - F_0 - \{u\}$ and $\mathrm{TN}^1_{n-1} - F_1 - \{v\}$ are connected. Since $2^{n-1} > 3n - 6 + 2 = 3n - 4$ for $n \ge 5$, there exists at least one crossed edge $(x, \bar{x})$ between $\mathrm{TN}^0_{n-1} - F_0 - \{u\}$ and $\mathrm{TN}^1_{n-1} - F_1 - \{v\}$, such that $x, \bar{x} \notin F$, so $\mathrm{TN}_n - F$ is connected (see Fig. 3).

Case 2 There exists a pair adjacent vertices $\{u, v\}$ of TN_n such that $N(u, v) \subset F$.

Since $|F \cap N(u, v)| \ge |N(u, v)| \ge 2n - 2$, we have $|F - N(u, v)| = |F| - |F \cap N(u, v)| \le 3n - 6 - (2n - 2) = n - 4$. By Lemma 5, $\mathrm{TN}_n - N[u, v]$ is connected. So the graph $\mathrm{TN}_n - N(u, v)$ has two components, one of which is $K_2 \cong \mathrm{TN}_n[u, v]$, the other is $\mathrm{TN}_n - F - \{u, v\}$.

4 Conditional Diagnosability of Twisted-Cube Connected Networks

By the method of Fan [8], it is easily obtained that the diagnosability of TN_n under the comparison model is n. This section will prove that the conditional diagnosability of TN_n is $3n - 5$ under the comparison model.

Let $u, v, w \in V(\mathrm{TN}_n)$, and $(u,\ w,\ v)$ be a path with length two. We set $A = N(u) \cup N(w) \cup N(v)$, $F_1 = A - \{w, v\}$, $F_2 = A - \{u, w\}$. It is easy to check that F_1 and F_2 are two conditional faulty sets, and F_1 and F_2 are indistinguishable. When $(u,\ w,\ v)$ is in a cycle of length four, we get $|F_1| = |F|_2 = 3n - 4$ and $|F_1 - F_2| = |F_2 - F_1| = 1$. Hence, we have the following result:

Theorem 1 $t_c(\mathrm{TN}_n) \leq 3n - 5$ *for* $n \geq 4$.

The following two lemmas are necessary in our context.

Lemma 7 *Let* F_1 *and* F_2 *be any two distinct conditional faulty subsets of* TN_n *with* $|F_1| \leq 3n - 5$, $|F_2| \leq 3n - 5$. *Denote by* H *the maximum component of* $\mathrm{TN}_n - F_1 \cap F_2$. *Then for every vertex* $u \in F_1 \Delta F_2$, $u \in H$.

Proof Without loss of generality, we assume that $u \in F_1 - F_2$. Since F_2 is a conditional faulty set, there is a vertex $v \in (\mathrm{TN}_n - F_2) - \{u\}$ such that $(u, v) \in E(\mathrm{TN}_n)$. Suppose that u is not a vertex of H. Then, v is not in H, so u and v are part of a small component in $\mathrm{TN}_n - F_1 \cap F_2$. Since F_1 and F_2 are distinct, $|F_1 \cap F_2| \leq 3n - 6$. So $\{u,\ v\}$ forms a component K_2 of $\mathrm{TN}_n - F_1 \cap F_2$ by Lemma 6, i.e., u is the unique neighbor of v in $\mathrm{TN}_n - F_1 \cap F_2$. This is a contradiction since F_1 is a conditional faulty set, but all the neighbors of v are faulty in $\mathrm{TN}_n - F_1$.

Lemma 8 [11] *Let* G *be a graph with* $\delta(G) \geq 2$, *and let* F_1 *and* F_2 *be any two distinct conditional faulty subsets of* G *with* $F_2 \subset F_1$. *Then* (F_1, F_2) *is a distinguishable conditional pair under the comparison diagnosis model.*

Theorem 2 *Let* F_1 *and* F_2 *be any two distinct conditional faulty subsets of* TN_n $(n \geq 7)$ *with* $|F_1| \leq 3n - 5$, $|F_2| \leq 3n - 5$. *Then* $(F_1,\ F_2)$ *is a distinguishable conditional pair under the comparison diagnosis model.*

Proof By Lemma 8, $(F_1,\ F_2)$ is a distinguishable conditional pair of TN_n if $F_2 \subset F_1$ or $F_1 \subset F_2$. Thus, we assume that $|F_1 - F_2| \geq 1$ and $|F_2 - F_1| \geq 1$. Let $A = F_1 \cap F_2$. Then, we have $|A| \leq 3n - 6$. Let H be the maximum component of $\mathrm{TN}_n - F_1 \cup F_2$. By Lemma 7, every vertex in $F_1 \Delta F_2$ is in H.

We claim that H has a vertex u outside $F_1 \cup F_2$ that has no neighbor in A. Since every vertex has degree n, vertex in A can have at most $n|A|$ neighbors in H. There are at most $2(3n - 5) - |A|$ vertices in $F_1 \cup F_2$ and at most two vertices of $\mathrm{TN}_n - A$ may not belong to H by Lemma 6. Since $|A| \leq 3n - 6$, we have

$$\begin{aligned} &2^n - n|A| - (2(3n-5) - |A|) - 2 \geq 2^n - (n-1)|A| - 2(3n-5) - 2 \\ &\geq 2^n - (3n-6)(n-1) - 2(3n-5) - 2 \\ &\geq 4 \text{ when } n \geq 7 \end{aligned}.$$

Thus, there must be vertices of H outside $F_1 \cup F_2$ have no neighbor in A. Let u be such a vertex.

If u has no neighbor in $F_1 \cup F_2$, then we can find a path of length at least 2 within H to a vertex v in $F_1 \cup F_2$. We may assume that v is the first vertex of $F_1 \Delta F_2$ on this path, and let q and w be the two vertices on this path immediately before v (we may have $u = q$), so q and w are not in $F_1 \cup F_2$. So the edges (q, w) and (w, v) show that (F_1, F_2) is a distinguishable conditional pair of TN_n by Lemma 1. Now assume that u has a neighbor in $F_1 \Delta F_2$. Then, since the degree of u is at least 3, and u has no neighbor in A, there are three possibilities:

1. u has two neighbors in $F_1 - F_2$;
2. u has two neighbors in $F_1 - F_2$; or
3. u has at least one neighbor outside $F_1 \cup F_2$.

In each subcase above, Lemma 1 implies that (F_1, F_2) is a distinguishable conditional pair of TN_n under the comparison diagnosis model, finishing the proof.

Theorem 3 $t_c(\text{TN}_n) = 3n - 5$. *for* $n \geq 7$

5 Conclusion

The issue of identifying faulty processors is important for the design of multi-processor interconnected systems, which are implementable with VLSI. The process of identifying all the faulty processors is the system-level diagnosis. This paper establishes the conditional fault diagnosability of twisted-cube connected networks. Since the conditional connectivity also has the same restriction as the conditional diagnosis, our intuition from this discussion is that the conditional connectivity can be used to bound the conditional diagnosability of some variants of hypercube and other network topologies. It is an attractive work to develop more different diagnosis model to expose the network reliability.

Acknowledgments This work was partly supported by the National Natural Science Foundation of China (No. 61072080), Natural Science Foundation of Fujian Province (Nos. 2013J01221, JA12073).

References

1. Preparata, F.P., Metze, G., Chien, R.T.: On the connection assignment problem of diagnosable systems. IEEE Trans. Comput. **16**, 848–854 (1967)
2. Malek, M.: A comparison connection assignment for diagnosis of multiprocessor systems. In: Proceedings of 7th International Symposium Computer Architecture, pp. 31–35 (1980)
3. Chwa, K.Y., Hakimi, S.L.: On fault identification in diagnosable system. IEEE Trans. Comput. **30**(6), 414–422 (1981)
4. Maeng, J., Malek, M.: A comparison connection assignment for self-diagnosis of multiprocessor systems. In: Proceedings of 11th International Fault-Tolerant Computing, pp. 173–175 (1981)
5. Sengupta, A., Dahbura, A.: On self-diagnosable multi processor systems: diagnosis by comparison approach. IEEE Trans. Comput. **41**(11), 1386–1396 (1992)
6. Wang, D.: Diagnosability of hypercubes and enhanced hypercubes under the comparison diagnosis model. IEEE Trans. Comput. **48**(12), 1369–1374 (1999)

7. Wang, D.: Diagnosability of enhanced hypercubes. IEEE Trans. Comput. **9**, 1054–1061 (1994)
8. Fan, J.: Diagnosability of crossed cubes under the comparison diagnosis model. IEEE Trans. Parallel Distrib. Syst. **13**(10), 1099–1104 (2002)
9. Yang, H., Yang, X.: A fast diagnosis algorithm for locally twisted cube multiprocessor systems under the MM* model. Comput. Math. Appl. **53**(6), 918–926 (2007)
10. Zheng, J., Latifi, S., Regentova, E., Luo, K., Wu, X.: Diagnosability of star graphs under the comparison diagnosis model. Inf. Process. Lett. **93**, 29–36 (2005)
11. Lin, C.-K., et al.: Conditional diagnosability of cayley graphs generalized by transposition tree under the comparison diagnosis model. J. Interconnection Netw. **9**(1), 83–97 (2008)
12. Lai, P.-L., et al.: Conditional diagnosability measure for large multiprocessor systems. IEEE Trans. Comput. **54**, 165–175 (2005)
13. Zhou, S., Chen, L., Xu, J.-M.: Conditional fault diagnosability of dual-cubes. Int. J. Found. Comput. Sci. **23**(8), 1729–1747 (2012)
14. Esfahanian, A.H.: Generalized measure of fault tolerance with application to n-cube networks. IEEE Trans. Comput. **38**, 1586–1691 (1989)
15. Latifi, S.: Combinatorial analysis of the fault diameter of the n-cube. IEEE Trans. Comput. **42**(1), 27–33 (1993)
16. Xu, J.-M., Deng, Z.: Wide diameter of Möbius cubes. J. Interconnection Netw. **6**(1), 51–62 (2005)
17. Yang, X., Evans, D.J., Megson, G.M.: The locally twisted cubes. Int. J. Comput. Math. **82**(4), 401–413 (2005)
18. Chang, C.-P., Wang, J.-N., Hsu, L.-H.: Topological properties of twisted cube. Inf. Sci. **113**, 147–167 (1999)
19. Wang, D., Zhao, L.: The twisted-cube connected networks. J. Comput. Sci. Technol. **14**(2), 181–187 (1999)
20. Hsu, G.-H., et al.: Conditional diagnosability of hypercubes under the comparison diagnosis model. J. Syst. Architect. **55**(2), 140–146 (2009)

Realization of Streaming Media Data Multicast Based on UDP

Zhenchuan Zhang and Zhanqun Lun

Abstract With wide usage and special needs of AV transmission services on the Internet, streaming media transport protocol and its technology is studied. In order to effectively resolve and control highly sensitive delay problem on the streaming media services and on carefully comparing TCP and UDP features, an UDP-based streaming media transmission multicast system is designed and implemented. By experiment, a delay time performance in different playout buffer (PB) length and different client numbers are achieved. The results demonstrated the real-time advantage of UDP on streaming media transmission.

Keywords UDP · Multicast · Streaming media · Socket · Playout buffer

1 Introduction

In early stage, Internet can only send text messages like files, E-mail, etc. With the emergence of the browser, text and image constitutes the main body of data performance on the Internet. Audio and video in the form of streaming media have experienced rapid development since mid-90s. Multimedia service from audio and video tends to be diversified [1–3]. Based on the traditional network transmission protocol, this paper is aiming at the characteristics of streaming media data transmission. The UDP protocol and technology were studied sharply. Streaming media data multicast system based on UDP were implemented through the network transmission software design.

Z. Zhang (✉) · Z. Lun
College of Information Science and Engineering, Northeastern University, Shenyang, China
e-mail: zhangzhenchuan@ise.neu.edu.cn

Z. Lun
e-mail: lunzhanqun@126.com

S. Patnaik and X. Li (eds.), *Proceedings of International Conference on Computer Science and Information Technology*, Advances in Intelligent Systems and Computing 255, DOI: 10.1007/978-81-322-1759-6_44,

2 Theory and Technology Base

UDP is a connectionless transport layer protocol of the OSI reference mode. UDP do not provide a datagram grouping, sorting, and assembling. It does not know if the messages arrive safely and integrated. UDP often supports data transmission which is in a better quality of the network and less sensitive to loss [2–4]. Being connectionless protocol, UDP has the advantage that resource consumption is small and processing speed is quick. So, it is used for audio and video with high delay sensitivity.

IP multicast technology is a TCP/IP network technology that allows one or more host to send a single data reporting to multiple hosts [5, 6]. As point to multipoint communication, multicast overcomes the drawback of unicast that causes system redundancy. And multicast makes up the radio that only communicates with subnet internal hosts and the lack of bandwidth.

Although the UDP protocol delay is small, it still affects transmission under the condition of unreliable network [7, 8]. In real-time applications based on UDP protocol, setting up a buffer in the client is used to reduce network delay. The received packet is firstly pressed into the buffer. When the buffer reached the expected number of package, it is decoded and played to reduce the effect of delay. This buffer is called to playout buffer (PB).

3 Pollution Resistance Network Coding Scheme Based on MAC

Designing and using an UDP-based streaming media transmission multicast system is based on network structure of the client and server. Due to the requirement that the client and server are able to stay connected no matter in any place and at any time, completing design should have two parts: TCP and UDP. In other words, the client joins IP multicast system using TCP, and the server and client use UDP to transmit and exchange streaming data.

Software development and operating platform is Windows XP, Linux "Redstar 2.0," and development tools Visual c++ 6.0.

3.1 Server Program

1. Use socket() to create an UDP socket, the second parameter is SOCK_DGRAM.
2. Initialize the sockaddr_in structure variable and assign a value.
3. Use the bind() to bind the top of the socket() and defined IP address and port.

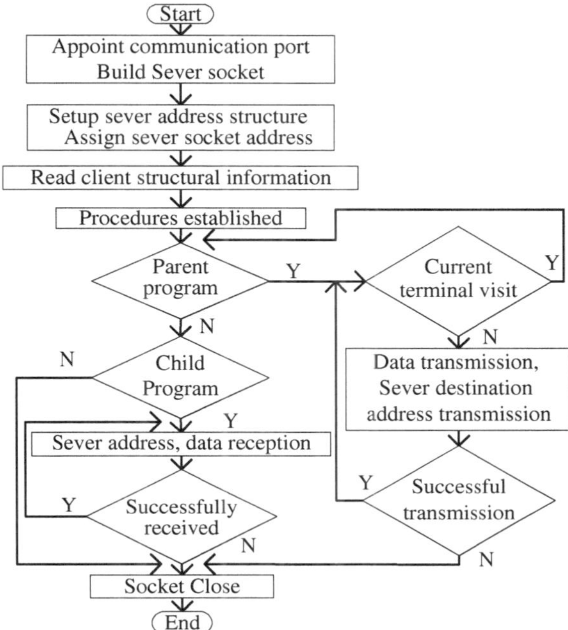

Fig. 1 Streaming socket UDP server program

4. Enter an infinite loop, use recvfrom() into a wait state until receiving the data that the client program sends. Then, send the feedback to the customer by handling received data. Here is that directly send the received data back to the customer program.

Streaming socket UDP server program process and streaming transmission process as shown in Figs. 1 and 2, respectively.

3.2 Client Program

1. Initialize the sockaddr_in structure variable and assign a value.
2. Use socket() to create an UDP socket, the second parameter is SOCK_DGRAM.
3. Use connect() to build the connection with the service program. UDP is non-connected, in fact also can be connected. Use UDP connections, the kernel can directly return error information to the user program to avoid that recvfrom() has been waiting due to not receiving data. Looks like, the client program do not respond.
4. Send data to the service program. Because for using an UDP connection, we can use the write() to replace the send to (). Here, the data read input from the user directly from the standard input.

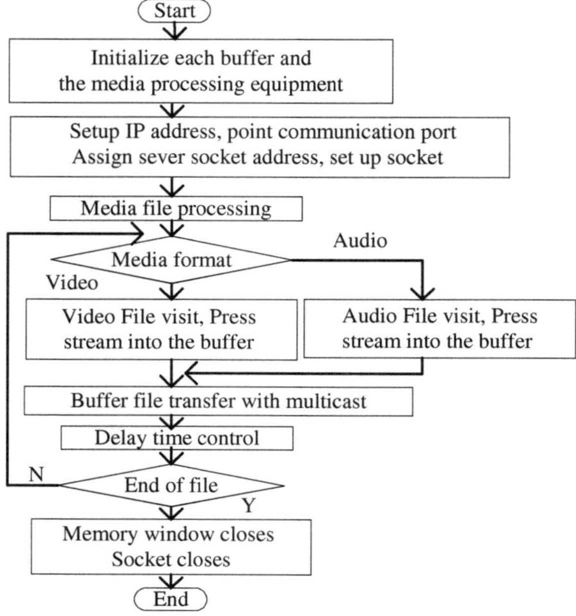

Fig. 2 Streaming transmission program

5. Receive the data that service program sends back. Also, use the read() to replace the recvfrom().
6. Process the received data. Here, output directly to the standard output.

Streaming socket client process and streaming receiving program process, respectively, as shown in Figs. 3 and 4.

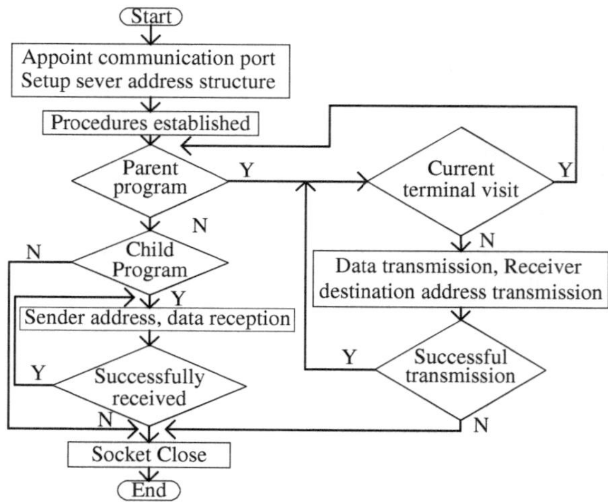

Fig. 3 UDP client program of stream socket

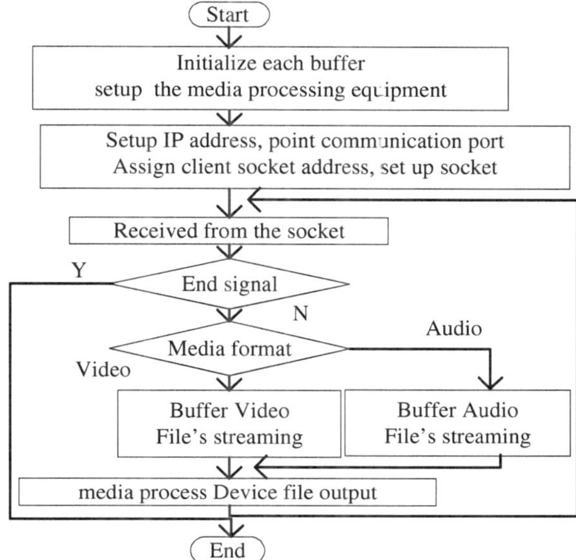

Fig. 4 Stream reception program

4 Experimental Test and Performance Analysis

The designed and implemented streaming media transmission multicast system, under the environment of 100 Base-T Ethernet, test audio and video transmission at the same time for different number of data sources client.

4.1 Time Delay Performance Under Different Playout Buffer Length

Set the number of concurrent client to 30. By changing the length of the buffer, obtain statistical results of delay time under different PB length.

As shown in Fig. 5, delay time decreases with the increase in the PB length. When the PB length is less than 32 kB, the buffer length impacts the delay time

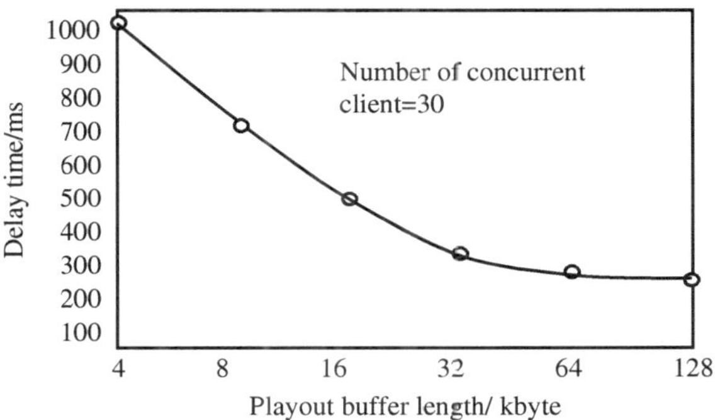

Fig. 5 Playout buffer length versus delay time

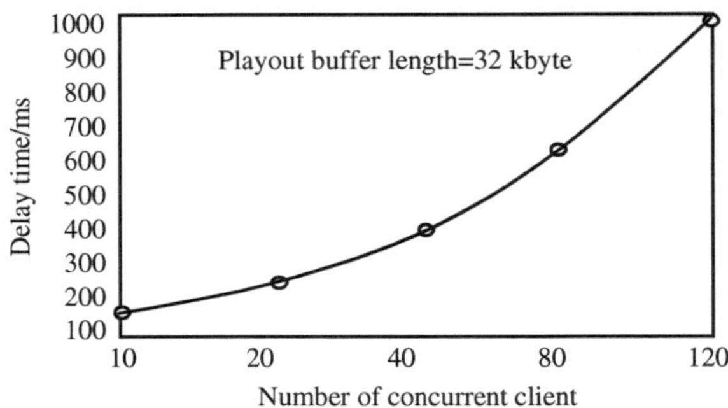

Fig. 6 Concurrent client numbers versus delay

more obviously. When the PB length is more than 32 kB, the tendency, the improvement of the delay time with the increase in PB length, tends to be slow. So, the reasonable PB length is 32 kB.

4.2 Time Delay Performance in Different Number of Concurrent Client

Set PB length as 32 kB. Change the number of concurrent client. Get statistical results of delay time under different client numbers as shown in the Fig. 6.

As shown in Fig. 6, delay time increases with the increase in the number of concurrent client, and the increasing speed of delay time is becoming sharper with the increase in the number of the clients. According to the 1,000 ms permissible delay, the system's number of concurrent clients is 120 or so.

5 Conclusion

By reasonable design, the IP multicast system based on UDP has been completed. The system supports the audio and video transmission at the same time in various formats under 100 Base-T Ethernet's environment. The system can support more than 100 remote data source client's visit at the same time. On the client, the time delay of network audio and video transmission is less than 1,000 ms. Experiments show that the system can provide audio and video multicast with high quality in current bandwidth.

References

1. Forouzan, B.A.: TCP/IP Protocol Suite 3rd. McGraw-Hill, New York (2007)
2. Fujimoto, K., Ata, S., Murata, M.: Adaptive playout buffer algorithm for enhancing perceived quality of streaming applications. Telecommun. Syst. **25**(3), 259–271 (2004)
3. Josevski, N., Chilamkurti, N.: A new TCP-friendly stabilised CSFQ mechanism for layered multicast. In: Proceedings of 14th IEEE International Conference on Networks, vol. 1, pp. 137–141. IEEE Computer Society, Piscataway (2006)
4. Lee, J., Jung, Y., Choe, Y.: Unequal error recovery scheme for multimedia streaming in application-level multicast. In: Proceedings of 7th International Conference on Computational Science, Part IV, pp. 668–675. Springer, Berlin (2007)
5. Kim, N., Kim, J.: A hybrid multicast connectivity solution for multi-party collaborative environments. Multimedia Tools Appl. **44**(1), 17–37 (2009)
6. Lorenz, M., Brunnett, G., Heinz, M.: Driving tiled displays with an extended chromium system based on stream cached multicast communication. Parallel Comput. **33**(6), 438–466 (2007)
7. Huang, N.-F., Chu, Y.-M., Chen, Y.-R.: Design of a P2P live multimedia streaming system with hybrid push and pull mechanisms. In: Proceedings of 2010 WRI International Conference on Communications and Mobile Computing, vol. 1, pp. 541–545. IEEE Computer Society, Piscataway (2010)
8. Han, S., Kim, N., Kim, J., Kim, J.: An extended media service framework for multiparty collaborative environments. In: Proceedings of IEEE International Conference on Computer Systems and Applications, pp. 814–817. IEEE Computer Society, Piscataway (2006)

Cross-Layer Design in HF Communications with the Consideration of IP Service Features

Yuan Jing, Guoce Huang, Jian-xin Guo, Yun-jun Qi and Li-yang Hou

Abstract So far, the high-frequency (HF) time-varying channel was always the development obstacle for HF data communication. A cross-layer system based on AMC and hybrid ARQ (HARQ) in HF IP communication was proposed to overcome the problem. In order to adapt the changing of the HF channel, the HARQ feedback message was used to transfer the modulation mode and coding mode which was decided by the receiver. And the IP frame average server time was made as the performance index of the system which could rise up the performance of IP service feature. The general closed-form expression of the IP frame average server time had been given. What was more, the advantages of the system have been proved in other queue performance about IP frame through theoretical analysis and simulation researches.

Keywords Component · Cross-layer design · High-frequency IP network · Average server time · Queue performance

1 Introduction

High-frequency (HF) communication is widely used in the field of navigation and aviation. HF IP network was developed with the constant progress of network technology and IP technology. At the beginning of this century, the HF IP network caused us attention. And the HF IP access gateway [1, 2] was introduced to consummate interconnection of HF IP network, and the implementation with cable, satellite, and other communication network under the IP protocol. But the HF communication under the influence of the ionosphere and the band width has

Y. Jing (✉) · G. Huang · J. Guo · Y. Qi · L. Hou
School of Information and Navigation, AFEU, Xi'an, China
e-mail: gmt_jingyuan@163.com

S. Patnaik and X. Li (eds.), *Proceedings of International Conference on Computer Science and Information Technology*, Advances in Intelligent Systems and Computing 255, DOI: 10.1007/978-81-322-1759-6_45,

certain limitation on both transmission quality and transmission rate, which would affect the quality of service (QoS) of IP business transmission in HF IP network.

This paper mainly focused on the cross-layer design based on AMC (adaptive modulation coding) technology and select repeat hybrid ARQ (HARQ) in HF IP access gateway. Researches and analysis were concerned about the IP data transmission performance in the gateway.

1.1 HF IP Network

In 2004 and 2006, North Atlantic Treaty Organization (NATO) has projected the AHFWAN66 (Allied High-Frequency Wide-Area Networking STANAG 5066) scene in military exercises of joined warrior interoperability individual (JWID) and coalition warrior interoperability individual (CWID) [2]. Figure 1 is a simplified schematic diagram for AHFWAN66.

HF IP network was based on IP protocol. HF communications connectivity of the network has been established. While current HF communication limited the efficiency of IP data transmission [3], HF IP access gateway (also named HF IP router [4, 5]) was mainly used to complete the IP protocol and HF protocol conversion include data transfer protocol and routing. With the gradual development of new HF data communication technology, HF IP network was developed from design to practical. This paper mainly studied the HF IP across-layer communication, and the key research was the performance of IP packet transmission in HF IP network.

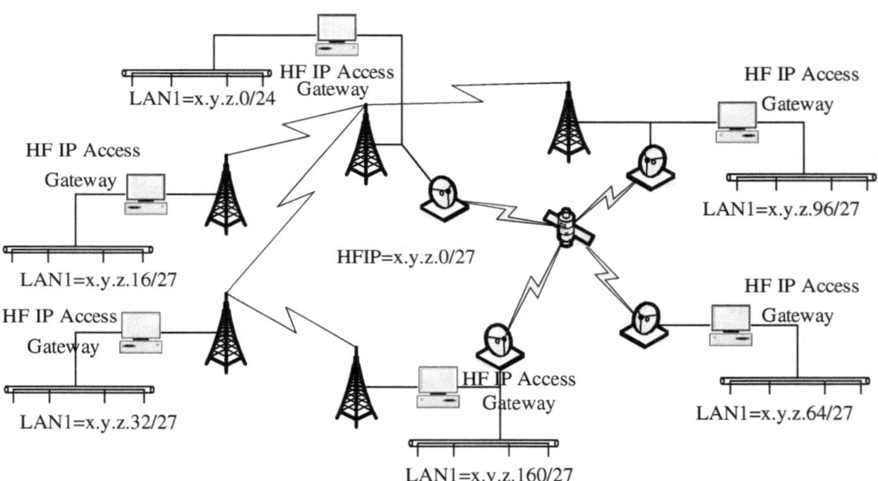

Fig. 1 AHFWAN66

1.2 AMC and HARQ

The principle of the AMC was adaptive changing of the coding and modulation method according to the current channel characteristics. The receiver would estimate (or predicate) the channel SNR and feedback the results through the feedback channel to inform the sender, meanwhile the sender according to the SNR to decide the next data which would transmitted and the transmission mode. Therefore, the SNR identification and the decision relationship between different transport modes were needed. According to the different transmission modes and the performance comparison, the best transport mode should be selected under a specific SNR.

HARQ mainly has three kinds, I-type HARQ, II-type HARQ, and III-type HARQ [5].

I-type HARQ is send after a group of channel coding and decoding error. If the receiver has got an error data which would be discarded directly would call for resending via a feedback channel. Grouping of exactly the same content sent twice. The error data could been retransmitted, as long as not reached the upper limit of data retransmission time.

In II-type HARQ, the receiver would store the wrong data and would be weighted combination with all the same received data, which could get a data of signal-to-noise ratio increase and then decode the data. Moreover at the first time, the data were made up with the parity bits and the information bits, and the retransmission packet only contain part of check bits which have not been send before. The receiver would receive the check bits and merger with the received data, than check and decoding.

While in III-Type of HARQ, the retransmission were made up with information bits and different check bits. The receiver would receive and decode the retransmission data, if it contains errors merged with the old data.

Here, I-type HARQ and the selective ARQ were used. There was more discussion for II-type HARQ and III-type HARQ as the same principle has been found in the actual system.

2 The Across-Layer System for HF Communication

2.1 System Introduction

HF channel is a time-varying channel. Changing data transmission mode based on channel characteristic could improve the communication performance of the system.

Firstly, some characteristics of the system would be expounded:

1. There was variety of judgment index for communication performance, such as grouping presented [7, 8], system throughput [9, 10], the bit error rate to judge the probability of interrupt [11]. In order to better study the influence of the IP data transmission for the HF across-layer communication system, the IP

packets average service time was used as the communication performance indicators. And there would be a detailed analysis in this paper.

2. HARQ and SR-ARQ mechanism were used. The feedback modulation mode selects low-order modulation mode, what is more the length of the feedback information is very short and error-free transmission. But the transmission delay should not be neglected.
3. Five kinds of modulation mode could be used, they were BPSK, QPSK, 8PSK, 16QAM, and 32QAM [12, 13]. And assumed that in a forward data frames contens L packets in spite of the modulation mode. That means the transmission frame structure was the same, and then, the different modulation modes have a different forward transmission time while the receiver uses coherent demodulation.
4. The receiver can estimate current SNR of the channel accurately. There were a lot of SNR estimation method [14], and the more commonly used was the PSAM [15] (pilot-symbol-aided modulation).

2.2 Modulation Signal Feature Analysis

There are five kinds of Modulation mode which were BPSK, QPSK, 8PSK, 16QAM, and 32QAM, and the modulation bandwidth is 3 kHz. Symbol rate is 300 symbols per second. Symbol rate was 300 symbols per second.

Figure 2 shows the relationship between SNR and average bit error rate, and different curves represented different modulation mode. Simulation results showed that a certain SNR circumstance, modulation with a higher-order corresponding to a higher bit error rate, but also a higher data rate which reflected in the shorter service time of IP packets. So different modulation modes have had an intersection point of communication performance under a certain SNR, which was the switching threshold of the standard for different modulation mode.

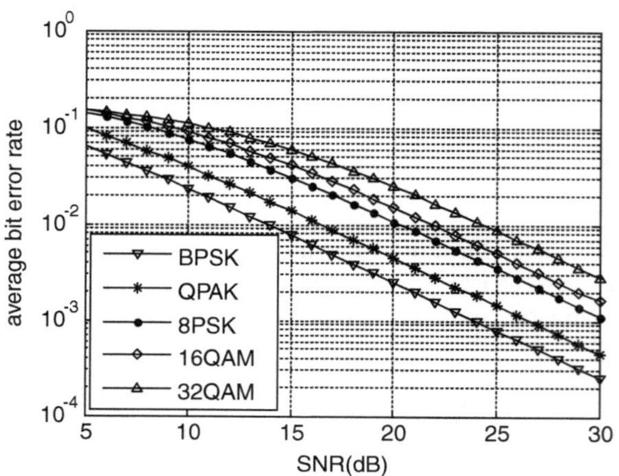

Fig. 2 The BER curve under different SNR for different modulation schemes

For MPSK and MQAM signals, the SNR and bit error rate exist the following relationship (specific results can be gained through the moment-generating function [16]).

MPSK:

$$P_b(\gamma_b) \approx \frac{2}{\log_2 M} Q\left(\sqrt{2\gamma_s} \sin\left(\frac{\pi}{M}\right)\right) \tag{1}$$

MQAM:

$$P_b(\gamma_b) \approx \frac{4}{\log_2 M} Q\left(\sqrt{\frac{3\gamma_s}{M-1}}\right) \tag{2}$$

For formula 1 and 2, the same symbol width brought same symbol SNR even for different modulation mode. But after formula conversion, different modulation modes have got a different bit SNR for the same symbol SNR. To do this in the following, calculations are symbol SNR without acknowledgment.

Q function is generally difficult to accurate calculation; here, moment-generating function was used to get the bit error rate accurately. For example with QPSK.

$$\begin{aligned} \bar{P}_{b,\mathrm{QPSK}}(\gamma_s) &\approx \int_0^\infty Q\left(\sqrt{\gamma_s}\right) p_{\gamma_s}(\gamma_s)\mathrm{d}\gamma_s \\ &= \frac{1}{\pi} \int_0^{3\pi/4} \left[\int_0^\infty \exp\left[\frac{-\gamma_s}{2\sin^2\phi}\right] p_{\gamma_s}(\gamma_s)\mathrm{d}\gamma_s\right] \mathrm{d}\phi \\ &= \frac{1}{\pi} \int_0^{3\pi/4} M_{\gamma_s}\left(\frac{-1}{2\sin^2\phi}\right)\mathrm{d}\phi \end{aligned} \tag{3}$$

In Rayleigh channel,

$$M_{\gamma_s}\left(\frac{-g}{\sin^2\phi}\right) = \left(1 + \frac{g\bar{\gamma}_s}{\sin^2\phi}\right)^{-1} \tag{4}$$

There was

$$\bar{P}_{b,\mathrm{QPSK}}(\gamma_s) = \frac{1}{\pi} \int_0^{3\pi/4} \left(1 + \frac{\bar{\gamma}_s}{2\sin^2\phi}\right)^{-1} \mathrm{d}\phi \tag{5}$$

Similarly, we could get the average bit error rate for other modulation mode. The average probability of successfully received data packets can be obtained under certain SNR.

$$\bar{P}_{d,i} = \left(1 - \bar{P}_{b,i}\right)^{N_{\mathrm{data}}} \tag{6}$$

3 IP Packet Average Service Time

The average service time of the IP packets can be a good characterization of service features for IP application which could reflect the transmission ability of HF IP network about IP packets.

3.1 Theory Analysis

First, the average service time for a single packet was researched. Assumed that interval time of the adjacent packets which need to be sent in the network obey the exponential distribution. Then, the characteristics of the packets average service time for different modulation mode were analyzed.

The $\bar{P}_{d,i}$ corresponded to the probability of the packet for receiving with no error under a special modulation mode. And the conditional probability of successfully received data packets could be calculated through formula 7.

$$\bar{P}'_{d,i}(\bar{\gamma}_s) = \int_{\gamma_{i-1}}^{\gamma_i} P_s(\gamma) p_{\bar{\gamma}_s}(\gamma) \mathrm{d}\gamma \tag{7}$$

The average service time refered to the time a single packet have been successfully received. And it also included the time for sending the feedback message. And n indicates the number of transmission until the packets has been correctly received. And relative to a certain fixed modulation mode, it can be represented as formula 8.

$$\begin{aligned} T_{\text{serve,data,i}} &= \sum_{n=1}^{\infty} n \cdot \bar{P}_{d,i} \cdot \left(1 - \bar{P}_{d,i}\right)^{n-1} \cdot \left(T_{\text{data,i}} + T_{\text{feedback}}\right) \\ &= \frac{T_{\text{data,i}} + T_{\text{feedback}}}{\bar{P}_{d,i}} \end{aligned} \tag{8}$$

As in cross-layer communication system, due to different channel conditions corresponded to different modulation mode, the selection of value $\bar{P}'_{d,i}$ was different as n changed.

And each modulation mode corresponded to a SNR range, so the probability of each SNR range could be calculated by the probability density function of SNR; then, the probability of different state could be calculated. Assuming that decision threshold respect to γ_1, γ_2, γ_3, and γ_4. And $f(\gamma)$ is the probability density function for γ, according to the ith modulation used in probability b_i.

$$b_i(\bar{\gamma}_s) = \int_{\gamma_i}^{\gamma_{i+1}} f_\gamma(\gamma, \bar{\gamma}_s)\mathrm{d}\gamma \tag{9}$$

So for cross-layer communication system, the average service time for packet can be represented as

$$\begin{aligned} T_{\text{serve,data}} &= \sum_{n=1}^{\infty}\sum_{i=1}^{5}\left(b_i \cdot n \cdot \bar{P}'_{d,i} \cdot \left(1 - \bar{P}'_{d,i}\right)^{n-1} \cdot \left(T_{\text{data,i}} + T_{\text{feedback}}\right)\right) \\ &= \sum_{i=1}^{5}\left(b_i \cdot \left(T_{\text{data,i}} + T_{\text{feedback}}\right) \sum_{n=1}^{\infty} n \cdot \bar{P}'_{d,i} \cdot \left(1 - \bar{P}'_{d,i}\right)^{n-1}\right) \\ &= \sum_{i=1}^{5}\left(\frac{b_i}{\bar{P}'_{d,i}} \cdot \left(T_{\text{data,i}} + T_{\text{feedback}}\right)\right) \end{aligned} \tag{10}$$

Because more than one packet would be sent in the process of interaction, a packet of service time shall not be less than the sum of a data transmission delay, a feedback delay, and the link cycle time:

$$\begin{cases} T_{\text{serve,data}} = \sum_{i=1}^{5}\left(\frac{b_i}{\bar{P}'_{d,i}} \cdot \left(T_{\text{data,i}} + T_{\text{feedback}}\right)\right) & T_{\text{serve,data}} \geq \min\left(T_{\text{data,i}} + T_{\text{feedback}}\right) \\ T_{\text{serve,data}} = \min\left(T_{\text{data,i}} + T_{\text{feedback}}\right) & T_{\text{serve,data}} \leq \min\left(T_{\text{data,i}} + T_{\text{feedback}}\right) \end{cases} \tag{11}$$

But relative to an IP frame (L_{ip} for the length of the IP packets, unit for bits, $T_{\text{serve,ip}}$ for service time), it was divided into multiple packets and then the IP frame service time was not linear summation of each packet service time, because in a $T_{\text{data,i}}$ time it contained L packets, and the probability of each packet for correct receiving was $\bar{P}_{di}$. Thus, the number of packets which could be received correctly in a transmission is $L_{\text{data,i}}$:

$$L_{\text{data,i}} = \mathrm{L} \cdot \bar{P}_{d,i} \tag{12}$$

For fixed modulation mode, for example in the ith state, the average service time of IP frames is as follows:

$$T_{\text{serve,ip,i}} = \frac{T_{\text{data,i}} + T_{\text{feedback}}}{L_{\text{data,i}}} \cdot \left\lceil \frac{L_{ip}}{L \cdot N_{\text{data}}} \right\rceil \tag{13}$$

$\lceil\,\rceil$ means the top integral function.

By the same token, in cross-layer communication system, the average service time of IP frames was as follows:

$$\begin{cases} L_{\text{data,i}} = L \cdot \bar{P}'_{d,i} \\ T_{\text{serve,ip}}(\gamma_s) = \left\lceil \frac{L_{ip}}{L \cdot N_{\text{data}}} \right\rceil \cdot \sum_{i=1}^{4} \left(b_i \cdot \frac{T_{\text{data,i}} + T_{\text{feedback}}}{L_{\text{data,i}}} \right) \end{cases} \tag{14}$$

Thus, we get the $T_{\text{serve,ip}}$ under average channel SNR of certain situations. This was mainly because in the calculation of b_i and $\bar{P}'_{d,i}$, the average SNR of the channel was used.

By formula 14, we also received the relational expression of SNR and average service time for IP frames.

3.2 Decision Threshold

The principle of modulation mode choosing was in particular signal SNR conditions to select the optimal modulation mode, to ensure that the IP frame on average service time optimal, which can be represented as

$$T_{\text{serve,ip}}(\gamma_s) = \min_{i=1}^{5} \left(T_{\text{serve,ip,i}}(\gamma_s) \right) \tag{15}$$

From formula 14, to learn $T_{\text{serve,ip}}(\gamma_s)$ related to γ_s was a monotone decreasing function, and the second derivative was less than 0. There was only one intersection point for different IP frames' average service time function under different modulation mode. So the following relationship can be launched:

$$T_{\text{serve,ip}}(\gamma_s) = \begin{cases} T_{\text{serve,ip,i}}(\gamma_s), & \gamma_s \in (-\infty, \gamma_1) \\ T_{\text{serve,ip,i}}(\gamma_s), & \gamma_s \in [\gamma_1, \gamma_2) \\ T_{\text{serve,ip,i}}(\gamma_s), & \gamma_s \in [\gamma_2, \gamma_3) \\ T_{\text{serve,ip,i}}(\gamma_s), & \gamma_s \in [\gamma_3, \gamma_4) \\ T_{\text{serve,ip,i}}(\gamma_s), & \gamma_s \in [\gamma_4, +\infty) \end{cases} \tag{16}$$

While $T_{\text{serve,ip,1}}(\gamma_1) = T_{\text{serve,ip,2}}(\gamma_1)$, $T_{\text{serve,ip,2}}(\gamma_2) = T_{\text{serve,ip,3}}(\gamma_2)$, $T_{\text{serve,ip,3}}(\gamma_3) = T_{\text{serve,ip,4}}(\gamma_3)$, $T_{\text{serve,ip,4}}(\gamma_4) = T_{\text{serve,ip,5}}(\gamma_4)$. Assuming that average number of packet in IP frames is $\bar{L}$.

By the preceding analysis, the different modulation mode of the condition probability for IP frame succeed receiving $\bar{P}'_{d,i}$ and IP frame average service time $T_{\text{serve,ip,i}}$ could be got through calculating bit error rate in particular modulation mode corresponded to γ_s.

It can be seen from Fig. 3, 8PSK scope is of almost no use and could be canceled. Only BPSK, QPSK, 16QAM, and 32QAM would be used in the cross-layer communication system. Also shows that in the same order number, power, and symbol rate, QAM modulation mode in the introducing of amplitude information when compared with the PSK modulation mode can reduce the bit error rate.

Table 1 The SNR threshold switching for different modulation methods

Modulation mode	γ_s SNR range (dB)
BPSK	$(-\infty, 20.2]$
QPSK	$(20.2, 25.7]$
16QAM	$(25.7, 31.4]$
32QAM	$(31.4, \infty]$

By calculation, we could get different modulation mode switching threshold (Table 1), also found that the probability of using 8PSK was very small, and can remove 8PSK modulation mode in optional modulation mode. And only four modulation modes (BPSK, QPSK, 16QAM, and 32QAM) were used. Also found in the simulation that significance of using 8PSK was not big (Fig. 3).

Calculate the intersection between $T_{\text{serve,ip,i}}$, that is, different modulation mode switching threshold. The threshold had relations with the packet length, the data frame length, the proportion of redundant information in data frames, feedback information length, and the transmission delay. So in the calculation, results of threshold value in different communication systems were not identical, but the calculation method was the same.

For particular SNR, there is a modulation mode can make the minimum average service time of IP frames. So the real-time SNR could be a basis for the modulation mode choosing, which could get the accurate judgment for modulation mode. In different SNR, the system could always use the most optimal transmission performance.

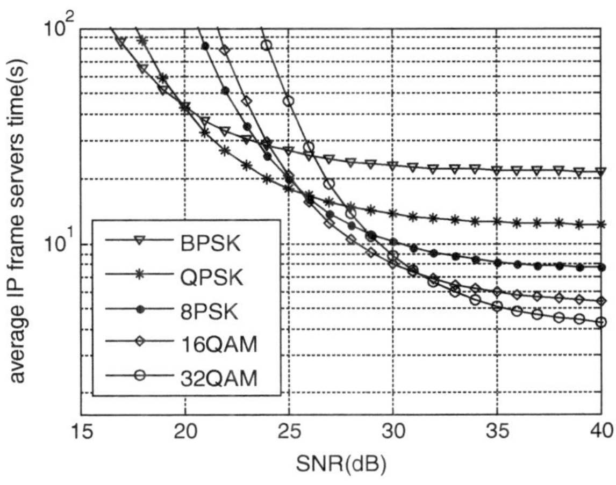

Fig. 3 The average IP frame servers time curve under different SNR for different modulation schemes

3.3 Simulation Research

Hypothesis 5 bytes of header information in the data frame, the data packet number L is 10, packet length is 35, and 32 bytes for useful information. What is more, the link cycle time is 1.2 s [17] and feedback transmission delay is 0.2 s.

In Fig. 4, the curve "CLD" refers to the cross-layer design of the HF IP communication system. By the simulation result shown in Fig. 4, the average service time for a single packet of the cross-layer design of the HF IP communication system introduced in this letter was always lower than the average service time of fixed modulation mode. Known from the theory analyzing that the average service time was mainly affected by two aspects of the bit error rate and bit rate, and the influence of the two effects on the group average service time was different under different channel conditions. Under the condition of low SNR (in the simulation result, the abscissa is average symbol SNR), the modulation mode select BPSK, select 32QAM conversely. The selection of dynamic modulation mode to balance the relationship between bit error rate and the bit rate could ensure the optimal packets average service time. Also see at the same time, decision threshold for cross-layer design of HF IP communication system was determined through optimal the average service time of IP frames and can achieve the same effect for packets on the average service time.

In the simulation, the average length of IP packets was 5,600 bits. IP packet was divided into different parts and then transmated in the HF channel and the transmission process of different packets were independent of each other, and the average service time was used by all packets transmated completely.

The simulation results could be seen in Fig. 5 after using cross-layer communication system. The average service time of IP frames was much lower than any of the fixed mode. Secondly, compared with Fig. 4 it can also be found that the curves of average service time for IP frames and packets have a similar trend.

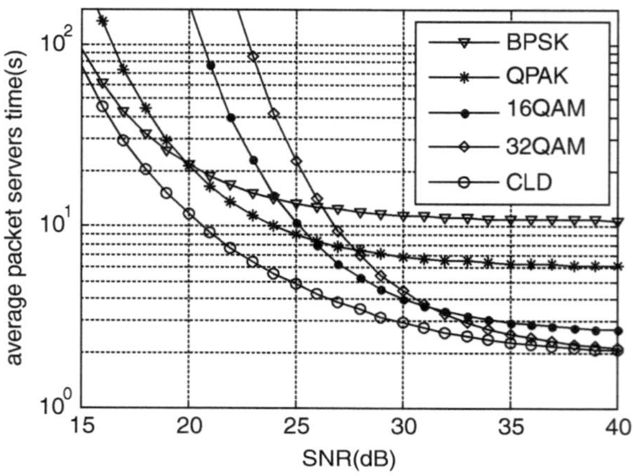

Fig. 4 The average packet servers time curve under different SNR for different modulation schemes

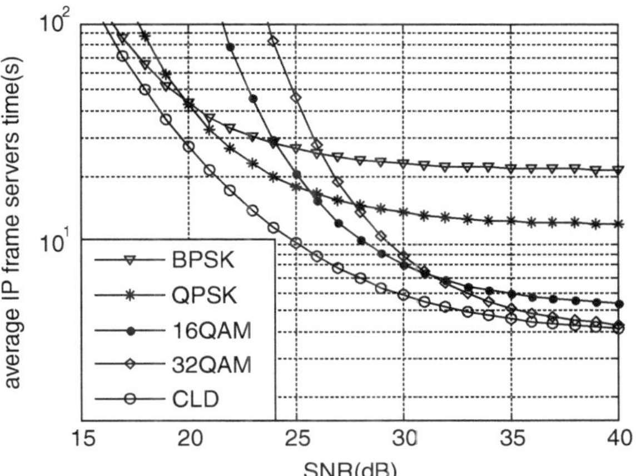

Fig. 5 The average IP frame servers time curve under different SNR for different modulation schemes

4 Other Relevant Characteristics Analysis

The service characteristics of IP frames are also represented in the average service waiting time and average waiting queue length, etc.; at the same time, it also had a direct relationship with local user IP packets generated in the quantity, the time interval, and average size.

The theoretical analysis and simulation research on the average service waiting time and average waiting queue length of IP frames in cross-layer design of the HF IP communication system would be done here.

4.1 The Average Service Waiting Time

Assume that HF IP network users send IP frames to obey the building process, which means the time interval between every two adjacent IP frames is obeying exponential distribution with parameters λ. Assume that the size of the IP frames also obey exponential distribution with parameter l. And the cache of the HF IP access gateway was infinite.

By the preceding analysis, we could know that the IP frames service time was not a standard normal process. So it could assume that IP packets waiting and transmission process was a M/G/1 queuing system.

The IP frame average service time $T_{\text{serve,ip}}$ was already obtained by theoretical analysis, and $T_{\text{wait,ip}}$ was used here to represent the IP frame average service waiting time.

In queuing problems, M/G/1/queuing process has these conclusions: $T_{\text{wait,ip}} = \frac{\lambda \cdot E[\chi^2]}{2(1-\rho)}$, $\rho = \frac{\lambda}{\mu}$.

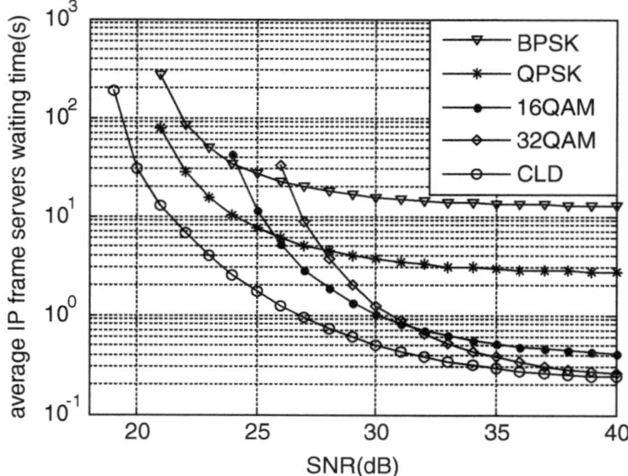

Fig. 6 The average IP frame servers waiting time curve under different SNR for different modulation schemes

χ represented the average service time for a single IP frame.

$$T_{\text{wait,ip}} = \frac{\lambda \cdot E\left[T_{\text{serve,ip}}^2\right]}{2\left(1 - \lambda \cdot E\left[T_{\text{serve,ip}}\right]\right)} \tag{17}$$

Choose IP packet arrival time interval of the simulation for 40 s, the average length of IP packets for 5,600 bit. It can be seen from Fig. 6, the cross-layer design of the HF IP communication system can ensure that cases of IP packets in different channel remain the optimal average service waiting time, especially when the SNR in the 18–35 dB, cross-layer design of the HF IP communication system corresponding to the IP packet waiting time is less than t any fixed mode. Using different modulation mode dynamically could ensure optimal average service waiting time. In Fig. 5 when the SNR less than 18 dB, the IP frame average service time increased obviously, the value of ρ was greater than 1, the average service time would be bigger than the time interval of IP frame arrival, the system average queue length gradually tended to infinity, and the IP frame average service waiting time would increase sharply.

4.2 Average Waiting Queue Length

The queueing process of IP frames in the HF IP access gateway met Little theorem and the average queue length was $\bar{N}_q$:

$$\bar{N}_q = \frac{\lambda^2 \cdot E\left[T_{\text{serve,ip}}{}^2\right]}{2(1 - \rho)} \tag{18}$$

Under the same conditions, the shorter average queue length showed a higher system processing capability, and vice system processing capability needed to be

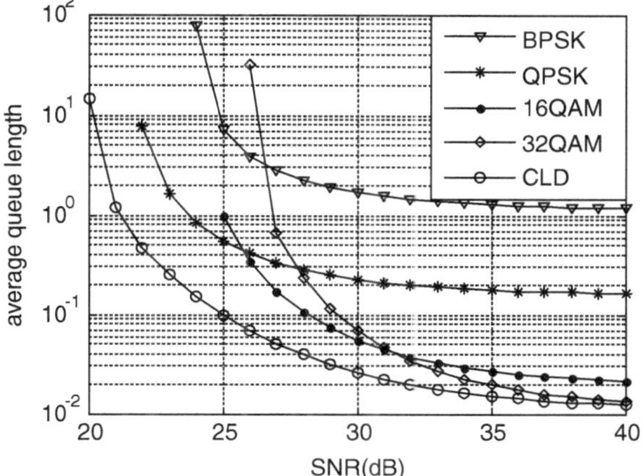

Fig. 7 The average queue length (the quantity of the IP frame) curve under different SNR for different modulation schemes

improved. By Fig. 7 and it could be seen in certain cases of λ and l, the average queue length of cross-layer design of the HF IP communication system was less than any other fix mode. When the channel condition is good, the high-order modulation mode should be used to improve data transfer rate, and with lower bit error rate, it could ensure low queue length. In bad channel conditions, the low-order modulation mode should be used to ensure that the data transmission with low bit error rate would not cause the queue length of unlimited growth. Also similar to Fig. 6, each kind of modulation mode has a minimum SNR, if the γ_s was lower than that SNR, the queue length would be infinite. After the text edit has been completed, the paper is ready for the template. Duplicate the template file by using the Save As command, and use the naming convention prescribed by your conference for the name of your paper. In this newly created file, highlight all of the contents and import your prepared text file. You are now ready to style your paper; use the scroll down window on the left of the MS Word Formatting toolbar.

5 Conclusion

A cross-layer design of the HF IP communication system was introduced in this letter. Theoretical calculation and simulation analyzes were made to find out the basic principle of the system. The IP frame average service time general expression was studied, and found decision threshold of different transport mode. Theoretical study and simulation results showed that the cross-layer design of the HF IP communication system can effectively optimize the IP frame queuing feature. In addition, the research method was also applicable to other adaptive technic.

Acknowledgments In this paper, funded by national natural science funds(61202490) and national defense science and technology key laboratory of funds(9140C020102110C0207).

References

1. Kneidel, T.A, Buchholz, H.: STANAG 5066-flexible software/hardware solution for the new NATO HF-communications protocol. In: Proceedings of IEEE Military Communications Conference, pp. 487–490 (2001)
2. Kallgren Donald, G., Smaal, J.W., Gerbrands, M., Andriesse, M.: An architecture for internet protocol over HF: allied high-frequency wide-area networking using STANAG 5066 (AHFWAN66). In: Proceedings of IEEE Military Communication Conference, pp. 102–114 (2005)
3. Isode Whitepapers. Why IP over HF Radio should be avoided [EB/OL]. 15 Feb 2012
4. http://www.isode.com/whitepapers/ip-over-stanag-5066.html
5. Kallgren Donald, G.: IP unicast/multicast operation over STANAG 5066. NATO 3G radio branch. In: Proceedings of IEEE, Military Communications Conference, vol. 10, pp. 501–505 (2001)
6. Ji, W., Zheng, B.: Enhanced cooperative packet retransmission employing joint cooperative diversity and symbol mapping diversity. J. Zhejiang Univ. Sci. A, **9**(8), 1090–1098 (2008)
7. Oien, G.E., Holm, H., Hole, K.J.: Impact of channel prediction on adaptive coded modulation performance in Rayleigh fading. IEEE Trans. Veh. **53**(3), 758–769 (2004)
8. Liu, Q., Zhou, S., Giannakis, G.B.: Queuing with adaptive modulation and coding over wireless links: cross-layer analysis and design. IEEE Trans. Wireless Commun. **4**(3), 1142–1153 (2005)
9. Harsini, J.S., Lahouti, F., Levorato, M., et al.: Analysis of non-cooperative and cooperative type II hybrid ARQ protocols with AMC over correlated fading channels. IEEE Trans. Wireless Commun. **10**(3), 877–889 (2011)
10. Al-zubi, R., Krunz, M.: Cross-layer design for efficient resource utilization in WiMedia UWB-based WAPANs. ACM Trans. Modeling Comput. Simul. **21**(1), 125–151 (2010)
11. Sun, Q., Tian, H., Dong, K., Zhang, P.: Cross layer scheduling for real-time traffic in multiuser MIMO-OFDMA systems. J. China Univ. Posts Telecommun. **16**(3), 24–29 (2009)
12. Guo, L., Yue, D.: Closed-loop MIMO-MRC cross-layer design scheme based on imperfect channel estimation information. Syst. Eng. Electron. **32**(3), 469–474 (2010)
13. MIL-STD-188-141B, Interoperability and performance standards for medium and high frequency radio systems. Department of Defense Interface Standard, 1999.3.1, Washington D C
14. MIL-STD-188-141C, Interoperability and performance standards for medium and high frequency radio systems. Department of Defense Interface Standard, 2011.7.25, Washington D C
15. Cai, X., Giannakis, G.B.: Adaptive PSAM accounting for channel estimation and prediction errors. IEEE Trans. Wireless Commun. **4**(1), 246–256 (2005)
16. Fu, H., Kam, P.Y.: A simple bit error probability analysis for square QAM in Rayleigh fading with channel estimation. IEEE Trans. Commun. **57**(8), 2193–2197 (2009)
17. Simon, M.K., Alouini, M.-S.: Digital Communication over Fading Channels: A Unified Approach to Performance Analysis. Wiley, New York (2000)
18. Johnson Eric, E., Balakrishnan, M., Tang, Z.: Impact of turnaround time on wireless MAC protocols. In: Proceedings of IEEE Military Communications Conference, pp. 375–381 (2003)

Study of Communication System for Transmission Corridor in the Smart Grid

Jiaquan Yang and Yangyang Song

Abstract To meet the requirements of online monitor for smart grid, a communication system in power grid transmission line is discussed. In this communication system, wireless mesh network is tightly coupling with backbone networks, such as Optical Fiber Composite Overhead Ground Wire (OPGW) Cable System or 2G/3G/4G Cellular Network. To achieve network topology fault tolerance and time synchronization, this system accommodates power grid transmission's linear topology to communication network's multi-path mesh topology. The simulation result indicates this kind of network can implement video monitoring and control every node while meeting the requirements of real time and reliability.

Keywords Smart grid · Transmission line · Wireless mesh network

1 Introduction

Online monitor of power grid transmission lines is the important part in smart grid. Transmission lines are often across long distance, located in remote area. While the current intelligent monitoring devices are lack of available remote communication channel, the OPGW optical fiber of the transmission line engineering

J. Yang (✉)
Department of Intelligent Power Distribution Network and Automation of Electric Power Systems Research and Application, Economic and Technological Development Zone, Yunnan Electric Power Research Institute, No.105, Yunda East Road, Kunming, Yunnan, China
e-mail: yjquan99@163.com

Y. Song
Department of Communication Engineering Institute, Chongqing University, No.174, Centre Street, Chongqing, China
e-mail: songyy5618@163.com

S. Patnaik and X. Li (eds.), *Proceedings of International Conference on Computer Science and Information Technology*, Advances in Intelligent Systems and Computing 255, DOI: 10.1007/978-81-322-1759-6_46,

implementation is difficult and 2G/3G/4G wireless public network do not cover all the monitor nodes. This paper proposes a hybrid communication method of transmission corridor.

In the intelligent online monitoring system, devices monitor and control the transmission lines in real time, converging the collected data through the uplink to the substation or higher-level monitoring center; substation sends state overhaul guide data information to the downlink intelligent monitoring devices. In order to transmit faster and error free, power transmission corridors need to build a broadband, real-time, reliability, and safety communication network [1, 2]. At the same time, the communication network should support the electric power business which is defined by the IEC61850 and IEC61970, so as to meet the intelligent monitoring system of geographic information system (GIS), fault temporal logic analysis, and other special requirements.

2 Related Research

The current power system has established the optical communication network using OPGW. Some researchers tried to use optical network to transmit data directly [3], while the optical network could not connect to all the communication section. In the literature of [4–7], the authors put forward using 2G/3G/4G technology as a way of data collection, but rent telecom network cannot ensure all node to be covered (especially important to monitor remote area). At the same time, the technology of the public communication network cannot meet the transmission corridor's requirements of real time, reliability, and security. Wireless sensor network composed of linear topological network to realize the data transmission [8] is proposed by other researchers, while in this network, the transmission distance, reliability, and bandwidth are limited. In the literature [9], the authors put forward a network communication mode that is a combination of optical and wireless networks, but it failed to reflect the geographical corridor cover characteristics of the transmission line. Some researchers suggested to use the network mode that is a combination of wireless ad hoc and optical networks [10]; it improved the topology reliability of the network, but it did not analyze the business reliability and real-time performance further.

Wireless mesh network (WMN) has the characteristics of broadband and convenient networking, suitable for the environment of transmission corridor. This paper carried out in the National High Technology Research and Development of China 863 key projects of smart grid.The project named "Internet of Things based on intelligent monitoring and life-cycle management of device in transmission lines and substations". Used the WMN that based on IEEE802.11 combining with the OPGW optical network and public land mobile network(PLMN), then the tightly coupling communication system is constituted. The reliability of communication network in this system is improved by dynamic topology. The reference clock for the intelligent terminal is provided by the entire network synchronization clock. The channel capacity and control transmission delay are improved by dynamic routing. The power consumption and interference of the equipment are reduced by using high antenna.

3 Communication System Study in Smart Grid Power Transmission Corridors

3.1 Three-Plane Network Mode

The communication system of smart grid power transmission corridors can be divided into three planes: optical network plane, wireless mesh network plane, and PLMN plane, as shown in Fig. 1. OPGW provides a reliable communication channel for the electric power remote scheduling on the transmission line, as well as provides backbone network for online monitoring devices.

Wireless mesh network converges the information of the transmission corridor multistage tower to the OPGW access point by wireless relay and then reduces OPGW open times. PLMN network act as the OPGW optical backbone network backup. WMN transfer the information to the communication backbone network through the 2G/3G/4G wireless cellular network.

The network fault tolerance design. Three-plane network provides diversified access and backbone network for information node to choose, which can constitute a multiple-ownership configuration to improve the reliability of the information, gathering access points. In order to overcome the unreliability of the network that brought by the multi-hop linear networking of the wireless relay equipment on the transmission tower, The topology of WMN in power corridor is the reticular logical topology, it provided redundant path for the multi-hop relay wireless network, as shown in Fig. 2.

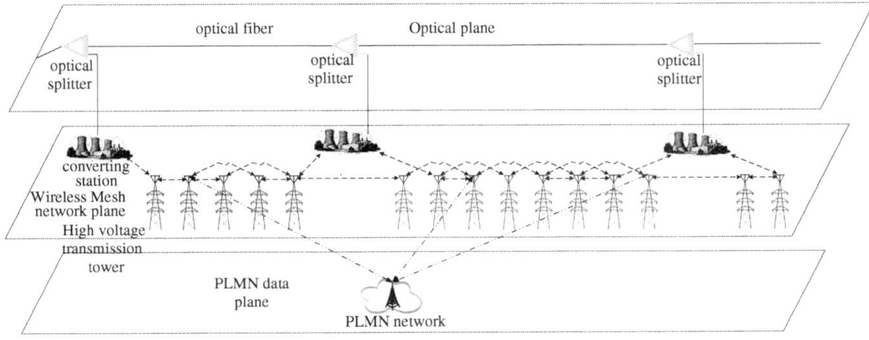

Fig. 1 Communication system in transmission corridors

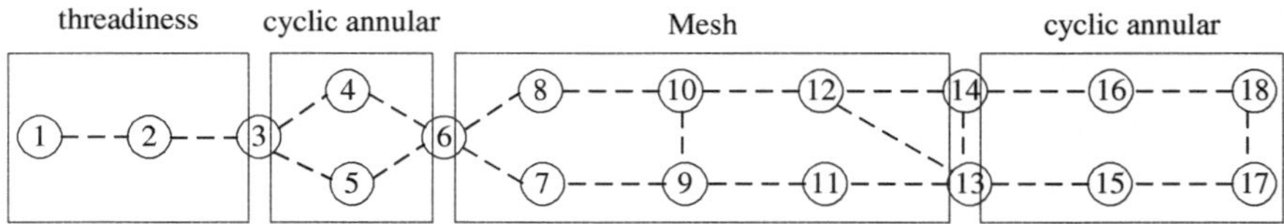

Fig. 2 Fault tolerance in communication network topology

Network control design. The network topology adopts IEEE802.11 equipment "point-to-point" and "point-to-multipoint" simple topology consisting of multiple independent basic service sets, and multi-hop WMN composes of extended service, OPGW optical network, and PLMN network, constituting a distributed system [11]. By simplification, BSS domain can avoid the hidden and exposed terminal problems in wireless mesh access protocol [12]. The network control zoning methods are shown in Fig. 3.

Network synchronization design. Smart business information terminals need communication network clock. BITS is a benchmark in a power system. WMN extracted reference clock from OPGW optical network and GPS as a backup reference clock. Therefore, WMN establishes a backbone network, setting the gateway node as the root node, using least series constitutes a dynamic clock tree, as shown in Fig. 4. In the synchronous tree, the node more near to the root has the higher level of the clock.

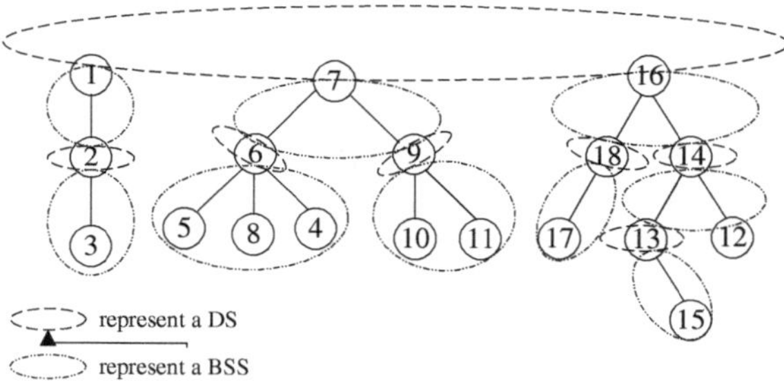

Fig. 3 The control area of communication network

	Synchronous tree 1	Synchronous tree 2	Synchronous tree 3
level 1 clock	1 (O)	7 (O)	16 (O)
level 2 clock	2 (M)	6 (M) 9 (M)	18 (M) (G) 14
Level 3 clock	3 (M)	(M) 5 (G) 8 (M) 4 (M) 10 (M) 11	17 (M) 13 (M) (M) 12
Level 4 clock			(P) 15

Fig. 4 The synchronization tree of communication network

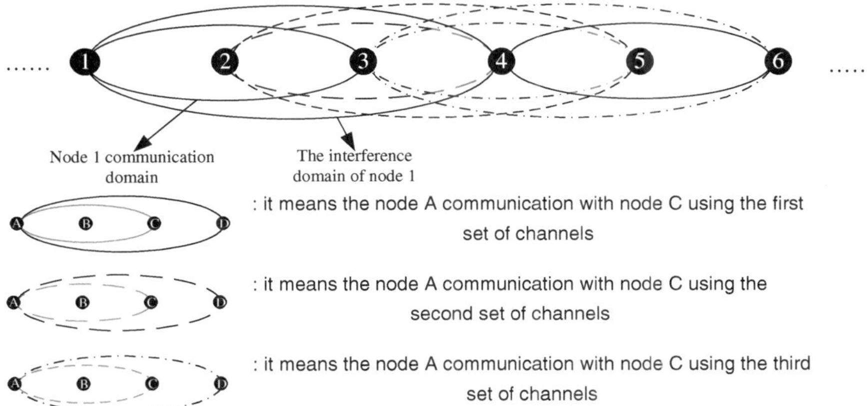

Fig. 5 The model of communication domain and interference domain

3.2 Key Technology in Wireless Mesh

WMN (working at 2.4 GHz or 5.8 GHz ISM spectrum) based on the IEEE802.11 and built multiple independent BSS service nodes need to realize the remote communication of several kilometers. For this, we need to consider power, wireless resources, protocol, and other limit conditions.

High-gain wireless technology. Long-span information transmission of the corridor needs to increase the transmission distance of the traditional IEEE802.11 equipment. It not only increases energy consumption by improving wireless transmission power, but also exceeds the limited transmission power. For example, IEEE802.11g equipment is limited to sending power less than 100 mw [13], and the working distance is 100 m. A long-distance communication requires transmission power exceeding the above regulations.

The information node of transmission corridor is linear distribution along the transmission tower, using high-gain antenna can improve the communication transmission distance and meet the provisions of the radio management committee. For example: if the sending and receiving side adopt the directional antenna that the gain is 20 dBi, and receiver sensitivity is -75 dBm, set the transmit power is 500 mw, then communication distance can be up to 5 km.

Wireless channel allocation. Wireless mesh network allocates wireless channel resources reasonably to reduce the interference between the channels. Due to wireless power limitation, the effective communication distance called communication domain, while beyond communication domain still can cause interference to other communication domain area called interference. WMN communication domain and interfering domain model of transmission corridor communication system are shown in Fig. 5.

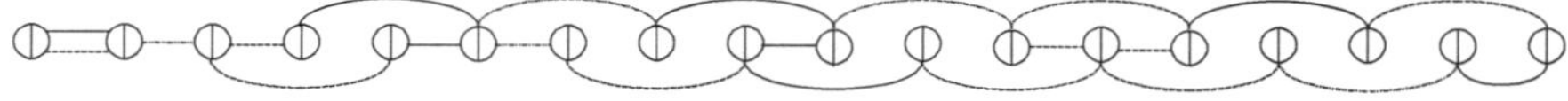

Means the transmission node with back-to-back directional antenna

Means the two sites communication with each other through the first set channel

Means the two sites communication with each other through the Second set channel

Means the two sites communication with each other through the Third set channel

Fig. 6 A radio resource allocation method

The principles of wireless channel allocation are the following: the maximum use of the given channel number and the communication port number, makes two nodes communicate on one node interference domain without interference, namely the communication domain will not overlap. Based on WMN topology and ensure that the nodes in WMN can communication, each node assign an independent channel to inhibition adjacent nodes interfere each other. For multiple nodes of a linear topology, you can use multiple communication interfaces between the two communication nodes, through the use of multiple redundant link can implement link level of resources redundancy, to improve the reliability of the network. Figure 6 shows the allocation way of the WMN when there are 3 set channels available.

QoS routing in WMN. Transmission system is a kind of control system in essence. It puts forward strict requirements for the real time and reliability of information transmission, such as online monitoring and control of power grid equipment. In this paper the communication network provides a comprehensive QoS guarantee measures that is provide QoS routing in WMN topology.

Routing is divided into two parts: the first is based on the QoS to establish and maintain the optimal path tree that node reaches the threshold and the second part is the node according to the optimal path of the tree and the QoS of different business, choose a different path transmit.

First of all, node periodically updates the entire wireless network link-state database, including delay, bandwidth, etc. After each update, node recalculates the optimal path tree under the restriction of business service quality. In addition, when the WMN is abnormal, the operation will be carried out immediately.

Node 11 uses the optical network as the business backbone network and uses the time delay and available bandwidth as a business service quality index of a link [value on the link represents the time delay and available bandwidth, such as value on the link (7, 8) is 1/4; it means the delay of the link is 1, the available bandwidth is 4] and suppose to get the link-state diagram as Fig. 7 shows.

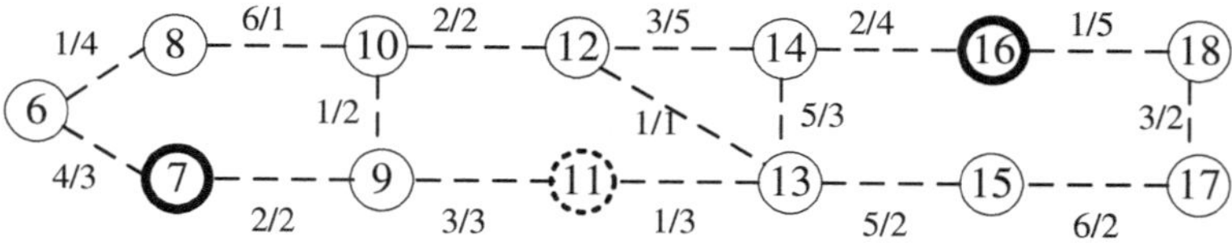

Fig. 7 Link status in part of the network

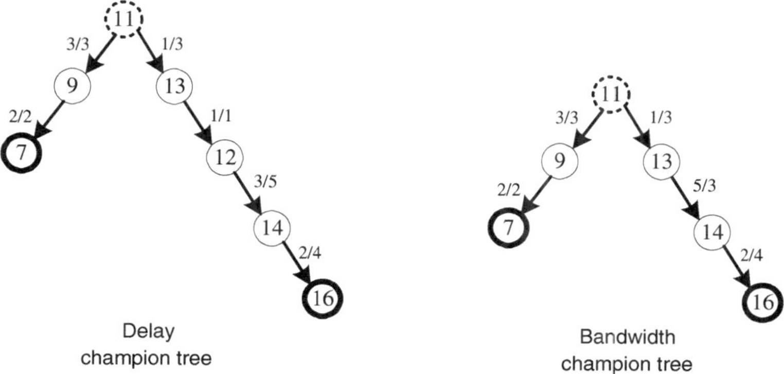

Fig. 8 Optimal path tree

We can obtain the optimal path tree from node 11 to gateway 7 and 16 by computing as shown in Fig. 8.

Then, the node chooses the best path from the optimal path tree based on the principle of optimal path for different internal business which has different QoS requirements. The figure shows the following:

The delay time between node 11 and node 7 is

$$D_{\text{path}(11,7)} = D_{(11,9)} + D_{(9,7)} = 5$$

The delay time between node 11 and node 16 is

$$D_{\text{path}(11,16)} = D_{(11,13)} + D_{(13,12)} + D_{(12,14)} + D_{(14,16)} = 7$$

The available bandwidth between node 11 and node 7 is

$$B_{\text{path}(11,7)} = \min\{B_{(11,9)}, B_{(9,7)}\} = 2$$

The available bandwidth between node 11 and node 16 is

$$B_{\text{path}(11,16)} = \min\{B_{(11,13)}, B_{(13,14)}, B_{(14,16)}\} = 3$$

And

$$D_{\text{path}(11,7)} < D_{\text{path}(11,16)}, \; B_{\text{path}(11,7)} < B_{\text{path}(11,16)}$$

Therefore, for the business that is sensitive to time delay, it can arrive in optical network through node 7 from node 11, while for the business that is sensitive to bandwidth, it can arrive in optical network through node 16 from node 11, which can realize the balanced transmission of network traffic.

Operation scheduling of WMN. In real-time communication scheduling, the business is divided into different priorities at the entrance of the information by the reliability and real time of information thus obtained distinguishing communication services. However, due to internal multi-hop transmission in WMN, for example,

some low-priority business gets the higher-priority service because of the short relay hop. So it needs to dynamically adjust the priority of information. For instance, the business of control has the highest priority (marked as 4), followed by the business of concentrated copy data (marked as 3); the lowest priority of business is monitor video (marked as 2). In WMN, according to the transmitted hop h can dynamic adjustment priority in multi-hops transmission and adopt different strategies for different bearing business.
Priority control = $\mathrm{Min}\{8,\ 4 + \mathrm{h}\}$, Priority reading = $\mathrm{Min}\{6,\ 3 + \mathrm{h}\}$, Priority video = 2. To the control and concentrated copy data, the priority adds 1 as the message jumps one hop, while it makes the priority of business of control higher than that of concentrated copy data to guarantee the business of control in WMN remains the same priority of general video business.

According to the priority weight and the limit of maximum tolerance delay, the node schedules the business of the largest priority first and schedules the business that of the earliest arrival deadline to ensure the real-time requirements of the business, which is based on the limit of maximum tolerance delay when the priority weights are the same.

4 Network Simulation

As we proposed a communication system and some key technology, a simulation with OPNET is finished to prove their feasibility. Some wireless parameter setting of the simulation is shown in Table 1.

Firstly, we compare the throughput in three scenarios below:

1. Point-to-point network in which distance between nodes is 10 m.
2. Point-to-point network in which distance between nodes is 5 km.
3. Mesh network in which the longest distance between nodes is 5 km.

Table 1 Wireless parameter in simulation

Spread Spectrum	Extended
Channel setting	Automatic
Data rate	54 Mbps
Sending power	0.05 w
Threshold of package receiving	−95
Threshold of package fragmentation	2,034 byte
RTS threshold	None
Long retransmission times	5
Buffer memory space	2,024,000 bit
Processing mode of large packet	Burst
The survival time of largest receives	0.5 s
SIFS	10 μs
Slot time	20 μs
DIFS	50 μs

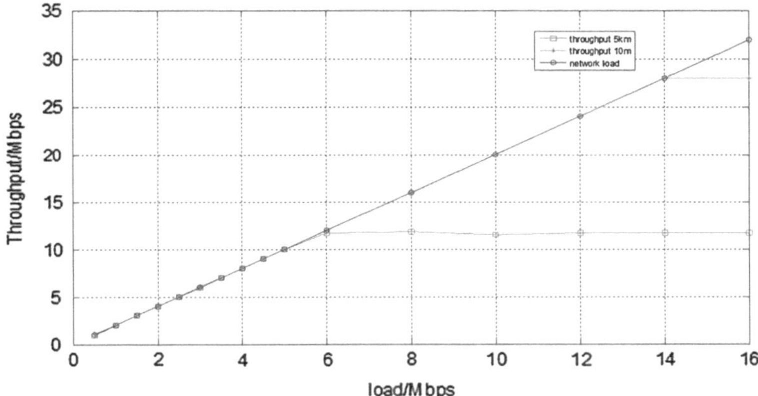

Fig. 9 Throughput of P2P network

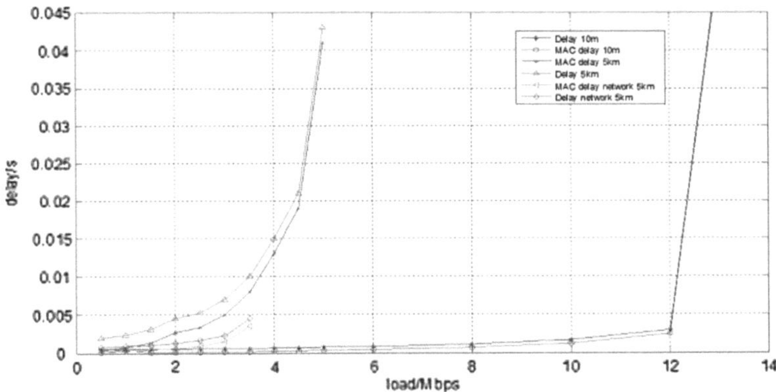

Fig. 10 Delay and MAC delay of the three scenario

The throughput is shown in Fig. 9, and the MAC delay is shown in Fig. 10. The throughput with a distance of 5 km is nearly half of the distance of 10 m. The increased distance decreases the probability of successful transmission of each node.

In Fig. 11, the network time delay has risen sharply when the load of node was 5 Mbps, and the network was unable to process all the data packets; then, part of the packets started to be discarded, and the network throughput reached its limits when the load of node was 6 Mbps, which can be known from the contrast of time delay.

Then follows the scenario of communication system in Fig. 1; the distance between nodes is a random number with a cap of 5 km.

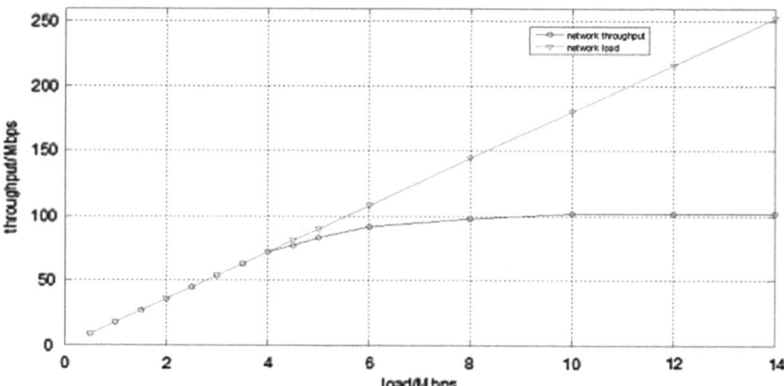

Fig. 11 Throughput of WMN

The comparison between the load and throughput in Fig. 11 shows that the ability of the sample network is nearly 100 MBps; the network can work properly only if the load of each node is less than 4 M; otherwise, some area of the network is going to congest. This is determined by the largest distance between the communication nodes. The throughput can meet the requirement of a standard video flow (2 Mbps).

5 Conclusion

This paper puts forward a kind of wireless communication system which is applicable to industrial information corridor and formed the three-plane communication network scheme that is tightly coupled by PLMN network, optical network, and WMN.

This communication system provided a way to levant the faults of link and topology and enhanced the reliability of the system and guaranteed reliability and real-time strategy for the business in view of the power system. Then, this paper presented simulation for the performance of this kind of network. The simulation result proved that mesh network can implement video monitoring and control every node while completely guaranteeing the real time and reliability.

Acknowledgments Foundation item: The National High Technology Research and Development of China (863 Program) (2011AA05120).

References

1. Sun, F., Lei, M., Yang, C., et al.: The construction of the IBM smart grid innovation operation management—new train of thought on the development of China power. The IBM global business services, Beijing (2006)
2. Liang, Z., Qiu, X., An, R.: The demand of smart grid for telecommunications. Telecommun. Electric Power Syst. **31**(215), 1–4 (2010) (in Chinese)
3. Zhang, L.: Networking solution for wind turbine generator control system in wind farm. Telecommun. Electric Power Syst. **33**(231), 138–142 (2012)
4. Xinbo, H., Jiabing, L., Xiangli, W., Huayu, Y.: On-line remote-monitoring system for transmission line insulator contamination based on the GPRS net. Autom. Electric Power Syst. **28**(21), 92–95 (2004) (in Chinese)
5. Z., Fu, Wu, B., Huang, X., Xin, Y.: Online monitoring system based on GSM network for transmission line fault. High Voltage Eng **33**(5), 69–72 (2007). (in Chinese)
6. Wang, Z., Song, G., Li, M.: The application of 3G communications technology in the acceptance of smart grid automation system. Telecommun. Electric Power Syst. **32**(228), 78–81 (2011) (in Chinese)
7. Cheng, P., Wang, L., Zhen, B., Wang, S.: Feasibility study of applying LTE to smart grid. IEEE Smart Grid Comm. 108–113
8. Wang, Y., Yin, X., You, D., Xu, T., Hua, H., Xiang, H.: A real-time monitoring and warning system for electric power facilities icing disaster based on wireless sensor network. Power Syst. Technol. **33**(7), 14–19 (2009). (in Chinese)
9. Aravinthan, V.: Wireless communication for smart grid applications at distribution level–feasibility and requirements. IEEE (2011)
10. Wang, X., Li, H., Cong, L.: A novel monitoring system for high voltage transmission lines based on wireless and optical communication technologies. Power Syst. Technol. **33**(18), 198–203 (2009) (in Chinese)
11. IEEE-SA Standards Board. IEEEStd 802.11™-2007. Wireless LAN medium access control (MAC) and physical layer (PHY) specifications. USA: IEEE (2007)
12. IEEE-SA Standards Board. IEEEStd 802.16j-2009, Part 16: Air interface for broadband wireless access systems amendment 1: multiple relay specification. USA: IEEE (2009)
13. IEEE-SA Standards Board. IEEEStd802.11g-2003, Part 11: Wireless LAN medium access control (MAC) and physical layer (PHY) specifications, Amendment 4: further higher data rate extension in the 2.4 GHz Band[S]. USA: IEEE (2003)

Network Selection Mechanism for Future Generation Networks Using Game Theory Model

C. P. Maheswaran and C. Helen Sulochana

Abstract Selecting an appropriate network to satisfy a service request in future generation mobile networks is quite recent. In heterogeneous network environments, diverse radio access technologies coexist, making it challenging to select the best network at any particular time. A Game Theory model is proposed, which specifically defines a game between the access networks themselves that compete in a non-cooperative manner to maximize their payoff. The concept of quality points in a game theoretical context, and a mathematical mechanism to select the best network is provided. Finally, it is also demonstrated that only the best network serves the service request of the user.

Keywords Heterogeneous networks · Network selection · Quality factor · Weighting coefficient

1 Introduction

Fourth-generation (4G) communication networks that allow subscribers to utilize networks having varying capabilities and dissimilar architectures have generated much interest with respect to the modeling, evaluation, and convergence of multiservice networks of various technologies [6]. As 4G networks incorporate interoperability, interworking, and convergence of diverse access, WCDMA and

C. P. Maheswaran (✉)
Department of Computer Science and Engineering, Noorul Islam Centre for Higher Education, Kumaracoil, Kanyakumari, TamilNadu, India
e-mail: maheswaran_ncp@yahoo.co.in

C. H. Sulochana
Department of Electronics and Communication Engineering, St. Xavier's Catholic College of Engineering, Chunkankadai, Kanyakumari, TamilNadu, India
e-mail: helenrajan@yahoo.com

S. Patnaik and X. Li (eds.), *Proceedings of International Conference on Computer Science and Information Technology*, Advances in Intelligent Systems and Computing 255, DOI: 10.1007/978-81-322-1759-6_47,

WLAN networks can be combined to form a converged communication networking framework, and Game Theory [2] can help in decision formulation of preferring a particular access network over another dissimilar access network for providing a required service.

Wireless access technologies and end-user terminals (smartphones, PDAs) offer rich connectivity, so that end-users will simultaneously be covered and connected by multiple access networks/technologies [2], creating new opportunities while posing novel challenges at different networking levels. On the network's side, when different networks operated by different and potentially competing actors coexist, resource allocation becomes challenging while attempting to develop effective strategies to allocate and dynamically manage radio resources, coupled with introducing effective network solutions to handle the vertical handover of end-users. Developing strategies to automatically select the "best" connectivity opportunity to match the user's quality-of-service (QoS) constraints becomes a challenge on the end-user's side.

The Game Theory approach for network selection when a single network is selected is analyzed, as network selection mechanism in a converged system involves decisions regarding available networks to a participating user that would best satisfy a specific service request upon activation, and during the whole duration of the session, resulting in either a single or a group of access networks.

A strong correlation between the twofold challenge of network resource allocation and user network selection exists. So, the 4G network environment is modeled using Game Theory, while introducing supplementary concepts and a diverse mechanism for network selection at the same time, allowing the best access network to serve the subscriber.

2 Related Work

Regarding network selection, Ref. [3] analyzes various metrics which force the selection phase after measuring the end-user's QoS level. Works based on modeling heterogeneous network environments using Game Theory [12] are limited, but have been extensively used in bandwidth allocation and pricing [8], modeling of peer-to-peer and ad hoc networks [4], and resource management [10], which has received exhaustive mention in [1].

Ormond et al. [11] propose Grey Relational Analysis and Analytic Hierarchy Processing to address selecting networks. Reference [1] uses a Game Theory model for 4G communication networks and sets up a game played in phases by assigning service requests to competing access networks. They also describe user utility and preferences which are related only to QoS parameters (delay, jitter, and packet loss), which is insufficient. Reference [7] presents algorithms for a utility-based model of heterogeneous networks that ensure user utility and help in distributing network loads evenly. Munasinghe et al. [9] explains the role of service platforms and terminals on network selection decisions. They favor a mobile terminal (MT)-controlled handover (HO) approach which might fail for handheld devices due to overhead signaling. Reference [5] identified differences between network-assisted and terminal-based network selection approaches and proposes a network-assisted selection mechanism (without an algorithm) compatible to ours. Existing literature do not explain the need for a balanced approach ensuring proportional impact of input factors on network selection decisions.

3 Modeling Network Selection as a Game

In heterogeneous networks, a set of requests needs to be disseminated to access networks, whose objective is to maximize payoff. It is user-centric, as decisions are made at the user's end and are based only on the user's satisfaction by the way their service requests are handled. Each access network also has its own internal admission mechanisms—the subset of service requests selected to be handled by a precise access network needs to be admitted into that network once they are elected.

In our Game Theory model, network selection is defined as a game between access networks in a heterogeneous system. As selfish players in the game, the access networks are non-cooperative decision-making entities, i.e., by rationally choosing the best policy, each one tries to maximize its own payoff.

The model of a finite strategic game is defined as having the following constituents:

1. A well-defined set of decision-making entities, i.e., the set of players.
2. A non-empty set of pure strategies or actions for each player.
3. A set of payoff functions for each player's possible strategies to a real number.

Consider a set of three service requests R (streaming video, surfing, and voice chat) that can be serviced by only one access point. We name it as the set of game resources that need to be distributed among competing players. The selection game is defined as a tuple including definitions for players, resources, each player strategies, and their payoffs.

Let's define the network selection game as $G = (N, R, (S_I)_{I \in N}, (V_{IJ})_{I \in N, j \in R})$ where N denotes the set of players (WCDMA, WLAN, WiMAX) between 1 and n, R denotes the set of service requests (set of game resources) between 1 and r, and S_i denotes the set of strategies for player i $i \in N$, where $s_i \in S$ is a non-empty set of strategies for the player i. Each pure strategy is the selection of a particular service request from the set R. V_{ij} denotes the payoff for each player i when choosing resource j, which is directly related to a user-generated preference value corresponding to that network. Section 4 further explains preference estimation and the resulting payoff function.

Fig. 1 Strategy space

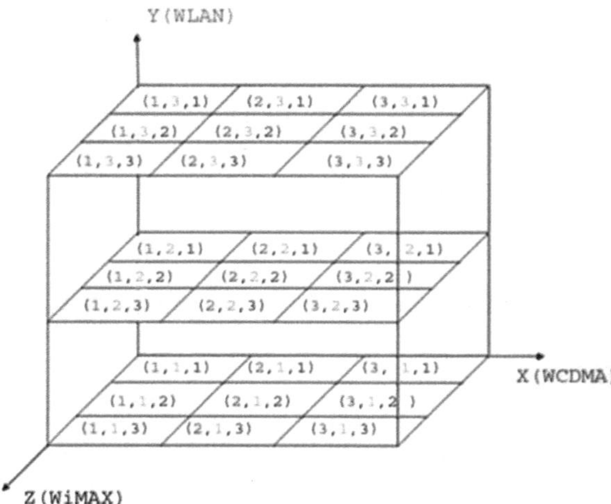

In our model, players (individual access networks) compete to win a service request. Here, WCDMA, WLAN, and WiMax are the X, Y, and Z axes, respectively. Each player (access network) chooses a strategy from a set of strategies (service requests) to achieve the highest payoff, which is depicted in Fig. 1.

For example, a strategy space coordinate (1,1,2) implies that both WCDMA and WLAN choose streaming video and WiMAX selects voice call. Generally, the three coordinate axes (access networks) and varying combinations of preferred strategies constitute the strategy space. In reality, the heterogeneous network environment constitutes more than the three access networks (players) mentioned above.

Since services are also classified as non-real time and real time, the actual heterogeneous network scenario is multidimensional with varying combinations of service requests. If m players (access networks) compete to serve n service requests, there will be n^m strategy space coordinates in the m-dimensional strategy space. We have considered three players (WCDMA, WLAN, WiMAX) to illustrate the possible multidimensional heterogeneous environment in a simple three-dimensional representation. So, our strategy space is three players and three service types (i.e., $m = 3$, $n = 3$).

4 Payoffs Calculation

Calculating payoffs of each competing access network depends on the choice of strategy of access networks and also on other factors (R1–R5) that have an associated condition at a particular time denoted by a character for clarity as shown below.

R1 The type of service: streaming video (A), internet surfing (B), and voice call (C)
R2 User preference: Cost (A) and Quality (B)
R3 Traffic state and signal strength of the network: bad (A), medium (B), and good (C)
R4 Speed of the user: High speed (A), Low speed (B), and Stable (C)
R5 Drainage rate of battery in each mode: (A), (B), (C), (D), (E), and (F)

The fifth input factor (R5) is kept dynamic. "A" implies that in a three-mode (WCDMA, WLAN, WiMAX)-enabled mobile device, WCDMA drains battery power the least, followed by WLAN and WiMAX, respectively. All the conditions are given below.

"A" means that WCDMA > WLAN > WiMAX	"B" means that WCDMA > WiMAX > WLAN
"C" means that WLAN > WCDMA > WiMAX	"D" means that WLAN > WiMAX > WCDMA
"E" means that WiMAX > WCDMA > WLAN	"F" means that *WiMAX > WLAN > WCDMA*

Input factors and associated conditions are collected by the core network as shown in Table 1.

A. Assignment of Quality Points

Quality points translate the relative "advantage" of one access network over another into an integer. For example, for high-speed users, WiMAX or WCDMA has higher-quality points, as WLANs have a smaller coverage area, causing an overwhelming number of handovers. Mappings from wireless parameters, user preferences, speed of users, power consumption of battery, and available bandwidth have quality points with defined thresholds. For example, a user speed of 80 km/h may map to a quality point of 0 for WLANs, as they are inefficient for a high-speed user. If the data transfer rate of an access network drops below a predefined threshold, it might map to a low-quality point for that network, making it less likely to win the service request. Information exchange across layers, design, and signaling ensures accurate mappings. As mentioned before, quality points and values assigned for all inputs in Table 1 need not necessarily be the same in reality.

B. Weighting Coefficients

We have considered five factors (R1–R5) affecting payoffs of the three competing access networks differently. We assign W_{R1}, W_{R2}, W_{R3}, W_{R4}, and W_{R5} as weighting coefficients of factors R1, R2, R3, R4, and R5, respectively. If R5 is five times more important in network selection than R1, we assign $W_{R5} = 25$ and $W_{R1} = 5$.

Table 1 Quality points based on service type, user preferences, and network state

Quality based on	Factor	Quality points		
		Q_{WCDMA}	Q_{WLAN}	Q_{WiMAX}
Service type (R1)	Video streaming	3	5	6
	Surfing	3	4	4
	Voice chat	7	4	4
User preference (R2)	Cost	3	7	4
	Quality	3	5	6
Signal strength and network state (R3)	$N\_t = \uparrow$ & SS $= \downarrow$	0	0	0
	$N_t = \leftrightarrow$ & SS $= \leftrightarrow$	7	3	5
	$N_t = \downarrow$ & SS $= \uparrow$	4	7	4

C. Equation for Calculating Payoff of the Networks

The total payoff of each network is the weighted sum of quality points it receives from each factor depending on specific conditions of the factors. The "payoff equation" providing the payoff of the competing access networks is provided below:

$$P_i = W_{\mathrm{R1}}Q_{\mathrm{R1}_i} + W_{\mathrm{R2}}Q_{\mathrm{R2}_i} + W_{\mathrm{R3}}Q_{\mathrm{R3}_i} + W_{\mathrm{R4}}Q_{\mathrm{R4}_i} + W_{\mathrm{R5}}Q_{\mathrm{R5}_i} \quad (1)$$

where P_i is the payoff of player i, W_{R1} to W_{R2} are weighting coefficients for factor R1 to R5, respectively, and Q_{R1_i} is the quality point that player i obtains from factor R1, depending on a specific condition. Equation (1) for three players can be written separately as follows:

$$P_1 = W_{\mathrm{R1}}Q_{\mathrm{R1}_1} + W_{\mathrm{R2}}Q_{\mathrm{R2}_1} + W_{\mathrm{R3}}Q_{\mathrm{R3}_1} + W_{\mathrm{R4}}Q_{\mathrm{R4}_1} + W_{\mathrm{R5}}Q_{\mathrm{R5}_1} \quad (2)$$

$$P_2 = W_{\mathrm{R1}}Q_{\mathrm{R1}_2} + W_{\mathrm{R2}}Q_{\mathrm{R2}_2} + W_{\mathrm{R3}}Q_{\mathrm{R3}_2} + W_{\mathrm{R4}}Q_{\mathrm{R4}_2} + W_{\mathrm{R5}}Q_{\mathrm{R5}_2} \quad (3)$$

$$P_3 = W_{\mathrm{R1}}Q_{\mathrm{R1}_3} + W_{\mathrm{R2}}Q_{\mathrm{R2}_3} + W_{\mathrm{R3}}Q_{\mathrm{R3}_3} + W_{\mathrm{R4}}Q_{\mathrm{R4}_3} + W_{\mathrm{R5}}Q_{\mathrm{R5}_3} \quad (4)$$

D. The Network Selection Decision and Procedure

Equations (2), (3), and (4) calculate each network's payoff, and the network with highest payoff serves a user's particular service request (strategy) if there are more than one.

We now illustrate the procedure for network selection mechanism.

Step 1: Get the input strategy (could be the service request that the access networks has chosen).
Step 2: Verify the given strategy with the existing one.
Step 3: Based on strategic decisions, the access networks can serve requests in case of differences.
Step 4: Otherwise obtain remaining policies.
Step 5: Based on the policies, assign quality points.
Step 6: Calculate total payoff of the access network using Eq. (1).
Step 7: Select the best network based for the user based on total payoff.

E. Maximizing the Payoffs

Here, players (access networks) try to maximize their payoffs. The three access networks reside according to their chosen strategies. An intelligent access network tries to maximize its payoff and moves to a different point in the strategy space, if, in its current position, it does not receive a higher payoff in winning the network selection race.

5 Network Selection Issue

A. Service and Access Provider's Role in Network Selection

Access network providers try to maximize their payoff and financial benefit by winning service requests. It provides hints to service/access providers to select a strategy from the strategy space, which provides a platform for intelligent network monitoring on part of network providers, which may be explored in future works. Thus, the provider can decide to serve those requests which would provide maximum benefit and cause least resource (i.e., bandwidth) consumption. We assume that the interconnection pricing will be the same for all cases, which would have no impact on selection strategy.

Network congestion [5] can degrade throughput and QoS which can be taken care of. The input R3 is based on state of the network (network traffic and congestion state) and signal strength. Quality points obtained from R3 by a particular access network will depend on network congestion which can knock out a congested network from the network selection game and connect the user to a better access network.

B. Proof of Concept

By providing a variety of inputs to a computer program, it precedes input information and decides which network will best serve the subscriber at a particular time. Although quite clear-cut, we presume that it will lay the basis for an intelligent agent, which will have many interfaces with a number of modules and other entities, thus ensuring that the user is connected to the best network. To select the best network, information exchange between different layers (network, link, and application layer) is required.

Assume a strategy space with three players and three service types (video streaming, surfing, and voice chat). We calculate payoffs of each competing access network, which depend on the choice of strategy and factors R1–R5, which is provided in Table 2.

In case 1, the three networks competed to win a service request for Internet surfing. The Input User Preference was A (cheaper service instead of high bandwidth and better quality). Input Traffic state and Signal Strength were signified for the three networks. The user's Input Speed was "A" (service request from high-speed user). Input Drainage rate was B (WCDMA had a better drainage rate of battery). Thus, WCDMA with the highest payoff (595) was eventually used for

Table 2 Network selection based on case 1 and case 2

	INPUT					OUTPUT			
	Type of service	User preference	Signal strength	Speed	Drainage rate	Payoff			Best Network
						WCDMA	WLAN	WiMax	
Case 1	B B B	A	A B A	A	B	595	545	525	WCDMA
Case 2	A A A	B	B B B	A	D	805	645	815	WiMax

Internet surfing. Similarly, in case 2, WiMAX had the highest payoff and served streaming video. This demonstrates that this new mechanism of network selection ensures those networks that have significant advantages for every factor or input will be chosen to serve the user.

6 Conclusion and Future Work

A novel Game Theory technique and strategy space-based modeling of a heterogeneous network environment, along with ideas for network selection ("quality points," "weighting factors") and "payoff equation," which ties up quality points and weighting factors in a transparent manner, has been introduced. A mathematical approach for network selection giving heterogeneous network users an opportunity to be "always best network connected" is discussed, taking into account the most influential factors for network selection from the user's perspective and providing intelligent network monitoring and resource allocation by including strategy space. Further investigation of quality points under various circumstances and the relationship between mobility, handoff issues, and network selection decision in heterogeneous networks will be examined in future works.

References

1. Altman, E., El Azouzi, R., Boulogne, T., Jimenez, T, Wynter, L.: A survey on networking games. Comput. Oper. Res. (2004)
2. Antoniou, J., Pitsillides, A.: 4G converged environment: modeling network selection as a game. In: Proceeding of IST mobile summit, Budapest, Hungary, (2007)
3. Bari, F., Leung V.: Service delivery over heterogeneous wireless systems: networks selection aspects. In: Proceeding of the 2006 international conference on wireless communications and mobile computing, Vancouver, British Columbia, Canada (2006)
4. Charilasa, D., Markakia, O., Nikitopoulos, D., Theologoua, M.: Packet switched network selection with the highest QoS in 4G networks. Comput. Netw. **52**(1), 248–258 (2008)
5. Da Silva L.A., Srivastava V.: Node participation in ad-hoc and peer to peer networks: a game-theoretic formulation. In: Workshop on games and emergent behaviour in distributed computing environments, Birmingham, UK, Sept 2004

6. Gavrilovska, L.M, Atanasovski, V.M.: Interoperability in future wireless communications system: a roadmap to 4G. Microw. Rev. **13**(1), 19–28 (2007)
7. Gazis, V., Houssos, N., Alonistioti, N., Merakos, L.: On the complexity of "always best connected" in 4G mobile networks. Vehicular Technology Conference (VTC 2003-Fall) 2003
8. Ho, C., Pingyl, F., Zhigang, C.: A utility-based network selection scheme for multiple services in heterogeneous networks. In: Proceeding of international conference on wireless networks, communications and mobile computing, Hawaii, USA (2005)
9. Munasinghe, K.S., Jamalipour, A., Vucetic, B.: Interworking between WLAN and 3G cellular networks: an IMS based architecture. In: Proceeding of Auswireless conference, Sydney, Australia (2006)
10. Myerson, R.: Game Theory: Analysis of Conflict. Harvard University Press, USA (1997)
11. Ormond, O., Murphy, J., Muntean, G.: Utility-based intelligent network selection in beyond 3G systems. In: Proceeding of IEEE ICC (2006)
12. Prasad, R.: Convergence paving a path towards 4G. In: Proceeding of International Workshop on Convergent Technologies (IWCT), Oulu, Finland (2005)

Research of Emergency Logistics Routing Optimization Based on Particle Swarm Optimization

Liyi Zhang, Yang Li, Teng Fei, Xi Chen and Guo Ting

Abstract Emergency logistics path optimization as a key link for today's emergency decision has not to be ignored. The particle swarm optimization (PSO) as originally conceived was to simulate the process of flocks of birds foraging, and subsequently, it is found that the PSO is the better optimization tool. In this paper, single-vehicle emergency logistics distribution routing optimization model of an extreme case is established, the PSO is used in solving the model, and its effectiveness is shown by simulation.

Keywords Particle swarm optimization (PSO) · Emergency logistics · Path optimization · Single-vehicle

1 Introduction

Emergency logistics path optimization as a key link for today's emergency decision has not to be ignored. Emergency logistics path optimization is that in the shortest possible time, relief supplies can be delivered to the affected point quickly and completely. Emergency logistics path optimization is different with the conventional path optimization. The emphasis of emergency logistics path optimization is the time not strong economy, while conventional path optimization shall focus on economy.

L. Zhang (✉) · T. Fei · X. Chen · G. Ting
Information Engineering College, Tianjin University of Commerce, Tianjin, China
e-mail: zhangliyi@tjcu.edu.cn

Y. Li
Economic College, Tianjin University of Commerce, Tianjin, China
e-mail: liyang@tjcu.edu.cn

S. Patnaik and X. Li (eds.), *Proceedings of International Conference on Computer Science and Information Technology*, Advances in Intelligent Systems and Computing 255, DOI: 10.1007/978-81-322-1759-6_48,

The PSO is originated in the simulation of a simple social system. The PSO as originally conceived was to simulate the process of flocks of birds foraging, and subsequently, it is found that the PSO is the better optimization tool [1]. First, single-vehicle emergency logistics distribution routing optimization model of an extreme case is established; second, the PSO is researched; third, the PSO is used in solving the model; and at last, its effectiveness is shown by simulation.

2 Emergency Logistics Path Optimization Model

2.1 Basic Problem Analysis

After disasters, the l affected areas request the delivery of relief supplies to the relief materials reserve center (RMRC). The RMRC has m' cars.

In this paper, it is assumed that the RMRC has only one car, namely $m' = 1$. In case that total demand for relief supplies of all the affected points is less than the vehicle load, the problem which presupposes that it can meet time requirements is transformed to TSP.

2.2 Model

In order to facilitate the model's establishment, the assumptions have been taken:

1. The total demand for relief supplies of all the affected points is less than the vehicle load;
2. The position of RMRC and the affected point is known;
3. The average velocity of vehicles is known, which indicates driving distance and time are proportional.

Basic constraints: Shortage relief supplies must arrive before require time of the disaster points. If it has not arrived, this path has been unable to pass because of the disaster; therefore, this path is to be given up, that is, it is required infinite time to pass through this path.

Due to the fact that emergency logistics have characteristics of weak economy, so the model of minimum time is established:

$$\min T = \sum_{i=0}^{l} \sum_{j=0}^{l} t_{ij} x_{ij} \tag{1}$$

$$t_{ij} = \frac{d_{ij}}{v} \tag{2}$$

$$T_i < t_{ei} \tag{3}$$

where l is the number of the affected points; t_{ij} is the time that vehicles drive from affected point i to point j; d_{ij} is the distance between point i and point j; v is the driving velocity of vehicles; T_i is the time that the car arrives at point i; and t_{ei} is the dead time of the distribution time for the affected points.

The RMRC will be numbered as 0, and affected points will be numbered from 1 to l. Define variable x_{ijk} as follows:

$$y_{ki} = \begin{cases} 1, & \text{point } i \text{ is serviced by vehicle } k \\ 0, & \text{other} \end{cases} \tag{4}$$

Formula (1) is target function of the model, which aims to shorten the time for the best. Formula (2) indicates the time that vehicles pass by affected points i, j. Formula (3) indicates the time demand of goods arriving at affected points.

3 The Basic Particle Swarm Optimization

3.1 Introduction of PSO

The particle swarm optimization (PSO), which is introduced in 1995 by Kennedy and Eberhart, is evolutionary computation technique based on swarm intelligence. Particle swarm algorithm is derived from the study of a flock of birds foraging behavior [2, 3, 4]. Flock of birds which's behavior is unpredictable in flight often sudden changes in direction, spreads out, and gatherers, but, the flock of birds can keep conformity overall and the most suitable distance between each other. It is found that a kind of social information sharing mechanism which provided the advantage to evolve the group in the biological groups is existed. This is the basis of formatting the PSO. Its search idea is that neighborhood of the better particle has the greater probability of adapting to a high-value solution, so the neighborhood, which is allocated more particles in the group optimal particle, can enhance search efficiency. Each particle also is avoided blind into a local optimum by other information simultaneously.

3.2 The Basic Particle Swarm Optimization

PSO is a group of random initialization particles, which has no volume and quality; each particle is regarded as a feasible solution of the optimization problem; good or bad about the particles is determined by the fitness function, which is set in advance. Each particle moved in the feasible solution space, and its direction and distance are determined by a speed variable. Usually, the particles will follow the current optimum particle and can search the optimal solution, which gets in the last after each generation. The particles will track the two extremes in every generation: One is optimal solution which is found by particle itself and the other is optimal solution which is found by a whole group.

Set that a group that consists of M particles can fly with a certain speed in D-dimensional search space. State properties of particle i in time t are set as follows:

Location: $x_i^t = (x_{i1}^t, x_{i2}^t, \ldots, x_{id}^t)^T$

$x_{id}^t \in [L_d, U_d], L_d, U_d$ are, respectively, lower and upper limits of the search space.

Speed: $v_i^t = (v_{i1}, v_{i2}, \ldots, v_{id})^T$

$v_{id}^t \in [v_{\min,\mathrm{d}}, v_{\max,\mathrm{d}}]$, $v_{\min}$, $v_{\max}$ are, respectively, minimum and maximum speed.

Personal best position: $p_i^t = (p_{i1}^t, p_{i2}^t, \ldots, p_{iD}^t)^T$

Global optimal position: $p_g^t = (p_{g1}^t, p_{g2}^t, \ldots, p_{gD}^t)^T$ where $1 \le d \le D$, $1 \le i \le M$.

Position obtained by the following formula in time $t + 1$ is as follows:

$$v_{id}^{t+1} = v_{id}^t + c_1 r_1 (p_{id}^t - x_{id}^t) + c_2 r_2 \left(p_{gd}^t - x_{id}^t\right) \tag{5}$$

$$x_{id}^{t+1} = x_{id}^t + v_{id}^{t+1} \tag{6}$$

where r_1 and r_2 are random numbers distributed uniformly in the interval (0, 1) and c_1 and c_2 are learning factors, usually $c_1 = c_2 = 2$.

Equation (1) is composed for three parts: the first part is inherited particle previous speed, it is said that particles which do the inertial motion according speed itself trust in their own state of motion; the second part is the cognitive part, it is said the thinking of the particle itself, policy decision which is to achieve the next step behavior with considering their past experiences is cognition which reflects an enhanced learning process; the third part is the social part, it is said information sharing and mutual cooperation among the particles. The particles in the search process remember their own experiences, besides considering the experience of its companions. When individual particle detects better experience of companion, it will adjust the adaptive cognitive processes for seeking a consistent.

3.3 Solving the Model

Step 1: Initialize the particle swarm parameters and set all kinds of the involved parameters in the basic particle swarm algorithm.

Step 2: Evaluate each particle and calculate the fitness value of each particle, namely each particle represents the time of the distribution.

Step 3: Compare the delivery time of each particle and its best position, if better, update as the current best position.

Step 4: Compare the delivery time of each particle and group's best position, if better, update as the current best position.

Step 5: Update the particle status. Profile the speed and position of the particles using Eqs. (5) and (6). If $v_i > v_{\max}$, set $v_{\max}$, and if $v_i < v_{\min}$, set $v_{\min}$.

Step 6: Checks whether it complies with end condition. If it has a good enough position or the number of iterations is in the maximum, and when the iteration is stopped, output the shortest delivery time. Otherwise, go to Step 2.

4 Simulation

Set the coordinates of the RMRC is (60,140) ,the given data is from Ref. [5]. The shortage relief supplies are shipped to the 12 affected points from the RMRC after public emergencies occur. The coordinate data and the end time of each affected point are given in Table 1. The speed of vehicle is 35 km/h. Distance of each affected point and from each affected point to the RMRC is calculated by the formula (7). The number of particle swarm is 100. The number of iterations is 100.

$$d_{ij} = \sqrt{(x_i - x_j)^2 + (y_i - y_j)^2}. \tag{7}$$

Figure 1 shows the distribution map of cities. Figure 2 shows optimal path plan, which is shown that the shortest delivery time to get should be distributed according to the plan. Figure 3 shows the procession for finding the optimal fitness value, that is, the algorithm training process.

As can be seen from the simulation, particle swarm algorithm has the following advantages:

Table 1 Experiment data

Affected point	Coordinate	End time
1	(31, 115)	19.5
2	(41, 37)	13.0
3	(49, 97)	17.5
4	(53, 121)	18.5
5	(93, 155)	5.0
6	(194, 67)	12.5
7	(95, 101)	5.5
8	(109, 102)	6.0
9	(44, 160)	20.8
10	(21, 55)	17.0
11	(109, 33)	14.0
12	(131, 89)	10.0

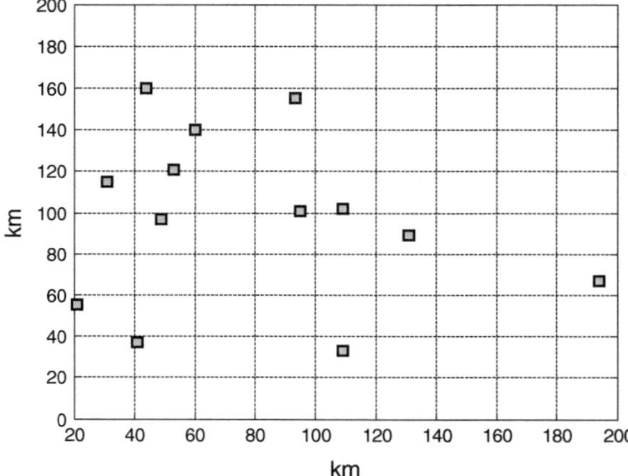

Fig. 1 Distribution map of cities

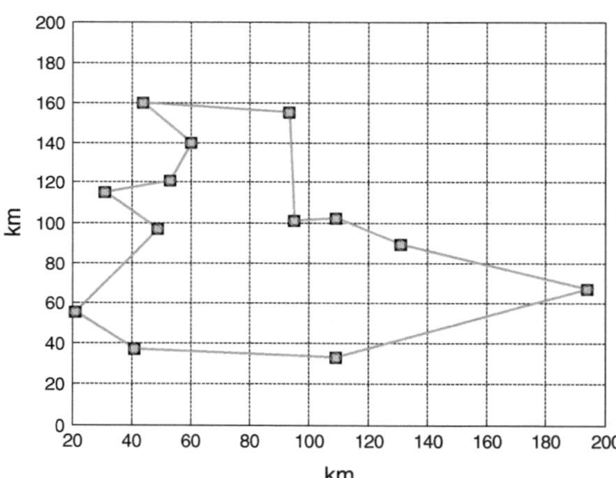

Fig. 2 Optimal path plan

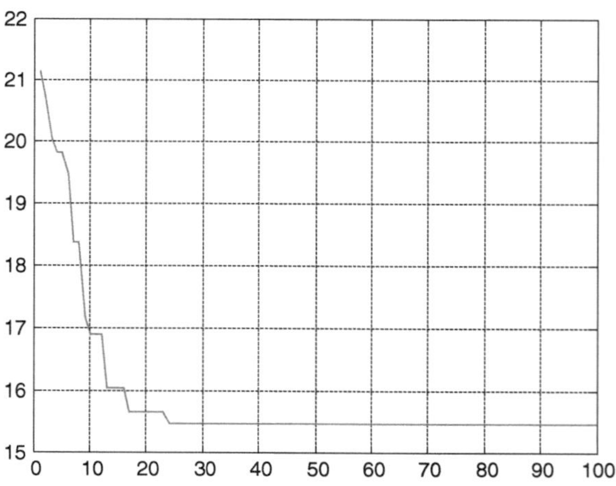

Fig. 3 The algorithm training process

1. The PSO, which has a strong global convergence and robustness, could carry out the cooperative search.
2. The PSO which reserved global search strategy is based the population.

5 Conclusion

In this paper, the PSO is used to solve the model of single-vehicle distribution of emergency logistics. From the simulation, it can be proved that the solving algorithm is effective and reflected that the algorithm has extensive application prospects. Study of the PSO is still in its infancy, and the solving multi-vehicle emergency logistics path optimization using the algorithm still needs to be explored in depth.

Acknowledgments This work is partially supported by the object of China Logistics Association (2012CSLKT027).

References

1. Eberhart, R.C., Kennedy, J.: A new optimizer using particles swarm theory. In: Proceeding sixth international symposium on micro machine and human science, Nagoya, Japan, p. 3943, IEEE Service Center, Piscataway, NJ (1995)
2. Kennedy, J., Eberhart, R.C.: Particle Swarm optimization. In: Proceedings of IEEE international conference on neural networks, vol. IV, pp. 1942–1948. IEEE service center, Piscataway, NJ (1995)
3. Cao, P., Chen, P., Liu, S.: Application of improved particle swarm optimization in TSP. Comput. Eng. **34**(11), 217–218 (2008)
4. Li, J., Guo, Y.: Optimization of Logistics Delivery Vehicle Scheduling Theory and Method. Beijing (2001)
5. Fei, T.: Research of ACO in the medical devices logistics distribution routing optimization. Taiyuan University of Technology, People's Republic of China (2010)

User Fairness-Based Adaptive Power Allocation in TD-LTE-A Downlink

Xuan-li Wu, Ming-xin Luo, Lu-kuan Sun and Nan-nan Fu

Abstract In TD-LTE-A system, the signal-to-leakage-and-noise ratio (SLNR) beamforming algorithm shows better performance in terms of sum capacity and average BER compared with other beamforming algorithms with moderate complexity. In order to further improve the performance of TD-LTE-A system, power-allocation scheme is analyzed based on the framework of SLNR beamforming and a user fairness-based adaptive power-allocation algorithm is proposed. Simulations show that the proposed algorithm can obtain the trade off between user fairness and system performance in TD-LTE-A downlink.

Keywords TD-LTE-A · Power allocation · User fairness · SLNR

1 Introduction

Since LTE-A has been accepted by IMT-A as one of the 4G standard, many new technologies have been introduced in LTE-A system to provide higher transmission data rate and spectrum efficiency. In LTE-A downlink, the non-codebook-based beamforming can reduce the interferences between different users, remove

---

X. Wu (✉) · M. Luo · L. Sun · N. Fu
Communication Research Center, Harbin Institute of Technology, Harbin, People's Republic of China
e-mail: xlwu2002@hit.edu.cn

M. Luo
e-mail: 822063769@qq.com

L. Sun
e-mail: nic81@foxmail.com

N. Fu
e-mail: betterfunan@126.com

S. Patnaik and X. Li (eds.), *Proceedings of International Conference on Computer Science and Information Technology*, Advances in Intelligent Systems and Computing 255, DOI: 10.1007/978-81-322-1759-6_49,

the requirement of feedback from the receiver end, decrease the power consumption, and improve the sum capacity of the system at the same time through the property of channel reciprocity in TDD mode. Hence, many researches have been done in the area of non-codebook-based beamforming in TD-LTE-A framework.

For single user beamforming, the singular value decomposition algorithm is used widely in transceivers because it can balance the trade off between complexity and system performance [1]. For multiuser beamforming, the optimal algorithm is dirty paper coding (DPC) in terms of channel capacity [2], however, due to the huge complexity, DPC algorithm is only used as an upper bound for performance comparison with other algorithms. In recent years, the algorithms with lower complexity have attracted more attention, such as zero forcing (ZF), minimum mean square error (MMSE), block diagonalization (BD) and signal-to-leakage-and-noise ratio (SLNR) [3–6]. In comparison, the performance of SLNR algorithm is very close to the optimal DPC algorithm in terms of channel capacity [6]. In this paper, the power allocation scheme is analyzed under the SLNR beamforming so that the obtained SLNR of different users can be used as the feedback parameter to improve the accuracy and performance of power allocation. In power allocation, although the water-filling algorithm is famous for its ability to obtain the optimal capacity, the user fairness is not considered. Hence, many modified algorithms were proposed to increase the user fairness through adaptive power allocation with respect to different parameters, such as channel norm, signal-to-interference-and-noise ratio (SINR), and SLNR [7]. However, the objective function of most existing literatures is only focused on the objective that all the users can be served with same transmission performance.

In this paper, an adaptive power-allocation algorithm is proposed based on SLNR beamforming considering different requirements of different users in TD-LTE-A downlink, and hence, the user fairness can be achieved with very little performance degradation in terms of channel capacity.

2 System Model of SLNR Beamforming in TD-LTE-A Downlink

In single cell TD-LTE-A system, only one eNodeB is considered with multiple users and the number of antennas used in transmitter and receiver are N_t, and N_r, respectively. The diagram of SLNR algorithm-based multiuser beamforming in TD-LTE-A downlink is shown in Fig. 1. In eNodeB, the channel state information is obtained by estimation of sounding reference signal (SRS), and then, the channel state information is used to obtain the SLNR of different users, finally, the power allocation is performed according to the results of beamforming at the transmitter.

In Fig. 1, the received signal in the kth UE can be expressed as:

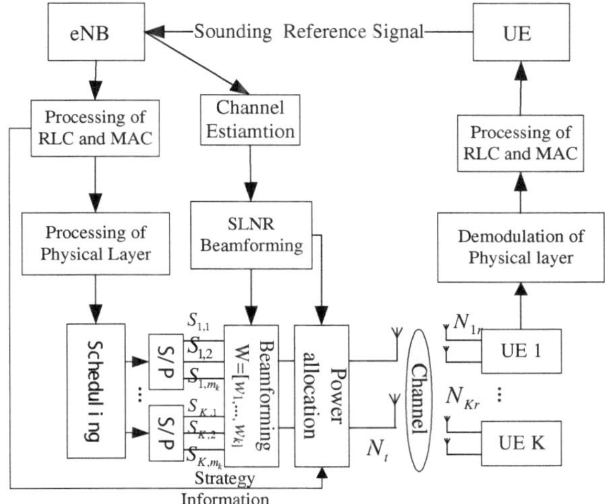

Fig. 1 SLNR algorithm-based multiuser beamforming in TD-LTE-A downlink algorithms

$$\mathbf{r}_k = \sqrt{p_k}\mathbf{H}_k\mathbf{w}_k\mathbf{S}_k + \sum_{j=1,j}^{K}\sqrt{p_j}\mathbf{H}_k\mathbf{w}_j\mathbf{S}_j + \mathbf{n}_k \tag{1}$$

where K is the total number of users, $\mathbf{H}_k$ is the matrix to show the channel gain from the kth user to eNodeB, $\mathbf{w}_k$ is the beamforming matrix for kth user, p_k is the power allocated to kth user, $\mathbf{n}_k$ is the normalized additive white Gaussian noise with the variance of σ^2, and $\mathbf{S}_k = [S_{1,1}, \ldots, S_{1,m_k}]^T$ is the transmitted signal of the kth user with normalized power, where m_{k} is the number of data streams of kth user.

The obtained SLNR of kth user can be expressed as:

$$\mathrm{SLNR}_k = \|\mathbf{H}_k\mathbf{w}_k\|_F^2 \Bigg/ \left(N_r\sigma^2 + \sum_{i=1,i\neq k}^{K}\|\mathbf{H}_i\mathbf{w}_k\|_F^2\right) \tag{2}$$

When using the SLNR-beamforming algorithm, the optimal beamforming matrix can be obtained using the following formulation with the restriction of $\mathrm{trace}\left(\mathbf{w}_k\mathbf{w}_k^{\mathrm{H}}\right) = N_t$:

$$\mathbf{w}_k = \arg\ \max(\mathrm{SLNR}_k) = \arg\ \max\frac{\mathrm{trace}\left(\mathbf{w}_k^{\mathbf{H}}\mathbf{H}_k^{\mathbf{H}}\mathbf{H}_k\mathbf{w}_k\right)}{\mathrm{trace}\left(\mathbf{w}_k^{\mathbf{H}}\left(\bar{\mathbf{H}}_k^{\mathbf{H}}\bar{\mathbf{H}}_k + N_r\sigma^2\mathbf{I}_{N_t}\right)\mathbf{w}_k\right)} \tag{3}$$

where $\bar{\mathbf{H}}_k = \left[\mathbf{H}_1^{\mathbf{H}}, \ldots, \mathbf{H}_{k-1}^{\mathbf{H}}, \mathbf{H}_{k+1}^{\mathbf{H}}, \ldots, \mathbf{H}_K^{\mathbf{H}}\right]^H$, $\mathbf{A}^{\mathrm{H}}$ is the conjugate transposed matrix of matrix $\mathbf{A}$, $\|\mathbf{A}\|_F^2$ is the Frobenius norm, and $\mathbf{I}_{N_t}$ is a unit matrix with order N_t.

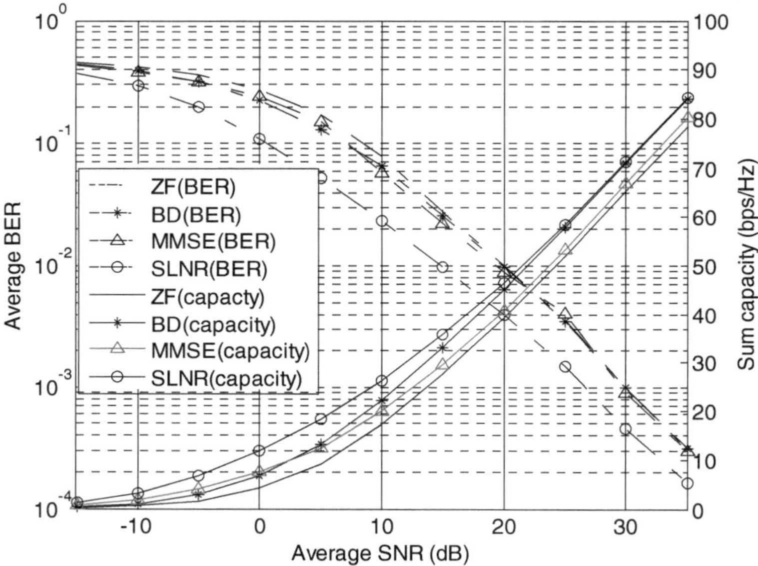

Fig. 2 Performance of different beamforming algorithms

Figure 2 shows the performance comparison of different beamforming algorithms used in TD-LTE-A downlink. Simulation results show that the SLNR-beamforming algorithm has a better performance in both average BER of all users and sum capacity than other conventionally used algorithms. As a result, power allocation can be more accurately based on SLNR beamforming.

3 User Fairness-Based Adaptive Power Allocation in SLNR Beamforming

The traditional water-filling-based power-allocation algorithm can provide the optimal capacity, however, the user fairness cannot be guaranteed because the users with poor channel condition will be assigned as less power regardless of the users' requirement. Although the modified algorithms can improve the user fairness, the difference of user requirement is neglected. Hence, a power-allocation algorithm is proposed to guarantee each user's requirement and provide a relatively high capacity as well.

Here, the scenario that all the traffics will be considered in the process of power allocation is considered, i.e., the power allocation of hybrid traffics is considered.

The SINR is used as the metric, and the requirements for SINR of different users are totally different. To satisfy different requirements of different users, the objective function can be formulated as (4) with the restriction that the received SINR of each users is bigger than the minimum SINR requirement.

$$\max_{1\le j\le K}\left\{\sum_{j=1}^{K}\log_2\left(\det\left(I_{N_r}+p_j\mathbf{w_j^H H_j^H H_j w_j inv}\left(\mathbf{N_r}\sigma^2\mathbf{I_{N_r}}+\sum_{\mathbf{k=1,k\neq j}}^{\mathbf{K}}\mathbf{p_k w_k^H H_j^H H_j w_k}\right)\right)\right)\right\}$$
$$s.t.\quad \mathrm{SINR}_j\ge \mathrm{SINR}_{j\,\min},\sum_{j=1}^{k}p_j=P_T \tag{4}$$

It is difficult to obtain the solution to (4), and hence, a simplified algorithm is proposed. The proposed algorithm can be divided into two steps, in the first step, the minimum required power for each user is calculated, and then, in the second step, the residual power is allocated to different users according to their requirement.

The first step can be expressed as the solution to the following problem:

$$\min\left\{\sum_{i=1}^{K}P_i\right\}\quad s.t.\quad SINR_i\ge SINR_{i\,\min} \tag{5}$$

The optimal solution to (5) can be obtained under the condition that $\mathrm{SINR}_i=\mathrm{SINR}_{i\ \min}$, i.e.,

$$\mathrm{SINR}_i=\left(p_i\mathbf{w}_i^{\mathbf{H}}\mathrm{H}_i^{\mathbf{H}}\mathbf{H}_i\mathbf{w}_i\right)\Big/\left(\sum_{j=1,j\neq i}^{K}p_j\mathbf{w}_j^{\mathbf{H}}\mathbf{H}_i^{\mathbf{H}}\mathbf{H}_i\mathbf{w}_j+\sigma^2\right)=\mathrm{SINR}_{i\,\min} \tag{6}$$

Then, the transmission power of ith user p_i can be calculated as:

$$p_i=\left(\sum_{j=1,j\neq i}^{K}p_j\mathbf{w}_j^{\mathbf{H}}\mathbf{H}_i^{\mathbf{H}}\mathbf{H}_i\mathbf{w}_j\mathrm{SINR}_{i\,\min}+\sigma^2\mathrm{SINR}_{i\,\min}\right)\Big/\left(\mathbf{w}_i^{\mathbf{H}}\mathbf{H}_i^{\mathbf{H}}\mathbf{H}_i\mathbf{w}_i\right) \tag{7}$$

According to the relationship between SLNR and SINR, and the weight obtained from the SLNR beamforming, the following expression can be obtained:

$$\mathbf{P}=\mathbf{BAP}+\sigma^2\boldsymbol{B\varphi} \tag{8}$$

where $\boldsymbol{\varphi}=[1,\ 1,\ldots,\ 1]^T$ is a vector with the dimension of $K\times 1$, $\mathbf{P}=[p_1,p_2,\ldots,p_k]^T$, and the nonzero elements of matrixes $\mathbf{A}$ and $\mathbf{B}$ are given as $A_{ij}=\mathbf{w}_j^H\mathbf{H}_i^{\mathbf{H}}\mathbf{H}_i\mathbf{w}_j$ when $i\neq j$ and $B_{ij}=\mathrm{SINR}_{i\,\min}/\left(\mathbf{w}_i^{\mathbf{H}}\mathbf{H}_i^{\mathbf{H}}\mathbf{H}_i\mathbf{w}_i\right)$ when $i=j$.

Hence, in the first step, the matrix for power allocation $\mathbf{P}^{(1)}$ can be calculated as:

$$\mathbf{P}^{(1)}=\sigma^2(\mathbf{I}-\mathbf{BA})^{-1}\boldsymbol{B\varphi} \tag{9}$$

In the second step, the allocated power to each user will not be smaller than the one obtained in (9), hence, the deduced SINR for ith user can be expressed as:

$$\begin{aligned}\mathrm{SINR}_i &= \left(\rho_i p_i^{(1)} \mathbf{w}_i^{\mathbf{H}} \mathbf{H}_i^{\mathbf{H}} \mathbf{H}_i \mathbf{w}_i\right) \Big/ \left(\sum_{j=1, j\neq i}^{K} \rho_j p_j^{(1)} \mathbf{w}_j^{\mathbf{H}} \mathbf{H}_i^{\mathbf{H}} \mathbf{H}_i \mathbf{w}_j + \sigma^2\right) \\ &= \left(p_i^{(1)} \mathbf{w}_i^{\mathbf{H}} \mathbf{H}_i^{\mathbf{H}} \mathbf{H}_i \mathbf{w}_i\right) \Big/ \left(\sum_{j=1, j\neq i}^{K} (\rho_j/\rho_i) p_j^{(1)} \mathbf{w}_j^{\mathbf{H}} \mathbf{H}_i^{\mathbf{H}} \mathbf{H}_i \mathbf{w}_j + \sigma^2/\rho_i\right)\end{aligned} \tag{10}$$

where $\rho_i, \rho_j \geq 1,\ \ i,j = 1,2,\ldots,K$ are the parameters to indicate the final power allocated to ith user.

Considering the requirement that $\mathrm{SINR}_i \geq \mathrm{SINR}_{i\,\min}$ for all the users, one possible solution for all the condition is to satisfy the following relationship:

$$\rho_j/\rho_i = 1, j \neq i, \quad i,j = 1,2,\ldots,K \tag{11}$$

Then ρ_i can be calculated as:

$$\rho_i = P_T \Big/ \sum_{i=1}^{K} p_i \quad i = 1,2,\ldots,K \tag{12}$$

Hence, the final matrix for power allocation can be obtained as follows:

$$\mathbf{P}_{\mathrm{all}} = \boldsymbol{\rho}\mathbf{P}^{(1)} = \left(P_T \Big/ \sum_{i=1}^{K} p_i\right)\mathbf{P}^{(1)} \tag{13}$$

where $\boldsymbol{\rho} = [\rho_1, \rho_2, \ldots, \rho_K]$ is a vector with the dimension of $1 \times K$.

4 Simulation Results

The simulations are based on the platform of TD-LTE-A downlink using SLNR as the beamforming algorithm. The number of antennas in eNodeB and UE are configured to be 8 and 1, respectively. The maximum transmission power in eNodeB is 8 dBW. The channel is flat fading channel, and the channel gain of each sub-channel obeys the complex Gaussian distribution with zero mean and unit variance. In order to verify the performance of the proposed algorithms, ideal channel estimation to SRS signal is assumed and the variance in additive Gaussian noise in each receiver end is set to be the same. Channel coding is not considered in this platform so that the real difference of different algorithms can be identified. The minimum required SINR of different uses are shown in the second row of Table 1.

To compare the performance of the proposed algorithms, the conventionally used water-filling algorithm [8] is given as a comparison, where water-filling algorithm shows the upper bound of sum capacity. The sum capacity and average BER performance are shown in Figs. 3 and 4, respectively.

Table 1 Minimum required SINR of different traffics

User number	1	2	3	4	5	6	7	8
$SINR_{min}$	0.8	1.2	1.6	2.0	2.4	2.8	3.2	3.6
SINR of proposed algorithm	1.0	1.4	1.9	2.4	2.9	3.4	3.9	4.4

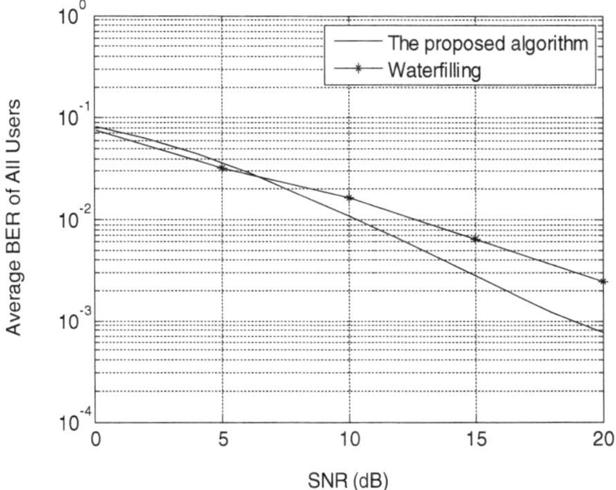

Fig. 3 Average BER performance

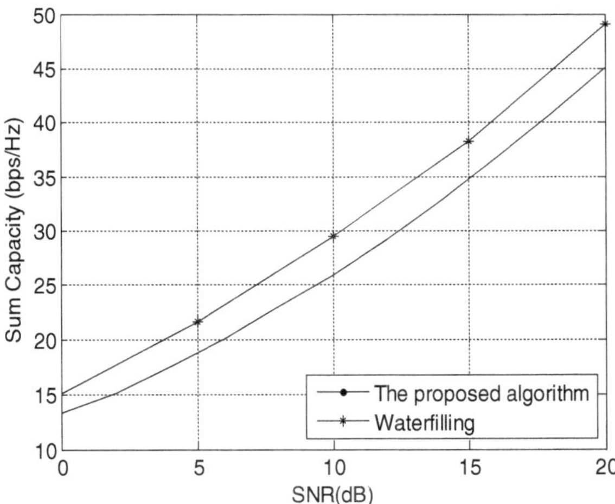

Fig. 4 Capacity performance

From Figs. 3 and 4, it can be seen that although water-filling algorithm can provide the highest sum capacity as it is designed to do so, its average BER performance is worse because the users with poor channel condition will be allocated with very limited power so that the SINR in the receiver end will be much lower. The obtained SINRs for the proposed algorithm are shown in the last row of Table 1. When using the proposed algorithm considering the difference of user requirement, all the SINR requirement of different users can be satisfied, and thus, the user fairness increased. In comparison, when using the water-filling algorithm, all the users have similar SINR value of about 2.83, and hence, the requirement of last two users cannot be satisfied.

5 Conclusions

In this paper, a power-allocation algorithm is proposed under the TD-LTE-A system using SLNR beamforming to satisfy different user requirements. Simulation results show that compared with the water-filling algorithm, the proposed algorithm can satisfy all the requirement of different users at the cost of very limited capacity loss, and thus guarantee the user fairness.

Acknowledgments This work is supported by the National Basic Research Program of China (973 Program) under grand no. 2013CB329003, Next Generation Wireless Mobile Communication Network of China under grant no. 2012ZX03001007-005, the Fundamental Research Funds for the Central Universities under grant no. HIT. NSRIF.2012020, Heilongjiang Post-doctoral Science-Research Foundation under grant no. LBH-Q12081, and China Scholarship Council.

References

1. Liu, L., Chen, R., Geirhofer, S., Sayana, K., Shi, Z., Zhou, Y.: Downlink MIMO in LTE-advanced: SU-MIMO vs. MU-MIMO. IEEE Commun. Mag. **50**(2), 140–147 (2012)
2. Wang, Y., Hur, S., Park, Y., Choi, J.: Efficient user selection algorithms for multiuser MIMO systems with zero-forcing dirty paper coding. J. Commun. Netw. **13**(3), 232–239 (2011)
3. Gao, X., Edfors, O., Rusek, F., et al.: Linear pre-coding performance in measured very-large MIMO channels. In: 2011 IEEE Vehicular Technology Conference (VTC Fall), pp. 1–5 (2011)
4. Lim, W.Y., Lei, Z.D.: Joint MMSE precoding & detection for interference alignment in MIMO X channels. In: 2012 IEEE 23rd International Symposium on Personal Indoor and Mobile Radio Communications (PIMRC), pp. 1651–1655 (2012)
5. Chen, R., Chen, Z., Andrews, J.G., Health, R.W.: Multimode transmission for multiuser MIMO systems with block diagonalization. IEEE Trans. Signal Process. **56**(7), 3294–3302 (2008)
6. Sadek, M., Tarighat, A., Sayed, A.H.: A leakage-based precoding scheme for downlink multi-user MIMO channels. IEEE Trans. Wireless Commun. **6**(5), 1711–1721 (2007)
7. Wang, J.J., Xie, X.Z., Zhang, Q.: Dynamic power allocation based on SLNR precoding for multi-user MIMO downlink. J. Chongqing Univ. Posts Telecommun. (Nat. Sci. Ed.) **20**(6), 643–645 (2008)
8. Palomar, D.P., Fonollosa, J.R.: Practical algorithms for a family of water filling solutions. IEEE Trans. Signal Process. **53**(2), 686–695 (2005)

Real-Time Compressive Tracking Based on Online Feature Selection

Zheng Mao, Jianjian Yuan, Zhenrong Wu, Jinsong Qu and Hongyan Li

Abstract As the projection matrix is only generated in the initial stage and kept constant in subsequent processing, so when the object is occluded or its appearance changes, this will result in drifting or tracking lost. To address this problem, this paper proposes a real-time compressive tracking algorithm based on online feature selection. First, the feature pools are constructed. Then, features with high confidence score are selected from the feature pool by a confidence evaluation strategy. These discriminating features and their corresponding confidences are integrated to construct a classifier. Finally, tracking processing is carried on by the classifier. Tracking performance of our algorithm compares with that of the original algorithm on several public testing video sequences. Our algorithm improves on the tracking accuracy and robustness; furthermore, the processing speed is approximately 25 frames per second. It meets the requirements of real-time tracking.

Keywords Online feature selection · Compressive sensing · Subregion features · Real-time tracking

1 Introduction

Object tracking plays great roles in computer vision. It has long been studied by many researchers. However, due to changes of illumination, pose, motion and occlusion, long-time and stable tracking is still a challenging problem.

Recently, the hotspot of tracking treats tracking problem as a binary classification problem. A model has been trained to separate the object from background which

Z. Mao (✉) · J. Yuan · Z. Wu · J. Qu · H. Li
Beijing University of Technology, Beijing, China
e-mail: maozheng@bjut.edu.cn

J. Yuan
e-mail: jianrobust@163.com

S. Patnaik and X. Li (eds.), *Proceedings of International Conference on Computer Science and Information Technology*, Advances in Intelligent Systems and Computing 255, DOI: 10.1007/978-81-322-1759-6_50,

shows promising results. This kind of tracking methods has much similarity with object detection, so they have been termed "tracking by detection." Grabner et al. [1] proposed a method of online feature selection based on Adaboost. The strong classier consists of several weighted weak classifiers. However, training samples used to train classifiers are cropped around the object position from consecutive frames. So, once some errors are generated in consecutive processing due to occlusion, illumination, and so on, it will result in drifting or tracking lost. Grabner et al. [2] proposed a semi-supervised learning approach. Its basic idea is how to label and unlabel training samples based on the model of data distribution. It mainly considers how to train and update a classifier with few labeled samples and a lot of unlabeled samples. This approach shows great results when the object leaves the field of views completely. But, it will affect the update of the classifier when the pose of the object changes continuously. Babenko et al. [3] proposed a tracking method based on multiple instances learning where training samples are presented in "bags" and labels are provided to the bags rather than to individual instance. Therefore, when tracking errors are generated, the correct training sample is found from the bags through online learning to update the classifier. However, when the object is heavily occluded, the errors generated by the occluded region still can result in drifting. In 2006, Candes and Donoho [4] formally proposed the concept of compressive sensing, and later this concept is introduced to computer vision [5]. Zhang et al. [6] proposed a real-time compressive tracking approach. Compressive features are extracted from a sparse project matrix, and then, probability distribution of positive and negative samples is used to construct classifier for tracking. However, there are two problems of this approach. First, feature descriptor is simple. When appearance texture or illumination changes severely, this will cause tracking lost or drifting. Second, the tracking features are determined by the projection matrix. But, the projection matrix keeps constant in consecutive processing. Therefore, tracking position may drift from precise object positioning due to the features extracted from region occluded.

The rest of paper is organized as follows. Section 2 reviews the original algorithm including features extraction and construction and update of the classifier. Section 3 introduces online feature selection mechanism into original algorithm. The main focus is how to evaluate the confidence of feature. Section 4 performs a lot of comparative experiments with the original algorithm.

2 Related Work

2.1 Compressive Features Extraction

If the random matrix satisfies Johnson–Lindenstrauss [8] lemma, the low-dimensional features can preserve almost all information of high-dimensional features. Compressive features are generated with the sparse matrix. The sparse matrix is defined as the following form:

$$r_{ij} = \sqrt{n} \times \begin{cases} 1 & p = 1/2n \\ 0 & p = 1 - 1/n \\ -1 & p = 1/2n \end{cases} \quad (1)$$

The tracking features of original algorithm are the weighted linear combination of haar features. However, the property of the tracking features is determined by the projection matrix. We can see that the entries of the matrix consist of 1 and −1 with probability 71 %. In this case, the tracking features describe the difference among image patches. It represents the texture property of the object. However, when the entries are all 1 or all −1, the tracking features represent the mean values of the image patch. The original algorithm's progress of feature extraction is shown in Fig. 1.

2.2 Classifier Construction and Update

To a random sample S, the compressive features are represented as $v = \{F_1, F_2, F_3 \ldots F_n\}^T$. Low-dimensional vectors are modeled with a Bayesian classifier as follows:

$$H(v) = \log\left(\frac{\prod_{i=1}^{n} p(v_i \mid y = 1)p(y = 1)}{\prod_{i=1}^{n} p(v_i \mid y = 0)p(y = 0)}\right) = \sum_{i=1}^{n} \log\left(\frac{p(v_i \mid y = 1)}{p(v_i \mid y = 0)}\right) \quad (2)$$

where a uniform prior is assumed that $p(y = 1) = p(y = 0)$, and y represents the label of the training samples. Diaconis and Freedmans showed that the random projections of high-dimensional random vectors are almost Gaussian [9]:

$$p(v_i \mid y = 1) \sim N(\mu_i^1, \sigma_i^1) \quad (3)$$

$$p(v_i \mid y = 0) \sim N(\mu_i^0, \sigma_i^0) \quad (4)$$

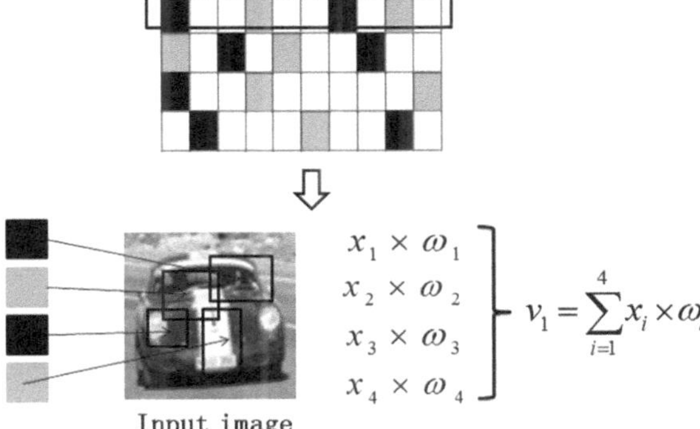

Fig. 1 Feature extraction of the original algorithm

The four parameters of model $(\mu_i^1, \sigma_i^1, \mu_i^0, \sigma_i^0)$ are updated incrementally. For more details see literature [6].

The projection matrix is only generated in initial frame and kept constant in consecutive processing. To address the two problems of the original approach, this paper proposed some strategy to improve the tracking performance. First, we generate two complementary projection matrices to generate tracking features. This can solve the drifting problem caused by changes of texture and illumination. Second, we adopt an online feature selection method to actively select features with more high confidence. This can weaken effects of features extracted from region occluded. The online selection mechanism can effectively improve the tracking performance.

3 Proposed Algorithm

3.1 Mechanism of Online Feature Selection

The main idea of online feature selection mechanism is how to select more confident features from the feature pool by a criterion. First, a feature pool is constructed by projecting high-dimensional vectors to low-dimensional vectors. Then, features are ranked by the confidence evaluation. Finally, the selected features and the corresponding confidences are integrated to construct or update the classifier (Fig. 2).

3.2 Construction of Features Pool

The features' property of original algorithm represents the texture property which is unstable. To make the system more stable, we generate another complementary projection matrix. Features generated by the two projection matrices have the complementary property. For more details see literature [7].

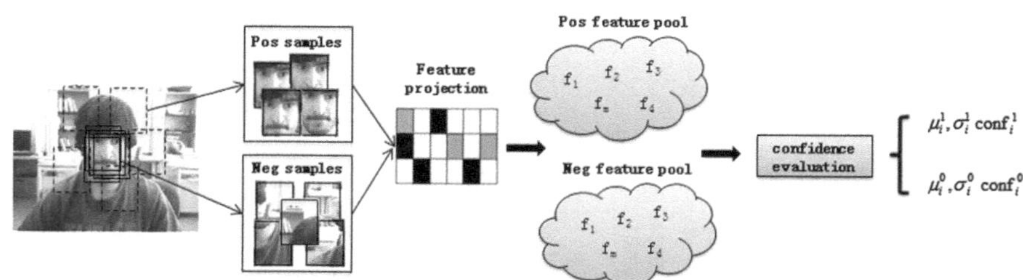

Fig. 2 Main components of online feature selection mechanism

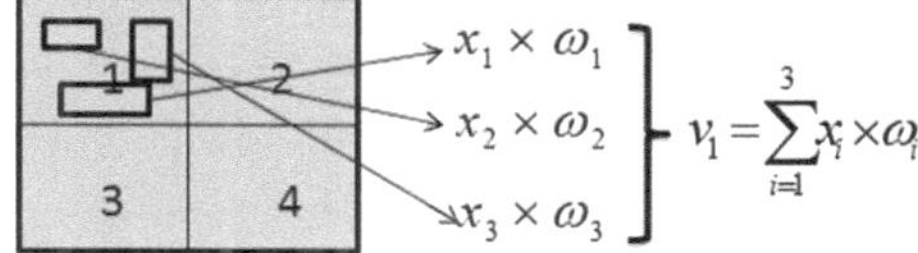

Fig. 3 Subregion feature extraction

Tracking features are represented by convolving a sample with a set of rectangle filters at different scales. However, the rectangle filters are randomly selected. So, if the rectangle filters convolve with patches from region occluded, the features can result in drifting. To address the problem, rectangle filter is no longer selected globally instead selected from a subregion [9]. Tracking features in the feature pool come from the four subregions.

The image region is divided into four subregions. The rectangle filters used to construct one feature are all from the same subregion. Therefore, when a subregion is occluded, features extracted from this subregion are no longer confident for the object. Features from the subregion occluded would not be selected to update classifier. The extraction of subregion features is shown as Fig. 3.

3.3 Evaluation of Feature Confidence

A criterion is needed to select features from features pool. This paper proposed a method based on Gaussian Kernel Density Estimation to evaluate feature confidence. The feature confidence is evaluated by the difference between means of positive samples. The Gaussian Kernel Density Estimation is defined as follows:

$$G(\mu_{k+1}) = e^{-||u_{k+1}-\mu_k||^2/(2\times\sigma_k^2)} \tag{5}$$

where μ_{k+1} is updated means of one feature distribution, and μ_k is the means of original feature distribution.

When features in positive feature pool are updated, all features are ranked by its corresponding confidence. We select N (smaller than the dimension of original feature vector) features to update the classifier. Although we construct positive and negative feature pool, the background changes all the time. It is meaningless to evaluate the confidence of negative features.

Considering the processing time and the stability, the number of features in feature pool is defined as N_o. However, the final dimension of feature vector is N_t. The selected features and its corresponding confidence are integrated to construct the classifier. The modified classifier is defined as follows:

$$H(v) = \prod_{i=1}^{n} G(v_i) \times \log\left(\frac{p(v_i\,|\,y=1)p(y=1)}{p(v_i\,|\,y=0)p(y=0)}\right) = \prod_{i=1}^{n} G(v_i) \times \log\left(\frac{p(v_i\,|\,y=1)}{p(v_i\,|\,y=0)}\right) \tag{6}$$

We can see that the feature with more high confidence will make more contribution to the final result. This can highlight the features with high confidence and weaken the features with low confidence. The main steps of our algorithm are summarized in Algorithm 1.

Algorithm 1. Online Feature Selection Compressive Tracking

Initial stage:

Sample two sets of image patches based on the initial tracking position defined artificially. N_o features are extracted to construct the feature pool. The initial classifier is constructed by features that selected from the feature pool randomly.

Tracking stage:

Input*:* *t*-th video frame

1. Sample a set of image patches, $D^{\gamma} = \{Z\|\| I(z) - I_{t-1}\| < \gamma\}$ where I_{t-1} is the tracking location at the $(t-1)$-th frame, and extract features with low dimensionality.

2. N_t confident features are selected by the evaluation strategy. Features and their corresponding confidence are integrated to construct the classifier. Use classifier *H* in (6) to each feature vector **v(z)** and find the tracking location with the maximal classifier response.

3. Sample two sets of image patches $D^{\alpha} = \{Z\|\| I(z) - I_{t-1}\| < \alpha\}$ and $D^{\zeta,\beta} = \{Z| \zeta < \| I(z) - I_{t-1}\| < \beta\}$ with $\alpha < \zeta < \beta$. Extract the features with these two sets of samples and update the parameters in the feature pool.

Output: Tracking location I_t and parameters in the feature pool.

4 Experiments and Discussion

The algorithm is implemented in Visual 2005 and OpenCV 2.4. The performance of our algorithm compares with original one on several publicly video sequences.

The target object in occluded face sequence in Fig. 4a undergoes large pose variation and heavy occlusion. The six figures in (a) show the two algorithms' tracking results at the 104-, 157-, 424-, 568-, 651-, and 709-th frame. We can see that when the face is not occluded, both algorithms can track the object precisely and stably. However, once the target object undergoes heavy occlusion (157-, 424-, 709-th frame), the comparison results show that our algorithm outperforms than the original one.

The target object in Panda video sequence in Fig. 4b undergoes large variation of appearance and partial occlusion. As figures shown in (b), they are tracking results of the 283-, 559-, 598-, 734-, 843-, and 921-th frame. As the target object walks around in the cage, it faces the pose variation and partial occlusion. The original algorithm drifts from the precise target position. A great number of tracking errors are accumulated at 734-, 843-, and 92-th frame. Our algorithm still can track the target object precisely.

Fig. 4 Tracking results of video sequences. (**a**) Occluded face. (**b**) Panda. (**c**) David

Table 1 The comparison of processing speed for two algorithms

Frames/s	Faceocc	Panda	David
CT	26	25	25
Our algorithm	25.5	26	24

For the David indoor video sequence shown in Fig. 4c, the illumination and pose of the object both change gradually. The figures shown in (c) are tracking results at 361-, 435-, 463-, 659-, 699-, and 758-th frame. If the target object undergoes the illumination and poses variation, this can affect the appearance model of target which can cause drifting or tracking lost.

As Table 1 shows that due to the construction of features pool, the number of features has increased. However, features are calculated based on integral histogram, and the final features used to track are selected from features pool, so the processing speed approximately reaches at 25 frames/s. It meets the requirements of real-time tracking.

References

1. Grabner, H., Grabner, M., Leordeanu, M.: Online selection of discriminative tracking features. PAMI **27**, 1631–1643 (2005)
2. Grabner, H., Leistner, C., Bischof, H.: Semi-supervised on-line booting. In: BMVC, pp. 47–56 (2006)
3. Babenko, B., Yang, M.H., Belongie, S.: Robust object tracking with online multiple instance learning. AMI **3**, 1619–1632 (2011)
4. Donoho, D.L.: Compressed sensing. Inf. Theory **52**(4), 1289–1306 (2006)
5. Li, H., Shen, C., Shi, Q.: Real-time visual tracking using compressive sensing. In: Computer Vision and Pattern Recognition (CVPR), pp. 1305–1312 (2011)

6. Zhang, K., Zhang, L., Yang, M.-H.: Real-time compressive tracking. In: ECCV (2012)
7. Zhu, Q.P., Yan, J., Zhang, H., Fan, C., Deng, D.: Real-time tracking using multiple features based on compressive sensing. Opt. Precis. Eng. **21**(2) (2013)
8. Achlioptas, D.: Database-friendly random projections: Johnson-Lindenstrauss with binary coins. J. Comput. Syst. Sci. **66**, 671–687 (2003)
9. Yan, J., Wu, M.Y.: Online boosting based target tracking under occlusion. Opt. Precis. Eng. **20**(2) (2012)

Part IV
Cloud Computing

A Sociology-Based Reputation Model for Cloud Service

Xiaoli Liu, Yujuan Quan, Weizhen Jiang and Zhenyu He

Abstract Reputation is the major factor of cloud service selection. In this paper, we first analyze the reputation based on social service. Then, we proposed a layered model with fuzzy estimation, in which the users' rating, anonymous rating, and experts' assessment are involved.

Keywords Reputation · Social ontology · Cloud computing

1 Introduction

Cloud computing is the use of computing resources (hardware and software) that are delivered as a service over a network [1]. Everything is a service in cloud computing. Service selection is becoming very critical for cloud security and stability. Reputation is the major factor for cloud selection.

Current research about reputation management is focused on the computation of various kinds of users' rating with different calculation model.

In this paper, we propose a research framework for reputation management of SaaS service. Current research about reputation management mainly focuses on the performance of a service, including the following parameters: correctness, response time, crash rate, and so on. However, we need more parameters for cloud service selection in practice, such as risk control, the compensation mechanism, price system, and so forth. A comprehensive reputation system is needed in service-oriented environment to balance the benefit of different entities. Service science technologies and sociology-related studies are introduced as support

X. Liu (✉) · Y. Quan · W. Jiang · Z. He
College of Information Science and Technology, Jinan University, Guangzhou, Guangdong, China
e-mail: txlliu@jnu.edu.cn

S. Patnaik and X. Li (eds.), *Proceedings of International Conference on Computer Science and Information Technology*, Advances in Intelligent Systems and Computing 255, DOI: 10.1007/978-81-322-1759-6_51,

theories of our research framework. Then, research method and implementation of our framework are presented in detail.

This paper is organized as the follows: Sect. 2 presents related works about reputation management of cloud service and research challenges; Sect. 3 presents the factors related to cloud reputation-based social service; the proposed reputation model is proposed in Sects. 4 and 5 concludes this paper.

2 Related Work

2.1 Research Challenges

Existing reputation systems are majorly in the domains of multiagent systems, web services, and online markets [2].

In [3] research, challenges related to cloud reputation and trust are discussed, including SLA specifications, open standards and accreditations, security measures, and so on.

2.2 Current Research

Current research about reputation management can be classified into three categories: evaluation procedure, parameters and reputation systems.

A useful reputation evaluation approach is introduced in [4]. It contains three steps: feedback checking, feedback adjustment, and feedback detection. The simulation result shows that this approach works well for service selection. A novel reputation management system for volunteer clouds is presented in [5]. The basic trust parameters in this system included performance, crashes rate, and correctness.

A survey of trust and reputation systems for online service provision is summarized in [6]. These reputation systems devote to the following factors: accuracy for long-term performance, weighting toward current behavior, robustness against attacks, and smoothness. An example is illustrated to show why users' rating is not enough for reputation evaluation [7].

3 Reputation Evaluation Factors

In this section, we analyze the different factors that affect the quality of cloud services based on the reputation mechanism of social service [8].

3.1 Performance Evaluation

The quality of service is the major factor of Internet services. In [9], the parameters reflecting the quality of Web services are presented in detail. The functional performance of service can be evaluated as the software. The major factors related to performance evaluation about cloud service include correctness, fault-tolerant rate, delay, clash rate, completeness, consistency, usability, and so on. We summarized these factors in Fig. 1.

3.2 Business Transparency

In society, business transparency is the major factor for reputation evaluation. For example, in politics, transparency is used as a means of holding public officials accountable and fighting corruption, and there is less opportunity for the authorities to abuse the system for any fraud. In cloud computing, the service providers have the permission to access clients' private data; it is requisite to promote transparency of business process. Transparent security could promote cloud providers to pay more attention to their security policies, design, practices, and so on [10].

3.3 Risk Management

Risks can be from platform, malicious users, normal users' misoperation, and providers. The different users have different level of risk tolerance. Computational risk management tactics should be established to identify, assess, and manage risks [11]. Service providers have the responsibility to tackle the various risks.

Fig. 1 Functional and non-functional factors of cloud service

Functional
• Correctness • Completeness • Consistency
Non-Functional
• Delay • Fault-tolerant rate • Clash rate • Usability

3.4 The Compensation

No perfect services exist. Compensation mechanism is needed to cope with the emergency of service misdelivery. Monetary compensation mechanisms should be constructed to protect the benefit of service users.

3.5 Price System

The pay-per-use pricing model is simple: It associates units (or units per time) with fixed price values. Dynamic pricing policies could achieve more economically efficient allocations and prices for high-value services [12].

4 Reputation Model

In this section, we give the details of our reputation model, including reputation calculation and parameter setting.

4.1 Reputation Model Overview

1. Rating

   Users' rating is the commonly used method for reputation evaluation in service science. Users' rating can reflect the actual quality of service with the users' experience. However, it is not enough because either some users' rating is prejudiced or some users may give wrong assessment due to providers' unfair policy.

   Anonymous rating is helpful for reputation evaluation in some ways. In [6], it pointed out that a way to avoid positive bias can consist of providing anonymous reviews. It can usually present specific assessment from a different view [6].

2. Assessment

   Part of users' rating is prejudiced. Professional assessment should be the major part of reputation calculation since it is objective and based on concrete estimation of different factors. The experts can analyze the different factors in Sect. 3 to sum up from 4 aspects: functional, non-functional, security performance, and price system.

3. The model

   The layered reputation model is presented in Fig. 2.

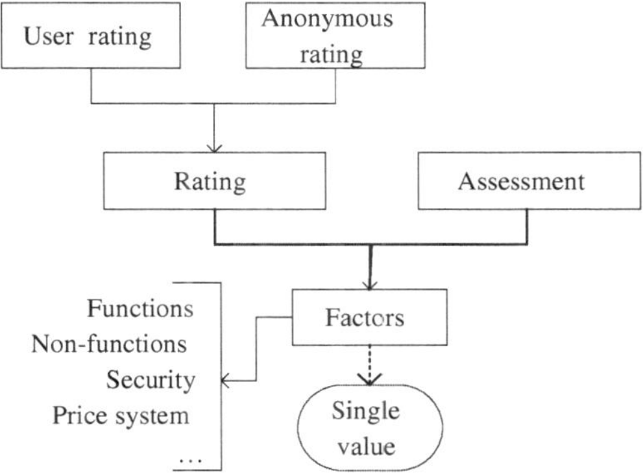

Fig. 2 The layered reputation model

4.2 Reputation Calculation

Fuzzy estimation is presented in [13]. At present, fuzzy estimation is used for reputation management. A model for trust management based on fuzzy logic is presented in [14]. Usually, rating value of existing online services varies from 1 to 5. Here, we use vector V_a to represent weights of different rating. $V_a = \langle a_1, a_2, a_3, a_4, a_5 \rangle$, while a_i represents the weight of "i star."

Fuzzy evaluation set $\tilde{C}_i(i = 1 \ldots 5)$ is from the users' evaluation about "i star." $\tilde{C}_i = \{c_{\text{if}}, c_{\text{in}}, c_{\text{is}}, c_{\text{ip}}\}$, c_{if} represents the proportion of function performance, c_{in} is for non-functional performance, c_{is} is for security performance, and c_{ip} is for price evaluation. The fuzzy matrix R is defined as $R = (\tilde{C}_1, \tilde{C}_2, \tilde{C}_3, \tilde{C}_4, \tilde{C}_5)^T$.

The evaluation can be calculated with the formula $\tilde{E} = V_a^{\circ} R$, while $\circ$ is the fuzzy matrix composite operator.

$$\tilde{E} = \langle a_1, a_2, a_3, a_4, a_5 \rangle^{\circ} \begin{bmatrix} c_{1f} & c_{1n} & c_{1s} & c_{1p} \\ c_{2f} & c_{2n} & c_{2s} & c_{2p} \\ c_{3f} & c_{3n} & c_{3s} & c_{3p} \\ c_{4f} & c_{4n} & c_{4s} & c_{4p} \\ c_{5f} & c_{5n} & c_{5s} & c_{5p} \end{bmatrix}$$

Professional evaluation is a vector $\tilde{E}$, $\tilde{E} = \langle e_f, e_n, e_s, e_p \rangle$; it can represent the reputation from the following four aspects: functional, non-functional, security performance, and price system.

In the same way, we can get the evaluation of users' rating $\tilde{U}$ and anonymous rating $\tilde{A}$. The integration evaluation of service reputation is $\tilde{R}$, while $\tilde{R} = \tilde{E} \times W_e + \tilde{U} \times W_u + \tilde{A} \times W_a$. The final reputation value is multioutput, which contains different factors: functions, non-functions, security, and price. The cloud users can select proper service with the different factor reputation value.

4.3 Parameter Setting

In the above model, the practical data are needed, including the evaluation from experts, users, and anonymous users. We need to do more experiment to train the data to get different weight value.

5 Conclusion

In this paper, we have proposed a research framework for reputation management of cloud service based on social ontology. The current research about cloud reputation and research challenges are discussed firstly. Then, we first analyze the reputation based on social service. Finally, we present a layered model with fuzzy estimation, in which the users' rating, anonymous rating, and experts' assessment are involved.

The framework components were designed to be generic for SaaS. For the validation of this reputation model, we need to do more work along with the research steps above. Future work mainly includes (1) experimentation about this model and (2) model improvement with more social information.

Acknowledgments This research is supported by "the Fundamental Research Funds for the Central Universities" (No.21612337), "Foundation for Distinguished Young Talents in Higher Education of Guangdong, China" (No. 2012LYM_0021), and "Fundamental Research Funds for the Central Universities" (No. 21613323).

References

1. Cloud computing, http://en.wikipedia.org/wiki/Cloud_computing. (2012)
2. Alnemr, R., Meinel, C.: From reputation models and systems to reputation ontologies. Trust Manage. V. pp. 98–116 (2011)
3. Habib, S.M., Ries, S., Muhlhauser, M.: Cloud computing landscape and research challenges regarding trust and reputation, pp. 410–415 (2010)
4. Wang, S.G., Sun, Q.B., Yang, F.C.: Reputation evaluation approach in web service selection (Chinese version). J. Softw. **23**(6), 1350–1367 (2012)
5. Muralidharan, S.P., Kumar, V.V.: A Novel Reputation Management System for Volunteer Clouds. pp. 1–5. IEEE (2012)
6. Josang, A., Ismail, R., Boyd, C.: A survey of trust and reputation systems for online service provision. Decis. Support Syst. **43**(2), 618–644 (2007)
7. Alnemr, R., Meinel, C.: Why Rating is not Enough: A Study on Online Reputation Systems. IEEE (2011)
8. Mezzetti, N.: A socially inspired reputation model. Public Key Infrastructure. pp. 605–605 (2004)
9. Malik, Z., Bouguettaya, A.: RATE Web: Reputation assessment for trust establishment among web services. VLDB J. **18**(4), 885–911 (2009)
10. Brodkin, J.: Gartner: Seven cloud-computing security risks. Infoworld. pp. 2–3 (2008)

11. Buyya, R., Yeo, C.S., Venugopal, S.: Market-Oriented Cloud Computing: Vision, Hype, and Reality for Delivering it Services as Computing Utilities. pp. 5–13 IEEE (2008)
12. Weinhardt, C., et al.: Business models in the service world. IT Prof. **11**(2), 28–33 (2009)
13. Chiu, S.L.: Fuzzy model identification based on cluster estimation. J. Intell. Fuzzy Syst. **2**(3), 267–278 (1994)
14. Supriya, M., Sangeeta, K., Patra, G.: Estimating trust value for cloud service providers using fuzzy logic. Int. J. Compu. Appl. **48**(19), 28–34 (2012)

Comparative Analysis and Simulation of Load Balancing Scheduling Algorithm Based on Cloud Resource

Tangang, Ranzhi Zhan, Shibo and Xindi

Abstract The high-performance large-scale cloud computing resource is the basic condition to achieve cloud computing services, on how the enormous resource scheduling is the successor to the problem of cloud computing services that must be addressed. We has carried out a comparative analysis for two kinds of cloud resource scheduling algorithm based on load balancing algorithm from core idea, algorithm complexity and the advantages and disadvantages and so on. And on this basis, we compare the difference in performance of each scheduling algorithm. We need to raise a more suitable algorithm for the cloud resource load balancing scheduling to meet the needs of practical application, in the case of the inconsistent of the mission requirements specification.

Keywords Cloud computing · Load balancing · Algorithm

Tangang (✉) · R. Zhan · Shibo
Chongqing Electric Power Information and Communication Branch Company, Chongqing, China
e-mail: tggump@126.com

R. Zhan
e-mail: cqzrz@sohu.com

Shibo
e-mail: shib@cq.sgcc.com.cn

Xindi
Information and Network Management Center, North China Electric Power University, Baoding, China
e-mail: xindi158@sina.cn

S. Patnaik and X. Li (eds.), *Proceedings of International Conference on Computer Science and Information Technology*, Advances in Intelligent Systems and Computing 255, DOI: 10.1007/978-81-322-1759-6_52,

1 Introduction

Cloud computing is a kind of Internet-based business and software service model provided by large institutions and can provide users with flexible, customizable virtual machine, virtual network, and virtual cluster. Cloud computing resources are on the space distribution but in essence is heterogeneous, and different management domains and organizations have different resource management strategies and access price model, and therefore, if we want to solve the scheduling problem of cloud resources, we must give full play to the role of cloud computing.

As processing power and capacity of each node is not balanced, the arrival patten of the job is not consistent, resulting a large number of idle resources and excessive load nodes, the problem of heterogeneous system under load balance need to be solved. For the heterogeneous system to obtain a good load balancing, we need to design a good load balancing task scheduler, and then, the key point depends on the load balancing scheduling algorithm. Based on the cloud resource load balancing scheduling problem, two commonly used methods are rotation scheduling algorithm (RR) and the minimum number of connections scheduling algorithm (LC) [1].

2 Load Balancing Algorithm

2.1 Rotation Scheduling Algorithm

The principle of the rotation scheduling algorithm is that assign the inner servers to the users, starting from 1, until n (number of internal server), and then restart cycle, where each scheduler i= (i + 1) mod n, and select the server i. The advantage of the algorithm is simplicity, and it does not need to record the current state of all connections, so it is a kind of stateless scheduling.

In system implementation, we introduced an additional condition, namely when the weight of the server is zero, it means that the server is not available and cannot be scheduled. The aim is to cut the server out of service (such as shielding server failure and system maintenance), consistent with the other weighted algorithms at the same time. So, the algorithm changes correspondingly, and the process of the algorithm is as follows:

RR algorithm process: Given a set of servers S = {S0 and S1,..., Sn−1}, an indicator variable i means the choice of the server of the last time, W(Si) means the weight of server Si. The variable i is initialized to n−1, where n > 0.

```
j = i;
do {
    j = (j + 1) mod n;
    if (W(Sj) > 0) {
        i = j;
        return Si;
        }
    } while (j ! = i);
return NULL;
```

It is assumed that all server processing performances are the same.

2.2 The Minimum Number of Connections Scheduling Algorithm

In fact, the time duration that the client request to the server for the cloud resource may be different from each time. With the extension of working hours, if simple RR algorithm is used, the number of the connection processes on the server may be significantly different, and this does not actually reach the real load balancing. The least-connection scheduling algorithm records each server which the internal need to load, recording the number of connections processed by the server. When there is a new service connection request, the current connection will assigned to the least number of servers. When there's a new connection request, the current request will be assigned to the server with the least connections,making the scheduling more realistic and the load more balanced.

The least-connection scheduling algorithm assigns the new connection request to the server with the least connections. The least number of connections scheduling is a kind of dynamic scheduling algorithm, and it estimates the load on the server through the current active number of connections of the server.

In system implementation, we also introduced that the weight of the server is zero and it means that the server is not available and not be scheduled; the process of the algorithm is as follows: Given a set of servers S = {S0 and S1,..., Sn−1}, W(Si) means the weight of server Si and C(Si) means the current number of connections of the server Si [2].

```
for (m = 0; m < n; m++) {
    if (W(Sm) > 0) {
    for (i = m + 1; i < n; i++) {
    if (W(Si) <= 0)
        continue;
    if (C(Si) < C(Sm))
    m = i;
```

```
    }
    return Sm;
  }
  }
return NULL;
```

When each server has the same processing performance, The Least-Connection Scheduling algorithm can assign the requests with a variable load to each server, making sure that all requests with a relatively long treatment time can not be sent to the same server.

3 The Comparative Analysis of Scheduling Algorithms

Round-robin scheduling algorithm would dispatch the requests to different servers in turn by the way of round robin. That is each time execute the rule i= (i + 1) mod n, and select the server i. The advantage of this algorithm was simplicity, without recording of the current status of all connections. So it was a stateless scheduling, or static scheduling, which is the special scheduling strategy that has been determined in advance irrespective of the runtime environment of system. The scheduling algorithm assumes that all servers' processing performances are same without considering the server's current number of connections and the response speed. So it is relatively simple and not suited for the heterogeneous hardware environment of servers, and this algorithm would lead to load imbalance among servers because of the variable request time.

Weighted round-robin scheduling algorithm could solve the heterogeneity of different servers. Weight represents the performance of the server, and its default value is 1. Assuming server A having a weight of 1 and server B having a weight of 2, it means that the processing performance of B is twice as A. Weighted round-robin scheduling algorithm assigns all requests to the allocated servers based on the weight and the round-robin method. The servers with high weight would receive connections firstly and would have more connections than those with lower weight, and the same weight servers would process the same number of connections. The scheduler satisfy the requests based on the different server processing power through the "Weighted Round-Robin" scheduling algorithm. This will ensure that the server with a high processing capability can satisfy more cloud resource requests. Scheduler can inquire automatically real server load and dynamically adjusts its weights.

Against with the RR algorithm, the least connection scheduling algorithm is dynamic. That so-called dynamic scheduling algorithm means load balancing system makes load scheduling in real time by using the full (or partial) information needed by each system node to run. It estimates the server load through the current active connections. Scheduler needs to record the number of connections for each server. When a request is scheduled on one server, the connection number is plus one; when a connection aborts or stopped, it is minus one. Such balancing algorithm is suited for request service, which needs long time to process. If the real servers in cloud data center have similar performance, the least connection scheduling algorithm can better balance the load. But this algorithm does not work well, when the processing capacity of each server is different, because cloud resource connection will enter into waiting state after processing request. Generally, the waiting time is of a few minutes. The connection is still occupied by the server's resources. That why this case happens. Server with high performance has processed the received connections, and connections will be in waiting state, while server with low performance is already busy with processing the received connections, but also continually receives new connection requests.

On the basis of least connection scheduling algorithm, weighted least connection scheduling algorithm allocates different weights for each server depending on its processing capacity to make the server accept same number of service requests corresponding to the weight. It is a superset of the least-connection scheduling, and each server shows their processing performance based on the corresponding weights. The default value of server is 1, and system administrator can set the server weight dynamically. When scheduling a new connection, weighted least connection scheduling should make server's established connections and weights proportional as much as possible. Scheduler can automatically inquire the load condition of real sever and dynamically adjust its weight.

Contrast load balancing scheduling algorithm based on cloud resources in the Table 1.

4 Algorithm Simulation

4.1 Process Design

Firstly, create objects with selected algorithm and then read the parameters in XML. Use these parameters to initialize the object and continuously perform the selected algorithm. Show the result, the number of requirements which are processed and the number of requirements which entered the server after dealing with all requirements [3]. When a requirement enters the server's queue, it is regarded as to suffer a network congestion, which leads to delay. When a delay occurs, the requirement is considered not being responded to. Finally, calculate the number of requirements which entered the queue and display its mean value on the page. Program flowchart is shown in Fig. 1.

Table 1 Comparative analysis of these four algorithms

Algorithm	Algorithm introduction	Complexity	Advantages	Disadvantages
RR	Distribute the connecting request to different servers According to the order in turn, to realize load balancing	Lower	The biggest advantage is simple and easy to practice	Not suitable for the occasion when the performance and mission requirement specifications vary in different servers
LC	The load balancer that records the server's connection numbers distributes users' requirements to the server which has the least connections	Lower	Smoothly distributes the requirements with high variable load to the servers	Not suitable when the performance varies in different servers
WRR	Using weight to indicate the performance of servers. After some time, the requirement number of each server tends to be the proportion of its weight	Lower	More suitable than the runner scheduling algorithm when the performance varies in different servers	Not suitable for the occasion when the mission requirement specifications vary in different servers
WRR	Using weight to indicate the performance of servers, distributing user requirements to the server which has the smaller ratio of the number of connections and weights	Higher	Better than the least connection scheduling algorithm	Still not suitable for the occasion when the mission requirement specifications vary in different servers

4.2 The Operating Results and the Analysis of the Simulation Algorithm

As can be seen from the test results, the performance of the runner scheduling algorithm is lower than the least connection scheduling algorithm which has fewer requirements to enter the queue [4]. The weighted algorithm's performance is better than the unweighted ones, while the difference is smaller when the weights are the same. The weighted algorithm's performance is better when the weights are different rather than same. And they can better distribute requirements to smoothly connect to different servers, thus reducing the length of server's response queue (Tables 2, 3).

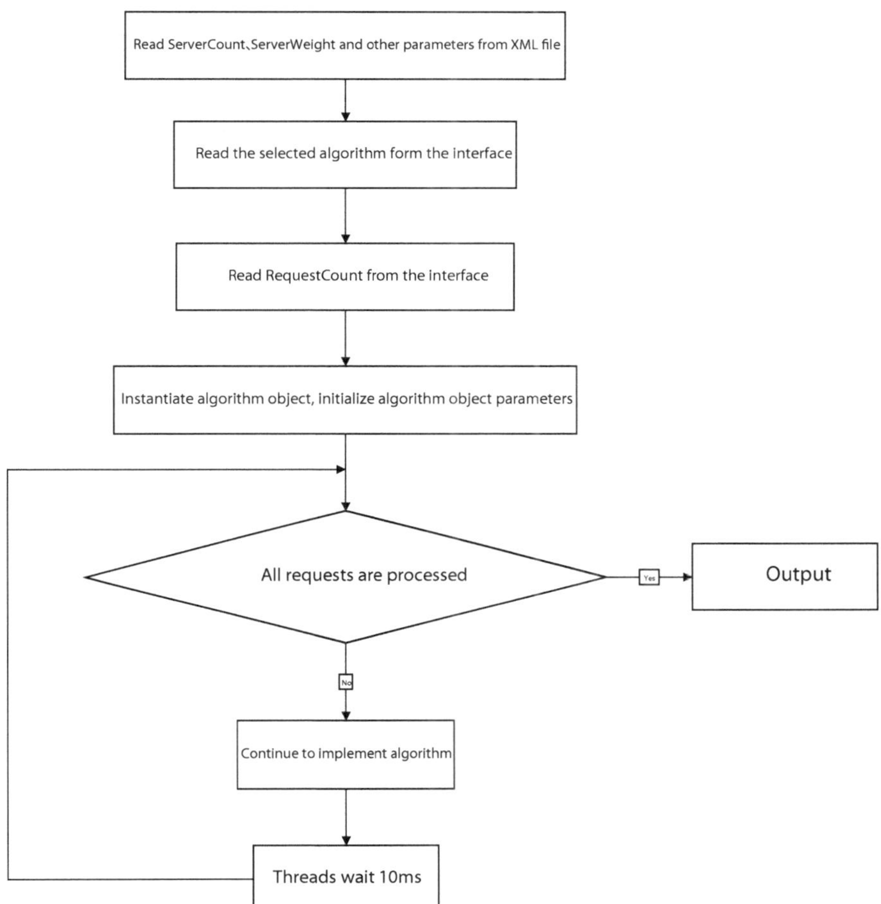

Fig. 1 Program flowchart

Table 2 ServerWeight is different

Algorithm	RequestCount	Average	ServerCount	ServerWeight
RR	1,000	397	3	16,8,4
WRR	1,000	307	3	16,8,4
LC	1,000	289	3	16,8,4
WLC	1,000	279	3	16,8,4

Table 3 ServerWeight is the same

Algorithm	RequestCount	Average	ServerCount	ServerWeight
RR	1,000	359	3	16,16,16
WRR	1,000	304	3	16,16,16
LC	1,000	221	3	16,16,16
WLC	1,000	226	3	16,16,16

5 Conclusion

After comparing two kinds of load balancing scheduling algorithms based on cloud resource, we can see that the runner scheduling algorithm is simple and easy to practice, but it is not suitable for the occasion when the performance and mission requirement specifications vary in different servers. The least connection scheduling algorithm can smoothly distribute the requirements with high variable load to the servers, but it fails to reach a good result when the performance varies in different servers. The corresponding weighted algorithms of the two have a little improvement, but the performance is still not good enough when the mission requirement specifications vary in different servers. This means that we still need to work on it to come up with better load balancing scheduling algorithms based on cloud resource [5].

References

1. Michael, A., Fox, A., Griffith, R., et al.: A view of cloud computing. Commun. Acm **53**(5), 50–58 (2010)
2. Saira, B., Muhammad, K.K.: Potential of cloud computing architecture. In: Informational Conference on Information and Communication Technologies (ICICT) 2011, Karachi, pp. 1–5 (2011)
3. Xu, B., Zhao, C., Hu, E., Hu, B.: Job scheduling algorithm based on Berger model in cloud environment. Adv. Eng. Softw. **42**(7), 419–425 (2011)
4. Fang, Y., Wang, F., Ge, J.: A task scheduling algorithm based on load balancing in cloud computing. In: Web Information Systems and Mining, pp. 271–277. Springer, Berlin (2010)
5. Pandey, S., Wu, L., Guru, S.M., Buyya, R.: A particle swarm optimization-based heuristic for scheduling workflow applications in cloud computing environments. In: 24th IEEE International Conference on Advanced Information Networking and Applications (AINA) 2010, pp. 400–407. (2010)

Max–Min Task Scheduling Algorithm for Load Balance in Cloud Computing

Yingchi Mao, Xi Chen and Xiaofang Li

Abstract In cloud computing, load balancing is required to distribute the dynamic local workload evenly across all the nodes. It helps to achieve a high user satisfaction and resource utilization by ensuring an efficient and fair allocation of every computing resource. Load balancing aids in minimizing resource consumption and avoids bottlenecks. Although many load balancing schemes have been presented, there is no scheme providing the elasticity in cloud computing. In this paper, a Max–Min task scheduling algorithm for load balance in the elastic cloud is proposed. To realize the load balancing, the proposed algorithm maintains a task status table to estimate the real-time load of virtual machines and the expected completion time of tasks, which can allocate the workload among nodes and realize the load balance. The extensive experiments demonstrate that the proposed Max–Min task scheduling algorithm can improve the resource utilization as well as reduce the respond time of tasks.

Keywords Elastic cloud · Load balance · Max–min algorithm · Task scheduling

1 Introduction

Cloud computing, a framework for enabling convenient and on demand network access to a shared pool of computing resources, is emerging as a new paradigm of large-scale distributed computing. A typical feature of cloud computing is its flexibility [1]. A cloud computing system with elastic features is called an elastic cloud. The load balancing in the cloud computing has become its key technology and hot research topic [2]. At present, the existing load balancing algorithms have

Y. Mao (✉) · X. Chen · X. Li
College of Computer and Information Engineering, Hohai University, Nanjing, China
e-mail: maoyingchi@gmail.com

S. Patnaik and X. Li (eds.), *Proceedings of International Conference on Computer Science and Information Technology*, Advances in Intelligent Systems and Computing 255, DOI: 10.1007/978-81-322-1759-6_53,

been proposed to meet the different goals considering the different constraints. For example, a QoS-based load balancing algorithm was presented to meet user service requirements under the SLA constraints [3, 4]. Against the rising cost of energy, during the establishment of green data centers, another load balancing algorithm was proposed to reduce energy consumption as well as improve the computing efficiency, which results in energy saving and environmental protection [5]. However, the above mentioned algorithms were not designed for general use, so their application fields are limited, resulting in the poor performance on other aspects. Moreover, they do not take full advantage of the elasticity in the cloud computing.

In order to improve the elasticity of cloud computing, a Max–Min task scheduling algorithm for load balancing in the cloud computing is proposed. The proposed algorithm maintains a task status table to estimate the real-time load of virtual machines and the expected completion time of tasks, which can realize the load balancing and improve the resource utilization. The experimental results show that Max–Min task scheduling algorithm can improve the resource utilization as well as reduce the respond time of tasks.

2 Related Work

Generally, a load balancing algorithm generally is divided into two types, static and dynamic. The static load balancing algorithm includes round robin (RR) and opportunistic load balancing (OLB). The RR algorithm allocates tasks to each node in turn, without considering the resource quantity of each server and the execution time of tasks. Similar to the RR algorithm, aOLB algorithm randomly allocates tasks to nodes available. On the other hand, dynamic algorithm considers the real-time workload and response time of nodes with a dynamic feedback mode. The dynamic load balancing algorithm includes minimum execution time (MET), minimum completion time (MCT), Min–Min and Max–Min algorithms. When a task arrives, a suitable node will be dynamically selected to perform the task execution based on the workload of each node.

However, there are limitations for those algorithms in cloud computing. The RR and random algorithms are simple to implement, but they cannot guarantee the stability of the load balancing effect due to the differences of each computing node and task. The MET and MCT algorithms take task differences into account, but these two algorithms are mainly for single task allocation and not for several tasks arriving in batches. The Max–Min and Min–Min algorithms can better allocate many tasks arriving in the same batch. For different application scenario and goals, many studies have improved the Max–Min and Min–Min algorithms.

The traditional Max–Min and Min–Min algorithms are for tasks arriving in the same batch. However, when the new tasks are arriving, the previous tasks are not completed. For this scenario, only simply using the Max–Min and Min–Min algorithms for task allocation cannot deal with the uncompleted tasks. The reasons are that the algorithm just allocates some overweight tasks to nodes with overload to execute, which leads the imbalanced workloads among nodes and reduces the system efficiency. Thus, the existing task scheduling algorithms cannot give a better solution under the constraints of the tasks arriving in the consecutive batches.

3 System Model

- Resource Model

A resource model is used to describe the processing capability of nodes. A unified resource model is established to describe the performance of node as a reference for task scheduling. The resource is defined as the computing capacity, storage and network transmission capacity in a time unit.

Definition 1 Node Resource: Resource index is used to quantify the resource available in a unit of time. The resource of Node i is represented with vector $\mathbf{K}_i$.

$$\mathbf{K}_i = [Kcpu_i, Kmem_i, Kdisk_i, Knet_i] \tag{1}$$

- Task Model

When a node executes a task, the task actually consumes the resources. A task's resource consumption index is in accordance with the resource index of the node.

Definition 2 Task Description: A task index is used to quantify the resource the task consumes. $\mathbf{J}_i$ is the resource vector task i consumes.

$$\mathbf{J}_i = [Jcpu_i, Jmem_i, Jdisk_i, Jnet_i] \tag{2}$$

Definition 3 Resource Weight: Resource weight is used to describe the weight of different load indexes. $\mathbf{W}$ is denoted as the resource weight vector.

$$\mathbf{W} = [w_1, w_2, w_3, w_4], \quad \sum_{i=1}^{4} w_i = 1 \tag{3}$$

In the different applications, the resource weight can be defined differently. For example, for a Web application $\mathbf{W} = [0.6, 0.05, 0.05, 0.3]$ can be used as default setting because CPU and network traffic are the key factors affecting the system performance.

Definition 4 Comprehensive Resource of Node: The weight sum of the resource vectors is calculated to describe the general performance of nodes. The comprehensive resource of node i is represented with R_i.

$$R_i = \mathbf{K}_i \times \mathbf{W}^T \tag{4}$$

Definition 5 Amount of Tasks: The weight sum of the resource vectors is calculated to evaluate the amount of tasks, which is denoted T_i as for the task i.

$$T_i = J_i \times \mathbf{W}^T \tag{5}$$

Definition 6 Execution Time: For the task i executed in a virtual machine j, suppose that the task i consumes all resource of the virtual machine j, then the time to complete the task i is the execution time of the task in the virtual machine j. H_i^j is the execution time of task i in the virtual machine j.

$$H_i^j = \frac{T_i}{R_j} \tag{6}$$

- Load Model

Definition 7 Node Load: To represent the load status of a node, the load of node i at the time t is denoted as L_i^t.

$$L_i^t = \frac{v_i \cdot length}{\Delta t} \tag{7}$$

Definition 8 Cluster Load: To calculate the loads of the cluster based on the average load of each node, the cluster load is denoted as L^t at the time t.

$$L^t = \frac{\sum L_i^t}{n}, \quad i \in Q_{\text{cluster}} \tag{8}$$

The prediction load P^t at the time t can be obtained with the load prediction algorithm. The cluster load L^t at the time t can be computed with the virtual status table. The cluster load at the time $t + 1$ is computed with P^t and L^t using the load prediction algorithm.

4 Max–Min Task Scheduling Algorithm

In this paper, a Max–Min task scheduling algorithm for the elastic cloud (ECMM) is proposed. Its main idea is to maintain an executing task status table and a virtual machine status table inside a load balancer. The task status table mainly includes task execution time, completion time and the latest update time. Meanwhile, the virtual machine status table contains the existing tasks in the virtual machine, the total execution time of tasks, the status of the virtual machine life cycle and the latest update time. When allocation tasks arriving in the same batch, it is first to select the task with the longest execution time (Max), calculate the estimated time of the tasks in each virtual machine with the virtual machine table, select the virtual machine with the shortest completion time (Min), and allocate the task to the relevant virtual machine. Meanwhile, it is to update the number of tasks and the total task execution time of the virtual machine in the virtual machine status table. The process cycles until all tasks are allocated.

4.1 Task Status Tables and Virtual Machine Status Tables

The core of the virtual machine task amount estimation algorithm is maintaining the executing task status table and the virtual machine status table. The executing task status table is set as set **S**. Every S_i in **S** is the status tuple of the task *i*.

$$S_i = \; < id, wm\_id, finish, total, last, submit > \tag{9}$$

where *id*, *wm_id*, *finish*, *total*, *last*, *submit* represent the task number, number of a virtual machine executing tasks, execution time of a completed task, task execution time, last update time and time when a task arrives, respectively.

The virtual machine status table is set as set **V** Each V_i in **V** is a status tuple of the virtual machine *j*. The description of the virtual machine status tuple is Eq. (10), where *vm_id*, *num*, *length*, *status*, *last* represent the virtual machine number, number of executing tasks, total remaining time of incomplete executing tasks, virtual machine lifecycle status, and last update time, respectively.

$$v_j = \; < vm\_id, num, length, status, last > \tag{10}$$

4.2 Update Algorithm of Task Status and Virtual Machine Status Tables

The update of a task status table mainly refers to removing completed tasks and updating the completion time of the incomplete task <S_i, *finish*>. The virtual machine status table comes from the task status statistics. After the task status table updates each time, the *num* and *length* of the virtual machine will be computed again according to the updated task status table.

Every the interval Δt, agent collects the task's status in the virtual machine *j* and sends back to the set of the executing tasks *E* and the set of the completed tasks *F*. **S** and **V** are updated with *E* and *F*. The detail of the update algorithm for task status table and virtual machine status table are as follows.

Step 1: Delete the completed tasks from **S** according to the set F sent back from the virtual machine j.

Step 2: For $\forall i \in E$, the task i has not been completed in the virtual machine. From the last task update time $S_i.last$ in the current time t, there are $v_j.num$ tasks executed. Considering the equal time sheet, Eq. (11) is used to update the task execution time for the task i.

$$S_i.finish = S_i.finish + \frac{t - S_i.last}{v_j.num} \tag{11}$$

If $S_i.finish > S_i.total$, it means that the estimation of the execution time for the task i is too short, thus, it needs to prolong the execution time. It sets to $S_i.total = 1.5 \times S_i.total$.

Step 3: For the task i, the last update time is set to $S_i.last = t$.

Step 4: After the step 2 and 3 are finished for all of tasks in the set E, the virtual machine status will be estimated. For $i \in \mathbf{S}$, the task i is allocated to the virtual machine j, that is $j = S_i.vm\_id$. Meanwhile, it initializes the status of all virtual machines, that is, $v_j.num = 0$, $v_j.length = 0$, $v_j \in \mathbf{V}$.

Step 5: Updating the status of the virtual machine j in the set $\mathbf{V}$. It sets $v_j.last = t$, $v_j.num = v_j.num + 1$, $v_j.length = v_j.length + S_i.finish$.

4.3 Task Allocation Algorithm

Definition 9 Expected Completion Time: For the task i executing in the virtual machine j, the actual completion time for task i in the virtual machine j is used as the expected completion time of task i, denoted as F_i^j.

Due to the processing of each task in parallel in each node, the executing task number *num*, and the incomplete time *length* can be obtained from the virtual machine status table. To simplify the computation, it is to execute *num* tasks in parallel with the length of *length*/*num* in one node of virtual machine. The expected completion time for the new incoming tasks can be calculated in the Eq. (12).

$$F_i^j = \begin{cases} (v_j.num + 1) \times H_i^j & H_i^j \le \frac{v_j.length}{v_j.num} \\ \left(\frac{v_j.num+1}{v_j.num}\right) \times v_j.length + \left(H_i^j - \frac{v_j.length}{v_j.num}\right) & H_i^j \le \frac{v_j.length}{v_j.num} \end{cases} \tag{12}$$

Definition 10 Cluster expected completion time: Cluster expected completion time is defined as the set of expected completion time of the task i in the virtual machine cluster, denoted as $\mathbf{F}_i$.

The details of the tasks allocation algorithm are as follows.

Step 1: Sort all of the tasks in the tasks set T based on the task's amount.
Step 2: Choose the task i with the largest task amount in T.
Step 3: Calculate $\mathbf{F}_i$, that is the cluster expected completion time for task i. Sort all F_i^j in $\mathbf{F}_i$ and select the virtual machine j with the shortest expected completion time F_i^j.
Step 4: Allocate the task i to the virtual machine j.
Step 5: Add the tuple $S_i = \left[i, j, 0, H_i^j, t, t\right]$ into the set $\mathbf{S}$.
Step 6: Update the virtual machine status v_j in $\mathbf{V}$, that is $v_j.last = t$, $v_j.num = v_j.num + 1$, $v_j.length = v_j.length + S_i.finish$. Remove the task i from the tasks set T.
Step 7: Repeat Step 2–6 until the tasks set T is empty.

5 Experiments Evaluation

5.1 Experimental Settings

In this paper, Cloud Sim, a simulation tool for cloud computing, is used to model and simulate task scheduling in the large-scale cloud computing. The virtual nodes, computing resource, energy consumption are modeled with Cloud Sim to evaluate the efficiency of load balancing for the proposed Max–Min task scheduling algorithm.

Since the tasks arrive in batches, the experiment need preprocesses the data set. For a given data set, Δt is used to divide time to a certain number of batches. The data between $t - \Delta t$ to t is set to the value when the data arriving at time t in the same batch. For the experimental results, the paper evaluates the performance of the algorithm with indexes such as task pending time, task response time rate, theoretical concurrency, and node resource utilization.

5.2 Experimental Results

The experiment is mainly used to verify the performance of the elastic cloud max–min (ECMM) task scheduling algorithm. The control group includes the RR and Max–Min algorithms.

For the experiment of 6 data sets in the Fig. 1, it can be seen that the ECMM task scheduling algorithm is better than the RR in terms of average task pending time. And the ECMM considers the current amount and number of tasks as well as the amount of arriving tasks and shortens the average task pending time by 21.4–30 % compared with the Max–Min algorithm.

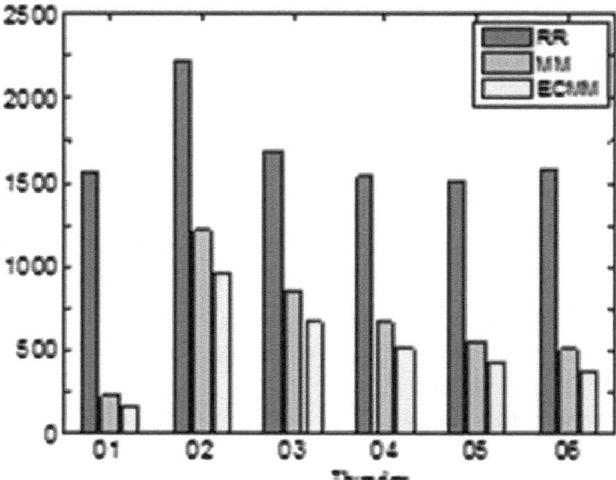

Fig. 1 Average task pending time

In terms of the task response time ratio, the Max–Min is the largest. This is because the Max–Min algorithm pursues the shortest task execution time. It leads to the accumulation of small tasks running on several nodes, so in these nodes, multiple small tasks are executed in concurrency, and the concurrency of tasks will increase significantly. In comparison, the RR algorithm allocates tasks to each node in turn, the number of tasks at each node is more balanced, and the concurrency of tasks is lower than the Max–Min algorithm. When the ECMM executes a task, the existing number of tasks of the node is also taken into account, so it is better than the RR algorithm without considering nodes in terms of the response time ratio, and its response time ratio drops 11–14 %, as shown in Fig. 2.

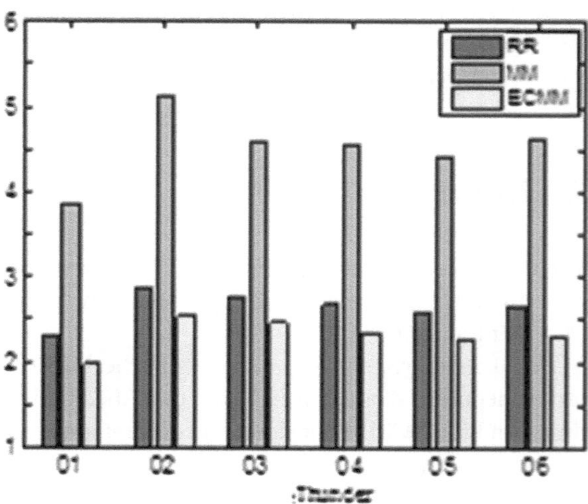

Fig. 2 Average task response time ratio

6 Conclusion

The paper introduces an elastic cloud task scheduling algorithm. It describes the traditional task scheduling algorithm and related researches. And according to the features of the cloud computing environment and tasks as well as the real environment, the algorithm takes reasonable simplification to create a resource model and a task model. For tasks arriving in consecutive batches, the algorithm designs and describes an improved Max–Min task scheduling algorithm.

Acknowledgments This research is partially supported by Fundamental Research Funds for the Central Universities 2013B06914; Key Laboratory of Geo-informatics of State Bureau of Surveying and Mapping 201005.

References

1. Data Center Knowledge [EB/OL]. http://www.datacenterknowledge.com
2. Chen, Q., Deng, Q.: Cloud computing and its key techniques. J. Comput. Appl. **29**(9), 2562–2567 (2009)
3. Gencay, E., Sinz, C., Kuchlin, W.: Towards SLA-based optimal workload distribution in SANs: NOMS 2008. In: IEEE/IFIP Network Operations and Management Symposium: Pervasive Management for Ubiquitous Networks and Services, Salvador, Bahia, Brazil, 7–11 April 2008. Institute of Electrical and Electronics Engineers Computer Society
4. Filin, S., Harada, H., Hasegawa, M., et al.: QoS-guaranteed load-balancing dynamic spectrum access algorithm. In: 2008 IEEE 19th International Symposium on Personal, Indoor and Mobile Radio Communications, PIMRC 2008, Poznan, Poland, 15–18 Sept 2008. Institute of Electrical and Electronics Engineers Inc
5. Lin, C., Tian, Y., Yao, M.: Green network and green evaluation: energy-saving scheme, model and evaluation. J. Comput. **2011**(4), 593–612 (2011)

CP-ABE Scheme with Revocation for Cloud Storage

Ning Pan, Lei Sun and Xiuqing Mao

Abstract Cloud Computing is the next-generation architecture of IT Enterprise because of its scalability and low cost. But it also raises security issues such as the security of data which is the major constraint to the development of cloud storage. To solve the safety issues in cloud storage services for data sharing characteristics, this paper proposed a new ciphertext policy attribute-based encryption (CP-ABE) scheme applied in cloud storage while shortening the length of the ciphertext and can be flexible to properties revocation. The security of the scheme is based on Decisional Bilinear Diffie-Hellman (DBDH) problem.

Keywords Cloud storage · ABE · Revocation · Proxy re-encryption

1 Introduction

With cloud computing booming, the security issue has been its major challenge. In recent years, there are more and more user data breaches which made it difficult for many companies to trust the privacy of data stored in the cloud. Therefore, the data security in cloud storage may impede its fast growth. With the numerous cloud storage encryption solution proposed as well as the deepening of the research, attribute-based encryption technology is becoming a research hotspot. Ciphertext policy attribute-based encryption (CP-ABE) is a technique in which user with secret key containing attributes is only able to decrypt the message if the attributes in the policy match with the attributes in secret key, which provides a new flexible

N. Pan (✉) · L. Sun · X. Mao
Zhengzhou Information Science and Technology Institute, Zhengzhou, China
e-mail: pan_ning1988@163.com

L. Sun
e-mail: sl0221@sina.com

S. Patnaik and X. Li (eds.), *Proceedings of International Conference on Computer Science and Information Technology*, Advances in Intelligent Systems and Computing 255, DOI: 10.1007/978-81-322-1759-6_54,

and efficient mechanism for realizing one-to-many encryption, and due to its flexible expressiveness, it is regarded as a promising tool for enforcing fine-grained access control over encrypted data.

Commercial interests is one of the root causes for the user to apply cloud computing, but when general cloud storage services are unable to meet users' security needs, they will turn to choose relatively expensive but more secure encryption cloud storage service. In CP-ABE, size of ciphertext and secret key will increase linearly with the number of attributes in policy; it will increase the transmission and users' cost; inevitably, there will be some system attributes revocation and user permissions change operation. In recently years, researchers have proposed a series of attribute encryption schemes. One of the efficient constructions of the CP-ABE with (t, n) threshold scheme can be found in the [1, 2]; the size of ciphertext in [1] is $n + O(1)$ and in [2] is $2(n - t) + O(1)$. In [3], authors proposed a constant length ciphertext scheme, in which the number of attributes in user's secret key is same as the number of attributes in policy. And in [4], authors proposed a scheme in which ciphertext remains constant in length, irrespective of the number of attributes, but not support attributes revocation.

Our Contributions To overcome the lack of the literature [4], we introduce the idea of tag mechanism proposed in [5] into our scheme to support attributes revocation on the basis of the combination of CP-ABE technology and proxy re-encryption technology and control the length of the ciphertext at the same time achieve high revocation of system attributes. Cloud storage service provider in the scheme is only responsible for the storage and proxy re-encryption of the cipher text, so do not worry about its own security problem of cloud storage service providers.

2 Security Assumptions

To enhance the security of model, many models assume that cloud storage provider (CSP) is not credible so that delegate all encryption calculation to clients. This so-called security is based on the user's own cost of computing power entirely at the expense of cloud computing itself a huge amount of computing power. So we assume that CSP is "honest but curious"; that is to say, CSP will follow our proposed protocol in general, but try to find out as much secret information as possible based on their inputs. Our scheme introduces a Third Party Auditor which is trusted by CSP and users and assumes that the Third Party Auditor and the CSP are not privately colluding to steal users' secret information stored, at the same time assumes the data communication channel between users, CSP, the party, and data users to be secured under existing security protocols such as SSL.

3 Security Model

In order to achieve the aforementioned objectives, we propose the model described in Fig. 1. The model is composed of the following components: authorization center, CSP, data owner (Owner), and data users (Users).

Authorization Center It generates the public, master secret key and version number for the system. The authorization center is responsible for users' keys issuing and revoking.

CSP It provides data outsourcing services and consists of data servers and data service manager.

Owner The owner who is registered to the trusted authority is considered as registered user. A registered user is responsible for defining attribute-based access policy and encrypting data under the predefined policy before storing at the cloud storage.

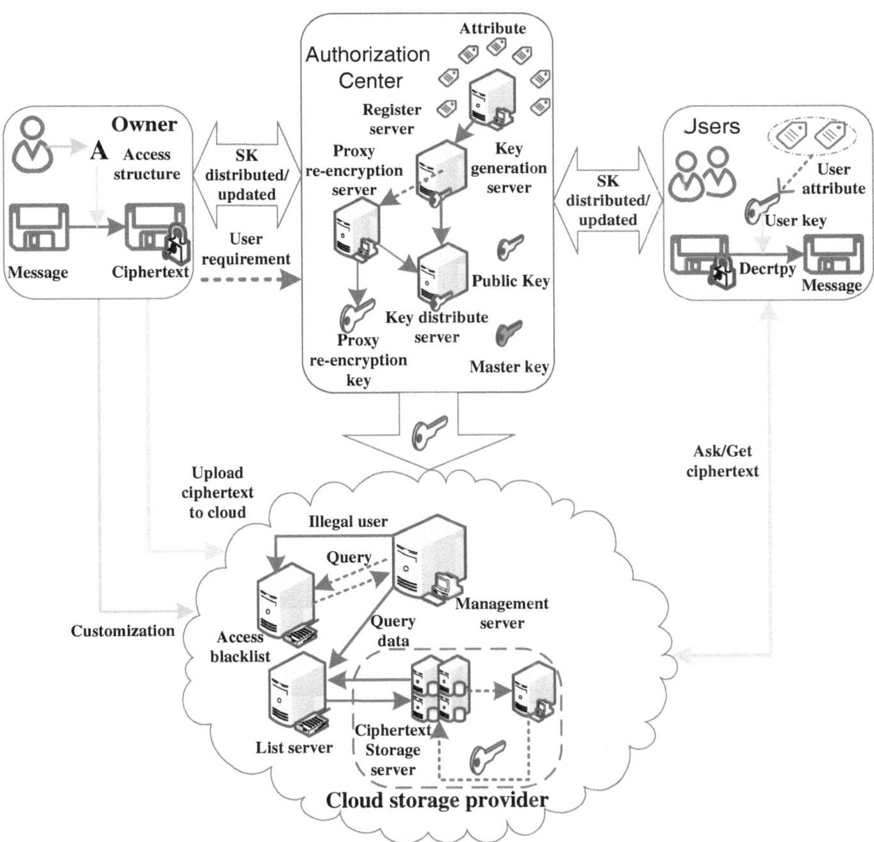

Fig. 1 Secure cloud storage model

Users The users download the data stored in cloud storage and decrypt it with their key. The scheme ensures that any data users can only decrypt the encrypted data if and only if they can successfully complete the access policy.

Attribute revocation

Authorization Center It generates the proxy re-encryption key and sends it to the CSP. The authorization center is responsible for users' keys and version number updating.

CSP It is responsible for the stored data re-recryption.

Users The users update their key through the authorization center and decrypt the re-encryption data.

If there exists the illegal user who is recognized by CSP or data owner, he will be put into the data access blacklist in which the users uploading data operation are shielded.

In our scheme, the data owner only needs to design the access structure and initial attribution encryption, which greatly reduces the computing resource requirements of the data owner. At the same time, all data are encrypted and stored, which ensures the CSP cannot get the stored information without illegal operation and then ensures the stored data's confidentiality and security.

In view of the large re-encryption calculation, we introduce the idea of lazy re-encryption namely that only when the stored data are accessed by the user that the data is updated. If the currently stored ciphertext version number is k, actually should be updated to the version $k + t$, the proxy re-encryption key is $\mathrm{CK}^{k+t} = \left(\prod_{j=0}^{t-1} \mathrm{ck}_1^{k+j}, \prod_{j=0}^{t-1} \mathrm{ck}_2^{k+j}, \ldots \prod_{j=0}^{t-1} \mathrm{ck}_n^{k+j}\right)$. Similarly, when the data user accesses the CSP, he contrasts the version number in the cipher-text and his own key at this time; if different, the user should get the latest version of the key from the authorization center.

4 Proposed Scheme

Z_p is a group of large prime order p. Groups G and G_1 are cyclic multiplicative groups of prime order p. Assume $\Omega = \{\mathrm{att}_1, \mathrm{att}_2, \ldots, \mathrm{att}_n\}$ be the set of all possible attributes in system and $A_i = \{v_{i,1}, v_{i,2}, \ldots, v_{i,n_i}\}$ be the set of all possible values for att_i, where $n_i = |A_i|$. Assume $L = [L_1, L_2, \ldots, L_n]$ be a set of attributes for user and $W = [W_1, W_2, \ldots, W_k]$ is an access structure. Here, $e : G \times G \to G_1$ is the admissible bilinear map function.

1. $Setup(1^k)$: input the security parameters k.

The authorization center selects a large prime number p, a bilinear group $e : G \times G \to G_1$ with order p whose generator is g, $h \in G$, $y \in Z_p$ and $t_{i,j} \in Z_p (i \in [1, n], j \in [1, n_i])$. It calculates $T_{i,j} = g^{t_{i,j}} (i \in [1, n], j \in [1, n_i])$, $Y = e(g, h)^y$ and initializes the version number ver $= 1$.

The authorization center releases the (ver, MPK) and saves MSK, where $\text{MPK} = \left(e, g, h, Y, T_{i,j}(i \in [1, n], j \in [1, n_i])\right)$, $\text{MSK} = \left(y, t_{i,j}(i \in [1, n], j \in [1, n_i])\right)$

2. *KenGen*(*MSK*, *L*): input the user attribute set L and the master key MSK.

The authorization center generates $r \in Z_p$ and calculates the SK of user as follows:

$$\text{SK}_L = \left(\text{ver}, h^{y+r}, \forall v_{i,j} \in L : D_{i,j} = \left(T_{i,j}\right)^r, g^r, L\right).$$

3. *Encrypt*(*MPK*, *M*, *W*): input message M, access structure W, and public parameters MPK.

Data owner selects $s \in Z_p$ and calculates CT = (ver, C_1, C_2, C_3, W), where $C_1 = MY^s$, $C_2 = g^s$ and $C_3 = \left(h^s, \forall v_{i,j} \in W : X_{i,j} = \left(T_{i,j}\right)^s\right)$.

4. $Decrypt(MPK, CT, SK_L)$: input public parameters MPK, ciphertext CT, and user's key SK_L.

Data users get the chipper text CT from CSP and check the version number in CT and SK_L, if different, update the private key from the authorization center. Assume AS $\subseteq L$ and AS $= W$. Therefore, after identifying the AS, users calculate the message M as follows:

$$\begin{aligned}
&\frac{C_1 e\left(g^r, h^s \prod_{v_{i,j} \in W} X_{i,j}\right)}{e\left(C_2, h^{y+r} \prod_{v_{i,j} \in \text{AS}} \left(T_{i,j}{}^r\right)\right)} \\
&= \frac{MY^s e\left(g^r, h^s \left(\prod_{v_{i,j} \in W} g^{t_{i,j}}\right)^s\right)}{e\left(g^s, h^{y+r} \prod_{v_{i,j} \in \text{AS}} g^{rt_{i,j}}\right)} \\
&= \frac{Me(g,h)^{ys} e(g^r, h^s, g^{sp})}{e(g^s, h^{y+r}) e(g^s, g^{rq})} \\
&= \frac{Me(g,h)^{ys} e(g,h)^{rs} e(g,g)^{rsp}}{e(g,h)^{ys} e(g,h)^{rs} e(g,g)^{rsp}} \\
&= M
\end{aligned}$$

Here, $p = \sum_{v_{i,j} \in W} t_{i,j}$, $q = \sum_{v_{i,j} \in \text{AS}} t_{i,j}$.

If system has attributes revocation, the authorization center generates the proxy re-encryption key and the corresponding users' key.

5. *Rekeygen*(*u*, *MPK*): input the set of attributes u which are to be updated and public parameters MPK.

If $v_{i,j} \in u$, the authorization center selects $t_i \in Z_p$ and calculates $\mathrm{ck}_i = t_i / t_{i,j}$; otherwise, $\mathrm{ck}_i = 1$. It sends the re-recryption key $\mathrm{ck} = \left(\mathrm{ver}, \forall v_{i,j} \in u, \mathrm{ck}_i\right)$ to CSP.

6. *ReEncrypt*(CT, ck, β): input the original ciphertext CT, re-recryption key ck, and the set of attribute β which is in the original access structure W and $\mathrm{ck} \neq 1$.

CSP checks the version number in the ciphertext CT and ck, if inconsistent directly outputs CT. Otherwise, if $v_{i,j} \in \beta$ and $v_{i,j} \in W$, CCS calculates $C_3' = \left(h^{sm}, \forall v_{i,j} \in W : X_{i,j}' = \left(T_{i,j}\right)^{\mathrm{sck}_i}\right)$ where $m = \sum_{\substack{v_{i,j} \in \beta \\ v_{i,j} \in W}} \mathrm{ck}_i$.

7. *Rekey*(SK_L, ck, β): input the user's key SK_L, re-recryption key ck, and the set of attribute β which is in the original access structure SK_L and $\mathrm{ck} \neq 1$.

The authorization center checks the version number in the key SK_L and ck, if inconsistent directly outputs $\mathrm{SK}_L' = \mathrm{SK}_L$. Otherwise, if $v_{i,j} \in \beta$ and $v_{i,j} \in L$, the center calculates $n = \sum_{\substack{v_{i,j} \in \beta \\ v_{i,j} \in L}} \mathrm{ck}_i, D_{i,j}' = \left(T_{i,j}\right)^{r\mathrm{ck}_i}$ and updates

$$\mathrm{SK'}_L = \left(\mathrm{ver} + 1, h^{y+rn}, \forall v_{i,j} \in L : D_{i,j} = \left(T_{i,j}\right)^{r\mathrm{ck}_i}, g^r, L\right).$$

Assume AS = L and AS = W. Therefore, after identifying the AS, users calculate the message M as follows:

$$\begin{aligned}
&\frac{C_1 e\left(g^r, h^{sm} \prod_{v_{i,j} \in W} X_{i,j}'\right)}{e\left(C_2, h^{y+rn} \prod_{vi,j \in \mathrm{AS}} \left(T_{i,j}{}^{r - rk_i}\right)\right)} \\
&= \frac{MY^s e\left(g^r, h^{sm} \left(\prod_{v_{i,j} \in W} g^{t_{i,j}}\right)^{\mathrm{sck}_i}\right)}{e\left(g^s, h^{y+\mathrm{rn}} \prod_{vi,j \in \mathrm{AS}} g^{r\mathrm{ck}_i t_{i,j}}\right)} \\
&= \frac{Me(g,h)^{ys} e(g^r, h^{sm} g^{sp})}{e(g^s, h^{y+rn}) e(g^s, g^{rq})} \\
&= \frac{Me(g,h)^{ys} e(g,h)^{rsm} e(g,g)^{rsp}}{e(g,h)^{ys} e(g,h)^{srn} e(g,g)^{rsq}} \\
&= M
\end{aligned}$$

Here, $p = \sum_{v_{i,j} \in W} t_{i,j} \mathrm{ck}_i$, $q = \sum_{v_{i,j} \in \mathrm{AS}} t_{i,j} \mathrm{ck}_i$, $n = \sum_{\substack{v_{i,j} \in \beta \\ v_{i,j} \in L}} \mathrm{ck}_i$, $m = \sum_{\substack{v_{i,j} \in \beta \\ v_{i,j} \in W}} \mathrm{ck}_i$.

5 Security Analysis

The proposed scheme satisfies the distinguishing ability of messages under the Decisional Bilinear Diffie-Hellman (DBDH) assumption.

Assume there exists a polynomial-time adversary A that can attack our scheme in the selective-set model with advantage ε. We build a simulator X that can play the DBDH game with advantage $\frac{\varepsilon}{2}$, where $N = \prod_{i=1}^{n} n_i$ which is the number of access structure. The simulation proceeds as follows:

The challenger generates $a, b, c, z \in_R Z_p$, $v \in_R \{0, 1\}$, and G with the generator g. If $v = 0$, the challenger sets $Z = e(g, g)^{abc}$; otherwise, it sets $(g, g^a, g^b, g^c, z) \in G \times G_1$. The challenger sends $(g, g^a, g^b, g^c, z) \in G \times G_1$ to X.

The challenger chooses the challenge access structure $W^* = [W_1^*, W_2^*, \ldots, W_k^*]$ and gives it to X.

The simulator X selects $u \in_R Z_p$ and calculates $Y = e(g^a, (g^b)^u) = e(g, g)^{ab}$ and $h = g^u$. For $t'_{i,j}\{i \in [1, n], j \in [1, n_i]\} \in_R Z_p$, if $v_{i,j} = W_i^*$, X calculates the private keys $t_{i,j}\{i \in [1, n], j \in [1, n_i]\} \in_R Z_p$ as $t_{i,j} = t'_{i,j}$ and the public keys $T_{i,j}\{i \in [1, n], j \in [1, n_i]\} \in_R Z_p$ as $T_{i,j} = g^{t'_{i,j}}$; otherwise, $t_{i,j} = bt'_{i,j}$ and $T_{i,j} = (g^b)^{t'_{i,j}}$. X sends $\mathrm{MPK} = (e, g, h, Y, T_{i,j}\{i \in [1, n], j \in [1, n_i]\})$ to the adversary A.

There exists $v_{i,j} = L_i$ and $v_{i,j} \neq W^*$ as $L \neq W^*$, so we can write $\sum_{v_{i,j} \in L} t_{i,j} = X_1 + bX_2$, where $X_1, X_2 \in Z_p$. Here, X_1 and X_2 can be represented as the sum of value $t'_{i,j}$. So X can calculate X_1 X_2. It selects $\beta \in_R Z_p$ and sets $r = \frac{\beta - ua}{X_2}$ and calculates

$$\mathrm{SK}_L = \left\{ g^{\frac{\beta}{X_2}} (g^a)^{-\frac{u}{X_2}}, (g^{ab})^u g^{\frac{\beta u}{X^2}} (g^a)^{-\frac{u^2}{X^2}}, \forall v_{i,j \in L} (g^{t_{i,j}})^{\frac{\beta}{X_2}} ((g)^a)^{t_{i,j}})^{-\frac{u}{X_2}} \right\}$$

where

$$(g^{ab})^u g^{\frac{\beta u}{X^2}} (g^a)^{-\frac{u^2}{X^2}} = (g^u)^{ab}\Big) (g^u)^{\frac{\beta - ua}{X_2}} = h^y h^r = h^{y+r},$$

$$g^{\frac{\beta}{X_2}} (g^a)^{-\frac{u}{X_2}} = g^{\frac{\beta - ua}{X_2}} = g^r \text{ and } (g^{t_{i,j}})^{\frac{\beta}{X_2}} ((g)^a)^{t_{i,j}})^{-\frac{u}{X_2}} = (g^{t_{i,j}})^r = (T_{i,j})^r.$$

So SK_L is a valid secret key. Adversary A identifies set $\mathrm{AS} \subseteq L$ and calculates $\sum_{v_{i,j \in \mathrm{AS}}} t_{i,j} = \sum_{v_{i,j \in W^*}} t_{i,j}$. If there exists $\mathrm{AS} \subseteq L$ such that $X_2 \bmod p = 0$ when $X_2 \bmod p = 0$, the probability is at most N^2/p.

The adversary A will submit two challenge messages M_0 and M_1 to the simulator X. The simulator flips a fair binary coin u and returns the ciphertext as $\mathrm{CT}^* = \langle C_1^*, C_2^*, C_3^*, W^* \rangle$ where $C_1^* = M_u Z_u$ $C_2^* = g^c$ and $C_3^* = h^s (g^c)^{\sum_{v_{i,j \in W^*}} t_{i,j}}$.

The adversary A will submit a guess u' of u. If $u' = u$, the simulator X will output 1; otherwise, it will output 0. There will be two cases: If $Z = e(g, g)^{abc}$, then the advantage of A is ε, so

$$\Pr\left[x \to 1 \middle| Z = e(g,g)^{abc}\right] = \Pr\left[u' = u \middle| Z = e(g,g)^{abc}\right] = 1/2 + \varepsilon.$$

If $Z = e(g,g)^z$, then A has no advantage to distinguish bit u, so

$$\Pr[x \to 0 | Z = e(g,g)^z] = \Pr[u' \neq u | Z = e(g,g)^z] = 1/2.$$

As above, the overall advantage of the simulator in the DBDH game is

$$\mathrm{Adv}_s = \frac{1}{2}\left(\Pr\left[x \to 1 \middle| Z = e(g,g)^{abc}\right] + \Pr[x \to 0 | Z = e(g,g)^z]\right) - \frac{1}{2} = \frac{\varepsilon}{2}.$$

6 Scheme Comparison

Our scheme is based on the literature [5], while ciphertext length is relatively small and the revocation of attribute. Table 1 is the comparison based on size of parameters with other schemes. Table 2 is the comparison based on size of parameters with other schemes. Here, N' is total number of attributes in the system and n is total number of attributes, $N' = \sum_{i=1}^{n} n_i$, where n_i is the number of possible values for attribute i, G_1, G_2, and G_T are bilinear groups, the notation $|G|$ shows the bit length of the element belonging to group G, the notations mG and mC_ε show the m times calculation over the group G and pairing operations, respectively, and x_1 and x_2 are the sets of attributes associated with ciphertext and secret key length.

Table 1 Size of parameters for ABE schemes

Scheme	MPK	MSK	SK	CT
2	$2\|G_1\| + \|G_T\|$	$\|G_1\|$	$(3+n)\|G_1\|$	$(1 + r_1 n)\|G_1\| + \|G_T\|$
4	$(2N' + 3)\|G_1\| + \|G_T\|$	$(N' + 1)\|Z_p\|$	$2\|G_1\|$	$2\|G_1\| + \|G_T\|$
7	$2n\|G_1\|$	$3\|Z_p\|$	$2n\|G_1\|$	$2n\|G_1\|$
This paper	$(4+n)\|G_1\|$	$\|Z_p\|$	$(2+n)\|G_1\|$	$(3+n)\|G_1\|$

Table 2 Computational time for each scheme

Scheme	Encrypt	Decrypt
2	$(1 + 3x_1 n)G_1 + 2G_T$	$(1 + n + x_1)C_\varepsilon + (3x_1 - 1)G_1 + 3G_T$
4	$(n+1)G_1 + 2G_T$	$2C_\varepsilon + 2G_T$
7	$(n+t+1)G_1$	$3C_\varepsilon + (x_2)G_T + O(n)a$
This paper	$(4+n)G_1$	$3C_\varepsilon + 2G_1$

7 Conclusion

In this paper, we propose a new CP-ABE scheme applied in cloud storage which achieves attribute revocation based on proxy re-encryption and tag mechanism. In addition, the scheme is based on DBDH problem and greatly reduces the storage overhead for the users. In future, we make this scheme for threshold ABE and add feature like the recipient's anonymity to increase the security.

References

1. Waters, B.: Ciphertext-policy attribute-based encryption: An expressive, efficient, and provably secure realization, p. 290. Cryptology ePrint report (2008)
2. Daza, V., Herranz, J., Morillo, P., R'afols, C.: Extended access structures and their cryptographic applications. Appl. Algebra Eng. Commun. Comput. 2008
3. Emura, K., Miyaji, A., Nomura, A., Omote, K., Soshi, M.: A ciphertext-policy attribute-based encryption scheme with constant ciphertext length. In: ISPEC 2009. LNCS, vol. 5451, pp. 13–23 (2009)
4. Doshi, N., Jinwala, D.: Constant Ciphertext Length in CP-ABE. arXiv.org 1208. 5991
5. Yu, SC., Wang, C., Ren, K., et al.: Attribute based data sharing with attribute revocation. In: Proceedings of Asian ACM Conference on Computer and Communications Security, pp. 261–270. ACM Press, New York (2010)
6. Sahai, A., Waters, B.: Fuzzy identity-based encryption. Advances in Cryptology—EUROCRYPT, pp. 457–473. Springer, Berlin (2005)
7. Herranz, J., Laguillaumie, F., R'afols, C.: Constant size ciphertexts in threshold attribute-based encryption. In: PKC 2010. LNCS 6056, pp. 19–34 (2010)
8. Wang, J., Zhang, M., Chen, Q.: An efficient attribute based encryption with attribute revocation. J. Comput. Appl. **32**(S1), 39–43 (2012)
9. Liu, F., Yang, M.: Ciphertext policy attribute based encryption scheme for cloud storage. Appl. Res. Comput. **29**(4), 1452–1456 (2012)
10. Feng, D., Zhang, M., Zhang, Y.: Study on cloud computing security. J. Softw. **22**(1), 71–83 (2011)
11. Feng, D.: Status quo and trend of cryptography. J. Chin. Inst. Commun. **23**(5), 18–26 (2002)

An Energy Efficiency Model Based on QoS in Cloud Computing

Xiaobo Cai and Xuejie Zhang

Abstract The energy usage of cloud computing systems is becoming an important consideration, especially for making the data center more energy efficient as a necessary task. Various methods for reducing the data centers' energy consumption have been investigated, but without paying much attention to the quality-of-service (QoS) expectations. In this paper, we proposed a simple model of energy efficiency–QoS (E-Q) aimed at capturing some key aspects of energy minimization while meeting the specified constraints on performance and/or QoS. Then, based on the proposed E-Q model, we give an algorithm which can enable the physical nodes to satisfy the performance's need for any set of jobs with a minimum-energy schedule and in the cloud. Experiment results demonstrate the effectiveness of the proposed algorithm compared to previous work, which can achieve objective optimization both on energy consumption and QoS.

Keywords Cloud computing · QoS · Green computing · Energy efficient · Job/tasks deployment

1 Introduction

Cloud computing offers utility-oriented IT services to users worldwide. Based on pay-as-you-go model, it enables hosting of pervasive applications from consumer, scientific, and business domains [1]. However, data centers hosting cloud

X. Cai · X. Zhang (✉)
School of Information Science and Engineering, Yunnan University, Kunming, China
e-mail: xjzhang@ynu.edu.cn

X. Cai
e-mail: 86828122@qq.com

X. Cai
College of Basic Science and Information Engineering,
Yunnan Agricultural University, Kunming, China

S. Patnaik and X. Li (eds.), *Proceedings of International Conference on Computer Science and Information Technology*, Advances in Intelligent Systems and Computing 255, DOI: 10.1007/978-81-322-1759-6_55,

applications consume huge amounts of electrical energy, contributing to high operational costs and carbon footprints to the environment [2]. Therefore, we need Green Cloud computing solutions to achieve not only the efficient processing and utilization of a computing infrastructure, but also to minimize energy consumptions cost.

Currently, the energy usage of cloud computing systems is as an important consideration. Various job/tasks allocation methods have been investigated in a cloud, focusing on reducing the data centers' energy consumption, especially for making the data center more energy efficient as a necessary task [3], without paying much attention to the QoS expectations. In [4], power management has been applied at the data center level; in this work, the main technique applied to minimize power consumption is concentrating the workload to the minimum of physical nodes and switching idle nodes off. Chase et al. [4] have considered the problem of energy-efficient management of homogeneous resources in Internet hosting centers. Beloglazov et al. [5] proposed energy-aware allocation heuristics providing data center resources to client applications in a way that improves energy efficiency of the data center, while delivering the negotiated quality of service. Quan et al. [6] proposed a way of saving energy for traditional data centers. The author focuses on the CPU usage rate to model the energy consumption of a server. Yao at [7] discussed the operating system can reduce the energy consumption by scheduling jobs appropriately. Goudarzi et al. [8] accounted for the energy cost associated with a client's VM running in a data center; a solution is proposed for SLA-aware resource allocation to minimize the total operational cost in the cloud computing system.

The theoretical study of speed scaling policies to manage energy was initiated in a seminal paper by Yao et al. [7], and we adopt their setting. The work in this paper proposed a mechanism considering both the energy consumption cost and the important performance indicators like system response time of QoS, focusing on trying to use smallest computing resource to process maximum number of tasks. In contrast to the discussed studies, we proposed a simple model of energy efficiency–QoS (E-Q) in which each job is to be executed between its arrival time and deadline by a variable speed processor with the SLAs' requirements of response delay. A precise definition of the model is given in Sect. 2. Moreover, we provide a more formal analysis of the minimum-energy scheduling problem. In Sect. 3, we give an algorithm that computes a maximum eq value for reducing energy consumption while meeting QoS requirements. In Sect. 4, we present the experiment and analysis of the resource allocation algorithm based on the proposed E-Q model, which enable the physical nodes use the minimum energy consumption while meeting the SLAs response delay requirements. Finally, we close with a discussion of some simulation results and open problems.

2 An Energy Efficiency Model Based on QoS

The theoretical study of speed scaling policies to manage energy was initiated in a seminal paper by Yao et al. [7], and we adopt their setting. Moreover, based on the quantitative QoS analysis, we define the eq. value as the QoS level of the physical node with 1 unit energy efficiency, and we propose an energy efficiency–QoS (E-Q) model to describe the data center's energy consumption and QoS level.

2.1 Energy Efficiency Cost Analyses

In [10], computer energy efficiency is defined as the amount of calculation of the CPU L(T) with 1 GHz frequency E(T) per second and may be expressed as Formula (1):

$$\eta(T) = \frac{L(T)}{E(T)}, \quad E(T) \neq 0 \tag{1}$$

In this paper, we define energy efficiency as the energy consumption cost per unit time, which describes the usage efficiency of the physical node more intuitively. We define the energy consumption cost based on the large data center cost function of Hamilton [9] as Formula (2).

$$c_i = (p \times E_{\text{server}} \times \text{PUE})(1 + q_i) \tag{2}$$

where p is the price of the unit electric energy. $E_{\text{server}} = p_{\text{idle}} + u_{\text{cpu}} + u_{\text{mem}} + u_{\text{disk}}$ is the energy consumption of the computing nodes provided by the cloud server. q_i is the QoS level of the ith physical node. Power usage effectiveness (PUE) denotes the energy efficiency of the data centers. In our work, specific to the physical node, the PUE approximately equals 1. Thus, the energy consumption cost of the ith physical node is presented in Formula (3).

$$c_i = (p \times E_{\text{node}})(1 + q_i) \tag{3}$$

The executed time of the jth job by the ith physical node is Δt_{ij}; based on the above analysis, we define the energy efficiency of the ith node which can be presented with the following Formula (4).

$$\eta(i) = \frac{c_i}{\sum_j \Delta t_{ij}}, \quad \Delta t \neq 0 \tag{4}$$

2.2 Energy Efficiency Cost Metrics Based on QoS Indicators

In this work, we use several QoS performance indicator metrics to evaluate the E-Q (energy efficiency–QoS) performance for cloud computing. In order to compare the energy efficiency of the system, we focus on the online services that are sensitive to response time and system service rate. Assume that the users' request rate under the cloud computing environment is a λ Poisson process, and each computing node can get the predicted average request rate as follows:

$$\lambda_i = \sum_{j=1}^{n} \lambda_{ij} \quad (5)$$

where λ_i is the predicted average request rate of the ith computing node. λ_{ij} is the required rate of the jth job on the ith computing node.$\overline{\mathrm{RT}}$ can be presented as the basic response time with following Formula (6).

$$\overline{\mathrm{RT}} = \sum_{i=1}^{n} \mathrm{rt}_i/n \quad (6)$$

where rt_i is the response time of the ith computing node, which is the interval time from the SLA agreement was signed to the users can use the computing resources. n is the number of the computing nodes with a cloud service provider.

We refer to $[a_j, b_j]$ as the interval of job j. A schedule is a pair $S = (s, \mathrm{job})$ of functions defined over $[t_0, t_1]$. $s(t) \geq 0$ is the processor speed at time t. job(t) defines the job being executed at time t (or idle if $s(t) = 0$). We require that $s(t)$ and job(t) are piecewise constants with finitely many discontinuities. A feasible schedule for instance J is a schedule δ that satisfies

$$\int_{a_i}^{b_i} s(t)\delta(\mathrm{job}(t), j)\mathrm{d}t = R_j \quad (7)$$

For all $j \in J$ (where $\delta(x, y)$ is 1 if $x = y$ and 0 otherwise). In other words, δ must give each job j the required number of cycles between its arrival time and deadline (with perhaps intermittent execution). Associated with each job j, its average rate requirement or density

$$d_j = \frac{R_j}{b_j - a_j} \quad (8)$$

We define a corresponding step function $d_j(t) = d_j$ for $t \in [a_j, b_j]$ and $d_j(t) = 0$ elsewhere. At any time t, the Average Rate Heuristic (AVR) sets the processor speed at

$$s(t) = \sum_{j} d_j(t) \quad (9)$$

We assume that the power energy consumed by the ith computing node per unit time is a convex function of the processor speed; the total energy consumed by a schedule δ is

$$E_i(\delta) = \int_{t_0}^{t_1} p(s(t))\,\mathrm{d}t \tag{10}$$

According to [8],

$$P(s) = s^2 \tag{11}$$

The goal of the scheduling problem is to find, for any given problem instance, a feasible schedule that minimizes $E_i(\delta)$.

We can use the earliest deadline policy to choose the available jobs. It is easy to see that the strategy yields a feasible schedule. q_i is the QoS of the ith node based on the response time. q_i can be presented with following Formula (12).

$$q_i = \overline{\mathrm{RT}}/\mathrm{rt}_i \tag{12}$$

In order to achieve the system utility specified by the SLA, the cloud service provider can adjust the service rate μ_i dynamically with the M/M/1 queue model.

$$\overline{\mathrm{RT}} = \sum_{i=1}^{n} \mathrm{rt}_i/n \tag{13}$$

$$q_i = \overline{\mathrm{RT}}/\mathrm{rt}_i \tag{14}$$

where μ_i is the service rate of the computing node, which is the service numbers of concurrent execution. We define the response time as an important performance index and assume T_D is the upper limit specified by the SLA agreements. The constraints on the response time of the rt_i may be expressed as follows:

$$\mathrm{rt}_i \leq T_D \tag{15}$$

Therefore, our goal is to maximize the global utility value. Besides the response time, we can consider the other QoS performance indexes like system availability, throughput, and system security to make the quantitative analysis, and we can get the q_i, as formalized in (16).

$$q_i = \beta_1\overline{\mathrm{RT}} + \beta_2\overline{\mathrm{SA}} + \beta_3\overline{\mathrm{TP}} + \beta_4\overline{\mathrm{SS}} + \ldots \tag{16}$$

where the parameter β is the different weight of the QoS indicators according to SLA agreements specifically specified.

2.3 An Energy Efficiency QoS Model

Using the model provided in the Sect. 2.2, the energy consumption cost of the ith computing node can be expressed as follows:

$$c_i = (p \times E_{\mathrm{node}})(1 + q_i) = (p \times \int_{t0}^{t1} P(s(t))dt)(1 + q_i) \tag{17}$$

Based on the energy efficiency Formula (5), we can get the Formula (18).

$$\eta(i) = \frac{c_i}{\sum_j \Delta t_{ij}} = \frac{\left(p \times \int_{t0}^{t1} P(s(t))dt\right)(1 + q_i)}{b_i - a_i} \tag{18}$$

We can define an integral of the power consumption function as shown in (19).

$$\begin{aligned} \mathrm{eq}_i &= \mathrm{QoS}/\mathrm{Energy} - \mathrm{Efficiency} \\ &= q_i/\eta(i) = q_i \times \sum_j \Delta t_{ij}/(p \times E_{\mathrm{node}})(1 + q_i) \end{aligned} \tag{19}$$

The formulation below describes the utility function for the ith cloud service provider:

$$U_i = \sum_{i=1}^{n} \mathrm{eq}_i \tag{20}$$

The goal of the scheduling problem is to find, for any physical node under the cloud computing environment, a feasible schedule that maximizes eq.

3 Deployment of Job/Tasks Algorithm for Cloud Computing Based on the E-Q Model

Currently, resource allocation in cloud computing aims to provide high performance while meeting quality-of-service (QoS) requirements specified by user via service level agreements (SLAs), without focusing on energy consumption cost. To solve it, we apply an algorithm of saving energy for traditional data centers, considering all the above features. In this section, the optimal heuristic for the foresaid problem is presented. We can compute the different energy consumption cost and eq value with different response time and processor speed. Based on the output, we can adjust the job/task deployment. Details of the QoS-based energy efficiency and eq maximization algorithm for short are presented below.

Program E-Q-driven-job/tasks-assignment (deployment of jobs, s,eq_{ij}, U_i,R_j).
{Inputs: job List,μ_i,λ_i,p,Δt,}
Set max EQ_{ij}=0 $U_i = 0$.

for each physical node i do
for each job j on the ith node do

```
if rt≤TD then compute eqij compute Ui
else adjust μi back to 4 compare the value of rt and TD
if eqij≥max EQ max EQij←eqij and Ui←eqij+Ui
else back to adjustss, δ (job (t), j)
end for
end for
store tuple (μi, s,eqij, δ(job(t), j)) in a list
```

The key issue of the improved job/tasks deployment algorithm is to find out the global optimum of the utility function by adjusting the service rate and the CPU speed of the physical node continuously.

4 Experiment Results and Analysis

As the targeted system is a generic cloud computing environment, it is essential to evaluate it on a large-scale virtualized data center infrastructure. However, it is extremely difficult to conduct repeatable large-scale experiments on a real infrastructure. Therefore, to ensure the repeatability of experiments, parameters are chosen based on Table 1 and the number of server types is set as Table 1.For each server type, an arbitrary number of job/tasks are placed in data centers. Processers in server types are selected from a set of Intel processors (e.g., Atom, *i*5, *i*7, and Xeon) with different cores, cache, power consumptions, and working frequencies. Energy cost is assumed to be 15 cents per KW h at all times. Request arrival rates of the client are chosen uniformly between 0.1 and 1 request per second. Model 1 denotes the traditional algorithm based on FIFO, and model 2 denotes the improved algorithm based on E-Q model. From the result in tables 2 and 3, we can clearly see that E-Q-driven algorithm finds out better solution with power consumption and the QoS level.

The experiment is done with two main goals. The first goal is studying the saving energy mechanism in different resource configuration scenarios with different service rate and different CPU speed of the physical nodes in the cloud computing. The second goal is checking the efficiency of the proposed algorithm with the traditional method. From the result in tables 2 and 3, we can see clearly that E-Q algorithm finds out better solution with much more eq value and utility value than the traditional FIFO does in any situation.

Table 1 Server configuration

Server type (i)	Total number	Nr.cores (m_i)	f_i (GHz)
1	1	1	2.2
2	1	2	2.0

Assume that execute time of the two jobs is 30 s and 20 s

Table 2 Simulation results of 2 job/tasks

Model	Node (i)	μ exchange	exchange	Job (j)	eq_{ij}	U_i
1	1	0.8	2.2	1	13.11	–
1	2	0.9	2.0	2	9.37	22.49
2	1	0.98	2.4	1	19.05	–
2	2	0.95	2.2	2	7.59	26.65

Assume that execute time of the five jobs is 30 s, 20 s,10 s,20 s, and 40 s

Table 3 Simulation results of 5 job/tasks

Model	Node (i)	μ exchange	exchange	Job (j)	eq_{ij}	U
1	1	0.8	2.2	1	13.11	–
1	1	0.8	2.2	2	44.27	–
1	1	0.8	2.2	3	354.19	–
1	2	0.9	2.0	4	75	–
1	2	0.9	2.0	5	9.37	495.95
2	1	0.9	2.2	4	52.88	–
2	1	0.95	1.9	5	9.84	–
2	2	0.85	2.0	1	45.48	–
2	2	0.90	2.2	2	52.88	–
2	2	0.88	1.9	3	538.05	699.13

5 Conclusion

In this paper, we have introduced a simplified model of variable speed processors with different response delay and analyze how the scheduling of jobs affects the overall power consumption cost and the QoS level. Moreover, we proposed the resource allocation algorithm based on E-Q-driven model presented in Sect. 3. Based on the results of this paper, it can be seen that considering the QoS indicators with effective job/tasks deployment can help to minimize the energy efficiency cost and maximize the QoS level.

The research work is planned to be followed by the development of the other QoS performance indicators like system availability, throughput, and security to be quantitative analysis; the E-Q model have yet to be investigated more fully.

Acknowledgments The work is supported by the National Natural Science Foundation of China (Grant No.: 61170222).

References

1. Beloglazov, A., Abawajy, J., Buyya, R.: Energy-aware resource allocation heuristics for efficient management of data centers for cloud computing. Future Gener. Comput. Syst. **28**(5), 755–768 (2012)
2. Srikantaiah, S, Kansal, A, Zhao, F.: Energy aware consolidation for cloud computing. In: Proceedings of the 2008 Conference on Power Aware Computing And Systems. USENIX Association, p. 10 (2008)
3. Pinheiro, E, Bianchini, R, Carrera, E V, et al. Load balancing and unbalancing for power and performance in cluster-based systems. In: Workshop on Compilers and Operating Systems for Low Power, vol. 180, pp. 182–195 (2001)
4. Chase, J.S., Anderson, D.C., Thakar, P.N., et al.: Managing energy and server resources in hosting centers. In: ACM SIGOPS Operating Systems Review, vol. 35(5), pp. 103–116 (2001)
5. Beloglazov, A, Buyya, R.: Energy efficient resource management in virtualized cloud data centers. In: Proceedings of the 2010 10th IEEE/ACM International Conference on Cluster, Cloud and Grid Computing. IEEE Computer Society, pp. 826–831 (2010)
6. Quan, D.M., Mezza, F., Sannenli, D., et al.: T-Alloc: a practical energy efficient resource allocation algorithm for traditional data centers. Future Gener. Comput. Syst. **28**(5), 791–800 (2012)
7. Yao, F., Demers, A., Shenker, S.: A scheduling model for reduced CPU energy. Foundations of Computer Science, 1995. In: Proceedings. 36th Annual Symposium on IEEE, pp. 374–382 (1995)
8. Goudarzi, H., Ghasemazar, M., Pedram, M.: SLA-based optimization of power and migration cost in cloud computing. Cluster, Cloud and Grid Computing (CCGrid), 2012 12th IEEE/ ACM International Symposium on. IEEE, pp. 172–179 (2012)
9. Cost of Power in Large-Scale Data Centers, Available at: http://perspectives.mvdirona.com/ 2008/11/28/CostOfPowerInLargeScaleDataCenters.aspx
10. Song, J., Li, T.T., et al.: Energy-Efficiency model and measuring approach for cloud computing. J Software **23**(2), 200–214 (2012)

Revenue-Sharing Contract in a Cloud Computing Service Supply Chain Under Asymmetric Information

Lingyun Wei and Shuo Qi

Abstract We aim to coordinate a three-staged cloud computing service supply chain under asymmetric information with revenue-sharing contract. The discussed supply chain consists of one application service provider (ASP), one application platform provider (APP), and one application infrastructure provider (AIP). AIP packages the computer capacity as services and supplies them to APP; APP integrates the application platform as services and supplies them to ASP; ASP sells the value-added application services to the market. ASP service system is modeled as an M/M/1 queueing system while the effects of congestion are considered. Moreover, the market is described as a price-sensitive random demand which follows the Poisson distribution. We discuss four situations: namely centralized control, decentralized control, revenue-sharing contract under symmetric information, and revenue-sharing contract under asymmetric information. When using revenue-sharing contract under asymmetric information, we find that the discussed supply chain can attain Pareto improvement.

Keywords Cloud computing · Supply chain · Revenue-sharing contract · Asymmetric information

1 Introduction

Cloud computing has been a hot topic in the recent years; however, most of the researchers focus on it from the technical aspects, such as cloud security or cloud storage, few of them from the operation and management aspects. Nowadays, a

L. Wei (✉) · S. Qi (✉)
School of Automation, Beijing University of Posts and Telecommunications, Beijing 100876, People's Republic of China
e-mail: weilingyun2010@sina.com

S. Qi
e-mail: qishuo05@sina.com

S. Patnaik and X. Li (eds.), *Proceedings of International Conference on Computer Science and Information Technology*, Advances in Intelligent Systems and Computing 255, DOI: 10.1007/978-81-322-1759-6_56,

company cannot succeed if he does not cooperate with upstream and downstream companies. The competition is no longer between different enterprises but between different supply chains. Hence, how to maximize the profits of the cloud computing service supply chain will have a significant impact on the cloud computing industry.

A cloud computing service supply chain generally includes three kinds of actors: application infrastructure providers (AIPs), application platform providers (APPs) and application service providers (ASPs). AIPs deliver the infrastructure including hardware (server, storage, and network) and associated software (operating systems virtualization technology, file system) as a service [1]. APPs deliver the platform as a service and offer the underlying services for application design, development, testing, deployment, and hosting over the application platform, which is deployed onto the cloud infrastructure [2]. ASPs deliver the software as a service and provide access to a software application over a network [2]. AIPs, such as Amazon EC2 and Windows Azure, make price based on the user's actual service time. APPs and ASPs, for example, Google App Engine and NetSuite ERP, charge a fixed fee per user by the month or year. The fixed fees charged are depended on the used computing capacity. Obviously, all of the actors in the cloud computing service supply chain hope the realization of an efficient supply chain; however, they are primarily concerned with optimizing their own objectives which often results in poor supply chain performance [3]. Therefore, supply chain excellence requires the coordination of disparate incentives [4]. Haluk and Hsing [5] study coordination strategies in a SaaS supply chain consisted of one application service provider and one application infrastructure provider. They propose that it is possible to create the right incentives so that the economically efficient outcome is also the Nash equilibrium. However, their model only discusses two supply chain members and does not take the supply chain contract into account [5].

Supply chain contract can be regarded as an effective method to achieve the coordination of a supply chain or to ensure the performance improvement [4]. Ilaria [6] aims to coordinate a supply chain with revenue-sharing contract which ensures the system efficiency to be achieved as well as the profits of all the supply chain members are improved. Palsule [7] developed a game theoretic model for revenue-sharing contracts in which the supply chain revenue is shared among the players. Unfortunately, those studies are based on the symmetric information and they do not take the asymmetric information into account. It is noteworthy that in many situations, some information is only privy to one party and the other party makes decisions with limited available information [8]. Therefore, the research on supply chain contracts under asymmetric information is more practical. Sinha [9] studies the coordination problem in a two-stage supply chain under uncertain cost information of the retailer. He designs a quantity discount contract to make the profit of the supply chain under asymmetric cost information as closely as the one under symmetric information. Kayis [10] studies the asymmetry information about suppliers' production costs and exploits the quantity discount and price-only contract to coordinate the supply chain. However, those researches mainly focus on the physical supply chain coordination under asymmetric information. Few efforts have been done on coordinating cloud computing service supply chain under asymmetric information.

In our study, we aim to coordinate a three-staged cloud computing service supply chain under asymmetric information with revenue-sharing contract. Four situations, including centralized control, decentralized control, revenue-sharing contract under symmetric information, revenue-sharing contract and under asymmetric information, are considered. Compared with previous studies, our research mainly has three improvements. (1) We use supply chain contract to coordinate a cloud computing service supply chain while asymmetry cost information is considered. (2) We discuss a three-staged cloud computing service supply chain, which consists of one AIP, one APP and one ASP. (3) The market demand is assumed to be a price-sensitive random demand which follows the Poisson distribution, and the ASP service system is modeled as an M/M/1 queueing system while the effects of congestion are considered.

In the following section, the basic model of the cloud computing service supply chain is introduced. In Sect. 3, we discuss four situations in the cloud computing service supply chain. Finally, a conclusion is provided.

2 The Model of Cloud Computing Service Supply Chain

We consider a three-staged cloud computing service supply chain which consists of risk neutral AIP, APP, and ASP. The goal of each member is to realize the profit maximization for itself. In order to provide application services for its customers, ASP acquires a computer capacity μ_S, which can serve μ_S users per unit time, between APP and APP charges w_S per unit of the capacity. Meanwhile, APP acquires a computer capacity μ_P between AIP and can serve μ_P users per unit time, and AIP charges w_P per unit time. Obviously, μ_S and μ_P could be different. However, under decentralized control, once ASP sets μ_S, APP will set $\mu_P = \mu_S = \mu$ as well to fully satisfy the market demand and maximize its profit. Moreover, ASP sells the value-added service at price p per unit customer to the market and faces a price-sensitive random market demand which obeys the Poisson distribution. λ is the customer's arrival rate which has a negative exponential relationship with price p. Assuming $V(\lambda)$ represents the total value of the whole supply chain associated with λ customers per unit of time, the marginal value of customer is $V'(\lambda) = \frac{k}{\sqrt{\lambda}}$ [11] where k is an arbitrary constant. We take the ASP service system as an M/M/1 queueing system. It means the market demand obeys the Poisson distribution; the per unit time computer capacity μ obeys the exponential distribution. There is only one service system. Let $T(\lambda, \mu)$ denotes the expected time each user spends in this system, including the actual service time and the waiting time. When this system is stable, $T(\lambda, \mu) = \frac{1}{\mu - \lambda}$ where $\mu > \lambda > 0$. Because of the queueing delay, customers incur a congestion cost which equals the delay cost parameter per transaction per unit of time v multiplied $T(\lambda, \mu)$ [11]. The marginal cost of customer is the sum of price p and the expected delay cost per customer. The market equilibrium is achieved when the marginal value equals the marginal cost per customer, namely $\frac{k}{\sqrt{\lambda}} = p + vT(\lambda, \mu)$ [11]. Then, one has $p = \frac{k}{\sqrt{\lambda}} - \frac{v}{\mu - \lambda}$. The task processing requirements per customer is a random r with mean $\bar{r}$ [12]. The system processing speed is a decision variable, m. Then, one has $\mu = \frac{m}{\bar{r}}$. To satisfy this system processing speed, AIP has a unit infrastructure service cost c_I; APP has a unit platform service cost c_P; ASP has a unit application service cost c_S. c is the unit service cost of the whole supply chain and $c = c_S + c_P + c_I$. The diseconomy of scale cost parameter related with AIP's management of service and infrastructure is described by e. This diseconomy of scale results from increasing costs of managing capacity and user access [13, 14] and rising complexity of the business model [15].

3 Four Situations of Cloud Computing Service Supply Chain

3.1 Centralized Control

In this situation, a single firm plays an integrated role of AIP, APP, and ASP. The supply chain can be coordinated and obtained the maximal profit. The per unit time expected profit of the whole supply chain $\mathrm{SP}_1(p, \lambda, m)$ can be described in Eq. 1.

$$\mathrm{SP}_1(p, \lambda, m) = p\lambda - cm - em^2 \tag{1}$$

Put $p = \frac{k}{\sqrt{\lambda}} - \frac{v}{\mu - \lambda}$ and $\mu = \frac{m}{\bar{r}}$ into Eq. 1, we can get the profit function $\mathrm{SP}_1(\lambda, m)$ in Eq. 2 which is only related with parameters λ and m.

$$\mathrm{SP}_1(\lambda, m) = k\sqrt{\lambda} - \frac{\lambda v \bar{r}}{m - \bar{r}\lambda} - cm - em^2 \tag{2}$$

To find the optimal per unit time supply chain's expected profit, first-order conditions require $\frac{\mathrm{dSP}_1(\lambda,m)}{\mathrm{d}\lambda} = 0$ and $\frac{\mathrm{dSP}_1(\lambda,m)}{\mathrm{d}m} = 0$ as follows:

$$\frac{\mathrm{dSP}_1(\lambda, m)}{\mathrm{d}\lambda} = \frac{k}{2\sqrt{\lambda}} - \frac{mv\bar{r}}{(m - \bar{r}\lambda)^2} = 0 \tag{3}$$

$$\frac{\mathrm{dSP}_1(\lambda, m)}{\mathrm{d}m} = \frac{\lambda v \bar{r}}{(m - \bar{r}\lambda)^2} - c - 2em = 0 \tag{4}$$

$\mathrm{SP}_1(\lambda, m)$ in Eq. 2 exists unique optimal solution if and only if $\frac{\mathrm{d}^2\mathrm{SP}_1(\lambda,m)}{\mathrm{d}\lambda^2} < 0, \frac{\mathrm{d}^2\mathrm{SP}_1(\lambda,m)}{\mathrm{d}m^2} < 0, \Delta_1 = \frac{\mathrm{d}^2\mathrm{SP}_1(\lambda,m)}{\mathrm{d}\lambda^2} * \frac{\mathrm{d}^2\mathrm{SP}_1(\lambda,m)}{\mathrm{d}m^2} - \frac{\mathrm{d}^2\mathrm{SP}_1(\lambda,m)}{\mathrm{d}\lambda\mathrm{d}m} * \frac{\mathrm{d}^2\mathrm{SP}_1(\lambda,m)}{\mathrm{d}\lambda\mathrm{d}m} > 0$. After some operations, we find that the optimal solution (λ^*, m^*), which can be obtained from Eqs. 3 and 4, exists if $\frac{2ek}{4\sqrt{\lambda^3}} + \frac{4emv\bar{r}^2}{(m-\bar{r}\lambda)^3} + \frac{2\lambda v\bar{r}k}{4\sqrt{\lambda^3}(m-\bar{r}\lambda)^3} - \frac{v^2m^2\bar{r}^2}{(m-\bar{r}\lambda)^6} - \frac{v^2\lambda^2\bar{r}^4}{(m-\bar{r}\lambda)^6} + \frac{2\lambda mv^2\bar{r}^3}{(m-\bar{r}\lambda)^6} > 0$.

3.2 Decentralized Control

In this situation, AIP, APP, and ASP independently make decisions to pursue optimal profits for themselves. ASP determines the optimal price p and the per unit time computer capacity μ, which purchase from APP. APP charges w_s per unit of the capacity. APP orders the equivalent computer capacity μ from AIP, who charges w_P per unit of actual service time. In the M/M/1 queueing system, the expected actual service time per customer is $\frac{1}{\mu}$ and the expected time each customer spends is $\frac{1}{\mu-\lambda}$. Therefore, in one unit of time, the actual service time equals $\frac{\mu-\lambda}{\mu}$, which does not include the waiting time. Let $\mathrm{HP}_2(\lambda,m), \mathrm{PP}_2(\lambda,m)$ and $\mathrm{AP}_2(p,\lambda,m)$ be the per unit time expected profits of AIP, APP, and ASP, respectively, and they can be described in Eqs. 5–7:

$$\mathrm{HP}_2(\lambda,m) = \frac{w_P(\mu-\lambda)}{\mu} - (c_I m + em^2) = \frac{w_P(m-\bar{r}\lambda)}{m} - \left(c_I m + em^2\right) \tag{5}$$

$$\mathrm{PP}_2(\lambda,m) = w_S m - \frac{w_P(\mu-\lambda)}{\mu} - c_P m = w_S m - \frac{w_P(m-\bar{r}\lambda)}{m} - c_P m \tag{6}$$

$$\mathrm{AP}_2(p,\lambda,m) = p\lambda - w_S m - c_s m \tag{7}$$

Put $p = \frac{k}{\sqrt{\lambda}} - \frac{v}{\mu-\lambda}$ and $\mu = \frac{m}{\bar{r}}$ into Eq. 7, we can get the profit function $\mathrm{AP}_2(\lambda,m)$ in Eq. 8, which is only related with parameters λ and m.

$$\mathrm{AP}_2(\lambda,m) = k\sqrt{\lambda} - \frac{\lambda v\bar{r}}{m-\bar{r}\lambda} - w_S m - c_s m \tag{8}$$

To find the optimal per unit time ASP's expected profit, first-order conditions require $\frac{\mathrm{dAP}_2(\lambda,m)}{\mathrm{d}\lambda} = 0$ and $\frac{\mathrm{dAP}_2(\lambda,m)}{\mathrm{d}\lambda} = 0$ as follows:

$$\frac{\mathrm{dAP}_2(\lambda,m)}{\mathrm{d}\lambda} = \frac{k}{2\sqrt{\lambda}} - \frac{mv\bar{r}}{(m-\bar{r}\lambda)^2} = 0 \tag{9}$$

$$\frac{\mathrm{dAP}_2(\lambda,m)}{\mathrm{d}m} = \frac{\lambda v\bar{r}}{(m-\bar{r}\lambda)^2} - w_S - c_s = 0 \tag{10}$$

$\mathrm{AP}_2(\lambda,m)$ in Eq. 8 exists unique optimal solution if and only if $\frac{\mathrm{d}^2\mathrm{AP}_2(\lambda,m)}{\mathrm{d}\lambda^2}<0, \frac{\mathrm{d}^2\mathrm{AP}_2(\lambda,m)}{\mathrm{d}m^2}<0, \Delta_2 = \frac{\mathrm{d}^2\mathrm{AP}_2(\lambda,m)}{\mathrm{d}\lambda^2} * \frac{\mathrm{d}^2\mathrm{AP}_2(\lambda,m)}{\mathrm{d}m^2} - \frac{\mathrm{d}^2\mathrm{AP}_2(\lambda,m)}{\mathrm{d}\lambda\mathrm{d}m} * \frac{\mathrm{d}^2\mathrm{AP}_2(\lambda,m)}{\mathrm{d}\lambda\mathrm{d}m} > 0$. After some operations, we can find that the optimal solution $\left(\lambda_2^*, m_2^*\right)$, which can be obtained from Eqs. 9 and 10, exists if $\frac{k(m-\bar{r}\lambda)^3}{2\sqrt{\lambda}} - v\bar{r}m^2 - v\lambda^2\bar{r}^3 + 2\lambda mv\bar{r}^2 > 0$.

3.3 Revenue-Sharing Contract Under Symmetric Information

In this situation, APP announces a lower price w_S to ASP and share a part of ASP's sale income. Let δ_S be the fraction of ASP's income that ASP keeps. Then, $(1-\delta_S)$ is the fraction of ASP's income that APP keeps. Meanwhile, AIP announces a lower price w_P to APP and share a part of APP's income. This income includes the shared sale income and the rent income from ASP. Let δ_p be the fraction of APP's income that APP keeps. Then, $(1-\delta_p)$ is the fraction of APP's income that AIP keeps. Let $\mathrm{HP}_3(p,\lambda,m), \mathrm{PP}_3(p,\lambda,m), \mathrm{AP}_3(p,\lambda,m)$ be the per unit time expected profits of AIP, APP, and ASP, respectively, and they can be described in Eqs. 11–13.

$$\mathrm{HP}_3(p,\lambda,m) = (1-\delta_p)[(1-\delta_S)p\lambda + w_S m] + \frac{w_p(m-\bar{r}\lambda)}{m} - c_I m - em^2 \quad (11)$$

$$\mathrm{PP}_3(p,\lambda,m) = \delta_p[(1-\delta_S)p\lambda + w_S m] - \frac{w_p(m-\bar{r}\lambda)}{m} - c_P m \quad (12)$$

$$\mathrm{AP}_3(p,\lambda,m) = \delta_S p\lambda - w_S m - c_s m \quad (13)$$

Put $p = \frac{k}{\sqrt{\lambda}} - \frac{v}{\mu-\lambda}$ and $\mu = \frac{m}{\bar{r}}$ into Eq. 13, we can get the profit function $\mathrm{AP}_3(\lambda,m)$ in Eq. 14, which is only related with parameters λ and m.

$$\mathrm{AP}_3(\lambda,m) = \delta_S k\sqrt{\lambda} - \frac{\delta_S\lambda v\bar{r}}{m-\bar{r}\lambda} - w_S m - c_s m \quad (14)$$

To find the optimal ASP's expected profit, first-order conditions require $\frac{\mathrm{dAP}_3(\lambda,m)}{\mathrm{d}\lambda} = 0$ and $\frac{\mathrm{dAP}_3(\lambda,m)}{\mathrm{d}m} = 0$ as follows:

$$\frac{\mathrm{dAP}_3(\lambda,m)}{\mathrm{d}\lambda} = \frac{\delta_S k}{2\sqrt{\lambda}} - \frac{\delta_S m v\bar{r}}{(m-\bar{r}\lambda)^2} = 0 \quad (15)$$

$$\frac{\mathrm{dAP}_3(\lambda,m)}{\mathrm{d}m} = \frac{\delta_S\lambda v\bar{r}}{(m-\bar{r}\lambda)^2} - w_S - c_s = 0 \quad (16)$$

$\mathrm{AP}_3(\lambda,m)$ in Eq. 14 exists unique optimal solution if and only if $\frac{\mathrm{d}^2\mathrm{AP}_3(\lambda,m)}{\mathrm{d}\lambda^2}<0, \frac{\mathrm{d}^2\mathrm{AP}_3(\lambda,m)}{\mathrm{d}m^2}<0, \Delta_3 = \frac{\mathrm{d}^2\mathrm{AP}_3(\lambda,m)}{\mathrm{d}\lambda^2} * \frac{\mathrm{d}^2\mathrm{AP}_3(\lambda,m)}{\mathrm{d}m^2} - \frac{\mathrm{d}^2\mathrm{AP}_3(\lambda,m)}{\mathrm{d}\lambda\mathrm{d}m} * \frac{\mathrm{d}^2\mathrm{AP}_3(\lambda,m)}{\mathrm{d}\lambda\mathrm{d}m} > 0$. After some operations, we can find that the optimal solution $(\lambda_3^*,\ m_3^*)$, which can be obtained from Eqs. 15 and 16, exists if $\frac{k(m-\bar{r}\lambda)^3}{2\sqrt{\lambda}} - v\bar{r}m^2 - v\lambda^2\bar{r}^3 + 2\lambda m v\bar{r}^2 > 0$. If $\lambda_3^* = \lambda^*, m_3^* = m_{3pp}^* = m^*$, namely $\frac{\mathrm{dAP}_3(\lambda,m)}{\mathrm{d}m} = \frac{\mathrm{dPP}_3(\lambda,m)}{\mathrm{d}m} = \frac{\mathrm{dSP}_1(\lambda,m)}{\mathrm{d}m} = 0$, $\frac{\mathrm{dAP}_3(\lambda,m)}{\mathrm{d}\lambda} = \frac{\mathrm{dSP}_1(\lambda,m)}{\mathrm{d}\lambda} = 0$, the revenue-sharing contract can coordinate the cloud computing service supply chain. m_{3pp}^* is APP's optimal solution of m and can maximize APP's profit in Eq. 12. $\frac{\mathrm{dPP}_3(w_S,m)}{\mathrm{d}m} = 0$ is described in Eq. 17. After some operations, we can realize that when $w_S = \delta_S(c + 2em) - c_s$, $w_p = \frac{m^2[\delta_p(1-\delta_S)(c+2em)+\delta_p w_S - c_P]}{\bar{r}\lambda}$, the cloud computing service supply chain is coordinated with revenue-sharing contract.

$$\frac{\mathrm{dPP}_3(\lambda, m)}{\mathrm{d}m} = \delta_p\left[\frac{(1-\delta_S)\lambda v\bar{r}}{(m-\bar{r}\lambda)^2} + w_S\right] - \frac{w_p\bar{r}\lambda}{m^2} - c_P = 0 \tag{17}$$

3.4 Revenue-Sharing Contract Under Asymmetric Information

In this situation, AIP cannot know the unit platform service cost c_P of APP, and APP cannot know the unit application service cost c_s of ASP. However, AIP has the prior probability distribution of c_P where $c_{p1} \le c_P \le c_{p2}$. APP has the prior probability distribution of c_s where $c_{s1} \le c_s \le c_{s2}$. APP designs a menu of contracts $(w_S(\widetilde{c_S}), \varphi_S(\widetilde{c_S}))$, which is related with $\widetilde{c_S}$, for ASP. $\widetilde{c_S}$ is the unit application service cost which ASP announces to APP. However, ASP does not need to report its true cost information and will report the one which can maximize its profit. So, APP should design a menu of contracts $(w_S(\widetilde{c_S}), \varphi_S(\widetilde{c_S}))$, which can make ASP obtain the optimal profit if it reports the true cost information. AIP designs a menu of contracts $\left(w_p\left(\widetilde{c_p}\right), \varphi_p\left(\widetilde{c_p}\right)\right)$, which is related with $\widetilde{c_p}$, for APP. $\widetilde{c_p}$ is the unit platform service cost which APP reports to AIP. However, APP does not need to report its true cost information and will report the one which can maximize its profit. So, AIP should design a menu of contracts $\left(w_p\left(\widetilde{c_p}\right), \varphi_p\left(\widetilde{c_p}\right)\right)$, which can make APP obtain the optimal profit if it reports the true cost information. Let $\mathrm{HP}_4(p, \lambda, m), \mathrm{PP}_4(p, \lambda, m), \mathrm{AP}_4(p, \lambda, m)$ be the per unit time expected profits of AIP, APP, and ASP, and they can be described as follows:

$$\begin{aligned}\mathrm{HP}_4(p, \lambda, m) = {} & \left(1 - \delta_p\left(\widetilde{c_p}\right)\right)\left[(1 - \delta_S(\widetilde{c_S}))p(\widetilde{c_S})\lambda(\widetilde{c_S}) + w_S(\widetilde{c_S})m(\widetilde{c_S})\right] \\ & + \frac{w_p\left(\widetilde{c_p}\right)\left(m\left(\widetilde{c_p}\right) - \bar{r}\lambda(\widetilde{c_S})\right)}{m\left(\widetilde{c_p}\right)} - c_I m\left(\widetilde{c_p}\right) - em\left(\widetilde{c_p}\right)^2\end{aligned} \tag{18}$$

$$\begin{aligned}\mathrm{PP}_4(p, \lambda, m) = {} & \delta_p\left(\widetilde{c_p}\right)\left[(1 - \delta_S(\widetilde{c_S}))p(\widetilde{c_S})\lambda(\widetilde{c_S}) + w_S(\widetilde{c_S})m(\widetilde{c_S})\right] \\ & - \frac{w_p\left(\widetilde{c_p}\right)\left(m\left(\widetilde{c_p}\right) - \bar{r}\lambda(\widetilde{c_S})\right)}{m\left(\widetilde{c_p}\right)} - c_P m\left(\widetilde{c_p}\right)\end{aligned} \tag{19}$$

$$\mathrm{AP}_4(p, \lambda, m) = \delta_S(\widetilde{c_S})p(\widetilde{c_S})\lambda(\widetilde{c_S}) - w_S(\widetilde{c_S})m(\widetilde{c_S}) - c_S m(\widetilde{c_S}) \tag{20}$$

If we denote $Z_S(c_S), Z_S(\widetilde{c_S}), Z_P(c_P)$ and $Z_P\left(\widetilde{c_p}\right)$ as Eqs. 21–24, the per unit time profit of APP and ASP can be described in Eqs. 25–26.

$$Z_S(c_S) = \delta_S(c_S)p(c_S)\lambda(c_S) - w(c_S)m(c_S) - c_S m(c_S) \tag{21}$$

$$Z_S(\widetilde{c_S}) = \delta_S(\widetilde{c_S})p(\widetilde{c_S})\lambda(\widetilde{c_S}) - w(\widetilde{c_S})m(\widetilde{c_S}) - \widetilde{c_S}m(\widetilde{c_S}) \tag{22}$$

$$\begin{aligned} Z_P(c_p) = \delta_p(c_p)[(1 - \delta_S(\widetilde{c_S}))p(\widetilde{c_S})\lambda(\widetilde{c_S}) + w_S(\widetilde{c_S})m(\widetilde{c_S})] \\ - \frac{w_p(c_p)(\mu(c_p) - \lambda(\widetilde{c_S}))}{\mu(c_p)} - c_P m(c_p) \end{aligned} \tag{23}$$

$$\begin{aligned} Z_P(\widetilde{c_p}) = \delta_p(\widetilde{c_p})[(1 - \delta_S(\widetilde{c_S}))p(\widetilde{c_S})\lambda(\widetilde{c_S}) + w_S(\widetilde{c_S})m(\widetilde{c_S})] \\ - \frac{w_p(\widetilde{c_p})(\mu(\widetilde{c_p}) - \lambda(\widetilde{c_S}))}{\mu(\widetilde{c_p})} - \widetilde{c_p}m(\widetilde{c_p}) \end{aligned} \tag{24}$$

$$PP_4(\widetilde{c_p}, c_P) = Z_P(\widetilde{c_p}) + (\widetilde{c_p} - c_p)m(\widetilde{c_p}) \tag{25}$$

$$AP_4(\widetilde{c_S}, c_S) = Z_S(\widetilde{c_S}) + (\widetilde{c_S} - c_S)m(\widetilde{c_S}) \tag{26}$$

As we mention above, AIP should design a menu of contracts $(w_p(\widetilde{c_P}), \varphi_p(\widetilde{c_P}))$, which can make APP obtain the optimal profit if APP reports its true cost information. APP should design a menu of contracts $(w_p(\widetilde{c_S}), \varphi_p(\widetilde{c_S}))$, which can make ASP obtain the optimal profit if ASP reports its true cost information. First-order conditions require $\frac{dAP_4(\widetilde{c_S}, c_S)}{d\widetilde{c_S}}|_{\widetilde{c_S}=c_S} = 0, \frac{dAP_4(\widetilde{c_P}, c_P)}{d\widetilde{c_P}}|_{\widetilde{c_P}=c_P} = 0$ in Eqs. 29 and 30. $\frac{dPP_4(\widetilde{c_P}, c_P)}{d\widetilde{c_P}} = 0$ and $\frac{dAP_4(\widetilde{c_S}, c_S)}{d\widetilde{c_S}} = 0$ are described in Eqs. 27 and 28 as follows:

$$\frac{dPP_4(\widetilde{c_P}, c_P)}{d\widetilde{c_P}} = \frac{dZ_P(\widetilde{c_P})}{d\widetilde{c_P}} + m(\widetilde{c_P}) + (\widetilde{c_P} - c_p)\frac{dm(\widetilde{c_P})}{d\widetilde{c_P}} = 0 \tag{27}$$

$$\frac{dAP_4(\widetilde{c_S}, c_S)}{d\widetilde{c_S}} = \frac{dZ_S(\widetilde{c_S})}{d\widetilde{c_S}} + m(\widetilde{c_S}) + (\widetilde{c_S} - c_S)\frac{dm(\widetilde{c_S})}{d\widetilde{c_S}} = 0 \tag{28}$$

$$\frac{dZ_P(c_p)}{dc_p} + m(c_p) = 0 \tag{29}$$

$$\frac{dZ_S(c_S)}{dc_S} + m(c_S) = 0 \tag{30}$$

From Eqs. 29 and 30, we can find that $Z_P(c_p)$ is a decreasing function of c_p, so when $c_p = c_{p2}$, APP has the minimum profit. Similarly, $Z_S(c_S)$ is a decreasing function of c_S, so when $c_S = c_{s2}$, ASP has the minimum profit. Let $Z_1 = Z_P(c_{p2}), Z_0 = Z_S(c_{s2})$, $Z_1, Z_0, Z_P(c_p), Z_S(c_S)$ can be described in Eqs. 31–34.

$$Z_1 = \delta_p[(1 - \delta_S)p\lambda + w_S m] - \frac{w_p(m - \bar{r}\lambda)}{m} - c_{P2}m \tag{31}$$

$$Z_0 = \delta_S p\lambda - w_S m - c_{s2}m \tag{32}$$

$$Z_P(c_p) = Z_1 + \int_{c_p}^{c_{p2}} m(c_p)\mathrm{d}c_p \tag{33}$$

$$Z_S(c_S) = Z_0 + \int_{c_s}^{c_{s2}} m(c_S)\mathrm{d}c_s \tag{34}$$

Z_1 and Z_0 are the reservation profit, which means that we should ensure the profit of APP and ASP under the revenue-sharing contract are not lower than the ones under the decentralized control. The per unit time profit of AIP can be described in Eq. 35, which equals the per unit time supply chain's profit minus the profit of APP and ASP.

$$\begin{aligned} \mathrm{HP}_4(\lambda, m) &= \int_{c_{p1}}^{c_{p2}} \int_{c_{s1}}^{c_{s2}} [Z(c_S, c_p) - Z_r(c_S) - Z_r(c_p)] f(c_s) f(c_p) \mathrm{d}c_s \mathrm{d}c_p \\ &= k\sqrt{\lambda} - \frac{\lambda v \bar{r}}{m - \bar{r}\lambda} - (c_I + c_{P2} + c_{s2})m - em^2 - Z_0 - Z_1 \end{aligned} \tag{35}$$

To find the optimal per unit time AIP's expected profit, first-order conditions require $\frac{\mathrm{dHP}_4(\lambda,m)}{\mathrm{d}\lambda} = 0$ and $\frac{\mathrm{dHP}_4(\lambda,m)}{\mathrm{d}\mu} = 0$ as follows:

$$\frac{\mathrm{dHP}_4(\lambda, m)}{\mathrm{d}\lambda} = \frac{k}{2\sqrt{\lambda}} - \frac{mv\bar{r}}{(m - \bar{r}\lambda)^2} = 0 \tag{36}$$

$$\frac{\mathrm{dHP}_4(\lambda, m)}{\mathrm{d}m} = \frac{\lambda v \bar{r}}{(m - \bar{r}\lambda)^2} - (c_I + c_{P2} + c_{s2}) - 2em = 0 \tag{37}$$

$\mathrm{HP}_4(\lambda, m)$ in Eq. 35 exists unique optimal solution if and only if $\frac{\mathrm{d}^2\mathrm{HP}_4(\lambda,m)}{\mathrm{d}\lambda^2} < 0, \frac{\mathrm{d}^2\mathrm{HP}_4(\lambda,m)}{\mathrm{d}m^2} < 0, \Delta_4 = \frac{\mathrm{d}^2\mathrm{HP}_4(\lambda,m)}{\mathrm{d}\lambda^2} * \frac{\mathrm{d}^2\mathrm{HP}_4(\lambda,m)}{\mathrm{d}m^2} - \frac{\mathrm{d}^2\mathrm{HP}_4(\lambda,m)}{\mathrm{d}\lambda\mathrm{d}m} * \frac{\mathrm{d}^2\mathrm{HP}_4(\lambda,m)}{\mathrm{d}\lambda\mathrm{d}m} > 0$. After some operations, we can find that the optimal solution $\left(\lambda_4^*,\ m_4^*\right)$ which can be obtained from Eqs. 36–37 exists if $\frac{2ek}{4\sqrt{\lambda^3}} + \frac{4emv\bar{r}^2}{(m-\bar{r}\lambda)^3} + \frac{2\lambda v\bar{r}k}{4\sqrt{\lambda^3}(m-\bar{r}\lambda)^3} - \frac{v^2m^2\bar{r}^2}{(m-\bar{r}\lambda)^6} - \frac{v^2\lambda^2\bar{r}^4}{(m-\bar{r}\lambda)^6} + \frac{2\lambda mv^2\bar{r}^3}{(m-\bar{r}\lambda)^6} > 0$.

Equations 15–17 should be also satisfied under asymmetric information to ensure ASP and APP obtain the optimal profit. According to Eqs. 15–17 and 36–37, the relationships between $w_s, w_p, \delta_s, \delta_p, c_s, c_P$ can be described as follows:

$$w_S = \delta_S(c_I + c_{P2} + c_{s2} + 2em) - c_s \tag{38}$$

$$w_p = \frac{\delta_p m^2 (1 - \delta_S)(w_s + c_s)}{\delta_s \bar{r}\lambda} + \frac{\delta_p w_s m^2 - c_P m^2}{\bar{r}\lambda} \tag{39}$$

As a conclusion, we can design a revenue-sharing contract based on Eqs. 31, 32, and 36–39 to improve the profits of AIP, APP, ASP, and the supply chain at the same time. Although the discussed supply chain cannot be coordinated with revenue-sharing contract under asymmetric information, it can attain Pareto improvement. Moreover, the reservation profit Z_1 and Z_0 can ensure the profits of APP and ASP in the revenue-sharing contract are not lower than the ones in the decentralized control.

4 Conclusions

In this paper, we aim to coordinate a three-staged cloud computing service supply chain under asymmetric information with revenue-sharing contract. The discussed supply chain includes one AIP, one APP, and one ASP. The market is characterized by a price-sensitive random demand, which obeys the Poisson distribution. And the ASP service system is modeled as an M/M/1 queueing system while the effects of congestion are considered. We take the asymmetric cost information into account and find that the discussed supply chain cannot be coordinated with revenue-sharing contract under asymmetric information. However, it can attain Pareto improvement.

Acknowledgments This work was supported by Beijing Natural Science Foundation (4122052).

References

1. Sushil, B., Leena, J., Sandeep, J.: Cloud computing: a study of infrastructure as a service (IAAS). Int. J. Eng. Inf. **2**, 60–63 (2010)
2. Vania, G., Pieter, B.: Adding value to the network: mobile operators' experiments with software-as-a-service and platform-as-a-service models. Telematics Inform. **28**, 12–21 (2011)
3. Cachon, G.P., Martin, A.: Supply chain coordination with revenue-sharing contracts: strengths and limitations. Manage. Sci. **51**, 30–44 (2005)
4. Cachon, G.P.: Supply chain coordination with contracts. Operation and Managements Science: Supply Chain Management. (2004)
5. Haluk, D., Hsing, K.C., Subhajyoti, B.: Coordination strategies in an SaaS supply chain. J. Manag. Inf. Syst. **26**, 119–143 (2010)
6. Ilaria, G., Pierpaolo, P.: Supply chain coordination by revenue sharing contracts. Int. J. Prod. Econ. **89**, 131–139 (2004)
7. Palsule, D., Omkar, D.: Supply chain coordination using revenue-dependent revenue sharing contracts. Omega-International J. Manag. Sci. **41**, 780–796 (2013)
8. Esmaeili, M., Zeephongsekul, P.: Seller–buyer models of supply chain management with an asymmetric information structure. Int. J. Prod. Econ. **123**, 146–154 (2010)
9. Sinha, S., Sarmah, S.P.: Buyer-vendor coordination through quantity discount policy under asymmetric cost information. In: 2007 IEEE International Conference on Industrial Engineering and Engineering Management. pp. 1558–1562. (2007)

10. Kayis, E., Erhun, F., Plambeck, E.: Delegation vs control of component procurement under asymmetric cost information and simple contracts. Manufact. Serv. Oper. Manag. **15**, 45–56 (2013)
11. Mendelson, H.: Pricing computer services: queuing effects. Commun. ACM **28**, 312–321 (1985)
12. Junwei, C., Kai, H., Keqin, L., Albert, Y.Z.: Optimal multiserver configuration for profit maximization in cloud computing. IEEE Trans. Parallel Distrib. Syst. **24**, 1087–1096 (2012)
13. Cotton, I.W.: Microeconomics and the market for computer services. Comput. Surv. **7**, 95–111 (1975)
14. Selwyn, L.L.: Economies of scale in computer usage: initial tests and implications for the computer utility. Ph.D. Dissertation. Sloan School of Management. (1970)
15. Rubens, P.: Infrastructure software vendors caught in a catch-22. ASPNews.com. (2001)

An Intelligent Storage Management System Based on Cloud Computing and Internet of Things

Jun Kang, Siqing Yin and Wenjun Meng

Abstract Cloud computing is an emerging model of network resource delivery and usage. Research how to create an operating platform based on cloud computing and Internet of Things (IoT), which can promote the development of IoT rapidly and satisfy the applied demand of IoT. In this paper, the two technologies, cloud computing and IoT, are introduced and analyzed. Then, an intelligent storage management system is designed combining of cloud computing and IoT. The designed system is divided into four layers: perception layer, network layer, service layer, and application layer. And the system's function modules and database design are also described. The system processes stronger applicability and expansion functions, and all of them can be extendedly applied to other intelligent management systems based on cloud calculating and IoT.

Keywords Cloud computing · IoT · Storage management

1 Introduction

Cloud computing is a new technology which appeared in recent years. In the cloud environment, due to virtualization technology, hardware and software resources are all integrated and shared efficiently. Cloud computing contains many

J. Kang (✉) · S. Yin
Software School, North University of China, Taiyuan, China
e-mail: kj_ty0015@sina.com

S. Yin
e-mail: yinsq@163.com

J. Kang · W. Meng
Mechanical Engineering College, Taiyuan University of Science and Technology, Taiyuan, China
e-mail: tyustmwj@126.com

S. Patnaik and X. Li (eds.), *Proceedings of International Conference on Computer Science and Information Technology*, Advances in Intelligent Systems and Computing 255, DOI: 10.1007/978-81-322-1759-6_57,

advantages, including a broad scale, multi-user concurrent operation, high scalability, high reliability, and ease of maintenance, which can be matched by no other technologies [1–3]. Nowadays, more and more researchers begin to study cloud computing technology and apply it to many industries including medical, insurance, traffic, energy, logistics, manufacture, education etc.

The concept of Internet of Things (IoT) was firstly presented by Kevin Ashton in 1999 and has been rapidly shifting from research to application. But until now, there is not a common architecture in IoT. In [4], Xu et al. design an intelligent fault prediction system based on IoT. The proposed structure is composed of four layers: sensor-monitoring layer, middleware-transmitting layer, prediction-application layer, and decision-feedback layer. The proposed system improves the working efficiency and intelligent level of fault prediction. In [5], Domingo provides an overview of IoT for people with disabilities. The proposed architecture of IoT can be divided into three layers: perception layer, network layer, and application layer.

Cloud computing and IoT have many respective advantages, and cloud computing can provide reliable technical support and services in the application of IoT; thus, how to combine the both technologies will bring the practical significance. In [6], Sun et al. designs a tailings dam monitoring and pre-alarm system (TDMPAS) based on IoT and cloud computing. The framework of the designed system consists of three layers as the sensor layer, the network layer, and the application layer. In [7], Fan et al. researches a DRAGON-lab platform and proposes an approach combining the IoT devices with confederation network in the DRAGON-lab based on the three-layer cloud structure. The new approach is essential for the next generation Internet.

In this paper, an intelligent storage management system is designed, which adopts the technology of IoT based on cloud service framework. First, the structure of three-layer model in IoT is introduced, and the disadvantage is analyzed. Second, the new four-layer model system is founded, adding the service layer. The new model is better meet the need of large scale of data processing and analyzing. Finally, an intelligent application of storage management system is presented as a case. The new model being designed in this paper provides a new view and approach for the use of cloud calculating and IoT.

2 Structure of IoT Based on Cloud Computing

2.1 Platform of Cloud Computing

The platform of cloud computing adopts layer structure [8], including three layers: the cloud service platform, the basic cloud platform, and the infrastructure platform, as illustrated in Fig. 1.

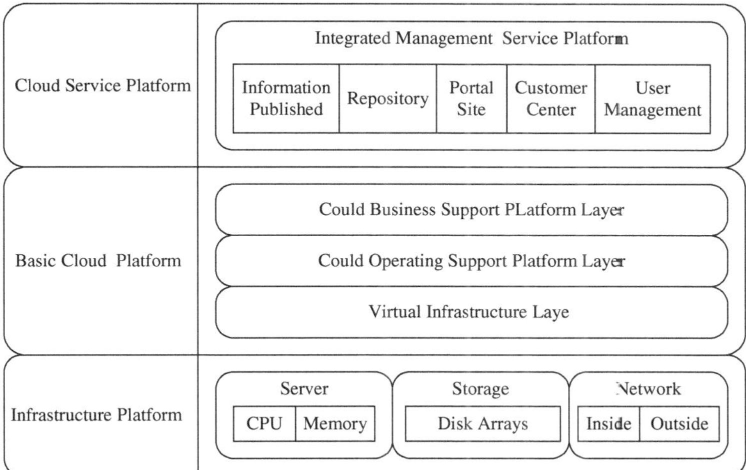

Fig. 1 The platform of cloud computing

- Cloud service platform: cloud service platform provides some functions such as information published, repository, portal site, customer center, user management etc. Besides, the platform provides interfaces of cloud support for applications.
- Basic cloud platform: basic cloud platform makes unified virtualization management of underlying hardware infrastructure and provides infrastructure and management service to run clouds, which includes the virtual infrastructure, could operating support platform, and could business support platform three layers.
- Infrastructure platform: infrastructure platform provides service for upper layers with various types of hardware.

2.2 Structure of IoT Based on Cloud Computing

The most representative model contains three layers: the perceptual layer, the transport layer, and the application layer [9–12].

- Perception layer: the perception layer formed by a variety of sensors and gateways, like the human nerve endings, is mainly used for automatically recognition and information collection.
- Transport layer: the transport layer is composed of a variety of network systems and cloud computing platform, which is responsible for transmission and process the information acquired from the sensing layer.
- Application layer: the application layer is the interface between the IoT and users, which is used to implement the types of intelligent application.

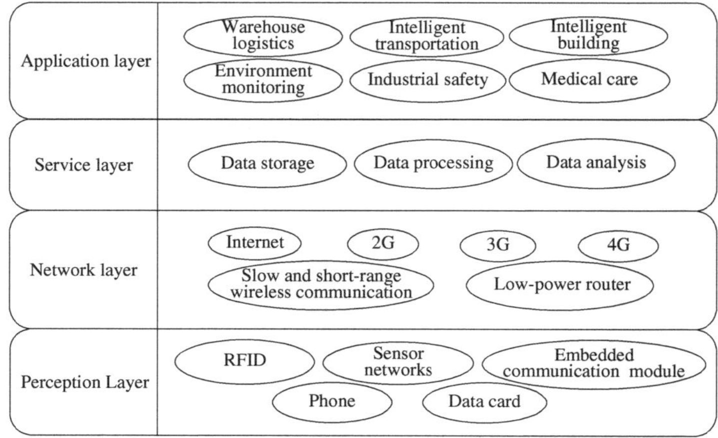

Fig. 2 The structure of IOT based on cloud computing

In this paper, a new four layers model is set up. The perceptual and application layer is the same as before. But the transport layer is broken down into two parts, the network layer and the service layer.

- Network layer: the network layer is only responsible for data transfer via different networks and Internet.
- Service layer: the service layer is responsible for data processing and analysis by data processing center based on cloud computing.

The structure of IoT based on cloud computing is shown in Fig. 2.

3 Intelligent Storage Management System

3.1 Structure of Intelligent Storage Management System

Intelligent storage management system is one kind of the applications of warehouse logistics. The system can improve the quality and efficiency of enterprises logistics, decrease inventory cost, and promote enterprise competitiveness. According to the characteristics and requirements of intelligent storage management system, union the model's features of IoT based on cloud computing, a four-layer structure of the system has been designed and is described as follows:

- First layer: the first layer is perception layer, which is composed of handheld devices with RFID sensor and wireless sensor networks. The user uses a handheld device to scan a product's electronic label and collect the product's information and then send the information to upper layer by wireless sensor networks.

- Second layer: the second layer is network layer. This layer includes intranet and Internet, and necessary network equipment such as switch, repeater, and hub. The information from the first layer is first transferred to intranet and then transferred to Internet.
- Third layer: the third layer is service layer, which is created based on cloud computing. The hardware contains disk arrays, sever cluster, management server, and necessary network equipment. The software includes cloud support platform software and cloud service platform software. The cloud support platform software could integrate different distributed heterogeneous computation resources, centralize monitoring and management resource, and schedule jobs automatically. The cloud service platform software has multiple functions, information published, repository, portal site, customer center, user management, etc.
- Forth layer: the forth layer is application layer, which provides application program as interfaces for end users.

3.2 Function Module of Intelligent Storage Management System

The system's function module is designed as shown in Fig. 3.

The system is divided into 6 modules: based information management, system setting, in-storage management, out-storage management, transfers management, and sale management. The functions of each module are as follows:

- Based information management: the major function of module mainly processes the based information, such as customer, product, location, department, device and staff information, and each of them includes operations of setting, browsing, inquiries, addition, modification, and deletion.
- System setting: the module includes parameter setting and user information setting.
- In-storage management: the module is one of the two primary functions in this system. In-storage is the process from applying for in-storage, splitting orders and assigning task, planning for in-storage, receiving, packing to putting in storage. In the process, the data collecting device is used to collect goods information to store in the database by RFID and barcode technology.
- Out-storage management: the module is one of the two primary functions in this system. Out-storage is the process from applying for out-storage, splitting orders and assigning task, planning for out-storage, picking, unpacking to remove from storage. Also, the data collecting device is used to collect goods information to store in the database by RFID and barcode technology.
- Transfers management: the module is responsible for the situation that the goods in the storage need to move to other location occasionally. Similarly, the goods information is used to collect by the data collecting device in the transfers' process.

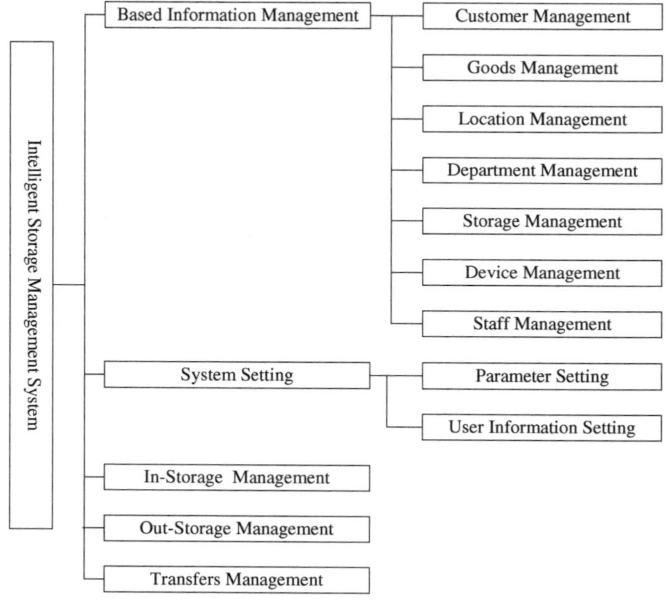

Fig. 3 The function module of intelligent storage management system

3.3 Database Designing of Intelligent Storage Management System

The database designing of intelligent storage management system is the most important part of system design, having complex structure and content. This section describes mainly part of the system's database by Entity-Relation (E-R) diagram, as shown in Fig. 4.

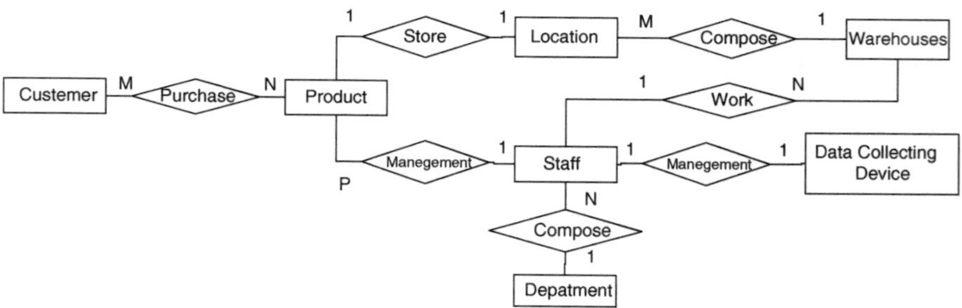

Fig. 4 The E-R diagram of intelligent storage management system

4 Conclusion

The rapid development of cloud computing and IoT provides a new research way for intelligent storage management system. IoT emphasizes on thing-to-thing connected with the Internet, and cloud computing provides processing power for the thing connected with cloud. The integration of two technologies can better satisfy the growing demand of information exchange.

In this paper, the features of cloud computing and IoT are analyzed and an intelligent storage management system is designed based on combining of cloud computing and IoT. The system has four layers, including perception layer, network layer, service layer, and application layer. In the service layer, the technology of cloud computing is used. The paper also described the function module and the database of the designed system. Through theoretical analysis and empirical research, this study found that the designed system has higher running efficiency, stronger security, great application values, and extended value.

Acknowledgments This work is supported partly by Research Project Supported by Shanxi Scholarship Council of China (Grant No. 2011-073), the Shanxi Teaching Reform Project (J2012060), and 2012' North University of China (NUC) Foundation.

References

1. Mell, P., Grance, T.: The NIST Definition of Cloud Computing. National Institute of Science and Technology, USA (2011)
2. Buyya, R., Broberg, J., Goscinski, A.: Cloud Computing: Principles and Paradigms. Wiley Press, USA (2010)
3. Quan, L., Wen-Wu, H., Jian-Hui, L.: Real-time service model based on Cloud Computing in the next generation internet. Int. J. Advancements Comput. Technol. **4**(9), 280–287 (2012)
4. Xiaoli, X., Chen, T., Minami, M.: Intelligent fault prediction system based on internet of things. Comput. Math. Appl. **64**(5), 833–839 (2012)
5. Domingo, M.C.: An overview of the Internet of Things for people with disabilities. J. Netw. Comput. Appl. **35**(2), 584–596 (2012)
6. Sun, E., Zhang, X., Li, Z.: The Internet of Things (IoT) and Cloud Computing (CC) based tailings dam monitoring and pre-alarm system in mines. Saf. Sci. **50**(4), 811–815 (2012)
7. Tong-rang, F., Feng, G., Xuan, Z., Xu, W.: Integration of IoT and DRAGON-lab in cloud environment. J. China Univ. Posts Telecommun. **19**(2), 87–91 (2012)
8. Winkler, V.J.R.: Securing the Cloud: Cloud Computer Security Techniques and Tactics. Elsevier Inc., Amsterdam (2011)
9. Atzori, L., Iera, A., Morabito, G.: The internet of things: a survey. Comput. Netw. **54**(15), 2787–2805 (2010)
10. Chen, C.-Y., Chao, H.-C., Tin-Yu, W., Fan, C.-I., Chen, J.-L., Hsu, J.-M.: IoT-IMS communication platform for future internet. Int. J. Adapt. Resilient Auton. Syst. **2**(4), 74–94 (2011)
11. Guo, B., Zhang, D., Wang, Z.: Living with internet of things, the emergence of embedded intelligence. In: 4th International Conference on Cyber, Physical and Social Computing, vol. 2, no.4, pp. 297–304, (2011)
12. Kun, G., Qin, W., Lifeng, X.: Controlling moving object in the internet of things. Int. J. Advancements in Comput. Technol. **4**(5), 83–90 (2012)

Research on Numerical Weather Prediction Based on Doppler Raw Radar Data and Cloud Model

Jianhua Du and Shenghong Wu

Abstract The new generation of Doppler weather radar base data transmission is always the assessment of key project in professional work and also is the necessity for many short-term weather forecast systems. Based on the quality control of radar data, we completed the provincial or regional networking of weather radar base data, which are able to provide radar echo data on the latitude/longitude/elevation grids in real time and generate some products, so as to provide a base for the application of radar data to meso-scale weather analysis, nowcasting, and hydrological research. Simulation results show that the method is applicable and effective.

Keywords Numerical weather prediction · Doppler raw radar data · Cloud model · Meteorological data

1 Introduction

Numerical weather prediction (NWP) is based on the actual situation of the atmosphere, in a certain initial and boundary conditions, through large-scale computer for numerical calculation, solving and describing weather evolution equations of fluid mechanics, and thermodynamics to predict the future of a certain period of time atmospheric motion and weather phenomena approach [1]. NWP

J. Du (✉) · S. Wu
Hainan Meteorological Service Centre, Haikou 570203 Hainan, China
e-mail: dudjh@163.com

J. Du
Qionghai Meteorological Service, Qionghai 571400 Hainan, China

S. Patnaik and X. Li (eds.), *Proceedings of International Conference on Computer Science and Information Technology*, Advances in Intelligent Systems and Computing 255, DOI: 10.1007/978-81-322-1759-6_58,

with the classic method for synoptic weather is different, it is a quantitative and objective forecasts [2]. The first requires the establishment of a numerical weather that better reflects the forecast period (short term, medium term) and the error is smaller than numerical prediction models to calculate steady and relatively fast arithmetic calculation. Secondly, NWP uses various means (conventional observations, radar, ship observations, satellite observations, etc.) to obtain meteorological data [3].

Quantitative and objective NWP prediction used or under atmospheric dynamics equations and the equations used in the same, that is, from the continuity equation, the thermodynamic equation, vapor equation, equation of state equations of motion and three equations posed [4]. Around the world, more than 30 countries and regions of the daily weather forecast numerical weather as the main method of production, in which many countries and regions in addition to making 2 days short NWP, but also about a week produced an interim value weather forecast. China began in 1955 to explore for NWP, in 1959 started on a computer NWP [5]; in 1969, the National Weather Service officially released short-term NWP, after the gradual improvement of NWP model and realized data entry, mapping, analysis automation, and forecasting output. Currently, in addition to completing the daily short-term NWP business is preparing an interim NWP.

Because the atmosphere is a kind of continuous motion scale continuum spectrum, so no matter how high-resolution mode, there are always close to or less than the grid length scale motions, not exactly in the pattern reflected, this movement called sub-grid process. Turbulence, convection, condensation, and radiative processes are included with sub-grid process. It has been used in NWP parametric approach to consider these processes, which uses large-scale variables to describe sub-grid scale processes on large statistical effect of movement. Although this method has achieved very good results, but there are still many unresolved issues. Parametric is not to be considered as a large-scale impact of small scale and feedback effects, lack of objective value of the parameter determination method for parametric model is too sensitive to such differences.

Since the radar is the only accurate description possible for convective-scale weather phenomena in the atmosphere observing system, how to make full use of Doppler radar data, to extract, the use of meaningful meteorological information to improve the initial field of numerical models to improve the model prediction capabilities, as increasingly central concern. In recent years, meteorologists have committed using high-resolution radar data through inversion, assimilation and nonadiabatic initialization method, the high-resolution mode radar observation information into the initial field, and the increase in the level of numerical prediction.

The rest of this paper is organized as follows: We study the Doppler raw radar data and cloud model in Sect. 2. Section 3 designs the NWPs optimal model. Section 4 discusses the experimental results. Finally, Sect. 5 concludes the paper and discusses some future research directions.

2 Doppler Raw Radar Data and Cloud Model

Doppler weather radar data is nowcasting using one of the highest frequency information, as nowcasting for demanding real-time data, while raw radar data quantity of data is large and high-temporal resolution, which made the data transmission a very high demand [6]. Doppler weather radar is raw radar weather detection, processing, generating, and displaying radar weather data application systems controlled by software radar. Radar data acquisition status controls interface generating and transmitting radio frequency (RF) pulse detecting meteorological targets acquisition and processing the reflected RF signal to a base obtained weather data (from the measured reflectivity, mean radial velocity, and the velocity spectrum width composition) sent RPGG system.

Cloud model with uncertain knowledge expressed with certainty in the stable, among the changes in the characteristics, reflects the basic principle of the evolution of species in nature [7]. For the cloud model C (Ex, En, HP), Ez can be expected through the cloud model Ex, entropy En, and hyper entropy three digits features to characterize an overall concept. Ex expectations reflected cloud droplet group's center of gravity position; entropy En reflected in the domain space can be the qualitative concept acceptable range, on the other hand, is also reflected in the spatial domain of the points to represent this qualitative concept probability, which means that the concept of cloud droplets appeared qualitatively randomness; hyper entropy He reflects the narrow domain of inter-language value that represents the uncertainty of all points cohesion, which cloud droplet condensation [8]. There are positive cloud generator and backward cloud generator. Cloud parameter is calculated as follows:

$$\begin{cases} E_x = (c_{\max} + c_{\min})/2 \\ E_n = (c_{\max} - c_{\min})/6 \\ H_e = k \end{cases} . \quad (1)$$

3 Numerical Weather Predictions Optimal Model

In recent years, along with a variety of large-scale meteorological equipment and meteorological data distribution network in a short time in the deepening of nowcasting applications, short-term nowcasting business to local probe data collected in real time, real-time automatic weather station data quality control, a variety of soundings time synchronization, data update frequency, data format, and data organization and put forward new demands.

Weather radar volume scan mode is the usual way of working, *X*-band weather radar volume scanning once for each observation will form an appropriate body scan documents. Body scan files are named according to VTByyyymmddhhxxzz.nn, VTB said that the body can scan mode, yyyymmddhh, respectively, observations indicate that the second body sweep the year, month, day, and hour, xxzz indicates that the second body scan observing the number of minutes and seconds, and nn represents the number of layers body scan [9]. Doppler raw radar data have unified data storage, each body stored as a separate file. SA/SB radar in the lower elevation of each scanned twice, in the base data, recorded as an elevation layer. VIL is a reflection of precipitation cloud system in a certain area of the vertical column bottom liquid water content in vivo distribution. Assuming meet MP raindrop distribution is calculated using the relation M_Z VIL value MVIL, the practical application of the following formula:

$$\begin{cases} a = \bar{y} - b\bar{x} \\ b = \left(n\sum_{i=1}^{n} x_i y_i - \sum_{i=1}^{n} x_i \times \sum_{i=1}^{n} y_i \right) \Big/ n\sum_{i=1}^{n} x_i^2 \end{cases}. \tag{2}$$

Since the vertical liquid water content is low to the high reflectivity calculated, so do not consider the impact of ground clutter, even small-scale VIL great value, all have a high level from the lower to the development of the system exists.

All data of this curve fitting according to different typhoons were fitted in different locations, each *Z–R* relationship analyze and calculate the respective changes in estimation error. Estimation error formula is as follows:

$$E = \sqrt{\frac{\sum_{i=1}^{n} (R_i - R_s)^2}{n}}. \tag{3}$$

In which, R_i is the first i time according to the relationship *Z–R* of the precipitation rate estimated; R_s is the sum of the time period measured ten times rainfall, E units of mm/h.

Weather radar detection system is an important means of precipitation, but also on strong convective weather monitoring and warning, one of the main tools in the monitoring hail, heavy precipitation, wind damage, and other severe convective weather has unique. The basic NEXRAD base reflectivity factor products include the average radial velocity and velocity spectrum width, on this basis, according to local rainfall and strong convective weather characteristics, through a certain algorithm, can be applied to local radar secondary products, which for a better grasp of the region and the strong convective weather precipitation occurrence and development of the law, a better job of disaster prevention and mitigation services and improve the ecological environment through artificial rainfall that all have more practical significance.

4 Experimental Test

Haikou and analyzed Doppler radar-based data, based on cloud model reflectivity, mean radial velocity, and spectrum width data, automatic station precipitation data and Hainan, local weather and climate characteristics and concepts of discrete jump process, the establishment of reflectivity factor and precipitation conceptual model, through great determination method for soft partition of the data set, the establishment of reflectivity and precipitation Boolean database, and then according to the given support and confidence of the soft field values associated with cloud-based knowledge mining algorithm to get the time, and reflectivity and precipitation association rules. With time, the conjunctive reflectivity as a rule antecedent to pieces after precipitation as a rule creates a rule generator, according to the estimation of digging out the rules, the establishment of quantitative precipitation system is for 0–3 h. 0–3 h is needed to achieve quantitative precipitation forecast products and systems operational.

Figure 1 is the 500 hPa geopotential height field (units: gpm). Figure 2 is the simulated rainfall at 4 km grid with radar data assimilation. Figure 3 is the sacterplot of the relation between reflectivity Z and precipitation rate R. Table 1 is the location of Hainan auto-precipitation stations and the relative radar. Table 2 is the relation of reflectivity rate and precipitation calculated by forms $Z = 50R^{1.35}$ and $Z = 250R^{1.42}$ (unit: mm/h).

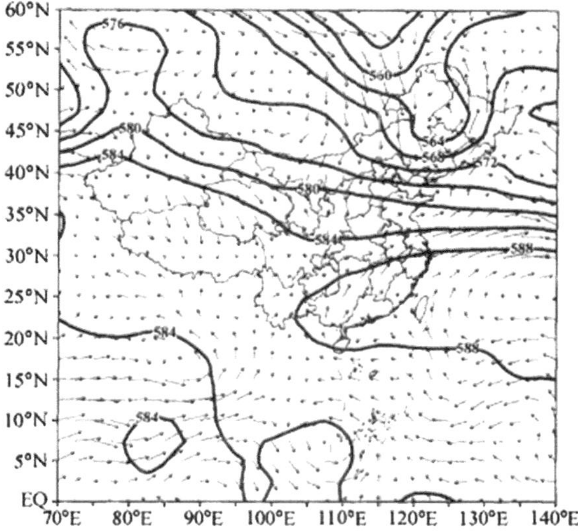

Fig. 1 500 hPa geopotential height field (units: gpm)

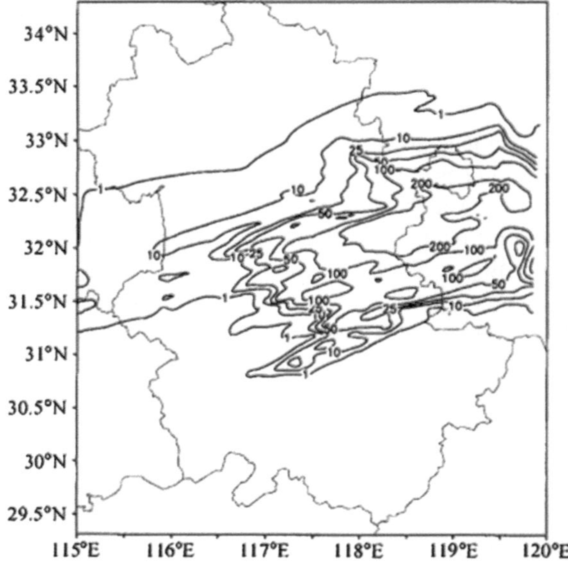

Fig. 2 Simulated rainfall at 4 km grid with radar data assimilation

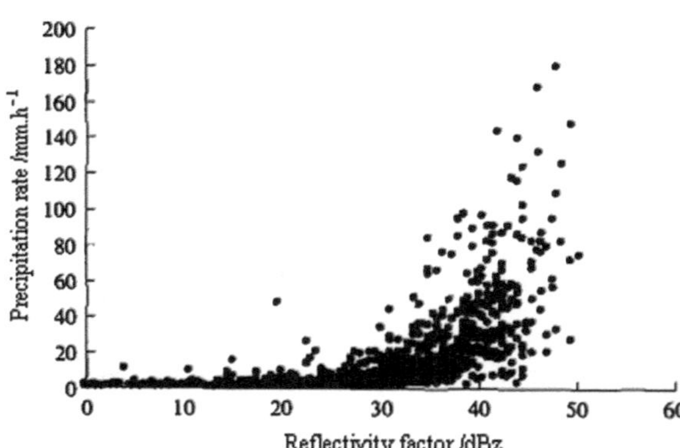

Fig. 3 The sactterplot of the relation between reflectivity *Z* and precipitation rate *R*

Table 1 Location of Hainan auto-precipitation stations and the relative radar

No.	Location	Longitude and latitude	Azimuth (°)	Distance/km
1	HanKou	20.02°N, 110.35°E	123.16	21.2
2	Qionghai	19.25°N, 110.46°E	122.30	9.85
3	Qiongshan	19.98°N, 110.33°E	123.09	26.3
4	Wenchang	19.61°N, 110.72°E	123.06	80.2
5	Dingan	19.68°N, 110.31°E	122.69	43.1
6	Wanning	18.8°N, 110.39°E	121.64	56.8
7	Tunchang	19.36°N, 110.1°E	122.05	32.6
8	Baisha	19.23°N, 109.44°E	121.18	54.6
9	Lingshui	18.48°N, 110.02°E	120.82	19.2
10	Changjiang	19.25°N, 109.03°E	120.77	10.6

Table 2 The relation of reflectivity rate and precipitation calculated by forms $Z = 50R^{1.35}$ and $Z = 250R^{1.42}$ (unit: mm/h)

Reflectivity factor/dBz	$Z = 50R^{1.35}$ precipitation rate	$Z = 250R^{1.42}$ precipitation rate
10	0.24	0.18
15	0.53	0.45
20	1.3	1.25
25	3.1	2.65
30	6.8	5.2
35	15.6	12.9
40	32.4	28.6
45	56.9	47.4
50	98.1	78.2

5 Conclusions

Atmospheric sounding of fast data collection and processing is the basis for meteorological operations and is the key short-term nowcasting one, this at home and abroad have done a lot of research, there are already many research results into business applications, which greatly promoted the meteorological operations modernization. This paper mainly achieved Doppler weather radar data service transmission design and implementation optimized for data compression and transmission processes. In this paper, a real-time automatic weather station data quality control, real-time data collection probe, radar network time synchronization, nowcasting data update frequency, nowcasting data format, and data file organization methods were analyzed.

Acknowledgments This research work was supported by the CMA Meteorological Observation Centre Focus Open Laboratory Projects (grant no. KLAS201110). The authors gratefully acknowledge the helpful comments and suggestions of the reviewers, which have improved the presentation.

References

1. Xu, Z., Xu, Y., Ge, W.: The impact of using radar and satellite data on meso-scale model numerical simulation. Scientia Meteorologica Sinica **22**(2), 167–173 (2002). (in Chinese)
2. He, K., Fan, Q., Li, K.: Z–R Relation with its application to typhoon precipitation in Zhoushan. J. Appl. Meteorol. Sci. **18**(4), 573–576 (2007)
3. Liang, J.J., Qin, A.K.: Comprehensive learning particle swarm optimizer for global optimization of multimodal functions. IEEE Trans. Evol. Comput. **10**(3), 281–295 (2006)
4. Sheng, C.Y., Pu, Y.F., Gao, S.T.: The effect of Chinese Doppler radar data on nowcasting of mesoscale model. Chin. J. Atmos. Sci. **30**(1), 93–106 (2006). (in Chinese)
5. Turpeinen, O.M.: Diabatic initialization of the Canadian regional finite element (REF) model using satellite data. Sensitivity to humidity enhancement, latent-heating profile and rain rates. Mon. Wea. Rev. **118**, 1396–1407 (1990)

6. Tong, M., Xue, M.: Ensemble Kalman filter assimilation of Doppler radar data with a compressible nonhydrostatic model: OSSE experiments. Mon. Wea. Rev. **133**, 1789–1807 (2005)
7. Li, D.Y., Liu, C.Y.: Study on the universality of the normal cloud model. Eng. Sci. **6**(8), 28–34 (2004)
8. Huang, H.S., Wang, R.C.: Subjective trust evaluation model based on membership cloud theory. J. Commun. **29**(4), 14–19 (2008)
9. Tuo, Y., Liang, H., Ma, S., et al.: A preliminary research on improving MM5 initial field using radar data. J. Nanjing Inst. Meteorol. **26**(5), 661–667 (2003)

Part V
Data Processing

Network Data Mining and Its New Technology to Explore

Jun-xi Liu

Abstract This paper investigates the network data mining and network information retrieval, differentiated on the basis of the general data mining on the characteristics of the network data mining, the type, and the latest technology.

Keywords Network data · Data mining · Mining techniques

1 The Basic Content of the Network Data Mining

1.1 Data Mining and Network Data Mining

Data mining is extracted from a large number of incomplete noise, fuzzy, or random data, but people do not know in advance useful information and knowledge. Network data mining is the application of data mining technology in the network information processing, data from the Web site to explore the relationships and rules. Each site on the Web is a data source, between each site and the organization is not the same, which constitute a large heterogeneous database environment. The network data mining not only takes advantage of all the general and standard database data mining technology, but also takes advantage of the characteristics of network data, using a special method.

J. Liu (✉)
Shanghai University of Political Science and Law, Shanghai, China
e-mail: liujunxi@163.com

S. Patnaik and X. Li (eds.), *Proceedings of International Conference on Computer Science and Information Technology*, Advances in Intelligent Systems and Computing 255, DOI: 10.1007/978-81-322-1759-6_59,

1.2 Network Data Mining and Network Information Retrieval Distinction

Network information retrieval system by the Robot, the index database, and query engine. The information gathering Robot WWW traverse, the discovery of new information as much as possible. The full-text indexing technology to collect information stored in the index database query engine receives and analyzes the user's query, traverse the index databases based on a relatively simple matching strategy, and finally submit the result set of addresses to the user. The network information retrieval system can only deal with the simple goal of a keyword; can not handle the complexity of the sample given by the user in the form of fuzzy target. Network data mining technology follows the Robot, full-text search, the outstanding achievements of the network information retrieval, integrated use of inductive learning, machine learning, statistical analysis methods and artificial intelligence, pattern recognition, a variety of techniques in the field of neural networks. Network data mining system and network information retrieval biggest difference is that it can in accordance with the requirements of the user-defined purpose of information search, according to the target feature information in the network or repository.

1.3 Network Data Mining Type

1. Web content mining
   Web content mining process to discover useful information from the network, data, documentation, access the object of the search engine in the web search. Many types of network information resources, from the perspective of a network of information sources, including Gopher, FTP, UseNet have hidden resources after the WWW form of private data of WWW information resources, database management, information systems and data, can not be indexed. From the terms of the form of the network resources, including text, images, audio, video and other forms of data.
2. Network structure mining
   The network structure mining, mining Web the potential link structure mode. This idea stems from the citation analysis, by analyzing a Web link and the number of links and object to create a Web link structure mode. This mode can be used for Web pages classified and can obtain information about the similarity between different pages and associate degrees. The network structure mining authority site to help users find related topics to the site, and you can find links to related topics.
3. Network usage mining
   Network usage mining is mainly used to understand the significance of network behavior data. Web content mining, network structure mining object is the

original online data network usage mining face of the second-hand data is extracted in the process of user and network interaction, including network server to access the records, proxy server logging browser logging, user profiles, and registration information, user sessions or transactions, user questions, style and so on.

1.4 The General Steps of the Network Data Mining

1. It is established in the target sample the target text selected by a user, as to extract the characteristics of the user information. This is the individual requirements of the user, based on the needs of users for data mining.
2. Establish a statistical dictionary to establish the main dictionary and thesaurus for feature extraction and word frequency statistics, contains the dictionary, and then according to the target word frequency distribution of the sample, the statistical dictionary feature vectors extracted mining target and calculated corresponding weight value. Often the feature item weights and the match threshold is the feedback feature items sample weights and matching threshold closely, so according to these sample data adjustment feature vector. The sample eigenvectors with these goals, you can use a traditional search engine technology and information collected.
3. Finally, the information is acquired by the feature vector, and the feature vector of the target sample is matched and will meet the threshold condition of the information presented to the user.

2 Network Data Mining Technology

Web data mining the primary technology solve semi-structured data source model and semi-structured data model query and integration issues. This must be a model to describe the data on the Web, while looking for a semi-structured data model is the key to solve the problem. In addition, you also need a semi-structured model extraction technique, semi-structured model of the technology that automatically extracted from the existing data. Can be seen as a semi-structured extensible markup language (XML) data model can easily corresponding attributes in the XML document describes the relational database, the implementation of precise query and model extraction. The use of XML, Web designers can create text and graphics, but also to build a multi-level document type definition, interdependent systems, data tree, metadata, hyperlinks, structure and style sheet.

2.1 Web Data XML to Uniquely Identify

XML, each database search software must understand how to build, because each database describing data formats are almost always different. Due to the presence of the integration problems of the data from different sources, it is now actually impossible to search for a variety of incompatible database. XML can be the structures of the data from different sources are easily combined. XML three-tier architecture is shown in Fig. 1.

Software agents can integrate data from back-end databases and other applications at the server of the middle layer. The data can then be sent to clients or other servers for further collection, processing, and distribution. XML-based data is self-describing data does not need to have internal description can be exchanged and processed. With the use of XML, users can easily carry out local computation, and processing XML format data is sent to the customer; the customer can use the application software to parse the data and for data editing and processing.

2.2 XML Applied to a Large Number of Computing Load Distribution in the Client

Customers can choose according to their needs and making a different application to handle the data, while the server is only required to issue the same XML file. Initiative in processing data to the client, the server just made the best possible accuracy of the data package into an XML file. XML's self-explanatory client to receive data at the same time understand the logical structure and meaning of the data, so that a wide range of common distributed computing become possible. This is also more conducive to meet stressed the needs of individual users in the network information data mining problems.

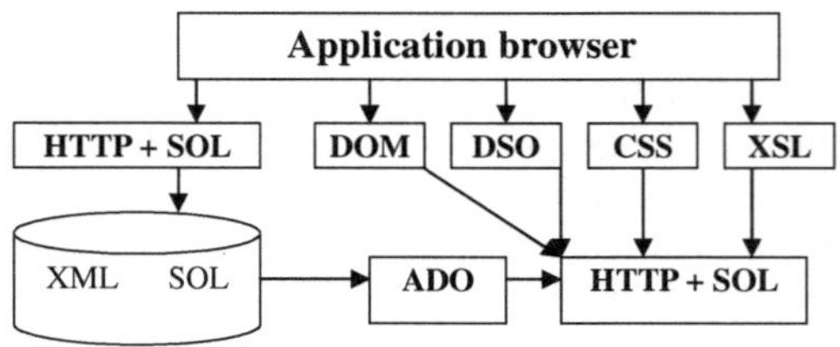

Fig. 1 The three-tier structure of XML

3 Network Data Mining Model Design

3.1 Network Data Mining Model Design

Characteristics: Heterogeneous network data are different from the general database data and are semi-structured. Each site is a data on the Internet Source; each data source has its own style, that is, information and organization of each site is different; the Internet is a huge heterogeneous database, unlike a traditional relational database. Data on the Internet is very complex, there is no uniform model describes the existence of these data have some degree of structural, but the readme level and complex interrelated, which is a non-fully structured data. Given these characteristics of network data, the data mining technology into the Internet, must be to do a certain amount of pre-processing work. Network based on data mining model is shown in Fig. 2.

Figure 2 is actually an improved model of traditional search engines; data mining technology is loaded into the search engine, enabling network data mining. XML information preprocessing module is a key link in the network data mining, but if follow conventional Spider or Robot heterogeneous network data collection, recombination, in accordance with a unified data structure, will be difficult to achieve because of the heavy workload. The emergence of XML has brought hope to resolve the problem of network data mining. XML is a simple, open, efficient, and scalable network markup language in line with international standards, its scalability and flexibility to be able to describe the data in different types of application software, which describe the data records collected pages. XML indexing of network data is a semi-structured data model, but the document describes the one-to-one correspondence with the attributes in the relational database, which can implement accurate query model extraction, complete integration of heterogeneous network data job.

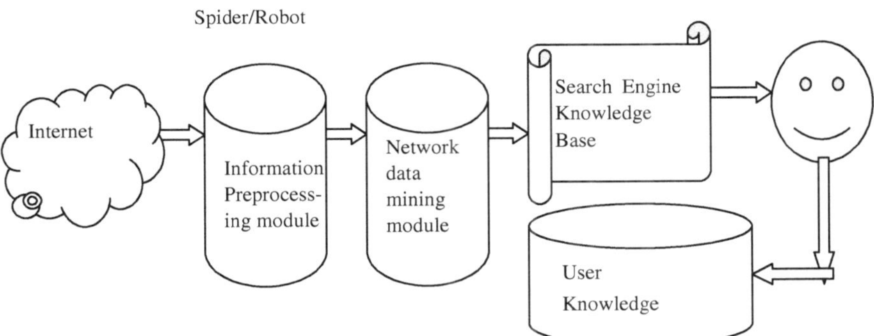

Fig. 2 Network data mining models

4 Network Data Mining Algorithms Support Intelligent Retrieval

4.1 Network Data Mining Support Intelligent Retrieval

Personalized information retrieval, or based on the content retrieval, and even knowledge retrieval, intelligent Web information retrieval system key is to know the user needs what, something with a high-quality (content related knowledge content) available to the user. Retrieve data mining of network intelligence support is reflected in a deep analysis of the information and networking source information the user needs to provide the key to intelligent retrieval necessary knowledge.

4.2 User Knowledge of Mining

Although with a specific user's information needs of the individual information, but the user base as a whole, the information needs of the user is random, for general information, the user needs analysis to a great deal of difficulty. Data mining from the overall situation, a rich, dynamic online query and analysis to understand the user's information needs. Questions online, the survey table, the system can obtain the information about the user's user name, user access to the IP address, the original user's occupation, age, hobbies and other information. Then, take certain mining rules (such as association rules, online analytical processing, etc.), these data fusion analysis, the result is the establishment of an information demand model for each user. And a full range of user needs information mining, and similar information needs of the user can be linked to implement "a minority" of the retrieval program. User knowledge mining has entered a practical stage mining tools, such as IBM's new DB2 UDB7.1 is an ideal knowledge of the user.

4.3 Network Knowledge Mining

The network knowledge mining is to find out the law of distribution of information in the vast amounts of data with extreme uncertainty, mining hidden information, and the formation of the model, to discover regularities knowledge. Network information distribution regularity is the correlation within the network information. The excavation of this correlation of network information is mainly reflected in two aspects: first, the Web content mining. Web content mining is a network source information in the form of text, images, audio, video, metadata, classification, clustering, and other forms of mining method to find useful information, and the information in the form of press meet some retrieval methods be the organization's process. Network structure mining—the network structure mining is established by analyzing the number as well as the object of a Web page link and be linked to the Web link structure mode. This mode can be used for Web classified and thus can obtain information about the similarity between different pages and associate degrees. Link structure model is conducive to intelligent navigation.

5 Conclusion

Data mining techniques to the development of the network resources, able to accelerate the development of intelligent retrieval. The results of data mining is the basis for intelligent retrieval, the intelligent retrieve the results can provide guidelines for data mining and clues. Currently, the network data mining data mining technology led products have been applied. Example, Net Perceptions developed Net Perception, can mine user information, and thus lay the foundation for the realization of personalized information services. If combined with machine learning, pattern recognition and other artificial intelligence techniques in the development of the endless network data resources, network data mining technology will be more perfect in practical applications.

With the rapid development of the Internet and information technology, network information resources have become the bottleneck of the further development of the Network Information Service. Network number dig digging a new branch in the data mining technology, it relates to the network technology, data mining technology, text processing, artificial intelligence technology, a powerful network data mining system is provided for the user's information gathering tools, the description of a new generation of network data language (XML) will be provided for the implementation of the network data mining great convenience.

References

1. Zhang, D.: A novel web usage mining approach for search engines. Comput. Netw. **39**, 303–310 (2002)
2. Hart, J.: Data Mining Concepts and Techniques. Morgan Kaufinann Press, San Francisco, pp. 435–449 (2001)
3. Beeferman, D., Berger, A.: Agglomerative clustering of a search engine query log. In: Proceedings of ACM KDD 2000, Boston, MA, USA
4. Greening, D.R.: Data mining on the web. Web Tech. **1**, 26–29 (2000)

Single-Layer Closed Contour Extraction from Craniofacial CT Data Using Curve Evolution

Kang Li, Guohua Geng and Shang Peng

Abstract Computer-aided craniofacial reconstruction refers to the process that recovers the appearance model of a face from the geometric characteristics of a skull model. One of the key problems is extracting a single-layer closed contour which wraps the whole area of complex skull structure in a CT slice accurately and effectively. This paper presented a single-layer closed contour extraction method based on curve evolution. Firstly, CT slices required to segment by threshold to separate the soft tissue and skull region and secondly initialize a single-layer closed curve just containing the skull region and evolving the curve using a novel differential equation until the curve fits snugly along the border of the skull region. The results of experiments indicate that this method can automatically process the skull CT slices and extract the single-layer outer contour of the skull region accurately and efficiently.

Keywords Craniofacial · Segmentation · Contour extraction · Curve evolution

1 Introduction

Contour extraction is a significant problem of image processing, pattern recognition, and computer vision [1]. The contour of an image is always one of the most important characteristics, which has a direct affection on the image understanding,

---

K. Li (✉) · G. Geng · S. Peng
School of Information Science and Technology, Northwest University, Xi'an, China
e-mail: likang@nwu.edu.cn

G. Geng
e-mail: ghgeng@nwu.edu.cn

S. Peng
e-mail: xdshangpeng@nwu.edu.cn

S. Patnaik and X. Li (eds.), *Proceedings of International Conference on Computer Science and Information Technology*, Advances in Intelligent Systems and Computing 255, DOI: 10.1007/978-81-322-1759-6_60,

image recognition, and other related fields. It is also an important step to build 3D models of face and skull from CT data for craniofacial reconstruction process [2]. In order to measure the depth of soft tissue accurately, we need to extract two different single-layer closed contours for warping bone and soft tissue. Most of classical methods for contour extraction convolve the image with an operator to detect the edges like Canny and Sobel [3]. The results of extracting contour of skull from CT data by classical methods always include several closed contours of different parts of skull, combined with a lot of redundant contours. In [4], a line scanning method is used to extract the outer contours of skull from the contours extracted by Sobel operator. This method cannot remove the redundant contours from inner components of skull and curves from each contour that still remains separated. Active contour is an image segmentation technique introduced by Kass et al. [5]. The active contour, or 'the snake,' behaves as an intelligent agent by changing its position and shape to minimize the energy function defined in order to segment an object in an image [5]. In [6], an improved active contour method is used to extract the outer contours. This method remove redundant contours inside the skull automatically, but can not extract the closed outer contour.

Various curve-evaluation-based models or partial differential equation (PDE)-based models have been proposed to extract edges of interest in medical image recent years [7, 8]. Curve evaluation is detailed in Sect. 2. Threshold-based image segmentation is performed to separate bone structure, and soft tissue is detailed in Sect. 3. Then, an initial curve is generated and evolved by a novel differential equation until the curve fits snugly along the border of the skull region, which is described in Sect. 4. Experimental results and analysis are presented in Sect. 5.

2 Curve Evolution Theory

There are several geometric properties that reflect the characteristics of a curve, such as curvature, smoothness measures, and curve length. They are represented by either explicit or implicit curve descriptor. The most important property is the curvature k that describes the rate of bending of the curve and the normal vector $\overrightarrow{N}$. Those properties also can be used to control the shape of curves.

The underlying principle of curve evolution is the evolution process of a simple closed curve whose points move in the direction of the normal with a prescribed velocity. Assume $C(p, t)$ describes the curve at time t, p is one of the parameters of the curve, like curve length. k describes the curvature and $\overrightarrow{N}$ as the normal vector. $C(p, 0)$ presents the initialized curve at $t = 0$. While $C(p, 0)$ is set up, the curve at $t\,(t \neq 0)$ can be achieved by evolving $C(p, 0)$ in the direction of the normal. The differential equation of curve evolution can be defined as follows:

$$\partial C(p, t)/\partial t = f\overrightarrow{N} \tag{1}$$

where $\partial C(p,t)/\partial t$ represents the evolution of curve over time and f is a function about the properties of the curve, which represents the speed evolving the curve. $\overrightarrow{N}$ represents the direction of evolving.

If f is set to a constant value, the curve will be evolving to the direction of normal in a fixed speed, equivalent to applying a constant force at each point of the curve. If f is a function of the curvature k, the speed of the curve evolving is relevant to the curvature. There are a lot of important curve evolving equations proposed with a different choice of function f. If $f = k$, it is the mean curvature motion equation:

$$\partial C(p, t)/\partial t = k\overrightarrow{N}. \tag{2}$$

If $f = k^{1/3}$, it is the affine-invariant geometric flow equation:

$$\frac{\partial C(p,t)}{\partial t} = k^{1/3}\overrightarrow{N}. \tag{3}$$

Equation 3 drives a closed non-convex curve to a smooth non-self intersection convex curve.

In this paper, we proposed a new curve evolving equation for the single-layer closed contour extraction of outer shape of skull and face from CT images, and we add several factors to control the shape of the curve to fit the borders of areas of interest and fill the holes and hollows they contained.

3 Threshold-Based Image Segmentation

We use threshold-based method to separate the bone and soft tissues on CT slices. This is a very popular method for medical image segmentation. We set up a CT data acquisition protocol to ensure the grayscale values of skull and soft tissue fixed to two well-separated ranges. The grayscale values remain stable between different acquisitions [9]. Each grayscale value in CT data is a 12-bit integer and varies from −2,048 to 2,048. According to the specification of data acquisition, value of soft tissue is between −200 and 350, that of bone is over 350, and that of fat and other tissues is below −200.

Using this scale to segment images, a pixel is considered to be the part of bone if grayscale value is over 350 and to be the part of soft tissues if grayscale value is between −200 and 350. Each CT slice can be divided into two different images: one is the bone part of skull separated from background and the other is all the soft tissue part out of bone and background. For convenient purpose, we set the value of area of interest to 255 and rest part to 0. The segment methods can be represented as given in Eq. 4:

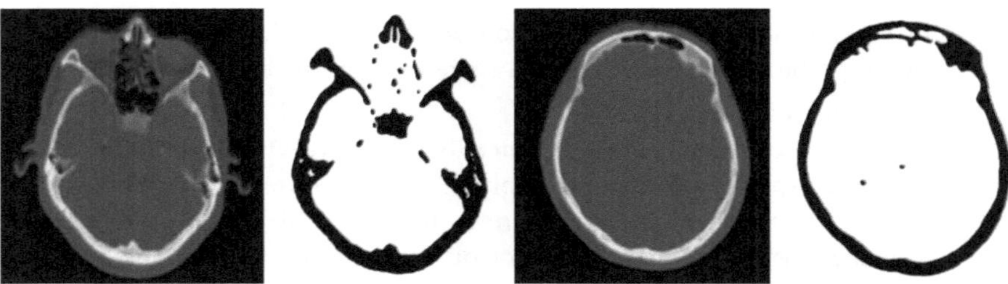

Fig. 1 Result of bone segmentation from craniofacial CT images

$$g'(i,j) = \begin{cases} 0, & g(i,j) < d \\ 255, & g(i,j) \geq d \end{cases} \quad i,j = 1,2,\ldots,n \tag{4}$$

where g represents the grayscale value of medical image, g' represents the grayscale value after segmentation, (i, j) represents the position of the pixel in the image, d is the threshold value, and the size of the image is $n \times n$. Result of segmentation is shown in Fig. 1.

4 Closed Outer Contour Extraction

The key step of this method is to extract the closed outer contour from the segmented bone images.

Up-to-date, there is no specified standard for craniofacial outer contour extraction [10]. We perform the extraction according to several rules to fulfill subsequent needs:

1. Contours must fit perfectly with the border of region of interest;
2. Contours must keep closed and non-self intersect, wrapping the region of interest include fill the holes they contains smoothly.

For those constraints, we extract the outer contours by evolving a curve to completely wrapping the region of interest and keep the contour smooth and closed. We use three factors to evolve the curve.

The first factor drags points on the curves to the inside direction of the curve, which are not fit along the border of the region of interest. We named this factor the attraction force. This force is a constant on all points of the curve and always points to the direction of the internal normal of the curve at a given point. As shown in Fig. 2, black area is a hollow bone region, the curve is a result of evolving, and the arrow that directs to the hollow represents the attraction force. This force constantly drags the point on the curve to the border of the region of interest until reaches the border or evolve stopped.

Fig. 2 Example of attraction force during evolving

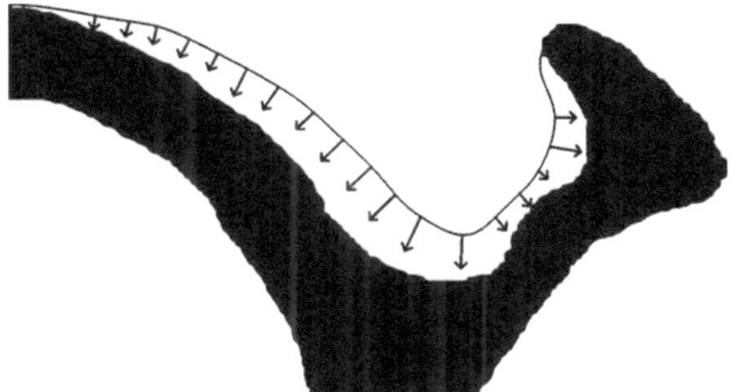

The second factor acts on those curve points which have any neighboring points that already fixed on the border of region of interest. This factor has the same direction with the attraction force and also a constant. It reinforces the attraction effect of the curve to the region of interest. We named this force adhesive force.

Those two forces make up the data extraction forces that extract the curve accurately evolving to the borders of the region of interest. But only those two factors are not sufficient enough to evolve the curve to wrap the region of interest. If the area includes a hole or a hollow, these two forces tend to drive the curve into the hole or the hollow and become more irregular. So we need the third factor to prevent this phenomenon.

When a curve evolves into a hole or a hollow under the effect of data extraction forces, its curvature at each point of the curve increases and becomes more irregular. The third factor we add tries to keep the curve regular. This factor is applied to all the points that not fixed on the border of region of interest. It has a direction of normal and the sign of curvature, and the value of the factor is proportional to the cube root of the curvature. We called this factor the curve regularization force. When this factor is applied with the attraction force and adhesive force, the curve will stop evolving at the balance position of those three factors and become accurate at the borders of the region of interest. The mathematical description of the curve evolution is given below:

$$\mathrm{d}C(p,\,t)/\mathrm{d}t = g'(C(p,\,t))\Big(p_1 + p_2(\mathrm{d}g'(C(p,t))/\mathrm{d}s^+ + \mathrm{d}g'(C(p,t))/ds^-) + p_3 k^{1/3}\Big)\overrightarrow{N(t)}. \quad (5)$$

This equation is an ordinary differential equation. In this equation, $C(p,\,t)$ represents the curve at time t and $\mathrm{d}C(p,\,t)/\mathrm{d}t$ represents the evolution of the curve over time. Left part of the equation describes the effect of three factors on the curve. g' represents the binary image after segmentation. $g'(C(p,\,t))$ inspects the forces acting on point $C(p,t)$ to determine whether to stop the forces acting on this point. If $C(p,t)$ is fixed on the border of the region of interest, this term equals 0 to cancel the effect of all the forces. Then, the point can be fixed on the given position of the region of interest. Otherwise, this term will equal 1. $\overrightarrow{N(t)}$ represents the internal normal of the curve at point $C(p,t)$ at time t. This term gives the curve evolution direction at each point. The evolution speed mainly depends on other parts of the equation.

The term p_1 represents the attraction force. It is a constant value. $\mathrm{d}g'(C(p,t))/\mathrm{d}s^+ + \mathrm{d}g'(C(p,t))/ds^-$ represents the adhesive force, and the terms $\mathrm{d}g'(C(p,t))/\mathrm{d}s^+$ and $\mathrm{d}g'(C(p,t))/\mathrm{d}s^-$ represent the derivatives of the image g' on the right and left of the curve point and along the curve. These derivatives are not equal to zero when the neighboring points on the right or left of this point have a different value to that point. In this term, this part generates an adhesive force which adds to the attraction force. If not, this part equals to zero. p_2 is a positive constant that determines the weight of the adhesive force. $k^{1/3}$ is the regularization force, k is the signed curvature of the curve, and p_3 is a constant which represents the regularization force.

During the experiment to extract the outer contour of the skull, all the constants are determined by experimental results. We set p_1 between 0.05 and 0.15, p_2 equals to 0.7, and p_3 equals to the maximum curvature reached by the curve.

5 Experimental Results

We use craniofacial CT images from 82 healthy persons as the raw data to test the curve evolution, and each set of CT image includes about 250 slices. First, we perform threshold-based image segmentation to separate the skull part out. Then, use the equations given in the 4th part of this paper to extract the closed outer contour of the skull part in each slice. The initial curve of each slice is the convex hull of the skull part. The significant result of each step is illustrated in Fig. 3.

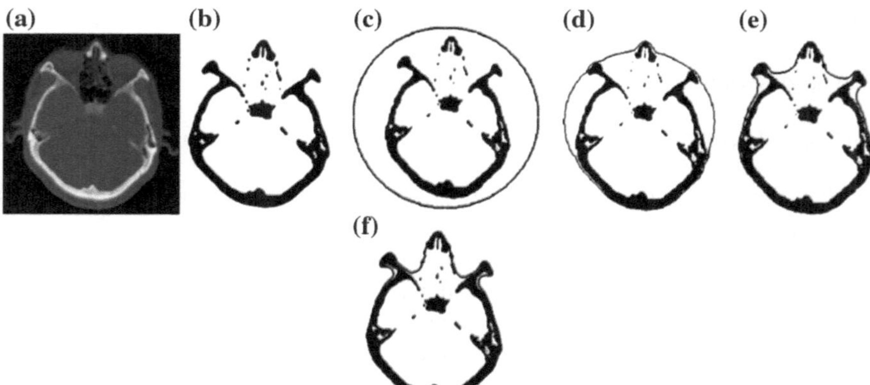

Fig. 3 Result of each steps in the curve evolution process. **a** A slice of CT images. **b** Segmentations of bone tissue. **c** Curve at $t = 0$ s. **d** Curve at $t = 1.8$ s. **e** Curve at $t = 3.5$ s. **f** Result of closed outer contour extraction

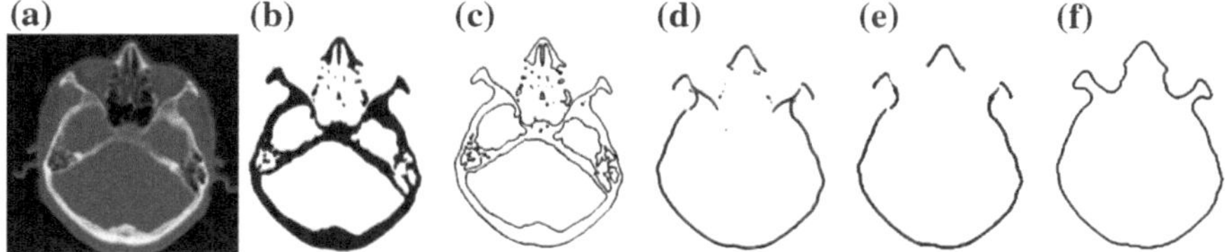

Fig. 4 Outer contour extraction by four different methods. **a** A CT slice of nose area. **b** Segmentations of bone tissue. **c** Result of Sobel operator. **d** Result of a line scanning method. **e** Result of snake method combined with line scanning method. **f** Result of curve evolution method presented in this paper

We set p1 equals 0.1 and p2 equals 0.7. Figure 3c is the initial curve at $t = 0$ s. It is convex circle that surrounds the bone tissue and has one pixel width. Figure 3d shows the curve after evolving for 1.8 s and several point of the curve already fixed on the border of the region of interest. Figure 3e shows the curve after evolving for 3.5 s. Most of points in the curve are already fixed on the border of region of interest and those points stopped evolving. Only the point over hollows and holes is still not fixed, and curve over hollows and holes cannot reflect the shape of hollows and holes. Those points keep on evolving to Fig. 3f. The result is perfectly fixed on the border of region of interest.

6 Discussion

We compared the contours extracted by several different methods with the result extracted by the methods of this paper. Figure 4a–b shows the CT slice of nose area and the segmented bone tissue image. Result extracted by Sobel operator is illustrated in Fig. 4c. The contours extracted are very complicated and different to form a single-layer outer contour. Figure 4d shows the result extracted by a line scanning method. The result is separated into several parts and has a lot of noise points. Figure 4e shows the result extracted by snakes combined with a line scanning method, and the outer contour is also separated into three parts. Figure 4f shows the result of curve evolution method presented in this paper. The extracted contour is a closed single-layer curve that perfectly fixed on the border of region of interest, and the nose area has a clean outer shape without noise points.

7 Conclusion

This paper focuses on the key problem that classical edge extraction methods usually come up with several separated closed contours of different parts of skull and hard to merge them to a single-layer contour. A single-layer closed contour extraction method based on curve evolution is presented in this paper. Craniofacial CT slices are segmented by threshold-based method to separate the bone tissue out. And then, evolves a single-layer closed curve which contains the skull region by a novel differential equation until the curve fits the border of the skull region perfectly. The results of experiments indicate that this method can automatically process the skull CT slices and extract the single-layer outer contour of the skull region accurately and efficiently. The results also well illustrate the shape of hollow and holes parts of the region of interest.

Acknowledgments This work was supported by Special Program for National Program on Key Basic Research Project of China (2011CB311802) and in part by the State Key Program of National Natural Science of China (Grant No. 60736008) and the National Natural Science Foundation of China (Grant No. 61172170).

References

1. Lakshmi, S., Sankaranarayanan, D.V.: A study of edge detection techniques for segmentation computing approaches. IJCA Special Issue on "Computer Aided Soft Computing Techniques for Imaging and Biomedical Applications" CASCT, pp. 35–40 (2010)
2. Tilotta, F., Richard, F., Glaunès, J.A., Berar, M., Gey, S., Verdeille, S., Gaudy, J.F.F.: Construction and analysis of a head CT-scan database for craniofacial reconstruction. Forensic Sci. Int. **191**, 112e1–112e12 (2009)
3. Raman, M., Himanshu, A.: Study and comparison of various image edge detection techniques. Int. J. Image Process. **3**(1), 1–11 (2009)
4. Song, L.: Research on the Information Extraction and 3D Reconstruction of Skull and Face Based on CT Data. Northwest University, Master Thesis (2009)
5. Kass, M., Within, A., Terzopoulos, D.: Snakes: active contour models. Int. J. Comput. Vision **1**(4), 321–331 (1988)
6. Wang, F., Geng, G.H., Feng, J.: Craniofacial reconstruction method based on snake model and ray method. Comput. Eng. **37**(2), 207–209 (2011)
7. Farzinfar, M., Xue, Z., Teoh, E.K.: A novel approach for curve evolution in segmentation of medical images. Comput. Med. Imaging Graph. **34**(5), 354–361 (2010)
8. Tannenbaum, A.: Three snippets of curve evolution theory in computer vision. Math. Comput. Model. **24**(5–6), 103–119 (1996)
9. Hounsfield, G.N.: Computerized transverse axial scanning(tomography): Part 1. Description of system. Br. J. Radiol. **46**, 1016–1022 (1973)
10. Tannenbaum, A.: Three snippets of curve evolution theory in computer vision. Math. Comput. Model. **24**(5–6), 103–119 (1996)

Distributed Data Platform System Based on Hadoop Platform

Jianwei Guo, Liping Du, Ying Li, Guifen Zhao and Jiang Jiya

Abstract Associated with the proposed rapid development of Web2.0, cloud computing, networking concepts, and technologies of the information age increasingly reflect the characteristics of its "big data." In order to exert the value of large-scale data, data mining technology in many areas of commercial, military, economic, and academic received more and more attention. At the same time, the huge scale of the data is a major challenge to the traditional data mining technology. A combination of data mining and cloud computing is becoming a trend in the industry rely on the robust-processing power provided by cloud computing and other distributed computing platform, and this kind of combination is constantly showing its strong advantages and potential.

Keywords Distributed computing · Hadoop

J. Guo (✉) · L. Du · Y. Li · G. Zhao · J. Jiya
Department of Information Technology, Beijing Municipal Institute of Science and Technology Information, Beijing 100044, China
e-mail: vipherovip@163.com

L. Du
e-mail: duliping_419@163.com

Y. Li
e-mail: shai_wang@hotmail.com

G. Zhao
e-mail: gfzh@hotmail.com

J. Jiya
e-mail: jiya_jiang@sina.com

S. Patnaik and X. Li (eds.), *Proceedings of International Conference on Computer Science and Information Technology*, Advances in Intelligent Systems and Computing 255, DOI: 10.1007/978-81-322-1759-6_61,

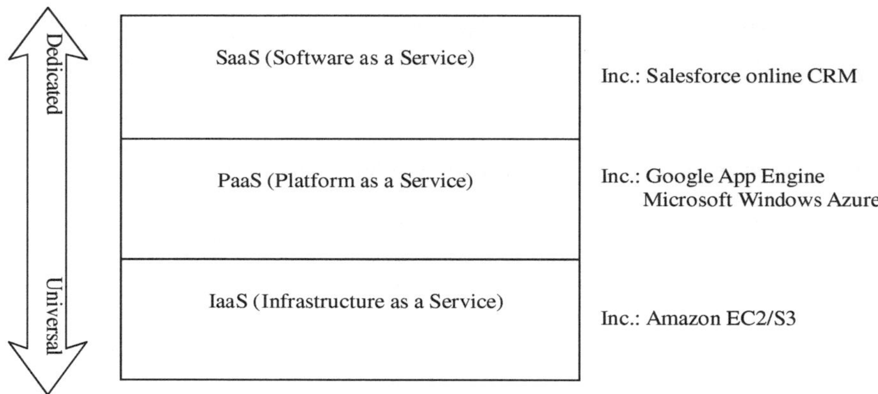

Fig. 1 The three major service types of cloud computing

1 Introduction

Cloud computing is generally regarded as a commercial computing model that uses the resource pool composed of a large number of computers for computation. This kind of resource pool is vividly called as "cloud" [1], from which the user can utilize the computing power, storage space, or information services according to requirements. The user of cloud computing can dynamically apply for some resources and submit various tasks to the cloud for autonomous management, operation, and maintenance by cloud services. The user and application developer can ignore the detailed distribution at the bottom layer and focus more on implementation of tasks to raise efficiency, reduce the costs, and promote technological innovation. The resource pool of cloud computing is also virtualized as computing and storage resources. Different resources can be dynamically allocated and organized according to demand, and the resources applied by the user can be recovered and reused by the system. This mode of operation can fully utilize the computing resources and improve the service quality.

Cloud computing is the development of parallel computing, distributed computing, and grid computing, or the commercial implementation of the computing concepts. It combines the concepts of virtualization and utility computing and provides a series of services from hardware to software. The three major service types of cloud computing are Infrastructure as a Service (IaaS), Platform as a Service (PaaS) and Software as a Service (SaaS), as shown in Fig. 1.

Hadoop [2] is an open source project of Apache foundation that provides the software framework for distributed computing environment, uses Hadoop Distributed File System (HDFS) and MapReduce parallel programming model as the core technology, integrates database, cloud computing management, machine learning, and other platforms, and gradually becomes the standard platform for application and research of cloud computing in the industrial and academic circles. Currently,

Hadoop is widely used in Facebook, Twitter, Yahoo!, and other famous companies and runs well in large-scale computer cluster with millions of computational nodes.

Hadoop was originated from the open source search engine Apache Nutch, one of the subprojects of Apache Lucene launched in 2002 [3]. In order to adapt to the sharp rise of data size, raise the data-handling capacity and ensure the search speed and accuracy of Nutch, an efficient distributed computation structure was in urgent need. In 2004, Google released the parallel data processing technology—MapReduce, one of its key technologies at the symposium on Operating System Design and Implementation, and published a thesis titled "MapReduce: Simplified Data Processing on Large Clusters [4]." At that time, Doug Cutting, the responsible person of Apache Nutch, saw the opportunity, lead his team to develop the open source MapReduce computation framework, combined it with the Nutch Distribution File System (NDFS), and integrated into the foundation platform of Nutch search engine. In February 2006, it was separated and became an independent project of Apache named as Hadoop. The core technologies of Hadoop are MapReduce parallel programming model and HDFS [5].

2 Structure of Hadoop Platform

The distributed system of Hadoop adopts a "Scale Out" mode to enhance the computing power. A comparative mode is called "Scale Up," which is represented by large stand-alone server. In the past decades, the development of computer and advancement in computing followed the Moore's law. With increase in data size, it was found that the problem of large-scale computing cannot be solved by relying on larger server only, but a new path should be opened, and more attention was paid to scaling out. In the case of Hadoop, scaling out means to organize low-end or commercial machines together and form a dedicated distribution system as shown in Fig. 2.

3 Distributed File System

3.1 Please Name Node and Data Node Architecture

Hadoop adopts the master/slave structure for distributed computation and storage, which includes two types of node, Name Node and Data Node. The two nodes play the master and slave roles, respectively in Fig. 3.

Name Node is at the master terminal of HDFS and plays the master role. Generally, there is only one Name Node in a Hadoop cluster. Name Node acts as the center of data management, but is not used as the hub of data transmission. It is

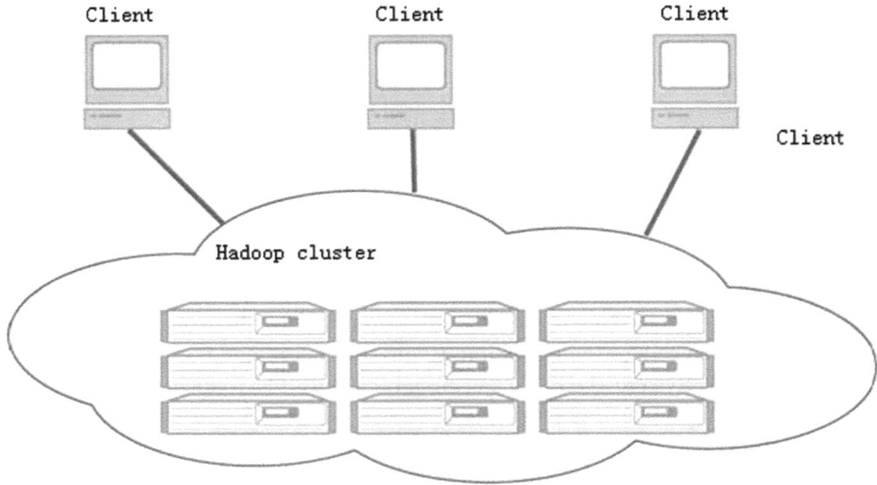

Fig. 2 Interactive mode of Hadoop cluster

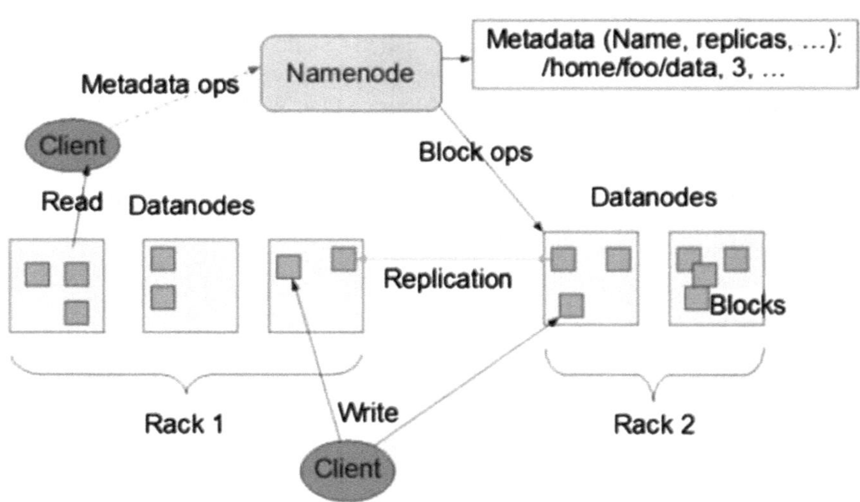

Fig. 3 HDFS architecture

used for management of the namespace of file system, storage of the directory structure of the whole file system and index node of files, and providing metadata service for HDFS.

3.2 Online Publication in Springer Link

All papers will be published in our digital library www.springerlink.com. Only subscribers to Springer's eBook packages or to the electronic book series are able to access the full text PDFs and references of our online publications. Metadata, abstracts, and author e-mail addresses are freely available for all users.

4 MapReduce Parallel Programming Model

MapReduce is an abstract programming model originally developed and used by Google, which can solve the problem of data-intensive operation under many big data environments and deliver the programs that are designed in distributed environment and ideal for parallel computing.

5 MapReduce Implementation

The part of key-value pair of intermediate data with sequence or marking information (defaults of authentication key) can usually be hashed, and the data will be transferred to the corresponding Reducer according to the result of hashing. Certainly, Hadoop will be a convenient method for definition of the allocation rule by the user. In the process of implementation, each data output by Mapper will be redirected on the network, and rule judgment of each data will be performed by Partitioner. Data with the same characteristics will be put into the same Reducer for processing. The process of transferring data into the correspnding Reducer is generally called shuffle as shown in Fig. 4.

6 MapReduce Framework

Each request for computation in the Hadoop cluster is called a job. Hadoop is used to complete the job in a distributed environment. Like HDFS of master/slave structure, MapReduce adopts the similar architecture composed of three types of server, JobTracker, TaskTracker, and JobClient. The master process in the MapReduce framework of Hadoop is a Hadoop called JobTracker, which is responsible for management of all jobs under the framework, and as a scheduling core, allocates tasks to various jobs as shown in Fig. 5.

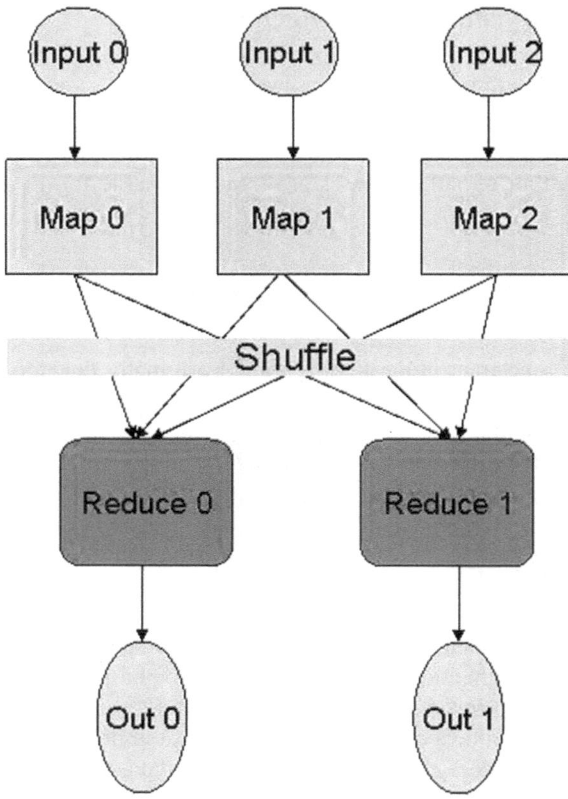

Fig. 4 Shuffle process of MapReduce

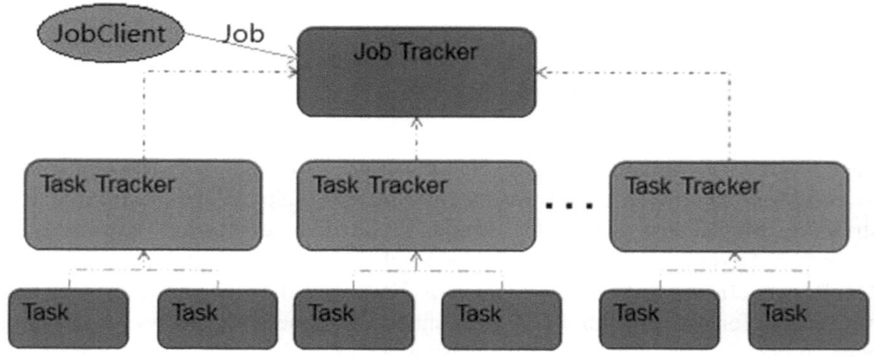

Fig. 5 Basic framework of Hadoop MapReduce

7 Conclusion

This paper has studied on the software architecture of Hadoop for building a distributed platform, focused on the core framework HDFS and the parallel programming model MapReduce, and introduced the design concept of HDFS and the basic method for file storage and management by practice. In particular, the parallel programming model MapReduce is systematically studied for understanding its powerful problem processing and implementation model, and an application template based on MapReduce is written for data mining.

References

1. Xu, Q., Wang, Z.: Cloud Computing: Application Development Practice, p. 12. China Machine Press, China (2011)
2. The Apache™ Hadoop® Project. http://hadoop.apache.org/
3. Lam, C.: Hadoop in Action. Manning Publications Co., New York (2011)
4. Dean, J., Ghemmawat, S.: MapReduce: simplified data processing on large clusters. In: Proceedings of the 6th Symposium on Operating System Design and Implementation, pp. 137–150. ACM Press, New York (2004)
5. Apache. Hadoop distributed file system, 2010-09-24. http://wiki.apache.org/hadoop/HDFS

Retraction: Developing Write-Back Caches and Information Retrieval Systems with EASEL

Mingqian Wang, Yingying Wang, Yueou Ren and Xi Zhao

Several conference proceedings have been infiltrated by fake submissions generated by the SCIgen computer program. Due to the fictional content the chapter "Developing Write-Back Caches and Information Retrieval Systems with EASEL" by "Mingqian Wang, Yingying Wang, Yueou Ren and Xi Zhao" has been retracted by the publisher. Measures are being taken to avoid similar breaches in the future.

S. Patnaik and X. Li (eds.), *Proceedings of International Conference on Computer Science and Information Technology*, Advances in Intelligent Systems and Computing 255, DOI: 10.1007/978-81-322-1759-6_62,

A Method of Archiving History Data Based on XML

Yan Zhang and Xu Luo

Abstract Analyzed radical problem of historical data archiving in data warehouse is presented, applying XML into archiving. Arithmetic of getting information about structuring XML pattern from data dictionary and arithmetic of structuring level direct graph were provided. The arithmetic of mapping from database structure to XML Schema based on level direct graph was implemented. In the meantime, the ways to ensure XML semantic integrity were provided. This schema can adapt to data analysis under MapReduce very well.

Keywords Data warehouse · Archiving · XML Schema · Directed graph · Mapping

1 Introduction

Archiving historical data in the data warehouse system refers to the early detail-level data, which generally store in relatively advanced relational database. However, for archiving historical data, availability, refactoring, and storage efficiency of data must be considered carefully. First, archiving historical data will grow fast in size and have to be stored for a long period. So a low-cost storage strategy should be

Y. Zhang (✉)
Professor, Teaching Department of Computer and Mathematical Foundation of Shenyang Normal University, Huanghe North Street 253, Huanggu District, Shenyang 110034 Liaoning, China
e-mail: Zhangyan_synu@126.com

X. Luo
Lecturer, Teaching Department of Computer and Mathematical Foundation of Shenyang Normal University, Huanghe North Street 253, Huanggu District, Shenyang 110034 Liaoning, China
e-mail: Luoxu_2002@sina.com

S. Patnaik and X. Li (eds.), *Proceedings of International Conference on Computer Science and Information Technology*, Advances in Intelligent Systems and Computing 255, DOI: 10.1007/978-81-322-1759-6_63,

adopted. Secondly, archiving historical data are old data instead of useless data. In the application of certain OLAP and DM, the archiving historical data play an irreplaceable role [1]. So it is advised to employ a storage technology that can guarantee the refactoring of the data. XML has its advantage of solving the problems above, which can be applied to archive the historical data [2].

- In the reuse of archiving historical data, the problem caused by differences in platform and system often exists. In this regard, the XML technology is independent of platform, which can improve the availability of archiving historical data.
- Relational database technology does not support semantic description well. So it is difficult to maintain the stability and transitivity. XML has the easily expressed features, which makes it to be an effective solution to the reconstructed data. XML can guarantee the integrity by dealing with differences in archiving data and current data.
- The MapReduce frame paralleled the tasks of data warehouse; it can process data sources with different type, XML for Analysis (XMLA) plays an important role in the course of receiving the request from user to calling computing tasks of MapReduce [3].
- For the increasing of data redundancy caused by data description demand in the same capacity condition, the space which XML document has taken up is larger. But, for the archiving historical data, the compressing effect of XML document is best and storage efficiency is highest [1].

2 Building XML Data Model

Archiving historical data are to be stored as XML documents. The key in guaranteeing the correctness and validity is XML Schema. In order to express the archiving database ensuring the associated data structure, three problems need to be solved. The first is how to construct relations among tables, which means how the relationships and referential integrity constraints between two entities are structured [3]. The second is how to form semantically correct, complete XML Schema with its entity integrity [4].

2.1 Building XML Data Model from the Data Dictionary Information Document

The data dictionary is a set of tables which store the database data. The information related to the XML data model mainly contains metadata for each table with its columns and index properties in the database, which are the basis for building XML data model. In archiving the historical database warehouse data,

SQL and similar tools are used to read the information out and form a programmable accessed document, which serves as a reference to the XML model [5]. The data document stores the required database names, table names, table filed tags with notes, and integrity constraints. In a teaching monitoring database (Instruction), for instance, message table of student (Students), message table of teacher (Teachers), message table of courses (Courses), message table of sore (Scores), message table of teaching evaluations (Evaluations), and message table of equipments running (Equipments), can form a metadata document. A XML data model of database with table students, table courses, and table scores can be structured, including field, data type, length, digits, null, primary key, external keys, and so on.

2.2 Build Directed Multi-Level XML Model Graphs

A database-centered directed graph $G = (V, E)$,where V is the finite set of vertices and E is the finite set of ordered two vertices in V, can be constructed based on a XML data model table. Here is an algorithm for building an XML model directed graph.

- Express the vertex set $V = \{V_1, V_2, \ldots\}$ on tables.
- Create the table set of vertices $V' = \{V_i, V_j, \ldots\}$ on the foreign key relationship table.
- Count the number of vertex n in set V', which indicates how many association references are supported currently, that is, the in-degree n.
- Express the vertex set $V(1)$ and $V(>1)$ when $n = a$ and $n > 1$, respectively, on the in-degree of the table. The in-degree of $n = 0$ means vertices $V(0) = V–V'$.
- Generate a vertex as the root whose name is the database name. Then, connect root vertex with vertex in $V(0)$. The root vertex is the end point, and the table field is the starting point. The root vertex is added to the set V.
- Construct vertex pairs in $V(0)$, $V(1)$, $V(>1)$ on foreign key tables. Connect vertex pairs with $V(i)$ as starting points and $V(j)$ as ending points where $i < j$, respectively.
- For any two nodes in V', V_i and V_j, if (V_i, V_j) and (V_j, V_i) are two directed edges, then generate a new table vertex, connecting V_i and V_j with V_{ij} as the starting point and V_i, V_j as the ending points, respectively. The newly generated table contains referenced attributes and properties. The referenced properties generate the compound keys in the table. The new vertex V_{ij} is added to set V after deleting the corresponding two vertices' table.

The XML model directed graph built by the algorithm has the following characteristics:

- There is one and only one vertex with in-degree 0, which is the root vertex.
- For any two nodes in V, V_i and V_j, no more directed edge (V_i, V_j) is here. This is called special level directed graph. Figure 1 shows the XML model based on the algorithm.

3 Generate Mapping XML Schema

It is possible to generate mapping XML Schema as it is similar to XML model level directed graph in structure. Here, an algorithm to generate an XML Schema based on XML model mapping is introduced.

- Make the root vertex in directed graphs map to the outermost element in the XML pattern.
- Make the one-degree vertex as child elements and nest them in its starting point to the edge element, which means they are in the parent vertex element.
- Make the vertex with in-degree more than one as child elements, nest them in the shortest ancestor vertex to which they are connected.

The XML level model in Fig. 1 can be mapped to XML Shema level structure in Fig. 2 via the algorithm above. A complement level structure including databases, data sheets, records, and fields can be got by the level structure based on XML level structure. The nest relationships are tables in databases, records in tables, and fields in records. A reference relationship exists in tables in the database. Detailed mapping rules are as follows:

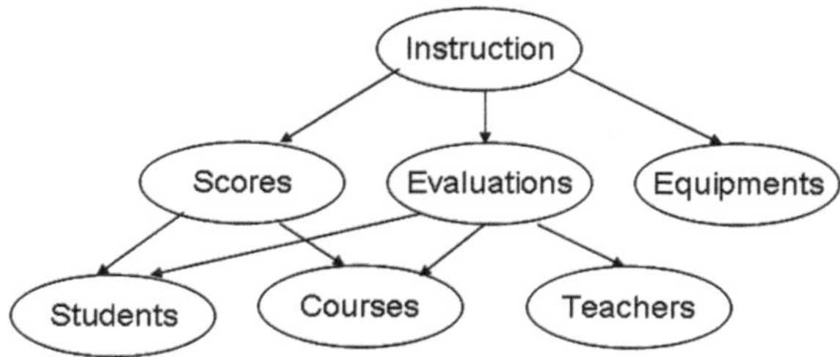

Fig. 1 XML pattern level directed graph

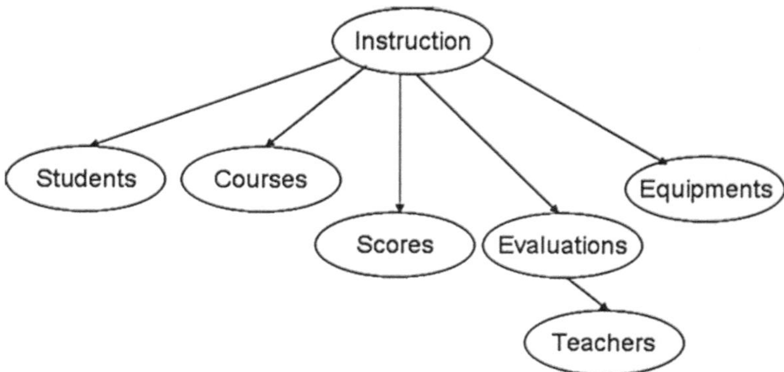

Fig. 2 XML Schema level structure

- Mapping the whole database to a root element, which is called the metaelement with the database name.
- A table element with the same name got from the plural form in the database maps the table elements to child elements [3].
- A record element is got by mapping each record to its child elements in each table. The name usually is in the single form.
- In each of the record element, further description of the elements of the various fields is contained. Under normal circumstances, elements as primary keys or foreign keys must be mapped to text field elements, with their field names. And for other fields, they can be mapped to elements or attributes [4]. However, to guarantee the semantic integrity of the archive historical data, the property is used to make some notes and instructions on the elements. So, it is recommended to map fields as elements with the element name and value mapping field names and attribute values and the relationship between modes, respectively. For field elements, to separate it from common elements, the primary key is defined as the key element and foreign key as type keyref. Meanwhile, the selector elements and field elements used to describe the positioning of the associated reference need to be declared.

4 To Guarantee the Semantic Integrity

In comparison with the XML, the database is weak in the semantic integrity. The advantages of XML can be used to improve the effectiveness and availability of historical data.

4.1 Description Properties of Supplement Physical Elements

In the database relationship mode, field name usually uses single English letters or combinations. Only designers and database administrators are aware of the real meanings. So the actual meaning must be described when archiving the historical data. For instance, in XML Schema, a description data type is added for each physical element (meta, table, record, field elements).

4.2 Description Child Elements for Supplementary Value Codes

Under normal circumstances, to enhance the running performance of the system, concise description and expression of practical code are often used. The code expression is the detailed description a unit or department generated within a certain period of time. As most usual database is with short period of validity and self-defined conventions, the ambiguity may not be a so serious issue. However, the complete description of the elements must be done for further use in archiving historical data. So in the XML in the archiving historical data model, elements with values as codes must be described using supplementary codes. The detail is as follows:

- Generate the set $A = \{a1, a2, \ldots\}$ with each element's actual meanings based on the decomposed code.
- For all $ai \in A$, form a two-column data table including code definitions and values, respectively. The definition means the actual meaning and declares as primary keys. The code value means corresponding values in the certain domain of concrete meanings.
- Expand the XML model level directed graph to form a reference relationship with code elements as starting points and ending points as pairs of vertices got from step two.
- Convert the code element primary keys from single structure to compound structure, consisting of primary keys of child elements and their supplementary child elements.

In order to guarantee the semantic integrity, the extended Student element XML model is shown in Fig. 3. The concrete XML Schema mapping method is the same as above. A repeated description is omitted.

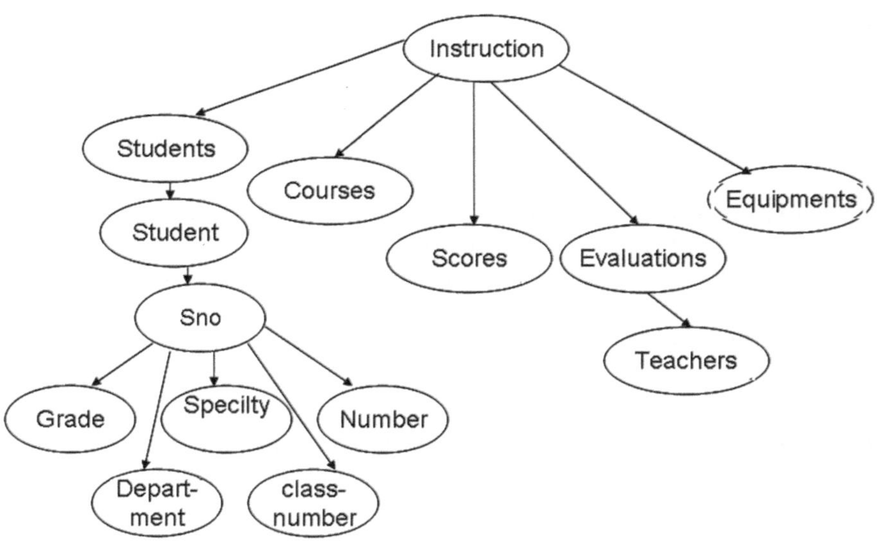

Fig. 3 Expanded XML Schema

Table 1 Comparison of data loading capacity and data loading time

Number of records	Loading capacity (M)		Loading time (ms)	
	DBF	XML	DBF	XML
164,000	18.3	21.6	107,516	99,633
410,000	45.8	53.9	212,239	222,289
574,000	64.2	75.5	336,101	352,803
820,000	91.7	107.9	443,966	438,840

5 Test Results

A test is done in the apache Hadoop release 1.0.4. Hadoop provides Stream Xml Record Reader class; it can be used by setting input type to Stream Input Format and setting attribute of stream.recordereader.class to org.apache. Hadoop. Streaming. Stream Xml Record Reader [6]. Test goals are the data loading capacity and data loading time. By comparing experimental results of the DBF document and the XML document, it is found that the space which XML document has taken up is a little larger. But, for the data loading time, there is no obvious superior or inadequate between the DBF document and the XML document. Comparison data are shown in Table 1. It can be inferred that the DBF document loading capacity and loading time will be greater than the XML document if more tables are in the database.

6 Conclusion

The archive of historical data of the data warehouse is an important part of the composition of the data warehouse system. Applying XML technology to the historical data archive is a proper solution to key problems in data warehouse. Using the technique above, mapping the data of historical structure in the database to XML Schema, a XML model with complete semantic integrity can be constructed. A XML document can be created based on the XML archiving technology. Finally, the availability and reconstruction of the data can be guaranteed and the value of the data can be improved.

References

1. Zhang, Y., Xia, X.-F., Yu, G.: Study of historical data archiving strategy in data warehouse. J. Shenyang Inst. Aeronaut. Eng. **21**(4), 65–67 (2004)
2. Zhang, Y., Xia, X.-F., Yu, G.: Realizing of historical data archiving in data warehouse based on XML. J. Shenyang Normal Univ. (Nat. Sci.) **23**(2), 166–168 (2005)
3. Li, Y.-F., Jia, S.-Y., Deng, S.-K., Han, Y.-Y.: MapReduce model based on tree structure. Comput. Technol. Dev. **23**(1), 32–45 (2012)

4. Qin, X.-P., Wang, H.-J., Du, X.-Y., Wang, S.: Big data analysis-competition and symbiosis of RDBMS and MapReduce. J. Softw. **30**(3), 102–105 (2013)
5. Wei, Y.-S., Zhang, F., Chen, X., Sun, Z.-L.: A optimization method for XML query based MapReduce. J. Microelectron. Comput. **30**(3), 102–105 (2013)
6. White, T.: Hadoop: The Definitive Guide, vol. 5, pp. 365–405. Publishing House of Tsinghua university, Beijing (2010)

Energy-Aware Resource Management and Green Energy Use for Large-Scale Datacenters: A Survey

Xiaoying Wang, Xiaojing Liu, Lihua Fan and Jianqiang Huang

Abstract As cloud computing gains a lot of attention that provides various service abilities, large-scale datacenters become dominant components of the cloud infrastructure. Huge energy consumption appears to be nonignorable leading to significant cost and also bad impacts on the global environment. How to efficiently manage the services while keeping energy consumption under control is an important problem. According to the analysis of prior representative literature, this paper makes an overview of energy-aware resource management approaches. The basic architecture of cloud datacenters and virtualization technology is introduced. Then, we conduct a survey of energy-aware approaches for adaptive resource management of such cloud environments. We also focused on some studies of renewable energy usage for green datacenters. Finally, the research problems are summarized and analyzed synthetically and possible future directions are given.

Keywords Cloud computing · Large-scale datacenters · Renewable energy · Energy-aware resource management

X. Wang (✉) · X. Liu · L. Fan · J. Huang
Department of Computer Technology and Applications,
Qinghai University, Xining 100086 Qinghai, China
e-mail: wxy_cta@qhu.edu.cn

X. Liu
e-mail: liuxj@qhu.edu.cn

L. Fan
e-mail: fanlh@qhu.edu.cn

J. Huang
e-mail: huangjq@qhu.edu.cn

S. Patnaik and X. Li (eds.), *Proceedings of International Conference on Computer Science and Information Technology*, Advances in Intelligent Systems and Computing 255, DOI: 10.1007/978-81-322-1759-6_64,

1 Introduction

In the cloud environment, the key infrastructure is usually comprised of large-scale datacenters, providing online computing services for thousands of millions of customers simultaneously. These datacenters own hundreds to thousands of heterogeneous server nodes, which could consume significant amount of energy. Furthermore, the ratio of energy cost can reach to more than 50 % compared with the total operational cost of the entire datacenter [1]. Each single year, more than 30 billion dollars are spent on dealing with the extra heat derived from massive enterprise services all over the world, even more than the money spent on buying new hardware and devices [2]. The problems it brought include not only the expensive maintenance and operational cost for datacenter providers, but also the pollution and bad impact on the global natural environment. One of the main reasons is that the power consumed by large-scale datacenters mainly comes from traditional manners [1], which use fuels and coals to generate power.

In order to reduce the total energy consumption of the entire datacenter by itself, a number of methods have been exploited and tried, including: (1) improvement of the chip manufacturing to reduce the hardware power consumption while keeping high performance; (2) server consolidation using virtualization techniques to reduce the number of active server nodes [3]; (3) rebuilding the heat dissipation systems, e.g., using new novel heat dissipation methods such as water cooling to reduce the power consumption. These approaches are all from similar point of view—saving and reducing. However, the basic function of large-scale datacenters is to provide services for high-performance computing and massive data processing, which would more or less limit the reduction amount of energy consumption. According to such considerations, a number of famous enterprises and IT service companies began to explore possible solutions of using renewable energy to provide the power supply for large-scale datacenters. For example, *Google* has been pondering a "floating datacenter" that could be powered and cooled by the ocean [4]; *Apple* planned to build a brand-new datacenter in Prineville, Oregon, and vowed to use "100 % renewable energy" [5]; *HP* also attempted to create a "Net-Zero" datacenter that requires no net energy from utility power grids [6].

Nevertheless, the generation of renewable energy is usually intermittent. For example, solar energy will be greatly impacted by the strength of the direct sunlight, which is usually high in daytime and low in nighttime. Hence, it is a great challenge to appropriately manage the resources of the datacenter and the workloads of the upper-level applications, in order to accurately control the energy consumption to match the fluctuation of the unstable incoming energy supply.

In this paper, we intend to review the related work about energy-aware resource management and the utilization of renewable energy in large-scale datacenters for cloud computing. The architecture and the execution mechanisms of such datacenters are first introduced, and then, we discuss some possible solutions to save energy while keeping system performance. The difficulties and challenges when leveraging the renewable energy are presented, and we will also review some existing approaches that aim to solve the intermittency and fluctuation features. After the comprehensive survey and relevant analysis, we will discuss about the possible future research directions based on the state of the art.

2 Overview of Large-Scale Green Datacenters

2.1 Architecture

The architecture of a typical large-scale datacenter with mixed energy supplies is shown in Fig. 1. The left half of the figure shows the supplying part of the whole system, which integrates the traditional grid utility and renewable energy. The automatic transfer switch (ATS) combines different energy supplies together and provides the energy to the datacenter. The right half of the figure shows the consumption part of the whole system. The functional equipments inside the datacenter consume energy for dealing with fluctuating incoming workloads. At the same time, some cooling units have to work in order to lower the temperature and guarantee the availability of the devices, which will consume considerable amount of power too.

2.2 Virtualization

Virtualization is an essential technology for cloud computing, which introduces a software abstraction layer between the hardware and the operating system with applications running upon it. Researches in labs and universities are developing approaches based on virtual machines to solve manageability, security, and portability problems [7, 8]. The ability of multiplexing hardware and server consolidation greatly facilitates the requirements of cloud computing.

Specifically, in a large-scale datacenter for cloud computing, virtual machines are usually deployed and used in a manner [9] as the architecture shown in Fig. 2. Multiple VMs can concurrently run applications based on different operating system environments on a single physical server in the datacenter. VMs can be

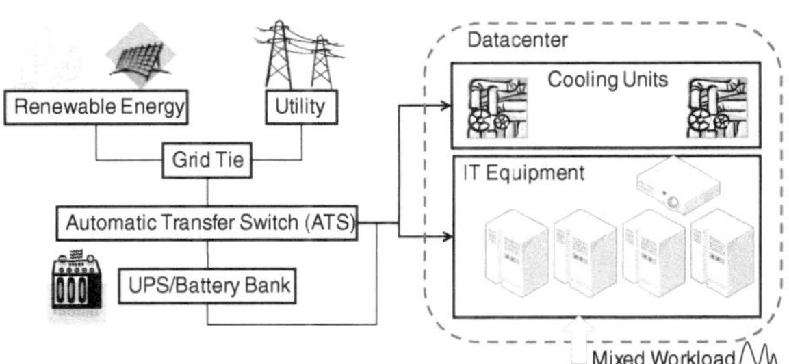

Fig. 1 Architecture of the large-scale green datacenter

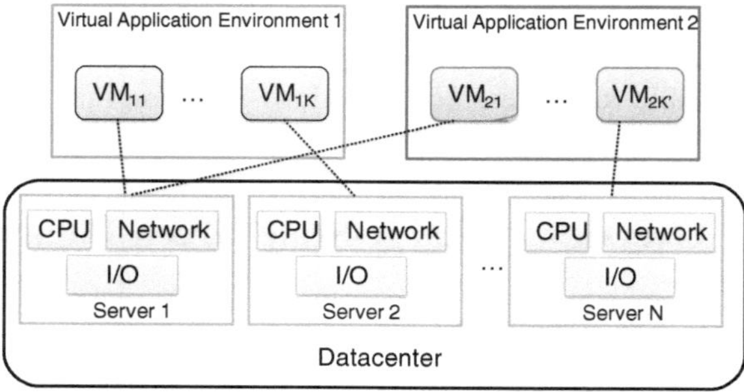

Fig. 2 Cloud infrastructure with virtual application environments

dynamically started and stopped according to incoming requests, providing flexibility of configuring various partitions of resources on the same physical machine according to different requirements of service requests [10].

3 Energy-Aware Resource Management

3.1 Switching On/Off Nodes

A straightforward consideration for saving energy is to properly shutdown some of the physical servers to reduce extra overhead, under a condition of compelling all of the workload of these servers outside. It requires the resource manager to efficiently analyze and redistribute loads and adjust resource allocation schemes. Some researchers have made efforts on such approaches. Chen et al. [11] characterized the unique properties, performance models, and power models of connection servers. Pinheiro et al. [12] developed systems that dynamically turn cluster nodes on and off to handle the load imposed on the system while at the same time save power under light load. In this way, power consumption of the whole system could be saved remarkably by turning some servers into off mode. However, the overhead and time latency of turning on/off nodes is also nonignorable, which might result in a delay in processing of workloads.

3.2 Dynamic Voltage and Frequency Scaling

Comparatively, another more finer-grained way to save energy for CPUs is to utilize the features of dynamic voltage scaling (DVS). Bohrer et al. [13] from *IBM Research* conducted a research on the impact of the workload variation in energy consumption and designed a simulator to quantify the benefits of dynamically scaling the processor voltage and frequency. Rajamony et al. [14], also from *IBM Research*, considered the energy saving issue from two different aspects and proposed independent voltage scaling (IVS) and coordinated voltage scaling

(CVS) policies. Sharma et al. [15] investigated adaptive algorithms for DVS in quality of services (QoS)-enabled web servers to minimize energy consumption subject to service delay constraints. Petrucci et al. [16] presented in their work a dynamic configuration approach for power optimization in virtualized server clusters, which leveraged a dynamic configuration model and outlined an algorithm to dynamically manage the virtual machines, with the purpose of controlling power consumption while meeting the performance requirements.

To sum up, using DVFS techniques to eliminate unnecessary waste for CPU power is a popular way in current researches. Since it supports fast switching with neglectable delay time, the workloads can be processed in time with low overhead.

3.3 Saving Cooling Energy Consumption

The previous subsections presented some existing approaches to reduce the power consumption of IT devices themselves. However, in large-scale datacenters, the energy consumption for heat dissipation and cooling also occupies a significant part of the total amount, even up to 50 % [17]. Hence, some researchers turned to focus on temperature-aware or thermal-aware resource management approaches.

Tang et al. [18] looked into the prospect of assigning the incoming tasks around the data center in such a way so as to make the inlet temperatures as even as possible, allowing for considerable cooling power savings. Pakbaznia et al. [19] presented a power and thermal management framework for datacenters where resources are dynamically provisioned to meet the required workload while ensuring that a maximum temperature threshold is met throughout the datacenter. Ahmad and Vijaykumar [20] proposed *PowerTrade* to trade off idle power and cooling power for each other and reduce the total power; and *SurgeGuard* to over-provision the number of active servers beyond that needed by the current loading so as to absorb future increases in the loading. Wang et al. [21] established an analytical model that describes datacenter resources with heat transfer properties and workloads with thermal features.

To sum up, this section reviews the researches in the area of resource management, task scheduling, load balancing that also partially considers energy consumption reduction. However, since the total energy consumption is significant for such large-scale datacenters, there is finally a limitation of the possible saved energy amount. Even if 10–20 % of the total energy consumption could be saved, the carbon emission will still be a huge amount.

4 Renewable Energy Use in Green Datacenters

New types of renewable energy such as solar, wind, and tidal bring advantages by their features including: sufficiency, cleanness, sustainability, nonpollution, and so on. This section summarizes some works about renewable energy use in green datacenters and illustrates some possible attempts.

4.1 Single Datacenter

Here, some researches about how to efficiently utilize renewable energy inside a single datacenter are first reviewed and summarized, as follows.

Deng et al. [22] proposed the concept of carbon-aware cloud applications, which treated carbon-heavy energy as a primary cost, provisioning a cloud instance only if its emission costs are justified by application-specific rules. Goiri et al. designed a framework called *GreenSlot* [23] that aims to schedule batch workloads, and another framework called *GreenHadoop* [24] that orients MapReduce-based tasks. Both are based on the prediction of the availability of renewable energy and try to maximize the utilization of available green energy by different scheduling strategies. Krioukov et al. [25] presented an energy agile cluster that is power proportional and exposes slack. Li et al. [26] proposed *iSwitch*, which switches between wind power and utility grid following renewable power variation characteristics, leverages existing system infrastructures. Arlitt et al. [6] from *HP Labs* introduced and designed a "Net-Zero energy" datacenter managed in a manner that uses on-site renewables to entirely offset the use of any nonrenewable energy from the grid.

4.2 Multiple Distributed Datacenters

Some big companies and enterprises usually establish multiple datacenters around different areas all over the world, powered by mixed green energy and utility grid, as shown in Fig. 3. Recently, there are also some works discussing the possibility of exploring the heterogeneousness of the distributed datacenters and co-scheduling the workload among multiple datacenters.

Stewart and Shen [27] outlined a research agenda for managing renewable in the datacenter, which compliments ongoing efforts to integrate renewables into the grid. Akoush et al. [28] introduced a design called "Free Lunch" that exploits otherwise wasted renewable energy by colocating datacenters with these remote energy sources and connecting them over a dedicated network. Chen et al. [29] proposed a holistic workload scheduling algorithm, called *Min Brown*, to minimize the brown energy consumption across multiple geographically distributed datacenters with renewable energy sources. Le et al. [30] sought to exploit geographically distributed datacenters that pay different and perhaps variable electricity prices, the benefit of different time zones and near sites that produce renewable electricity. Heddeghem et al. [31] looked at the feasibility of globally distributing a number of these renewable sources for powering already distributed datacenters and provided a mathematical model for calculating the carbon footprint. Li et al. [32] proposed a collaborative cost optimization framework by coupling utilities with datacenters via dynamic pricing.

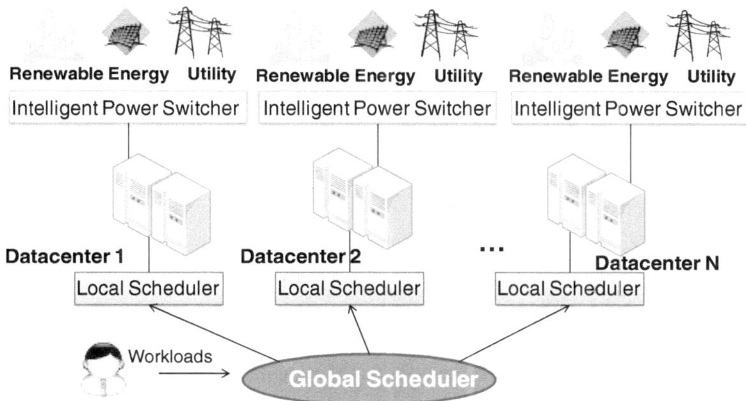

Fig. 3 Architecture of geographically distributed datacenters

To sum up, the above works considered to utilize the benefit of geographically distributed locations, different time zones, and prices to schedule and dispatch loads onto multiple datacenters.

5 Conclusion and Future Directions

In this paper, we reviewed relevant research works about energy-aware resource management approaches inside large-scale datacenters for cloud computing. From the comprehensive survey, we found that although there have been a number of researches starting to explore the energy-efficient management issues, the relevant study of renewable energy usage is still preliminary. Considering that the generation process of renewable energy is usually intermittent and random, we intend to discuss some possible future research directions as follows: (1) It is necessary to study on how to incorporate energy-related metrics into the optimization objectives. (2) A holistic framework has to be established, which could describe the relationships between job scheduling and IT business power consumption, resource allocation and power consumption, transaction and cooling power consumption, and so on. (3) The looseness and slackness of the dominant workload inside the datacenter should be further exploited, which could facilitate the adjustment of resource allocation toward the requirements of varying power supply strength. (4) Since the utilization of all nodes in the datacenter is usually unbalanced, the resource-controlling approach should be aware of temperature and locations. How to design location-aware and thermal-aware strategies is still an important open issue that needs to studied.

Acknowledgments This research is funded in part by National Natural Science Foundation of China (No. 61363019, No. 60963005) and Tsinghua—Tencent Joint Laboratory for Internet Innovation Technology (No. 2011-1).

References

1. Le, K., Bilgir, O., Bianchini, R., et al.: Managing the cost, energy consumption, and carbon footprint of internet services. In: Proceedings of ACM SIGMETRICS Performance Evaluation Review, p. 357–358. ACM (2010)
2. Bianchini, R., Rajamony, R.: Power and energy management for server systems. Computer **37**(11), 68–76 (2004)
3. Uddin, M., Rahman, A.A.: Server consolidation: an approach to make data centers energy efficient and green. Int. J. Sci. Eng. Res. **1**(1), 1–7 (2010)
4. LaMonica, M.: Google files patent for wave-powered floating data center. Available from http://news.cnet.com/8301-11128_3-10034753-54.html (2008)
5. Rogoway, M.: Apple outlines 'green' energy plans for Prineville data center. http://www.oregonlive.com/silicon-forest/index.ssf/2013/03/apple_outlines_green_energy_pl.html (2013)
6. Arlitt, M., Bash, C., Blagodurov, S., et al.: Towards the design and operation of net-zero energy data centers. In: Proceedings of IEEE ITherm2012, pp. 552–561. IEEE (2012)
7. Zhao, M., Figueiredo, R.J.: Experimental study of virtual machine migration in support of reservation of cluster resources. In: Proceedings of Experimental Study of Virtual Machine Migration, pp. 1–8. ACM (2007)
8. Sundararaj, A.I., Sanghi, M., Lange, J.R., et al.: Hardness of approximation and greedy algorithms for the adaptation problem in virtual environments. In: Proceedings of ICAC '06, pp. 291–292. IEEE (2006)
9. He, L., Zou, D., Zhang, Z., et al.: Developing resource consolidation frameworks for moldable virtual machines in clouds. Future Gener. Comput. Syst. (2012). doi: 10.1016/j.future.2012.05.015 (in press)
10. Beloglazov, A., Abawajy, J., Buyya, R.: Energy-aware resource allocation heuristics for efficient management of data centers for cloud computing. Future Gener. Comput. Syst. **28**(5), 755–768 (2012)
11. Chen, G., He, W., Liu, J., et al.: Energy-aware server provisioning and load dispatching for connection-intensive internet services. In: Proceedings of the 5th USENIX Symposium on Networked Systems Design and Implementation, pp. 337–350. NSDI (2008)
12. Pinheiro, E., Bianchini, R., Carrera, E.V., et al.: Load balancing and unbalancing for power and performance in cluster-based systems. In: Workshop on Compilers and Operating Systems for Low Power, pp. 182–195. Barcelona, Spain (2001)
13. Bohrer, P., Elnozahy, E., Keller, T., et al.: The case for power management in web servers. Graybill, R., Melhem, R. (eds.) Power Aware Computing. Springer, New York (2002)
14. Rajamony, R., Elnozahy, M., Kistler, M.: Energy-efficient server clusters. In: Proceedings of the Second Workshop on Power Aware Computing Systems. Springer (2002)
15. Sharma, V., Thomas, A., Abdelzaher, T., et al.: Power-aware QoS management in web servers. In: Proceedings of RTSS'03, p. 63. IEEE (2003)
16. Petrucci, V., Loques, O., Niteroi, B., et al.: Dynamic configuration support for power-aware virtualized server clusters. In: Proceedings of 21th Euromicro Conference on Real-Time Systems. Ireland, (2009)
17. Sawyer, R.: Calculating total power requirements for data centers. White Paper, American Power Conversion, (2004)

18. Tang, Q., Gupta, S., Varsamopoulos, G.: Thermal-aware task scheduling for data centers through minimizing heat recirculation. In: Proceedings of IEEE Cluster Computing, pp. 129–138. IEEE (2007)
19. Pakbaznia, E., Ghasemazar, M., Pedram, M.: Temperature-aware dynamic resource provisioning in a power-optimized datacenter. In: Proceedings of Design, Automation and Test in Europe Conference and Exhibition (DATE), pp. 124–129. European Design and Automation Association (2010)
20. Ahmad, F., Vijaykumar, T.: Joint optimization of idle and cooling power in data centers while maintaining response time. In: Proceedings of the Fifteenth Edition of ASPLOS on Architectural Support for Programming Languages And Operating Systems, pp. 243–256. ACM (2010)
21. Wang, L., Khan, S.U., Dayal, J.: Thermal aware workload placement with task-temperature profiles in a data center. J. Supercomput. **61**(3), 780–803 (2012)
22. Deng, N., Stewart, C., Gmach, D., et al.: Policy and mechanism for carbon-aware cloud applications. In: Proceedings of NOMS2012, pp. 590–594. IEEE (2012)
23. Goiri, Í., Beauchea, R., Le, K., et al.: GreenSlot: scheduling energy consumption in green datacenters. In: Proceedings of 2011 International Conference for High Performance Computing, Networking, Storage and Analysis, p. 20. ACM (2011)
24. Goiri, Í., Le, K., Nguyen, T.D., et al.: GreenHadoop: leveraging green energy in data-processing frameworks. In: Proceedings of the 7th ACM European Conference on Computer Systems, pp. 57–70. ACM (2012)
25. Krioukov, A., Alspaugh, S., Mohan, P., et al.: Design and evaluation of an energy agile computing cluster. Technical Report UCB/EECS-2012-13, University of California, Berkeley, (2012)
26. Li, C., Qouneh, A., Li, T.: iswitch: coordinating and optimizing renewable energy powered server clusters. In: Proceedings of 39th Annual International Symposium on Computer Architecture (ISCA), pp. 512–523. IEEE (2012)
27. Stewart, C., Shen, K.: Some joules are more precious than others: managing renewable energy in the datacenter. In: Workshop on Power Aware Computing and Systems, 2009
28. Akoush, S., Sohan, R., Rice, A., et al.: Free lunch: exploiting renewable energy for computing. In: Proceedings of HotOS 2011, pp. 17–17. USENIX (2011)
29. Chen, C., He, B., Tang, X.: Green-aware workload scheduling in geographically distributed data centers. In: Proceedings of CloudCom 2012, pp. 82–89. IEEE (2012)
30. Le, K., Bianchini, R., Martonosi, M., et al.: Cost-and energy-aware load distribution across data centers. In: Proceedings of HotPower (2009)
31. Van Heddeghem, W., Vereecken, W., Colle, D., et al.: Distributed computing for carbon footprint reduction by exploiting low-footprint energy availability. Future Gener. Comput. Syst. **28**(2), 405–414 (2012)
32. Li, Y., Chiu, D., Liu, C., et al.: Towards dynamic pricing-based collaborative optimizations for green data centers. In: Second International Workshop on Data Management in the Cloud (DMC), pp. 272–278. IEEE (2013)

Storage and Accessing Small Files Based on HDFS

Yingchi Mao and Wei Min

Abstract Hadoop distributed file system (HDFS) becomes a representative cloud platform, benefiting from its reliable, scalable and low-cost storage capability. Unfortunately, HDFS does not perform well for huge number of small files because massive small files imposed heavy burden on NameNode of HDFS. This paper introduces an optimized scheme, structured index file merging (SIFM), using two-level index file and structure metadata storage, to reduce the I/O operations and improve the access efficiency. Extensive experiments demonstrate that the proposed SIFM can effectively achieve better performance in terms of storing and accessing for huge number of small files on HDFS, compared with native HDFS and Hadoop Archive (HAR).

Keywords Small file · Storage an access · HDFS · Structured index files storage

1 Introduction

With the rapid development of Internet, the amount of data growing exponentially, there have been appeared many large server architecture such as data centers and cloud computing. In large data processing, the Google's GFS provide an effective way to handling large files [1]. Hadoop is composed of one NameNode and DataNodes as architecture components. NameNode stores all the metadata in main memory. A large number of small files take significant impact on the metadata performance of Hadoop distributed file system (HDFS) and become the bottleneck for handling metadata requests of massive small files.

Y. Mao (✉) · W. Min
College of Computer and Information Engineering, Hohai University, Nanjing, China
e-mail: maoyingchi@gmail.com

S. Patnaik and X. Li (eds.), *Proceedings of International Conference on Computer Science and Information Technology*, Advances in Intelligent Systems and Computing 255, DOI: 10.1007/978-81-322-1759-6_65,

HDFS is designed for storing the large files, and therefore, it suffers performance efficiency in dealing with small files [2]. In fact, many current systems in the area of energy, climatology, biology, social networks, e-Business, and e-Learning contain huge amounts of small files [3, 4]. For example, National energy research scientific computing center stored over 13 million files in 2007, 99 % of which were less than 64 MB and 43 % of files were less than 64 kB [5, 6]. Therefore, storing and accessing a large number of small files face a great challenge to HDFS. The reason is that memory of NameNode is highly consumed by huge number of files, and no optimization scheme is provided to improve the access efficiency.

To improve the access performance on HDFS, the efficiency problem of reading and writing a large number of small files is analyzed. Based on the analysis of small files, an optimized scheme, structured index file merging (SIFM), is proposed for HDFS to reduce the memory consumption of NameNode and to improve the reading efficiency of small files. In SIFM, the correlations between small files and directory structure of data are comprehensively considered to assist the small files to be merged into large files and generate index files. Distributed storage architecture is used in index files management. In addition, SIFM adopts data prefetching and caching mechanism to improve the access performance.

2 Related Work

In recent years, research on small file optimization for HDFS has attracted significant attention. Hadoop Archive (HAR), Sequence File, and MapFile are typical general solutions to small file optimization.

HAR packs a number of small files into large HDFS blocks so that the original files can be accessed in parallel transparently and efficiently without expanding the files. It contains metadata files and data files [6]. The file data are stored in multiple part files, which are indexed for keeping the original separation of data intact. The metadata files can record the original directory information and the file states. HAR can reduce the memory consumption of NameNode, but it has some problems. Creating an archive generates a copy of original files, which brings extra burden on disk space. There is no mechanism to improve the access efficiency.

A sequence file is a flat file consisting of binary key-value pairs. It uses filename as the key and file contents as the value. You can write a program to put small files into one single sequence file and process the small files using MapReduce operating on the sequence file [2]. In addition, it provides compression and decompression at the block level and record level. However, there are problems of sequence file. It only supports append operation and does not provide update/delete operation for a specific key. If you look up a specific key in the sequence file, the whole file has to be viewed, which results in the low access performance.

A MapFile is a type of sorted sequence file with an index to permit lookups by key. It contains two files, a data file and a smaller index file. The index file stores all the key-value pairs sorted by key in the map. The index file stores key location the keys' location and the location is the offset where the first record containing this key is located [7]. The index file is read entirely into memory, so the index file should be kept itself small. Different from the Sequence File, when looking up a specific key in a MapFile, it needs not to search the whole data file. However, similarly with Sequence File, MapFile only supports append operation for a specific key.

3 Small Files Storage and Accessing Scheme

In this paper, we proposed an optimization scheme, SIFM, to improve the storage and access efficiency for small files on HDFS. The core ideas of SIFM include (1) file correlations are considered when merging files, which reduce the seek time and delay in reading files. (2) Structured distributed architecture for storing the metadata files is applied to reduce the seek operations of requested files. (3) Considering the access locality in inter-block on DataNodes, prefetching and caching strategy is used to reduce the access time when reading number of small files.

3.1 File Merging Strategy

File filtering criteria To effectively deal with the small files in HDFS, the first important issue is to identify the cutoff point between large and small files.

Many applications consist of a large number of small files. For example, in biology, the human genome generates up to 30 million files averaging 190 kB [8]. Sloan Digital Sky Survey hosted 20 million images with average size of less than 1 MB [3]. According to the fact from the real applications, in this paper, we treat the size of files smaller than 1 MB as small files.

When storing a small file, according to the small file filtering criteria, the client checks whether the size of file is larger than 1 MB. If it is a large file, it will be stored using the native HDFS method. If it is a small file, the proposed file merging operation will be executed.

Structured index file creation NameNode only maintains the metadata of merged files and the relevant index files are created for each original small file to indicate its offset and length in a merged file. In SIFM, a structured index file is built for each merged file and is loaded in the memory of NameNode. A structured index file is composed of two index sets, small file index and merged file index.

- A small file index is used to indicate the id, name, and length of a small file. SF_flag indicates its validation. Since the most frequent operations on small files index is queries by file ids, indexes are sorted by file id.
- Merged file index is built for each merge file, which indicates the offset and length for each original small file in it. Beside them, because a merged file may occupy multiple blocks, the merged file index indicates the block id where the

merged file is stored, id of the merged file, name of the merged file. Similarly, MF_flag indicates the validation of the merged file. Using merged file index, it can be convenient to analyze the location of a small file in the read process.

File merging operation File merging operations are carried out in HDFS clients, which merge related small files into a large merged file. NameNode only maintains the metadata of merged files and does store the original small files; thus, file merging can reduces the number of files that need to be managed by NameNode. When writing a small file, if it is a small file, the proposed file merging operation will be carried out. Otherwise, it will use the native HDFS method. The details of file merging operations are as follows:

Step 1: Preparation procedure. The number of small files is computed and the size of the small file is also calculated. The small index files will be created in Step 2.
Step 2: Creation the structured index file. If the size of current small file is less than the available size of one HDFS block, the small file index is created and the SF_flag is set to TRUE. The offset and length of the small file are calculated, and the merged file index is updated. If current HDFS block cannot provide enough space to store the small file, the remaining small space in HDFS block will be abandoned. The small file will be written from the first place of the next new HDFS block. Meanwhile, the small file index is created, and the merged file index is created based on the offset and length of the small file.
Step 3: Small file merging. According to the offset and length of each file in the merged file, files are merged into the merged in turn.

3.2 Metadata Files Storage

In HDFS, metadata of small files store the mapping information from the small file to the merged file. To reduce the memory consumption of NameNode, and improve the accessing performance on HDFS, it needs to optimize the metadata management. In SIFM, the structured P2P architecture, Chord, is adopted to store and manage the metadata file.

Metadata of small files are stored in NameNode with the key-value pairs. In metadata, SF_id is the only one identifier based on the filename and directory, which is as key stored in NameNode. SF_name and MF_name denote the name of original small file and the merged file, respectively. The offset represents the offset of small file in HDFS block. MF_length indicates the length of the merged file. MF_flag indicates the validation of the merged file. SF_name, MF_name, offset, MF_length, MF_flag are created value via the SHA-1 algorithm. Based on the hashed value, the corresponding metadata can be stored in the node of cluster.

When reading one small file, utilizing the Chord routing mechanism, NameNode can quickly locate the metadata in NameNode and seek the correspond small file based on the mapping information in the metadata.

3.3 Prefetching and Caching Files

Prefetching and caching are widely used approaches of accessing optimization [1]. Prefetching can avoid disk I/O cost and reduce the response time by considering the access locality and fetching data into cache before they are requested. In SIFM, prefetching and caching strategy is used to improve access efficiency, which includes metadata caching, index fetching, and data file fetching.

- Metadata caching. When a small file is requested, it is mapped to a merged big file to get the metadata of the big file from NameNode. If the metadata of the merged big file have cached, the client can directly obtain the metadata from cache.
- Index prefetching. Based on the metadata of the merged big file, the client should identify the connected blocks to obtain the requested file. However, NameNode still calculate the offset and length of the requested file from the index file. If the index file has been cached from DataNode, accessing small files, which belong to the same merged big file, only run the I/O operations without calculation.
- Data file prefetching. For accessing the files in a merged big file, it is necessary to exploit the access locality among files in a merged file. After a requested file is returned to the client, the correlated files are to be prefetched based on the access locality among files in a merged file.

4 Experiments Evaluation

In this section, in order to demonstrate that the proposed optimization scheme can improve the storage and access efficiencies of small files, extensive experiments on the different number of small files are conducted, and their results compared with the ones of native HDFS, Sequence File, and HAR.

4.1 Experimental Settings

The experimental platform is built on a cluster with five nodes. One node, which is HP server, acts as NameNode. It has 4 Intel Xeon CPU (2.6 GHz), 8 GB memory and 800 GB memory. The other four nodes, which are DELL PC, act as DataNodes. Each of them has 2 Intel CPU (2.6 GHz), 2 GB memory and 500 GB disk. All nodes are interconnected with 100 Mbps Ethernet network. In each node, Fedora 10 is installed. Hadoop version is 0.20.2, and Java version is 1.7.0. The number of replicas is set to 3, and HDFS block size is 64 MB by default.

To evaluate the storage and access efficiency of the proposed scheme, 80,000 files are selected as the dataset. The distribution of file sizes is shown in Fig. 1. The

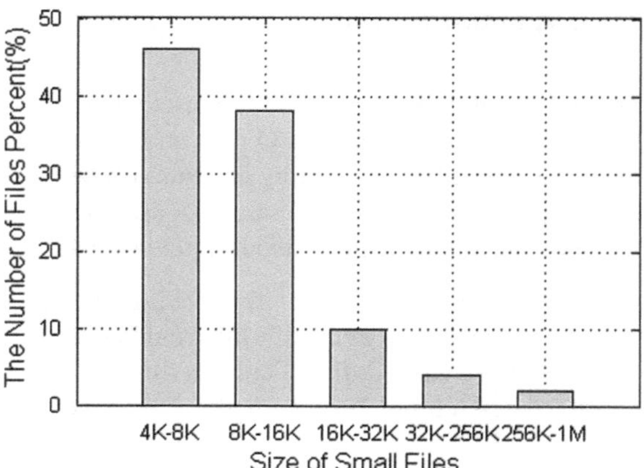

Fig. 1 Distribution of file size

file sizes in the dataset range from 4 kB to 1 MB, and files whose seizes are less than 32 kB account for 95 % of the total files. In this paper, 1 MB is taken as the cutoff point between large and small files. All files of dataset are small files.

4.2 Experimental Results

Storage Efficiency Storage operation is performed on SIFM, native HDFS, and HAR. In the storage operation, we evaluate the storage time and memories usage while uploading 1,000, 5,000, 10,000, 20,000, 40,000, and 80,000 small files, respectively, to an empty HDFS. Figure 2 shows that the time consumption for storage increases as the number of files increase. For native HDFS, HAR, and SIFM, when writing 1,000 small files, the storage time are 450, 313, and 151 s, respectively. SIFM greatly outperforms native HDFS and HAR. The reason is that SIFM adopts the merging scheme to reduce the I/O operation between NameNode and DataNode.

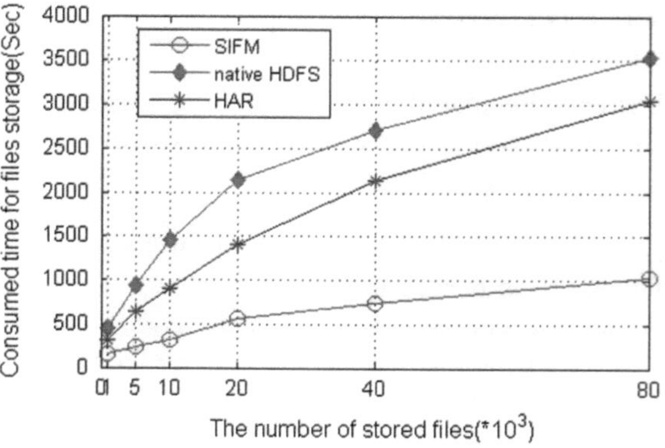

Fig. 2 Comparison under time for file storage

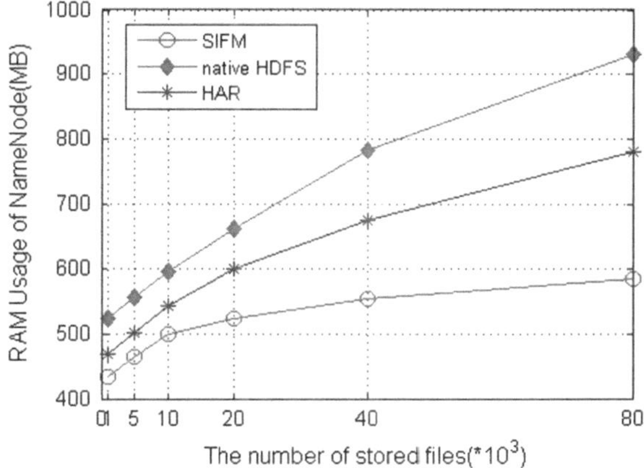

Fig. 3 Memory usage of NameNode

Memory Usage For native HDFS, HAR, and SIFM, the consumed memories of NameNode are measured when storing small files. The results are shown in Fig. 3. As expected, due to their file archiving and merging facilities, SIFM can consume less memories of NameNode than native HDFS and HAR. When storing 80,000 small files, the storage efficiency increases up to 42 % and 26 %, respectively. In addition, SIFM also has advantage on the memories usage of DataNode. Because of its structure index file strategy, SIFM consumes little more memory compared with HAR and much less memory than native HDFS. Although SIFM results in the extra overhead on the DataNode, the whole performance of SIFM is better than that of HAR (Fig. 4).

File Access Efficiency Figures 5 and 6 illustrate the access time by applying SIFM, native HDFS, HAR, and Sequence File in random files and sequence files, respectively. When reading 8,000 random files, the SIFM scheme can reduce the access time by 47, 52, and 33 %, compared with native HDFS, HAR, and Sequence File, respectively. Moreover, with the increase in the number of files,

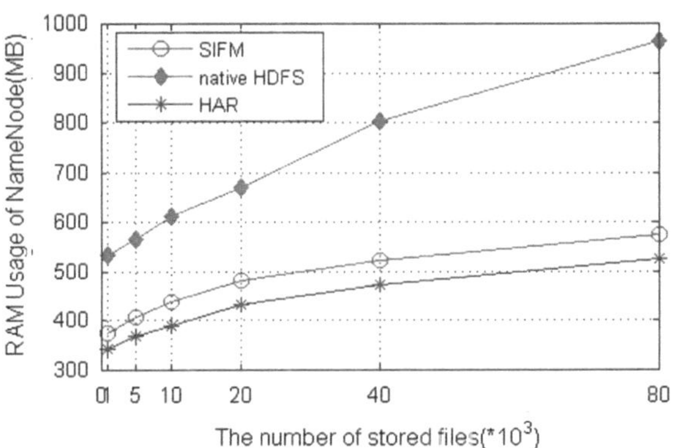

Fig. 4 Memory usage of DataNode

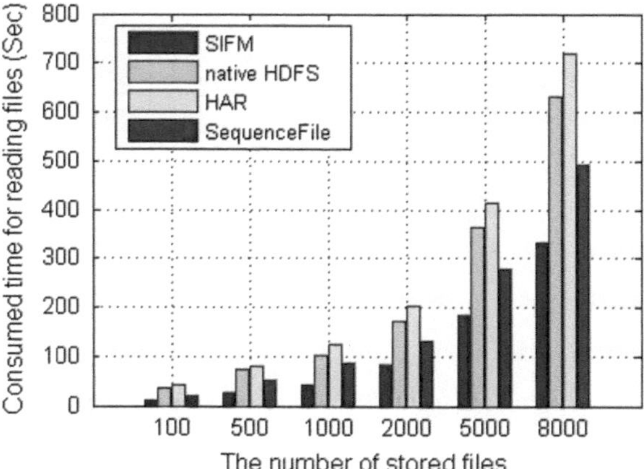

Fig. 5 Access time for random files

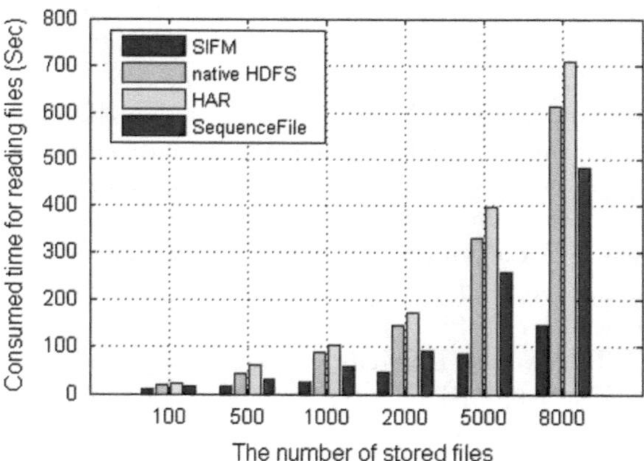

Fig. 6 Access time for Sequence files

from 100 to 8,000, the access time of all of schemes also increases. The reason is that the increase in the number of files can increase the seek operations on DataNodes, which can result in the higher access latency. In addition, reading random files and sequence files, the access time of the proposed SIFM scheme has some difference. For SIFM, reading random files has better access efficiency than reading sequence files. Reading sequence files can reduce the I/O operations between NameNode and DataNodes, which result in the improvement of access performance.

5 Conclusion

In this paper, the optimized scheme, SIFM, is proposed to effectively improve the storage performance, which outperforms HAR and Sequence File. As for access efficiency, SIFM reduce the access time by 50–80 %. The improvement on access efficiency benefits from three aspects: (1) File correlations are considered when merging files, which reduce the seek time and delay in reading files. (2) Structured distributed architecture for storing the metadata files is applied to reduce the seek operations of requested files. (3) Considering the access locality in inter-block on DataNodes, prefetching and caching strategy is used to reduce the access time when reading huge number of small files.

Acknowledgments This research is partially supported by the National Key Technology Research and Development Program of the Ministry of Science and Technology of China under Grant No. 2013BAB06B04 and Fundamental Research Funds for the Central Universities 2013B06914.

References

1. Liu, X., Han, J., Zhong, Y., Han, C., He, X.: Implementing WebGIS on Hadoop: a case study of improving small file I/O performance on HDFS. In: Proceedings. of IEEE International Conference on Cluster Computing, 1–8 (2009)
2. White, T.: Hadoop: The Definitive Guide. Yahoo Press, Sebastopol (2010)
3. Neilsen, E.: The sloan digital sky survey data archive server. omput. Sci. Eng. **10**(1), 13–17 (2008)
4. Shen, C., Lu, W., Wu, J., Wei, B.: A digital library architecture supporting massive small files and efficient replica maintenance, In: Proceedings of the 10th annual joint conference on digital libraries, ACM, pp. 391–394 (2010)
5. NERSC: Petascale and data and storage institute NERSC file system statistics (2007)
6. Mackey, G., Sehrish, S., Wang, J.: Improving metadata management for small files in HDFS. In: IEEE International Conference on Cluster computing 2009, J. Venner, Pro Hadoop, pp. 1–4. Springer (2009)
7. Min, L., Yokota, H.: Comparing Hadoop and fat-btree based access method for small file I/O , WAIM 2010, LNCS 6184, pp.182–193. (2010)
8. Bonfield, J.K., and Staden, R.: ZTR: A new format for DNA sequence trace data, Bioinformatics. **18**(1), 3–10, (2002)

Performance Analysis of EDCA Under Light, Heavy, and Variable Data Sources

A. Anitha and J. Jaya Kumari

Abstract With Enhanced Distributed Channel Access (EDCA), high-priority traffic has a higher chance of being sent than low-priority traffic. With the shorter contention window (CW) and shorter Arbitration Inter-Frame Space (AIFS), the high-priority stations wait for a less time before sending its packet than the low-priority stations. This paper deals with the integration of EDCA with light, heavy, and variable network loads in WLANs. This work analyzes the EDCA method of WLANs using light, heavy, and variable packet sizes and is simulated using ns-2.28 simulator. The simulation results show that high throughput is achieved for the saturated data sources and minimum delay for the unsaturated data sources. The results for the variable data loads increase the throughput for high data loads with high-traffic sources and obtained a minimum delay for minimum data load with low-traffic sources.

Keywords TXOP · Saturated source · Unsaturated sources · AIFS

1 Introduction

In this paper, the behavior of light sources (saturated), heavy sources (unsaturated), and variable network loads are analyzed using the Enhanced Distributed Channel Access (EDCA) method in IEEE 802.11e WLAN. With the increasing demand of wireless services, users expect better QoS and performance. The better QoS can be

A. Anitha (✉)
CSE Department, Noorul Islam Centre for Higher Education, Kumaracoil, India
e-mail: anidathi@yahoo.co.in

J. Jaya Kumari
ECE Department, Noorul Islam Centre for Higher Education, Kumaracoil, India
e-mail: jkumaribharat@yahoo.com

S. Patnaik and X. Li (eds.), *Proceedings of International Conference on Computer Science and Information Technology*, Advances in Intelligent Systems and Computing 255, DOI: 10.1007/978-81-322-1759-6_66,

achieved by the EDCA method [1], provided by the IEEE 802.11e standard with the medium access scheme (MAC). The performance of the network in correspondence with the user service is determined by the QoS. The QoS is set to achieve how the information is being carried along the network. This has to be delivered in a better manner. The traffic delay is reduced using the MAC protocol. The traffic that the user can be expected in recent days is the delay in transmission of audio and video data. The handling of real-time traffic is a challenging task.

The IEEE 802.11 standard is widely deployed to be the medium for transmission. Distributed coordination function (DCF) is one of the transmission modes, which uses CSMA/CA to communicate between two stations. A new Task Group E has been enhanced in the WLAN to support the real-time traffic in an efficient manner. 802.11e is the standard with two features: enhanced DCF (EDCF) and hybrid coordination function (HCF) [2]. These two modes are mainly designed to support the infrastructure and the traffic in the user network to reduce the delay of packets.

The saturation delay [3] and the throughput in WLAN are increased to obtain a better QoS. This is achieved by using the EDCA method. EDCA is an Enhanced Distributed Channel Access, which is present inside the data link layer.

1.1 Distributed Coordination Function

The DCF is one of the transmission modes deployed in IEEE 802.11 standard. All these nodes are used to contend for sharing the same medium. Due to this occurrence, collision is unavoidable. To overcome this situation, the back-off time is introduced by DCF and the back-off time calculation is given in Eq. 1

$$\text{Back-off Time} = \text{rand}(0, \text{CW}) \times \text{slottime} \tag{1}$$

If station A wants to transmit data to station B, station A listens to the channel. If the channel is busy, it waits until the channel becomes idle. After detecting the idle channel, station A further waits for a DIFS period and invokes the back-off procedure. The value of contention window (CW) is doubled if collision occurs.

1.2 EDCA

EDCA is used as the fundamental access mechanism for the MAC layer in IEEE 802.11e. EDCA supports priority service in which the higher-priority nodes get higher chance than the lower-priority nodes. Access categories (AC) are also known as traffic classes. Each station can run a maximum of four access categories, AC_0, AC_1, AC_2, and AC_3 where AC_3 has the highest priority. The frames generated by each station can be mapped to any of these access categories. Each access category can run the back-off process separately for the frame transmission. The adjustable parameters for the back-off process are as follows: CW_{min}, CW_{max}, Arbitration Inter-Frame Space (AIFS), and TXOPLimit.

EDCA supports prioritized service which improves multimedia transmissions such as voice, video transmissions. To guarantee QoS required by the real-time applications, EDCA has a principal mechanism called TXOP.

1.3 TXOP

Transmission opportunity is a mechanism which allows multiple consecutive frame exchanges without back-off. TXOP reduces the overhead of the contention time during the frame transmission. The TXOP in which it can transmit as many frames as possible during the maximum duration of the allotted time interval is called TXOPLimit. TXOPLimit is the particular duration in which a station gets the right to transmit. TXOP can be obtained through contention in EDCA. Using this mechanism, the packets can be transmitted as bursts for transmitting the video frames.

During the TXOP periods, the frames from the same packets are transmitted consecutively without any back-off, and each frame is separated by the SIFS interval. On the reception of each frame by the receiver node, after an SIFS time interval, corresponding acknowledgment frame is send back to the sender node.

1.4 TXOP Mechanism

- If a sender node wishes to transmit a data frame, senses the channel before transmission.
- If the channel is idle for DIFS amount of time, then the node gains the channel access.
- The sender node sends an RTS frame, and with an SIFS interval, the CTS frame is transmitted by the receiver node.
- On the reception of the CTS frame, the sender node waits for an SIFS interval and transmits a frame to the receiver node.
- The receiver node on the successful reception of the data frame transmits the corresponding acknowledgment to the sender node.
- After receiving the acknowledgment, the sender node sends the next frame to the receiver node.
- The process continues till the TXOP Limit expires.
- After the TXOPLimit, the sender node waits for DIFS time interval and enters the back-off phase.

A node can end its transmission if there are no frames to send by an access category or if there occurs any transmission failure or if there is any lack of space for transmitting the frame or the corresponding acknowledgment. The transmission overhead is also caused in the TXOP mechanism. The transmission overhead is shared by all the frames, which leads to higher performance.

2 Related Work

Misic et al. [4] proposed a non-saturation method which works only for the single-hop network. The parameters CW, TXOPLimit, and AIFS are considered. In this work, error-prone channel condition is assumed and the external collision is considered. This work achieved a 20 % throughput increase. Engelstad et al. [5] proposed an approach for predicting mean queuing delay in IEEE 802.11e EDCA. The priority scheme with no RTS is considered. Five different stations with no topology and each station used all four access categories, which leads to virtual collision. Inan et al. [6] proposed the back-off priority scheme in IEEE 802.11e standard. Here, the parameters are minimum back-off size, back-off window-increasing factor, and retransmission limit.

Xiao [7] proposed HCF access method. Here, the contention-based EDCA and contention-free HCF controlled channel access are combined to provide QoS stations with prioritized access. In this paper, EDCA defines the prioritized CSMA/CA mechanism. This method achieved accurate average throughput in the saturation situation.

Tao et al. [8] proposed EDCA which supports up to eight priorities in a station, which are mapped into four different access categories. This paper assumes that all the nodes have the same parameters. The network operates under the saturation condition (i.e., heavy load condition). Inan et al. [6] proposed an EDCA method which analyzed the accuracy of varying traffic loads, EDCA parameters, and MAC layer buffer space. This model predicted the EDCA performance and traffic load range from a lightly loaded non-saturated channel to a heavily congested saturated medium. But this model assumed only one AC per station due to space limitation.

3 Proposed Method

This work analyzes the performance of the WLAN by varying the data loads [4] than using fixed loads in the EDCA method of IEEE 802.11e standard. Then, the analysis is carried out by using light data source and heavy data source using the EDCA method [9]. Light data source means non-real-time data source which includes text data. Heavy data source means real-time data which includes audio and videos. Saturated data source [3] is created by creating CBR data source over UDP agent. Unsaturated data source is created using FTP over TCP agent. The minimum node created is 10 and is increased up to 50. For each set of nodes, the throughput and delay are measured for both saturated and unsaturated data. From the analysis, the throughput of the saturated data leads higher than the unsaturated data, but the delay for the unsaturated data is minimum than that of saturated data.

For analyzing the EDCA method with the variable data load, network of first 20 nodes is created and the packet size is given as 1024 bits and the throughput and delay are measured. Then, the packet size is increased and the parameters are measured. In this method, the access categories are assigned with the default parameter values. From the simulation results, high throughput values are obtained for the increasing packet sizes with high nodes in the network. Minimum delay is obtained for the small packet size with minimum node in the network. The drop increases with the increase in the nodes and packet size.

4 Results and Discussion

4.1 Performance Analysis

The performance is evaluated using ns-2.28 simulator. To make the nodes to communicate with each other, we have investigated the transition between all the traffic classes. The four traffic classes are placed upon different set of nodes. These nodes are able to communicate with each other, and each node is able to send frames from TXOP allocation to randomly chosen destination. By this approach, we have evaluated network performance for four traffic classes under TXOP allocations. The presence of EDCA gave good performance result by using variable loads. Good performance is being achieved through EDCA. To analyze the effect of mobility, pause time was varied from 0s (high mobility) to 100s (low mobility). First, the number of nodes is taken as 20 and is increased to 30 and the analysis is carried out. To study the effect of traffic load on the network, various numbers of maximum connections were set up between the nodes with the traffic rate of 4 packets per second, where each packet size was 512 bytes. A set of traffic generation files are created, which corresponds to different traffic to be generated.

First, the simulation setup is done for the variable network load analysis using the EDCA method. The EDCA with four different access categories with the default parameters is used. First 20 nodes are created and then the 30 nodes. These nodes are able to communicate with each other, and each node is able to send frames from TXOP allocation to randomly chosen destination. AODV is the routing protocol used in this scenario. To create a traffic connection file, the type of traffic connection, the number of nodes, and maximum number of connections are set up between them. A random seed is used for the CBR connections.

To analyze the QoS, the parameters measured are throughput and delay for each set of nodes with variable packet sizes. First, the throughput is measured using 1024 bits. 30 nodes gave a 4 % increase in throughput than the 20 nodes. Then, the

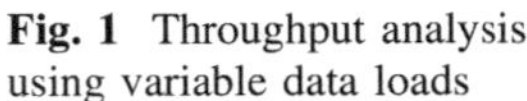

Fig. 1 Throughput analysis using variable data loads

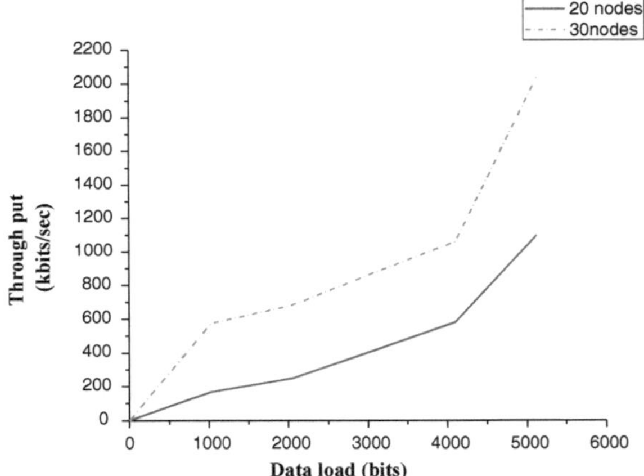

packet size is increased to 2048 bits, which too gave a high throughput value than the 20 nodes. The packet size is increased to 3072 bits, which too gave the same result. Then, the size is increased to 4096 bits and then to 5120 bits. Same as the previous results obtained, the increase in packet sizes gave a high increase in throughput for the 30 nodes than the 20 nodes. When the analysis is carried out on the basis of increasing packet size, the results show that the increase in packet sizes leads to an increase in throughput and is shown in Fig. 1.

The high throughput is obtained for the high packet size with more number of nodes. Here, we obtained high throughput value for 5120 bits and the number of nodes in the network is 30. Under moderate loads, TXOP feature is triggered first with the low-priority traffic classes since their back-off time is the longest. From moderate to high load, TXOP allocation increases for all traffic classes until maximum values are reached. The results obtained gave a high throughput value than using fixed packet sizes for the different set of nodes in the network. Then, for the same set of nodes and the packet sizes, the end-to-end delay analysis is conducted. The delay analysis results show that increase in packet sizes increases the delay. For first 20 nodes, the analysis is conducted with 1024 bits, which gave the minimum delay, and by gradually increasing the packet size, the delay too increased gradually, and for the maximum packet size used (5120 bits), a high delay value is obtained.

Then, for the same packet sizes, the nodes are increased to 30 and the delay is analyzed. From the results, delay increased tremendously than the 20 nodes and reached to a high delay value. From the analysis, we studied that the increase in packet sizes and the nodes increases the delay and is shown in Fig. 2.

Then, the analysis is conducted with the saturated and unsaturated data sources by increasing the nodes. First 10 nodes are used, and then, the number of nodes is increased gradually to 20 then to 30, 40, and at last to 50 nodes. First, for different sets of nodes, the saturated data are given and the result shows that as the number of nodes increases, the throughput too increased gradually. Then, for unsaturated

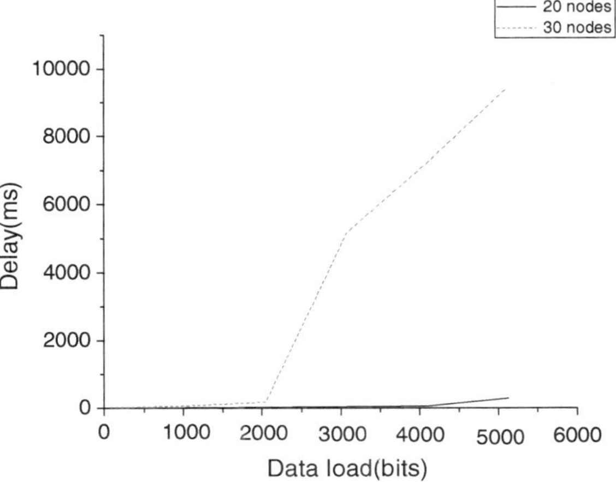

Fig. 2 Delay analysis using variable data loads

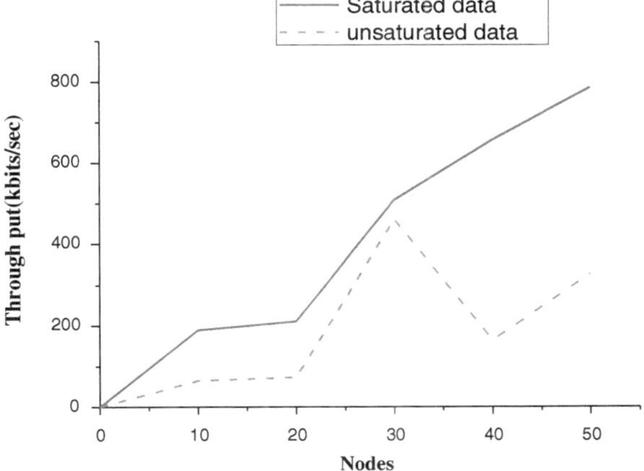

Fig. 3 Throughput analysis of saturated and unsaturated data

data, the results show that the throughput value obtained is less than the saturated one. But as the number of nodes increases, the throughput value too increased and these results are shown in Fig. 3. Then, the delay analysis is conducted for the same set of nodes and the data. The results show that for saturated data, the delay obtained for 10 nodes is less than the 20 nodes and obtained the same 10 nodes' delay value for the 30 nodes. The delay started increasing when the nodes are increased to 40 and 50. The unsaturated data obtained much lesser delay than the saturated one. But as the number of nodes is increased, the delay too increased and the results are shown in Fig. 4.

From the above analysis, it shows that the high throughput is achieved for the saturated data source and the minimum delay for the unsaturated data source with more number of nodes in the network.

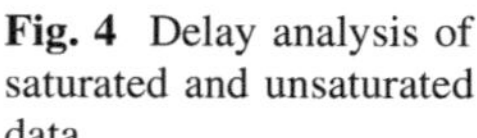

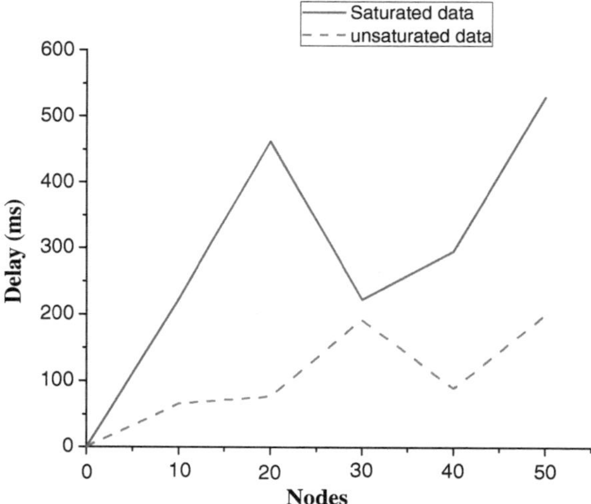

Fig. 4 Delay analysis of saturated and unsaturated data

5 Conclusion

This paper analyzes the saturated, unsaturated, and variable data load performance of IEEE 802.11e EDCA. Using this model, the fair service differentiation is investigated by means of EDCA parameter sets: CWmin, CWmax, AIFS, and TXOPLimit. This work is simulated using the ns-2.28 simulator. The proposed scheme gained high throughput for the saturated sources and low access delay for the unsaturated sources than the existing DCF model. The results for the variable data loads increase the throughput for high data loads with high-traffic sources and obtained a minimum delay for minimum data load with low-traffic sources. The performance of the EDCA method can be improved by altering the values of the EDCA category parameter.

References

1. Wu, H., Wang, X., Zhang, Q., Shen, X.: IEEE 802.11 e enhanced distributed channel access (EDCA) throughput analysis. In: Proceedings of IEEE International Conference on Communications, ICC'06, pp. 223–228. IEEE (2006)
2. Hour, B., Hameed, S.: Enhancement of IEEE802.11e WLAN through real time simulation study. Int. Arab J. Inf. Technol. **6**(4), 371–377 (2009)
3. Xiong, L., Mao, G.: Saturated throughput analysis of IEEE 802.11e EDCA. Comput. Netw. **51**, 3048–3068 (2007)
4. Misic, J., Rashwand, S., Misic, V.B.: Analysis of impact of TXOP allocation on IEEE 802.11 e EDCA under variable network load. IEEE Trans. Parallel Distrib. Syst. **23**(5), 785–799 (2012)
5. Engelstad, P.E., Østerbø, O.N.: Queueing delay analysis of IEEE 802.11e EDCA. In: Proceedings of IFIP WONS. IEEE (2007)

6. Inan, I., Keceli, F., Ayanoglu, E.: Modeling the 802.11e enhanced distributed function. In: Proceedings of IEEE Global Telecommunications Conference (Globecom'07), vol. 2, pp. 2546–2551. IEEE (2007)
7. Xiao, Y.: Performance analysis of priority schemes for IEEE 802.11 IEEE 802.11e wireless LANs. IEEE Trans. Wireless Commun. **4**(4), 1506–1516 (2005)
8. Tao, Z., Panwar, S.: Throughput and delay analysis for the IEEE 802.11 e enhanced distributed channel access. IEEE Trans. Commun. **54**(4), 596–603 (2006)
9. Nguyen, S.H.: Performance analysis of IEEE 802.11 WLANs saturation and unsaturated sources. IEEE Trans. Veh. Technol. **61**(1), 333–345 (2012)

Part VI
Algorithms

Possibilistic C-means Algorithm Based on Collaborative Optimization

Jing Zang and Chenghua Li

Abstract In this paper, a new possibilistic C-means (PCM) clustering algorithm is proposed based on particle swarm optimization (PSO) and simulated annealing (SA). Two optimizing algorithms automatically define the centers and number of clustering by different searching mechanisms in each sub-swarms and collaborative interaction among sub-swarms, which provides the optimal initialization for the new algorithm's good clustering result within a bid to solve the problem that PCM algorithm is very sensitive to initialization and parameter. The new PCM has the strong ability in the global searching and has less sensitivity to initialization and less possibility of stagnation in local optimum. Furthermore, the PSO-SA defines the centers and numbers of clustering automatically. Its advantages lie in the fact that it can not only improve the clustering performance but also has better accuracy and robustness, avoiding the problem of coincident clusters.

Keywords Collaborative optimization · Possibilistic C-means clustering · Particle swarm optimization · Simulated annealing

1 Introduction

Possibilistic C-means clustering (PCM) is proposed by Krishnapuram and Keller [1], in a bid to improve fuzzy C-means clustering (FCM), effectively eliminate the influence of noise on clustering results. But, PCM is very sensitive to the choice of initialization and parameter [1, 2].

J. Zang (✉)
Information Science and Engineering College, Shenyang Ligong University, Shenyang 110159, China
e-mail: zang_jing@163.com

C. Li
Mechanical Engineering College, Shenyang Ligong University, Shenyang 110159, China
e-mail: chenghuali2000@yahoo.com

S. Patnaik and X. Li (eds.), *Proceedings of International Conference on Computer Science and Information Technology*, Advances in Intelligent Systems and Computing 255, DOI: 10.1007/978-81-322-1759-6_67,

Particle swarm optimization (PSO) is inspired by the behavioral regularities of birds, fish, and human society. PSO [3] is an optimized technique based on swarm intelligence, with an ability of effectively solving problem, but easily trapping into local extreme point in actuality, slower convergence in later evolution; Simulated annealing (SA) optimization [4] algorithm has the strong ability in jumping out of local optimal solution, so SA is applied to PSO in a bid to foster strengths and circumvent weaknesses.

In view of the advantages of two kinds of optimization algorithm, this paper develops PCM based on collaborative action of two kinds of optimization algorithm, with improving the accuracy of PCM clustering results.

2 Possibilistic C-means

For a data collection $X = (x_1, x_2, \ldots, x_n)$, C is the category number of data collection, $(A_1, A_2, \ldots, A_c)$ is the possible subset of X, U is the similar possibilistic partition matrix, the clustering center of each prototype is $(v_1, v_2, \ldots, v_c)$, $u_k(x_i)$ is the membership grade of x_i in class A_k, (that is u_{ik}), and the objective function of PCM can be described as follows:

$$\begin{gathered} 0 \leq u_{ij} \leq 1, \sum_{j=1}^{N} \mu_{ij} > 0 \quad (i = 1, 2, \ldots, C, \; j = 1, 2, \ldots, N) \\ J_m(U, V; X) = \sum_{i=1}^{C} \sum_{j=1}^{N} (\mu_{ij})^m (d(x_j, v_i))^2 + \sum_{i=1}^{C} \eta_i \sum_{j=i}^{N} (1 - \mu_{ij})^m \end{gathered} \tag{1}$$

$$v_i = \sum_{j=1}^{N} \mu_{ij}^m x_j \Big/ \sum_{j=1}^{N} \mu_{ij}^m, \quad i = 1, 2, \ldots, C, \; j = 1, 2, \ldots, N \tag{2}$$

$$\mu_{ij} = \left(1 + \left(\frac{d_{ij}^2}{\eta_i} \right)^{1/(m-1)} \right)^{-1} \tag{3}$$

$$\eta_{\mathrm{i}} = K \left(\sum_{j=1}^{N} \mu_{ij}^m * d^2(x_j, v_i) \Big/ \sum_{j=1}^{N} \mu_{ij}^m \right) \tag{4}$$

$m \in [1, \infty)$ is weighted index of controlling the possibilistic degree of clustering, where η_i denotes a scale parameter.

PCM aims to find the best class U, which could make J_m minimum.

3 Particle Swarm Collaborative Optimization Based on SA

3.1 PSO Algorithm

PSO is a population-based searching process, with each individual in swarm moving toward the good position based on its fitness to environment, and being regarded as a particle without volume in D-dimensional space and the position of each particle is regarded as the potential solution on optimizing problem. During their flight, each particle dynamically adjusts its position and velocity according to its and its partner's experiences. For given x a particle position and v its corresponding flight velocity in a search space, the ith particle is represented as $X_i = (X_{i1}, X_{i2}, \ldots, X_{id})$ in the D-dimensional search space, the best previous position of the ith particle is recorded and represented as $P_i = (p_{i1}, p_{i2}, \ldots, p_{iD})$, and the index of the best particle among all the particles in the group is represented by $P_g = \left(p_{g1}, p_{g2}, \ldots, p_{gD}\right)$. The flight velocity for particle i is represented as $V_i = (v_{i1}, v_{i2}, \ldots, v_{iD})$. The modified velocity and position of each particle can be calculated by the current velocity and the distance from p_i to p_g as shown in the following formula as [5, 6]:

$$v_{id}^{k+1} = w \times v_{id}^{k} + c_1 \times \text{rand}() \times \left(p_{id} - x_{id}^{k}\right) + c_2 \times \text{rand}() \times \left(p_{gd} - x_{id}^{k}\right) \quad (5)$$

$$x_{id}^{k+1} = x_{id}^{k} + v_{id}^{k+1} \quad (6)$$

where w is the inertia weight and generally sets up to vary linearly from 0.9 to 0.2; c_1 is the cognitive acceleration parameter; c_2 is the social acceleration parameter; rand is uniformly distributed in the range [0, 1]. PSO has shortcomings, such as easily trapping into the local extreme value point, while SA algorithm has strong ability in jumping out of local optimal solution; therefore, this paper combined two kinds of algorithm for getting the initial clustering parameters of PCM algorithm, which results in success in clustering analysis without supervision.

3.2 SA Algorithm

SA algorithm [7] belongs to the global optimization algorithm. SA's thought is that with the dropping randomly of temperature, it searches for the optimal solution of the objective function in the solution space, and it finally tends to the global optimal solution using characteristic of mutation on probability when it traps into a local optimum.

3.3 Particle Swarm Collaborative Optimization Based on SA

The main idea of SA and PSO collaborative optimization [8–11] strategy is to achieve global optimization through synergy of the different sub-swarms, with the prerequisite of dividing the whole swarm into s ($s \geq 2$) sub-swarms.

Step 1: For the first sub-swarm, SA is used to search.
Step 2: For the second to the ($s - 1$)th sub-swarm, as an independent particle swarm, respectively, update velocity and position of particles in sub-swarm based on the optimal position it has found, using large inertial factor to ensure the comprehensive in searching range, as well as introducing variation factor in order to improve the particle's space exploration ability of the book sub-swarm.
Step 3: For the sth sub-swarm, update velocity and position of particles in their group based on the optimal position the whole population has searched, as well as introducing variation factor to ensure the fast convergence of the algorithm and to avoid the local optimum.

4 Possibilistic C-means Algorithm Based on Collaborative Optimization

The core of the PCM algorithm is the certainty of the clustering centers. Each particle of PSO algorithm represents a choice of a clustering center. If the size of the swarm is N, it means there are N clustering ways. We define the fitness function for the valuation of each particle in the following formulas:

$$f(x_i) = 1/J_m(u, A) + 1 \tag{7}$$

The better the clustering effect is, then the smaller $J_m(u, A)$ is, and the higher the fitness of individual is. The value of the fitness for every particle indicates the quality of the cluster effect based on the choice of the clustering center. Therefore, the choice of the global and the individual extremum in the PSO algorithm means every clustering center should be based on the fitness function [12].

Termination conditions: satisfy the maximum algebra, or $J_m(u, A)$ is less than the given number, or the matrix U will never change, that is $\left|u_{ij}^{k} - u_{ij}^{k-1}\right| \leq \varepsilon$.

4.1 Solving Process of the Algorithm

Step 1: Determine the fuzzy index m, the number of classes C, the size of swarm N, and the number of subgroup s. Set the relevant parameters of the algorithm, such as the original position, velocity for each particle, inertia weight, and initial temperature of annealing T_0; coefficient of cooling temperature C; iterations N; pbest i is the present position of every particle; gbest is the best position of all particles in the present swarm.

Step 2: The first sub-swarm uses the SA algorithm to search the optimal solution and updates the individual and global extremum of the particle using the state-producing function with the additional disturbance.

step 3: Compute the membership u_{ij} for every clustering center, calculate the fitness of every particle $f\_pbest(i, k)$, and the optimal fitness of every subgroup $f\_gbest(i, k)$ by formula (7), and then update the global optimal position with the optimal position of the sth sub-swarm.

Step 4: For the top $(s - 1)$ sub-swarms, determine whether the variation on all their particles is needed to jump out of the local optimal value and keep the global with the gathering degree and the steady degree probability of particles.

$$mut\_p(i, k) = f\_gbest(i, k) + \alpha \times (f\_avg(i, k) - f\_gbest(i, k)), a \in [0, 2] \quad (8)$$

Step 5: Compare each fitness of the whole swarm ($fgbest(i, k)$) with its best position pbest. If it is better, it is regarded as the present best position.

Step 6: For the top s sub-swarms, respectively, compare the optimal fitness of every particle in sub-swarm with the fitness of the best position gbest what the sub-swarm has experienced. If it is better, $f\_gbest(i, k)$ is regarded as the global best position of the present sub-swarm. Finally, we replace the global optimal position with that of the sth sub-swarm

Step 7: The first sub-swarm searches the optimal solution with the SA algorithm and updates the individual and global extremum of the particle by the state-producing function with the additional disturbance. The particles of the rest sub-swarms, respectively, update the velocity and position by formulas (5) and (6). (The master sub-swarm and the sub-swarm, respectively, have the different inertia weight cognition and social acceleration coefficients.);

Step 8: If the termination condition is satisfied, output the solution; If not, return to step 2.

4.2 Simulation Experiment

The first experiment uses the famous wine sample data to test. The data includes 160 samples and every sample represents the four-dimension vector of four features. The clustering class is four. The samples of every cluster are 40. The fuzzy index m is 2.0. The original clustering center of every cluster gets three ones of 160 samples. The initial cluster uses the SA algorithm to set the parameters as follows: initial temperature is $T = 1{,}000{,}000$, coefficient of cooling temperature C is 0.5, and c_1 and c_2 are equal to 2. The maximum iteration is 50. The

Table 1 The experimental result based on the wine dataset

Algorithms	The clustering accuracy (%)
PCM	67.52
PCM on PSO	91.17
PCM based on the hybrid collaborative optimization	94.63

Table 2 The clustering center of three algorithms

Algorithms	The clustering center	ΔV
PCM	(4.6385, −0.0068)	4.64E−1
	(2.9876, 0.0496)	
	(1.3127, 0.0133)	
PCM on PSO	(1.0361, −0.0496)	1.36E−1
	(2.9398, 0.0394)	
	(4.9930, −0.0520)	
PCM based on the hybrid collaborative optimization	(1.0621, −0.0101)	9.89E−2
	(4.9348, 0.0126)	
	(2.9706, 0.0750)	

experimental result is indicated in the Table 1. The result of Table 1 shows that the cluster accuracy of PCM and PCM algorithm based on PSO is inferior to PCM algorithm based on the hybrid cooperative optimization; the experimental result is indicated in the Table 1.

The second experiment tests the robustness. The formula is as follows:

$$\Delta V = \|V_I - V_{\text{ideal}}\| = \sqrt{\sum_{i=1}^{i=c} \sum_{j-1}^{j-c} \left(v_{ij} - v_{\text{ideal},i,j}\right)^2} \quad (9)$$

The dataset is equal to the Zhang and Yeung [13]. V_I represents the clustering center matrix of clustering algorithm. V_{ideal} represents the ideal clustering center, ΔV is bigger, and then, the result of clustering is worse. The experimental result indicates in the Table 2.

5 Conclusions

In this article, SA is introduced into PSO in a bid to improve the particles of different sub-swarms exploring the new space and to avoid PSO trapping into the local extremum efficiently with the iterative process of particle mutation and SA. The article examined PCM based on collaborative optimization, with verification of its advantages through the experiments, such as the less sensitivity to the initial parameters, the improved robustness, clustering analysis without supervision, and success in avoidance of coincident clusters, and it can be used in image processing, data mining, and other classification problem, including the unclassified machine parts, seeds classification, and malfunction diagnosis.

Acknowledgments This research is partly supported by Research on Case-based Reasoning in Precision Spade Punch Planter Design Theory and Method (Project No. 51075282) from the National Natural Science Fund, China.

References

1. Krishnapuram, R., Keller, J.: A possibilistic approach to clustering. IEEE Trans. FS **1**(2), 98–110 (1993)
2. Bezdek, J.C.: Pattern Recognition with Fuzzy Objective Function Algorithms, pp. 95–107. Plenum Press, New York (1981)
3. YouRui, H.: Intelligent Optimization Algorithm and Application, pp. 85–97. National Defence Industry Press, Beijing (2008). (in Chinese)
4. Vaz, A., Ismael, F., Pereira Ana, I.P.N et al.: Particle swarm and simulated annealing for multi-global optimization. WSEAS Trans. Inf. Sci. Appl. **2**(5), 534–539 (2005)
5. Kathiravan, R., Ganguli, R.: Strength design of composite beam using gradient and particle swarm optimization. Compos. Struct. **81**(4), 471–479 (2007)
6. Zhang, J., Liu, S., Zhang, X.: Improved particle swarm algorithm. Comput. Eng. Des. **28**(17), 4215–4216 (2007)
7. Gao, Y., Xie, S.: Particle swarm cooperative optimization based on the simulated annealing. Comput. Eng. Appl. **40**(1), 47–50 (2004). (in Chinese)
8. Liang, J.J., Suganthan, P.N.: Dynamic multi-swarm particle swarm optimizer. In: Proceedings of IEEE International Swarm Intelligence Symposium, vol. 3(4), pp. 124–129 (2005)
9. Iwamatsu, M.: Locating all global minima using multi-species particle swarm optimizer the inertia weight and the constriction factor variants. In: Proceedings of 2006 IEEE Congress on Evolutionary Computation, pp. 816–822. Vancouver (2006)
10. Seo, J.H., Im, C.H., et al.: Multimodal function optimization based on particle swarm optimization. IEEE Trans. Magneties **2**(4), 1095–1098 (2006)
11. Timm, H., Borgelt, C., Doring, C., Kruse, R.: An extension to possibilistic fuzzy cluster analysis. Fuzzy Sets Syst. **147**(1), 3–16 (2004)
12. Gao, Y., Wang, X., Lu, X., Yin, Y.: The study of PCM clustering algorithm of PSO. Comput. Simul. **27**(9), 177–180 (2010)
13. Zhang, J.S., Yeung, Y.W.: Improved possibilistic c-means clustering algorithms. IEEE Trans Fuzzy Syst. **12**(2), 209–217 (2004)

Infrared Small Target Tracking Algorithm Based on Fusion Feature Matching and Mean Shift Correction

Rui Li, Xinsheng Huang, Ruitao Lu and Lurong Shen

Abstract IR small target tracking has a very important significance for guidance and targeting early warning, but infrared image has low signal-to-noise ratio and less information. So, an infrared small target tracking algorithm based on fusion match and Mean Shift correction is proposed in this paper. The fusion feature is used to ensure that the target has a clear distinction with background or interference targets. Mean Shift correction gradually corrects the match result to the local brightest point. Experiment results show that the algorithm has high tracking precision and low false tracking, and it is better than original Mean Shift algorithm and cross-correlation algorithm.

Keywords Infrared image · Fusion feature · Matching · Mean shift · Correction

1 Introduction

IR small target tracking is a difficult problem in the field of video surveillance, guidance, and targeting early warning, because in IR sequence, the gray scale of infrared image is severely compressed, infrared image has low signal-to-noise ratio

R. Li (✉) · X. Huang · R. Lu · L. Shen
College of Mechatronics and Automation, National University of Defense Technology, Changsha 410073 Hunan, People's Republic of China
e-mail: lirui600@sina.com.cn

X. Huang
e-mail: huangxinsheng@sina.com.cn

R. Lu
e-mail: luruitao@sina.com.cn

L. Shen
e-mail: shenlurong@sina.com.cn

S. Patnaik and X. Li (eds.), *Proceedings of International Conference on Computer Science and Information Technology*, Advances in Intelligent Systems and Computing 255, DOI: 10.1007/978-81-322-1759-6_68,

and less information, and the visual effect of infrared image is poor because of the imaging mechanism. Therefore, IR small target tracking becomes a very challenging problem in the field of pattern recognition and computer vision [1–4].

Tracking algorithm based on matching is a classic tracking method that has been widely used because of its simple principle. Generally, matching method algorithms are divided into two categories: One is feature-based matching and the other is matching based on gray-level information. The main idea of feature-based matching is searching and matching the image gray matrix in image with template gray matrix using some kind of metrics [5–7]. Single feature is difficult to do match work under the complex scenes; feature fusion matching has become a trend [8].

Exhaustive search is a non-inspired blind search process; every point correlation coefficients were calculated. Exhuastive search cost most of the time in image matching process. Hill-climbing method is one of the most simple heuristic search algorithms. The searching direction is the direction of the steepest ascent, so it is possible to achieve the peak [9].

Feature matching tends to consume plenty of time, so it is usually used as a rough track in tracking problem and then it corrects the feature matching result as the last result. This paper uses the basic principle of Mean Shift to correct feature matching result.

Mean Shift concept is proposed by Fukunaga and Hostetler [10] on the paper "The Estimation of the Gradient of a Density Function." Later, Comaniciu introduced Mean shift algorithm into the field of target tracking, greatly reducing the amount of computation of the tracking algorithm [11, 12]. Recent research on Mean Shift for IR small target tracking either a combination with other methods [1, 3, 13] or the use of new similarity metric [14].

This paper proposed an infrared small target tracking algorithm based on feature fusion matching and Mean Shift correction. Firstly, extract the weighted gradient histogram and weighted gray histogram as template. Secondly, search the best match point in searching area of next frame by hill-climbing method. Finally, use Mean Shift drift to the local brightest point.

2 Fusion Feature Matching

Infrared small target cannot completely reflect the shape of the small target because of the limitations of imaging equipment. In IR sequence, the same small targets in different frames have a lot of differences in each frame, as shown in Fig. 1.

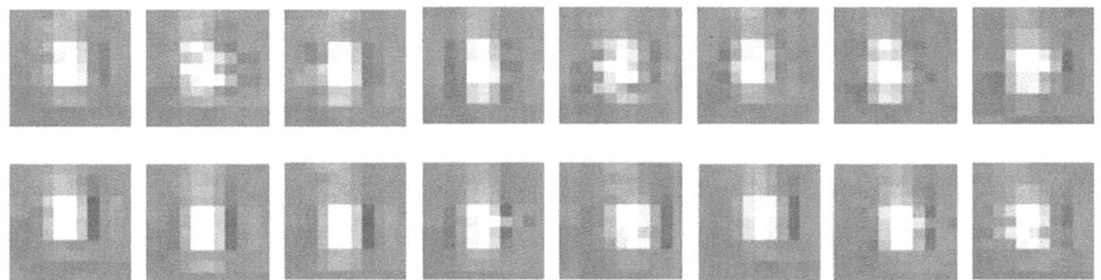

Fig. 1 The same infrared small target in different frames (11 by 11 pixels)

2.1 Histograms of Oriented Gradients

Histograms of oriented gradients (HOG) [15] mainly present target's contour information. The basic formula to calculate HOG is as follows:

$$\begin{cases} \Delta x = \mathrm{I}(x+1,y) - \mathrm{I}(x-1,y) \\ \Delta y = \mathrm{I}(x,y+1) - \mathrm{I}(x,y-1) \\ \theta(x,y) = \arctan\left(\frac{\Delta y}{\Delta x}\right) \\ m(x,y) = \sqrt{\Delta x^2 + \Delta y^2} \end{cases} \tag{1}$$

Because of the low signal-to-noise ratio in infrared images, target contours have a lot of differences in different frames. So, Epanechnikov weighted module is introduced in calculating HOG (Fig. 2). In order to decrease the effect of noise, Gaussian frequency domain filtering is introduced before calculating HOG.

2.2 Gray-Level Histogram

Gray-level histogram is the probability of occurrence of a certain gray. It is calculated using statistic gray-level distribution of the target area. The Epanechnikov weighted module is also introduced to ensure the reliability of the data of the point near center point.

Fig. 2 Oriented gradients (*left*) and Epanechnikov kernel (*right*)

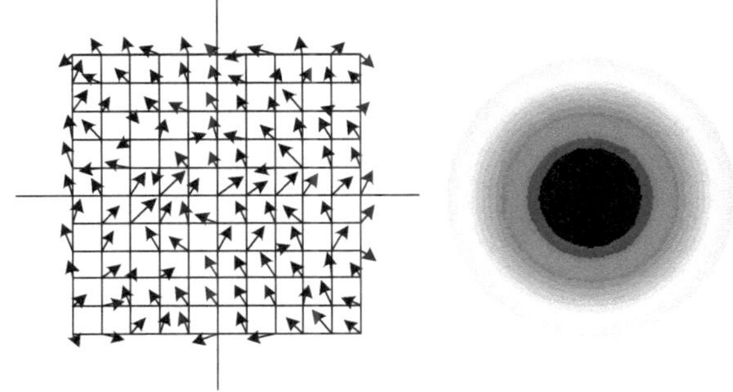

2.3 Gradient and Gray-Level Fusion

By taking the advantages of both the gradient and gray-level information, the fusion information of gradient and gray scale is used to do match work. The effect of fusion information is shown in Fig. 3.

This paper introduced the hill-climbing algorithm to quickly search the best match point. Euclidean distance is chosed for measuring the difference between two feature vectors in this paper. Initial climb points' selection rules are as follows:

$$\begin{cases} x_i = x - \frac{w}{2} + \frac{i}{3}w & i = 1,\ 2 \\ y_i = y - \frac{h}{2} + \frac{i}{3}h & i = 1,\ 2 \end{cases} \tag{2}$$

$(x,\ y)$ is the center point of the target in the previous frame, w is the width of the window of the climbing area, and h is the height of the window of the climbing area. The four initial climbing point coordinates are (x_1, y_1), (x_1, y_2), (x_2, y_1), (x_2, y_2), as shown in Fig. 4.

The search process is summarized as follows: Get the best match point in the neighbor of current point; if the current point is not the best match point, then the best match point is the new current point; otherwise, climbing is over. Hill-climbing method is easy to fall into local optimum value. This problem could be maked up by increasing the number of initial climb point. Therefore, we use the average four initial climb points as mentioned above.

3 Mean Shift Correction

Given n sample points $x_i(i = 1,\ 2, \ldots, n)$ in the d-dimensional space R^d, the Mean Shift Vector at point x is as follows:

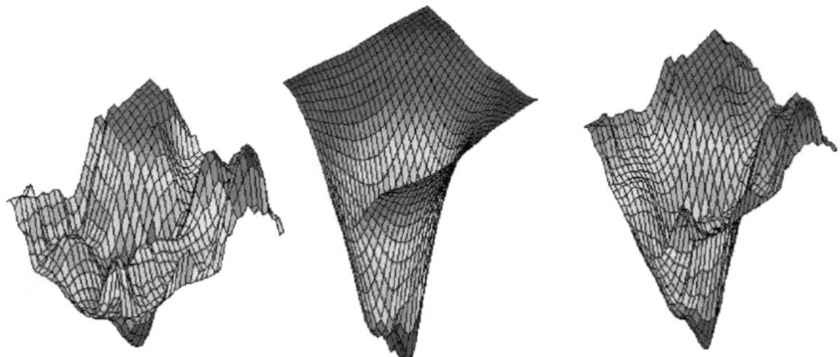

Fig. 3 Correlation surface of HOG match (*left*), gray-level match (*middle*), and fusion information match (*right*) of same target region in the same image

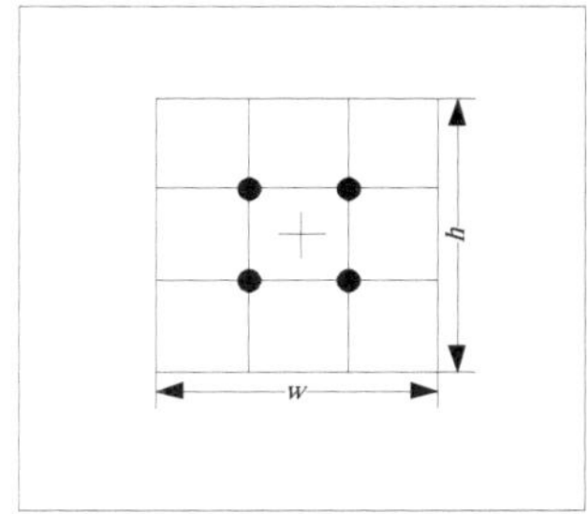

Fig. 4 Four initial climbing points (*black dot*)

$$M_h(x) = \frac{1}{k}\sum\nolimits_{x_i \in S_h} (x_i - x) \tag{3}$$

In Formula 3, S_h is a set of points y that satisfy Formula 4 in a circle of radius 12.

$$S_h(x) = \left\{ y : (y - x)^T (y - x) \leq h^2 \right\} \tag{4}$$

In Formula 3, the k present k points out of n sample points satisfy Formula 4.

As shown in Fig. 5, the average offset M_h (x) points to the area where most of sample points are located, that is, the probability density function of the gradient direction.

The brightness center of the target area is calculated as follows:

$$m_{pq} = \sum\nolimits_{x, y \in S_h} x^p y^q I(x, y) \tag{5}$$

$$C(x, y) = (Cx, Cy) = \left(\frac{m_{10}}{m_{00}}, \frac{m_{01}}{m_{00}} \right) \tag{6}$$

where $I(x, y)$ is the gray level of the image coordinate point (x, y) and S_h is circle used to statistic fusion feature. Cx and Cy denote the two coordinate values of S_h.

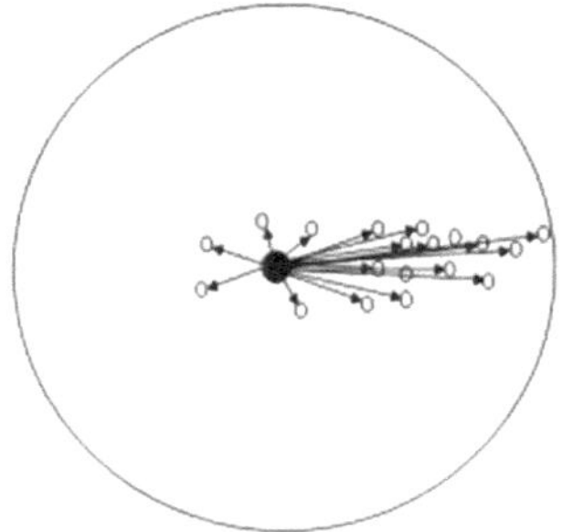

Fig. 5 Mean shift vector

Mean Shift correction algorithm performs the following steps, until the end conditions are satisfied:

(1) Calculate $C(x, y)$;
(2) $C(x, y)$ is assigned to (x, y);
(3) $\|C(x, y) - (x, y)\| < \varepsilon$, the end of the cycle; if not, continue to perform (1).

4 Experiment Results

The radius of Gaussian frequency domain filter in this paper is 40. In experiments, the size of climbing area is 15 by 15 pixels. An average of four initial climbing points is mentioned in 2.3.

In order to evaluate the performance of the proposed algorithm, four IR sequence images (ir1, ir2, ir3, and ir4) are used. The four videos were taken with a same infrared camera, and the size of image is 480×640 pixels.

Figure 6 shows the tracking error of different algorithms. The three algorithms are original Mean Shift algorithm, the cross-correlation algorithm, and proposed algorithm in this paper.

Experiments show that original Mean Shift algorithm has poor performance on infrared small target tracking. Cross-correlation algorithm performance is good, when there is no interference, as shown in Fig. 6d. But it cannot well distinguish target and interference, especially when the interference is much like the target, as shown in Fig. 6b. The proposed algorithm has better performance on all of four IR sequence images than the other two algorithms.

Table 1 shows the average tracking error of the three algorithms.

Figure 7 shows image tracking results of three algorithms. Red, green, and blue circles represent the image tracking result of original Mean Shift algorithm, cross-correlation algorithm, and proposed algorithm, respectively. In ir2(86) and ir2(105), the result of original Mean Shift does not show up because the tracking results are far away from the real target, which is shown in Fig. 6b. The first frame of each sequence images shows the target Fig. 7.

In the experiment, if the tracking error is bigger than the radius of statistic circle, then we consider the result as false tracking result. For example, in ir1, the radius of statistic circle is 5 pixels, and then, the tracking errors that are bigger than 5 pixels are false tracking results. The false tracking ratios of three algorithms in four IR sequence images are shown in Table 2.

In ir4, false tracking ratio of proposed algorithm is 0 %, which means 100 % true tracking ratio. Cross-correlation algorithm also performs well because there are only a dozen frames in which target is close to the interference. In ir2, the false

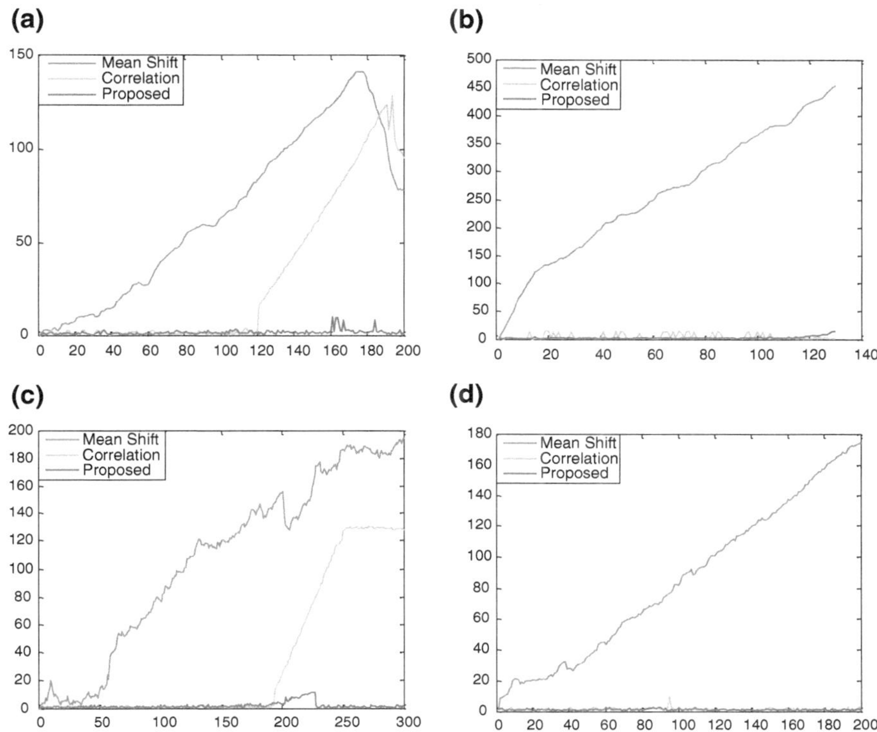

Fig. 6 Tracking error of **a** ir1, **b** ir2, **c** ir3, and **d** ir4

Table 1 Average tracking error

IR sequences	Methods		
	Mean shift (pixel)	Cross-correlation (pixel)	Proposed (pixel)
ir1	67.7645	29.5157	1.4935
ir2	258.8314	3.7462	2.3674
ir3	107.1899	35.0645	1.8014
ir4	84.8799	1.1853	0.8829

tracking ratio of proposed algorithm is 9.23 %, and it is the highest ratio of the three IR sequence images. The reason is that in the last 12 frames in ir2, the camera angle and the target shape change greatly, and the proposed algorithm does not consider these changes.

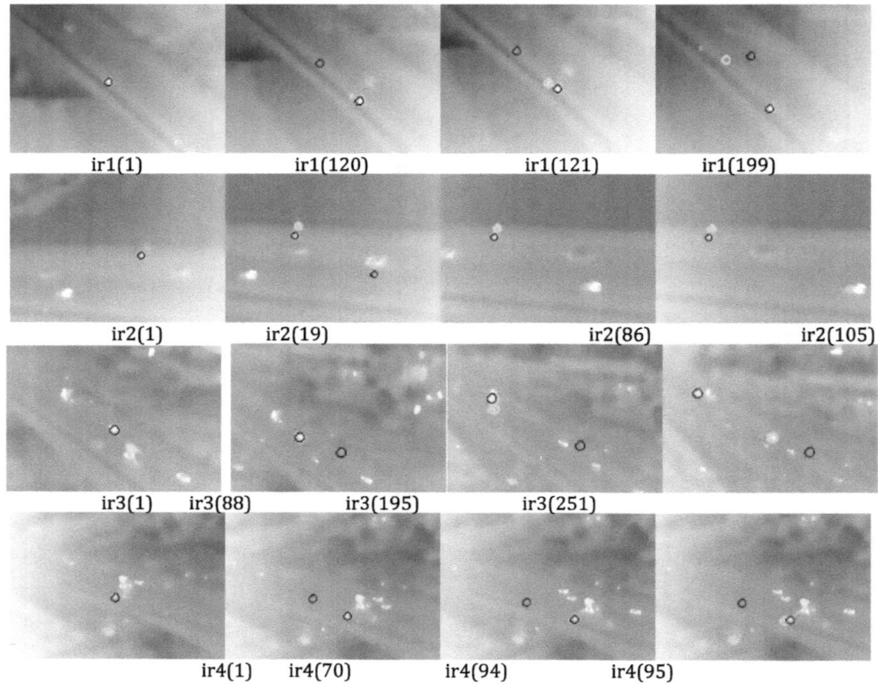

Fig. 7 Image tracking results (*numbers in brackets are the number of frames*)

Table 2 False tracking ratio

IR sequences	Methods		
	Mean shift (%)	Cross-correlation (%)	Proposed (%)
ir1	94.50	40.00	2.50
ir2	99.23	21.54	9.23
ir3	94.33	35.67	7.67
ir4	99.50	0.50	0.00

5 Conclusion

This paper proposed an infrared small target tracking algorithm based on fusion match and Mean Shift correction. The fusion feature is used to ensure that the target has a clear distinction with background or interference targets. Mean Shift correction gradually corrects the match result to the local brightest point. Experiment results show that the algorithm has high tracking precision and low false tracking, and it is better than original Mean Shift algorithm and cross-correlation algorithm.

References

1. Wang, Z., Hou, Q., Hao, Ling: Improved infrared target-tracking algorithm based on mean shift. Appl. Opt. **51**(21), 5051–5059 (2012)
2. Shaik, J., Iftekharuddin, K.M.: Detection and tracking of targets in infrared images using Bayesian techniques. Opt. Laser Technol. **41**, 832–842 (2009)
3. Shuang, Z., yu-ping, Q.: Mean shift algorithm apply for infrared imaging tracking. In: AASRI Conference on Computational Intelligence and Bioinformatics, pp. 52–57 (2012)
4. Li, H., Wei, Y., Li, L., Tang, Y.Y.: Infrared moving target detection and tracking based on tensor locality preserving projection. Infrared Phys. Technol. **53**, 77–78 (2010)
5. Wu, D., Gong, J., Wu, H., Tian, J.: New algorithm of image rotation matching based on feature points. Comput. Sci. **36**(12), 248–250 (2009). 262
6. Yongfang, G., Ming, Y., Yicai, S.: Study on an improved robust algorithm for feature point matching. In: International Conference on Medical Physics and Biomedical Engineering, vol. 33, pp. 1810–1816 (2012)
7. Zitová, B., Flusser, J.: Image registration methods: a survey. Image Vis. Comput. **21**, 977–1000 (2003)
8. Fu, Y., Cao, L., Guo, G., Huang, T.S.: Multiple feature fusion by subspace learning. In: International Conference on Image and Video Retrieval, pp. 127–134 (2008)
9. Yan, Y., Wang, Y., Huang, X.: Fast image matching based on hill-climbing. Sci. Technol. Rev. **20**, 72–75 (2008)
10. Fukunaga, K., Hostetler, L.D.: The estimation of the gradient of a density function. IEEE Trans. Inf. Theory **21**, 32–40 (1975)
11. Comaniciu, D., Meer, P.: Mean shift analysis and application. In: Proceeding of the International Conference on Computer Vision, pp. 1197–1204 (1999)
12. Comaniciu, D., Ramesh, V., Meer, P.: Real-time tracking of non-rigid objects using mean shift. Comput. Vision Pattern Recognit **2**, 142–149 (2000)
13. Zhou, H., Yuan, Y., Shi, C.: Object tracking using SIFT features and mean shift. Comput. Vis. Image Underst. **113**, 345–352 (2009)
14. Leichter, I.: Mean shift trackers with cross-bin metrics. IEEE Trans. Pattern Anal. Mach. Intell. **34**(4), 695–706 (2012)
15. Dalal, N., Triggs, B.: Histograms of oriented gradients for human detection. In: IEEE Conference on Computer Vision and Pattern Recognition, pp. 886–893 (2005)

A New Threshold-Constrained IFT Algorithm for Segmenting IC Defects

Honghua Cao, Junping Wang and Guangyan Zhang

Abstract In integrated circuit (IC) manufacturing, the defect is a major factor in affecting the IC functional yield. The defect separated from the defect image can be used to describe the defect features. Therefore, defect segmentation is extremely important in layout optimization. To speed up the execution efficiency in defect segmentation, this paper proposes a new threshold-constrained image forest transform (IFT) algorithm. Firstly, the image is mapped to a weighted graph, on which seed pixels are selected by an interactive mode. Secondly, every pixel is assigned to an optimum cost path proceeding from the seed pixels based on improved growing criteria. Finally, the aggregation of entire path forms an IFT tree. Experiments show that our proposed algorithm can not only achieve quite better effect in the IC defect segmentation, but also reduce the search ranges and improve the execution efficiency compared with the traditional IFT, region-growing algorithm, and Otsu threshold algorithm.

Keywords IC · Redundancy material defect · Threshold-constrained IFT · IFT tree

H. Cao (✉) · J. Wang · G. Zhang
School of Telecommunications Engineering, Xidian University, Xi'an 710071, China
e-mail: honghcao@163.com

J. Wang
e-mail: jpwang@mail.xidian.edu.cn

G. Zhang
e-mail: zh_gy_358@126.com

J. Wang
Microelectronics Institute, Xidian University, Xi'an 710071, China

S. Patnaik and X. Li (eds.), *Proceedings of International Conference on Computer Science and Information Technology*, Advances in Intelligent Systems and Computing 255, DOI: 10.1007/978-81-322-1759-6_69,

1 Introduction

In integrated circuit (IC) manufacturing, functional yield loss is closely related to defects. Therefore, the segmentation of defect is extremely important in improving the manufacturing yield and the reliability of circuits [1–3]. The shapes of defect involved in the lithography process are varied. So we should segment the defect effectively and determine the origins, the diameter distribution, and the shape of those defects precisely for giving a theoretical basis to the layout optimization. There are many kinds of image segmentation methods. However, only a few of them are applied in IC real defect segmentation. Thus, the application of image forest transform (IFT) algorithm in IC real defect segmentation has become very worthwhile research.

The IFT is a booming image processing technique in image segmentation [4, 5]. Nowadays, the research of interactively IFT has become the interest and focus of the image processing. For example, Falcão and Bergo [6] realized a volume segmentation interactively with differential image foresting transforms (DIFT) in medical images which achieved substantial efficiency gain; Spina et al. [7] proposes an unified framework for fast interactive segmentation of natural images using the IFT, which completes natural object extraction more effectively.

Based on the analysis of existing theory, this paper proposes a threshold-constrained IFT according to the characteristics of IC defect image and the request of image segmentation. Firstly, the image is mapped to a graph, on which seed pixels are selected with an interactive mode. Then, the threshold limitations are added in the path cost calculation of the growing criteria, and every pixel is assigned to an optimum cost path by computing the path cost of each adjacent pixel, which proceeds from the seed pixels. Finally, the aggregation of entire path is an IFT tree and the graph is mapped to image again. This completes the defect segmentation. Through the comparison with the general region-growing algorithm and commonly used threshold segmentation algorithm, the experiments show that our algorithm can not only achieve quite better effect in the IC real defect segmentation, but also reduce the search ranges of the traditional algorithm and improve the execution efficiency.

2 IC Real Defect

There are many factors that can produce defects in the process of IC manufacturing. Defect can be divided roughly into global defects and local defects. The redundancy material defects belonging to local defects have an important role in blocking light, which led to various IC structural defects in the lithography and etching. This influence accounts for more than 60 % in functional yield loss [8].

Therefore, when considering the prediction and improvement of IC yield, the effect of redundancy material defects must be taken into account. This paper

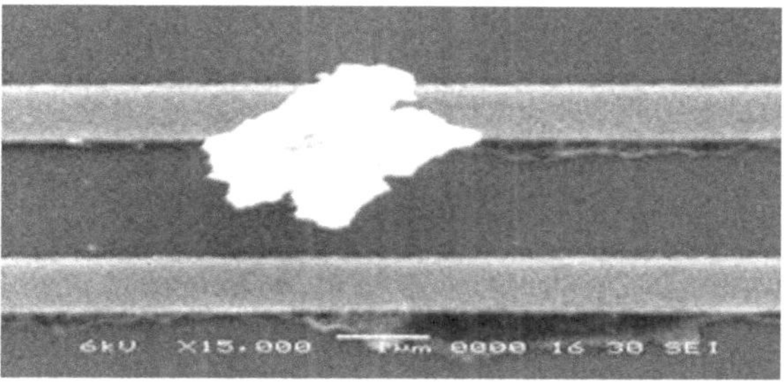

Fig. 1 Redundancy material defects left on the layout

focuses on the segmentation of the redundancy material defects (hereinafter referred to as IC real defect). Figure 1 shows a defect sample picked at random from several samples of real defects. Through analyzing the theory, general defect layout has the following characteristics: (1) Image format is gray scale or true color. (2) There are noises, interconnecting wires, defects, and backgrounds in the figure. (3) Compared with normal layout images, defects that can cause circuit failures are existed in the layout image. They are just what we need to segment.

3 Threshold-Constrained IFT Algorithm for Segmenting IC Defects

IFT is a framework for image segmentation, where the image is interpreted to a graph. It designs the growth criteria based on the connectivity of the graph. The essence of this algorithm is that image is mapped to graph firstly, and then, a marked graph is obtained by the calculation of the optimum-path forest in the graph, and finally, the marked graph is converted into the needed image [9–13]. This algorithm has a lower misclassification rate in the image segmentation and is little affected by noise. However, the speed of the algorithm remains improved. Based on the analysis of existing IFT algorithm, a threshold is added in the calculation of the path cost to reduce the search range and improve the execution efficiency. According to actual application demand of the IFT, it is important to select appropriate path cost function and seed pixel set.

3.1 The Selection of Seed Pixels

Seed pixels determine the starting position for the search of the path. We can find different target areas from different seed pixels. The forest formed by the seed pixel set needs to be consistent with the segmentation result what we expected to achieve. At the same time, it should ensure that a better segmentation is obtained with the smallest number of seed pixels. In this paper, the interactive model is applied to select seed pixel in the IC real defect segmentation, and the seed pixel must be selected from the inner regions of the target so as to get accurate and comprehensive information of the defect.

3.2 The Path Cost Function of Threshold-Constrained IFT

Threshold-constrained IFT defines an optimum-path forest in the graph. Each node of the forest corresponds to an image pixel, the connectivity of each node is represented by the relation of the adjacent between the pixels (4-neighborhood or 8-neighborhood, etc.), and the path cost is determined by a specific path cost function. Usually, the path cost function depends on local properties such as the color, gradient, and the pixel position of the image.

In finding the optimum path, path cost function was used to calculate the cost of each node. By calculating the cost of this node from each path to it, the minimum cost path to this node is obtained. In general, path cost function is a function of arc weight. This paper improves the traditional algorithm and uses the following path cost function:

$$f_{\mathrm{sum}}(<t>) = h(t) \tag{1}$$

$$f_{\mathrm{sum}}(<\pi> \bullet <s,\, t>) = \begin{cases} f_{\mathrm{sum}}(<\pi>) + w(s,\, t) \\ +\infty \end{cases} \tag{2}$$

$$\begin{array}{c} w(s,\, t) < T \\ w(s,\, t) > T \end{array} \tag{3}$$

where $(s,\, t) \in A$, s is the starting node of path π and t is the ending node. $h(t)$ is a fixed handicap cost for any paths which start at pixel t. $w(s,\, t)$ is a non-negative weight that is assigned to the arc $(s,\, t)$. The path cost function is required to be monotonically $f(<\pi> \bullet <s,\, t>) - f(\pi) \geq 0$; In this paper, the arc weight is defined as the absolute value of gray difference of two nodes corresponding to two adjacent pixels, and the additive path cost function f_{sum} is adopted.

Through testing the IC real defect segmentation, threshold T takes the average gray value of the entire image. Generally, internal gray distribution of IC real defect is uniform and large gray jumps do not exist. The proposed algorithm will not traverse all the nodes of the image during the search process. It only calculate parts of pixels which have a difference with seed pixels below the threshold. Its essence is to reduce the search area and improve the execution efficiency.

The realization steps of the threshold-constrained IFT are as follows: Firstly, the image is mapped to an edge-weighted graph. Secondly, the root of forest is extracted from the given pixels. Then, based on our improved path cost function, threshold-constrained IFT assigns to each node the mark of the closest seed pixel, which is determined by the minimum cost path from the set of seed pixels. Finally, all paths constitute a directed forest, which forms the segmented image expected.

3.3 Threshold-Constrained IFT Algorithm Based on the FIFO Priority of Queue

The steps of threshold-constrained IFT algorithm based on the FIFO priority of queue structure are given as follows:

Input: An image = (I, *I*); an adjacency relation $A \subset I \times I$; and a monotone increasing path cost function *f*; seed pixel set *R*

Output: path cost map *C*; label map *L*; an optimum-path forest *P*

Auxiliary Data Structures: priority queue *Q* and cost variable cst.

---

1. Compute threshold *T*, that is the average gray of the entire image;
2. For each $t \in I$, set $L(t) \leftarrow 0$, $C(t) \leftarrow f(<t>)$;
3. For all seed nodes $r \in R$, set $C(r) \leftarrow 0$, $L(r) \leftarrow 1$, insert *r* in *Q*;
4. while(!QueueEmpty(*Q*)):

4.1. Sort *Q* based on *C*(*s*) from smallest to largest, remove *s* from *Q* such that *C*(*s*) is minimum;

4.2. For each node *t* such that $(s, t) \in A$:

4.2.1. Compute $w(s, t)$, if $w(s, t) < T$, then cst $= f_{\text{new}}(f_{\text{sum}}(s) \bullet w(s, t))$; otherwise, cst $= +\infty$;

4.2.2. if cst $\leq C$, Set $C(t) \leftarrow$ cst, $L(t) \leftarrow L(s)$;

4.2.2.1. if $t \notin Q$, insert *t* in *Q* based on *C*(*t*);

4.2.2.2. if $t \notin Q$, remove *t* from *Q*, then insert *t* in *Q* based on *C*(*t*) and update the location of *t* in *Q*;

3.4 The Application of Threshold-Constrained IFT in IC Defect Segmentation

To validate the effectiveness and adaptability of the proposed algorithm, the experiments are conducted in Visual C++ programming on a dual-core 3.3 GHz PC with 4-GB memory. Our algorithm is also tested on many standard layout images with different defect types from a lot of different sources in our experiments.

In this paper, the seed pixels are selected by manual scribing in the IC real defect segmentation. Those points on the line are regarded as seed pixels. Figure 2 shows a gray layout image randomly selected with a single defect, and black parts of which in the defect regional center are seed pixels by manual scribing. All pixels that satisfy the growing criteria around the seed pixels are contained. Then, an IFT tree that included all defect pixels is formed, as shown in Fig. 2(b). More test results are shown in Fig. 2. Figure 2(c) shows the defect segmented by traditional IFT algorithm. Figure 2(d) shows the defect segmented by region-growing algorithm proposed by Verma et al. [14]. The defect in Fig. 2(e) is obtained by commonly used Otsu threshold segmentation algorithm [15].

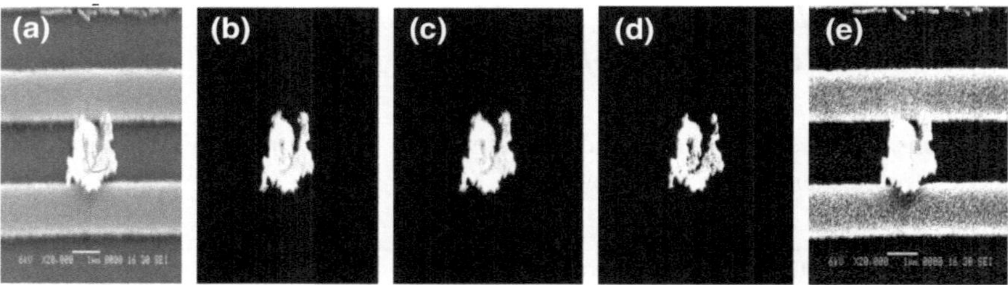

Fig. 2 The compared results of image with one defect segmented by four different methods. **a** Traditional image with manual scribing pixels. **b** The result by threshold-constrained IFT. **c** The result by traditional IFT. **d** The result by region-growing algorithm. **e** The result by Otsu threshold segmentation

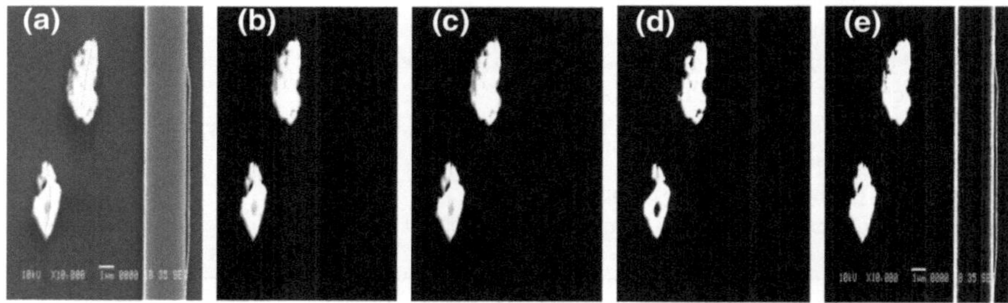

Fig. 3 The compared results of image with more defects segmented by four different methods. **a** Traditional image with manual scribing pixels. **b** The result by threshold-constrained IFT. **c** The result by traditional IFT. **d** The result by region-growing algorithm. **e** The result by Otsu threshold segmentation

For a layout image with many defects, the selection of seed pixels by manual scribing also has the advantage that all defects can be segmented, as long as the seed pixels are selected completely. So, human interaction is clearly a significant factor for the segmentation result. Figure 3 shows the segmentation results of the layout image with more defects. Every part of Fig. 3 has the same meaning with Fig. 2.

The evaluation indicators of IC real defect segmentation result have to meet two aspects. One is that there is a clear boundary between defect and noise, interconnects or nets of the layout. And the edge of defect must be segmented precisely. Another is that the inner regions of the defect segmented must be uniform without too many holes. To the IC defect segmentation results, threshold-constrained IFT and traditional IFT all have better performance compared to the region-growing and Otsu threshold segmentation in the segmentation of IC real defects. They are also less affected by noise and can segment the edge part of defect precisely.

In the meantime, through numbers of validations of many standard layout images, our proposed method reduces the time complexity and has a greater improvement compared to the traditional IFT. Table 1 shows the compared results

Table 1 The compared results of execution efficiency between threshold-constrained IFT and traditional IFT

Time(s)	Image 1	Image 2	Image 3	Image 4	Image 5	Image 6
Traditional IFT	64.50	96.23	2.87	32.10	0.23	97.58
Proposed IFT	21.59	46.34	2.11	7.14	0.17	60.12

of execution efficiency between threshold-constrained IFT and traditional IFT, where image 1 to image 6 are layout images with defects selected randomly.

From what we have mentioned above, we can clearly see that threshold-constrained IFT not only has the highest reliability, most accurate segmentation results which can segment the texture and depict every detail of the defect, but also has a lower time complexity.

4 Conclusion and Future Work

In the field of image processing, the segmented defect can be used to analyze its size and structure properties, which provide the basis for the research of defect model. Based on the threshold-constrained IFT, the segmentation of IC defects can be completed effectively. Our proposed algorithm is superior to the traditional IFT, the region-growing algorithm, and the Otsu threshold algorithm with a lower false recognition rate and a lower time complexity.

However, for locating the position of defect and realizing segmentation effectively, the seed pixels' selection is completed by manual work. Although the execution time of the proposed algorithm is short, it takes more time when human selects the seed pixels before executing the algorithm. Thus, when processing a number of images or images with many defects, the entire process is quite time-consuming. Therefore, as future work, we intend to find a quick and effective method to select the seed pixels automatically according to the properties of the images based on the threshold-constrained IFT.

Acknowledgments This work was supported in part by the National Natural Science Foundation of China under Grant 61173088, in part by the Science and Technology Program of Xi'an, China, under Grant CX1248⑤, and in part by the 111 Project under Grant B08038.

References

1. Sundareswaran, S., Maziasz, R., Rozenfeld, V., Sotnikov, M.,Konstantin, M.: A sensitivity-aware methodology to improve cell layouts for DFM guidelines. In: 12th International Symposium on Quality Electronic Design (ISQED), pp. 431–436. IEEE Press, Santa Clara (2011)

2. Bickford, J.P., Hibbeler, J.D., Mueller, D., Peyer, S., Kumar, V.S.: Optimizing product yield using manufacturing defect weights. In: 23rd Annual SEMI on Advanced Semiconductor Manufacturing Conference (ASMC), pp. 16–20. IEEE Press, Saratoga Springs (2012)
3. Tam, W.C., Blanton, S.: To DFM or not to DFM? In: 48th ACM/EDAC/IEEE on Design Automation Conference (DAC), pp. 65–70. IEEE Press, New York (2011)
4. Malmberg, F., Lindblad, J., Nyström, I.: Sub-pixel segmentation with the image foresting transform. In: IWCIA 2009, LNCS, vol. 5852, pp. 201–211. Springer, Heidelberg (2009)
5. Tao, W.B., Chang, F.: Interactively multiphase image segmentation based on variational formulation and graph cuts. Pattern Recogn. **43**, 3208–3218 (2010)
6. Falcao, A.X., Bergo, F.P.G.: Interactive volume segmentation with differential image foresting transforms. Med. Imaging **23**, 1100–1108 (2004)
7. Spina, T.V., Montoya-Zegarra, J.A., Falcao, A.X., Miranda, P.A.V.: Fast interactive segmentation of natural images using the image foresting transform. In: 16th International Conference Digital Signal Processing, pp. 1–8. IEEE Press, Santorini-Hellas (2009)
8. Wang, J.P., Hao, Y., Jing, M.E.: Skeleton extraction of IC real defects using morphology. J. Xidian Univ. **32**, 206–209 (2005)
9. Miranda, P.A.V., Falcao, A.X., Ruppert, G.C.S.: How to complete any segmentation process interactively via image foresting transform. In: 23rd SIBGRAPI—Conference on Graphics, Patterns and Images, pp. 309–316. IEEE Press, Gramado (2010)
10. Miranda, P.A.V., Falcao, A.X., Ruppert, G.C.S., Cappabianco, F.A.M.: How to fix any 3D segmentation interactively via image forest transform and its use in MRI brain segmentation. In: 2011 IEEE International Symposium on Biomedical Imaging: From Nano to Macro, pp. 2031–2035. IEEE Press, Chicago (2011)
11. Malmberg, F.: Image foresting transform: On-the-fly computation of segmentation boundaries. In: SCIA 2011. LNCS, vol. 6688, pp. 616–624. Springer, Heidelberg (2011)
12. Cappabianco, F.A.M., Araujo, G., Falcao, A.X.: The image forest transform architecture. In: ICFPT 2007. International Conference Field-Programmable Technology, pp. 137–144. IEEE Press, Kitakyushu (2007)
13. Falcao, A.X., Stolfi, J., de Alencar Lotufo, R.: The image foresting transform: Theory, algorithms, and applications. Pattern Anal. Mach. Intell. **26**, 19–29 (2004)
14. Verma, O.P., Hanmandlu, M., Susan, S., Kulkarni, M., Jain, P.K.: A simple single seeded region growing algorithm for color image segmentation using adaptive thresholding. In: 2011 International Conference on Communication Systems and Network Technologies (CSNT), pp. 500–503. IEEE Press, Katra (2011)
15. Zhu, Q.D., Jing, L.Q., Bi, R.S.: Exploration and improvement of Otsu threshold segmentation algorithm. In: 8th World Congress on Intelligent Control and Automation (WCICA), pp. 6183–6188. IEEE Press, Jinan (2010)

Based on Difference Evolutionary Video Feature Classification of Video Watermarking Algorithm

Chengzhong Yang and Xiaoshi Zheng

Abstract According to the video features, using the method of evolution for video image classification processing, suitable for video performance and human visual characteristics of video watermarking, puts forward using a kind of Difference Evolution algorithm (DE) for video image classification processing of video watermarking algorithm. All the key frame numbers are converted to a binary matrix after data transformation; at last, the binary image is produced. From this, the watermarking image reflects the video's features, so it is effective to resist the watermark copy attack.

Keywords HVS · Difference Evolution algorithm · Video characteristics · Texture attribute

1 Introduction

Video is viewed as a composition of images, where different areas of every frame have different properties. The property can be viewed as the composition of frame property and image property. If the watermarking strength is the same in every area, the watermark will probably affect the quality of the video. The factors affecting the human eye's feeling are mainly composed of the following three features: brightness features, region activities, and edge features. So, we should embed the different watermarking strength in different areas according to these properties.

C. Yang (✉)
College of Information Engineering, Guizhou Mizu University,
Guiyang, 550025 Guizhou, China
e-mail: loocy_me@126.com

X. Zheng
Shandong Computer Science Center, Jinan, 250014 Shandong, China
e-mail: zhengxs@keylab.net

S. Patnaik and X. Li (eds.), *Proceedings of International Conference on Computer Science and Information Technology*, Advances in Intelligent Systems and Computing 255, DOI: 10.1007/978-81-322-1759-6_70,

This paper is about a new method based on video characteristics, using the evolutionary computation algorithm for video image classification. First give some theory about the use of the video property in video watermark and then, according to the new classification standard for image area, about the watermark embedding and extraction. Here, we use the Difference Evolution algorithm (DE) for image area. At last, we do some simulation tests to verify our algorithm.

2 The Video Watermarking Principle of Considering the Characteristics

The DC coefficient expresses image block's mean by brightness; it reflects the basic information of the image block. According to the difference of the adjacent frame, we can detect video frame whether there is a big change; besides most of the Signal energy is concentrated in the alternating current coefficient, DCT exchange coefficient can determine the image block whether they contain the detailed information.

2.1 Figures and Tables

It is essential that all illustrations are as clear and as legible as possible. Vector graphics—instead of rasterized images—should be used for diagrams and schemas whenever possible. Please check whether the lines in line drawings are not interrupted and have a constant width. Grids and details within the figures must be clearly legible and may not be written one on top of the other. Line drawings are to have a resolution of at least 800 dpi (preferably 1,200 dpi). The lettering in figures should not use font sizes smaller than 6 pt ($\sim$2 mm character height). Figures are to be numbered and should have a caption which should always be positioned under the figures, in contrast to the caption belonging to a table, which should always appear above the table. Captions are set in 9-point type. If they are short, they are centered between the margins. Longer captions, covering more than one line, are justified (Figs. 1 and 2 show examples). Captions that do not constitute a full sentence do not have a period. Text fragments that are fewer than four lines should not appear at the top or bottom of pages, following a table or figure. In such cases, it is better to set the figures right at the top or right at the bottom of the page. A figure should never be placed in the middle of a paragraph. If screenshots are necessary, please make sure that the essential content is clear to the reader.

Remark 1 In the printed volumes, illustrations are generally black and white (halftones), and only in exceptional cases, and if the author or the conference organization is prepared to cover the extra costs involved, are those where the colored pictures are accepted. Colored pictures are welcomed in the electronic

Fig. 1 Original frame and original watermark

Fig. 2 The frame embedded the watermark and image watermark extracted using HVS

version free of charge. If you send colored figures that are to be printed in black and white, please make sure that they really are also legible in black and white. Some colors show up very poorly when printed in black and white.

2.2 Formulas

Displayed equations or formulas are centered and set on a separate line (with an extra line or half line space above and below). Displayed expressions should be numbered for reference. The numbers should be consecutive within the contribution, with numbers enclosed in parentheses and set on the right margin. Please do not include section counters in the numbering.

2.3 Footnotes

The superscript numeral used to refer to a footnote that appears in the text either directly after the word to be discussed or—in relation to a phrase or a sentence—

following the punctuation mark (comma, semicolon, or period). For remarks pertaining to the title or the authors' names, in the header of a paper, symbols should be used instead of a numbers (see first page of this document). Please note that no footnotes may be included in the abstract.

3 Differential Evolution Algorithm Profile

3.1 Copyright Form

Difference Evolution algorithm (DE) is a group based on difference evolutionary algorithm for solving continuous global optimization problems. The algorithm is by the USA at the University of California, Berkeley, Rainer Storn and Kenneth Price in 1995 and is put forward for solving the Chebyshev polynomial.

DE is an evolutionary algorithm based on real number encoding, and its basic idea is to use the current population of individual difference to restructure the variation among populations and then use the intermediate species and father generation populations to directly get new species, then use new species and the direct competition with his father generation populations for the next generation of population. According to different types of mutation and hybridization, Rainer Storn and Kenneth Price advocated that the differences of ten kinds of evolution pattern thus formed the different difference evolution algorithm. The general form of difference evolution algorithm is a DE/*x*/*y*/*z*. Among them, *x* is an individual that restructures with difference vector; It can be any parent individuals, also can be a certain individual or the current best individuals; *y* is the number of difference vector when restructuring; *z* represent miscellaneous fork model, divided into bin mode and exp mode. exp represents index of hybrid mode, namely the hybrid first fixed in a hybrid start bit, after crossing from the start bit, in turn, the hybrid mode is a continuous change in several location of adjacent genes. Bin represents binomial hybrid mode, namely the hybrid is not fixed in position, but to decide the source of the component in $V_i(t)$ is completely by hybrid probability CR. This article will use two modes: "DE/best/1/exp" and "DE/best/1/bin". "DE/best/1/bin" model of DE algorithm is described as follows [1]:

$t \leftarrow 0$; //t is evolution generation
Initialization population is $\{ X_i(t) | 1 \leq \mathrm{i} \leq \mathrm{NP}\}$; //NP is the size of the population
Evaluation of population and the best saved individuals, recorded as $X_{\mathrm{best}}(t)$;
Do
for $i = 1$ to NP do
$m \leftarrow \lfloor \mathrm{rand} \times D \rfloor$;//rand $\in [0, 1)$, D is the dimensions of the solution space, or the length of the individual chromosomes
for $j = 1$ to D do
In [0, 1], produce a random decimal r
if ($r \leq$ CR or $j = m$) then // Dim Pc as double CR $\in [0, 1]$

$$V_i^j(t) \leftarrow X_{\text{best}}^j(t) + F^*(X_a^j(t) - X_b^j(t));$$

//$F \in [0, 2]$, $a \neq b \neq$ best
else $V_i^j(t) \leftarrow X_i^j(t)$;
next j
if $\text{fit}(V_i(t)) > \text{fit}(X_i(t))$ then
//fit(x) as the fitness evaluation function

$$X_i(t+1) \leftarrow V_i(t)$$

else $X_i(t+1) \leftarrow X_i(t)$
next i
Evaluation of population and the best saved individuals should remember $X_{\text{best}}(t)$;
$t \leftarrow t + 1$;
until satisfying the termination conditions;
output $X_{\text{best}}(t)$ and $\text{fit}(X_{\text{best}}(t))$.

4 The Classification of Image Region

First of all, we should determine the position of the watermark embedding. In the DCT domain, if we use DC coefficient embedding digital watermark, this will be very easy to be blocked; but also because of the embedded information content is too small and difficult to identify, in the high frequency coefficient embedding watermark, we can ensure that the watermark is well hidden, but compression, filtering, and adding noise can easily remove watermark; Low-frequency coefficient gathered the most of the image's signal energy, so they have enough robustness, and its value is very big, hidden effect is better, so we can find the watermark is embedded in 2–4 line 2–4 columns.

DCT coefficient of energy can be used to determine the image block that contains detailed information. The experimental results show that [2] in high-brightness area, the sensitivity of the human eye is relatively lower, while in a dark area, the sensitivity of the human eye is higher. In the grain area, the sensitivity of the human eye is higher than the edge of no edge area, but below the smooth area. Thus, we will be dividing a visualization into three areas: smooth area, edge area, and grain area.

P value indicates the change of area. The *P* value more in the stable region; when *P* is oppositely bigger, then it is in the grain area; when *P* is very large, it is in the edge area. *P* artificial bee colony algorithm evaluation function definition is given by:

$$P = \lg\left(\sum_{k=1}^{9} \sqrt{X_k}\right) \tag{1}$$

where X_k is the DCT domain exchange coefficient value. According to the brightness, visualization is divided into sensitive area, general sensitive area, and not sensitive area. They are dependent on the definition of B value [3]:

$$B = \frac{\sum_{m=1}^{3} \sum_{n=1}^{3} (A(m, n) - 128)^2}{\beta} \tag{2}$$

Among them, (m, n) is point (m, n) in 3×3 piece of a pixel value, and β is elasticity factor. In this chapter, $\beta = 70,000$. We can get nine area recorded as $P = (P1, P2, \ldots, P9)$.

5 Embedding Watermark

Zheng et al. [4] put forward a kind of image watermarking algorithm, which can meet the video watermarking using HVS requirements, and so, we chose this algorithm. But, we also chose the jpeg lossy compression resistance against the effect of better additional low-frequency coefficient to eliminate watermarking spatial correlation to extract better watermark. The steps are as follows:

Step 1: Obtain watermarking information and then implement scrambling formula (3)

$$\begin{pmatrix} x' \\ y' \end{pmatrix} = \begin{pmatrix} 1 & 2 \\ 1 & 1 \end{pmatrix} \begin{pmatrix} x \\ y \end{pmatrix} (\bmod N) \tag{3}$$

Step 2: Reading video information, according to the random mark and test target sample “conclusion” [3, 5], random selection of 70 % of all frame sequence embedded to this conclusion shows that we can extract watermark in the random and frame probability up to 99.7 %, and in a single-frame watermark, this method not only damages, but also can resist proposed attack. We need every frame of the gray image, so we get Y channel by changing the RGB model, to YUV model.

Step 3: Put Y channel into 8×8 pieces, each is the piece of executive DCT transformation.

Step 4: Calculation of each piece of P, B value, according to human visual characteristics for one kind of watermark strength, modifies 2–4 line 2–4 series of value as follows:

$$\begin{cases} T(x,y) + a^{*}K1(x), & \text{if } (w(i,j) = 1) \\ T(x,y) + a^{*}K2(x), & \text{if } (w(i,j) = 0) \end{cases} \tag{4}$$

Here, $T(x, y)$ is 2–4 line 2–4 the average value of a column; $K1$ and $K2$ consists of two one-dimensional homogeneous distribution of randomly related number sequence generation; and a is watermarking strength.

Step 5: For each piece, perform the discrete cosine transform.
Step 6: Change the transformation model of the model.

6 Extracting Watermark

Step 1: Random selection and frame change the RGB model of the transformation model in each frame and then point for channel 8 × 8 pieces, each is the piece of executive CDT transformation.
Step 2: Calculation 2–4 line 2–4 column and $K1$, $K2$ mark value pk2 railway, the relationship between the value of the watermark extraction is as follows:

$$w(i,j)^{*} = \begin{cases} 1, & \text{if } (\text{pk1} > \text{pk2}) \\ 0, & \text{if } (\text{pk1} < \text{pk2}) \end{cases} \tag{5}$$

Step 3: $\mathbf{w}(\mathbf{i},\mathbf{j})^{*}$ performs the scrambling transformation.

7 Simulation Experiment

We choose a video named 1.avi (frame number: 352). The platform is the size of the watermark vc6.0 (20).

First of all, we in the video embedding watermark are not at the same time using stable power 30 and 8 (30 is the biggest strength settings and is the lower limit value), based on HVS embedding watermark. Strength $a = \{12, 10, 8, 18, 13, 11, 30, 25, 15\}$. We also random selection for some frame embedding watermark, and calculation of the PSNR value. For the result see Table 2. We can also see more intuitive results: random frame image, as shown in Figs. 1, 2, 3, and 4 is according to the diagrams in Tables 1, 3, 4, and 5.

Fig. 3 The frame embedded the watermark and the extracted with the watermarking strength 8 without HVS

Fig. 4 The frame embedded the watermark and the extracted with the watermark strength 30 without HVS

Table 1 Nine divided area ("Bri" represents brightness, "Tex" represents texture)

	Insensitivity	General sensitivity	Sensitivity
Smooth	*P*1	*P*2	*P*3
Edge	*P*4	*P*5	*P*6
Texture	*P*7	*P*8	*P*9

Table 2 The value of PSNR ("\"means the frame is not embedded with the watermark; PSNR1 shows the value of PSNR using HVS, PSNR2, PSNR3 values of PSNR using strength 30, 8 without HVS separately)

Random frame No.	4	27	144	177	63	327	56	155
PSNR1	39.807	\	39.185	39.156	37.328	41.5679	40.975	42.757
PSNR2	36.328	33.162	\	33.672	36.995	28.892	32.789	33.478
PSNR3	\	42.957	42.255	38.794	\	38.995	42.986	40.976

Table 3 The result of Gaussian noise using HVS

Random frame No.	31	102	1	56	12	21	211	5
Extracted watermark								
NC value	0.9949	0.7625	0.9935	0.7850	0.9850	0.9925	0.5325	0.9750

Table 4 The result of Gaussian noise using watermarking strength 8 without HVS

Random frame No.	29	34	115	68	132	72	254	28
Extracted watermark								
NC value	0.5720	0.6225	0.5450	0.5375	0.9323	0.7425	0.9275	0.5402

Table 5 The result suffering rotation 20C attack

Random frame No.	4	6	25	82	124	131	117	331
Extracted watermark								
NC value	0.5150	0.4975	0.4975	0.8250	0.7425	0.5803	0.6225	0.5451

8 Conclusion

This paper brings the HVS in video watermarking and video classification adapted to the evolution method. According to the analysis of experimental results for improving the video quality at the same time, to ensure the robustness, this algorithm can resist inter-frame attack, cut attack, the reorganization of the attack against (such as noise attack, filtering attack, resample, etc.) and jpeg compression attack. After the attacks, we still can extract clear watermark. But, this algorithm can effectively resist geometric attacks (such as rotation). In addition, we try to use evolution method, adaptive to choose embedded position to improve the algorithm, but the relating to the function for selection, still we need to do further research and get better results.

Acknowledgments This work was supported by the Guizhou Science and Technology Foundation Project (No. LKM [2011] 28).

References

1. Li, H., Lui, K.: Artificial colony algorithm and bee algorithm and difference evolution algorithm in numerical optimization of the comparative (in Chinese)
2. Xu, Y., Cong, J.: A new blind image watermarking algorithm in DCT domain. J. Comput. Eng. Appl. **2**, 47–49 (2004)
3. Qi, H., Zheng, D., Zhao, J.: Human visual system based adaptive digital image watermarking. J. Sig. Proc. **88**(l), 174–188 (2007)
4. Zheng, X., Zhao, Y., Li, N.: Randomly marking and detecting of target samples and its application in video digital watermark. (China: CN200610055l298)
5. Zhou, Z., Liu, J.-Y., Ma, L.: Video content analysis technique. Comput. Eng. Des. China **29**, 1766–1769 (2008)

Removal Algorithm for Redundant Video Frames Based on Clustering

Xian Zhong and Jingling Yuan

Abstract For the problem of a lot of redundant video frames existing in the massive image and video, a new removal algorithm RRFC for redundant video frame based on the classic *k*-means clustering algorithm but keeping the frame order by means of calculating the dissimilarity between frames was proposed in this paper. The experimental results show that removal rate is generally about 15 %, which verify the effectiveness of the algorithm. The algorithm provides the support for the subsequent key frame extraction, semantic feature modeling, and semantic retrieval.

Keywords Redundant video frames · Removal algorithm · Clustering algorithm · Dissimilarity alpha

1 Introduction

With the "Big Data" era coming, massive videos and images of the Internet and monitoring are increasing and video retrieval is also facing bottlenecks and challenges. Since massive video is not all valid, how to remove redundant video frames and extract useful video frames has become a key issue [1, 2]. Because of their successive orderliness, video frames contain a chronological sequence of uptake of useful information such as task, object, scene, and event. But there is also a large number of redundant or useless video such as only the background of the video when no vehicle tunnel monitoring midnight or almost no motion video of the Plaza Square crowd in camera shot [6, 8].

X. Zhong
Computer School, Huazhong University of Technology, Wuhan 430074, China
e-mail: cloudy0902@hotmail.com

J. Yuan (✉)
Computer School, Wuhan University of Technology, Wuhan 430070, China
e-mail: yuanjingling@126.com

S. Patnaik and X. Li (eds.), *Proceedings of International Conference on Computer Science and Information Technology*, Advances in Intelligent Systems and Computing 255, DOI: 10.1007/978-81-322-1759-6_71,

In order to facilitate subsequent video key frame extraction, we study clustering algorithm-based method for removing redundant video frames [1–4]. Our work will provide support for the efficient semantic retrieval, video retrieval, and mobile visual search and benefit to save memory and improve retrieval efficiency.

2 Dissimilarity Measurement of Video Frames

In the intelligent transportation, safe city, smart city, and other large projects are mostly deployed in oriented camera, that is, usually the direction is not adjusted and background is fixed. The color histogram differences can be compared to distinguish moving objects in the background in addition to day and night, the weather changes, and other specific circumstances. So the clustering based on dissimilarity measurement of color histogram is considered to extract useful video frames and remove redundant frames [5–8]. For color histogram of each frame, the 16 binary HSV color histogram is employed, in which eight bits represent shades, four bits represent saturation, and the remaining four represent value.

The color difference between frame *fi* and frame *fj* is computed by the following formula (1),

$$\mathrm{Coldist}(i,j) = \frac{\sum_{h=1}^{8}\sum_{s=1}^{4}\sum_{v=1}^{4}\min(\mathrm{HSV}_{i(h,s,v)},\mathrm{HSV}_{j(h,s,v)})}{N} \qquad (1)$$
$$\mathrm{Coldist}(i,j) \in [0,1]$$

$\mathrm{HSV}_i(h,s,v)$ represents the *fi* HSV color histogram, and N represents the total pixel number of any frame. The formula (1) is also used to calculate the subsequent key frame extraction fidelity.

On the basis of dissimilarity measure, a multipolar threshold can be set on the video frame sequence for clustering such as k-means algorithm. The formula 1 can be optimized further, for example, the three values of h, s, and v can be, respectively, calculated relative to the dissimilarity then integrated.

3 Removal for Redundant Video Frames Based on Clustering

The traditional global k-means clustering algorithm in many fields has been widely used, but it is not fully applicable to the removal of redundant frames clustering video frame, because the video frame sequence is ordered; clustering even after removal of redundant frames must keep the original order. Therefore, our algorithm for removing redundant frames takes a previous video frame sequence as the basis for classification according to the traditional clustering algorithm, in order to ensure the orderliness and coherence of video frames.

Fig. 1 Definition of class and data structure

```
Initialize State:
Video Frame Sequence ←Frame;  //Define the sequence of video frames
Dense ←β ;// Definition clustering density β
Cluster←SubCluster;// Defined subclass SubCluster
  Sum of SubCluster ←sum;// Defined the number of subclass
  Typedef struct Frame
{Color Histogram ←HSV;
// HSV indicates color histogram of  the current frame
    Frame* next;   //next is the next frame
    int N;   //N is the total number of pixel
  }f;
Coldist(f(i),f(j))←coldist;
// Definition of the color difference between ith frame and jth frame
  Typedef struct Clusters
  {
    Color Histogram← center;
    //center is the center point of the current subclass
    Frame* next ;
    int f_count;
    //f_count represents the video frame number of current subclass
  }SubCluster;
End State
```

Suppose there are M-frame frame sequence, $F = \{f1, f2, \ldots, fi, \ldots, fm\}$; the dissimilarity between frames is given in the previous formula (1). Let B be cluster density.

In order to express frame orderliness, one-way linked list structure is used to represent the node and subclass. Each node f is one frame of the video, and each subclass must maintain the original sequence of video frames for the video frame clustering. Classes and data structures are defined in Fig. 1.

The key three steps of improved clustering algorithm remove redundant frame based on clustering (RRFC) are as follows:

1. Initialization, let the first frame as the frame in the first subclass, input video frame sequence and cluster density β, the number of this subclass frames is 1, and this subclass' center point $f =$ HSV1. The pseudo-code is shown in Fig. 2.
2. Compute dissimilarity and clustering, starting from the second frame, select the next frame *f*_next orderly, cluster the video frame, and calculate centroid

Fig. 2 Initialization

```
Input Frame, β ;
// Input sequence of video frames, cluster density  β
Get(f);
  // Get the first frame of a video sequence as the current frame f
New Frame  f_head;
// Create the initial frame  of original series
    f_head.next = f;// Point the current frame
    sum = 0;  //   The number of subclass is initialized 0
    New SubCluster  sub(sum+1); //    Build one subclass
    sub(1).next = f;
      // The current frame is the first frame of the first subclass
    sub(1).center = f.HSV;
    //the center point is the HSV of the current frame
    sub(1).f_count=1; // The number of frames is  1
    sum++ ;
      f = f.next; // Point to next frame
    sub(1).next -->next = Null;
```

```
While(f != Null)
{ // While the current frame is empty, the loop ends
    f_head.next = f;
Boolean succ = false;
// Determine whether the current frame belongs to a sub-class successfully
   // calculate Coldist(f,Sub(i)) between the current frame and ith sub-class
     For i = 1 to sum
       float temp_sum=0;
       coldist(f,sub(i)) = 0;// coldist is initialized 0
       For (h=1 to 8 & s,v=1 to 4)
            temp_sum = temp_sum + min(f.HSV(h,s,v), sub(i).center(h,s,v));
       End For
       coldist(f,sub(i)) = 1 - temp_sum/N;
// When clustering dissimilarity is less than the density  β, then the current
frame is attributed to the i-th sub-class, and recalculate the centroid
       If (coldist(f,sub(i)) <= β)
       {//Put the current frame into the ith subclass
         f_head.next = f.next;
         f.next = sub(i).next;
         sub(i).next = f;
         succ = true;
         sub(i).center =(f.HSV+
         sub(i).center*sub(i).f_count)/(sub(i).f_count+1);
         sub(i).f_count++;
         f= f_head.next;
         break;// Clustering ends, jump loop
       }
     End For
     If (succ == false)
{//Traverse all subclass, when the current frame doesn't belong to one subclass,
   build new subclass
       New SubCluster sub(sum+1); // Build a subclass sub(sum+1)
       sub(sum+1).next = f;
        // The first frame of the(sum+1) is the current frame
        sub(sum+1).center = f.HSV;
        //The cent point is HSV of the current frame
       sub(sum+1).f_count=1; // The frame count is1
       f = f.next;
       sub(sum+1).next-->next = Null;
       sum++;
     }
     }
```

Fig. 3 Clustering based on dissimilarity

dissimilarity Coldist(*f*, Sub(*i*)) between the current frame and the existing subclass. The pseudo-code is shown in Fig. 3.

3. Output in reverse order: When the video frame is complete, output each subclass. When the current frame is classified, the new classified video frame is put in the classified list header. The original order of the video frame of each subclass is reverse order, so each subclass will be output in reverse order for ensuring the orderly clustering.

After the end of the algorithm, select the first, middle, and the last frame of each generated subclass as the representative of the whole; if the frame number is less than three, then all should be taken; thus, the process of removing redundant frames is completed.

RRFC algorithmic complexity is polynomial time; the results are highly intuitive and understandable.

Table 1 Removing results of reluctant frames

	Number of removed redundant frames	The rest number of the frames	Removal rate (%)
Network video (3,470 frames)	510	2,960	14.7
Network video (2,967 frames)	456	2,511	15.4
Tunnel video (3,918 frames)	966	2,952	24.7
Square video (2,239 frames)	188	2,051	8.4

4 Experimental Results

We selected a number of network video and other video frame sequence to remove redundant frames; the following results are shown in Table 1.

From the Table 1, we can see that some video removed redundant frames more and some video removed less redundant frames, mainly in different places and different behavioral characteristics of a moving object, which result in the removal number. Removed features are about 15 % of the original characteristics of the video frame; the video effect remains unchanged. The experiment result is very meaningful for the subsequent key frame extraction, semantic feature modeling, and semantic retrieval.

5 Conclusion

In this paper, in order to solve the redundant frames in massive video, on the basis of classical clustering algorithm, we propose a new removal algorithm RRFC for redundant video frame. The core of the algorithm is to compute dissimilarity between frames but keeping the frame order. Experimental results demonstrate the effectiveness of the algorithm.

Acknowledgments The research work was supported by National Natural Science Foundation of China under Grant No.61303029 and National Key Technology R&D Program 2012BAH89F00.

References

1. Gu, J., Wu, H., Zhu, H.: Data set enrichment research summary. Comput. Appl. Softw. **29**(10), 211–215 (2012)
2. Duan, L., Huang, T., Alex, C.K., Gao, W.: Technology bottlenecks and challenges of moving visual search. China Comput. Fed. Commun. **8**(12), 8–14 (2012)
3. Sara, M., Affendey, L.S., Mustapha, N., et al.: An integrated semantic-based approach in concept based video retrieval. Multimedia Tools Appl. **64**(11), 77–95 (2013)
4. Theodoridis, S., Koutroumbas, K.: Pattern Recognition. Sciences Press, Beijing (2006). (The latest version is 2010,Electronic Industry Press)

5. Hand, D., Mannila, H., Smyth, P.: Principles of Data Mining. Machine Press, Beijing (2003)
6. Popoola, O.P., Wang, K.: Video-based abnormal human behavior recognition—A review. IEEE Trans. Syst. Man Cybern. Part C Appl. Rev. **42**(6), 865–878 (2012)
7. Metze, F., Ding, D., Younessian, E., et al.: Beyond audio and video retrieval: Topic-oriented multimedia summarization. Int. J. Multimedia Inf. Retrieval **2**(2), 131–144 (2013)
8. Vijayakumar, V., Nedunchezhian, R.: A study on video data mining. Int. J. Multimedia Inf. Retrieval **1**(3), 153–172 (2012)

Research on Distributed Data Mining System Based on Hadoop Platform

Jianwei Guo, Ying Li, Liping Du, Guifen Zhao and Jiya Jiang

Abstracts Distributed systems, represented by Hadoop, are becoming an essential component of large-scale mining system. Therefore, this paper is to complete a data mining task in the Hadoop distributed system, whose main purpose is to build a distributed cluster computing environment by Hadoop and perform data mining tasks in the environment. The paper studies the Hadoop system structure and acquires an in-depth understanding on the distributed file system HDFS and the principle and implementation of MapReduce parallel programming model. We achieve a systemic control of the data mining process, apply the traditional data mining algorithms to MapReduce programming model, research the implementation of data mining algorithms on Hadoop platform, and mainly analyze the execution efficiency and scalability.

Keywords Data mining · *K*-means

J. Guo (✉) · Y. Li · L. Du · G. Zhao · J. Jiang
Department of Information Technology, Beijing Municipal Institute of Science and Technology Information, Beijing 100044, China
e-mail: vipherovip@163.com

Y. Li
e-mail: duliping_419@163.com

L. Du
e-mail: shai_wang@hotmail.com

G. Zhao
e-mail: gfzh@hotmail.com

J. Jiang
e-mail: jiya_jiang@sina.com

S. Patnaik and X. Li (eds.), *Proceedings of International Conference on Computer Science and Information Technology*, Advances in Intelligent Systems and Computing 255, DOI: 10.1007/978-81-322-1759-6_72,

1 Introduction

The characteristics of big data are usually defined as "4V" by the industry, that is [1]: Volume, Variety, Value, and Velocity.

Data mining technology [2] is a natural outcome of the development of information technology and has evolved into an important data analysis tool with increasingly outstanding features. The data "mining" is able to discover important data pattern, extract "interesting" knowledge and thus convert the data grave into knowledge "gold," which makes great contributions to business decisions, knowledge base, military, medical research and other fields. To address the current increasingly prominent problem known as "rich data, scarce info" faced by the information industry, it seems more urgent to deepen the research on data mining technology.

2 Introduction of Data Mining

Data mining [3] is a technology that finds valuable information from large data through analysis. Due to the existence and wide application of large-scale data, it is in urgent need to make the best of these data for the industrial sector, information sector and even the academic sector. Therefore, it is known in many industries as the Knowledge Discovery in Databases (KDD for short) or a fundamental step in the process of knowledge discovery in database [4]. The KDD process falls into three steps: (1) Data preparation. The data preprocessing before the data mining may include cleanup of noise data, integration of data sources, selection of targeted data, and conversion and unification of data format, so as to meet the requirements for data mining. (2) Data mining. It is the most important step, where the intellectual algorithm is applied to analyze data and extract data patterns according to the requirements. (3) Result evaluation and explanation. In the evaluation phase, the extracted data patterns are effectively analyzed to recognize the real "interesting" patterns from all the extracted results and exclude meaningless patterns according to a certain measure of interest. And the explanation phase mainly relates to the interaction between the system and users and focuses on the presentation of mining results as shown in Fig. 1.

3 Basic Tasks of Data Mining

The task of data mining determines the direction of data mining work and guides the algorithm to find the targeted data patterns. The data mining steps vary according to different mining tasks and different application areas. Generally speaking, in line with the features of mining work, data mining tasks can be

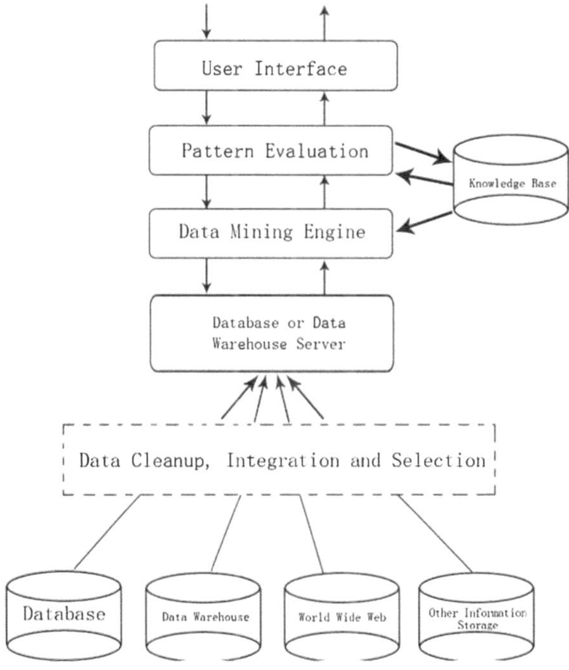

Fig. 1 Systemic structure of data mining

divided into two categories, the descriptive mining and the predictive mining. The descriptive mining task focuses on discovering the general characteristics of data and describing the existing data in the database, including data summary and search for data relationship and data type etc. The predictive data mining mainly makes deduction from current data and makes prediction for more data or new data.

According to different tasks of data mining, the mining work will get different data patterns as a result. The data mining system should have the capability to extract different kinds of data patterns to meet different task requirements and user demands and allow users to participate in the data mining process to instruct and focus on the mining of interesting data patterns. The traditional data mining tasks and the data patterns that they are able to discover can be subdivided into the following six types:

(1) Data description and visualization
(2) Correlation grouping and correlation analysis
(3) Classification and prediction
(4) Clustering
(5) Complex data mining
(6) Data evolution analysis.

4 Data Clustering Algorithm

Clustering refers to the process that divides the whole data set into multi-level subsets and groups together similar data items. In this way, the members in the same subset bear certain similar properties. Different from classification method, the clustering algorithm focuses more on automacity, which automatically groups the unlabeled data into data clusters. The process mainly relies on data properties and data interrelationship.

The clustering process of data mining can be defined in the following form:

A given database D, whose set of data entries is $X = \{x_1, x_2, x_3, \ldots, x_n\}$, among which $x_i(i = 1, 2, 3, \ldots n)$ represents a certain data entry and n refers to the number of data entries. The clustering of a certain set X is to divide X into m subsets $C_1, C_2, C_3, \ldots C_m$ and make them meet the following conditions:

(1) $C_{i \neq \Box}$, $i = 1, 2, \ldots, m$
(2) $\bigcup_{i=1}^{m} C_i = X$
(3) $C_i \cap C_j = \emptyset$, $i \neq j$, $i, j = 1, 2, 3 \ldots, m$.

5 *K*-Means Algorithm

The *K*-means algorithm is the most commonly used and the most typical partitioning cluster method, regarded as one of the top ten classical data mining algorithms. The basic principle of *K*-means is to cluster all the data orientated around the center point *K* in the space and the data objects choose the nearest center point as the temporary category when being clustered. In the process of successive iterations, the values of clustering centers are constantly updated and the data are continuously reallocated, until conditions are met to obtain the better clustering results.

That is, C_{ki} should be represented by the mean value of all data objects of a cluster. Since each process of iteration is to take the minimum value of current cost, the overall cost of all clusters will continue to decrease (or remain constant) but not increase, which guarantees that the *K*-means will finally achieve a minimum value. However, the *K*-means algorithm cannot ensure to find the global optimal solution. Usually, the achieved minimum value is locally optimal, which even cannot be guaranteed to be, but only close to the optimal solution. However, the results are very effective in reality, that is why *K*-means has its obvious advantages. This method significantly reduces the complexity of iterative calculation and improves the computational efficiency.

The execution process of *K*-means algorithm is shown as Fig. 2:

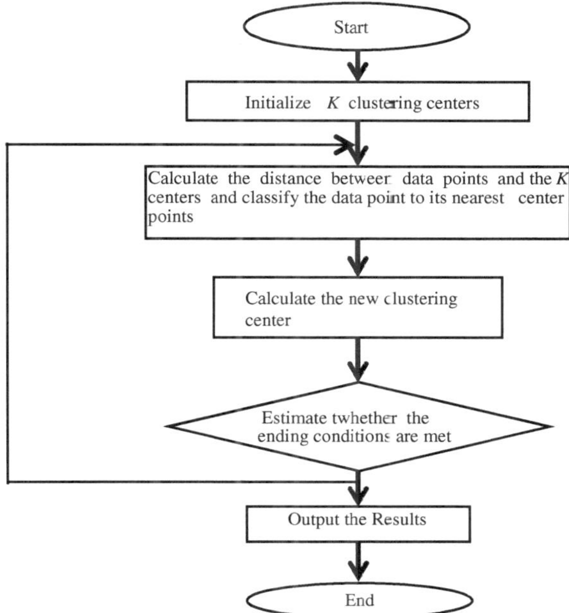

Fig. 2 The basic process of *K*-means algorithm

6 Distributed *K*-Means Process

According to the process of traditional *K*-means algorithm, we can find that the main procedure is represented by iterative operation. The operations conducted in all procedures are identical and to some degree discrete and the information exchange between two iterative operations mainly occurs in the update of *K* center vectors. Therefore, the iterative process can be isolated out independently so as to make a complete MapReduce operation to perform an iterative computation and finally re-perform the MapReduce process to achieve the purpose of iterative operation.

In the Reduce Function, the last part of iterative operation of *K*-means algorithm, that is, the updating of cluster center vector has been completed. The input of Reduce Function is the key-value pair <Cluster, List<Item>>, among which the key represents the category of cluster to be updated, while the value List<Item> refers to the collected data subsets belonging to the cluster. Each Item represents the sum of several data objects and at the same time records the number of collected data objects.

The operation and execution of Reduce Function is similar to that of Combine Function, which is to go through the Item list, extract the content vector of each Item for accumulation to obtain the vector *V* and at the same time accumulate the count variable of each Item and record the sum *C*. After the accumulation, we should calculate the average value of vectors *V*/*C* according to the definition of the center point of data cluster. The final result can be considered as the center position of current data cluster. Before the end of the Reduce Function, all *K*-updated cluster centers should be written into the hardware and overwrite the original documents of center points.

The codes of Reduce Function are shown in the following:

public void reduce(IntWritable key, Iterator < Item > values,

```
                  OutputCollector<Text,    Text>    output,
Reporter reporter)
                  throws IOException {
          System.out.println("reduce: "+key);
          Item centerItem = new Item();
          centerItem.category = key.get();
          while(values.hasNext()){
              Item item = values.next();
              centerItem.addVector(item);
          }
          centerItem.averageVector();
          newCenterList.add(centerItem);
    }
```

After writing the codes of MapReduce process, we should fulfill the mission statement and parameter configuration to perform the *K*-means algorithm in Hadoop environment. Therefore, we should complete the core "run" method of MapReduce framework, which is known as the "Driver" of MapReduce. The task of Driver is to exemplify, configure, and transfer a JobConf job named by the job configuration object and start the MapReduce job on the cluster by communicating with JobTracker. The JobConf object should include all configuration parameters that are required to keep the job running. In the Driver, we need to customize the basic parameters of MapReduce job, including the input paths of clustering data sets and the results and the earlier implemented Mapper category and Reduce category. Moreover, we should reset the default job attributes according to the format and procedure of input data and output data. Here, we use the TextInput-Format as the input format of data sets. Once completed and transferred to JobClient, the JobConf object is regarded as the overall planning of the job and becomes the blueprint for how to run the job. The contents of "run" Function in Driver that is paralleled with *K*-means algorithm is shown as follows:

```
    public int run(String[] arg) throws Exception {
        Configuration conf = getConf();
        // Add Distributed Cache
        DistributedCache.addCacheFile(nwe
  Path(conf.get("cachePath")).toUri(), conf);
        JobConf job = new JobConf(conf, MyJob.class);
        Path in = new Path(arg[0]);
        // Specify the Input Path
        FileInputFormat.setInputPaths(job, in);
        job.setJobName("PK-means!");
        // Configuration Classes
        job.setMapperClass(MapClass.class);
        job.setCombinerClass(Combine.class);
        job.setReducerClass(ReduceClass.class);
        // Configure the Input Formats
        job.setInputFormat(TextInputFormat.class);
        job.setOutputFormat(NullOutputFormat.class);
        job.setOutputKeyClass(IntWritable.class);
        job.setOutputValueClass(Item.class);
        // Submit the Job
        JobClient.runJob(job);
        return 0;
    }
```

7 Summary

In this paper, we systemically study the basic task and process of data mining and conduct in-depth analysis on *K*-means clustering algorithm. Based on MapReduce programming model, the parallel implementation of *K*-means is proposed and achieved according to the characteristics of distributed computing environment.

We use multiple ordinary PC host machines to build distributed clusters and carry out experiments on cluster data mining based on artificially generated data sets of different scales. By comparing and testing different data and computer cluster scales, we have proven the feasibility of data mining tasks in Hadoop distributed computing environment and validated the computing advantages and great potentials of the program by analyzing the speedup ratio, efficiency, data scalability, and cluster extendibility.

References

1. Xu, Z.: Big data: The impending data revolution, Guangxi Normal University, Guangxi (2012)
2. Han, J., Kamber, M.: Data Mining: Concepts and Techniques, 2nd edn. Morgan Kaufmann, San Francisco (2006)
3. Chow, J.: Redpoll:A machine learning library based on hadoop, CS Department, Jinan University, Guangzhou (2010)
4. Qin, G., Li, Q.: Knowledge acquisition and discovery based on data mining. Comput. Eng. (2003)

Novel Algorithms for Scan-Conversion of Conic Sections

Xin Chen, Lianqiang Niu, Chao Song and Zhaoming Li

Abstract Novel algorithms for the generation of conic sections whose axes are aligned to the coordinate axes are proposed in this paper. The algorithms directly calculate difference between the expressions satisfied by two adjacent points lied on curves, and then, a kind of iteration relation including residuals is established. Furthermore, according to the criterion of nearest to curve, a residual in 1/2 is employed to obtain an integer decision variable. The analyses and experimental results show that in the case of the ellipse, our algorithm is faster than basic mid-point algorithm and Agathos algorithm because constant term accumulated in every loop is eliminated, but its idea and derivation are at least as simple as basic mid-point algorithm. Moreover, our algorithm does not set erroneous pixels at region boundaries, and anti-aliasing is easily incorporated.

Keywords Conic sections · Scan-conversion · Anti-aliasing

X. Chen (✉) · C. Song
School of Science, Shenyang University of Technology, Shenyang, China
e-mail: 491079175@QQ.com

C. Song
e-mail: 75719224@qq.com

L. Niu
School of Information Science and Engineering, Shenyang University of Technology, Shenyang, China
e-mail: niulq@sut.edu.cn

Z. Li
Network Center, Changchun Vocational Institute of Technology, Changchun, China
e-mail: lzm@cvit.com.cn

S. Patnaik and X. Li (eds.), *Proceedings of International Conference on Computer Science and Information Technology*, Advances in Intelligent Systems and Computing 255, DOI: 10.1007/978-81-322-1759-6_73,

1 Introduction

Conic sections (ellipse, hyperbola, and parabola) are important computer graphics primitives, and how to scan-convert them has therefore received considerable research attention. Several authors have proposed algorithms for the scan-conversion of ellipses and other conic sections [1–8]. Among these, Bresenham's algorithm is the most typical two-point integer scan-conversion algorithm of line and circle [1]. The mid-point ellipse algorithm proposed by Van Aken [2, 3] takes advantage of the position of middle point to simplify the derived process that was used by Bresenham. Pitteway [4] described an algorithm for drawing a general conic section based on the mid-point method. McIlroy [6] extended the proof of best-fit accuracy of Bresenham's algorithm to the more general case in which the square of the circle's radius is an integer. Kappel [7] and Fellner and Helmberg [8] proposed some advanced algorithms depending on the mid-point method and gave some deeper discussions about how to generate ellipse correctly as well as gave different criteria to change from a region to another. Agathos, Theoharis, and Boehm proposed an algorithm based on Bresenham-like method but combined mid-point technique which can perform correct region transitions [9]. Several other techniques to speed up the scan-conversion of arcs and ellipses are reported in [10–18].

In essence, a progress to generate conic sections depends on distances between grid points and real curve to select candidate pixels. Unfortunately, these distances are floating points. Most of the presented algorithms do their best to avoid them and have proposed some efficient techniques like difference between two squares of the residual, sign of the residua of mid-point, scaling, and so on. Agathos algorithm generates conic sections by combined two-point method and residual.

In this paper, we will describe a method to draw conic sections based on two residuals. The new algorithm is easy to be derived, and the values of the decision variable may be exploited for anti-aliasing with Wu method [15]. Another merit of this method is that const item accumulated in the other algorithms is eliminated, so the drawing speed is improved.

2 Derivation of the Ellipse-Generating Algorithm

Due to the four-way symmetry of an ellipse, we need only to consider the first quarter region, and other regions may be obtained with symmetry. Equation of a standard ellipse which is centered at the origin of the 2D Cartesian space is:

$$F(x,y) = b^2x^2 + a^2y^2 - a^2b^2 = 0. \tag{1}$$

The first quadrant of the ellipse is separated into two parts called Region 1 and Region 2 by a point D on the curve where $dy/dx = -1$, shown in Fig. 1. In Region 1, the axis of major movement is X, and in Region 2, it is Y. For Region 1, if point

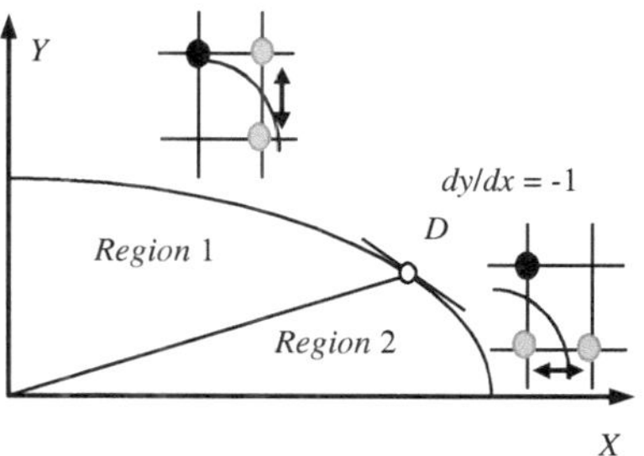

Fig. 1 Two regions of the first quarter ellipse

(x_i, y_i) has been plotted in ith step, next iteration step we always have $x_{i+1} = x_i + 1$ and need to decide if a y-stepping should occur, i.e., $y_{i+1} = y_i$ or $y_{i+1} = y_i - 1$. For the Region 2, next iteration step we always have $y_{i+1} = y_i - 1$ and need to decide if $x_{i+1} = x_i$ or $x_{i+1} = x_i + 1$.

2.1 Generating Region 1

Assuming that the ellipse is generated in a clockwise manner starting from the point (x_0, y_0), where $x_0 = 0$, $y_0 = b$. δ_{i+1} is defined as the distance from point (x_{i+1}, y_i) to real ellipse (Noting that δ_{i+1} is allowed to be negative), $i \geq 0$. For the $(i + 1)$th and $(i + 2)$th selections, point $(x_{i+1}, y_i - \delta_{i+1})$ and point $(x_{i+2}, y_{i+1} - \delta_{i+2})$ all lie on the ellipse (shown in Fig. 2) and meet the ellipse Eq. (1), we obtain:

$$\begin{cases} b^2x_{i+1}^2 + a^2(y_i - \delta_{i+1})^2 - a^2b^2 = 0 \\ b^2x_{i+2}^2 + a^2(y_{i+1} - \delta_{i+2})^2 - a^2b^2 = 0 \end{cases}.$$

Thus, we have

$$2a^2y_{i+1}\delta_{i+2} - a^2\delta_{i+2}^2 = 2a^2y_i\delta_{i+1} - a^2\delta_{i+1}^2 + b^2(x_{i+2}^2 - x_{i+1}^2) + a^2(y_{i+1}^2 - y_i^2).$$

Let

$$d_{i+1} = 2a^2y_i\delta_{i+1} - a^2\delta_{i+1}^2 \tag{2}$$

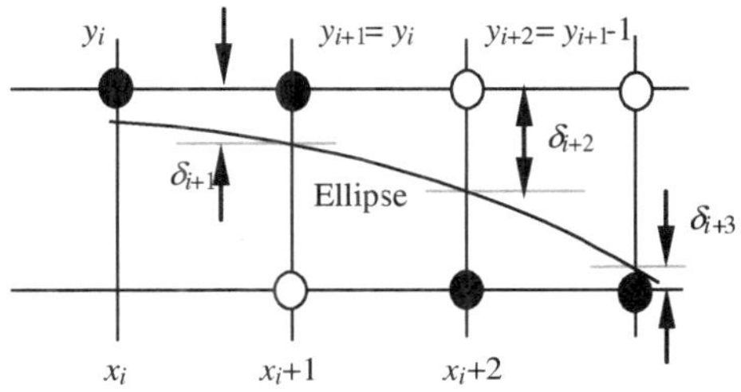

Fig. 2 Point selection of $(i + 1)$th step in Region 1

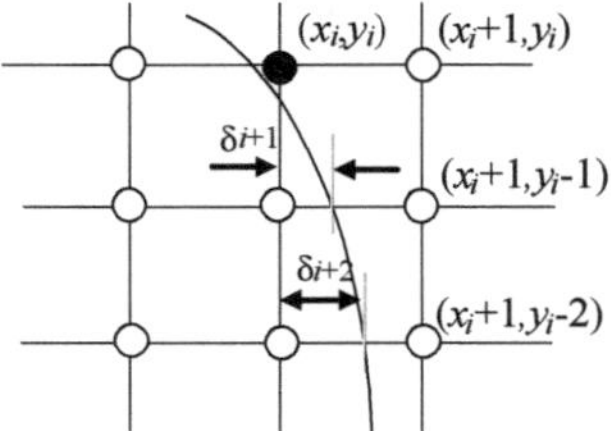

Fig. 3 Point selection of $(i + 1)$th step in Region 2

Due to $x_{i+2} = x_{i+1} + 1$, we obtain $d_{i+2} = d_{i+1} + b^2(2x_{i+1} + 1) + a^2(y_{i+1}^2 - y_i^2)$. If $\delta_{i+1} < 0.5$, we should select y_i in $(i + 1)$th step, i.e., $y_{i+1} = y_i$ and have

$$d_{i+2} = d_{i+1} + b^2(2x_{i+1} + 1) \tag{3}$$

If $\delta_{i+1} \geq 0.5$, we should select $y_i - 1$ in $(i + 1)$th step, i.e., $y_{i+1} = y_i - 1$ and have

$$d_{i+2} = d_{i+1} + b^2(2x_{i+1} + 1) - a^2(2y_{i+1} + 1) \tag{4}$$

Substituting $\delta_{i+1} = 0.5$ in formula (2), we gain

$$d_{i+1} = a^2 y_i - a^2/4 \tag{5}$$

Since $\frac{d(d_{i+1})}{d(\delta_{i+1})} = 2a^2(y_i - \delta_{i+1})$, we know that d_{i+1} is a monotonically increasing function of δ_{i+1} (Seeing Sect. 3.1). Thus, $\delta_{i+1} < 0.5$ is equivalent to $d_{i+1} < a^2 y_i - a^2/4$, and $\delta_{i+1} \geq 0.5$ is equivalent to $d_{i+1} \geq a^2 y_i - a^2/4$. Considering that point $(1, r - \delta_1)$ lies on the ellipse and meets the ellipse equation, we substitute the point's coordinate in (1) and get the initial value of iteration as follows:

$$d_1 = 2a^2 r\delta_1 - a^2\delta_1^2 = b^2 + a^2r^2 - a^2b^2 \tag{6}$$

Noting that decision variable d is an integer, we can replace $a^2/4$ in (5) by its integer part or regard $4d$ as a decision variable.

2.2 Generating Region 2

For Region 2, we have $y_{i+1} = y_i - 1$ in every iteration step. Assume that point (x_D, y_D) is the last point of the Region 1, and let δ_{i+1} is the distance from point (x_i, y_{i+1}) to the ellipse, where $i \geq D$. For the $(i + 1)$th and $(i + 2)$th selections, points $(x_i + \delta_{i+1}, y_{i+1})$ and $(x_{i+1}+\delta_{i+2}, y_{i+2})$ all lie on the ellipse, shown in Fig. 3 and meet the ellipse equation, we obtain:

$$\begin{cases} b^2(x_i + \delta_{i+1})^2 + a^2y_{i+1}^2 - a^2b^2 = 0 \\ b^2(x_{i+1} + \delta_{i+2})^2 + a^2y_{i+2}^2 - a^2b^2 = 0. \end{cases}$$

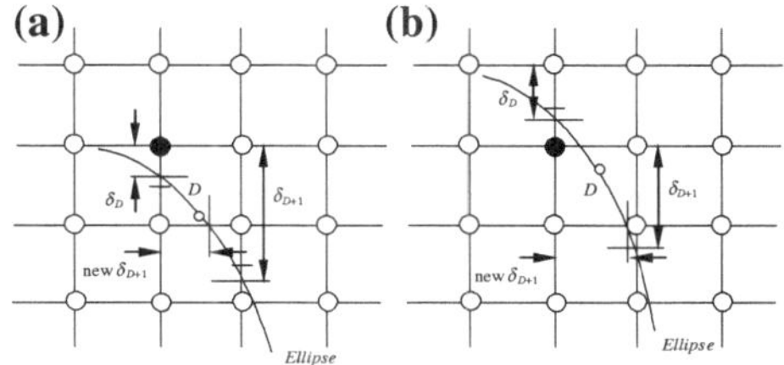

Fig. 4 Finding out the point D. (**a**) Case of $\delta_D < 1/2$ and $\delta_{D+1} \geq 3/2$, (**b**) Case of $\delta_D \geq 1/2$ and $\delta_{D+1} \geq \delta_D$

Let

$$d_{i+1} = 2b^2 x_i \delta_{i+1} + b^2 \delta_{i+1}^2 \tag{7}$$

We gain $d_{i+2} = d_{i+1} + b^2(x_i^2 - x_{i+1}^2) + a^2(y_{i+1}^2 - y_{i+2}^2)$. If $\delta_{i+1} < 0.5$, we have $x_{i+1} = x_i$ and

$$d_{i+2} = d_{i+1} + a^2(2y_{i+2} + 1) \tag{8}$$

If $\delta_{i+1} \geq 0.5$, we have $x_{i+1} = x_i + 1$, and

$$d_{i+2} = d_{i+1} - b^2(2x_{i+1} - 1) + a^2(2y_{i+2} + 1) \tag{9}$$

Substituting $\delta_{i+1} = 0.5$ in expression (7), we have

$$d_{i+1} = b^2 x_i + b^2/4 \tag{10}$$

Similarly, we know that $\delta_{i+1} < 0.5$ is equivalent to $d_{i+1} < b^2 x_i + b^2/4$ while $\delta_{i+1} \geq 0.5$ is equivalent to $d_{i+1} \geq b^2 x_i + b^2/4$.

Since the point $(x_D + \delta_{D+1}, y_{D+1})$ lies on the ellipse and meets its equation, substituting its coordinates in the expression (1), we obtain $b^2(x_D + \delta_{D+1})^2 + a^2 y_{D+1}^2 - a^2 b^2 = 0$. Thus, the initial value of iteration in Region 2 is obtained as follows:

$$d_{D+1} = a^2 b^2 - b^2 x_D^2 - a^2 y_{D+1}^2 \tag{11}$$

The constant $b^2/4$ in (10) can be replaced by its integer part.

2.3 Deciding the Region Transition Criterion

According to the definition of δ_i in our algorithm, after the point (x_D, y_D) has been selected, only two possible cases exist if we consider about relative positions of the ellipse and grid points and their residua δ_i, shown in Fig. 4, and we can gain simplified transition criterion as follows: (1) If $\delta_D < 1/2$, we will test the condition $\delta_{D+1} \geq 3/2$ and (2) If $\delta_D \geq 1/2$, we will test whether $d_{D+1} \geq d_D$.

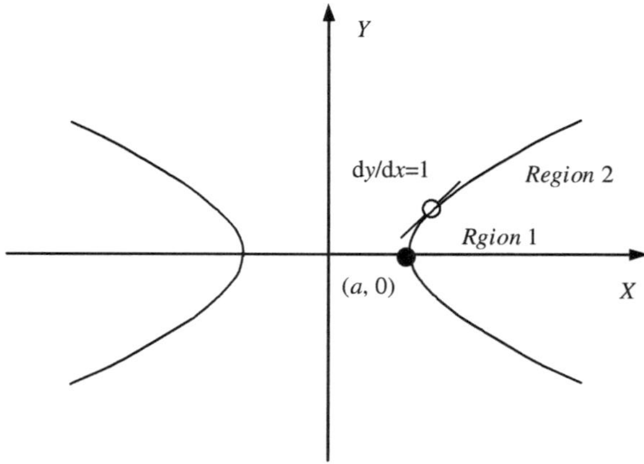

Fig. 5 Two regions of a quarter hyperbola

Besides, a logical operation takes more time than a sign test. When the algorithm is coded, logical operation (5) and (11) should be converted into sign test. For instance, for the expression $d_{i+1} < (a^2y_i - a^2/4)$, we can regard $d_{i+1}-(a^2y_i - a^2/4)$ as a new decision variable d_{i+1}. As a result, we replace the logical operation $d_{i+1} < (a^2y_i - a^2/4)$ in the algorithm by $d_{i+1} < 0$. In addition, d_{i+1} can be regarded as pixel intensity measured on a scale of 0 to $4a^2y_i - 2a^2 - a^2/2$, so we can convert it into an anti-aliasing algorithm directly [19].

Two main differences between our algorithm and other algorithms are: (1) Midpoint algorithm and Agathos algorithm calculate new decision variable d with iteration formula $d = d + \Delta(x, y) + c$, where $\Delta(x, y)$ is dx or $dx - dy$. The iteration formula in our algorithm can be implemented with $d = d + \Delta(x, y)$. The subtraction operation about constant term is eliminated in our algorithm. (2) Transition criterion is tested only for y-stepping rather than for every iteration step in our algorithm. The analysis result shows that the computation consumed by our algorithm is about 60 % of that consumed by mid-point algorithm (or Agathos algorithm).

3 The Hyperbola- and Parabola-Generating Algorithms

In a similar manner to the ellipse, we can derive incremental error expressions for scan-conversion of the hyperbola and parabola.

3.1 The Hyperbola-Generating Algorithm

Assuming that a hyperbola centers at (0, 0) and is symmetric about the X and Y axes, it is shown in Fig. 5 and is defined by the equation:

$$F(x,y) = b^2x^2 - a^2y^2 - a^2b^2 = 0 \tag{12}$$

If $a > b$, we can see that the hyperbola has two regions, and they are separated by the point where the slope of tangent dy/dx is 1. The major axis of movement in Region 1 is Y and in Region 2 is X. If $a \le b$, Region 2 will disappear. We consider here only the case of $a > b$.

- Generating Region 1

We assume that the hyperbola is generated in a clockwise manner starting from the point (x_0, y_0), where $x_0 = a$, $y_0 = 0$. Let δ_{i+1} be the distance from point (x_i, y_{i+1}) to real curve, where $i \ge 0$. For $(i + 1)$th step and $(i + 2)$th step selections, due to point $(x_i + \delta_{i+1}, y_{i+1})$ and $(x_{i+1}+\delta_{i+1}, y_{i+2})$ lie on the hyperbola and satisfy the Eq. (13), we have

$$\begin{cases} b^2(x_i + \delta_{i+1})^2 - a^2y_{i+1}^2 - a^2b^2 = 0 \\ b^2(x_{i+1} + \delta_{i+2})^2 - a^2y_{i+2}^2 - a^2b^2 = 0. \end{cases}$$

Let $d_{i+1} = 2b^2x_i\delta_{i+1} + b^2\delta_{i+1}^2$, we have $d_{i+2} = d_{i+1} + b^2(x_i^2 - x_{i+1}^2) + a^2(y_{i+2}^2 - y_{i+1}^2)$. If $\delta_{i+1} < 0.5$, then $x_{i+1} = x_i$. We gain $d_{i+2} = d_{i+1} + a^2(2y_{i+1} + 1)$. If $\delta_{i+1} \ge 0.5$, an x-stepping will occur, i.e., $x_{i+1} = x_i + 1$. Thus, we have $d_{i+2} = d_{i+1} - b^2(2x_{i+1} - 1) + a^2(2y_{i+1} + 1)$. Substituting $\delta_{i+1} = 0.5$ in the expression of d_{i+1}, integer decision variable is obtained by $d_{i+1} = b^2x_i + b^2/4$.

Since d_{i+1} is a monotonically increasing function about δ_{i+1}, $\delta_{i+1} < 0.5$ is equivalent to $d_{i+1} < b^2x_i + b^2/4$, and $\delta_{i+1} \ge 0.5$ is equivalent to $d_{i+1} \ge b^2x_i + b^2/4$. Considering that point $(x_0 + \delta_1, y_1)$ lie on the hyperbola and meets its equation, we substitute the point coordinate in the equation and gain $b^2(x_0 + \delta_1)^2 - a^2y_1^2 - a^2b^2 = 0$. Thus, the initial value of iteration is $d_1 = 2b^2x_0\delta_1 + b^2\delta_1^2 = a^2$.

- Generating Region 2

Let δ_{i+1} be the distance from point (x_{i+1}, y_i) to the hyperbola. For $(i + 1)$th step and $(i + 2)$ selections, where $i \ge D$, due to point $(x_{i+1}, y_i + \delta_{i+1})$ and $(x_{i+2}, y_{i+1}+\delta_{i+2})$ meet the hyperbola equation. Let $d_{i+1} = 2a^2y_i\delta_{i+1} + a^2\delta_{i+1}^2$, we gain $d_{i+2} = d_{i+1} + b^2(2x_{i+1} + 1) + a^2(y_i^2 - y_{i+1}^2)$. If $\delta_{i+1} < 0.5$, only an x-stepping will occur, i.e., $y_{i+1} = y_i$, then we gain $d_{i+2} = d_{i+1} + b^2(2x_{i+1} + 1)$. If $\delta_{i+1} \ge 0.5$, we have $y_{i+1} = y_i + 1$, and $d_{i+2} = d_{i+1} + b^2(2x_{i+1} + 1) - a^2(2y_{i+1} - 1)$.

In a same manner, we know that $\delta_{i+1} < 0.5$ is equivalent to $d_{i+1} < a^2y_i + a^2/4$ and $\delta_{i+1} \ge 0.5$ is equivalent to $d_{i+1} \ge a^2y_i + a^2/4$.

Assuming that point (x_D, y_D) is the last point of Region 1 and considering that point $(x_{D+1}, y_D - \delta_{D+1})$ lies on the hyperbola and satisfies its equation, we gain $b^2x_{D+1}^2 - a^2(y_D + \delta_{D+1})^2 - a^2b^2 = 0$. Thus, the initial value of iteration in Region 2 is $d_{D+1} = b^2x_{D+1}^2 - a^2y_D^2 - a^2b^2$.

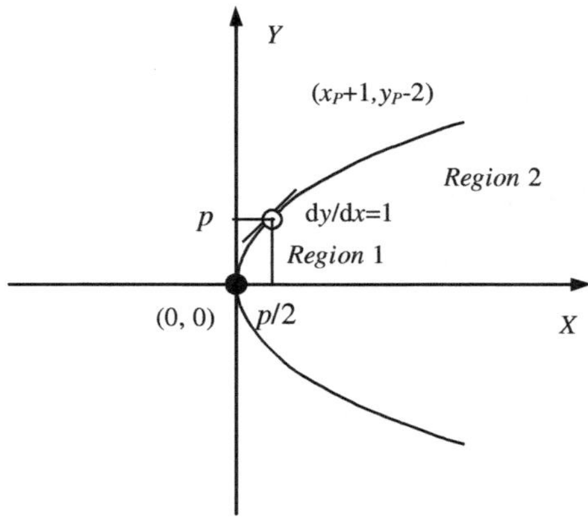

Fig. 6 Two regions of a quarter parabola

3.2 The Parabola-Generating Algorithm

Assuming that a parabola centers at (0, 0) and is symmetric about the *X* axis. It is shown in Fig. 6 and is defined by the equation:

$$F(x, y) = y^2 - 2px = 0 \tag{13}$$

The slope *dy*/*dx* of tangent of the parabola at (*p*/2, *p*) is 1, so it is the region change point. The major axis of movement in Region 1 (from (0, 0) to (*p*/2, *p*)) is *Y* while it is *X* in Region 2 (the other). We consider here only the case of $p > 0$ and describe the process briefly as follows:

For generating Region 1, let δ_{i+1} is the distance from point (x_i, y_{i+1}) to the parabola and $d_{i+1} = 2p\delta_{i+1}$. From $d_{i+2} = d_{i+1} + (y_{i+2}^2 - y_{i+1}^2) + 2p(x_i - x_{i+1})$, we have $x_{i+1} = x_i$ and $d_{i+2} = d_{i+1} + 2y_{i+2} - 1$ for $\delta_{i+1} < 0.5$, and we have $x_{i+1} = x_i + 1$ and $d_{i+2} = d_{i+1} + 2y_{i+2} - 1 - 2p$, for $\delta_{i+1} \geq 0.5$, where $\delta_{i+1} < 0.5$ is equivalent to $d_{i+1} < p$. The initial value is $d_1 = 1$.

For generating Region 2, let δ_{i+1} is the distance from point (x_{i+1}, y_i) to the parabola, and $d_{i+1} = 2y_i\delta_{i+1} + \delta_{i+1}^2$. For ($i$ + 1)th and (i + 2)th selections, where $i \geq D$, we get $d_{i+2} = d_{i+1} + (y_i^2 - y_{i+1}^2) + 2p(x_{i+2} - x_{i+1})$. If $\delta_{i+1} < 0.5$, i.e., $d_{i+1} < y_i + 1/4$, we select $y_{i+1} = y_i$ and $d_{i+2} = d_{i+1} + 2p$. Otherwise, we have $y_{i+1} = y_i + 1$ and $d_{i+2} = d_{i+1} - (2y_{i+1} - 1) + 2p$. Assuming that ($x_{D+1}$, $y_D + \delta_{D+1}$) is the last point of Region 1, if p is an odd, we get $d_D = p$ and $d_{D+1} = 3p - 1$. Otherwise, we get $d_D = 0$ and $d_{D+1} = 2p - 1$.

4 Conclusion

In this paper, a new method for the generation of conic sections by means of residuals is described. This method directly calculates differences between the expressions satisfied by two adjacent points lied on curves, and according to the criterion of nearest to curve, a kind of integer decision variable and its iteration relation is established. We think that its idea and derivation are at least as simple as basic mid-point algorithm. The analyses and experimental results show that in the case of the ellipse, our algorithm is faster than basic mid-point algorithm and Agathos algorithm. Moreover, new algorithms do not set erroneous pixels at region boundaries, and anti-aliasing is easily incorporated. Furthermore, we have extended this method to the generation of general conic sections and think that it would become a kind of effective technique to scan-conversion of curves.

References

1. Bresenham, J.E.: A linear algorithm for incremental digital display of circular arcs. Commun. Assoc. Comput. Mach. **20**, 100–106 (1977)
2. Van Aken, J.R.: An efficient ellipse-drawing algorithm. CG&A **4**, 24–35 (1984)
3. Van Aken, J.R., Novak, M.: Curve-drawing algorithms for raster display. ACM Trans. Graph. **4**, 147–169 (1985)
4. Pitteway, M.L.V.: Algorithm for drawing ellipse or hyperbolae with a digital plotter. Comput. J. **10**, 282–289 (1985)
5. Foley, J.D.: Computer Graphics Principles and Practice. Addison Wesley Publishing Company, Reading Massachusetts (1990)
6. McIlroy, M.D.: Getting raster ellipses right. ACM Trans. Graph. **11**, 259–275 (1992)
7. Kappel, M.R.: An ellipse-drawing algorithm for raster displays. In: Earnshaw, R. (ed.) Fundamental Algorithms for Computer Graphics, NATO ASI Series. Springer, Berlin (1985)
8. Fellner, D.W., Helmberg, C.: Robust rendering of general ellipses and elliptical arcs. ACM Trans. Graph. **12**, 251–276 (1993)
9. Agathos, A., Theoharis, T., Boehm, A.: Efficient integer algorithms for the generation of conic sections. Comput. & Graph. **22**, 621–628 (1998)
10. Wu, X., Rokne, J.G.: Double-step incremental generation of lines and circles. Comput. Vis., Graph. Image Process. **37**, 331–344 (1987)
11. Cheng, J., Lu, G., Tan, J.: A fast arc drawing algorithm. Software J. **13**, 2275–2280 (2002)
12. Niu, L., Xue, J., Zhu, T.: A run-length algorithm for fast circle drawing. J. Shenyang Univ. Technol. **32**, 411–416 (2010)
13. Liu, Y., Shi, J.: Double-Step Circle Drawing Algorithm with and without Grey Scale. J. Comput. Aided Des. Comput. Graph. **17**, 34–41 (2005)
14. Wu, X., Rokne, J.G.: Double-step generation of ellipse. CG&A **9**, 376–388 (1989)
15. Hsu, S.Y., Chow, L.R., Liu, C.H.: A new approach for the generation of circles. Comput. Graph. Forum **12**, 105–109 (1993)
16. Liu, K., Hou, B.H.: Double-step incremental generation of ellipse and its hardware implementation. J. Comput. Aid. Des. & Comput. Graph. **15**, 393–396 (2003)
17. Yao, C., Rokne, J.G.: Hybrid scan-conversion of circles. IEEE Trans. Comput. Vis. Graph. **1**, 31–318 (1995)
18. Yao, C., Rokne, J.G.: Run-length slice algorithms for the scan-conversion of ellipses. Comput. & Graph. **22**, 463–477 (1998)
19. Wu, X.: An efficient antialiasing technique. Computer Graphics, Proceedings of SIGGRAPH'91, vol. 25. pp. 143–152 (1991)

Stereo Matching Algorithms with Different Cost Aggregation

Kelin Ning, Xiaoying Zhang and Yue Ming

Abstract Stereo matching is one of the most active research fields in computer vision. The paper introduces the categories and the performance index of stereo matching and introduces three high-speed and state-of-the-art stereo matching algorithms with different cost aggregation: fast bilateral stereo (FBS), binary stereo matching (BSM), and a non-local cost aggregation method (NLCA). By comparing the performance in terms of both quality and speed, we concluded that FSB deals with the effects of noise well; BSM is suitable for embedded devices and has a good performance with radiometric differences; NLCA combines the efficiency with the accuracy of state-of-the-art algorithms.

Keywords Stereo match · Disparity image · Matching accuracy · Matching speed

1 Introduction

Stereo match which is to gain the depth or the disparity map from two images describing the same scene is a research focus in computer vision. It has been widely applied to visual navigation, target detection, three-dimensional reconstruction, and other field.

Scharstei and Szeliski [1] developed a taxonomy and categorization scheme of stereo algorithms. Stereo algorithms generally perform the following four steps: (1) matching cost computation; (2) cost (support) aggregation; (3) disparity computation/optimization; and (4) disparity refinement. And Scharstei and Szeliski [1] separated the stereo algorithms into two classes: local algorithms and global

K. Ning (✉) · X. Zhang · Y. Ming
Beijing Key Laboratory of Work Safety Intelligent Monitoring, School of Electronic Engineering, Beijing University of Posts and Telecommunications, Beijing 100876, People's Republic of China
e-mail: nkl256@163.com

S. Patnaik and X. Li (eds.), *Proceedings of International Conference on Computer Science and Information Technology*, Advances in Intelligent Systems and Computing 255, DOI: 10.1007/978-81-322-1759-6_74,

algorithms. Local algorithms usually focus on steps 1 and 2, while global algorithms always seek a disparity assignment data (step 3) that minimizes a global cost function and often skip the aggregation step (step 2).

The evaluation indexes of stereo matching algorithm are mainly matching accuracy and matching speed. Matching accuracy is decided by the algorithm computational framework itself. The Middlebury Web site [2] collected the state-of-the-art algorithms and ranked the accuracy according to the test results on the Middlebury benchmark. So far, various advanced stereo matching algorithms are available to a very high accuracy, but greatly different in matching speed. With the increase in demand for real-time stereo matching, scholars have focused on improving the execution efficiency of the algorithm, which is difficult to have a consistent evaluation standard because of the different implementation methods and hardware platforms.

There are two ways to improve the matching speed. One is to reduce the computational complexity of matching algorithm. De-Maeztu et al. [3] proposed a linear stereo matching which is highly scalable for higher-resolution stereo pairs. Rhemann et al. [4] developed a fast cost-volume filtering method based on a linear model. The other way is to improve the hardware computational speed, especially using the floating-point arithmetic and powerful GPU with paralled computing ability for stereo matching. Mei et al. [5] built an accurate stereo matching system on graphics hardware by adopting an AD-census cost measure.

This paper mainly introduces three kinds of state-of-the-art algorithms with lower computational complexity: fast bilateral stereo (FBS), binary stereo matching (BSM), and a non-local cost aggregation method (NLCA). Each algorithm is proposed with a special cost aggregation method. We have conducted experiments to demonstrate the performance and advantages of the three algorithms from matching accuracy and speed.

2 Algorithms

2.1 Fast Bilateral Stereo

Mattoccia et al. [6] proposed a cost aggregation strategy for stereo correspondence based on joint bilateral filtering which is referred to as FBS. Bilateral filtering is a non-iterative feature-preserving image smoothing technique [7] and is widely used in computer vision. It independently enforces a spatial filter and a range filter which considered the geometric and color differences. For each point p of the given image, the value is a weighted convolution with points q_i in its neighbor space $S(p)$ with the center point p:

$$\hat{I}(p) = \frac{\sum_{q_i \in S(p)} W_S(p, q_i) \cdot W_C(I(p), I(q_i)) \cdot I(q_i)}{\sum_{q_i \in S(p)} W_S(p, q_i) \cdot W_C(I(p), I(q_i))}, \quad (1)$$

where $W_S(p, q_i)$ and $W_C(I(p), I(q_i))$ are the weighting functions of the space and color distances of the point p and q_i. The two functions are independent, and the closer to the central point p in space and color, the greater the weight is. Typically, the weights W_S and W_C are assigned according to Gaussian functions with variance γ_S and γ_C, respectively.

The cost aggregation strategy of FBS came from the method adaptive weight (AW) [8]. In the AW approach, the weight assigned to each point is computed by two independent bilateral filtering in the reference and target image. The cost of correspondence $C(p_r, p_t)$ between pixel p_r and p_t can be expressed as

$$C(p_r, p_t) = \frac{\sum_{q_r \in S(p_r), q_t \in S(p_t)} W(p_r, q_r) \cdot W(p_t, q_t) \cdot \mathrm{TAD}(q_r, q_t)}{\sum_{q_r \in S(p_r), q_t \in S(p_t)} W(p_r, q_r) \cdot W(p_t, q_t)}, \quad (2)$$

where $W(p_r, q_r)$ means the weight of geometric and color differences, $TAD(q_r, q_t)$ means the truncated absolute differences (TAD) scores with a truncation value T to control the limit of the mating cost.

To deal with noise in the regions with similar color intensity, Mattoccia et al. proposed a simple but effective noise regularization stage for the range filter. Given two points $p_r \in I_r$, $p_t \in I_t$, and the associated support windows $S(p_r)$ and $S(p_t)$ of the size $W \times W$, they partition the support windows into $\frac{W}{w} \times \frac{W}{w}$ regular blocks.

For each block $b_r(u, v)$ in the support window $S(p_r)$ of the reference image, the weight is

$$W_C(I_r(p_r), \bar{I}_r(b_r(u, v))) = \exp\left(-\frac{\|I_r(p_r) - \bar{I}_r(b_r(u, v))\|}{\gamma_c}\right), \quad (3)$$

with $\bar{I}_r(b_r(u, v))$ representing the average value of pixels within block $b_r(u, v)$.

By using small w, the algorithms reduce the variance within each block as well as the computational complexity of the step cost aggregation. Especially if $w = 1$, the FSB computational framework is equivalent to adaptive support-weight algorithm [8].

2.2 Binary Stereo Matching

Zhang et al. [9] proposed a novel cost computation and aggregation approach for stereo matching named BSM. Completely different with traditional local methods, they introduced brief descriptor [10] into cost computation. The brief descriptor $B(x)$ is a brief computation for each point x in the input image pair which is defined as

$$B(x) = \sum_{1 \le i \le n} 2^{i-1} \tau(p_i, q_i), \tag{4}$$

where $\langle p_1, q_1 \rangle$, $\langle p_2, q_2 \rangle \cdots\cdots \langle p_n, q_n \rangle$ are n pairs of pixels sampled by an isotropic Gaussian distribution in a $S \times S$ with the center point x. $\tau(p_i, q_i)$ is a binary function defined as

$$\tau(p_i, q_i) = \begin{cases} 1 : I(p_i) > I(q_i) \\ 0 : I(p_i) \le I(q_i) \end{cases}, \tag{5}$$

with $I(x)$ representing the intensity of pixel x. The cost volume is constructed as

$$C(x, d) = \ ||B(x)\ XOR\ B(x_d)||_1, \tag{6}$$

where x_d is the corresponding pixel of x in the other image. *XOR* is a bitwise XOR operation. Then, the disparity d of the point x is computed as

$$d = \arg\min_{d \in D_d} C(x, d). \tag{7}$$

To deal with the edge-fattening problem, they proposed a novel cost aggregation method by introducing a binary string which they call binary mask. In the step of refinement, they proposed a voting-based depth refinement method which uses a bilateral filter to handle occluded area and random errors due to mismatch.

2.3 Non-local Cost Aggregation Method

Yang [11] proposed a non-local method by suggesting aggregation cost on a tree structure. He treats a guidance image (typically the reference image) as a connected, undirected graph and derives a minimum spanning tree (MST) from this graph. The vertices are the pixels in the image, and the edges are all the edges between the neighboring pixels. All the vertices are connected, and the sum of the weights is minimum after the edges with large weights are removed.

In the method, the weight between two neighboring pixels s and r is

$$w(s, r) = w(r, s) = |I(s) - I(r)|, \tag{8}$$

where $I(s)$ is the intensity of pixel s. The similarity between two pixels is decided by the shortest distance in the MST. $D(p, q)$ denotes the distance which is the sum of the weights of all the connected edges between pixels p and q, and the similarity can be denoted as

$$S(p, q) = S(q, p) = \exp\left(-\frac{D(p, q)}{\sigma}\right), \tag{9}$$

where σ is an empirical constant to adjust the similarity. Then, the aggregation from the MST structure is

$$C_d^A(p) = \sum_q S(p,q)C_d(q) = \sum_q \exp\left(-\frac{D(p,q)}{\sigma}\right)C_d(q), \tag{10}$$

with $C_d(q)$ representing the matching cost and $C_d^A(p)$ representing the aggregated cost at disparity d.

Normal local algorithms require a specified or an automatically detected window; each pixel supports the window whose center is this pixel, and the support weight outside the window is 0. In contrast, the aggregation cost of each pixel in NLCA algorithm is superimposed by all other pixels in the image. Therefore, compared to the previous methods, the NLCA algorithm gives a more natural image.

The greatest advantage of NLCA algorithm is the low computational complexity. From the properties of the MST, an aggregation cost computational formula of each node is

$$\begin{aligned} C_d^A(v) &= C_d^{A\uparrow}(v) + S(P(v),v) \cdot [C_d^A(P(v)) - S(v,P(v)) \cdot C_d^{A\uparrow}(v)] \\ &= S(P(v),v) \cdot C_d^A(P(v)) + [1 - S^2(v,P(v))] \cdot C_d^{A\uparrow}(v), \end{aligned} \tag{11}$$

where $C_d^{A\uparrow}(v)$ represents the aggregated cost of node v and its child nodes, $P(v)$ represents the parent node of node v and $C_d^{A\uparrow}(v) = C_d^A(v)$ if v is the root node. Because $S(v,P(v)$ and $1 - S^2(v,P(v))$ can be precomputed, only 2 addition/subtraction and 3 multiplication operations are required for each pixel at each disparity level to compute the aggregation cost, which is low computational complexity.

3 Experimental Results

This section shows the experimental results of three algorithms on the Middlebury benchmark. All experiments were conducted with one core and one process on a 3.2-GHz Core I7 CPU.

Figure 1 shows the ground truth and the disparity maps with images from top to bottom: ground truth, disparity maps computed by algorithm FBS, BSM, and NLCA. Table 1 shows the accuracy for the considered algorithms. From the disparity maps, we concluded that the results are all similar to the ground truth, and from the accuracy table, we noticed that BSM and NLCA have a higher robustness and less bad pixels than FBS.

Table 2 shows the comparison of matching speed on the image Tsukuba including the cost computation and aggregation computation. For easy comparison, FBS used $W = 45$ with $\text{FBS}_{45}(3)$ representing $w = 3$ while $\text{FBS}_{45}(5)$ representing $w = 5$. In contrast, BSM takes more time than FBS and NLCA especially for preprocess. The total time for $\text{FBS}_{45}(3)$ and NLCA is only about 1 s with high accuracy, while NLCA shows the best result during the cost aggregation. At the

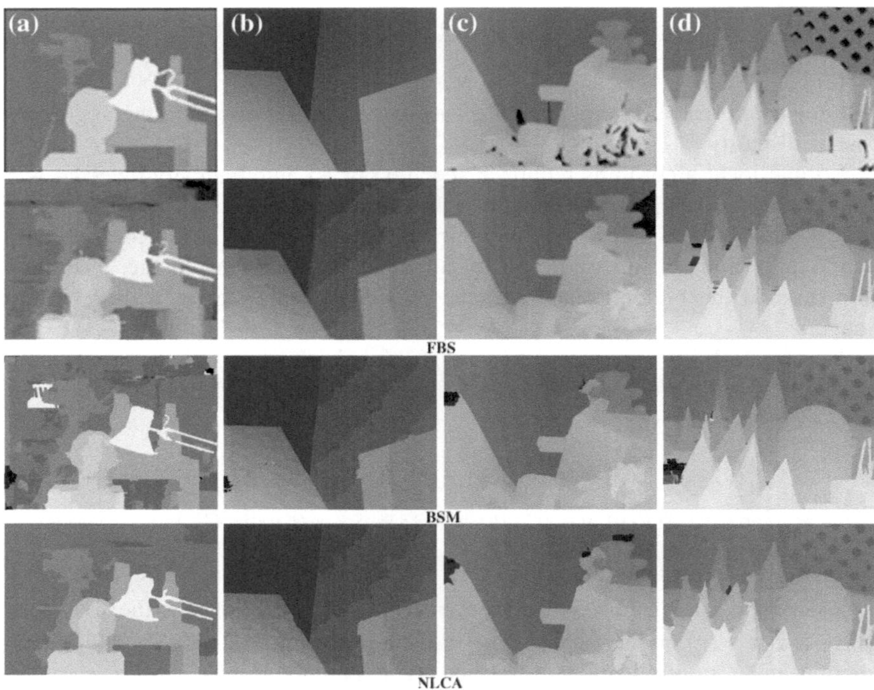

Fig. 1 Stereo results on the Middlebury benchmark. **a** Tsukuba, **b** Venus, **c** Teddy, and **d** Cone

Table 1 Stereo evaluation on the Middlebury benchmark

	Tsukuba NOCC ALL DISC	Venus NOCC ALL DISC	Teddy NOCC ALL DISC	Cones NOCC ALL DISC	Percentage of bad pixels
FBS	2.38 2.80 10.4	0.34 0.92 4.55	9.83 15.3 20.3	3.10 9.31 8.59	7.31
BSM	3.08 3.38 7.80	0.26 0.42 2.03	5.74 8.95 14.8	2.34 8.79 6.80	5.42
NLCA	1.47 1.85 7.88	0.25 0.42 2.60	6.01 11.6 14.3	2.87 8.45 8.10	5.48

same time, we can conclude that with small growth of w, the aggregation computation time decreases a lot.

4 Conclusions

With the comparison and principle analysis of the three state-of-the-art algorithms, we can conclude that FSB deals with the effects of noise well with a controllable speed, BSM takes more matching time than FBS and NLCA, but it is suitable for embedded devices and has a good performance with radiometric differences, and

Table 2 Comparison of matching speed on the image Tsukuba

Method	Cost computation time(s)	Aggregation computation time(s)	Total time(s)
$FBS_{45}(3)$	0.018	1.083	1.101
$FBS_{45}(5)$	0.018	0.426	0.444
BSM	45.73	4.34	50.07
NLCA	0.664	0.280	0.944

NLCA performs the best combining the efficiency with the accuracy of state-of-the-art algorithms with a good robustness.

Stereo matching algorithms are maturing with the continuous improvement of accuracy and efficiency, but processing by CPU is hard to meet the requirement of both the real-time images and the high accuracy. GPU implementation with powerful floating-decimal arithmetic for stereo match is a development direction in the field of computer vision.

Acknowledgments The work presented in this paper was supported by the National Natural Science Foundation of China (Grants No. NSFC-61170176), Fund for the Doctoral Program of Higher Education of China (Grants No. 20120005110002), National Great Science Specific Project (Grants No. 2011ZX0300200301, 2012ZX03005008), and Beijing Municipal Commission of Education Build Together Project.

References

1. Scharstei, D., Szeliski, R.: A taxonomy and evaluation of dense two-frame stereo correspondence algorithms. Int. J. Comput. Vis. (S0920–5691) **47**(1/3), 7–42 (2002)
2. Scharstein, D., Szeliski, R.: Middlebury stereo evaluation, version 2. http://vision.middlebury.edu/stereo/eval/ (2013)
3. De-Maeztu, L., Mattoccia, S., Villanueva, A., Cabeza, R.: Linear stereo matching. ICCV (2011)
4. Rhemann, C., Hosni, A., Bleyer, M., Rother, C., Gelautz, M.: Fast cost-volume filtering for visual correspondence and beyond. CVPR (2011)
5. Mei, X., Sun, X., Zhou, M., Jiao, S., Wang, H., Zhang, X.: On building an accurate stereo matching system on graphics hardware. GPUCV (2011)
6. Mattoccia, S., Giardino, S., Gambini, A.: Accurate and efficient cost aggregation strategy for stereo correspondence based on approximated joint bilateral filtering. ACCV (2009)
7. Tomasi, C., Manduchi, R.L.: Bilateral filtering for gray and color images. ICCV (1998)
8. Yoon, K.-J., Kweon, I.-S.: Adaptive support-weight approach for correspondence search. PAMI **28**(4), 650–656 (2006)
9. Zhang, K., Li, J., Li, Y., Hu, W., Sun, L., Yang, S.: Binary stereo matching. ICPR (2012)
10. Calonder, M., Lepetit, V., Strecha, C., Fua, P.: BRIEF: binary robust independent elementary features. In: European Conference on Computer Vision, pp. 778–792. (2010)
11. Yang, Q.: A non-local cost aggregation method for stereo matching. CVPR (2012)

An Improved Algorithm for Mining Association Rules Based on Partitioned and Compressed Association Graph

Hao Jiang, Yabo He and Wei Wan

Abstract Mining association rules have been one of the research focuses in data mining; an improved algorithm based on graph is presented for discovering various types of association rules in this paper. We introduce the original algorithm based on graph; on that basis, we improve it according to the idea of partition and compression. Firstly, the association graph constructed is partitioned into many association subgraphs. Secondly, we compress these subgraphs and mine frequent itemsets respectively which will not influence each other in different parts. In the end, the improved algorithm is compared with the original one in performance.

Keywords Association rules · Frequent itemsets · Association graph · Partitioned · Compressed

1 Introduction

There are many algorithms [1] of mining frequent itemsets on association rules; the most representative are Apriori algorithm [2] and FP-tree algorithm [3]. Apriori algorithm is a basic algorithm to mine frequent itemsets, using a kind of iterative algorithm by searching layer by layer, and k-itemsets are used to generate $k + 1$ itemset. The FP-tree algorithm which only scans the database twice does not need to generate candidate itemsets; it adopts a divide-and-conquer strategy. However, the transaction database is generally relatively large. For counting the support in each iterative process, the cost of scanning the database every time is extremely high. To solve the problem, we can start from two aspects: reducing the generation of candidate itemsets and the amount of data scanned in the process of each iteration.

H. Jiang · Y. He (✉) · W. Wan
School of Computer Science and Engineering, Southeast University, Nanjing, Jiangsu, China
e-mail: helloyabo@gmail.com

S. Patnaik and X. Li (eds.), *Proceedings of International Conference on Computer Science and Information Technology*, Advances in Intelligent Systems and Computing 255, DOI: 10.1007/978-81-322-1759-6_75,

Paper [4] proposes a graph-based approach to generate frequent itemsets of association rules from a large database of customer transactions. This approach scans the database once to construct an association graph and mine large itemsets. This paper makes a further improvement on the basis of algorithm introduced in Paper [4], and our experimental results prove the validity of the efficiency about these improvements. Moreover, relevant experimental analysis has also been provided in this paper.

2 A Graph-Based Algorithm for Mining Association Rules

2.1 Basic Idea of DLG Algorithm

The graph-based mining algorithm, referred to as the DLG algorithm [4, 5], constructs an association graph to reflect association between itemsets and then traverses the association graph to generate frequent itemsets. The DLG algorithm uses a vertical data format for mining frequent itemsets [6], in which data are represented by the symbol of {item:tid_set}, and it consists of the following three stages:

1. The first stage: generating frequent 1-itemset. The phase generates frequent a-itemset and records relevant information. After scanning the database once, we number items and calculate their supports to create a bit vector, whose length is the total number of transaction in database.
2. The second stage: creating association graph. The stage generates frequent 2-itemsets by using the logical AND operator (can be denoted as the notation "$\wedge$") among bit vector which is created in the first stage.
3. The third stage: generating frequent $K + 1$-itemsets ($K >= 2$). The stage generates larger frequent itemsets based on the second stage than before.

2.2 Efficiency Analysis of DLG Algorithm

The DLG algorithm has some advantages, compared with other association rule mining algorithms such as Apriori algorithm and FP-tree algorithm. It only needs to scan the database once; you can construct the transaction matrix, greatly reducing the I/O times and improving efficiency. However, the DLG algorithm has the similar problem too. The common problem is that they all exist in iterative process from frequent itemsets to candidate itemsets and from candidate itemsets to frequent itemsets, and therefore, it generates inevitably redundant candidate itemsets. In addition, the DLG algorithm is less efficient in the face of some huge transaction databases.

3 Improved Algorithm Based on DLG

3.1 Theoretical Basis

An improved algorithm based on partitioned and compressed association graph is presented in this paper. We will introduce three theorems, which are the theoretical basis of our algorithm, used in association graph before describing the algorithm.

Theorem 1 *In the association graph, assume that the nodes are numbered sequentially. We partition the graph by the following properties: according to the order of smallest to largest in node number, the current node and those nodes, whose number are larger than the current one, directly connected to the former will be divided into the same part. So, it will generate many independent and disjoint sub-association graphs. Then mine frequent itemsets in each sub-association graph respectively. At last the union set from total frequent itemsets of all parts is equal to all frequent itemsets generated by the original association graph.*

Proof This proof can be divided into two steps.

First Proof: We will prove that the set of frequent itemsets generated by all parts is a subset of the set of all frequent itemsets generated by the original association graph.

We assume that the original association graph, G for short, is divided into n sub-association graphs, such as $G_1, G_2, \ldots, G_n$, and the frequent itemsets generated by association graph G can be named P. Based on the algorithm of DLG, the frequent items obtained by sub-association graph must be a subset of association graph G, that is, $P_1 \subset P$, $P_2 \subset P$, ..., $P_n \subset P$. In addition, $P_1 \cup P_2 \cup \cdots \cup P_n \subset P$ can work according to the property of set. So the conclusion is correct.

Second Proof: We will prove that the set of all frequent itemsets generated by original association graph is a subset of the set of frequent itemsets generated by all parts.

We use mathematical reductio ad absurdum to prove this problem.

Assume that there exists a frequent itemset, I for short, generated by the original association graph, which is not in union set of frequent itemsets generated by all parts. Therefore, there must exist one such scenario: One item A of I is only from one sub-association graph, and another item B of I only comes from another sub-association graph. Based on the nature that subset of frequent itemsets must be frequent itemsets, we know that there must be an edge between A and B. Besides, according to the division process, A and B, adjacent to each other, are necessarily divided into the same subgraph, and this leads to a contradiction. Hence, the total frequent itemsets of all parts contain all frequent itemsets generated by the original association graph.

Based on the nature of set, set A will be equal to set B if A is a subset of B and B is a subset of A at the same time. In conclusion, the Theorem 1 is correct according to the above two proofs.

Theorem 2 *If the degree of some node is less than $k - 1$, the node can not be extended into frequent k-itemsets when generating k-itemsets in association graph. Then the node together with its adjacent edges can be removed from the association graph, and thus this association graph is compressed. The proof of this theorem can be seen in paper [7].*

Theorem 3 *When extending the edge of the known frequent k-itemset in association graph, if there is no edge between a item node of the frequent k-itemset and the node to be extended, it should not be extended. This extended $k + 1$ -itemset does necessarily not belong to the frequent itemsets so that we can tailor the extended candidate itemsets. The proof of this theorem can be seen in paper [8] where the author improved the DLG algorithm depending on this theorem.*

3.2 PCAG Algorithm Description

The improved algorithm based on partitioned and compressed association graph is described as follows:

1. Scan the database to construct the transaction matrix, generate all frequent 2-itemsets by using the logical AND operator among BVi, and construct association graph with these frequent 2-itemsets;
2. Divide this graph in accordance with Theorem 1 into all sub-association graphs.
3. Mine frequent itemsets in each sub-association graph.
4. Extend frequent $k - 1$-itemsets into k-itemsets for each sub-association graph as follows: We first calculate the degrees of each node; the node together with its adjacent edge will be removed from the sub-association graph if its degree is less than $k - 1$ according to Theorem 2. Then, we apply Theorem 3 while extending frequent $k - 1$-itemsets.
5. k increases by one. If the nodes of sub-association graph are less than k, the algorithm is finished, otherwise turn to 4.

Now, the PCAG algorithm pseudo-code, except the description of the first step, is presented in the Tables 1, 2 and 3.

The paper improves the DLG algorithm based on its shortcomings that generate many redundant candidate itemsets and is not suited to the circumstances with large number of transactions. We first divide the association graph to reduce the complexity, especially in high volumes of customer transactions. Two improved solutions are adopted when we extend candidate itemsets for each sub-association graph. One of the two solutions is to reduce the size of association graph and the other is that extended condition of the itemsets is limited, thereby helping to improve the generation efficiency of the frequent itemsets.

Table 1 PCAG algorithm

```
Algorithm 1: PCAG Algorithm;
Input: Database Transaction Set:D, minimum support rate minisup;
Output: All frequent k-itemsets in D (K >= 2).
MiningFrequentItems(){
   AssociationGraphPartition(G,{Gi}); //Partition
   For each Gi{ //Generate the frequent itemsets for each sub-association graph
     K=2;
     AssociationGraphCompression(Gi); /*Compression and Mining frequent
     itemsets*/
   }
   MergeAllFrequentItemset(); /*combine all frequent itemsets of all sub-association
graph */
   }
```

3.3 Experimental Results and Analysis

This section is to compare the efficiency of the DLG algorithm and PCAG algorithm. Figure 1 shows the relative execution time for the two algorithms under different transaction counts, assuming that the minimum support threshold, minisup for short, is 2 %, and Fig. 2 shows the relative execution time for the two algorithms under different minisup, assuming that the number of transaction is 3,000. From the two tables, we know that both in different transaction counts and minisup, the PCAG algorithm has better performance.

Table 2 Partition algorithm of PCAG

```
Algorithm 2: Partition Algorithm of PCAG;
Input: association graph G
Output: all sub-association graphs Gi.
AssociationGraphPartition(G, {Gi})
{
   Initialize({Gi}); // Gi initialization
   For each Node∈ G{//According to Theorem 1 (G: Association Graph)
      Gi .node=Gi.node ∪ Node[i];
      while(ExistEdge(Node[i],Node[j])) {/* for all nodes whose number is larger than
      Node[i] and that is adjacent to it. */
          Gi .node=Gi.node ∪ Node[j]; //get the nodes of sub-association graph
      }
      For each Node∈ Gi{//get the edges of sub-association graph
          If(ExistEdge(Node[k],Node[j])){
             Gi.edge = Gi.egde ∪ Edge(Node[k],Node[j]);
          }
      }
   }
}// AssociationGraphPartition(G, {Gi})
```

Table 3 Compression algorithm of PCAG

```
Algorithm 3: Compression Algorithm of PCAG
Input: sub-association graph Gi;
Output: all frequent itemsets generated by sub-association graph Gi (K >= 2).
Method:
AssociationGraphCompression(Gi)
{
    K_FrequentToK+1_Frequent(Gi); /*Generate frequent k+1-itemsets Based on
frequent k-itemsets*/
    K++;
}
K_FrequentToK+1_Frequent(Gi){
    While Gi.node >= K {
        For each Node∈ Gi{
         If Node.degree < K{//according to Theorem 2
             Delete Node;
             For each Node∈ Gi {/*Remove edges related to the node*/
                If(ExistEdge(Node, otherNode) or ExistEdge(otherNode, Node)) {
                Delete Edge(Node, otherNode) or Edge(otherNode, Node);
                }
             }//for
         }//if
        }//for
    }//while
}// K_FrequentToK+1_Frequent(Gi)
```

Fig. 1 shows that the number of transactions has less effect on PCAG algorithm than DLG algorithm. In other words, compared with DLG algorithm, the larger the transaction database is, the better the efficiency of PCAG algorithm is. In Fig. 2, in

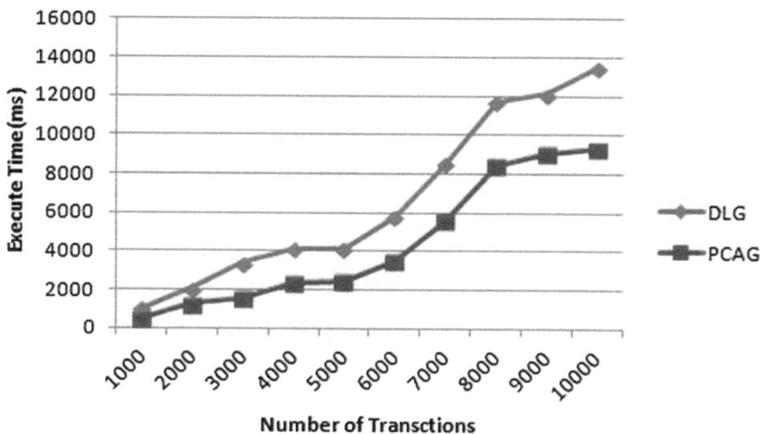

Fig. 1 Efficiency of different transaction counts with 2 % minisup

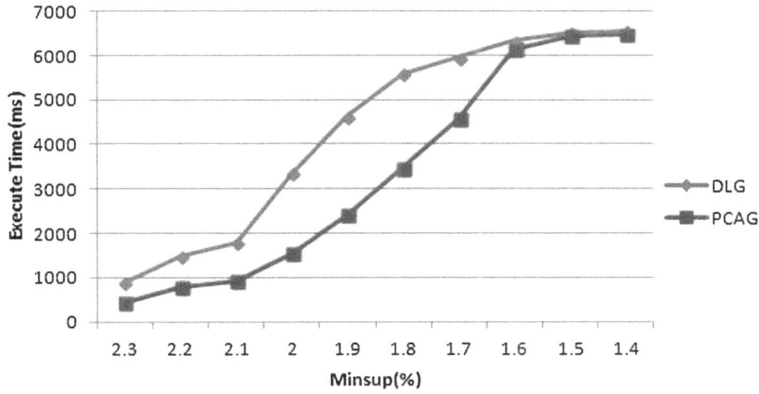

Fig. 2 Efficiency of different minisup with 3,000 transactions

general, with the changing minisup, the PCAG algorithm has higher performance than DLG algorithm on the basis of the same transaction database.

In contrast to DLG algorithm, the PCAG algorithm has improved greatly in efficiency as shown in Figs. 1 and 2.

4 Conclusion and Future Work

Some feasible improvement thoughts on the basis of association rule mining algorithm called DLG are presented in this paper. On the one hand, we introduce the idea of partition and come up with a method to divide and conquer the association graph. This method is particularly applicable to association rule mining in the large-scale transaction database and makes up for disadvantages of the DLG algorithm. On the other hand, we propose a method to compress the associated graph. Combined with the property of Apriori, we impose the further restrictions on the extension of the frequent itemsets. All these methods together improve the efficiency of generating frequent itemsets. Obviously, the PCAG algorithm also has its defects. The improvement effect is not obvious when the minisup is too small. Because a dense association graph created by too many frequent itemsets will decrease the performance of PCAG algorithm when the minisup is too small.

From the experimental results, we know that the efficiency of the PCAG algorithm decreases slowly with the increase in the transaction database size; in other words, the number of transactions has less effect on PCAG algorithm. Taking into account these natures of algorithm, we can mine parallelly frequent itemsets of each sub-association graph when dealing with large transaction databases in our future study, because all sub-association graphs are independent of each other in the process of mining. Besides, we also attempt to apply the PCAG algorithm in other fields except database.

References

1. Agrawal, R., Imielinski, T., Swami, A.: Mining association rules between sets of items in large databases. ACM SIGMOD **22**(2), 207–216 (1993)
2. Agrawal, R., Srikant, R.: Fast algorithm for mining association rules. IBM Research Report RJ9839 (1994)
3. Han, J., Pei, J., Yin, Y.: Mining frequent patterns without candidate generation. In: Proceedings of ACM SIGMOD International Conference on Management of Data, pp. 1–12 (2000)
4. Yen, S.-J., Chen, A.L.P.: A graph-based approach for discovering various types of association rules. IEEE Trans. Knowl. Data Eng. **13**(5), 839–845 (2001)
5. Yen S.J., Chen, A.L.P.: An efficient approach to discovering knowledge from large databases. In: Proceedings of the IEEE/ACM International Conference on Parallel and Distributed Information Systems, pp. 8–18 (1996)
6. Han, J., Kamber, M.: Data Mining Concepts and Techniques, 2nd edn. China Machine Press, Beijing (2007)
7. Wan, W.: Research and Application: Algorithm of Mining Associtton Rules Based on Graph. SouthEast University, Nanjing (2012)
8. Huang, H.: Revised algorithm of mining association rules based on graph. Comput. Digit. Eng. **37**(12), 38–42 (2009)

Research on Spectral Reflectance Reconstruction Algorithms for Munsell Color Card

Xin Jing, Tianxin Yue and Li Zheng

Abstract *XYZ* chrominance parameters of color image have some limitations for the characterization of the color information of the image. Spectral reflectance can fully and accurately reflect the color information. It is critical to obtain accurate spectral reflectance using spectral reconstruction. In the paper, mathematical models have been constructed using Wiener estimation algorithm and Principal component analysis algorithm by the given spectral reflectance in 1269 Munsell and tristimulus values in D65 illumination, to find the relationship between spectral reflectance and tristimulus values by mathematical models. The average value 0.0522 of the root mean square error (RMSE) criterion and the mean value 0.9886 of goodness-of-fit coefficient (GFC) which have been obtained by Wiener estimation algorithm are better than the results of the Principal component analysis algorithm.

Keywords Munsell color card · Wiener estimation · Principal component analysis · Tristimulus values · Spectral reconstruction

1 Introduction

Color science is a booming discipline in recent decades, which is an essential subject in color TV, color printing, color image processing, textile, paint, lights, and other industries [1, 2]. Modern optical theory thinks that the light emitted from

X. Jing (✉) · L. Zheng
School of science, Shenyang Jianzhu University, Shenyang, China
e-mail: jingxin@sjzu.edu.cn

L. Zheng
e-mail: zhengli_shunv@sina.com

T. Yue
Faculty of Information and Control Engineering, Shenyang Jianzhu University, Shenyang, China
e-mail: yuetianxin331@163.com

S. Patnaik and X. Li (eds.), *Proceedings of International Conference on Computer Science and Information Technology*, Advances in Intelligent Systems and Computing 255, DOI: 10.1007/978-81-322-1759-6_76,

Table 1 Comparison of the spectral reconstruction accuracy

	Wiener estimation		Principal component analysis	
	RMSE	GFC	RMSE	GFC
a	0.0475	0.9906	0.0460	0.9911
b	0.0568	0.9866	0.0810	0.9723
Average	0.0522	0.9886	0.0635	0.9817

the light source through the reflection surface of the object into the eye and the color information on the surface is formed on the retina of the human eye, then the color information through the nerve center of the optic nerve to the brain, and finally, the information in the brain for processing, handling, and forming color perception [3]. Thus, the surface of the color is formed to have the following three main parts, namely a light source, the reflective surface, and the human eye. Where the light source characteristics are described by the "spectral power distribution function," the reflective surface characteristics are described by the "surface spectral reflectance function," the human eye feature is usually used "standard colorimetric observer" to describe. In this paper, Wiener estimation algorithm and Principal component analysis algorithm have been used to form mathematical models according to stochastic samples for given 1269 Munsell spectral reflectance and tristimulus values in D65 illumination. In addition, it uses root mean square error (RMSE) and goodness-of-fit coefficient (GFC) to obtain spectral reconstruction accuracy. From Table 1, it can be seen that the average values of the RMSE and the average values of GFC are, respectively, 0.0522 and 0.9886, better than the results of Principal component analysis algorithm.

2 CIE Tristimulus Values

Tristimulus values by the formula

$$
\begin{aligned}
X &= K\int_{\lambda} R(\lambda)I(\lambda)\bar{x}(\lambda)\mathrm{d}\lambda \\
Y &= K\int_{\lambda} R(\lambda)I(\lambda)\bar{y}(\lambda)\mathrm{d}\lambda \\
Z &= K\int_{\lambda} R(\lambda)I(\lambda)\bar{z}(\lambda)\mathrm{d}\lambda
\end{aligned}
\tag{1}
$$

Formula (1) uses the manner of integral to obtain the relationship between tristimulus values and the spectral reflectance of the surface of the object. In addition, the use of the vectors and matrices is a common form. In this form,

$$v = Hr = KUIr \tag{2}$$

where v is the vector of tristimulus values, i.e., $(X, Y, Z)^T$, r is the spectral reflectance vector and H is transformation matrix of the v and r vectors. Matrix H can be represented as follows:

$$H = KUI \tag{3}$$

In formula (3), $U = [\bar{x}(\lambda), \bar{y}(\lambda), \bar{z}(\lambda)]^{\mathrm{T}}$, K is the scale factor, $\bar{x}(\lambda), \bar{y}(\lambda)$ and $\bar{z}(\lambda)$ are the CIE1931 standard colorimetric observer color-matching functions. Matrix I (light source spectral power distribution) is a diagonal matrix.

3 Wiener Estimation and Principal Component Analysis Models

3.1 Wiener Estimation Model

Using Wiener estimation model to estimate the spectral reflectance of the surface is a good method [4, 5]. The method minimizes the mean square error between original spectral reflectance of and spectral reflectance reconstructed through sample training:

$$E\left\{\|r - \hat{r}\|^2\right\} \to \min \tag{4}$$

where r is a column vector which corresponds to the original spectral reflectance and $\hat{r}$ is a column vector which corresponds to the reconstructed spectral reflectance .

Assuming

$$v = Hr \tag{5}$$

$$\hat{r} = Gv \tag{6}$$

where v represents the tristimulus values vector, H represents the transformation matrix, and G represents the estimated matrix.

By the matrix knowledge we know,

$$\|r - \hat{r}\|^2 = trace\left\{(r - \hat{r})(r - \hat{r})^{\mathrm{T}}\right\}. \tag{7}$$

Suppose $f(G) = E\left\{\|r - \hat{r}\|^2\right\}$, solve its derivation,

$$\begin{aligned}\frac{\partial f(G)}{\partial G} &= \frac{\partial \mathrm{trace}}{\partial G}\left\{E\left[rr^{\mathrm{T}} - r\hat{r}^{\mathrm{T}} - \hat{r}r^{\mathrm{T}} + \hat{r}\hat{r}^{\mathrm{T}}\right]\right\} \\ &= \frac{\partial \mathrm{trace}}{\partial G}\left\{E\left[rr^{\mathrm{T}} - r(GHr)^{\mathrm{T}} - (GHr)r^{\mathrm{T}} + (GHr)(GHr)^{\mathrm{T}}\right]\right\} = 0\end{aligned}$$

Therefore,

$$G = E(rr^{T}H^{T})E(Hrr^{T}H^{T})^{-1} = E(rv^{T})E(vv^{T})^{-1} \tag{8}$$

where G represents estimation matrix, $E(rv^{T})$ is the cross-correlation matrix between vectors r and v, and matrix $E(vv^{T})$ is the autocorrelation matrix of vector v. So formula (6) is converted to

$$\hat{r} = E(rv^{T})E(vv^{T})^{-1}v. \tag{9}$$

The formula (2) can be changed to Eq. (9), the model has been obtained as follows:

$$\hat{r} = E\left(r(KUIr)^{T}\right)E\left((KUIr)(KUIr)^{T}\right)^{-1}v. \tag{10}$$

Using Eq. (10) which gives the Wiener estimation model and the known sample tristimulus values, spectral reflectance of the sample can be estimated.

3.2 Principal Component Analysis Model

Assuming spectral reflectance r is a n-dimensional column vector, then the vector group R can be composed of q samples of spectral reflectance to meet $R = \{r_1, r_2, \ldots, r_q\}$. By statistical analysis and linear operation, we can obtain k basic feature vectors $E = [e_1, e_2, \ldots, e_k]$ in the collection R, $k < n$. So any spectral reflectance of a sample can be estimated by the following formula:

$$r = EA \tag{11}$$

where $A = [a_1, a_2, \ldots, a_k]^{T}$ is the appropriate scale factor which is main element.

Obviously, through Principal component analysis (PCA) [6], it puts the n-dimensional spectral reflectance solving problems into solving k-dimensional principal component in order to realize reduction of spatial dimension and computational complexity. Typically, average vector r_0 of all the spectral reflectance vectors can be calculated in vector group R, and then, it is to use PCA to represent the difference between the spectral reflectance of the target vector and the average vector, i.e.,

$$r = r_0 + EA. \tag{12}$$

Using a linear model, the formula (12) into Eq. (2), the primary element coefficient vector can be expressed as

$$\tilde{A} = (KUIE)^{-1}(v - KUIr_0). \tag{13}$$

Then, $\tilde{A}$ substituting into Eq. (12). Thus, PCA model is as follows:

$$r = r_0 + E(KUIE)^{-1}(v - KUIr_0). \tag{14}$$

Using the command in MATLAB,

```
[pc,score,latent,tsquare] = princomp();
```

The first three eigenvalues are 12.6726, 2.1843, and 0.5681 from the pc, and the contribution ratio of 98.73 % is up to standard. So E can be determined in formula (14) selecting the corresponding three eigenvectors. Further r can be obtained.

4 Experimental Procedure and Results

In the experiment, the data are selected from the Munsell Laboratory experiments of 1269 Munsell spectral reflectance as test samples, the measured spectral wavelength range of 380–780 nm, steps of 1 nm, so that the spectral reflectance ratio is a vector composed of $N = 401$. In addition, the choice of training and test samples is stochastic in accordance with a and b situations. There a represents a group of the first half of Munsell spectral reflectance data as training samples, all of 1269 sets of data as the test sample. Also b represents a group of the latter half of Munsell spectral reflectance data as training samples, all of 1269 sets of data as the test sample. It shows that degree of curve fitting for reconstruction of spectral reflectance with the original using Wiener estimation and Principal component analysis algorithm from Fig. 1 which is obtained in a way. Clearly, dashed line obtained by Principal component analysis algorithm is close to original spectrum. Also it looks that degree of curve fitting for reconstruction of spectral reflectance with the original using Wiener estimation and Principal component analysis algorithm from Fig. 2 obtained in b mode. Obviously, dotted line obtained by Wiener estimation is close to original spectrum. Where curve represents original

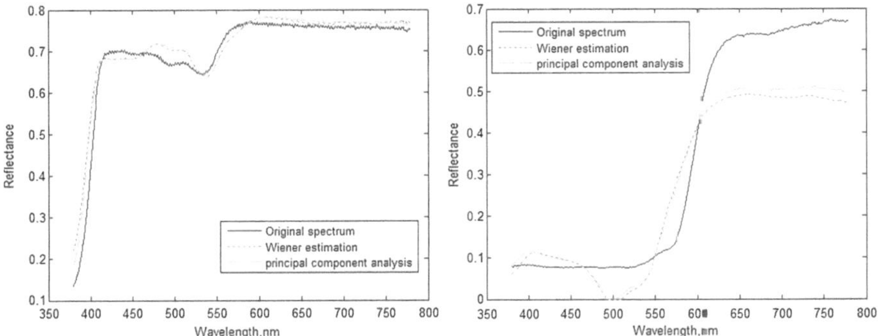

Fig. 1 Estimated average spectra for a situation

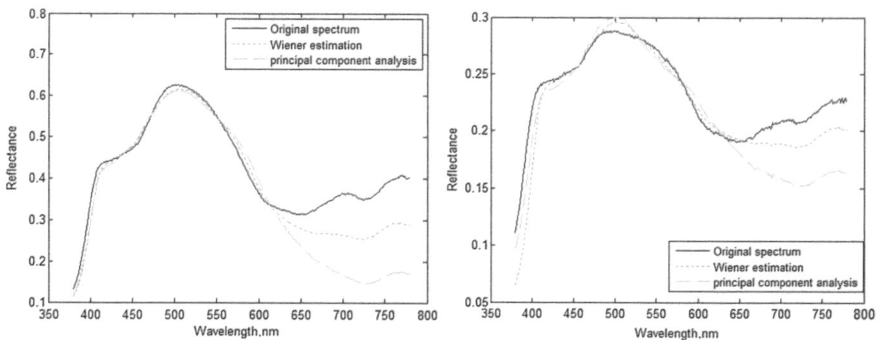

Fig. 2 Estimated average spectra for *b* situation

spectrum, dotted line represents Wiener estimation algorithm for reconstruction of spectral reflectance and dashed line indicates spectral reflectance obtained by Principal component analysis algorithm.

In addition, we use RMSE [7] and GFC [8] as spectral reflectance reconstruction accuracy of the evaluation criteria. Let original spectral reflectance data set $R(r_1, r_2, \ldots r_n)$ and reconstructed spectral reflectance data set $\hat{R}(\hat{r}_1, \hat{r}_2, \ldots \hat{r}_n)$.

RMSE can assess physical differences between original spectral reflectance and reconstructed spectral reflectance, and its expression is as follows:

$$\text{RMSE} = \sqrt{\frac{1}{n}\sum_{i=1}^{n}(r_i - \hat{r}_i)^2}. \tag{15}$$

GFC is based on Schwartz inequality and expressed as follows:

$$\text{GFC} = \frac{\sum_{\lambda=380}^{780}\hat{r}(\lambda)r(\lambda)}{\left(\sum_{\lambda=380}^{780}\hat{r}(\lambda)^2\right)^{1/2}\left(\sum_{\lambda=380}^{780}r(\lambda)^2\right)^{1/2}}. \tag{16}$$

Table 1 gives spectral reconstruction accuracy measured by Wiener estimation model and Principal component analysis model in various training and testing samples. One can see that the spectral reconstruction accuracy measured by two models in a mode is better than *b* mode. In addition, 0.0522 that the average RMSE obtained by Wiener estimation model is smaller than 0.0635 that one obtained by Principal component analysis model, while 0.9886 that the average GFC obtained by Wiener estimation model is bigger than 0.9817 that one obtained by Principal component analysis model. The changes that in Principal component analysis model RMSE and GFC are, respectively, 0.0810 and 0.9723 in *b* mode with 0.0460 and 0.9911 in a mode larger. In the Wiener estimation model, RMSE and GFC have smaller changes in two ways. So Wiener estimation model is better for spectral reflectance reconstruction.

5 Conclusion

In the paper, it uses Wiener estimation algorithm and Principal component analysis algorithm to form mathematical models in given 1269 Munsell spectral reflectance and corresponding to tristimulus values in D65 illumination, to find the relationship between spectral reflectance and tristimulus values by mathematical models. From Table 1, the distinction of accuracy can be seen from comparing with two kinds of models. The average result obtained by Wiener estimation is 0.0522, while the average value of RMSE is 0.0635 obtained by Principal component analysis method. Also, the average value of GFC is 0.9886 obtained by Wiener estimation while the same value is 0.9817 obtained by Principal component analysis. The simulation illustrates that Wiener estimation algorithm is superior to Principal component analysis method in the experiments for spectral reflectance reconstruction.

Acknowledgments The paper has been supported by Natural Science Foundation of Liaoning Province (2013020013).

References

1. Yi, Y.H., Liu, J.H., Gao, R.Y., Su, H.: Study on color characterization method of color inkjet printer. China Printing Packag. Study **4**(2), 17–23 (2012)
2. Zhang, X.D., Wang, Q., Wang, Y.: The XYZLMS interim connection space for spectral image compression and reproduction. Opt. Lett. **37**(24), 5097–5099 (2012)
3. Li, J., Wang H.W., Chen, G.X.: Study on key technologies of multi-spectral color reproduction. In: Proceedings of SPIE International Symposium on Multispectral Image Processing and Pattern Recognition, SPIE, Bellingham, USA, 800622-1-800622-7 (2011)
4. Urban, P., Rosen, M.R., Berns, R.S.: Spectral image reconstruction using an edge preserving spatio-spectral Wiener estimation. J. Opt. Soc. Am. A. **26**(8), 1865–1875 (2009)
5. Bochko, V.: Spectral color imaging system for estimating spectral reflectance of paint. J. Imaging Sci. Technol. **51**(1), 70–78 (2007)
6. Lehtonen, J., Parkkinen, J., Jaaskelainen, T., Kamshilin, A.: Principal component and sampling analysis of color spectra. Optical Rev. **2**(16), 81–90 (2009)
7. Stigell, P., Miyata, K., Hauta-Kasari, M.: Wiener estimation method in estimating of spectral reflectance from RGB images. Pattern Recognit. Image Anal. **17**(2), 233–242 (2007)
8. Imai, F.H., Rosen, M.R., Berns, R.S.: Comparative study of metrics for spectral match quality, CGIV 2002. In: The First European Conference on Colour Graphics, Imaging, and Vision, pp. 492–496 (2002)

Waveform Design Based on Water-Filling Algorithm

Bin Wang, Jinkuan Wang, Fengming Xin and Yuhuan Wang

Abstract Waveform design is an important problem in radar system design. In this paper, water-filling algorithm is used to solve the problem of waveform design based on mutual information. Two cases that no clutter exists and clutter exists are considered. Then, the optimal water-filling waveform can be obtained, respectively. The simulation results show that the optimized transmitted waveform allocates as more energy into different modes of target response as possible and obtains more information of target from echo in both cases. Finally, the whole paper is summarized.

Keywords Waveform design · Clutter · Water-filling

1 Introduction

Cognitive radar is a new concept in the research field of radar. It is proposed by research team of Simon Haykin in 2006 [1–3]. Different from traditional radar, cognitive radar continuously learns about the environment through experience gained from interactions of the receiver with the environment, the transmitter adjusts its illumination of the environment in an intelligent manner and the whole radar system constitutes a closed-loop dynamic system.

It is very important for the transmitted waveform to be adaptively designed, and many excellent works on waveform optimization have been done. In [4], the authors present illumination waveforms matched to stochastic targets in the presence of signal-dependent interference. In [5], the author applies information theory to design radar waveforms in channel-noise-only case and develops the

B. Wang (✉) · J. Wang · F. Xin · Y. Wang
EOSA Institute, Northeastern University at Qinhuangdao,
NO. 143, Taishan Road, Qinhuangdao, Hebei, China
e-mail: wangbinneu@qq.com

S. Patnaik and X. Li (eds.), *Proceedings of International Conference on Computer Science and Information Technology*, Advances in Intelligent Systems and Computing 255, DOI: 10.1007/978-81-322-1759-6_77,

optimal waveforms for target detection and information extraction. In [6], the authors propose a method to employ waveform agility to improve the detection of low-RCS targets on the ocean surface that present low signal-to-clutter ratios due to high sea states and low grazing angles. In [7], the authors address the problem of radar waveform design for target identification and classification. Both the ordinary radar with a single transmitter and receiver and the recently proposed MIMO radar are considered. In [8], the authors investigate the use of information theory to generate the optimum waveform matched to a Gaussian-distributed target ensemble with known spectral variance in the presence of signal-dependent clutter. In [9], the authors describe the optimization of an information theoretic criterion for radar waveform design.

In this paper, we consider waveform design problem in two cases that no clutter exists and clutter exists

2 Signal Model Description

The block diagram in Fig. 1 represents the signal model in signal-dependent interference. The meanings of the symbols are as follows:

$x(t)$: A finite-energy waveform with duration T.
$g(t)$: A extended target $(\sigma_G^2(f))$.
$c(t)$: A zero-mean complex random process $(\sigma_C^2(f))$.
$n(t)$: The zero-mean receiver noise process $(P_n(f))$.
T_g: The time duration where most of the target impulse's energy resides.

We assume that the transmitted signal has no significant energy outside the frequency interval $w = [-W, W]$. Since $z(t)$ and $d(t)$ are the response of a linear time-invariant system to the transmitted signal, neither do they have significant energy outside the frequency interval $w = [-W, W]$.

Let $y(t)$ be the received signal given by

$$y(t) = z(t) + d(t) + n(t). \tag{1}$$

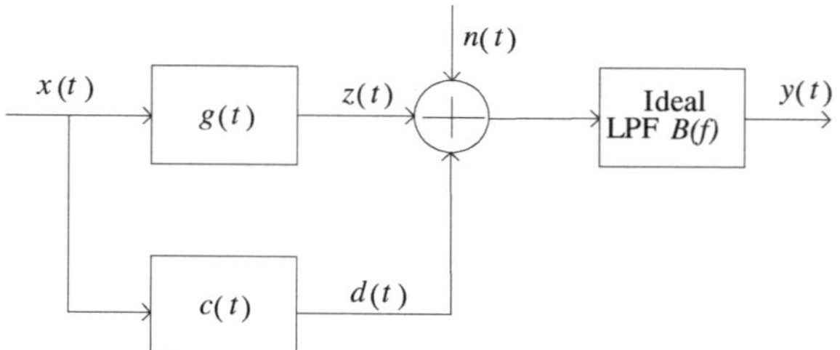

Fig. 1 Signal model in signal-dependent interference

We note that $x(t)$ is a deterministic waveform. It is explicitly denoted in $I(y(t); g(t)|x(t))$ because the mutual information is a function of $x(t)$, and we are interested in finding those functions $x(t)$ that maximize $I(y(t); g(t)|x(t))$ under constrains on their energy and bandwidth.

Bell proposed the derivation of the mutual information in the channel-noise-only case and derived the information-based waveform solution. According to his theory, we provide the derivation of the mutual information in the presence of signal-dependent clutter.

Assume we have a channel with input Z, a zero-mean Gaussian random variable with variance σ_Z^2, the clutter D, a zero-mean Gaussian random variable with variance σ_D^2 and additive zero-mean Gaussian noise N with variance σ_N^2. The mutual information $I(Y; Z)$ between Y and Z is

$$I(Y; Z) = H(Y) - H(Y|Z) = \frac{1}{2}\ln\left(1 + \frac{\sigma_Z^2}{\sigma_N^2 + \sigma_D^2}\right). \tag{2}$$

The mutual information between each sample Z_m of $\widehat{z}_k(t)$ and the corresponding sample Y_m of $\widehat{y}_k(t)$ is

$$I(Y_m; Z_m) = \frac{1}{2}\ln\left[1 + \frac{2|X(f_k)|^2\sigma_G^2(f_k)}{T_y\left\{P_n(f_k) + 2|X(f_k)|^2\sigma_C^2(f_k)\right\}}\right]. \tag{3}$$

The mutual information between the random target impulse response and the received radar waveform is

$$I(y(t); g(t)|x(t)) = T_y\int_W \ln\left(1 + \frac{2|X(f)|^2\sigma_G^2(f)}{T_y\left\{P_n(f) + 2|X(f)|^2\sigma_C^2(f)\right\}}\right)\mathrm{d}f. \tag{4}$$

3 Water-Filling Algorithm

As the energy of transmitted signal is limited, meantime major energy of transmitted signal is ensured to distribute in $\omega = [0, W]$. So energy constraint is

$$\int_W |X(f)|^2\mathrm{d}f \le E_x \tag{5}$$

where E_x is the total energy of transmitted signal and W is bandwidth of transmitted signal. So the optimization problem can be viewed as

$$\begin{aligned} &\max\ T_y \int_W \ln\left(1+\frac{2|X(f)|^2\sigma_G^2(f)}{T_y\left\{P_n(f)+2|X(f)|^2\sigma_C^2(f)\right\}}\right)\mathrm{d}f \\ &\text{s.t.}\ \int_W |X(f)|^2\mathrm{d}f - E_x \le 0 \end{aligned} \tag{6}$$

When KKT condition is satisfied, the optimized waveform can be obtained

$$|X(f)|^2 = -R(f) + \sqrt{R^2(f) + S(f)(A - D(f))} \tag{7}$$

where

$$D(f) = \frac{T_y P_n(f)}{2\sigma_G^2(f)} \tag{8}$$

$$R(f) = \frac{P_n(f)\left(2T_y\sigma_C^2(f)+\sigma_G^2(f)\right)}{4\sigma_C^2(f)\left(T_y\sigma_C^2(f)+\sigma_G^2(f)\right)} \tag{9}$$

$$S(f) = \frac{P_n(f)\sigma_G^2(f)}{2\sigma_C^2(f)\left(T_y\sigma_C^2(f)+\sigma_G^2(f)\right)}. \tag{10}$$

As power spectral density of transmitted signal is nonnegative, so optimized waveform can be expressed as

$$|X(f)|^2 = \max\left[0, -R(f) + \sqrt{R^2(f) + S(f)(A - D(f))}\right] \tag{11}$$

where A can be obtained by the following energy constraint

$$E_x \ge \int_W \max\left[0, -R(f) + \sqrt{R^2(f) + S(f)(A - \mathrm{D}(f))}\right] df. \tag{12}$$

When the background has only noise, which means $\sigma_C^2(f)=0$. The mutual information of echo and target is

$$I(y(t); g(t)|x(t)) = T_y \int_W \ln\left(1 + \frac{2|X(f)|^2\sigma_G^2(f)}{T_y P_n(f)}\right)\mathrm{d}f. \tag{13}$$

Using Lagrange multiplier method, we can get

$$L\left(|X(f)|^2, \lambda\right) = -T_y \int_W \ln\left(1 + \frac{2|X(f)|^2\sigma_G^2(f)}{T_y P_n(f)}\right)\mathrm{d}f + \lambda\left(\int_W |X(f)|^2\mathrm{d}f - E_x\right). \tag{14}$$

So

$$|X(f)|^2 = \max\left[0, A - \frac{P_n(f)T_y}{2\sigma_G^2(f)}\right]. \tag{15}$$

A is a constant that influences the energy constraint of transmitted signal. It can be seen that the waveform tends to put more energy to the larger band of $\sigma_G^2(f)/P_n(f)T_y$.

When clutter exists, using Lagrange multiplier method, we can get

$$L\left(|X(f)|^2, \lambda\right) = -T_y \int_W \ln\left(1 + \frac{2|X(f)|^2\sigma_G^2(f)}{T_y\left\{2|X(f)|^2\sigma_C^2(f) + P_n(f)\right\}}\right)\mathrm{d}f + \lambda\left(\int_W |X(f)|^2\mathrm{d}f - E_x\right). \tag{16}$$

After solving the problem, we can get the optimized waveform

$$|X(f)|^2 = \max\left[0, -R(f) + \sqrt{R^2(f) + S(f)(A - D(f))}\right]. \tag{17}$$

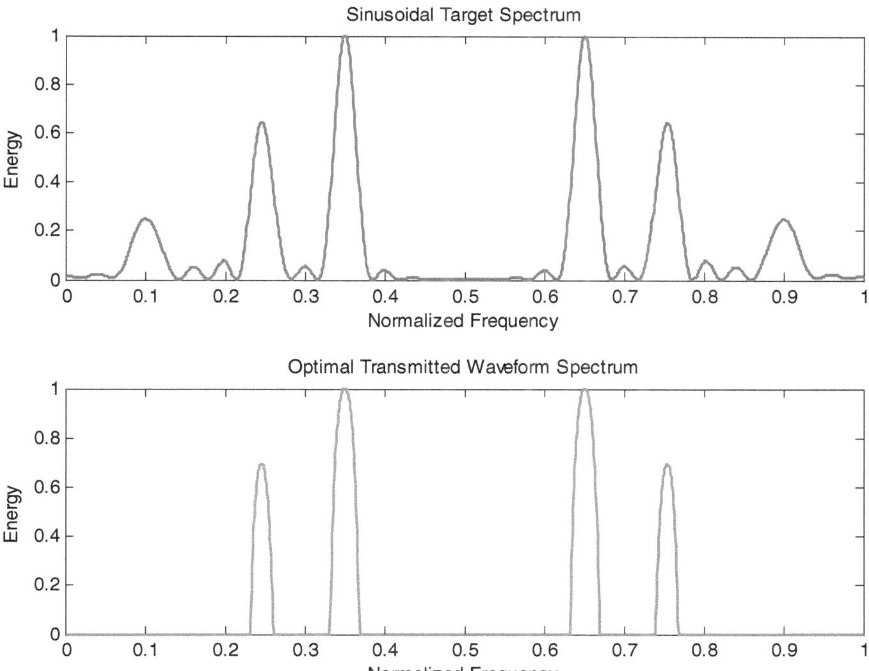

Fig. 2 Spectrum of sinusoidal target and optimal transmitted waveform

4 Simulations

In order to facilitate comparison, energy spectral of optimized transmitted waveform and margin of extended targets' impulse response is normalized. Simulation conditions are as follows. Signal energy is $E_x = 1$, signal frequency is $f \in [0, 1]$, sampling frequency is 2, and power spectral density of noise is $P_n(f) = 0.1$.

Figure 2 is spectrum of sinusoidal target and optimal transmitted waveform. It shows that in the situation there is no clutter, the optimized transmitted waveform allocates as more energy to spectral peak of target response as possible. As various modes of target may contain important information of target, so optimized transmitted signal allocates as more energy into different modes of target response as possible and obtains more information of target from echo.

Figure 3 is spectrum of target and clutter. Figure 4 is optimal transmitted waveform spectrum. It can be seen that the optimized waveform allocates the energy to the whole bandwidth. When PSD of clutter is small or 0 in certain band, the transmitted waveform will allocate more energy to this band. However, in the band, clutter is strong or target is weaker than clutter, energy spectrum of transmitted waveform will be reduced.

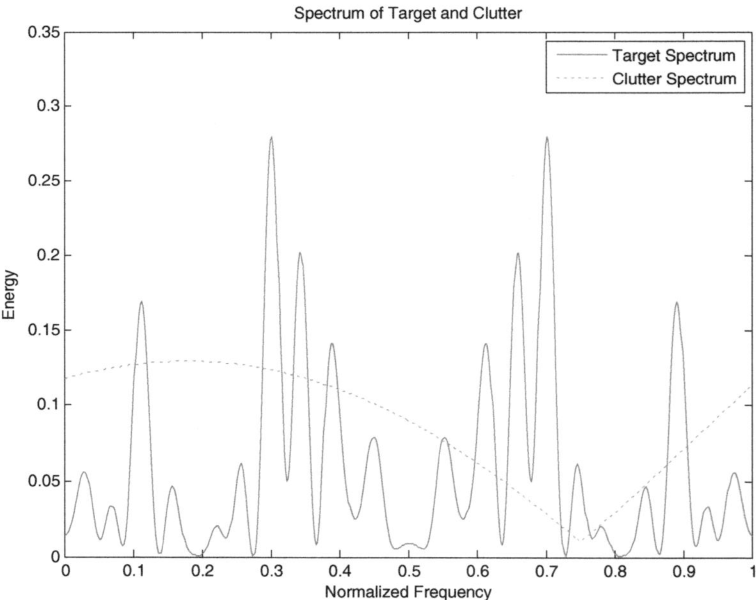

Fig. 3 Spectrum of target and clutter

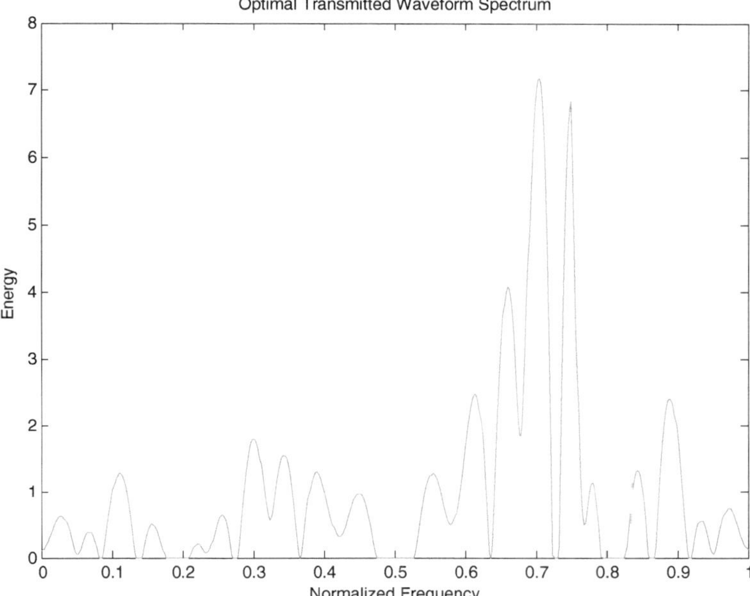

Fig. 4 Optimal transmitted waveform spectrum

5 Conclusions

In this paper, we use water-filling method to obtain optimized water-filling waveform based on mutual information. We consider two cases that no clutter exists and clutter exists. The simulation results show that the optimized transmitted signal allocates as more energy into different modes of target response as possible and obtains more information of target from echo in both cases that no clutter exists and clutter exists. This method can reduce the uncertainty of target through maximizing the mutual information of echo and target.

Acknowledgments This work was supported by the National Natural Science Foundation of China (No. 61004052 and No. 61104005), the Natural Science Foundation of Hebei Province (No. F2013501075), the Fundamental Research Funds for the Central Universities (No. N110323005), and the Doctoral Scientific Research Foundation of Liaoning Province (No. 20131030).

References

1. Haykin, S.: Cognitive radar: a way of the future. IEEE Signal Process. Mag. **23**(1), 30–40 (2006)
2. Haykin, S.: Cognition is the key to the next generation of radar systems. In: Proceedings of Digital Signal Processing Workshop and 5th IEEE Signal Processing Education Workshop (DSP/SPE 2009), pp. 463–467. IEEE Press (2009)

3. Haykin, S., Xue, Y.B., Davidson, T.: Optimal waveform design for cognitive radar. In: 42nd Asilomar Conference on Signals, Systems and Computers, Pacific Grove, CA, pp. 3–7 (2008)
4. Romero, R.A., Goodman, N.A.: Waveform design in single-dependent interference and application to target recognition with multiple transmissions. IET Radar Sonar Navig. **3**(4), 328–340 (2009)
5. Bell, M.R.: Information theory and radar waveform. IEEE Trans. Inf. Theory **39**(5), 1578–1597 (1993)
6. Sira, S.P., Cochran, D.: Adaptive waveform design for improved detection of low-RCS targets in heavy sea clutter. IEEE J. Sel. Top. Signal Process. **1**(1), 56–66 (2007)
7. Yang, Y., Blum, R.S.: MIMO radar waveform design based on mutual information and minimum mean-square error estimation. IEEE Trans. Aerosp. Electron. Syst. **43**(1), 330–343 (2007)
8. Romero, R., Goodman, N.A.: Information-theoretic matched waveform in signal dependent interference. In: IEEE Radar Conference, pp. 1–6. IEEE Press, (2008)
9. Leshem, A., Naparstek, O., Nehorai, A.: Information theoretic adaptive radar waveform design for multiple extended targets. IEEE J. Sel. Top. Signal Process. **1**(1), 42–55 (2007)

The Artificial Fish Swarm Algorithm to Solve Traveling Salesman Problem

Teng Fei, Liyi Zhang, Yang Li, Yulong Yang and Fang Wang

Abstract The artificial swarm algorithm, which has characteristics of global search, quick convergence speed, and efficient search, is an intelligent search algorithm based on modern elicitation method. In this paper, the solving of TSP by the artificial fish swarm algorithm is researched deeply, and steps of TSP is analyzed. The validity of the presented approach is demonstrated by simulation example.

Keywords Artificial fish swarm algorithm · Traveling salesman problem · Combination optimal problem

1 Introduction

Traveling salesman problem (TSP) belonged to the NP complete [1–3], which's complexity with the size of the input is an exponential function of growth, is one of a typical combinatorial optimization problem with many engineering applications and important theoretical research value. Because of simple description of the problem and board application in a variety of optimization problems, the TSP gripped many investigators to research for a long time.

The artificial fish swarm algorithm (AFSA) is brought by Dr. Xiaolei Li who introduced a new type model to search by way of the research on characteristics of

T. Fei · L. Zhang (✉) · Y. Yang · F. Wang
Information Engineering College, Tianjin University of Commerce, Tianjin, China
e-mail: fei_8825@163.com

T. Fei
e-mail: feiteng@tjcu.edu.cn

Y. Li
Economic College, Tianjin University of Commerce, Tianjin, China

S. Patnaik and X. Li (eds.), *Proceedings of International Conference on Computer Science and Information Technology*, Advances in Intelligent Systems and Computing 255, DOI: 10.1007/978-81-322-1759-6_78,

fish behavior and application in the model of animal commune in 2002 [4]. With the understanding and study of the algorithm, the AFSA has been widely applied in many fields of communications, digital and image processing, neural networks, data mining, and combinatorial optimization.

In this paper, the solving of TSP by the AFSA is researched deeply, and steps of TSP is analyzed. The validity of the presented approach is demonstrated by simulation example.

2 Basic Artificial Fish Swarm Algorithm

The AFSA is an optimization model of the self-governing body based on the animal behavior. The optimization is realized by foraging, cluster, and rear-end behavior that is imitated by the constructed artificial fish. The model of artificial fish is structured by multiparallel channel structure using the behavior in literature [5].

2.1 Related Definitions

The total number of artificial fish is N, and its individual state is $X = (x_1, x_2, \ldots, x_n)$, where x_i $(i = 1, 2, \ldots n)$ is variable to be optimized. The longest moving step of artificial fish is Step, perceived distance of the artificial fish is Visual, congestion factor is δ, the distance of the artificial fish i, j is $d_{ij} = |x_i - x_j|$, food concentration of artificial fish in the current location is $Y = f(x)$, where the objective function value is Y and the number of attempts is Try_number.

2.2 Basic Behavioral Description

Foraging behavior Foraging behavior is a basic behavior of artificial fish, which is activity of tending to the food. Artificial fish choices is tended though perceiving the account of food or concentration in the water by the vision or palate.

The current state of artificial fish is x_i and select a state x_j randomly in its perception

$$x_j = x_i + \text{Visual} \cdot \text{Rand}() \tag{1}$$

where Rand() is random number in the range between 0 and 1.

If $Y_i < Y_j$, the step forward must be made to this direction

$$x_i^{t+1} = x_i^t + \frac{x_j - x_i^t}{\|x_j - x_i^t\|} \cdot \text{Step} \cdot \text{Rand}(). \tag{2}$$

Otherwise, select status x_j randomly again, judge whether the forward condition is satisfied, and repeat Try_number times; if forward condition is still dissatisfied, select step randomly

$$x_i^{t+1} = x_i^t + \text{Visual} \cdot \text{Rand}(). \tag{3}$$

Huddling behavior In nature, the fish will naturally cluster in order to ensure the survival of groups and avoid danger in swimming. AFSA provides that each fish should move to the center of the neighboring partners as much as possible and not be overcrowding.

The current state of artificial fish is x_i. Search for number of partners n_f and central location x_c. If $Y_c/n_f > \delta Y_i$, partner center has more food and is not crowed, so, the forwarding step must be made to the location of the partner center

$$x_i^{t+1} = x_i^t + \frac{x_c - x_i^t}{\|x_c - x_i^t\|} \cdot \text{Step} \cdot \text{Rand}(). \tag{4}$$

Otherwise, perform foraging behavior.

Following behavior In the swimming process, the neighboring partners will quickly reach the food point following one fish which finds the food. Rear-end behavior is a chase behavior, which has the highest fitness, to the nearby artificial fish, also a process of advancing to near-optimal partner.

The current state of artificial fish is x_i. Search the partner which maximum value is Y_j in the current neighborhood is x_j. If $Y_j/n_f > \delta Y_i$, the state of x_j has a higher concentration of food and not crowded around, therefore, the forwarding step must be made to x_j

$$x_i^{t+1} = x_i^t + \frac{x_j - x_i^t}{\|x_j - x_i^t\|} \cdot \text{Step} \cdot \text{Rand}(). \tag{5}$$

Otherwise, perform foraging behavior.

Random behavior Random behavior is to choose a state randomly and then move to the direction which is selected. Random behavior is a default behavior of foraging behavior.

Bulletin board Bulletin board is used to record the individual state of the optimum artificial fish and food concentration of artificial fish position. In the optimization process for each artificial fish, the own state and billboard status after per action once can be tested. If the own state is better than the status of the bulletin board, translate the billboard status about the own state, in order that the bulletin board could record the history optimal state.

Behavior assessment Behavior assessment is use to reactive the way of the artificial fish autonomy behavior. For TSP, it is able to perform cluster and rear-end behavior, then, to evaluate the value of actions, to select the optimal behavior which's value is to execute. The default behavior is the foraging behavior.

3 Solving TSP by AFSA

3.1 The Model of TSP

Set that there are n cities, d_{ij} is the distance between city i and city j, and the model of TSP is

$$\min \sum_{i \neq j} d_{ij} x_{ij} \tag{6}$$

S.T.

$$\sum_{j=1}^{n} x_{ij} = 1 \quad i = 1, 2, \ldots, n \tag{7}$$

$$\sum_{i=1}^{n} xi_j = 1 \quad j = 1, 2, \ldots, n \tag{8}$$

$$\sum_{i,j \subseteq s} x_{ij} \leq |s| - 1,\ 2 \leq |s| \leq n - 2,\ s \subset \{1, 2, \ldots, n\} \tag{9}$$

$$x_{ij} \in \{0, 1\}\ i, j = 1, \ldots, n, \quad i \neq j \tag{10}$$

3.2 Steps of Solving

The steps of solving the traveling salesmen problem are as follows:

Step 1: Algorithm initialization. Get the number of cities and specific coor-dinate of each city, and set up the distance matrix between the city. The parameters of the total number of artificial fish is N, the maximum number of iterations NC, perceived distance of the artificial fish Visual, the longest moving step of artificial fish Step, and congestion factor δ.
Step 2: If the recent iterations is Times $= 0$, artificial fish individual is able to be generated, so that, the initial fish stocks is to be formatted.
Step 3: Artificial fish selects the optimal behavior to do in foraging behavior, rear-end behavior, and cluster behavior.
Step 4: The artificial fish tests its own state and billboard status after every action.
Step 5: Check whether Times is the maximum iterative times. If it is reached, display the state in bulletin board. If it is not reached, Times $+ 1$, turn to Step 3.

4 Algorithm Testing

Using MATLAB simulation, basic AFSA parameters are set as follows: the total number of artificial fish is $N = 10$, the maximum number of iterations is NC = 200, perceived distance of the artificial fish is Visual = 6, and congestion factor is $\delta = 0.8$. The TSP of 22 cities whose coordinates are shown in Table 1 is solved using AFSA.

Figure 1 shows the best routing of solving the TSP using AFSA, and it is shown below:

1——11——10——8——6——3——2——7——9——4——5——12——14——20——22——18——21——19——16——13——15——17——1.

Figure 2 shows curve of optimal value iteration. It is shown that the optimal value is 284.3044 and the total time is 0.351 s.

Solving TSP by using AFSA, the results showed that:

1. AFSA has a faster convergence rate and can be converged to a feasible solution in a relatively short period of time. It is used to solve the actual requirements. For the simulation of the above, convergence time of the artificial fish swarm is about 3.051 s. For less precision occasions, satisfactory solution can be quickly obtained.
2. Simulation of the above can be seen, as long as the implementation of the foraging behavior can converge to the optimal solution. The foraging behavior laid the foundation of AFSA, huddling behavior intensified the stability the algorithm's convergence the rear-end behavior had muscled the rapid and global convergence of the algorithm, Behavior evaluation provided effective protection for convergence and stability of the algorithm.
3. AFSA has many advantages, but it still has some flaws. With the increase in the number of artificial fish, AFSA may cause an increase in the calculated amount of growth and computation time.

Table 1 Coordinates of cities

No.	Coordinate	No.	Coordinate
1	(145, 215)	12	(128, 231)
2	(151, 264)	13	(156, 217)
3	(159, 261)	14	(129, 214)
4	(130, 254)	15	(146, 208)
5	(128, 252)	16	(164, 208)
6	(163, 247)	17	(141, 206)
7	(146, 246)	18	(147, 193)
8	(161, 242)	19	(164, 193)
9	(142, 239)	20	(129, 189)
10	(163, 236)	21	(155, 185)
11	(148, 232)	22	(139, 182)

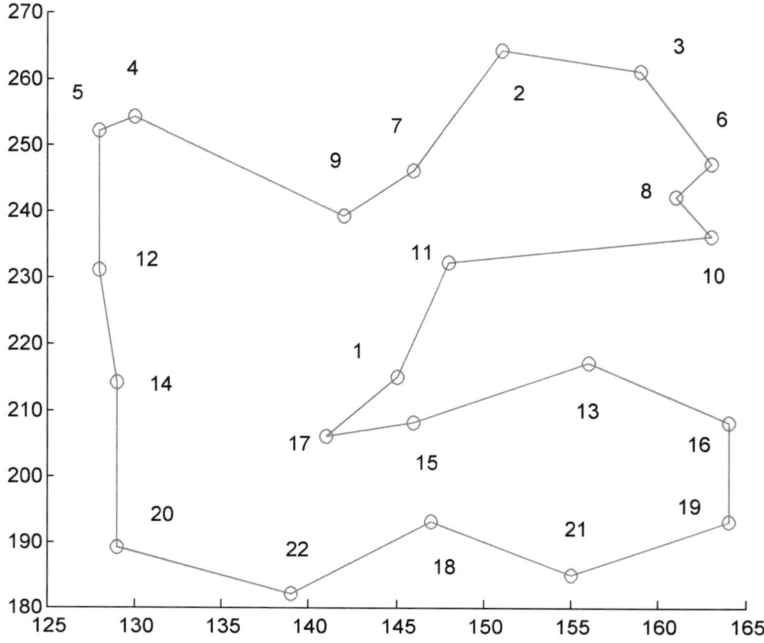

Fig. 1 The routing of the best

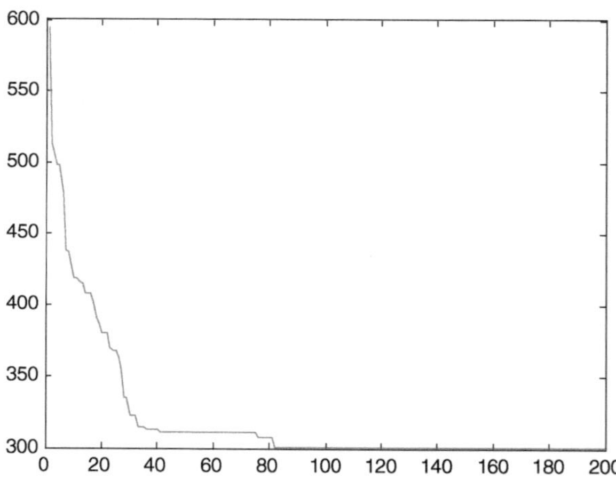

Fig. 2 Curve of optimal value iteration

5 Conclusion

In the paper, TSP is solved by researching the AFSA. The simulation proves that the algorithm is effective and has characteristics of fast convergence in the early part, finding the local optimization quickly. The AFSA, which needs a further profound study, is still in the beginning stages of research.

Acknowledgments This work is partially supported by the object of China Logistics Association (2012CSLKT027).

References

1. Li, Y., Fan, J., Zhang, Q.: Solving TSP with improved artificial fish swarm algorithm. J. Shijiazhuang Tiedao Univ. (Nat. Sci.) **24**(2), 103–110 (2011)
2. Baraglia, R., Hidalgo, J.I., Perego, R.: A hybrid heuristic for the traveling salesman problem. IEEE Trans. Evol. Comput. **5**(6), 613–622 (2001)
3. Zhu, M.H., She, X.Y.: Improved artificial fish school algorithm to solve traveling salesman problem. Appl. Res. Comput. **27**(10), 3734–3736 (2010)
4. Xiaolei, L., Jixin, Q.: Artificial fish swarm algorithm optimization: mode of top-down. In: Process Systems Engineering Conference Proceedings (2001)
5. Li, X.L., Shao, Z., Qian, J.X.: An optimizing method based on autonomous animates: fish-swarm algorithm. Inst. Syst. Eng. **22**(11), 32–38 (2002)

Multi-Sub-Swarm PSO Algorithm for Multimodal Function Optimization

Yanwei Chang and Guofang Yu

Abstract Inspired by individuals' clustering to sub-swarm through learning between individuals and between sub-swarms, we propose a new algorithm called dynamic multi-sub-swarm particle swarm optimization (MSSPSO) algorithm for multimodal function with multiple extreme points. In the evolutionary process, the initial particles, that are separately one sub-swarm, merge into bigger sub-swarms by calculating a series of dynamic parameters, such as swarm distance, degree of approximation, distance ratio and position accuracy. Simulation results show that, in single-peak function optimization, MSSPSO algorithm is feasible but search speed is not superior to the PSO algorithm, while in a multimodal function optimization, MSSPSO algorithm is more effective than PSO algorithm, which cannot locate the required number of extreme points.

Keywords Multimodal function optimization · Multi-sub-swarm PSO · Dynamic parameters · Degree of approximation · Position accuracy

1 Introduction

Optimization problem is ubiquitous in our society, some of which about mathematics, engineering, social, economic and management, requires not only to find the global optimal solution in feasible region, and often also to search

Y. Chang (✉)
School of Mechanical and Electrical Engineering, Jiangsu Normal University, Xuzhou 221116 Jiangsu, China
e-mail: changyw@jsnu.edu.cn

G. Yu
School of Information and Electrical Engineering, China University of Mining and Technology, Xuzhou 221116 Jiangsu, China
e-mail: gfyu@cumt.edu.cn

S. Patnaik and X. Li (eds.), *Proceedings of International Conference on Computer Science and Information Technology*, Advances in Intelligent Systems and Computing 255, DOI: 10.1007/978-81-322-1759-6_79,

multiple global optimal solution and meaningful local optimal solution for a variety of options for policy-makers in practice. Such problems are generally known as multi-peaks function optimization or multimodal function optimization problems, for example, linear programming problem, neural network ensemble training, optimal control law design, complex systems parameters, and structure identification.

How to construct an optimization algorithm that can search all global optimal solutions and many local optimal solutions has become an ongoing area of research [1, 2].

2 Multimodal Function Optimization Problems

Multimodal optimization mathematical description [3–5] is as follows:

Given a nonempty set S as the search space, and f: $S \mapsto R$ is the objective function to be optimized.

Definition 1 If $x^* \in S$, for $\forall x \in S$, then

$$f(x) \le f(x^*) \tag{1}$$

So x^* is global optimal solution of the maximization problem $(\max_{x \in S} f(x))$, $f(x^*)$ is a global optimum.

Definition 2 If $x^* \in S$, and $\exists\, \delta > 0$, for $\forall x \in S$ and $\|x - x^*\| < \delta$, then

$$f(x) \le f(x^*) \tag{2}$$

So x^* is local optimal solution of the maximization problem $(\max_{x \in S} f(x)), f(x^*)$ is a local optimum.

For these problems above, if and only if there exists a $x^* \in S$ so that formula (1) holds, then $f(x)$ is a single-peak function, and if there exists different $x_1, x_2, \ldots, x_m \in S$, so that $f(x_i)(i = 1, 2, \ldots, m)$ are the global optimum value and local optimum value, then $f(x)$ is called multimodal function.

For multimodal optimization problem, the traditional optimization methods mainly include derivative-based analytic methods and numerical optimization methods [6], all of which have a strong restriction of the objective function, such as continuity, differentiability, so there are many difficulties to solve complex multimodal optimization problem with discontinuous or nondifferentiable, non-convex objective function.

Swarm intelligence algorithms [7, 8] are heuristic, so they do not subject to the conditions above. Swarm intelligence methods provide a new way to solve the problem of multimodal function optimization, and get a lot of achievements, e.g., genetic algorithms [1, 9–12], immune algorithm [13–16], and particle swarm optimization (PSO) algorithm [2, 17–20] studied in this paper.

However, there are still two problems to be solved urgently in solving multi-modal function optimization problem with the swarm intelligence as follows:

1. The algorithms cannot run without the peak distance and the number of peaks.
2. The algorithms lack an effective method of measuring the swarm diversity, so the species easy to fall into premature.

PSO [17] is a swarm intelligence optimization algorithms [9, 10] based on the theory of stochastic optimization. For the two problems above, a new algorithm is proposed to solve the first problem effectively by dynamically forming sub-swarm on base of the information of particles and between the particles. The second type of problem will be described in another paper.

3 Dynamic Multi-Sub-Swarm PSO

3.1 Basic Concept of Dynamic Parameters

Let a given search space is D dimensional space, and let $X_i^D(t)$, $V_i^D(t)$, $\mathrm{Pb}_i^D(t)$ and $\mathrm{Gb}_i^D(t)$ is the position, velocity, individual history local optimum, and global optimum of the ith particle in the tth generation.

Particle swarm according to Eqs. (3) and (4) to update their velocity and position:

$$V_i^D(t+1) = W_i^D(t)V_i^D(t) + C_1R_1\left(\mathrm{Pb}_i^D(t) - X_i^D(t)\right) + C_2R_2\left(\mathrm{Pb}_i^D(t) - X_i^D(t)\right) \quad (3)$$

$$X_i^D(t+1) = X_i^D(t) + V_i^D(t) \quad (4)$$

To distinct with notations of PSO, the best position of the ith sub-swarm sb_i is used instead of gb in the Eqs. (3) and (4). For writing convenience, superscript D is negligible, and the generation t is denoted by superscript, and then the update formula of the kth particle of the ith sub-swarm is as follows:

$$v_{i,k}^{t+1} = w_{i,k}^t v_{i,k}^t + c_1r_1\left(\mathrm{pb}_{i,k}^t - x_{i,k}^t\right) + c_2r_2\left(\mathrm{sb}_{i,k}^t - x_{i,k}^t\right) \quad (5)$$

$$x_{i,k}^{t+1} = x_{i,k}^t - v_{i,k}^{t+1}. \quad (6)$$

Definition 1 The t-generation distance between the sub-swarm i and j is the distance between the ith sub-swarm optimum and the jth sub-swarm optimum. Denoted as follows:

$$\Delta sb_{i,j}^t = \left\|sb_i(t) - sb_j(t)\right\| \quad (7)$$

As the particles oscillate viciously in the evolutionary process, so a relatively stable sub-swarm optimal value is chosen as measure. Sub-swarm optimal values may be from the particles position of this evolutionary generation or of the past generation.

Definition 2 The distance between the kth particle of sub-swarm i and of sub-swarm j is the distance between the kth particle optimum in sub-swarm i and the jth sub-swarm optimum, denoted as follows:

$$\Delta ps_{i,j}^{k}(t) = \left\| ps_i^k(t) - sb_j(t) \right\| \tag{8}$$

Definition 3 The t-generation n-times degree of approximation between the sub-swarm i and j is the ratio of the t-generation and $t + n$-generation distance between the sub-swarm i and j. Denoted as follows:

$$\pi_{i,j}^{n}(t) = \begin{cases} \frac{\Delta sb_{i,j}^{t} - \Delta sb_{i,j}^{t+n}}{\Delta sb_{i,j}^{t} + \Delta sb_{i,j}^{t+n}} & \Delta sb_{i,j}^{t} + \Delta sb_{i,j}^{t+n} \neq 0 \\ 0 & \Delta sb_{i,j}^{t} + \Delta sb_{i,j}^{t+n} = 0 \end{cases} \tag{9}$$

where $\pi_{i,j}^{n}(t) \leq 1$ and $\pi_{i,j}^{n}(t) = 0$ expressed that the $t + n$-generation and t-generation distance of two sub-swarms is constant. Degree of approximation represents approaching speed of two sub-swarms in the evolutionary process, i.e., the greater the degree of approximation of two sub-swarm is, the faster the two sub-swarms in the evolutionary process approaches, v.v.

Definition 4 The t-generation distance ratio of sub-swarm i and j is as follows:

$$\lambda_{i,j}^{t} = \frac{\Delta sb_{i,j}^{t}}{\Delta_{\max}^{t}} \tag{10}$$

Definition 5 The t-generation position accuracy of sub-swarm i and j is as follows:

$$\beta_{i,j}^{t} = \frac{\Delta_{\max}^{t} - \Delta sb_{i,j}^{t}}{\Delta_{\max}^{t}} \tag{11}$$

where $0 \leq \lambda_{i,j}^{t} \leq 1$. The smaller the value of $\lambda_{i,j}^{t}$ is and the greater the value of $\beta_{i,j}^{t}$ is, the closer the centers of two sub-swarms is. Position accuracy represents relative positional relationship of any two sub-swarms and also discloses the distribution of all sub-swarms in the swarm and diversity of sub-swarms.

3.2 Algorithm Flow

Dynamic parameters are the measurement of the relationship between the sub-swarms in evolutionary process, so the multi-sub-swarm PSO algorithm process is described as follows:

Step 1 Initialize algorithm parameters, i.e., inertia weight w, w_{max}, and w_{min}, learning factor c_1 and c_2, sub-swarm degree of approximation threshold θ_π, the distance ratio threshold θ_λ, the error threshold ε, the maximum number of iterations t_{max}.
Step 2 Initialize particles and sub-swarms that include $v_i(0)$, $x_i(0)$, min $f(x)$, $pb(0)$, $sb_i(0)$, $\pi_{i,j}(0)$, $\lambda_{i,j}^0$, the distance matrix Δ, and the fitness of each particle. Each particle is a sub-swarm.
Step 3 Update the speed and position of the particle swarms according to the algorithm update Eqs. (5) and (6).
Step 4 Calculate the fitness of each particle and update $pb(t)$ and $sb_i(t)$.
Step 4.1 Update the distance matrix Δ and calculate $\pi_{i,j}^n(t)$ and $\lambda_{i,j}^t$ of the optimal particle for each sub-swarm.
Step 4.2 Update sub-swarms (merge sub-swarms) on comparing with $\pi_{i,j}^n(t)$, $\lambda_{i,j}^t$, θ_π, and θ_λ of every sub-swarm and then update the number of sub-swarms K and $\mathrm{sb}_i(t)$.
Step 5 Go to Step 8 if $t > t_{max}$.
Step 6 Stop evolution of the sub-swarm in which there is no merge operation and $|sb_i(t-1) - sb_i(t)/sb_i(t)| < \varepsilon$.
Step 7 Ends if all sub-swarms have stopped evolving or else go to Step 3.
Step 8 Ends.

4 Multi-Sub-Swarm PSO Algorithm Simulation

4.1 Test Functions and Parameter Selection

For testing the convergence of MSSPSO algorithm, a nonlinear symmetric singlet Sphere function and a multiplet Schwefel function with many different close peaks are selected as the test function.

1. Sphere function

$$f_1(x) = \sum_{i=1}^{30} x_i^2 \tag{12}$$

where $-100 \le x_i \le 100$.

This function is used for testing the performance of sub-swarms merging.

2. Schwefel function

$$f_2(x) = \sum_{i=1}^{30} \left(x_i \sin\left(\sqrt{|x_i|}\right)\right) \tag{13}$$

where $-500 \le x_i \le 500$.

Table 1 Selection parameters in MSSPSO and PSO

Parameters	PSO	MSSPSO
$v_{\max}$ (sphere)	90	5
$v_{\max}$ (Schwefel)	400	10
$w_{\max}$	0.9	0.9
$w_{\min}$	0.4	0.4
c_1	1.3	1.5
c_2	1.3	1.3
θ_π	–	0.3
θ_λ	–	0.1
The number of particles	30	100

This function is used for testing the convergence of sub-swarms.

The selection of the initial velocity of MSSPSO algorithm should be smaller speed to increase the local search capability than of the standard PSO algorithm to increase the exploration capability. The selection of parameters is shown as in Table 1.

We select 10-dimensional Sphere function and the two-dimensional Schwefel function for the experiments, which evolve 3,000 generations and 200 generations to search target to be minimum value and maximum value separately. For each function, we have conducted 20 experiments, in one of which the sub-swarms convergence are shown as follows:

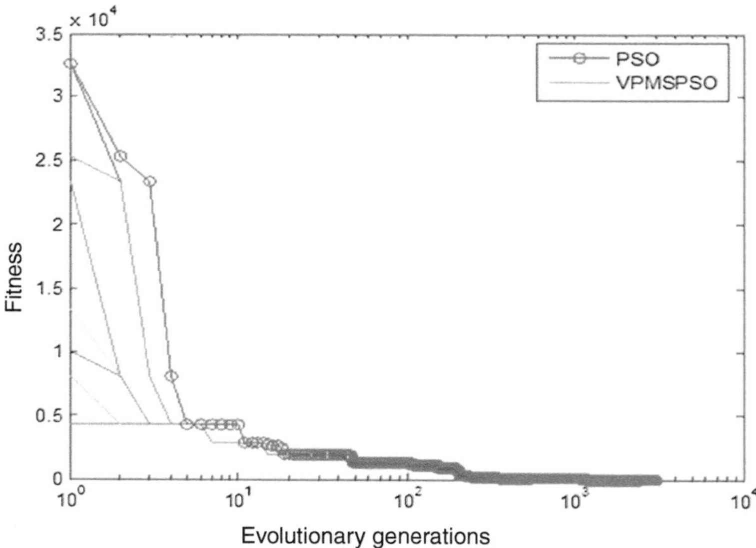

Fig. 1 Convergence *curve* of Sphere function using PSO and MSSPSO

4.2 Analysis of Simulation Results

Figure 1 shows that with evolution process, the sub-swarms, which formed in the beginning of MSSPSO algorithm, gradually merged into one final swarm and converges the optimal value, while the only one swarm, which exists in PSO algorithm also converges to the optimal value from the beginning to the end.

For multimodal function search, Fig. 2 shows that with the number of iterations increases, the sub-swarms of MSSPSO algorithm gradually merged into a finite number of sub-swarms and converges to its corresponding extreme value, while Fig. 3 shows that the swarm of PSO algorithm in the 20 experiments, two of which converges as the curve 1 or as the curve 2, i.e., did not find the optimal value, can only search one extreme value.

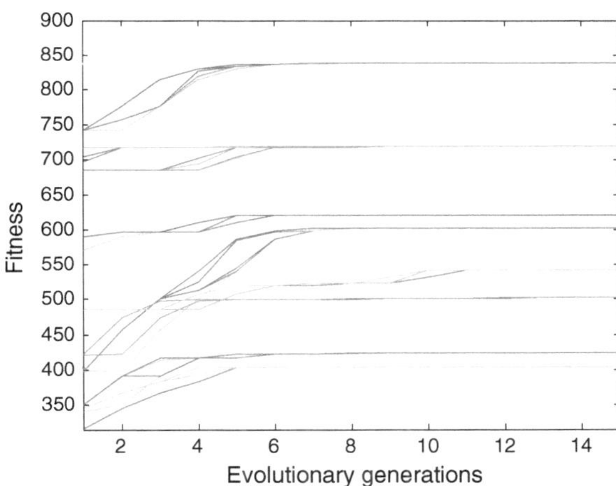

Fig. 2 Convergence *curve* of Schwefel function using MSSPSO

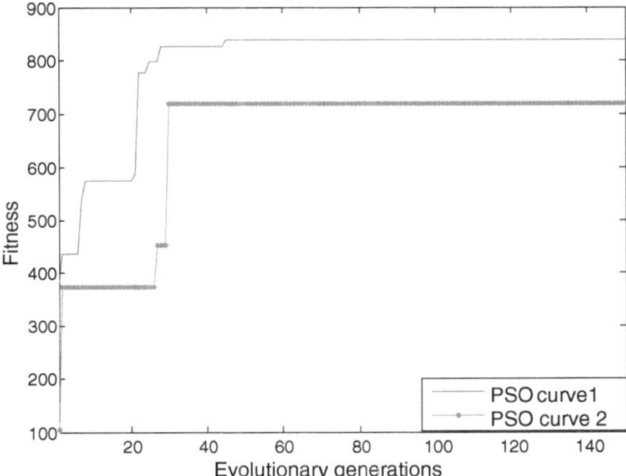

Fig. 3 Two convergence *curves* of Schwefel function using PSO

5 Conclusions

Under the inspiration of individuals' gathering to form biological communities through exchange information between each other, a dynamically multi-sub-swarm PSO algorithm is proposed in this paper. The initial particles themselves are separately one sub-swarm and merge into bigger sub-swarms with the search process, in accordance with the exchange of information between sub-swarms. Evolutionary topology of every sub-swarm is local neighborhood topology, so each sub-swarm has a rich diversity and a strong learning ability. The simulation results show that the MSSPSO algorithm is more effective than the PSO algorithm for multimodal function problems though its convergence rate is not faster than PSO algorithm for a single-peak problem.

References

1. Li, M., Kou, J.: Coordinate multi-population genetic algorithm for multi-modal function optimization. Acta Automatica Sin. **04**, 18–22 (2002). (cnki:ISSN:0254-4156.0.2002-04-004)
2. Barrera, J., Coello, C.: A review of particle swarm optimization methods used for multimodal optimization. In: Lim, C.P., Jain, L.C., Dehuri S. (eds.) Innovations in Swarm Intelligence, vol. 248, pp. 9–37. Springer, Berlin (2009) (doi:10.1007/978-3-642-04225-6_2)
3. Jia, Hongwei: Application of regional two-stage evolutionary algorithm to multimodal function optimization. J. Jimei Univ. Nat. Sci. **03**, 67–69 (2005). (cnki:ISSN:1007-7405.0. 2005-03-006)
4. Guo, Y.: Co-evolutionary algorithm with dynamic population size model, theory and applications. Ph.D. Dissertation, University of Science and Technology of China (2008)
5. Törn, A., Žilinskas, A.: Global optimization. In: Lecture Notes in Computer Science. Springer, Berlin (1989)
6. Himmelblau, D.M.: Applied Nonlinear Programming. McGraw-Hill, New York (1972)
7. Eberhart, R., Shi, Y., Kennedy, J.: Swarm Intelligence. USA: The Morgan Kaufmann Series in Artificial Intelligence (2001)
8. Bonabeau, E., Dorigo, M., Theraulaz, G.: Swarm Intelligence: From Natural to Artificial Systems. Oxford University Press, Inc., Oxford (1999)
9. Goldberg, D.E., Richardson, J.: Genetic algorithms with sharing for multimodal function optimization. In: Proceedings of the Second International Conference on Genetic Algorithms on Genetic Algorithms and their Application, Cambridge, Massachusetts, USA. pp. 41–49 (1987)
10. Beasley, D., Bull, D.R., Martin, R.R.: A sequential niche technique for multimodal function optimization. Evol. Comput. **1**(2), 101–125 (1993)
11. Zhang, G., Yu, L., Shao, Q., Feng, Y.: A Clustering based GA for multimodal optimization in uneven search space. The Sixth World Congress on Intelligent Control and Automation, Dalian, China. pp. 3134–3138 (2006). (doi:10.1109/WCICA.2006.1712944)
12. Yu, X., Wang, Z.: Improved sequential niche genetic algorithm for multimodal optimization. J. Tsinghua Univ. (Sci. Technol.) **03**, 705–709 (2001). (cnki:ISSN:1000-0054.0.2001-03-004)
13. De Castro, L.N., Timmis, J.: An artificial immune network for multimodal function optimization. In: Proceedings of the 2002 Congress on Evolutionary Computation, Honolulu, HI,USA. pp. 699–704 (2002). (doi:10.1109/CEC.2002.1007011)
14. De Castro, L.N., Von Zuben, F.J.: Learning and optimization using the clonal selection principle. IEEE Trans. Evol. Comput. **6**(3), 239–251 (2002)

15. Luo, Y., Li, R., Zhang, W.: Multimodal functions parallel optimization algorithm based on immune mechanism. Acta Simulata Syst Sin. **02**, 164–168 (2005). (cnki:ISSN:1004-731X.0. 2005-02-00I)
16. Xu, X., Zhu, J.: Immune algorithm for multi modal function optimization. J. Zhejiang Univ. (Eng. Sci.) **05**, 18–22 (2004). (cnki: ISSN:1008-973X.0.2004-05-003)
17. Kennedy, J., Eberhart, R.C. Particle swarm optimization. In: Proceedings of IEEE International Conference on Neural Networks, Piscataway (1995)
18. Seo, J.H., Im, C.H., Heo, C.G., Kim, J.K., Jung, H.K., Lee, C.G.: Multimodal function optimization based on particle swarm optimization. IEEE Trans. Magn. **42**(4), 1095–1098 (2006)
19. Ozcan, E., Yilmaz, M.: Particle swarms for multimodal optimization. In: Lecture Notes Computer Science, vol. 4431, pp. 366–375 (2007)
20. Yang, S.Q., Xu, W.B., Sun, J.: A modified niching particle swarm optimization algorithm for multimodal function. Jisuanji Yingyong/J. Comput. Appl. **27**(5), 1191–1193 (2007)

Part VII
Artificial Intelligence

Reliable License Plate Recognition by Cascade Classifier Ensemble

Bailing Zhang, Hao Pan, Yang Li and Longfei Xu

Abstract License Plate Recognition (LPR) is used in various security and traffic applications. This paper introduces a LPR system using morphological operations and edge detection for plate localization and characters segmentation. To emphasize the importance of classification reliability that is essential for reducing the cost caused by incorrect decisions, a cascaded classification system is designed, which consists of two modules, i.e., local mean *k*-nearest neighbor and one-versus-all support vector machine, each with reject option controlled by a properly defined reliability parameter. The impact of using the proposed cascade scheme is evaluated in terms of the trade-off between the rejection rate and classification accuracy. Experimental results confirm the effectiveness of the proposed system.

Keywords License plate recognition · Local mean *k*-nearest neighbor · Classification confidence

1 Introduction

License plate recognition (LPR) is a technology that uses a camera to identify vehicles by their license plates. Usually, a LPR system consists of three steps: license plate detection, character segmentation, and character recognition. Despite the progress of LPR technology, there are still many challenges to be met in order

B. Zhang (✉)
Xi'an Jiaotong-Liverpool University, Dushu Lake Higher Education Town, Suzhou Industrial Park, Suzhou, China
e-mail: bailing.zhang@xjtlu.edu.cn

H. Pan · Y. Li · L. Xu
Department of Computer Science & Software Engineering, Xi'an Jiaotong-Liverpool University, Suzhou, China

S. Patnaik and X. Li (eds.), *Proceedings of International Conference on Computer Science and Information Technology*, Advances in Intelligent Systems and Computing 255, DOI: 10.1007/978-81-322-1759-6_80,

to reach a reliable LPR system adapted to the variability of the environment. The real-world difficulties may result from outdoor imaging conditions such as a complex scene, bad weather conditions, low contrast, blurring, and viewpoint changes.

In this paper, a reliable LPR method will be presented. We emphasize the classification step of the LPR system and propose a reliable recognition algorithm with reject option. Classification with reject option has been a research issue attracting much interests recently. It is often a practical choice when the cost of incorrect classification is high. By refusing to classify the uncertain cases, the excessive misclassifications could be curbed. Such an enhancement of reliability, however, is at the expense either of a manual handling of rejected samples, or of their automatic processing by a more accurate but also computationally more costly classifier. Therefore, a trade-off between the accuracy attainable on non-rejected samples and the amount of rejections is often required [1, 2].

In the framework of the minimum risk theory, Chow [3] defined the optimal classification rule with the reject option in terms of posterior probabilities of different classes. In the applications where the posterior probabilities are unknown or difficult to find, some other scoring indices have to be used instead. On the other hand, methods more complicated than a simple rejection threshold on a single classifier have been explored, with their effectiveness demonstrated on various applications [4]. To achieve a working reliable LPR classification, we designed a cascade classifier system. Cascade classifiers have been extensively investigated for visual object detection [5] and different pattern classification tasks [6–8].

Most state-of-the-art classification schemes are based on discriminative classifiers, with support vector machines (SVMs) as the most commonly used. Another kind of widely used classification algorithms are distance-based like *k*NN. To benefit from the advantages of the distance-based classifier and the SVM methods, we developed a two-stage classifier which combines the two. In the first stage, the fast distance-based classifier, called local mean *k*NN [9], is used. However, before assigning a vector to the nearest class, the decision for this class has to pass a reliability test. If the test is passed, the classification is completed after the first stage and we have basically a more reliable distance-based decision. If it fails, the second stage of the classification is triggered and the vector is reclassified using the more reliable one-versus-all SVM classifier.

2 License Plate Localization and Character Segmentation

The overall license plate localization processes can be illustrated in Fig. 1. As the first step, vehicle detection is implemented by an improved Adaboost algorithm. Then, license plate is located with proper edge extraction algorithm and morphological processing. The rationale of a cascade of boosted classifiers proposed in [10] is that generative and discriminative classifiers are complementary in enhancing detection performance. To improve the vehicle detection performance,

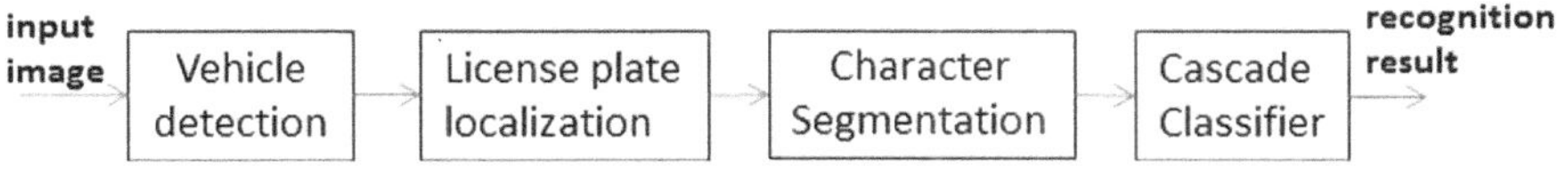

Fig. 1 The overall license plate recognition process system

images will be scaled down from high-resolution images (1,024 × 1, 360) to 128 × 128, 256 × 256 and 512 × 512, respectively. Then, Haar and HOG [11] features of each scaled image will be extracted and passed to the detector. The cascade AdaBoost classifier will generate hypothesis images delineated by rectangular boxes, which indicate potential vehicle objects. These hypothesis images will undergo scaling up operation with proportions of 2, 4, and 8 times. The bounding boxes of these amplified hypothesis images will be mapped back onto the image with size 512 × 512, followed up by a thresholding operation, which disregard the small hypothesis images of size less than the threshold.

The detected vehicle image will be converted into gray level, followed by some morphological operations to reduce irrelevant information such as the environmental pixels around the vehicle. A special morphological operator called top-hat transformation is applied, which emphasize the region of interests (ROI) containing the number plate while suppressing noises around the ROI. With the help of structuring elements, top-hat transformation can stress regions smaller than the specified structuring element while ignoring those which are bigger. The potential number plate regions can then be localized by appropriate edge statistical analysis, mainly in three steps. Pixels are first combined to form lines and lines are merged into regions. Then, an optimal region among the candidates is selected, following certain prescribed criteria.

The character segmentation is simply based on projection. Initially, the binary image of a number plate is projected vertically, the peaks of which indicate the potential locations of the characters. With some priori knowledge about the layout of characters on Chinese license plate, some criteria delineating the minimum width and height of a valid character can be applied to eliminate noises. Figure 2 shows the segmentation result of the number plate from Fig. 3, with each character of size 24 × 12.

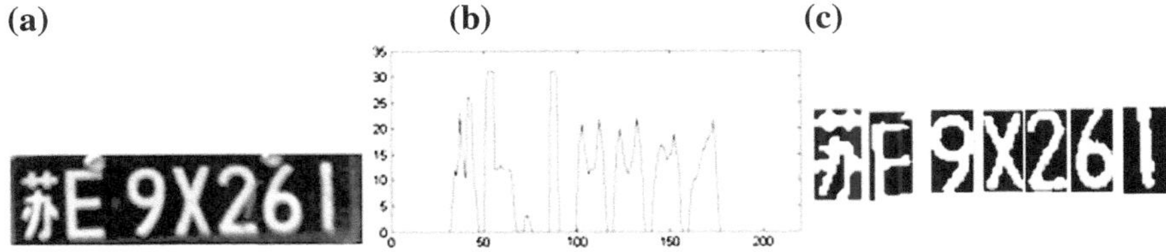

Fig. 2 Character segmentations by vertical projection. (**a**) Localized license plate. (**b**) Vertical projection. (**c**) Segmented characters

Fig. 3 **a** A raw image. **b** Detected vehicle. **c** Processed with top-hat transformation. **d** Localized license plate

3 Reliable License Plate Recognition

Two situations may contribute to low classification reliability. The first situation, denoted as type *a*, is the presence of outliers. The second situation, denoted as type *b*, corresponds to the classification errors due to the test samples near the decision boundaries thus class scores of two (or more) are nearly equal. To distinguish between these two unreliable classifications, we follow the train of thought in [1, 2] to define two reliability parameters, Ψ_a and Ψ_b, which quantify the reliability corresponding to the above two situations. By combination of Ψ_a and Ψ_b, an inclusive parameter Ψ can be defined as the comprehensive reliability measure. A conservative choice is $\Psi = \min\{\Psi_a, \Psi_b\}$ [2].

3.1 Reliability Parameter for Local Mean-Based Nonparametric Classifier

The k-NN method is a simple approach, which generally achieves good results when the available number of prototypes is large. Mitani and Hamamoto have proposed a local mean-based nonparametric classifier [9], which not only overcomes the influence of the outliers but also can achieve better performance especially in small training size situations.

Given a sample set $X^i = \{x^i_j | j = 1, \ldots, N_i\}$ from class ω_i, where N_i is the number of training samples from class ω_i. The detailed steps of local mean-based classifier can be referred to [9]. Denote O_i as the minimum distance between x and the set of the local means from different classes. The smallest distance O_{win} is

$$O_{\text{win}} = \min_i O_i = \min_i d(c_i, x) \tag{1}$$

where c_i is the local mean of *i*th class. The reliability probability Ψ_a can be defined as

$$\Psi_a = 1 - \frac{O_{\text{win}}}{O_{\max}} \tag{2}$$

the value of O_{max} is the highest among the values of O_{win} obtained for samples taken from a set disjoint from both the training set and test set.

The reliability parameter Ψ_b must be a function of both O_{win} and O_{2win} where O_{2win} is the distance between x and the local mean with the second smallest distance:

$$\Psi_b = 1 - \frac{O_{win}}{O_{2win}} \tag{3}$$

The classification reliability for the local mean kNN classifier is given by

$$\Psi = \min\{\Psi_a, \Psi_b\} \tag{4}$$

3.2 Reliability Parameter for One-Versus-All Support Vector Machine

SVMs were originally proposed for binary classification. For multi-class classification problems, the one-versus-all (OVA) [12] is the earliest implementation. It is expected that if, during training, a sample belonging to the k-th class is presented to all of the SVMs, the kth SVM will assume a value equal to 1, while all the other outputs will assume a value equal to -1. An effective definition of the reliability parameter Ψ_a can be defined as

$$\Psi_a = O_{win} \tag{5}$$

where O_{win} is the output of the winner SVM. In this way, the nearer to 0 the value of O_{win} the less reliable the classification is considered.

Samples of type b typically generate output vectors with two or more elements having similar values. Thus, for a given O_{win}, the higher the difference between O_{win} and O_{2win} (the output of the SVM having the highest value after the winner), the safer the decision of attributing the sample to the winning class. Consequently, a suitable parameter for evaluating the reliability of these situations can be

$$\Psi_b = O_{win} - O_{2win} \tag{6}$$

In conclusion, the classification reliability of the OVA classifier can be measured by

$$\begin{aligned} \Psi &= \min\{\Psi_a, \Psi_b\} = \min\{O_{win}, O_{win} - O_{2win}\} \\ &= O_{win} - O_{2win} = \Psi_b \end{aligned} \tag{7}$$

3.3 A Two-Stage Classification System for Reliable License Plate Recognition

To improve the system reliability, we proposed a two-stage classifier ensemble. At all stages, a pattern can be either classified or rejected. The decision of a classifier is rejected if its reliability parameter falls below a pre-defined threshold. Rejected patterns are fed into the next stage. Usually, the first stage represents coarser decision. The second stage need only operate on the surviving inputs from the previous stage.

4 Experiment

A set of 3,734 segmented characters was created with the technology introduced in Sect. 2. In the first experiment, the classification accuracy and the rejection rates for the local mean *k*NN and OVA SVM were independently studied, by varying

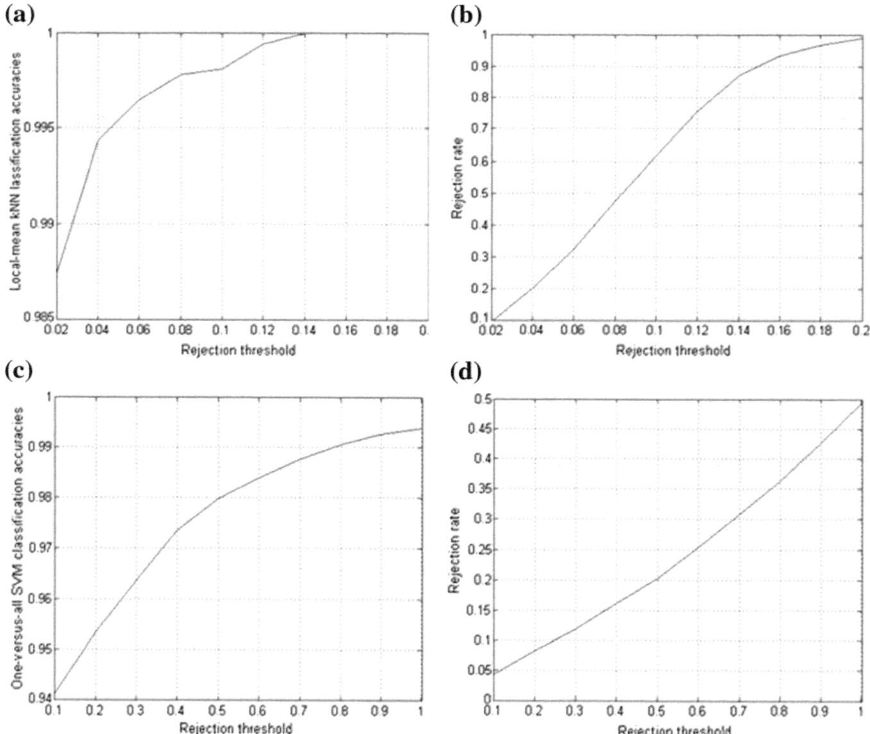

Fig. 4 Local mean kNN (*top*): **a** Accuracies versus reliability threshold. **b** Rejection rates versus reliability threshold. OVA SVM (*bottom*): **c** Accuracies versus reliability threshold. **d** Rejection rates versus reliability threshold

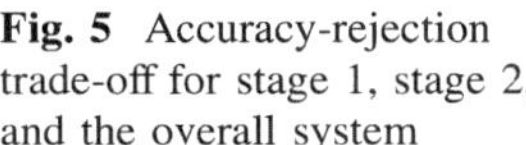

Fig. 5 Accuracy-rejection trade-off for stage 1, stage 2, and the overall system

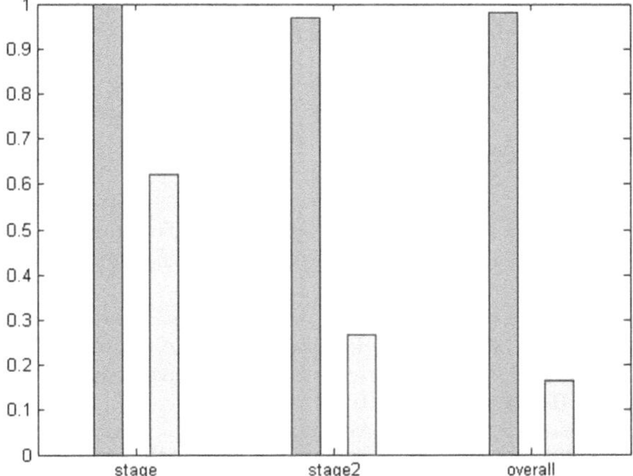

the corresponding thresholds of reliability parameters. For the local mean *k*NN, the threshold value η_T was changed from 0.02 to 0.2. For each η_T, the characters were randomly split into training and testing sets, with 80 % of the samples for training and 20 % for the test set. The accuracies and rejection rates were recorded and then averaged from the 100 runs. To simplify our discussion, the number of neighbors was fixed as 10, and variation of this parameter does not bring obvious performance change. The dependence of the accuracies and rejection rates on the reliability thresholds are illustrated in Fig. 4. At threshold of 0.1, the trade-off between accuracy and rejection rate is 99.9 and 80 %. This means that the 20 % of testing samples could be accepted for classification with high confidence. The less confident 80 % samples will be passed to the next stage.

The above experiment was similarly conducted for the OVA SVM. In Fig. 4c and d, the dependence of classification accuracy and the rejection rate on the reliability threshold value η_T were illustrated. At the threshold of 0.5, the accuracy reaches 98 % while the rejection rate equals to 20 %. For the overall cascaded classifier, the accuracy-rejection trade-off is given in Fig. 5, which confirms that a high classification accuracy is attained at a cost of much reduced rejection rate.

5 Conclusion

In this paper, a two-stage classifier system was proposed with rejection strategies for LPR. Rather than simply pursuing classification accuracy, we emphasized reject option in order to minimize the cost of misclassifications while secure high classification reliability. Reliability parameters of local mean *k*-NN and one-versus-all SVM have been proposed, and their use in reliable classifier design has been discussed. The proposed classification system can offer a highly reliable recognition performance for the segmented characters from license plate.

Acknowledgments The project is supported by Suzhou Municipal Science and Technology Foundation Grants SS201109 and SYG201140.

References

1. Cordella, L.P., Foggia, P., Sansone, C., Tortorella, F., Vento, M.: Reliability parameters to improve combination strategies in multi-expert systems. Pattern Anal. Appl. **2**(3), 205–214 (1999)
2. Sansone, C., Tortorella, F., Vento, M.: A Classification reliability driven reject rule for multi-expert systems. Int. J. Pattern Recognit. Artif. Intell. **15**(6), 885–904 (2001)
3. Chow, C.K.: On optimum recognition error and reject tradeoff. IEEE Trans. Inf. Theory. **IT-16**(1), 41–46 (1970)
4. Zhang, B.: Reliable classification of vehicle types based on cascade classifier ensembles. IEEE Trans. Intell. Transp. Syst. **14**(1), 322–332 (2013)
5. Viola, P., Jones, M.: Robust real-time face detection. Int. J. Comput. Vision **57**(2), 137–154 (2004)
6. Li, M., Sethi, I.K.: Confidence-based classifier design. Pattern Recognit. **39**, 1230–1240 (2006)
7. Pudil, P., Novovicova, J., Blaha, S., and Kittler, J.: Multistage pattern recognition with rejection option. In: Proceedings of 11th International Conference Pattern Recognition, Vol. B, pp. 92–95, (1992)
8. Kaynak, C., Alpaydin, E.: Multistage cascading of multiple classifiers: one mans noise is another mans data. In: Proceedings of 17th International Conference Machine Learning, pp. 455–462 (2000)
9. Mitania, Y., Hamamoto, Y.: A local mean-based nonparametric classifier. Pattern Recogn. Lett. **27**(10), 1151–1159 (2006)
10. Negri, P., Clady, X., Hanif, S., Prevost, L.: A cascade of boosted generative and discriminative classifiers for vehicle detection. EURASIP J. Adv. Sig. Process. **2008**, Article ID 782432 doi:10.1155/2008/782432 (2008)
11. Dalal, N., Triggs, B.: Histograms of oriented gradients for human detection. In: IEEE Computer Society Conference on Computer Vision and Pattern Recognition (CVPR 2005), pp. 886–893, San Diego, CA (2005)
12. Vapnik, V.: Statistical learning theory. Wiley, New York (1998)

Risk Prediction of Heart Disease Based on Swarm Optimized Neural Network

R. Chitra and V. Seenivasagam

Abstract Heart disease (HD) remains the biggest cause of deaths worldwide. This shows the importance of HD prediction at the early stage. In this paper, multi-layer feedforward neural network (MLFFNN) optimized with particle swarm optimization (PSO) is adopted for HD prediction at the early stage using the patientâ€™s medical record. The network parameters considered for optimization are the number of hidden neurons, momentum factor, and learning rate. The efficiency of the PSO optimized neural network (PSONN) is calculated using the records collected from standard Cleveland database and Real time clinical dataset. The results show the proposed system can predict the likelihood of HD patients in a more efficient and accurate way.

Keywords Particle swarm optimization · Heart disease prediction · Momentum factor · PSO optimized neural network

1 Introduction

Cardiovascular diseases remain the biggest cause of deaths worldwide over the last two decades. The diagnosis of heart disease (HD) in most cases depends on a complex combination of clinical and pathological data. An intelligent HD prediction system built with the aid of data mining technique like decision trees, naÃve

R. Chitra (✉)
Department of Computer Science and Engineering, Noorul Islam Centre for Higher Education, Kanyakumari, India
e-mail: jesi_chit@yahoo.co.in

V. Seenivasagam
Department of Computer Science and Engineering, National Engineering College, Kovilpatti, India
e-mail: yespee1094@yahoo.com

S. Patnaik and X. Li (eds.), *Proceedings of International Conference on Computer Science and Information Technology*, Advances in Intelligent Systems and Computing 255, DOI: 10.1007/978-81-322-1759-6_81,

Bayes, and artificial neural network was proposed by Palaniappan and Awang [1]. The result illustrated the peculiar strength of each of the methodologies in comprehending the objectives of the specified mining objectives. Srinivas et al. [2] applied data mining techniques to predict heart attack. Based on the calculated significant weightage, the frequent patterns having value greater than a predefined threshold were chosen for the valuable prediction of heart attack.

Yao and Liu [3] proposed evolutionary feedforward neural network with neuro-genetic approach for optimization of a trading agent. Carlos Ordonez used association rules to improve HD prediction. Association rules were applied on a real dataset, contacting medical records of patient with HD, and the risk factors were identified. Sulaiman et al. [4] proposed hybrid multi-layer feedforward neural network (MLFFNN) technique for predicting the output from a grid-connected photovoltaic system. The classification of biomedical dataset was done with hybrid genetic optimized back-propagation network by Kuok et al. [5] and Karegowda et al. [6] also proposed genetic optimized back-propagation network for classification. According to Song and Gu [7], PSO algorithm has been promised to solve many optimization problems. Lee et al. [8] have used PSO and GA for excess return evaluation in stock market. Based on their experiment, it is proven that PSO algorithm is better compared to GA. In this paper, an intelligent HD prediction system is proposed using an optimized feed forward neural network. In order to improve the performance of the feed forward neural network, its parameters are tuned using PSO algorithm, and it is discussed in the rest of the paper.

2 Materials and Methodology

This section describes the dataset and proposed methodology for HD prediction. The inputs are normalized; minâ€“max normalization techniques are used for normalization, and the normalized attributes are given as the input to the PSONN.

2.1 Heart Disease Dataset

HD dataset is obtained from UCI (University of California, Irvine CA) centre for machine learning and intelligent systems. The data collected from 270 patients are used for proposed work, and it contains 13 risk factors. HD risk factors never occur in isolation, but they are correlated with each other. The digitized data have 150 normal and 120 abnormal cases. In the selected dataset, class 0 specifies the absence of heart attack and class 1 specifies the presence of HD. Normally, direct support clinical decision-making is the intention behind the design of a clinical decision support system, and it presents patient-specific assessments or recommendations produced using the characteristics of individual patients to clinicians for consideration. The 13 attributes considered are age, gender, blood pressure, cholesterol, fasting blood sugar, chest pain type, maximum heart rate, ST segment slope in ECG, number of major vessels colored in angiogram, and thallium test value.

2.2 Multi-Layer Feed Forward Neural Network

Artificial neural networks (ANN) can be used to tackle the said problem of prediction in medical dataset involving multiple inputs. A MLFFNN is a three-layer network with input unit $x_i = \{x_1, x_2, \ldots, x_n\}$, hidden layer $h_j = \{h_1, h_2, \ldots, h_n\}$, and output layer $y_k = \{y_1, y_2, \ldots, y_n\}$. A MLFFNN consists of a layer of input units, one or more layers of hidden units, and one output layer of units. Each connection between nodes has a weight associated with it. In addition, there is a special weight (w) given as $w_{ij} = \{w_1, w_2, \ldots, w_n\}$, where n denotes the number of neurons in the hidden layer, which feeds into every node at the hidden layer and a special weight (z) given as $z_{jk} = \{z_1, z_2, \ldots, z_n\}$ that feeds into every node at the output layer. These weights are called the bias and set the thresholding values for the nodes. Each hidden node calculates the weighted sum of its inputs and applies a thresholding function to determine the output of the hidden node. The thresholding function applied at the hidden node is a sigmoid activation function. The back-propagation (BP) algorithm is a commonly used learning algorithm for training ANN. The network is first initialized by setting up all its weights to be small random numbers between [0, +1]. The 13 inputs are applied, and the output is calculated. The actual output (t) obtained is compared with the target output (y), and the error E is calculated. This error is then used mathematically to change the weights in such a way that the error will get minimized. The process is repeated again and again until the error is minimal. In the proposed work, mean square error function defined in Eq. (1) is used for training.

$$E = \frac{1}{n}\sum_{i=1}^{n} (y_i - t_i)^2 \tag{1}$$

The weights are updated using Eq. (2)

$$\Delta w_{ij}(t+1) = -\eta \frac{\partial E}{\partial w_{ij}} + \alpha \Delta w_{ij}(t) \tag{2}$$

In Eq. (2), Î± is the momentum factor and Î· is the learning rate.

The momentum factor Î± allows for momentum in weight adjustment. The magnitude of the persistence is controlled by the momentum factor. Momentum in the BP algorithm can be helpful in speeding the convergence and avoiding local minima. Learning rate Î· is a training parameter that controls the size of weight and bias changes when learning the selection of a learning rate is of critical importance in finding the true global minimum of the error. Increasing number of hidden

Table 1 Optimizing parameters selected in FFNN

Parameter	Range
Number of hidden neurons in hidden layer	6â€“25
Momentum factor (Î±)	0.9â€“0.931
Learning rate (Î·)	0.01â€“0.114

neurons increases processing power and system flexibility. But the complexity and cost of the system also increase depending on the number of hidden neurons. Range of optimizing parameters used is given in Table 1.

2.3 Particle Swarm Optimization

Particle swarm optimization (PSO) is a population-based stochastic optimization technique developed by Eberhart and Kennedy [9]. PSO does not require gradient information of the objective function under consideration. PSO can reach the global optimum value with less iteration. In PSO, each particle in the population has a velocity $v_i(t)$, which enables it to fly through the problem space. Therefore, each particle is represented by a position $x_i(t)$ and a velocity vector. Dimensions of position and velocity vectors are defined by the number of decision variables in the optimization problem. Modification of the position of a particle is performed by using its previous position information and its current velocity.

$$v_i(t+1) = wv_i(t) + c_1\,\text{rand}_1 + c_2\,\text{rand}_2(\text{Gbest}_i - x_i(t)) \tag{3}$$

$$x_i(t+1) = x_i(t) + v_i(t+1) \tag{4}$$

where $v_i(t)$ is velocity of particle i at iteration t, $x_i(t)$ is current position of the particle i at iteration t, Pbest_i is personal best of particle i, Gbest_i is best position in the neighborhood, rand is random number between 0 and 1, w is weighting function, c_1 is cognition learning rate, and c_2 is social learning rate

The PSONN was mainly developed to search for the optimal training parameters, i.e., the number of neurons in the hidden layer, the learning rate, the momentum rate, the transfer function in the hidden layer, and the learning algorithm. These training parameters are also known as the decision variables for the optimization task. The objective function for the optimization process was to minimize the MSE during training. PSO is chosen and applied in feed forward neural network to enhance the learning process in terms of convergence rate and classification accuracy.

3 Heart Disease Prediction System Using PSONN

The input data to PSONN need to be normalized as NN only work with data represented by numbers in the range between 0.001 and 0.999. Three-layer PSONN was used in this study for model calibration. The number of input neurons depends on the input attributes, and there is only one neuron in the output layer. In the proposed work, 13 input neurons and one output neuron for prediction of HD are used. The optimal training parameters considered are the number of neurons in the hidden layer, momentum rate and the learning rate, the transfer function in the hidden layer, and the learning algorithm.

The PSONN algorithm for intelligent heart attack prediction is given below

```
For each particle
    Initialize particle for NN problem
End
Do
    For each particle
        Calculate fitness value (feedforward error or MSE
in If the fitness value is better than the best fitness value
(pBest) in history
 Then set current value as the new pBestEnd
Choose the particle with the best fitness value of all the
particles as the gBest
For each particle
        Calculate particle velocity
        Update particle position
End
```

The PSO parameters and c_1, c_2 are the acceleration constants which are initially set to 2, and r_1 and r_2 are random integers set to a range between [0, 1]. The initial population is randomly selected, and the population size is set to 25. The total number of iterations is chosen as 200. With these parameters, the number of hidden layer neurons, momentum rate and the learning rate, the transfer function in the hidden layer, and the learning algorithm are optimized. The whole process is repeated until the generation reaches 200 or the minimum fitness is achieved.

4 Experimental Results

In this work, we analyze the use of the PSO algorithm and the cooperative variant with the weight decay mechanism for neural network training, aiming better generalization performances in HD prediction. For evaluating these algorithms, we apply them to benchmark classification problems of the medical field. The training set and testing set accuracy are calculated and analyzed for the standard HD dataset. In the standard dataset, 270 data are used for training and the network is also tested with same data. The performance metrics accuracy, sensitivity, and specificity of training and testing cases of both dataset are given in Table 2. The performance metrics obtained in the proposed system for standard dataset are compared with SVM classifier and feed forward neural network classifier and are shown in Fig. 1.

ROC curve is a graphical plot created by plotting false positive rate versus true negative rate, and it is used to analyze the performance of the classifier. The ROC curve for SVM, feed forward neural network, and PSO optimized neural network (PSONN) is shown in Fig. 2. From Fig. 1, it is observed that the PSONN has significantly good performance compared with all other classifiers.

The ROC curve depends on the true positive rate and false positive rate. From this curve, it is shown that the performance of the PSONN-based prediction system has shown a significant improvement on classifying the diseased and non-diseased patterns. When analyzing these graphs, the PSONN shows a better performance in terms of accuracy, sensitivity, and specificity. A true positive rate of 0.889 is obtained with a false positive rate of 0.0987 which is achieved by the proposed classifier. The sensitivity and specificity achieved by the PSONN are 88.98 and 90.13 for standard dataset with an overall accuracy of 89.6, which is higher than the other two well-known classifiers. This increase is because of the fact that the neural network structure is optimized with the PSO algorithm. The BP algorithm primarily depends on the initial parameter setting, and the performance of the classifier depends on these parameters. From Fig. 1, it is noticed that FFNN has an improved performance than the SVM; this is due to the fact that SVM does not depend on the

Table 2 Training and testing performance metrics

Performance metric	PSONN	
	Training set	Testing set
Accuracy	88.4	90.8
Sensitivity	87.54	90.42
Specificity	89.62	90.64

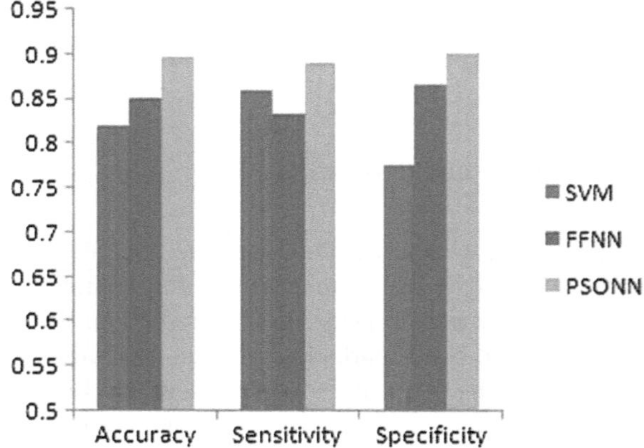

Fig. 1 Performance analysis for Cleveland dataset

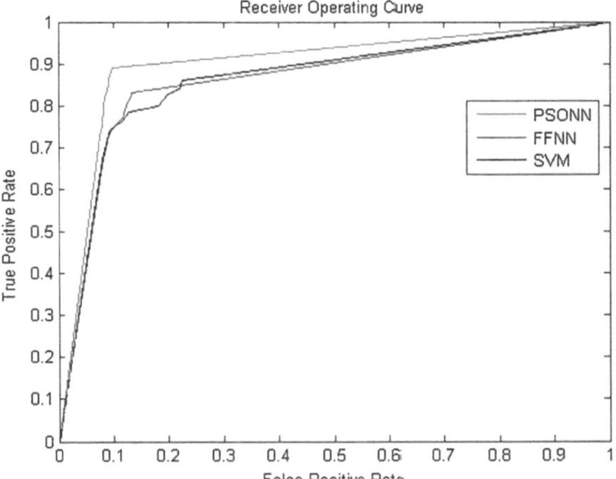

Fig. 2 ROC plot for different classifiers

generalization error. Although FFNN produces a good classification accuracy for prediction of HDs, this classifier selects the initial parameter setting manually, which will take a considerable time to classify the patterns. Hence, FFNN and SVM will not be suitable for the physicians to analyze and predict the diseased and non-diseased patterns. The results of the experiments show that an improvement of 4 % accuracy has been achieved by applying optimized neural network classifiers. This implies that the optimized neural network is one of the desirable classifiers, which can be used as an aid for the physicians to predict the HDs.

5 Conclusion

In this paper, we have proposed an adaptive intelligent mechanism for HD prediction using the profiles collected from the patients. PSONN adopted global and local optimization of the network parameters within the specified range. By exploiting this distinct feature of the PSONN, a computerized prediction algorithm is developed that is not only accurate but is also computationally efficient for heart attack prediction. With the proper optimization using PSO, the method can thus evolve an optimum number of hidden units within an architecture space. The results of proposed system have achieved better accuracy than most of the existing algorithms.

References

1. Palaniappan, S., Awang, R.: Intelligent heart disease prediction system using data mining techniques. In: IEEE, pp. 108–115 (2008)
2. Srinivas, K., Rani, B.K., Govrdhan, A.: Applications of data mining techniques in healthcare and prediction of heart attacks. Int. J. Comput. Sci. Eng. **2**(2), 250–255 (2010)

3. Yao, X., Liu, Y.: Towards designing artificial neural networks by evolution. Appl. Math. Comput. **91**, 83–90 (1998)
4. Sulaiman, S.I., Rahman, T.K.A., Musirin, I.: A genetic algorithm-based hybrid multi-layer feedforward NeuNetwork for predicting grid-connected photovoltaic system output. In: IACSIT Hong Kong Conferences, IPCSIT, vol. 25 (2012)
5. Kuok, K.K., Harun, S., Shamsuddin, S.M.: Particle swarm optimization feedforward neural network for hourly rainfall runoff modeling in Bedup Basin, Malaysia. Int. J. Civ. Environ. Eng. **9**(10) 2010
6. Karegowda, A.G., Manjunath, A.S., Jayaram, M.A.: Application of genetic algorithm optimized neural network connection weights for medical Diagnosis of pima Indians diabetes. Int. J. Soft Comput. **2**(2), 15–23 (2011)
7. Song, M.P., Gu, G.C.: Research on particle swarm optimization: a review. In: IEEE, pp. 2236–2241 (2004)
8. Lee, J.S., Lee, S., Chang, S. Ahn, B.H.: A Comparison of GA and PSO for Excess Return Evaluation in Stock Markets, pp. 221–230. Springer, Berlin (2005)
9. Kennedy, J., Eberhart, R.C.: Particle swarm optimization. In: Proceedings of IEEE International Conference on Neural Networks, vol. 4, pp. 1942–1948 (1995)

A Comparative Study on Speaker Gender Identification Using MFCC and Statistical Learning Methods

Hanguang Xiao

Abstract In this study, we built databases for mandarin speeches under quiet and noisy environments, respectively. After using mel-frequency cepstrum coefficient (MFCC) to extract feature vectors for the speech records, we performed speaker gender identification using three statistical learning methods: K-nearest neighbor (KNN), probabilistic neural network (PNN), and support vector machine (SVM) and analyzed the influences of frame size, normalization, and noise on the identification result. The experiment showed that (1) the best appropriate frame size is 2,048; (2) feature normalization increased the whole accuracy about 3 %; (3) the accuracies of SVM are highest than those of KNN and PNN, which reached 100, 97.8 and 95.8 % accuracies in the quiet, noise, and hybrid database.

Keywords Gender identification · Feature extraction · Mel-frequency cepstrum coefficients · Support vector machine

1 Introduction

Speaker gender identification is to identify the gender of the speaker through speech signals of the speaker; it is an important branch of speech recognition studies. Gender identification is a pre-classification technique for speaker recognition, it could reduce the complexity of the speech recognition studies and improve the accuracy of speech recognition systems [1, 2]. In a human–computer interaction system, gender identification could help the system to decide what kind of voice response to use according to the gender of the speaker, so that the system could be more friendly and humanized. Computers could perform automatic

H. Xiao (✉)
School of Optoelectronic Information, Chongqing University of Technology, Chongqing, China
e-mail: simenxiao1211@163.com

S. Patnaik and X. Li (eds.), *Proceedings of International Conference on Computer Science and Information Technology*, Advances in Intelligent Systems and Computing 255, DOI: 10.1007/978-81-322-1759-6_82,

gender identification to short speech signals (for example, signals of only scores of milliseconds) [3–5].

Common approach for gender identification is to analyze the pitch of the speech and use the pitch determination standards for gender identification, and it works well [4]. However, when it comes to large amount of speakers, the gender identification with only single feature and criterion has low accuracy. Mel-frequency cepstrum coefficients (MFCC) could reflect the differences in time domain and frequency domain spaces, also performs stably in noisy and spectral distortion situations, it has been widely applied to speaker recognition area [6–8].

In speech recognition, framing and windowing technique is commonly used for the extraction of feature vectors. For the determination of frame size, it is more reasonable to decide manually first according to the short-term stability of the speech, and then determine the frame size by the self-feedback of the recognition [6, 9–12]. Therefore, it is necessary to study the influence of frame size on recognition accuracy. In addition, the feature vectors extracted by MFCC is not normalized, and most studies would use the MFCC feature vectors directly for classifying training and recognition; although those studies received positive results, normalization is a processing technique worth discussion from the perspective of improving the recognition results [13].

Currently, there are a few commonly used statistical learning methods, including K-nearest neighbor (KNN) [14], probabilistic neural network (PNN) [15], and support vector machine (SVM) [16]. KNN and PNN are two mature, simple, and effective classification methods. SVM is a supervised statistical learning method proposed by Vapnik and his coworkers based on structural risk minimization (SRM) [16]; it is considered the model for statistics and learning with small sample. Since SVM could find the global optimal solution without determining the class condition probability density and prior probability of each class, and has great generalization ability, it has been applied to various areas. These areas include text classification, handwritten numeral recognition, air quality forecasting, tumor and cancer diagnosis, microarray gene expression data analysis, drug design, protein–protein interaction prediction, and protein structure and function prediction [16, 17].

In this paper, we propose to use MFCC to extract the gender feature vectors of the speaker, and use several statistical learning methods for training, recognition, and comparison; the influences of frame size, normalization, and noise on the classification result are also discussed.

2 Classification Principle of Support Vector Machine

Support vector machine was proposed by Vapnik and coworkers [16, 18] based on the SRM principle from statistical learning theory and has been widely used in real world for classification and regression. In linearly separable cases, the training of SVM is to construct a hyperplane H

$$W \cdot P + b = 0 \tag{1}$$

that separates two different classes of feature vectors with a maximum margin. The hyperplane is constructed by finding another vector W and a parameter b that minimizes $\|W\|$ and satisfies the following conditions:

$$W \cdot P_i + b \geq 0, \quad (\text{positive}, y = +1) \tag{2}$$

$$W \cdot P_i + b < 0, \quad (\text{negative}, y = -1) \tag{3}$$

where P_i represents each vector of training set and y is the class label of the vector P_i, and $\|W\|$ is the Euclidean norm of W. After the determination of W and b, a given vector P_j can be classified by:

$$\text{sign}\,(W \cdot P_i + b). \tag{4}$$

In nonlinearly separable cases, SVM maps the input vectors into a high-dimensional feature space using a kernel function $K(P_i, P_j)$. Thus, the nonlinearly separable problem is transformed to be a linearly separable problem in this high-dimensional space. In this case, the decision function of classification is changed into:

$$\text{sign}\left(\sum_{i=1}^{l} \alpha_i^0 y_i K(P, P_i) + b\right) \tag{5}$$

where l is the total number of all samples in training set. The coefficients α_i^0 and b are determined by maximizing the following Lagrangian expression:

$$\sum_{i=1}^{l} \alpha_i - \frac{1}{2} \sum_{i=1}^{l} \sum_{j=1}^{l} \alpha_i \alpha_j y_i y_j K(P_i, P_j) \tag{6}$$

under conditions:

$$\alpha_i \geq 0 \text{ and } \sum_{i=1}^{l} \alpha_i y_i = 0. \tag{7}$$

The Gaussian kernel has been used widely in different classification studies. The expression of the Gaussian kernel function is:

$$K(P_i, P_j) = \exp\left(\frac{-\|P_i - P_j\|^2}{2\sigma^2}\right). \tag{8}$$

According to the value of (4) or (5), one can classify the query vector P_j into the positive or negative group. If the value is +1, it belongs to the positive group, otherwise the negative group.

3 Experiment and Analysis

3.1 Database and Preprocessing of Speech Signals

In this study, we recorded the speeches of 43 male speakers and 35 female speakers in quiet offices with professional digital voice recorders. The speakers were asked to talk about a topic in ordinary speed and tone, and the content is not restricted. Each speaker had a recording time of 30 s, and all speech data were collected with the sampling frequency of 32 kHz. After the recording, 78 audio files were obtained. To reduce the redundancy of the speech data, the files were resampled with the frequency of 11,025 Hz, i.e., down sampled. The resampled audio files were saved as wave files in database Mic.

To study the influence of noisy background on identification result, a database of speeches in noisy environments were also set-up. For this database, the speech data were collected without live recording. We searched mandarin speech files with all types of background noises on the Internet, and selected 91 and 44 representative male and female speakers from the searching result. While playing these speech files, we recorded the speech data with professional digital recorders and then processed using the same sampling and timing settings with the database Mic, finally the database Web was obtained.

The nonspeech data need to be removed from the recorded speeches. The common method for this is average energy and average zero-crossing rate checking. Average energy and average zero-crossing rate are the most basic time domain features, which is used in this paper.

After preprocessing, a speech is framed with overlap as Fig. 1. In order to avoid Gibbs' effect while performing fast Fourier transform (FFT), smoothing filtering is needed for [1, 2, …, *N*]. Hamming window is a commonly used smoothing filter, its expressions are as Eqs. (9) and (10). The window size is 512, 1,024, 2,046, and 4,096, respectively, for discussing its effect on the accuracy of SVM.

$$W_i = 0.54 - 0.46\cos\left(2\pi\frac{i}{L}\right), \quad i = 0, 1, \ldots, L-1 \tag{9}$$

$$\mathbf{X}_{ji} = \mathbf{X}_{ji} W_i, \quad j = 1, 2, \ldots, N;\ i = 0, 1, \ldots, L-1. \tag{10}$$

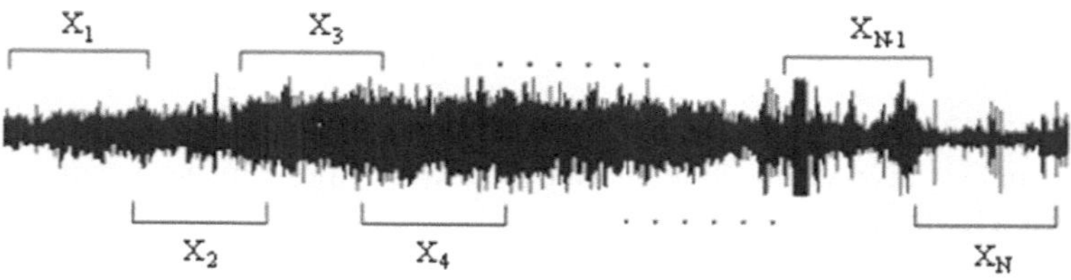

Fig. 1 Cut out short-term time series into frames [$\mathbf{X}_1$, $\mathbf{X}_2$, …, $\mathbf{X}_N$]

3.2 Feature Extraction of MFCC

Let $\boldsymbol{x}_j$ indicate one frame of a pretreated speech signal, perform FFT to $\boldsymbol{x}_j$, obtain cepstrum coefficients, and find the spectrum coefficient $\boldsymbol{f}_j$, then use triangle filter set scaled by Mel-frequency cepstrum features to perform filtering.

The triangle filter set is as shown in Fig. 2. Each triangle can be used as a band-pass filter with a center frequency and an upper and down cutoff frequency. Center frequency is the perception center of human ears in a frequency range, and upper and down frequency is the perception range of human ears in this frequency range. The number of filters is even at the lower frequency range; with the increase of frequency, the number of filters decreases exponentially. The shape of the filters could be triangle, Hamming, or Hanning. In this study, we chose the commonly used triangle filter and decided the number of filters as 24.

Perform additive filtering to the spectrum coefficient $\boldsymbol{f}_j$. In the same triangle, multiply the spectrum and corresponding amplitude of the triangle and then summate to obtain 24 coefficients m_j. Calculate the logarithm of these 24 coefficients and perform DCT transform to get the MFCCs as below:

$$C_i = \sqrt{\frac{2}{k}} \sum_{j=1}^{k} \ln(m_j) \cos|\pi \times i/k(j - 0.5)| \tag{11}$$

where k is the number of the triangle filters, m_j is the output of the jth filter, and C_i is the ith $(i = 1, 2, 3, \ldots, k)$ component of MFCC. Then, the MFCC features of training and testing set were normalized into −1 to +1.

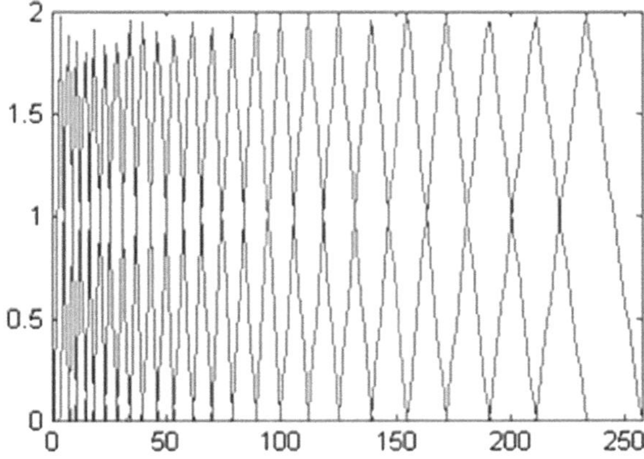

Fig. 2 The bank of triangle filters

3.3 Training and Recognition

After the preprocessing of databases Mic, Web, and Mic-Web, extract the feature vectors of the first 15 s of the speech files using MFCC feature extraction method to compose the training vector set. Extract the feature vectors of the second 15 s of the speech files to compose the test vector set and perform normalization. Set the number of nearest neighbors as 3 for KNN method. Identify the optimal value of gamma parameter g for PNN using grid-searching method. For SVM, choose radial basis function as the core function, set C as 10,000, and optimize g through grid searching.

Set the feature vectors for male and female speakers as positive samples and negative samples, respectively. During the test, input all test samples of a speaker into KNN, PNN, and SVM obtain the positive/negative status of the samples and compare with the actual positive/negative status. If the number of samples correctly identified accounts for over 50 % of the total number of the samples, the gender identification for the speaker is correct. Use the same approach for gender identification of the test sample set of other speakers, calculate the number of speakers been correctly recognized, and obtain the overall accuracy.

Let true positive (TP) indicate the number of correctly recognized male speakers, false negative (FN) indicate the number of male speakers incorrectly recognized as female speakers, true negative (TN) indicate the number of correctly recognized female speakers, and false positive (FP) indicate the number of female speakers incorrectly recognized as male speakers, then the identification accuracy could be calculated by the following formula:

$$Q_p = \mathrm{TP}/(\mathrm{TP} + \mathrm{FN}) \tag{12}$$

$$Q_n = \mathrm{TN}/(\mathrm{TN} + \mathrm{FP}) \tag{13}$$

$$Q = (\mathrm{TP} + \mathrm{TN})/(\mathrm{TP} + \mathrm{FN} + \mathrm{TN} + \mathrm{FP}). \tag{14}$$

Tables 1 and 2 present the identification accuracies of KNN, PNN, and SVM for test sample sets of Mic, Web, and Mic-Web before and after normalization, respectively, when the frame size is 512. Table 3 presents the identification accuracies of KNN, PNN, and SVM for test sample sets of Mic, Web, and Mic-Web after normalization with different frame sizes. It can be seen in Table 1 that the identification accuracies for database Mic are much higher than those for database Web. There are two reasons for such situation: noises would influence the identification, and the database Web contains speech data of more speakers, which increased the difficulty of identification. The latter reason is also reflected by that the identification accuracies for database Mic-Web are much lower than that of databases Mic and Web. It can be seen in Table 2 that normalization could increase the identification accuracies of KNN, PNN, and SVM by 2–3 %, the identification accuracies of PNN and SVM even reached 100 %. When the frame size is changed from 512 to 1,024, 2,048, and 4,096 gradually, the identification

Table 1 Identification accuracies of KNN, PNN, and SVM for the test sets of databases Mic, Web, and Mic-Web

Database	Classifier	TP	FN	TN	FP	Q_p (%)	Q_n (%)	Q (%)
Mic	KNN	43	0	32	3	100	91.4	96.2
	PNN	43	0	33	2	100	94.3	97.4
	SVM	43	0	34	1	100	97.1	**98.7**
Web	KNN	90	1	23	21	98.9	52.3	83.7
	PNN	88	3	32	12	96.7	72.7	88.9
	SVM	91	0	37	7	100	84.1	**94.8**
Mic-Web	KNN	133	1	54	25	99.3	68.4	87.8
	PNN	133	1	63	16	99.3	79.7	92.0
	SVM	134	0	64	15	100	81.0	**93.0**

Notes: The bold is the best accuracy for a certain database.

Table 2 Identification accuracies of KNN, PNN, and SVM for the test sets of normalized databases Mic, Web, and Mic-Web

Database	Classifier	TP	FN	TN	FP	Q_p (%)	Q_n (%)	Q (%)
Mic	KNN	43	0	34	1	100	97.1	98.7
	PNN	43	0	35	0	100	100	**100**
	SVM	43	0	35	0	100	100	**100**
Web	KNN	88	3	26	18	96.7	59.1	84.4
	PNN	86	5	37	7	94.5	84.1	91.1
	SVM	91	0	41	3	100	93. 2	**97.8**
Mic-Web	KNN	133	1	57	22	99.3	72.2	89.2
	PNN	133	1	65	14	99.3	82.3	93.0
	SVM	134	0	70	9	100	88.6	**95.8**

Notes: The bold is the best accuracy for a certain database.

Table 3 Recognition accuracies of KNN, PNN, and SVM for the test sets of the normalized database Mic-Web with different frame sizes

Frame size	Classifier	TP	FN	TN	FP	Q_p (%)	Q_n (%)	Q (%)
1,024	KNN	132	2	61	18	98.5	77.2	90.6
	PNN	131	3	69	10	97.8	87.3	93.9
	SVM	134	0	73	6	100	92.4	97.2
2,048	KNN	131	3	63	16	97.8	79.7	91.1
	PNN	131	3	72	7	97.8	91.1	95.3
	SVM	134	0	75	4	100	94.9	**98.1**
4,096	KNN	131	3	61	18	97.8	77.2	90.1
	PNN	132	2	67	12	98.5	84.8	93.4
	SVM	134	0	74	5	100	93.7	97.7

Notes: The bold is the best accuracy for different frame size and classifier.

accuracies of KNN, PNN, and SVM all rose and then dropped; and when the frame size is 2,048, the identification accuracies of the three methods reached the highest, 91.1, 95.3, and 98.1 %, respectively. These accuracies increased by 2–3 % than the identification accuracies of KNN, PNN, and SVM for the normalized database Mic-Web. Seen from Tables 1, 2, and 3, SVM has the highest identification accuracies, followed by PNN. In addition, it can be seen from Q_p and Q_n that the identification accuracies for male speakers are much higher than that for female speakers, indicating that males have more concentrated speech features, whereas females have more diverse speech features.

4 Conclusions

In this study, we set up two independent speech databases and one integrated speech database, performed gender feature vector extraction to the speech data using MFCC, and conducted training and test with statistical learning methods. The experiment showed that SVM has better gender identification accuracies for speeches in each database than PNN and KNN. Although the identification accuracies of SVM significantly reduced when the noise and data volume increased, appropriate frame size and normalization method could improve the accuracies. The gender identification for male speakers is much more accurate than that for female speakers, indicating that female speakers have more diverse speech features, and it is worth investigation to use other feature extraction methods to improve the gender identification accuracy for female speakers.

Acknowledgments This work is supported by science and technology research projects of Chongqing city board of education (No. KJ120817).

References

1. Zhang, H.D., Li, J.W.: Speech recognition based on CHMM classified by gender identification. Comput. Eng. Appl. **43**, 187–189 (2007)
2. Deng, Y., Welde, O.G.: Gender identification using HMM. Comput. Eng. Appl. **40**, 74–75 (2004)
3. Li, M., Han, K.J., Narayanan, S.: Automatic speaker age and gender recognition using acoustic and prosodic level information fusion. Comput. Speech Lang. **27**, 151–167 (2013)
4. Sorokin, V.N., Makarov, I.S.: Gender recognition from vocal source. Acoust. Phys. **54**, 571–578 (2008)
5. Xiao, H.G., He, W.: Gender recognition of speaker based on MFCC and SVM. J. Chongqing Univ. **32**, 770–774 (2009)
6. Bekios-Calfa, J., Buenaposada, J.M., Baumela, L.: Revisiting linear discriminant techniques in gender recognition. IEEE Trans. Pattern Anal. Mach. Intell. **33**, 858–864 (2011)
7. Huang, T., Yang, Y., Wu, Z.: Combining MFCC and pitch to enhance the performance of the gender recognition. In: 8th international conference on signal processing, **1**(4), 787–790 (2006)
8. Kang, S.I., Chang, J.H.: Discriminative weight training-based optimally weighted MFCC for gender identification. IEICE Electron. Express **6**, 1374–1379 (2009)

9. Bocklet, T., Bauer, A.J., Burkhardt, G.F.: Age and gender recognition for telephone applications based on GMM supervectors and support vector machines. In: 2008 IEEE International Conference on Acoustics, Speech and Signal Processing, vols. 1–12, pp. 1605–1608. New York, IEEE (2008)
10. Sahidullah, M., Saha, G.: A novel windowing technique for efficient computation of MFCC for speaker recognition. IEEE Signal Process. Lett. **20**, 149–152 (2013)
11. Zou, M.: A novel feature extraction methods for speaker recognition. In: Zhao, M.T., Sha, J.P. (eds.) Communications and Information Processing, Pt 1. vol. 288, pp. 713–722. Springer, Berlin (2012)
12. Sapijaszko, G.I., Mikhael, W.B.: An overview of recent window based feature extraction algorithms for speaker recognition. In: Garimella, A., Purdy, C.C. (eds.) 2012 IEEE 55th International Midwest Symposium on Circuits and Systems, pp. 880–883 (2012)
13. Xiao, H.G., Cai, C.Z.: A comparison study of normalization of feature vector. Comput. Eng. Appl. **45**, 117–119 (2009)
14. Cover, T., Hart, P.: Nearest neighbor pattern classification. IEEE Trans. Inf. Theor. **13**, 21–27 (1967)
15. Fe, S.D.: Probabilistic neural networks. Neural Networks **3**, 109–118 (1990)
16. Vapnik, V.: The Nature of Statistical Learning Theory. Springer, New York (1995)
17. Xiao, H.G., Cai, C.Z., Liao, K.J.: Recognition of military vehicles by using acoustic and seismic signals. Syst. Eng. Theor. Pract. **26**, 108–113 (2006)
18. Cortes, C., Vapnik, V.: Support vector machine. Mach. Learn. **20**, 273–297 (1995)

Design and Implementation of Interpreting Engine Under the Generation Platform

Shisheng Zhu, Mingming Zhou and Haitao Xiao

Abstract This paper proposes an enterprise resource planning (ERP) system generation platform, which is suitable for micro and small business. This generation platform based on Eclipse plug-in technology and extracts services from enterprise' business process and function according to the SOA standard. By using the business process modeling environment and the page editing environment provided by the platform, the user can model the enterprise business process and design the interface of ERP application system respectively. Then they can be delivered to the interpretation engine for explanation. This interpretation engine can generate an ERP Web application system rapidly, which satisfies different requirements from enterprises. It contributes to reduce the software developing cycle, and improve the capability of the system to fit the various enterprise operation situations. Interpreting engine is the core of generation platform because it ensures process's proper operation and the right jumps within the pages. It also deploys and releases the ERP application system to the Web application server rapidly and provides an important guarantee for the realization of the generation platform.

Keywords Generation platform · Interpreting engine · SOA · Process modeling

1 Introduction

Enterprise resource planning (ERP) system contains not only advanced management ideology, but also information technology, including software engineering, network technology, database technology, etc. Along with the unceasing informationization of enterprises, more and more small enterprises have adopted ERP system to enhance their competitiveness. However, due to the mass number, small

S. Zhu (✉) · M. Zhou · H. Xiao
Department of Computer, Shantou University, Shantou, 515063 Guangdong, China
e-mail: sszhu@stu.edu.cn

S. Patnaik and X. Li (eds.), *Proceedings of International Conference on Computer Science and Information Technology*, Advances in Intelligent Systems and Computing 255, DOI: 10.1007/978-81-322-1759-6_83,

scale, weak strength, less budget of the implementation of ERP invest, frequently adjusted management mode, and the existing ERP system function module is fixed and the scalability is not high, these lead to extremely high failure rate of the implement of ERP [1]. Therefore, in order to solve these problems, reduce the cost of ERP system development, and shorten development cycle times [2], this paper proposes a specifical ERP system generation platform with low cost for small enterprises.

Interpreting engine under the generation platform is in the core position. It provides the most important guarantee for the platform. The traditional interpreting engine can only achieve a description of commercial activities in the abstract high-level languages (such as IBM WSFL, Microsoft XLANG) analysis. To some extent, it can meet the requirements of business process of rapid changes in demand, but in practice, due to the lack of interaction with the engine, a graphical modeling tool and a perfect, concise symbols for process modeling, the development of the software which make use of this kind of engine is not flexible and the market popularization is also very hard. BPEL is a language for the automation of business process specification. It is based on XML. BPEL process can use a standardized way to interact within Web services. These processes can be implemented in any BPEL compliant platform or product. Currently, it has been generally recognized [3] in industry. The interpreting engine of this paper meets BPEL specifications. It is the extension of the traditional interpreting engine. Simultaneously, it not only realizes the interaction with interface by linking up all the deployment through an Ant Bus, but also realizes the automatic deployment and releasing of ERP application system.

2 About the Generation Platform

The generation platform adopted SOA, BPEL, and other new technologies. It can set up application of Web ERP system quickly according to the personalized needs of enterprises. What's more, the generated Web system is easy to maintenance, modify, and expand. Moreover, along with the successful application of the platform, it will eventually accumulate various functional service components. As a kind of resources, these service components can be constantly added, updated, and reused according to the requirement. Besides, the platform can greatly reduce the cost of the development of small enterprises information system and shorten the development cycle.

The generation platform uses the plug-in technology to design and divide the software architecture. The platform function modules are designed into separate Eclipse plug-in [4]. Considering the micro- and small enterprises need variety, strong personality, we use a page editing environment [5] and ERP-BPEL (a business process modeling tool based on Eclipse plug-in technology) for visual business process modeling [6, 7]. Interpreting engine is used to analyze BPEL process and generates code to achieve the development of enterprise ERP system.

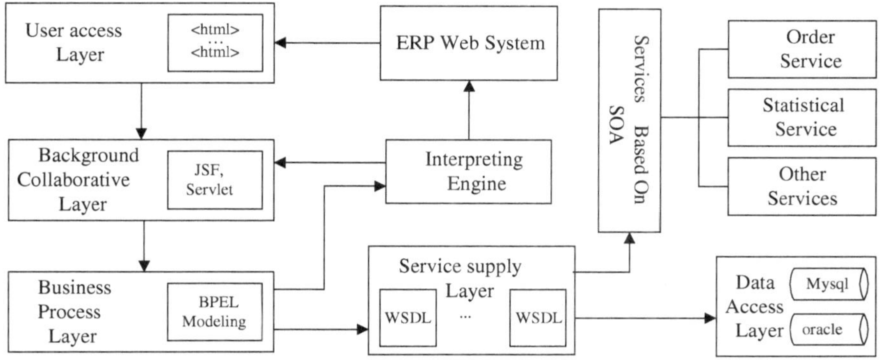

Fig. 1 The system framework of the generation platform

The generation platform architecture is divided into user access layer, background collaborative layer, business process layer, service supply layer, and data access layer which as shown in Fig. 1.

User access layer mainly uses the technology of HTML, Ajax, and so on. It can display the ERP system interface to the user as a Webpage and interact with the background collaborative layer through HTML and Ajax technology. Background collaborative layer mainly uses the technology of JSP and Servlet. It accepts the request and data from the user access layer to handle preliminary and then deliver to business process. Business process layer (display to the platform users as controls) receives the request which is distributed by background collaborative layer, calls the controls corresponding business process, and then calls the WEB services which described by WSDL files that are provided by service supply layer. Data access layer provides access to various databases. The following parts of this paper will introduce the interpreting engine architecture design and implementation under the platform in details.

3 Design and Implementation of Interpreting Engine

Interpreting engine as the core module in generation platform mainly completes two aspects functions, the first is to deploy the BPEL generated by the business process modeling module (business process layer) and other files, then providing a runtime for these files [8]. The second is to get Web application output after interpretation parsing using Web page design code as input. So the engine not only parses BPEL files, but also parses Web page design code and finally releases in the form of a Web application. According to its features, the engine is divided into process interpreting module and page interaction module. As there are many small functional nodes during the interpreting procedure, such as engine start, file loader, and file compiler. These functional nodes are executed in order. The design uses

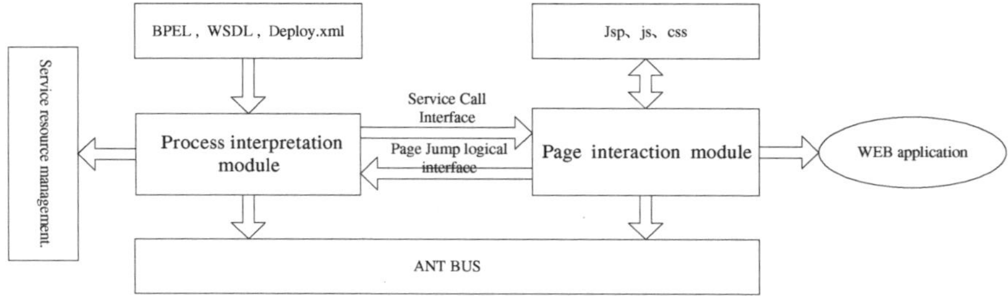

Fig. 2 The interpreting engine frame structure

Ant scripts to realize it. All functional modules are hanged on an Ant bus. Interpreting engine acts as a set of Ant scripts. The interpreting engine frame structure is shown in Fig. 2.

3.1 Process Interpreting Module

Process interpreting module is actually a BPEL interpreting engine, which initiates a business process instance according to the defined BPEL file, then completes the driver and execution of the entire business process according to jump condition, call target, and return value. Overall, the BPEL engine must address the following three questions:

Process scheduling problem It must be ensured that the engine is able to deal with the complex structure of the business process, such as serial, parallel, branch, polymerization, and be ensured that processes run from one node to another in this complex structure. There are a lot of methods to implement engine scheduling mechanism, many of which using token scheduling mechanism affected by Petri Net and others like traversal loop algorithms method.

Issues of process execution When processes run to a node, some problems are needed to resolve: whether to execute the node or not? How to execute the node? How to deal with the exception of executing process? How to save the status of the running node? etc. For these problems, some of the complex process engine often depends on constraints and changes of "the process instance state" or "activity instance state" to deal with. So sometimes a process interpreting engine is called "state machine."

Process instance object For each process instance, an entire set of process instance objects to describe the status and results of the running process instance are needed. Usually, process instance object contains the state or control information of the process instance. Normally, one or more activities exist in a process. Usually for these activities, the entire process instance's state information and control information is visible. But for the entire process, these activities' status and

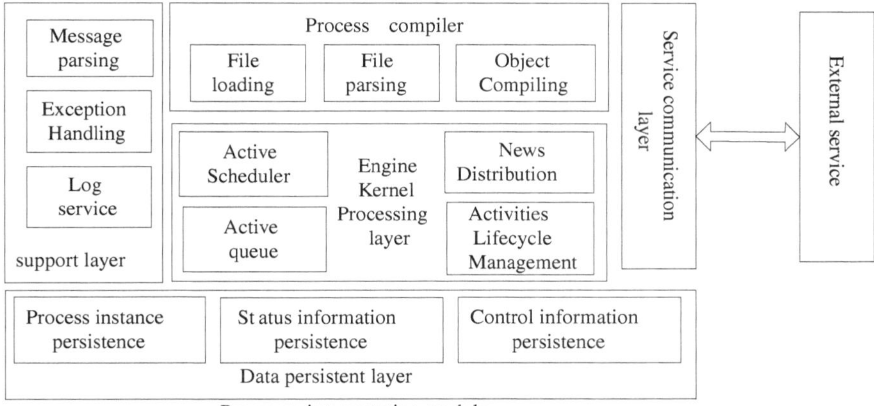

Fig. 3 Architecture diagram of process interpreting modules

control information are not visible, similar to the relationship between global variables and local variables in programming languages.

According to business process templates defined by BPEL modeling tools on the generation platform, the most basic task of the BPEL engine are to initiate a business process instance, and to complete the driver and execution of the entire business process according to the logical judgment and the jump condition in the business process instance, as well as intermediate variables, executing information of each jump and return value. After preliminary summary of the last chapter of the process interpreting module function features, the entire process interpreting engine is divided into five parts by different levels of functions, and the layers are: process compiler, runtime support layer, service communication layer, the engine kernel processing layer, and data persistence layer. Architecture diagram is shown in Fig. 3.

Process compiler It mainly does some preprocessing works before running BPEL file, including file loading, parsing, and compiling. File loading puts BPEL files which will be used in modeling environment and related with WSDL files into a specific folder, which will view the fold periodically, interpret the new files when added, and then generate Java object according to data types of WSDL, XML schema referenced by BPEL files. It will also generate BPEL controller (a servlet).

Support layer It provides some complementary tools for running process interpreting module to support process instance running, including system exception handling, log service, message parsing, and reconstruction. Exception handling handles all kinds of exception in the process of running interpreting engine, caused by internal logical of the business processes, system errors, network exceptions, etc. Log service mainly records some operations and timing of a process instance in its life cycle. During system running, many different types message will generated, such as HTTP message and SOAP message.

The interpreting engine defines a generic message format. The parsing and reconstruction of messages are to encapsulate all kinds of messages into a standard format.

Engine kernel processing layer Because of the engine kernel processing layer mainly focusing on "meet the basic running," we try to strip external service needed by engine and engine kernel. Simplifying kernel structure by placing some operations in periphery (running support layer) and only including the most basic objects and service, the scheduling and running mechanisms is used to solve the process running in engine kernel. It will not only make the entire engine hierarchy more clear, but also commodiously do some functional extension of engine by providing some extension interfaces in the engine kernel. At present, the engine kernel mainly handles some actives scheduling problems and system messages distributes.

Data persistence layer Its main task is to complete a variety of persistent operating logic decision, which is an important part in the interpreting engine, and to be responsible for the persistence of the BPEL. Instance of the template and various information generated during the BPEL process instance running, the layer will prevent the memory data loss during running process, which causes process instance run abnormally.

Service communication layer In a BPEL process, completion of a lot of actives needs to call external services. For BPEL itself, it is a Web service. To call external services must handle the communications in these services, which refer to message communication, service addressing, codes generation between client and server-side problems. This design uses Axis2 to solve the above problems. Axis2 is a technology framework (Architecture) of Web Service implementation, also a SOAP engine which encapsulates the processing of SOAP messages, provides a framework to process SOAP messages. The user can extend it for each service or operating. Users can model different message exchange patterns (MEP) based on the framework, which provides Web service deployment and client interface to call Web service functions.

3.2 Page Interaction Module

Page interaction module makes the engine interact with page designers, combined with the previously generated Java code to complete generating the page logic layer Servlet, and finally deploying and releasing Web applications (ERP system).

Page interactive process The interpreting engine has two interactive process with the page altogether. One of which is to tell user how to use the modeling BPEL according to BPEL server code after user finishes modeling in business process modeling environment. This design provides a defined *.html file whose format defines as (server-name. do{action parameters, parameter} {return type}) to the interface, and another of which is that the interface tells BPEL how service page jumps, after user uses BPEL in page editing environment. This design

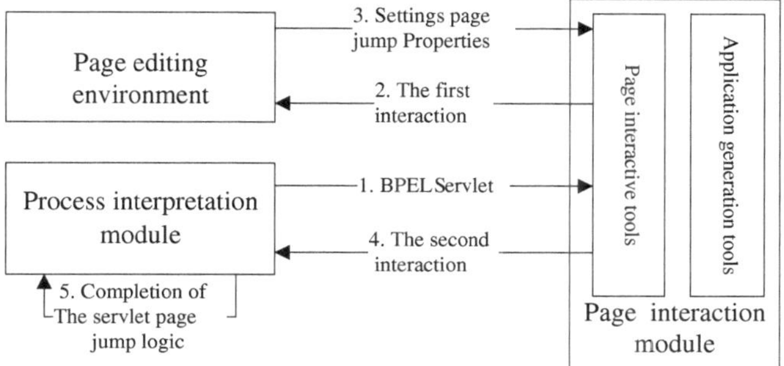

Fig. 4 The interaction between process interpreting and interface

provides another defined *.do file whose format defines as action-name #page (jump to) # [post | get] # additional-information. The specific process of interaction is shown in Fig. 4.

Application generation process It deploys the designed files including BPEL process files, interface design files and other relevant files after user finish system design in the platform, and finally generates available Web applications. In these processes, many relatively independent function modules involve, which must be able to be execute in the order in accordance to guarantee correct generation of Web applications. To resolve the execution order problem, XML scripts written by Ant are used in the design, which concatenates all kinds of small functional modules together as a bus (ANT script contents are omitted), instead of hard-coding in a Java file directly.

4 Case Practice

To prove the correctness and feasibility of design of the interpreting engine, a case is given to explain how interpreting works. A commerce application can be quickly generated through the platform for a toy factory. Due to limited space, only the most simple login system functions are introduced. First generating modeling file—login.bpel using modeling tools supported by business process layer in the platform, then the interpreting engine parses the file to generate corresponding server-side code—loginServiceprocessServlet.java and loginService.html file used by interface, whose contents are as follows:

```
(loginServiceprocess.do{arg0,arg1,}{loginService.Login
Response})
```

When designing the system login page, the related information of jump logic is set on "Jump to Url" attributes of the corresponding controls, with dragging controllable components panel. Finally the information is packed into a *.do file.

For example, the corresponding file in Login Service is LoginServiceprocess.do, which includes the following contents:

```
LoginServiceprocess.do#index.jsp#post#
```

Login Service process Servlet only include service call code at the beginning. It's content will be the jump logic of login Service process Servlet. Other interface design in the platform follows the same steps above mentioned. All the code in the project will be packaged and released through the Ant scripts in the system to generate the final Web application system. Confined to the length of the thesis and Chinese display, the operation effect figure of generating ERP application system is for omissions.

5 Conclusion

Through the generation platform, enterprises can quickly construct enterprise ERP application system which conforms to the individual needs. Running the generation system code turns out that the design scheme of the interpreting engine is correct, feasible, and suitable for model explanation processing that meets small enterprise requirements. In this design, the interpreting engine has achieved to analyze front desk interface. It is a breakthrough when compared with the traditional business process engine. But there are some problems in it. It is lack of a visualization user tool. It does not support the cross-platform application and artificial activities in the process. These are the tasks we will take on in the future.

Acknowledgment This work is supported by Shantou University innovation team project ITC12001.

References

1. Wang, J., Xiao, J.: The cause analysis and suggestions for the high failure rate of ERP. Chin. New Technol. Prod. **24**, 205 (2010)
2. Yu, X., Hong, Y.: Design and realization of the eclipse-based database model editor. Micro Comput. inf. **26**(3), 146–148 (2010)
3. Alves, A., Arkin, A.: Web services business process execution language version 2.0.OASIS Standard, 11 April 2007
4. Feng, X., Cui, K., Shen, J.: Research and implementation of the application framework oriented the plugin. Comput. Eng. Appl. **45**(10), 89–91 (2009)
5. Lu, J., Yang, D., Wang, Y.: Research of graphics editor technology based on the GEF. Value Engineering, March 2011
6. Pan, C., Pan, A., Zhou, X.: Research and implementation of process modeling based on the BPEL2.0 standard business. Comput. Eng. Des. **31**(21), 4607–4609 (2010)
7. Juric, M.B.: WSDL and BPEL extension for event driven architecture. Inf. Softw. Technol. **52**(10), 1023–1043 (2010)
8. Huang, L., Yao, F.: Research on web service combination technology under Apache ODE environment. Comput. Technol. Dev. **07**, 98–104 (2011)

The Application of Versioning Technology in XBRL Taxonomy Engineering

Ding Wang, Qilu Cao, Huang Min and Ying Wang

Abstract Versioning is the core of software maintenance in the software engineering. In this paper, versioning is applied to e**X**tensible **B**usiness **R**eporting **L**anguage (XBRL), which is the latest technology used in processing accounting information. Based on the parsing of the XBRL versioning specification, the XBRL versioning system is built in order to save costs and reduce errors. The system has a great practical value for the taxonomy creators and users.

Keywords XBRL · Taxonomy · Versioning · Software maintenance

1 Introduction

XBRL stands for e**X**tensible **B**usiness **R**eporting **L**anguage. It is one of a family of "XML" languages, which is becoming a standard means of communicating information between businesses and on the Internet. XBRL greatly increases the speed of handling of financial data, reduces the chance of error, and permits automatic checking of information.

D. Wang · Q. Cao
Information Technology Department, University of International Relations,
Beijing, China
e-mail: wangding09b@mails.ucas.ac.cn

Q. Cao
e-mail: Luke_cao29@hotmail.com

D. Wang · H. Min (✉) · Y. Wang
Engineering and Information Technology College, University of Chinese
Academy of Sciences, Beijing, China
e-mail: huangm@ucas.ac.cn

Y. Wang
e-mail: ywang@ucas.ac.cn

S. Patnaik and X. Li (eds.), *Proceedings of International Conference on Computer Science and Information Technology*, Advances in Intelligent Systems and Computing 255, DOI: 10.1007/978-81-322-1759-6_84,

XBRL allows report preparers to place electronic tags on specific content (graphs, numbers, text, etc.) in their reports by using an existing "XBRL taxonomy" so that it can be easily understood and processed by computers. The taxonomy simply means a list of tags organized into a single set. A taxonomy consists of the core part which is a schema (or more schemas) and linkbases. The schema is the part that contains definitions of elements, whereas linkbases provide relationships between them.

Up to now, dozens of government departments and organizations released their taxonomies. The most influential taxonomies are undoubtedly the International Financial Reporting Standards (IFRS) taxonomy and the US Generally Accepted Accounting Principles (GAAP) taxonomy. In recent years, both of them are released as a new version every year, especially the IFRS taxonomy; it has released 14 versions since 2002. As accounting rules change and as industries and companies change, the taxonomy definitions and items must evolve with them. The taxonomies must be maintained and developed. And as market places evolve, there will always be a need for new taxonomies that help define a business information process.

Under this condition, the versioning technology has a broad application space in the entire life cycle of taxonomy. However, around the world, the application of versioning in taxonomy is only in the preliminary stage. As a result, there is no versioning tool with comprehensive functions.

With parsing the XBRL versioning specification, a XBRL versioning system is built in this paper. The user can input the old and new versions of taxonomies into the system, and then, the comparison is done automatically. With the minimum degree of human involvement, the advantage of the system is obvious, saving costs and reducing errors.

2 Versioning in XBRL Taxonomy Engineering

XBRL taxonomy engineering is the application of systematic, formal, quantifiable approach to the design, building, usage, and maintenance of XBRL taxonomies, that is, application of software engineering to taxonomies [1].

The development cycle from software engineering is used to categorize the phases of XBRL taxonomy development. Six phases are set out as follows: planning and analysis, design, building, testing, publication and recognition, and maintenance. The phases are defined in the taxonomy life cycle, and the proposed ordering and feedback loops allow a manageable and controllable taxonomy development, leading to more quantifiable XBRL projects [2].

Taxonomy versioning is a part of the maintenance phase. The experience derived from developed projects indicates that taxonomy maintenance requires taxonomy developers to monitor the taxonomy usage and track the issues and bugs that users experience with the taxonomy. Feedback from the taxonomy usage and maintenance phase is used in the next version of the information model and improves consequently following taxonomy release. The taxonomy creator can transfer the changes between two versions by the versioning report in order to minimize the impact of version update. Approaches from software versioning system provide a starting point for taxonomy versioning [3]. XBRL International is currently working on a standardized solution to XBRL taxonomy versioning that can enhance the maintenance of the taxonomies from the users' perspective.

With numerous projects worldwide facing the same requirement for a standardized methodology for documentation of changes between taxonomies, the XBRL International developed the XBRL versioning specification [4], which provides an overview of how the XBRL versioning work. The XBRL versioning specification defines XML syntax for a versioning report that is designed to allow taxonomy creators to communicate information about the changes between two taxonomy versions in a structured format that will minimize the impact of the update on the users of the taxonomy.

The most important result of taxonomy versioning is versioning report, which contains the technical differences and the semantic differences. The goal of the versioning report is to facilitate XBRL instance users or consumers in adapting existing systems to accommodate different taxonomies and XBRL instance structures.

3 XBRL Versioning System

In theory, every change of taxonomy version update can be captured by the software. The objective of taxonomy versioning is the process of version update lowering the level of artificial participation; ideal case is completely without human involvement. A good versioning system can automatically accomplish the comparison of two versions, with cost saving and error reducing.

Currently, most XBRL taxonomy editing software, like UB matrix system [5], does not contain the version management functions. Arelle [6] is an open source tool made by Herman Fische. This tool is designed for version management only; its functions include the following: rendering taxonomy, comparing two different versions of taxonomies, and producing versioning reports. However, the functions of this tool is relatively simple; it can only handle element adding, deleting, and renaming and producing versioning report in XBRL format; it cannot produce versioning report in Excel table format.

In this paper, the XBRL versioning system is built, which is the first versioning application software on XBRL taxonomy in China. Figure 1 depicts the function framework of the system. The users log in the system through the portal and select new and old versions to import. The contents needed to compare is read by the input module. After processing by the base processing module, the element processing module, and the relationship processing module, a versioning report is generated by the output module.

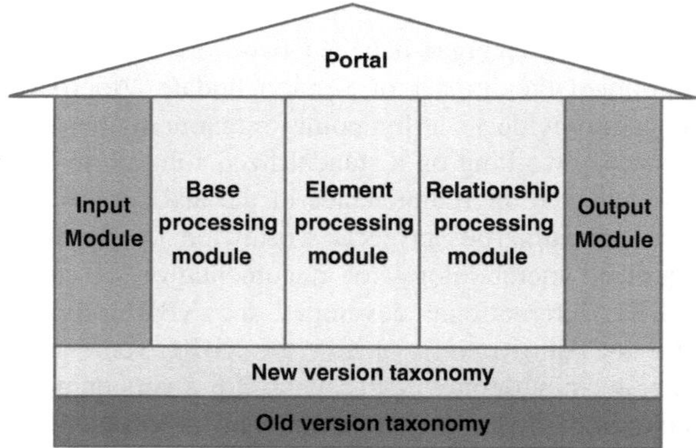

Fig. 1 The function framework of XBRL versioning system

The main function module of the system is divided according to XBRL versioning specification. The system accomplished the functions of Base module, Concept Basic module, Concept Extended module, and Relationship Sets module. The base processing module corresponds to Base module. The element processing module corresponds to Concept Basic module and Concept Extended module. The relationship processing module corresponds to Relationship Sets module.

Figure 2 depicts the comparison process of changing in elements and their attributes between the old and new version taxonomies. The elements in the old version are compared with the new version in order to record the elements removed; then, the type of element is determined, calling different processes according to different types; finally, elements in the new version are compared with the old version in order to record the elements added.

The base processing module corresponds to Base module, used to handle the change in namespace, the addition, removal, or renaming of a role between two versions.

The element processing module corresponds to Concept Basic module and Concept Extended module, used to handle the addition, removal, or renaming of an element, a label, or a reference, the change in element attribute like data type, period type, label, or reference, and the change in the content of a tuple between two versions. Concept Basic module and Concept Extended module have a very close relationship, so both modules are combined in a processing module.

The relationship processing module corresponds to Relationship Sets module, used to handle the changes in the presentation linkbase, calculation linkbase, and structure of dimensional syntax in the definition linkbase between two versions.

The input module contains eight reading functions, used to read all kinds of namespace, roles, elements, attributes, tuples, and linkbases.

The output module contains twenty-one writing functions, used to write all kinds of changes in the versioning report.

The portal of the system provides users with a user-friendly interface. Users can select two versions to import and choose comparison, generation of versioning

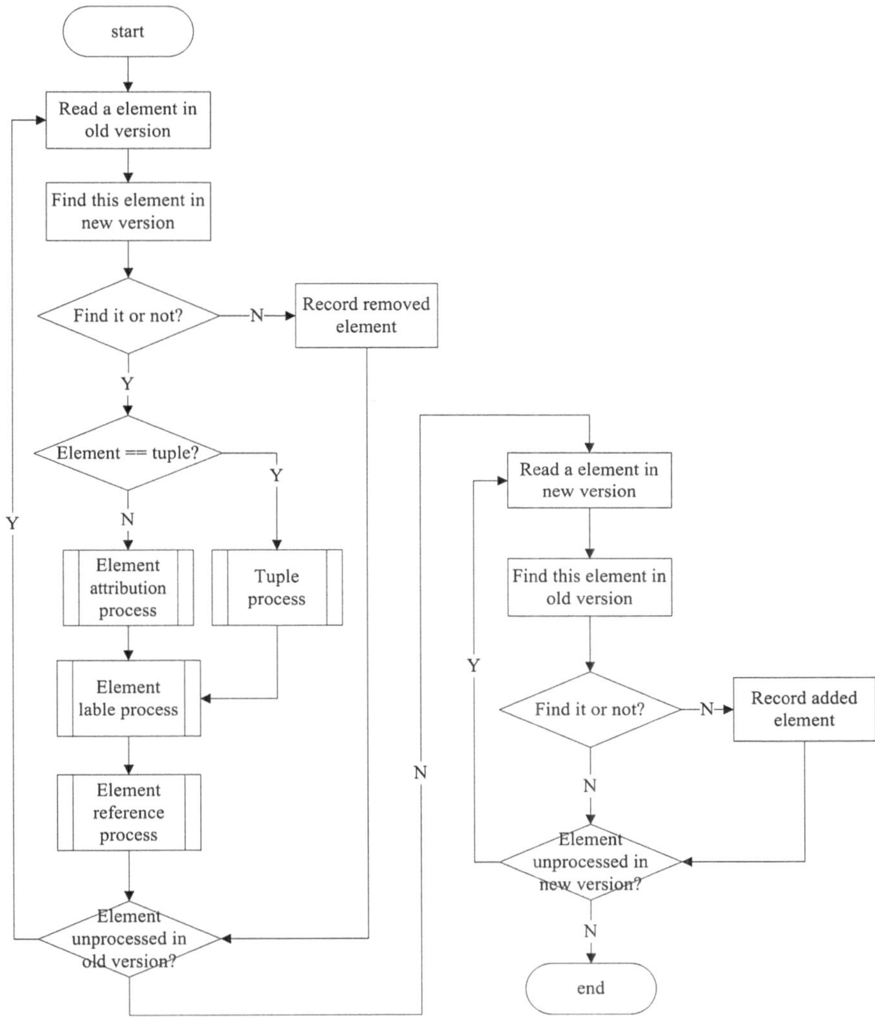

Fig. 2 The comparison process of elements between the old and new version taxonomies

report, or generation of tables. The function of comparison is rendering the result of the comparison on the current page.

The function of generation of versioning report is generating a versioning report in XBRL format, in other words, a document written in XBRL code. The function of generation of tables is generating a versioning report in Excel table format.

4 Conclusions

Now, the versioning technology has a broad application space in the world. However, the application of versioning in taxonomy is only in a preliminary stage. Under the consideration of the situation that there is no versioning tool with comprehensive functions, the XBRL versioning system is built in this paper. The function of the system is divided according to XBRL versioning specification.

The user can input the old and new versions of taxonomies into the system, and then, the comparison is done automatically. With the minimum degree of human involvement, the advantage of the system is obvious, saving costs and reducing errors.

Acknowledgments This work was financially supported by National Training Programs of Innovation and Entrepreneurship for Undergraduates of University of International Relations.

References

1. Hoffman, C.: Financial Reporting Using XBRL: IFRS and US GAAP Edition. Economic Science Press, Beijing (2006)
2. Debreceny, R.: XBRL for Interactive Data. Springer, Berlin (2009)
3. Shiping, L.: XBRL Practical Case Analysis. Economic Science Press, Beijing (2010)
4. XBRL International. Overview of Versioning 1.0. http://www.xbrl.org/WGN/versioning-overview/WGN-2011-05-11/versioning-overview-WGN-WGN-2011-05-11.html (2011). Accessed 2011
5. UBmatrix. XBRL Processing Engine. http://www.edgr.com/active.aspx?contentID=4 (2012). Accessed 2012
6. Arelle Work Group. Arelle. http://arelle.org/ (2012). Accessed 2012

Intelligent Automotive Fault Diagnosis Platform Based on ARM and Linux

Yanli Hou, Shenglong Huang and Duonian Yu

Abstract According to the requirements of modernization and informatization for the new generation of medium and heavy truck, we raise a corresponding design and solution and complete the work of hardware and software designing and realization independently. Firstly, I analyze the background and status of the proposed scheme; on this basis, we discuss the necessity of the hardware, software, and the related technology. Then, I describe the details of the structure and principles for the hardware. Meanwhile, the information processing process is described, and I explain the specific implementation of software via detailed description of CAN message received and classified threads. Finally, we get the corresponding conclusions through the functional verification and performance diagnosis.

Keywords CAN · Automotive fault · Diagnosis platform · ARM · Linux

1 Introduction

The foundation of our country's automobile industry is relatively weak, and there is a great gap between the foreign medium and heavy off-road trucks R&D capability and ours. But the long-distance road transport vehicles and field

Y. Hou (✉)
Center of Computer Fundamental Education, Jilin University, Changchun, Jilin, China
e-mail: hyl@jlu.edu.cn

D. Yu
College of Automotive Engineering, Jilin University, Changchun, Jilin, China
e-mail: yudn@jlu.edu.cn

S. Huang
R&D Center, China FAW Group Corporation, Changchun, Jilin, China
e-mail: hslhero@163.com

S. Patnaik and X. Li (eds.), *Proceedings of International Conference on Computer Science and Information Technology*, Advances in Intelligent Systems and Computing 255, DOI: 10.1007/978-81-322-1759-6_85,

transport vehicles are an integral part of vehicle in modern society, now. The demands of its modernization has forced us to raise the R&D level of medium and heavy trucks. At the same time, because of foreign technology blockade and restrictions of medium and heavy trucks, we must design and innovate the medium and heavy trucks off our own bat, and only in this way, can make our R&D technology continue to evolve, and gradually mature.

Because of the harsh environment of medium and heavy trucks, it must have a high environmental adaptability, and has the ability to troubleshoot quickly, especially in the field during long-distance transport. Secondly, truck after a fault cannot be repaired in the 4S shop such as an ordinary vehicle. Because of this, the modernization, informatization, and self-diagnostic capabilities of medium and heavy trucks are particularly important. The driver or attendant not only needs to keep abreast of the vehicle driving status, fault and other information, but also needs to real-time monitor the automotive fault, warn early, and rapid troubleshooting.

We proposed the design ideas of the intelligent automotive fault diagnosis platform based on ARM and Linux for the special requirements. The platform includes vehicle information, fault information, voice prompts, data storage, export, and other functions. Thus, facilitating driver's access to the vehicle's status and fault information, and to help maintenance personnel to locate the point of fault quickly, to accommodate the special circumstances of the field long-distance transport.

2 System Design

Intelligent automotive fault diagnosis platform is based on ARM9 processor, adopt embedded Linux operating system, I achieved most functions use the embedded C, C++, and QT language. The platform receiving the vehicle status messages and fault messages via CAN bus display to the screen real time. Meanwhile, the platform identifies the point of fault quickly according to the vehicle status and fault information, prompts the driver or technical personnel via voice, the screen or other forms, and gives professional repair advice.

Shown in Fig. 1, the platform is divided into 12 independent functional modules, such as: power supply, CAN-bus accessing, driving and fault information extraction, fault diagnosis, voice alarm, repair recommendations, user identification and rights management, user guide, data export, system setting, display, and human–computer interaction. It will reduce the difficulty of realization through analyzing the data flow and control flow between modules.

The driving and fault information extraction module includes the CAN message reception, CAN message parsing, message information store into the queue, driving status and fault code extraction, and several other steps.

Meanwhile, the platform includes a vehicle troubleshooting and repair expert database that established a group of technicians possessed very rich experience in automotive research and development, and maintenance staff had a long-term

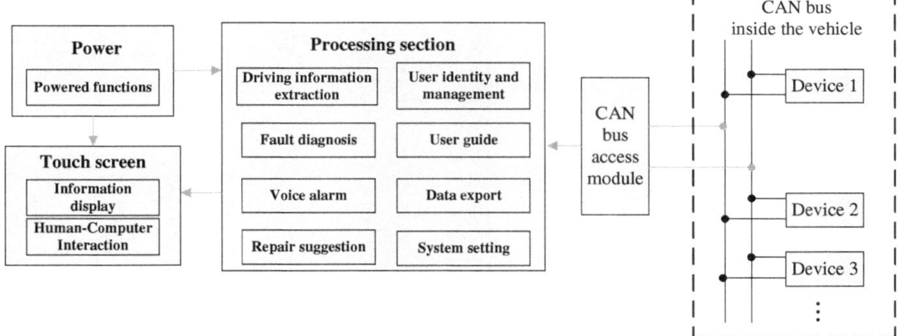

Fig. 1 Block diagram of automotive fault diagnosis platform

experience in the real vehicle repair. The expert database gives the appropriate information and maintenance recommendations. It meets the troubleshooting needs in long-distance field transport.

The platform can also record the vehicle's driving and fault information within the last three months, and exported all of the data, provide the basic data for assessing the truck's performance and reliability. Display and interactive program is realized using QT language. Users can get all information by reading the screen and control the machine via the touch screen and a virtual keyboard.

3 Hardware Architecture

In consideration of the needs of real-time information processing, we adopt Samsung ARM9-based 32-bit processor S3C2440, a high performance and low-power processor. Its interface is rich, can be connected to the CAN message reception module, and is widely used in mobile phones, automotive electronics, intelligent machinery, and other fields [1]. And we use a common CAN controller SJA1000, an independent controller that is developed by PHILIPS. SJA1000 can receive and transmit standard and extended packet [2].

Shown in Fig. 2, SJA1000's RD and WR pin connects to S3C2440's RD and WR pin, respectively, SJA1000's RST pin connects to S3C2440's RST pin, INT pin connects to S3C2440's external interrupt pin (EINT).

4 Software Design

We divide the software into two parts, one is information processing program, the other is interactive program.

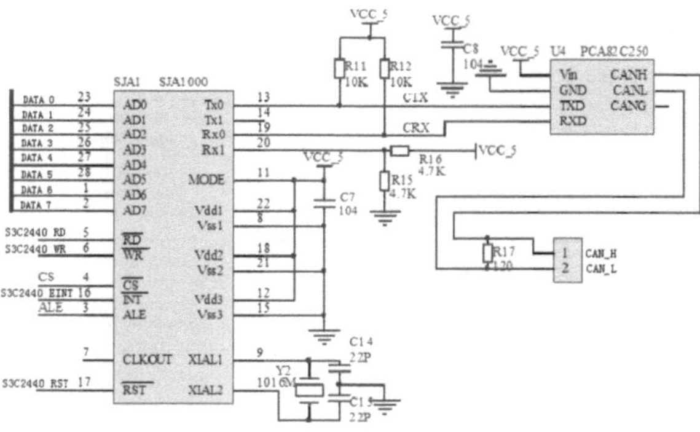

Fig. 2 Connection diagram between S3C2440 and SJA1000

4.1 Information Processing Program

Because CAN message reception and processing speed mismatch, if we execute received CAN messages, CAN packet diagnosis, extract information, treatment, and display threads orderly, it will result in some messages loss; in order to solve this problem, we use the message queue mechanism between each functional module. Namely the module M1 processes and stores the data into the message queue MQ1, the next module M2 gets the data from MQ1, then processes and stores the data into an other message queue MQ2.

The header file of message queue:

```
#include<sys/types.h>
#include<sys/msg.h>
#inlcude <sys/ipc.h>
```

The function prototype of creating a message queue:

```
int msgget(key_t key, int msgflg);
```

The function prototype of storing message queue:

```
int msgsnd(int msgid, const void* msg_ptr, int msg_sz, int msgflg);
```

The function prototype of reading message queue:

```
ssize_t msgrcv(int msgid, void* ptr, size_t msg_sz, long msgtype, int msgflg);
```

The function prototype of modifying the message queue attributes and deleting the message queue:

```
int msgctl(int msqid, int cmd,struct msqid_ds *buf);
```

Meanwhile, in order to process the function modules parallel and real time, we use multi-thread technology in our work, also avoid the time and resources overhead brought by switching processes.

The header file of thread:

```
#include<pthread.h>
```

The function prototype of creating a thread:

```
int pthread_create(pthread_t* thread, pthread_attr_t*
attr, void*(*start_rtn)(void), void* arg);
```

The function prototype of terminating a thread:

```
void pthread_exit(void* rval_ptr);
```

Figure 3 is the information processing flowchart of the platform. Now, taking CAN message reception and classifying thread is an example to describe the implementation details of information processing.

Figure 4 is a flowchart of CAN message received and classified thread. We customize CAN application layer transport protocol according to the CAN-bus communication protocol and SAE J1939 protocols. We use single packet to transmit the vehicle status information and divide fault information message into two kinds of single-packet and multi-packets transmissions; we broadcast messages to inform the receiving node the number of packets when multi-packets transmission. Table 1 is fault message single-packet and multi-packet CANID.

This thread contains two timers, One is the CAN network monitor timer (timer_10 s), when the platform is not able to receive the CAN messages within 10 s after powered gives a tip to driver. Another is the fault information packets receive time-out timer (timer_750 ms), that is if receive node is not able to receive all fault data packets within 750 ms, then gives a tip and discards this fault data messages.

The following illustrates the process of processing vehicle condition information. Firstly, CAN message reception and classified thread receive vehicle condition information and store it into a message queue according to the packet ID (0X0CFE6C17). Then, driving information about parsing thread gets the data from the tail of the message queue and parses the data frame. Assuming that the data frame is 0X FF FF FF FF FF FF 3C 00, because the second byte is 0X3C = 60,

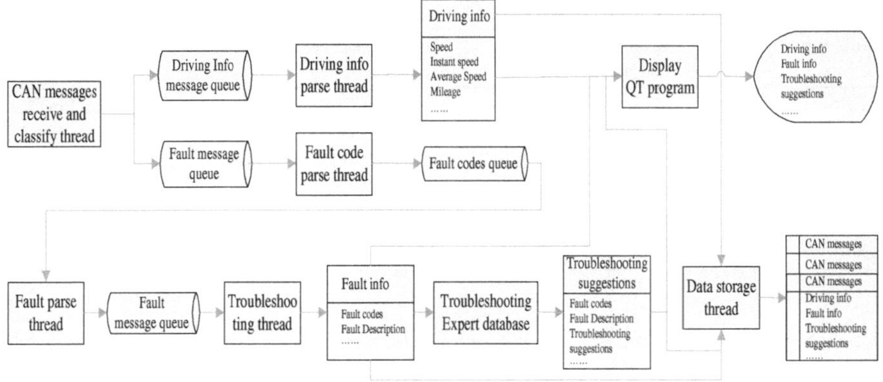

Fig. 3 Information-processing flowchart

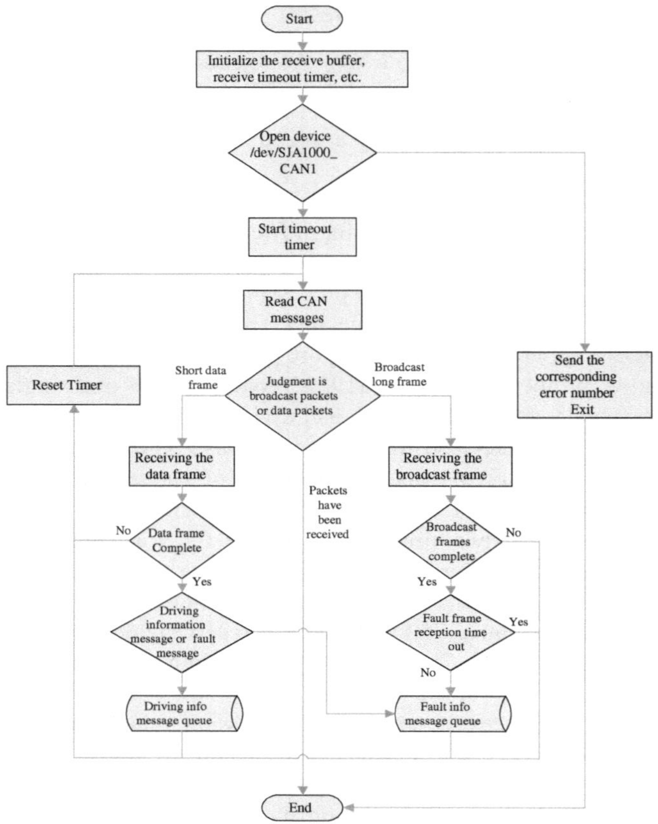

Fig. 4 Flowchart of CAN message reception and classifying thread

Table 1 Fault information message ID

Fault type	Single-packet CANID	Multi-packet broadcast packets CANID	Multi-packet data packets CANID
ABS	0X18FECA0B	0X1CECFF0B	0X1CEBFF0B
AMT	0X18FECA03	0X1CECFF03	0X1CEBFF03
EMS	0X18FECA00	0X1CECFF00	0X1CEBFF00
ComfortCAN1	0X18FECA31	0X1CECFF31	0X1CEBFF31
ComfortCAN2	0X18FECA21	0X1CECFF21	0X1CEBFF21
ComfortCAN3	0X18FECA E5	0X1CECFF E5	0X1CEBFF E5

we can get the real-time speed of truck as 60 km/h. The real-time vehicle information is displayed on the screen and stored into a structure that is used to analyze vehicle's breakdown and performance.

4.2 Display and Interactive Program

In order to adapt the ARM platform and embedded Linux application environment, we use the excellent cross-platform language Qt/Embedded to develop display and interactive interface, it provides a complete set of class libraries, including multi-threaded programing, databases, serial and parallel communications, etc. [3]. It reduces the difficulty in the development of work greatly [3].

Firstly, I achieve the Q widget base class for ensuring the unified type of the various windows, the other windows are derived from the base class. I implement the virtual keyboard for input, where users can operate the platform and set system using it.

User guidance and the troubleshooting expert database contain a large number of images for describing the operation and processing in detail. So we use an embedded browser (Arora). I cross-compile Arora, port it to the platform by following steps:

```
git clone git://github.com/Arora/arora.git
cd arora
qmake ''CONFIG-=debug'' --r
make
```

5 Test and Results Diagnosis

Since it is difficult to test diagnostic capabilities of the platform in a real truck quickly, I built a laboratory-integrated test environment to simulate extreme operating conditions of an actual vehicle, test, and verify the platform.

In this paper, I took vehicle condition test as an example to illustrate the functional test. Vehicle condition test includes the following aspects:

1. Normal condition test. Test each module under normal operating condition.
2. Electromagnetic interference environment test. Test each module in the case of weak and strong electromagnetic interference environment.
3. CAN-bus environment instability test. Test real-time display and packets loss in the case of CAN-bus debt ratio and size of the packet sending interval instability.
4. Durability test. Test the platform's function and performance in a situation of long-distance field transport.

Table 2 is an AMT fault description used during the test, fault diagnosis and processing program describe the fault code according to this table.

Table 3 is an ABS fault information and troubleshooting suggestions. Fault message parsing thread and troubleshooting expert database give the trouble-shooting suggestions following this table.

Table 2 AMT-fault description

Pcode	Fault code	Fault description
E0F62700	P1C45	Transmission neutral fault
E1F62700	P1C46	Starter gear fault
E2F62700	P1C47	Clutch cannot be kept separate
E7F62700	P1C48	Pressure is insufficient
EBF62700	P1C49	Unknown fault type
F0F62700	P1C4D	Clutch-combined fault
F2F62700	P1C4E	Clutch separate fault

Table 3 ABS-fault information and troubleshooting suggestions

Pcode	Fault description	Troubleshooting suggestions
02031500	Left-ring sensor, Incorrect tire	Check the tire circumference and the ring-gear teeth
05031500	Left-ring sensor, open circuit	Check the sensor and sensor wiring, and connectors
0A031500	Left-ring sensor, speed interrupt	Adjusting the gap between the sensor and the ring gear, push the sensor close to the ring gear. Inspect the bearing clearance and ring-gear deflection. Check sensor wiring and connectors, and prevent intermittent connection
0C031500	Left-ring sensor, frequency too high	Check sensor wiring and connectors, and prevent intermittent connection. Check whether there is squeezing in the brake. If not, replace the electrical components
01031600	Right front-ring-gear sensor, sensor gap	Adjusting the gap between the sensor and the ring gear, push the sensor close to the ring gear. Check the bearing clearance and ring-gear deflection
0C031600	Right front-ring-gear sensor, frequency too high	Check sensor wiring and connectors, and prevent intermittent connection. Check whether there is squeezing of brake sound

Figure 5 is a fault description and repair advices given by the troubleshooting expert database, using Arora browser. Finally, we complete the high and low temperature, humidity, and durability tests following the GB. In the above tests, automotive fault diagnosis platform meets the design requirements.

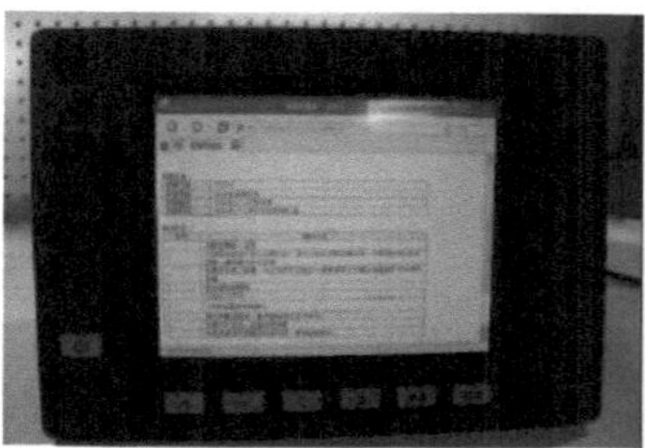

Fig. 5 Arora and repair suggestions

6 Conclusions

The platform has small size, lightweight, intelligent voice alarm, convenient interactive, and other characteristics can be embedded in the truck's center console. It monitors the fault information of the truck, helps maintenance personnel to locate the point of fault quickly, and guides the repair process. The platform raises the level of informatization, survivability, and reliability of medium and heavy trucks, contributing our strength for the modernization, electronic, and innovation of China's new generation of medium and heavy-duty trucks.

References

1. Yang, S., Zhang, J., Shi, Y.: ARM Embedded Linux System Development Techniques Explain. Electronic Industry Press, Beijing (2009)
2. Rao, Y., Zou, J., Wang, J., Zheng, Y.: Fieldbus CAN Principle and Application Technology. Beijing University of Aeronautics and Astronautics Press, Beijing (2007)
3. Jasmin Blanchette, Mark Summerfield: C++ GUI Qt 4 Programming. Electronic Industry Press, Beijing. (Yan Fengxin translated) (2008)

Simultaneous Object Tracking and Classification for Traffic Surveillance

Julfa Tuty and Bailing Zhang

Abstract Object tracking is the problem of estimating the positions of moving objects in image sequences, which is significant in various applications. In traffic surveillance, the tasks of tracking and recognition of moving objects are often inseparable and the accuracy and reliability of a surveillance system can be generally enhanced by integrating them. In this paper, we proposed a traffic surveillance system that features of classification of pedestrian and vehicle types while tracking, which works well in challenging real-word conditions. The object tracking is implemented by the Mean Shift and object classification is implemented with several different classification algorithms including k-nearest neighborhood (kNN), support vector machine (SVM), multi-layer perceptron (MLP), and random forest (RF), with high classification accuracies.

Keywords Traffic surveillance · Object tracking · Object classification · Mean shift

1 Introduction

Visual object tracking is an important research topic in computer vision. Its main purpose is to extract the desired information from a particular environment. Many important applications such as video surveillance, vehicle navigation, robotics, and video summarization require object tracking. For example, the trajectory of a vehicle committed in illegal acts, such as breaking the traffic rules, can be recorded and analyzed from tracking [1–3]. Despite the progresses made, tracking in an outdoor environment is still a challenging task [2].

J. Tuty · B. Zhang (✉)
Department of Computer Science and Software Engineering, Xi'an Jiaotong-Liverpool University, Suzhou, China
e-mail: bailing.zhang@xjtlu.edu.cn

S. Patnaik and X. Li (eds.), *Proceedings of International Conference on Computer Science and Information Technology*, Advances in Intelligent Systems and Computing 255, DOI: 10.1007/978-81-322-1759-6_86,

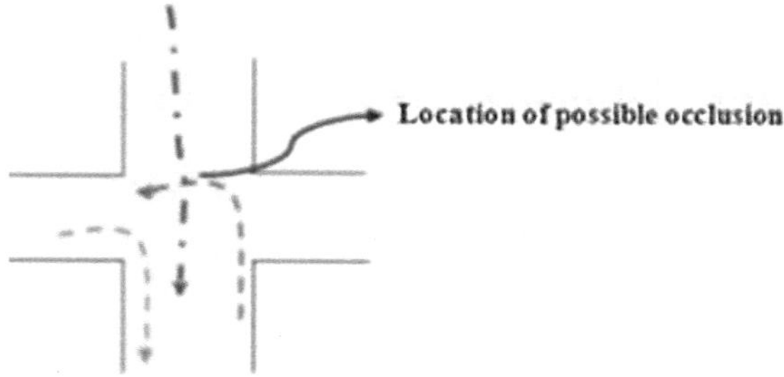

Fig. 1 Possible occlusion in the intersecting road

On the other hand, object classification has often been performed separately from object tracking module in many applications. In recent years, an increasing number of studies attempted to execute these two tasks together [4–6]. In traffic surveillance, it is particularly pertinent to simultaneously track and classify moving objects on road, such as pedestrians and different types of vehicles. In [7] the authors implemented a system by jointly estimating the posterior distribution. Such an integrated system is expected to be reliable in various adverse conditions. Considering the dynamic, complex outdoor environment, however, integration of tracking and classification still has many challenges, with common obstacles including occlusion, scale variations, and changing direction [8].

With the problems described above, this paper designs an efficient tracking and classification system, with low quality surveillance video as the main target. While tracking generates the trajectory of the objects of interest from a video stream, classification can infer different properties of a particular object for different high-level tasks. To address the real-time applicability of such a system, speed and cost are the main concern. Accordingly, we choose the efficient Mean Shift algorithm to implement tracking. Simple yet competent classification algorithms, including k-nearest neighbor (kNN), support vector machine (SVM), multi-layer perceptron (MLP), and random forest (RF), were experimentally compared, with satisfactory performance reported on a real-traffic surveillance video.

2 Object Tracking by Mean Shift Algorithm for Traffic Surveillance

The objective of video tracking is to associate target object in several consecutive video frames. To implement tracking for a video surveillance system, there are many issues to be taken into account. The most common factors that may critically influence many video processing tasks include occlusions, changing orientation, and changing speed. Particularly, object occlusion is the prime reason for the failure of many tracking algorithms. Generally, there are two types of occlusions, i.e., partial occlusion and complete occlusions. Partial occlusion occurs when part of the target object is occluded by other objects (Fig. 1).

Among the existed algorithms, the Mean Shift determines the next position of the target object using a confidence map. It finds highest density area inside a bounding box. While the Kalman Filter is often considered as a more reliable technique for tracking, it is not always appropriate in every scenario, particularly when the objects are subjective to changes in its moving direction as is the case we addressed for the surveillance of traffic at intersections.

Mean Shift is an iterative method usually starts with an initial estimate. A predefined kernel function determines the weight of nearby points for re-estimation of the mean. Typically, we use the Gaussian kernel on the distance to the current estimate, $K(x_i - x) = e^{c|x_i - x|^2}$. The weighted mean of the density in the window is then determined by

$$m(x) = \frac{\sum_{x_i \in N(x)} K(x_i - x)x_i}{\sum_{x_i \in N(x)} K(x_i - x)} \quad (1)$$

where $N(x)$ is the neighborhood of x, a set of points for which $K(x) \neq 0$. The Mean Shift algorithm then sets $x \leftarrow m(x)$, and repeats the estimation until $m(x)$ converges. Mean Shift usually converges in static distributions. Kernel density estimation is often applied in which the size of the kernels is varied depending upon either the location of the samples or the location of the test point. A histogram is a simple nonparametric estimation of a probability distribution. By analyzing the peak of a confidence map from the previous objects surroundings, Mean Shift assigns the new location of that object in the next frame based on the computed result. The most attractive advantages of Mean Shift algorithm include its low-computational cost and high-processing speed.

3 Classification Methods Compared

3.1 k-nearest Neighbor

kNN classifies object based on the minimal distance to training examples by a majority vote of its neighbors [9]. Specifically, the object is assigned to the most common class among its *k*-nearest neighbors. A satisfactory performance of kNN algorithm often requires a large number of training data set.

3.2 Multi-Layer Perceptron

In neural network, MLP consists of multiple layers of nodes, i.e., the input layer, single or multiple hidden layer, and an output layer. An MLP classifier is usually trained by the error backpropagation algorithm.

Fig. 2 Illustration of the simple tracking process (**a**), example of the tracked vehicle (**b**) and example of partial occlusion (**c**)

3.3 Support Vector Machine

Given labeled training data, SVM will generate an optimal hyperplane to categorize new examples. Intuitively, the operation of the SVM algorithm is based on finding the hyperplane that gives the largest minimum distance to the training examples. And the optimal separating hyperplane maximizes the margin of the training data.

3.4 Random Forest

RF consists of many decision trees combined as one collection. While creating a node, only a fraction of randomly chosen features are used among the available features. In order to obtain good classification accuracy, low bias and low correlation between the constituent trees are essential. To get low bias, trees are grown to maximum depth. To get low correlation, randomization, by selecting random subset features to split anode and of training samples to build a tree, is applied [10].

4 Experiment

To test the performance of the Mean Shift algorithm, several experiments have been conducted. Figure 2a shows the simple procedure of tracking. A target object is first manually selected and marked by a bounding box. Then, Mean Shift will automatically calculate the position of object in subsequent frames.

In object tracking, a common negative factor that might lead to failure is the background noise. As shown in the figure above, tracked vehicle is passing a crossing line, which may disturb the prediction of the next position of the object. The result from the Mean Shift shows the robustness in such a condition, as shown in Fig. 2b.

Fig. 3 Changing speed, direction, and orientation of a motorcycle

Fig. 4 Small objects that move away, i.e., changing to smaller scale

As a frequent occurred difficult situation in a video tracking, two objects may become close to each other and even occlude. As can be seen in Fig. 2c where a person on bike was occluded by a car, Mean Shift can produce satisfactory results in such a crowded area in the low-resolution video. It is also obvious from Fig. 2c that when one tracked object moving closer to the other object, the tracking window might shift to the second object.

As the objects move in an intersection area, orientation of a tracked object may be subjective to frequent changes. As illustrated in Fig. 3, Mean Shift generates robust tracking results. Figure 4 demonstrates another situation that a moving motorcycle that changes its speed and moves away from the central video frame, which can be successfully tracked.

To verify the classification performance, a collection of detected objects was prepared. Some of the samples are illustrated in Fig. 5. A common holdout procedure was followed, with 80 % of data for training the classifiers and 20 % for testing. There are two experiments separately carried out for the classification. First, experiment was conducted to compare the two categories of data (i.e., vehicle and human). Comparison was made for kNN, RF, MLP, and SVM. In this experiment, vehicle consists of car, bus, and van while human class consists of pedestrian, bike, and motor. From Fig. 6, we can find that kNN and SVM reached almost 100 % accuracy while RF and MLP are slightly below 100 %.

The second experiment was conducted for three categories of vehicle types including SUV, bus, and van. These three categories of vehicle types are easily confused due to their similar appearance. The comparison results are provided in Fig. 7. From Fig. 7, it is obvious that SVM outperforms the other classifiers, with the accuracy rate of 97.41 %. In contrast, RF does not perform as well, which only

Fig. 5 Samples of the collected data

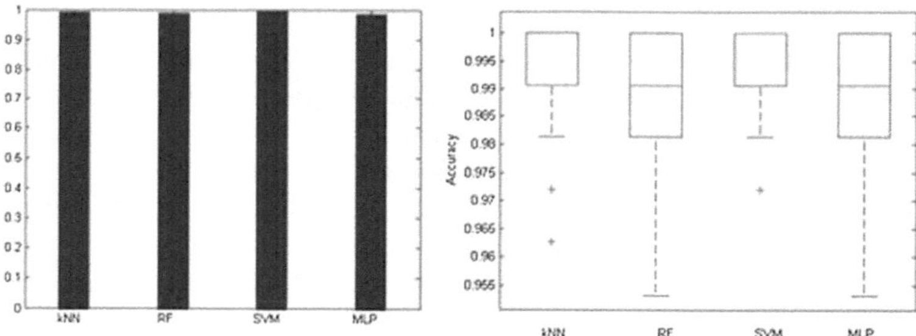

Fig. 6 Classification accuracies for the people and vehicles

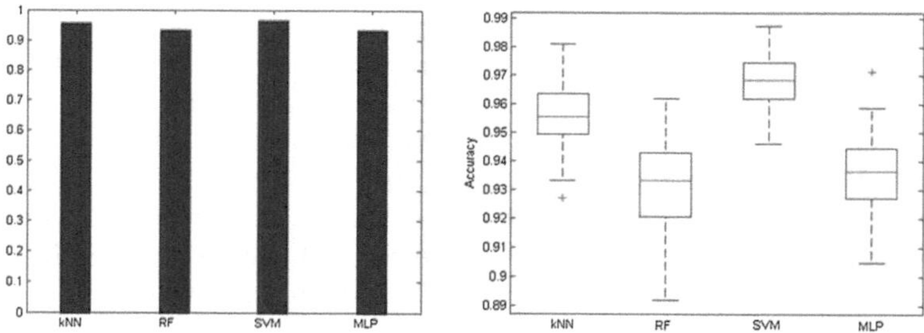

Fig. 7 Classification accuracies for SUV, van, and bus

reaches 93.62 %. On the other hand, kNN can still be regarded as the suitable classifier for the simultaneous tracking and classification system due to the trade-off between accuracy and computational cost.

5 Conclusion

In this paper, we proposed an integrative approach to traffic surveillance, which jointly applies moving object tracking and classification. Comparing with some other proposed methods, the advantages of our approach include the simple yet efficient Mean Shift algorithm for real-time application, together with an appropriate classification algorithm to successfully deliver a reliable information about the moving object. Experiments on real-life surveillance videos verified the satisfactory performance of the proposed approach.

Acknowledgments The project is supported by Suzhou Municipal Science and Technology Foundation grants SS201109 and SYG201140.

References

1. Salhi, A., Jammoussi, A.Y.: Object tracking system using Camshift, Meanshift and Kalman filter. World Acad. Sci., Eng. Tech. **64**, 674–679 (2012)
2. Suliman, C., Cruceru, C., Moldoveanu, F.: Kalman filter based tracking in a video surveillance system. Adv. Electr. Comput. Eng. **10**(2), 30–34 (2010)
3. Rasid, L.N., Suandi, S.A.: Versatile object tracking standard database for security surveillance. In: 10th International Conference on Information Science, Signal Processing and their Applications (ISSPA), pp. 782–785, Kuala Lumpur (2010)
4. Hsieh, J.W., Yu, S.H., Chen, Y.S., Hu, W.F.: Automatic traffic surveillance system for vehicle tracking and classification. IEEE Trans. Intell. Tranp. Syst. **7**(2), 175–186 (2006)
5. Morris, B., Trivedi, M.: Robust classification and tracking of vehicles in traffic video streams. In: Intelligent Transportation Systems Conference (ITSC'06), pp. 1078–1083, Toronto, Ont., Sept 2006
6. Lin, H.Y., Wei, J.Y.: A street scene surveillance system for moving object detection, tracking and classification. In: Proceedings of 2007 IEEE Intelligent Vehicles Symposium, pp. 1077–1082, Istanbul, Turkey, 13–15 June 2007
7. Hampapur, A., Brown, L., Connell, J., Ekin, A., Haas, N., Lu, M., Merkl, H., Pankanti, S., Senior, A., Shu, C.F., Tian, Y.L.: Smart video surveillance: exploring the concept of multiscale spatiotemporal tracking. IEEE Signal Processing Magazine, pp. 38–51, March 2005
8. Li, L.Z., Zhang, Y.B., Liu, Z.G.: Research on detection and tracking of moving target in intelligent video surveillance. In: International Conference on Computer Science and Electronics Engineering (ICCSEE), pp. 477–481, Hangzhou, China (2012)
9. Silva, L.A., Hernandez, E.D.M.: A SOM combined with KNN for classification task. In: Proceedings of International Joint Conference on Neural Networks, pp. 2368–2373, San Jose, California, USA, 31 July–5 Aug 2011
10. Ali, S.S.M., Joshi, N., George, B., Vanajakshi, L.: Application of random forest algorithm to classify vehicles detected by a multiple inductive loop system. In: International IEEE Conference on Intelligent Transportation Systems, pp. 491–495, Anchorage, Alaska, USA, 16–19 Sept 2012

Part VIII
Computer Applications

Optimization of the Translation of Labeled Transition Systems to Kripke Structures

Long Zhang, Wanxia Qu and Yang Guo

Abstract Labeled transition systems (LTSs) and Kripke structures (KSs) are the two most prominent semantic models in concurrency theory. Firstly, the models described by LTSs are usually translated to KSs. Then, an ad hoc model checker is used to verify the properties of the models. The performance of the embedded algorithm and the size of KS have severely confined the verification of LTS models. To address the problem in the translation of LTSs into KSs, a new and optimized algorithm was presented in this study. The new algorithm creates a smaller-scaled KS with a slight increase in time and space cost, thereby relieving the state explosion problem significantly.

Keywords Labeled transition system · Kripke structure · Model checking

1 Introduction

Kripke structures (KSs) and labeled transition systems (LTSs) are the two most prominent semantic models in concurrency theory. Many studies conducted in the past 20 years [1–4] show that LTSs and KSs are equivalent in terms of expression power. The KS is mainly used in the field of model checking. Firstly, the KS describes a system and then verifies whether the system is satisfied with the properties described by temporal logics. The LTS is an action-based model that is very suitable for describing reactive systems. The LTS is commonly used to

L. Zhang (✉) · W. Qu · Y. Guo
College of Computer, National University of Defense Technology, Changsha 410073, China
e-mail: longzhang.cav@gmail.com

W. Qu
e-mail: quwanxia@nudt.edu.cn

Y. Guo
e-mail: guoyang@nudt.edu.cn

S. Patnaik and X. Li (eds.), *Proceedings of International Conference on Computer Science and Information Technology*, Advances in Intelligent Systems and Computing 255, DOI: 10.1007/978-81-322-1759-6_87,

describe the behavior of the actual system. To verify the systems described by the LTS, the LTS must be translated to a KS, and a mature model checker must be used for the verification process. The performance of the translation algorithm and the size of the KS have severely confined the verification of LTS models.

The basic translation method, referred to in the present paper as **KS**, was firmly established by De Nicola and Vaandrager in [1]. In [2], the method was further improved and expanded, and the silent action was added to represent the unobserved action. Such enhancements were made to ensure that the KS produced by the conversion is a fair structure. The size of the KS produced by De Nicola's method is represented by $|S| + |T|$, where $|S|$ is the state number of LTS, and $|T|$ is the number of non-silent actions. With the increase in system size, the verification of LTS is faced with the state space explosion problem. The symbolic method [5, 6] has been used to improve system verification scale, but not to reduce the state space. To increase the coverage estimation of the verification process, a system described by FSM was translated to KS by Xu et al. [7]. As of this writing, minimization techniques are the main research directions. Reniers and Willemse [3] showed that bisimilarity or stuttering equivalence minimization in one domain can be obtained by implementing minimization in the other domain. Rob [4] extended their results by proving that the same property holds for similarity. However, [3, 4] mainly focused on minimization techniques, with the translation algorithm the same as that in [1]. As mentioned in [4], translating LTSs to KSs can be done in many ways, and finding the best algorithm will be beneficial.

In this paper, we propose a new algorithm to translate LTSs into KSs. Compared with the **KS**, the new algorithm generates a smaller-scaled KS. The KS generated by the proposed algorithm is equal to the KS generated by the **KS**. In the proposed algorithm, some redundant states are deleted, greatly reducing the state space of the KS model. Some randomly generated LTSs are processed to evaluate the efficiency of the algorithm. A parameterized cache coherence protocol described by an LTS is also processed to generate a KS that will be verified by ad hoc model checkers. The experimental results show that the new algorithm creates a smaller-scaled KS with a slight increase in time and space cost, thereby relieving the state explosion problem significantly.

The rest of the paper is organized as follows. In Sect. 2, we introduce the preliminaries. In Sect. 3, a new translation algorithm is proposed. Section 4 analyzes the experimental results. Section 5 finishes the paper with conclusions and possible future work.

2 Preliminaries

In this section, we formally introduce the computational models of LTS's and KS's and the definitions used for the proofs in Sect. 3.

Definition 1 (Labeled Transition System) An LTS is a three-tuple $\langle S, \mathrm{Act}, \rightarrow \rangle$, where

- S is a set of states.
- Act is a finite, non-empty set of visible actions.
- $\rightarrow \subseteq S \times (\text{Act} \cup \tau) \times S$ is the transition relation. A three-tuple $(s, a, t) \in \rightarrow$ is called a transition, where $s, t \in S$ and $a \in \text{Act}$.

We write $s \xrightarrow{a} t$ whenever $(s, a, t) \in \rightarrow$. Note that in the definition of the LTS, a special constant τ is assumed outside the alphabet of the set of actions Act. This constant is used to represent the so-called silent steps in the transition system. The size of an LTS $T = \langle S, \text{Act}, \rightarrow \rangle$ is defined as the number of states it has. Hence, given another LTS $T' = \langle S', \text{Act}', \rightarrow' \rangle$, $|T| \leq |T'|$ if and only if $|S| \leq |S'|$.

Definition 2 (Kripke Structure) A KS is a four-tuple $\langle S, \text{AP}, \rightarrow, L \rangle$, where

- S is a set of states.
- AP is a set of atomic propositions.
- $\rightarrow \subseteq S \times S$ is a total transition relation. An element $(s, t) \in \rightarrow$ is called a transition, where $s, t \in S$.

AP is a finite, non-empty set. In lieu of the convention for KS, we write $s \rightarrow t$ whenever $(s, t) \in \rightarrow$, that is, a transition occurs from state s to t. The size of a KS $K = \langle S, \text{AP}, \rightarrow, L \rangle$ is defined as the number of states it has. Hence, given another KS $K' = \langle S', \text{AP}', \rightarrow', L' \rangle$, $|K| \leq |K'|$ if and only if $|S| \leq |S'|$.

The inputs of some model checkers can be described merely as "fair" structures, that is, KS's in which all paths are of infinite length. To solve this problem, a special constant τ is introduced in the paper [2]. The constant τ used to present silent actions is not in Act. The silent actions are unobserved to all LTSs. An ordinary LTS is extended from finite paths to infinite paths by adding τ loops to the last state. The KS converted from the extended LTS satisfies the fairness constraint.

Definition 3 (Path) Let $K = \langle S, \text{AP}, \rightarrow, L \rangle$ be a KS. A path starting in state $s \in S$ is an infinite sequence $s_0 s_1 \ldots$, such that $(s_i, s_{i+1}) \in \rightarrow$ for all i's, and $s = s_0$. The set of all paths starting in s is denoted as Path(s).

Definition 4 (Trace) Let $K = \langle S, \text{AP}, \rightarrow, L \rangle$ be a KS. Let $\pi = s_0 s_1 \ldots$ be a path starting in s_0. The trace of π, denoted as Trace(π), is the infinite sequence $L(s_0)L(s_1)\ldots$.

Definition 5 ($\perp$ Equivalence) Let $K_1 = \langle S_1, \text{AP}_1, \rightarrow_1, L_1 \rangle$ and $K_2 = \langle S_2, \text{AP}_2, \rightarrow_2, L_2 \rangle$ be two KSs. For any $s \in S_1, t \in S_2$, if Trace(s) = Trace(t) without consideration of special label $\perp$, then KS K_1 and K_2 are$\perp$ equivalent.

3 Model Translation Algorithm

We first describe the original translation from LTSs to KSs referred to as **KS**. The new algorithm called the optimized **KS** (**OKS**) is then presented in detail.

3.1 Original Model Translation: KS

Definition 6 (KS) Let $T = \langle S, \text{Act}, \rightarrow\rangle$ be an LTS. The translation **KS**: LTS $\rightarrow$ KS is formally defined as $\mathbf{KS}(\text{T}) = \langle S', \text{AP}, \rightarrow', L\rangle$, where

- $S' = S \cup \{(s, a, t) \in \rightarrow | a \neq \tau\}$.
- $AP = \text{Act} \cup \{\perp\}$, where $\perp \notin \text{Act}$.
- $\rightarrow'$ is the least relation satisfying:

$$\frac{}{s \rightarrow (s, a, t)} \quad \frac{}{(s, a, t) \rightarrow t} \quad \frac{s \xrightarrow{\tau} t}{s \rightarrow t}$$

- $\forall s \in S, L(s) = \{\perp\}$, and $L((s, a, t)) = \{a\}$.

In this translation, the fresh symbol $\perp$ is used to label the states from the LTS. Let S_d be the subset of S that has no successors. The size of the sets of states and the transitions of the KS produced by **KS** is given by the following formula:

$$|S'| = \begin{cases} n + m - u & \text{if } d = 0 \\ n + m - u + 1 & \text{if } d \neq 0 \end{cases}, \tag{1}$$

$$| \rightarrow' | = \begin{cases} 2m - u & \text{if } d = 0 \\ 2m - u + 1 + d & \text{if } d \neq 0 \end{cases}, \tag{2}$$

where $n = |S|, d = |S_d|, m = | \rightarrow |$, and u is the number of unobservable transitions.

The basic idea of **KS** can be described by Fig. 1. The left LTS generates the right KS using **KS**, as shown in the figure. Each visible transition leads to an additional state that is labeled with that of the original transition symbol. A large number of states are labeled with $\perp$, indicating that the state is an invisible action. As a KS can be considered as a non-deterministic automaton, the states labeled with $\perp$ are redundant in the actual systems described by the LTS.

3.2 New Model Translation Algorithm: OKS

This paper proposes a new algorithm called the OKS for translating LTSs into KSs. This new algorithm deals with each state in an LTS one by one and groups the incoming transitions by the action tag. The current state is split into a number

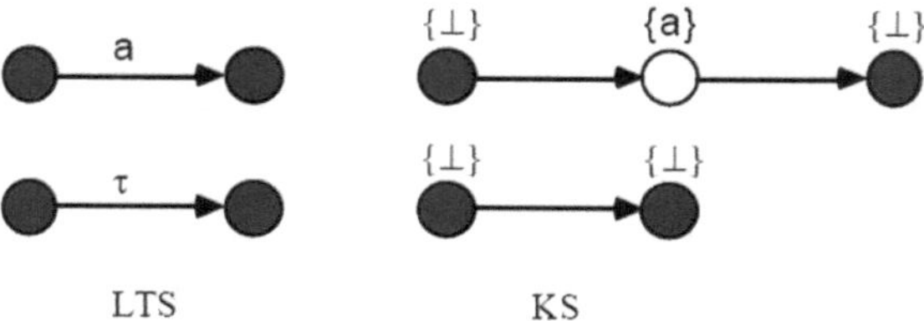

Fig. 1 KS translation

of sub-states equal to the number of groups. Each sub-state is labeled with the action tag, which is used to group transitions, and then the symbol on the transitions is removed to obtain the final KS.

The new algorithm described with a pseudocode is shown as *Algorithm 1*. The input of ***OKS*** is an LTS $T = \langle S, \text{Act}, \rightarrow \rangle$, and the output of ***OKS*** is a KS ***OKS*** $(T) = \langle S', \text{AP}, \rightarrow', L \rangle$. The entire algorithm can be divided into four steps:

1. Initialization
2. Splitting of state
3. Addition of transition
4. Construction of the set of atomic proposition

Algorithm 1 OKS

Input: an LTS $T =< S, Act, \rightarrow>$

Output: an KS **OKS**$(T) =< S', AP, \rightarrow', L >$

```
    /*Initialization*/
 1: S' := {∅}, AP := {∅}, →' := {∅}, L := {∅}
    /*Splitting of state*/
 2: for all s ∈ S do
 3:   for all a ∈ Act do
 4:     if ∃s', (s', a, s) ∈→ then
 5:       S' := S' ∪ {(s', a, s)}
 6:     end if
 7:   end for
 8: end for
    /*Addition of transition*/
 9: for all s ∈ S' do
10:   for all q ∈ S' and q ≠ s do
11:     if ∃a ∈ Act and ((parent(q), a, parent(s)) ∈ →)
        then
12:       →' := →' ∪ (q, a, s)
13:       L(s) := a
14:     end if
15:   end for
16: end for
    /*Construction of the set of atomic proposition*/
17: AP = Act
```

Step 1 Initialization. During initialization, the set S', AP, $\rightarrow'$, and L are assigned as $\{\varnothing\}$.

Step 2 Splitting of state. The code on lines 2–8 completes the splitting of states. The current state s is split by the incoming transitions. If the number of different actions is m, the state s is split into m sub-states, which are then added to S'. To add the transition quickly in Step 3, the parent state should record all information about the child nodes. All states are stored in an incoming format, and the state is split using the recorded information. An auxiliary function parent(s) is used to find the parent state of $s \in S'$. We use a hash table to store the information about the state splitting. An access to the function *parent*() is equivalent to a table lookup process.

Step 3: Addition of transition. The code on lines 2–8 completes the function of the addition of transition to the new KS. We use the auxiliary function *parent*() proposed in Step 2. For each $s, p \in S'$, if there exists $a \in$ Act, and (parent(p), a, parent(s)) $\in \rightarrow$, then we add the new transition (p, s) to $\rightarrow'$. As every parent node has a record of the information for the sub-states, the loop lines 10–15 can be completed in constant times.

Step 4: Construction of the set of atomic proposition. AP is directly assigned as Act.

The model checkers are equipped with elegant temporal logics such as the computation tree logic (CTL) and the CTL*, which are used to describe properties of systems. A new action-based logic, called the action-based CTL (ACTL) proposed by De Nicola and Vaandrager, is used to describe the properties of LTSs. In relation to the mapping from LTSs to KSs, a mapping from the ACTL formula to the CTL formula is also included in [1].

The correctness of **OKS** is stated by Theorems 1 and 2.

Theorem 1 (**KS** consistency) *Let* $T = <S, Act, \rightarrow>$ *be an LTS,* **KS** $(T) = \langle S', \mathrm{AP}, \rightarrow', L\rangle$ *be a KS, and* $q \in S$*; let* Φ *be an ACTL formula.* $T, q| = \Phi$ *if and only if* **KS**(*T*), **KS**(*q*)| = **KS**(Φ).

Proof The complete proof is reported in [1]. □

Theorem 2 (Model Equivalence) *Let* $T = <S, Act, \rightarrow>$ *be an LTS.* **KS**(*T*) *and* **OKS**(*T*) *are* $\perp$ *equivalent.*

Proof Let **KS**$(T) = \langle S_1, \mathrm{AP}_1, \rightarrow_1, L_1\rangle$ and **OKS**(T) $= \langle S_2, AP_2, \rightarrow_2, L_2\rangle$. For all $s \in S_1$ on one side, Trace(s) $= L^1(s_0)L^1(s_1)L^1(s_2)\ldots$, where $s = s_0$. Based on the definition of **KS**, we know that all even states are labeled with $\perp$, and all odd states are labeled with $a_i \in$ Act. All $\perp$ labels in Trace(s) are removed, then Trace(s) $= a_0a_1a_2\ldots$. Given the definition of **OKS**, we can find a state $s' \in S_2$ so that $\mathrm{Trace}(s') = a_0a_1a_2\ldots$. For all $s \in S_2$ on the other side, Trace(s) $= L^2(s_0)L^2(s_1)L^2(s_2)\ldots$, where $s = s_0$. Based on the definition of **OKS**, we know that all states are labeled with $a_i \in$ Act. A $\perp$ label is added in every state $s_i \in S_2$, then $\mathrm{Trace}(s) = \perp a_0 \perp a_1 \perp a_2 \perp\ldots$. Given the definition of **KS**, we can find a state $s' \in S_1$ so that $\mathrm{Trace}(s') = \perp a_0 \perp a_1 \perp a_2 \perp\ldots$

In summary, **KS**(*T*) and **OKS**(T) are $\perp$ equivalent. □

By Theorems 1 and 2, we know that **OKS** can ensure the consistency in the translation.

3.3 Complexity Analysis

The complexity analysis mainly involves Steps 2 and 3 in **OKS**. Step 2 consists of two layers of for-loops, and this step lasts for $O(|S| \times |\mathrm{Act}|)$. If the information of the incoming transition is stored, then this step lasts for $O(|S|)$. Step 3 also consists of two layers of for-loops. The loop body of the inner loop is the function parent(), which can be completed in $O(|S|)$. In the worst case, every state is split into $|Act|$ sub-state, and the longest time is $O(|S|^3 \times |\mathrm{Act}|^2)$. However, if the information about state splitting is stored, the loop can be completed in $O(|S| \times |\mathrm{Act}|)$. In summary, the final time complexity of **OKS** with optimization is $O(|S| \times |Act|)$.

4 Experiments

We implemented the **OKS** algorithm to some randomly generated LTSs and multiprocessor cache coherence protocols described by an LTS. The results were also compared with those of the original **KS**. All experiments were conducted on a PC (running Windows 7 and Microsoft Visual Studio 2008) with a 3.3-GHz Intel Core processor and 4 Gb of available main memory.

The experimental results are shown in Table 1. For randomly generated LTSs, the size of the KS produced by the **OKS** was smaller than the one produced by the **KS**, and the average state space was reduced by up to 30 %. As dividing states and adding transitions are more complex, the computation time overhead of the **OKS** was two to four times larger than that of the **KS**. To improve the execution speed, the cost of space was increased within the permissible range.

The execution time of the **KS** and the **OKS** increased with the increase in the transition number for the same-sized state number. The complexity analysis of **OKS** in Sect. 3 showed that adding transitions is time consuming. We effectively controlled the time cost in linear growth by optimizing the algorithm.

A parameterized cache coherence protocol described by an LTS [8] was used as the input data. MESI protocol is a well-known cache coherence protocol, and the subscript presents the number of nodes. For example, MESI_{10} represents an LTS that is the asynchronous synthesis of 10 nodes. Moreover, the **OKS** generated smaller KSs than the **KS** did in parameterized cache coherence protocol models. The **OKS** showed significant advantages in reducing the size of a KS in an LTS with more transitions and less action symbols. As more transitions were organized

Table 1 Experimental result

Source	LTS		KS		OKS	
	$\|S\|$	$\|\rightarrow\|$	$\|S\|$	Time (s)	$\|S\|$	Time (s)
Random	256	585	841	0.02	545	0.05
Random	4,096	15,457	19,553	0.45	13,769	1.33
Random	16,384	231,843	248,227	5.04	153,523	16.15
Random	4,096	31,703	35,799	0.76	28,167	2.96
Random	4,096	64,994	69,090	1.34	57,603	6.05
$MESI_8$	272	71,816	72,088	1.12	15,105	2.36
$MESI_9$	530	276139	276669	4.24	33380	7.49
$MESI_{10}$	1044	1079506	1080550	16.48	73080	24.37

in the same group, the time needed to divide the states and add transitions was reduced. The experimental results showed that the new algorithm has great value in practice.

5 Conclusion

LTSs and KSs are the two most prominent semantic models in concurrency theory. Hence, investigating the equivalent transformation between the two models is valuable. Based on the work of Nicola et al, we proposed and optimized a new translation algorithm to translate LTSs to KSs. The new algorithm **OKS** creates smaller-scaled KSs and relieves the state explosion problem significantly. The **OKS** can complete model transformation efficiently and can be used to verify the cache coherence protocols.

For our future work, we plan to improve the **OKS**. To meet the needs of large-scale validation, we will combine the minimization techniques to reduce the state space. Expanding the algorithm to generate a KS with fair paths is also a research direction.

Acknowledgments This work was supported by the National Nature Science Foundation of China (Nos. 61070036 61133007).

References

1. De Nicola, R., Vaandrager, F.W.: Action Versus State Based Logics for Transition Systems. Semantics of Systems of Concurrent Process LNCS, vol 469, pp. 407–419. Springer, Berlin (1990)
2. De Nicola, R., Fantechi, A., Gnesi, S., Ristori, G.: An Action-Based Framework for Verifying Logical and Behavioural Properties of Concurrent Systems. Computer Aided Verification LNCS, vol. 575, pp. 37–47. Springer, Berlin (1992)
3. Michel, A.R., Tim, A.C.W.: Folk Theorems on the Correspondence between State-Based and Event-Based Systems. In: SOFSEM2011: Theory and Practice of Computer Science, LNCS 6543, pp. 494–505. Springer, Berlin (2011)
4. Rob, S.: Correspondence between Kripke structures and labeled transition systems for model minimization. Technische Universiteit Eindhoven (2011)
5. Fantechi, A., Gnesi, S., Mazzanti, F., Pugliese, R., Tronci, E.: A symbolic model checker for ACTL. In: Proceedings of the International Workshop on Current Trends in Applied Formal Method(FM-TRENDS 98). LNCS 1641, pp. 228–242. Springer, Berlin (1999)
6. Pecheur, C., Raimondi, F.: Symbolic Model Checking of Logics with Actions. Model Checking and Artificial Intelligent, LNCS, vol. 4428, pp. 113–128. Springer, Berlin (2007)
7. Xu, S., Kimura, S., Horikawa, K., Tsuchiya, T.: Transition-based coverage estimation for symbolic model checking. In: Proceedings of the 2006 Asia and South Pacific Design Automation Conference, pp. 1–6 (2006)
8. Guo, Y., Qu, W., Zhang, L., Xu, W.: State space reduction in modeling checking parameterized cache coherence protocol by two-dimensional abstraction. J. Supercomput. doi:10.1007/s11227-012-0755-0

Retrieving Software Component by Incidence Matrix of Digraph

Chunxia Yang, Yinghui Wang and Hongtao Wang

Abstract Component retrieval is important to improve software productivity in the field of component-based software development (CBSD). Since the implementation of component is universally encapsulated, and outward published is a set of interfaces, the information implied in component interface attracts more and more attention in component retrieval system. In this paper, interface automaton is defined in formal to capture the operation signatures and the invocation sequence of the operations for retrieval. And three kinds of matching models are developed to satisfy exact or approximate matching according to the information retriever can give. Then, the matching is implemented based on incidence matrix of digraph corresponding to interface automaton and an example is illustrated. Since the implementation mainly involved manipulations of digraph incidence matrix except semantic matching, it is more suitable for computer to process.

Keywords Component retrieval · Interface automaton · Incidence matrix

---

C. Yang (✉) · Y. Wang
School of Computer Science and Engineering, Xi'an University of Technology, Xi'an, People's Republic of China
e-mail: ycxxaut@163.com

Y. Wang
e-mail: wyh_925@163.com

C. Yang
College of Electronics and Information, Xi'an Polytechnic University, Xi'an, People's Republic of China

H. Wang
School of Printing and Packaging Engineering, Xi'an University of Technology, Xi'an, People's Republic of China
e-mail: whtxaut@163.com

S. Patnaik and X. Li (eds.), *Proceedings of International Conference on Computer Science and Information Technology*, Advances in Intelligent Systems and Computing 255, DOI: 10.1007/978-81-322-1759-6_88,

1 Introduction

Component retrieval is an important issue in the field of component-based software development (CBSD). A good component retrieval method can effectively help users find the appropriate components from component repository, so as to improve the efficiency of software development. At present, various component retrieval methods have been put forward; however, none of them has a wide range of applications. Thus, further exploration on this issue is still needed. As to component, it is universally accepted that a component should conform to and provide the physical realization of a set of interfaces. As the only channel for component to interact with its environment, interface describes the behavior of the component to a great extent. And the information provided by interface is used as retrieval content for component retrieval.

Earlier, static behavior information declared in component interface, such as operations and their types, is employed for component retrieval. The representative methods are signature matching [1, 2] and specification matching [3, 4]. Component signature is the union of the operations' signatures it declares, and component specification includes the precondition and post-condition for every operation besides its signature. Signature matching will act well if the retriever knows in advance the component signature. Specification matching usually has good results due to its underlying mathematical rigor; however, the high cost caused by formal expression and the equivalence proof limits its application.

Later, dynamic behavior information implied in interface, such as the change of the type of operations and the range of variables that happened after the component is run, is captured for component retrieval. Component is described by a set of tuples and each tuple represents a characteristic input–output transformation of a component [5]. Mili et al. [6] use a pair (S, R) to describe the specification of a component, where S is the space of the variables that the component defines on, R is a relation on S and describes the change of space before input and after output. The components in repository are constructed as lattices by the refinement relationship of R for a certain S. And component retrieving is implemented by the lattice fixed by S and the vertexes fixed by the refinement relationship of the lattice according to R.

The other type of dynamic behavior information implied in component interface is the invocation sequence of the operations declared in interface [7, 8]. In [7], a specification of business component is described in two levels: one is the business-operation signature, including input business data types, output business data types, and the taxonomy of business operations; the other is the invocation sequence of the operations, including sequence relationship and concurrent relationship between business operations. The weighted sum of these two similarity degrees concludes the similarity of the two components. In [8], component interface is described as an automaton; and retrievers' requirement is expressed by a flow chart that is transformed into an automaton later. Component matching is found by the inclusion relationship among the digraphs corresponding to automatons, which is different from the similarity comparison of [7]. The design comes from the assumption that the operations and their invocation order given in requirements are assured by retriever, and they should exist in the candidate component.

Comparatively speaking, the digraph corresponding to automaton is more intuitive and complete to describe the behavior information implied in component interface than other forms discussed above. Although the digraph corresponding to automaton is used in [8], the formal definition for automaton is not given clearly. Meanwhile, component matching is implemented by matching of the route sets that does not make full use of the characteristic of digraph. In this paper, the formal definition of interface automaton is given firstly, and then three kinds of matching model is presented, then an improved implementation of them is developed by employing incidence matrix of the digraph that is analyzed to be lower time cost than the method proposed in [8], and an example is showed at last to demonstrate feasibility of the implementation at last.

2 Redefined Interface Automaton

Automaton model is usually used to discuss component composition [9]. Since the operations information in component interface is highly abstracted in origin definition of interface automaton, it should be redefined for the purpose of component retrieval. Here, the modified interface automaton used for component retrieval is defined.

Definition 1 An interface specification of a component is a deterministic finite automaton $M := <V, v_0, \Sigma, T>$, where

V is a set of states;
$v_0 \in V$ is a initial state;
$\Sigma = \{\delta | \delta = \langle ?/!, \text{Opname}, \text{OutType}, \text{InType} \rangle\}$ is a set of operation signatures declared in the interface, and each operation is a four tuple, in which the symbol "?/!" denotes the call direction of the operations, in the other words, it indicates that the operation is a request function or a provide function, Opname denotes the name of the operation, OutType denotes the type of output, and InType denotes a set of types of input parameters in order;

$T = \{v_i \times \delta \times v_j \mid v_i, v_j \in V\}$ is a set of steps.

An interface automaton can be represented as a digraph, where the states, steps, and operations of the automation correspond to the vertexes, directed edges, and labels of edges of the digraph, respectively. An example of interface automaton is given in Fig. 1. According to [8], an example of retriever's requirements is shown in Fig. 2.

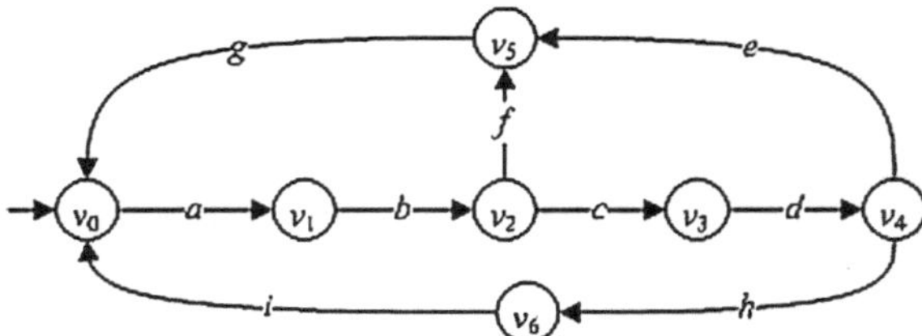

Fig. 1 Interface automaton M

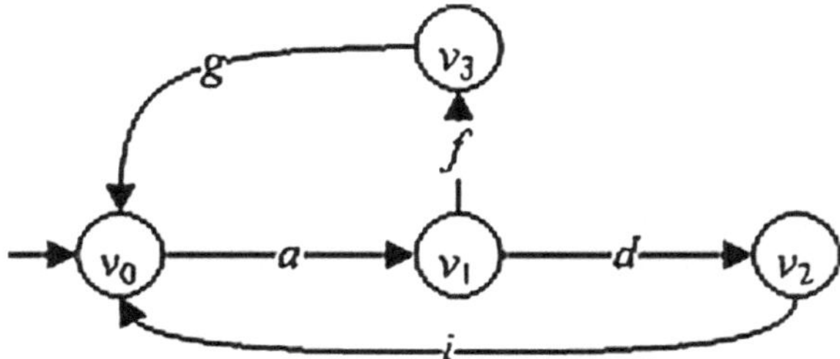

Fig. 2 Query automaton Q

We specify $|\Sigma|$ as the size of M. user cannot always describe the requirements comprehensively due to the underlying complexity of requirements. Thus, users are demanded to describe no more than the behavior features he/she assures. Therefore, $|Q| \leq |M|$ always holds.

Definition 2 In an interface automaton, if there exists some steps that constitute a continuous walk, e.g., $v_i \times \delta_i \times v_j$, $v_j \times \delta_j \times v_l, \ldots,$ $v_m \times \delta_m \times v_n$, then the operations of the previous step is called an ancestor of the operation of the following step, or the latter is called a junior of the former, i.e., we call operation δ_i an ancestor of operation δ_j or operation δ_j a junior of operation δ_i, denoted by $\delta_i = \text{ancestor}(\delta_j)$ or $\delta_j = \text{junior}(\delta_i)$, respectively.

The ancestor–junior relationship is transitive. And we specify that the operation first arrived from v_0 is an ancestor of the rest operations in a circle of an interface automaton.

3 Matching of Query and Interface Automaton

Definition 3 If there exists a mapping $f : Q \to M$ satisfying the following three conditions, then f is called a strong constraint matching (SCM for short), and M is called a SCM component of Q:

(1) $f(\delta_i) = f(\delta_j) \Rightarrow \delta_i = \delta_j,\ \delta_i,\ \delta_j \in \Sigma_Q$;
(2) $\delta_i \equiv f(\delta_i)$, where the meanings of "$\equiv$" is shown as follows: (a) The call direction of δ_i is same to that of $f(\delta_i)$; (b) distance (Opname, f(Opname)) $\leq w$, i.e., the semantic of Opname is same to that of f (Opname), and w is the semantic similarity threshold set by retriever; (c) The output type of δ_i, i.e.,

OutTpye, is same to that of $f(\delta_i)$; (d) The number and the input types of parameters of δ_i, i.e., InType, is same to that of $f(\delta_i)$, respectively;
(3) $\delta_i = \text{ancestor}(\delta_j) \Rightarrow f(\delta_i) = \text{ancestor}(f(\delta_j))$.

Condition (1) states that the mapping f is an injective mapping, which ensures that there is a one-to-one corresponding relationship between operations of Q and related operations of M. Condition (2) gives the meaning of operation signature matching, in which the semantic matching of operation names is preferable to exact string matching. Condition (3) means that the invocation sequence of operations in Q should be kept in M.

Definition 4 If there exists a mapping f: $Q \rightarrow M$ satisfying the following three conditions, then f is called a strong constraint approximate matching (SCAM), and M is called a SCAM component of Q:

(1) $f(\delta_i) = f(\delta_j) \Rightarrow \delta_i = \delta_j$, δ_i, $\delta_j \in \Sigma_Q$;
(2) $\delta_i \approx f(\delta_i)$, some tuples of δ_i can be neglected, especially, the elements of InType can be neglected completely or partially. And for the given tuples of δ_i, "$\approx$" means that (a) the call direction of δ_i is same to that of $f(\delta_i)$; (b) distance (Opname, f (Opname)) $\leq w$, whose meaning is same to above; (c) The OutType of δ_i is same to that of $f(\delta_i)$; (d) Each of the input types listed in InType of δ_i has a consistent items in that of $f(\delta_i)$;
(3) $\delta_i = \text{ancestor}(\delta_j) \Rightarrow f(\delta_i) = \text{ancestor}(f(\delta_j))$.

Definition 5 If there exists a mapping f: $Q \rightarrow M$ satisfying the following three conditions, then f is called a weak constraint approximate matching (WCAM), and M is called a WCAM component of Q:

(1) $f(\delta_i) = f(\delta_j) \Rightarrow \delta_i = \delta_j$, δ_i, $\delta_j \in \text{sub}(\Sigma_Q)$, where $\text{sub}(\Sigma_Q)$ is a subset of Σ_Q, and the percentage of $|\text{sub}(\Sigma_Q)|/|\Sigma_Q|$ is set by retriever;
(2) $\delta_i \approx f(\delta_i)$, the meanings of "$\approx$" is same to condition (2) of Definition 4;
(3) $\delta_i = \text{ancestor}(\delta_j) \Rightarrow f(\delta_i) = \text{ancestor}(f(\delta_j))$.

Obviously, the conditions (1) and (2) of the above three kinds of matching are weakening in order. The loosing matching strategy is benefit for retriever to give the query flexibly based on his/her assurance of the need. Meanwhile, condition (3) is required to be kept for all of the three kinds of matching, which is due to the invocation sequence is perceived by retriever according to its application environment and is supposed to be necessary for the retrieved component.

4 Matching Method

Hereafter, the interface automaton describing user's query is called query automaton Q, and the interface automaton in repository is called component automaton M, and Dig (Q) and Dig (M) denote the digraphs corresponding to them, respectively. In this study, the decision process of M matching with Q is considered as the decision process of Dig (M) including Dig (Q). And we define the digraph–inclusion relationship in Definition 6.

Definition 6 Given two digraphs Dig(Q) = $<V_Q, E_Q, L_Q>$ and Dig(M) = $<V_M, E_M, L_M>$, where V_t, E_t and L_t ($t \in \{M, Q\}$) are the vertex set, the directed edge set, and the label set of edges, respectively. If there is an injective mapping from L_Q (or sub(L_Q)) to L_M, such that for every $l \in L_Q$ (or $l \in$ sub(L_Q)), there is only one $l' \in L_M$ satisfying l matches with l', meanwhile, for any two mapping pairs (l_1, l'_1) and (l_2, l'_2), if l_1 = ancestor(l_2), then there must be l'_1 = ancestor(l'_2), we say Dig(M) including Dig(Q). And, we say $l' \in L_M$ is a matching edge of $l \in L_Q$ (or $l \in$ sub(L_Q)) if there is an edge l matching with l'; or else l' is called an irrelevant edge.

All the three kinds of interface automaton matching can be realized by the digraph–inclusion relationship. Specifically, label matching is corresponding to operation signature matching and "sub(L_Q)" in Definition 6 is derived from "sub(Σ_Q)" in Definition 5. As the labels of the edges are different from each other, we use the name labels to denote the directed edges.

Theorem *Component automaton M matching with query automaton Q can be reduced to the inclusion relationship between their digraphs* Dig(M) *and* Dig(Q), *i.e., if* Dig(M) *includes* Dig(Q), *then component automaton M matches with query automaton Q.*

Proof The theorem establishes obviously.

The work makes use of incidence matrix of digraph to determine whether the inclusion relationship exists between two digraphs. But, we do not intend to discuss the semantic matching of Opname and type matching in detail, which can be conceived and optimized by other deliberate considerations. The paper puts emphasis on the rest of matching matters by SCM, SCAM and WCAM. The whole decision process of Dig(M) including Dig(Q) is divided into four steps in detail. And the incidence matrix of them is denoted as M_m and Q_m respectively.

Step 1: According to the kind of matching chose by retriever (SCM, SCAM, or WCAM), the operation signature matching is carried out. If there is one-to-one corresponding relationship between the edges set (or subset, in WCAM) of Q_m and the edges subset of M_m, go to Step 2. Or else, the matching is failure.

Step 2: In M_m, all irrelevant edges and the columns fixed by them are deleted, and the connected relationship of them is extended to the matching edges. Then, the matching edges are arranged in the same order as their pre-images appearing in Q_m.

The connected relationship of irrelevant edges is dealt with as follows. For an irrelevant edge u in M_m, there must be a value "1" and a value "−1" in the column fixed by u, let the locations of them be M_m [v_1, u] and M_m [v_2, u], respectively.

Firstly, the "-1" is found in the row fixed by v_1 (there is at least a "-1" in the row since the digraph is connected), let the location be M_m [v_1, w]; then the value "0" located in M_m [v_2, w] is modified with "-1," i.e., the connected relationship of u is extended.

The inherent implication of above process is outlined as follows. Firstly, for an edge u in M_m, the locations of "1" and "-1" in the column fixed by u indicate the start point and the end point of u. For example, let the locations of the "1" and the "-1" be M_m [v_1, u] and M_m [v_2, u], it means that the start point of u is v_1 and the end point is v_2. "The "-1" is found in the row fixed by v_1, let the location be M_m [v_1, w]," which means the end point of w is v_1. Let the start point of w be v, so the relation between of u and w is $v \xrightarrow{w} v_1 \xrightarrow{u} v_2$. Since u is an irrelevant edge, in order to delete u and keep the connected feature from v to v_2, the value of M_m [v_2, w] is modified with "-1" which result in $v \xrightarrow{w} v_1 \xrightarrow{w} v_2$, and the connected feature of u is extended to w.

We delete the edge u and the column fixed by it. Analogously, all the other irrelevant edges are deleted and the connected feature of them is kept at the same time. Some situations that may arise in the above process are discussed as follows. Let u be an irrelevant edge, and the locations of "1" and "-1" in the column fixed by u are M_m [v_1, u] and M_m [v_2, u] respectively.

(1) There is only one "-1" in the row fixed by v_1, and its location is M_m [v_1, w], but the value of location M_m [v_2, w] is "1" (which means u and w construct a circle in Dig(M)). Then, the value "1" remains unchanged and u is deleted.
(2) There is a "-1" in the row fixed by v_1, and its location is M_m [v_1, w], but the value of location M_m [v_2, w] is already "-1" (which always results from the forgone deletion of the irrelevant edges). Here, the value "-1" remains unchanged and u is deleted.
(3) There are more than one "-1" in the row fixed by v_1, and the values in corresponding locations of the row fixed by v_2 is various. Here, all "0" in the row are modified with "-1" and "1," and "-1" remain unchanged. Then u is deleted. For example, suppose there are three "-1," and let the locations of them be M_m [v_1, w], M_m [v_1, x], and M_m [v_1, y], respectively. The values in M_m [v_2, w], M_m [v_2, x], and M_m [v_2, y] are "1," "-1," and "0," respectively. Here, values "1" and "-1" remain unchanged and value "0" is modified with "-1," then u is deleted.

Step 3: The corresponding vertex in M_m is found for each vertex in Q_m, and they are arranged in the same order. If the correspondence relationship cannot be established or there is a conflict during the establishing process, the matching is failure.

The corresponding vertex in M_m is found for each vertex in Q_m as follows. For an edge u in Q_m and its matching edge u' in M_m, let the locations of "1" in the columns fixed by u and u' are Q_m [v_n, u] and M_m [v_{m}, u'], respectively, then v_m is the corresponding vertex in M_m to vertex v_n in Q_m, denoting as v'_n for understanding easily. Once the corresponding relationship is established, other vertexes in M_m can no more be denoted as v'_n during the later process.

The inherent implication of the above process is explained as follows. When the connected relationship of irrelevant edge is extended to a matching edge, the end point of the matching edge is changed but the start point is unchanged. Therefore, the corresponding relationship between vertexes in Q_m and that in M_m is established by start points of a pair of matching edges.

The following situation may arise during the above process. Suppose edges u' and w' in M_m match with edges u and w in Q_m respectively, and the locations of "1" in the columns fixed by u and w are Q_m [v_n, u] and Q_m [v_n, w], i.e., both the start point of u and w are v_n. Suppose the locations of "1" in the columns fixed by u' and w' are M_m [v_{m}, u'] and M_m [v_k, w'], respectively. If u is processed before w in Q_m, then the corresponding relationship between v_n and v_m is established, and v_m is modified with v'_n. However, according to the principle outlined above, v_k corresponds to v_n and should be modified with v'_n, too. Thus, the only thing we do is to add all the value in the row fixed by v_k to the corresponding values in the row fixed by v'_n, i.e., vertex v_m and v_k are combined as vertex v'_n. Therefore, the start point of u' and w' is same, which conforms to that of u and w.

Step 4: While the corresponding vertexes and the matching edges in M_m are arranged in the same order as they appear in Q_m, the matching edges and corresponding vertexes of M_m construct a sub-matrix that has a same order of Q_m. Comparing the sub-matrix with Q_m, if for any nonzero value in Q_m, there is a same value in the corresponding location of the sub-matrix, then we conclude the ancestor–junior relationship of operations in Dig(Q) is kept in Dig(M). Furthermore, we conclude that digraph Dig(M) includes digraph Dig(Q). Or else, the matching is failure.

The matching method in this work reduces the time cost needed in [8], in which component retrieval is implemented by matching a set of routes with the other. It is time costly, because the route in component digraph is matched one by one to find the matching route for a route in user's digraph, which is time costly, and the involving repeated semantic matching is time costly. With incidence matrix, the matching is resolved by semantic matching once and some digital replacements, and the latter is more suitable for computer.

5 Examples

Here, a simple example is given to illustrate the component-matching process. M_m and Q_m are incidence matrixes of automation shown in Figs. 1 and 2. The lowercase letters on the head of columns of M_m and Q_m stand for edges in digraph (operation signatures in automation), and the same lowercase letters stand for the matching edges.

$$M_m = \begin{array}{c} \\ v_0 \\ v_1 \\ v_2 \\ v_3 \\ v_4 \\ v_5 \\ v_6 \end{array} \begin{array}{c} \begin{array}{ccccccccc} a & b & c & d & e & f & g & h & i \end{array} \\ \begin{bmatrix} 1 & 0 & 0 & 0 & 0 & 0 & -1 & 0 & -1 \\ -1 & 1 & 0 & 0 & 0 & 0 & 0 & 0 & 0 \\ 0 & -1 & 1 & 0 & 0 & 1 & 0 & 0 & 0 \\ 0 & 0 & -1 & 1 & 0 & 0 & 0 & 0 & 0 \\ 0 & 0 & 0 & -1 & 1 & 0 & 0 & 1 & 0 \\ 0 & 0 & 0 & 0 & -1 & -1 & 1 & 0 & 0 \\ 0 & 0 & 0 & 0 & 0 & 0 & 0 & -1 & 1 \end{bmatrix} \end{array}$$

$$Q_m = \begin{array}{c} \\ v_0 \\ v_1 \\ v_2 \\ v_3 \end{array} \begin{array}{c} \begin{array}{ccccc} a & d & g & f & i \end{array} \\ \begin{bmatrix} 1 & 0 & -1 & 0 & -1 \\ -1 & 1 & 0 & 1 & 0 \\ 0 & -1 & 0 & 0 & 1 \\ 0 & 0 & 1 & -1 & 0 \end{bmatrix} \end{array}$$

Firstly, suppose the kind of matching retriever chose is WCAM and the percentage of $|\text{sub}(\Sigma_Q)|/|\Sigma_Q|$ set by retriever is 80 %. We found $|\text{sub}(\Sigma_Q)|/|\Sigma_Q| = 100\ \% > 80\ \%$, then go to Step 2.

Then, the connected feature of irrelevant edges b, c, e, and h is extended and then these edges are deleted in M_m, (M'_m) is given as the intermediate result after deleting b, and the final result is got as M'_m.

In the next step, corresponding relationship between vertexes is established between Q_m and M_m', and the intermediate and the final result are (M_m'') and M_m'' respectively.

$$(M'_m) = \begin{array}{c} \\ v_0 \\ v_1 \\ v_2 \\ v_3 \\ v_4 \\ v_5 \\ v_6 \end{array} \begin{array}{c} \begin{array}{cccccccc} a & c & d & e & f & g & h & i \end{array} \\ \begin{bmatrix} 1 & 0 & 0 & 0 & 0 & -1 & 0 & -1 \\ -1 & 0 & 0 & 0 & 0 & 0 & 0 & 0 \\ -1 & 1 & 0 & 0 & 1 & 0 & 0 & 0 \\ 0 & -1 & 1 & 0 & 0 & 0 & 0 & 0 \\ 0 & 0 & -1 & 1 & 0 & 0 & 1 & 0 \\ 0 & 0 & 0 & -1 & -1 & 1 & 0 & 0 \\ 0 & 0 & 0 & 0 & 0 & 0 & -1 & 1 \end{bmatrix} \end{array} \quad M'_m = \begin{array}{c} \\ v_0 \\ v_1 \\ v_2 \\ v_3 \\ v_4 \\ v_5 \\ v_6 \end{array} \begin{array}{c} \begin{array}{ccccc} a & d & g & f & i \end{array} \\ \begin{bmatrix} 1 & 0 & -1 & 0 & -1 \\ -1 & 0 & 0 & 0 & 0 \\ -1 & 0 & 0 & 1 & 0 \\ -1 & 1 & 0 & -1 & 0 \\ 0 & -1 & 0 & 0 & 0 \\ 0 & -1 & 1 & -1 & 0 \\ 0 & -1 & 0 & 0 & 1 \end{bmatrix} \end{array}$$

$$(M''_m) = \begin{array}{c} \\ v_0 \to v'_0 \\ v_1 \\ v_2 \to (v'_1) \\ v_3 \to v'_1 \\ v_4 \\ v_5 \to v'_3 \\ v_6 \to v'_2 \end{array} \begin{array}{c} \begin{array}{ccccc} a & d & g & f & i \end{array} \\ \begin{bmatrix} 1 & 0 & -1 & 0 & -1 \\ -1 & 0 & 0 & 0 & 0 \\ -1 & 0 & 0 & 1 & 0 \\ -1 & 1 & 0 & -1 & 0 \\ 0 & -1 & 0 & 0 & 0 \\ 0 & -1 & 1 & -1 & 0 \\ 0 & -1 & 0 & 0 & 1 \end{bmatrix} \end{array} \quad M''_m = \begin{array}{c} \\ v'_0 \\ v'_1 \\ v'_2 \\ v'_3 \\ v_1 \\ v_4 \end{array} \begin{array}{c} \begin{array}{ccccc} a & d & g & f & i \end{array} \\ \begin{bmatrix} 1 & 0 & -1 & 0 & -1 \\ -1 & 1 & 0 & 1 & 0 \\ 0 & -1 & 0 & 0 & 1 \\ 0 & -1 & 1 & -1 & 0 \\ -1 & 0 & 0 & 0 & 0 \\ 0 & -1 & 0 & 0 & 0 \end{bmatrix} \end{array}$$

Finally, for each nonzero value in Q_m, a same value is found in the corresponding location of M''_m, then we conclude Dig(M) includes Dig(Q), i.e., component interface M matches with query automaton Q.

The conclusion can be confirmed by digraphs in Figs. 1 and 2 directly, which states the correctness of the method. And more complex matching of components can also be dealt with our method.

6 Conclusions

This paper developed a new matching method for component retrieval system. With the definition of interface automaton, the interface information including operation signatures and operation invocation sequence is described and three matching models are developed. Moreover, a well-designed implementation of the matching method is proposed based on the incidence matrix of digraphs corresponding to automaton. The method is analyzed to be superior to the method presented in the work of [8] in the view of time cost and it is more suitable for computer than the others since it is mainly digital manipulations except once semantic matching. Though the method given in this study has a rigorous foundation of mathematics, it does not influence the achievement of approximate results.

Acknowledgments This work was supported by the National Natural Science Foundation of China (No. 61100009), Shaanxi Province Major Project of Innovation of Science and Technology (No. 2009ZKC02-08), Shaanxi Province Department of Education Industrialization Training Project (No. 09JC08), and Shaanxi Technology Committee Industrial Public Relation Project (No. 2011K06-35).

References

1. Zaremski, A.M., Wing, J.M.: Signature matching: A tool for using software libraries. ACM Trans. Softw. Eng. Methodol **4**, 146–170 (1995)
2. Zaremski, A.M., Wing, J.M.: Specification matching of software components. ACM Trans. Softw. Eng. Methodol **6**, 335–369 (1997)
3. Zaremski, A.M., Wing, J.M.: Signature matching: A key to reuse. In: Proceedings of 1st ACM SIGSOFT Symposium on the Foundations of Software Engineering, pp. 182–190. Los Angeles, 08–10 Dec 1993
4. Hemer, D., Lindsay, P.: Specification-based retrieval strategies for module reuse. In: Proceedings of Australian Software Engineering Conference, pp. 235–243. Canberra, 27–28 Aug 2001
5. Mittermeir, R.T., Pozewaunig, H.: Classifying components by behavioral abstraction. In: Proceedings of 4th Joint Conference on Information Sciences, pp. 547–550. North Carolina, 23–28 Oct 1998
6. Mili, R., Mili, A., Mittermeir, R.T.: Storing and retrieving software components: A refinement based system. IEEE Trans. Softw. Eng. **23**, 445–460 (1997)
7. Meng, F.C., Zhan, D.C., Xu, X.F.: A specification-based approach for retrieval of reusable business component for software reuse. World acad. Sci. Eng. Technol. **15**, 240–247 (2006)
8. Wang, Y.H., Yang, C.H.X.: A perfect design of component retrieval system. Information **15**, 1687–1704 (2012)
9. Alfaro de, L., Henzinger, T.A.: Interface automata. In: Proceedings of the Joint 8th European Software Engineering Conference and the 9th ACM SIGSOFT Symposium on the Foundations of Software Engineering, pp. 109–120. Vienna, 10–14 Sept 2001

Functional Dependencies and Lossless Decompositions of Uncertain XML Datasets

Ping Yan, Teng Lv, Weimin He and Xiuzhen Wang

Abstract With the increase in uncertain data in many new applications, such as sensor network, data integration, web extraction, etc., uncertainty of both relational databases and Extensible Markup Language (XML) datasets have attracted high-research interests in recent years. As functional dependencies (FDs) are critical and necessary to schema design in relational databases and XML datasets, it is also significant to study the FDs and their applications in lossless decompositions of uncertain XML datasets. This paper first proposed three new kinds of FDs for uncertain XML datasets based on tree-tuple model, then three lossless decomposition methods based on the proposed three FDs, respectively, are given to decompose an XML dataset losslessly.

Keywords Uncertainty · XML · Functional dependency · Lossless decomposition · Tree-tuple model

P. Yan
School of Science, Anhui Agricultural University, Hefei 230036, China
e-mail: want2fly2002@163.com

T. Lv (✉) · X. Wang
Teaching and Research Section of Computer, Army Officer Academy, Hefei 230031, China
e-mail: LT0410@163.com

X. Wang
e-mail: orange1204@163.com

W. He
Department of Computing and New Media Technologies, University of Wisconsin-Stevens Point, Stevens Point, WI 54481, USA
e-mail: whe@uwsp.edu

S. Patnaik and X. Li (eds.), *Proceedings of International Conference on Computer Science and Information Technology*, Advances in Intelligent Systems and Computing 255, DOI: 10.1007/978-81-322-1759-6_89,

1 Introduction

Extensible Markup Language (XML) has become the de facto standard of data exchange and is widely used in many fields. With the increase in applications such as data integration, web extraction, sensor networks, etc., XML datasets may be obtained from heterogeneous data sources and are not always deterministic. In such cases, XML datasets may contain uncertain data for the same attribute or element due to different data sources, information extraction, approximate query, and data measurement units. Due to uncertainty, functional dependencies (FDs) of uncertain XML datasets are much more complicated than the counterparts of relational databases and deterministic XML datasets. FDs of uncertain XML datasets are critical to lossless decomposition, which is a major concentration of the paper.

1.1 Related Work

Although there has been a lot of significant work in FDs for relational databases and deterministic XML datasets, none of them can be directly applied to uncertain XML datasets due to the different structures between XML and relational database or uncertainty in uncertain XML.

For relational databases. FDs are thoroughly studied for several decades [1, 2]. Reference [3] proposed a concept of FDs in relational databases that can deal with slight variations in data values. Reference [4] proposed the conditional FDs to detect and correct data inconsistency. It is obvious that the techniques of relational databases cannot be directly applied to XML due to the significant difference in structure between XML documents and relational databases.

For deterministic XML datasets. FDs are also thoroughly studied in recent years. There are two major approaches to define FDs in XML, i.e., path-based approach and sub-tree/sub-graph-based approach. In path-based approach [5–11], an XML dataset is represented by a tree structure and some paths of the tree with their values are used in defining XML FDs. In sub-tree/sub-graph-based approach [12, 13], FDs of XML datasets are defined by sub-graph or sub-tree in XML datasets. A sub-graph or a sub-tree is a set of paths of XML datasets. As an extension of sub-tree/sub-graph-based approach, References [14, 15] considered XML FDs with some constraint condition. All the above XML FDs cannot deal with uncertainty in XML datasets. Reference [16] proposed an approach to discover a set of minimal XML Conditional Functional Dependencies (XCFDs) from a given XML instance to improve data consistency. An XCFD extends the traditional XML Functional Dependency (XFD) by incorporating conditions into XFD specifications. But XCFDs cannot deal with uncertainty in XML datasets, either.

For uncertain relational databases. Reference [17] proposed the probabilistic FDs for probabilistic relational databases that are associated with a likelihood of the traditional deterministic FDs. Reference [18] proposed some kinds of FDs for probabilistic relational databases, such as Probabilistic Approximate Functional Dependencies (pAFD), Conditional Probabilistic Functional Dependencies (CpFD), and Conditional Probabilistic Approximate Functional Dependencies (CpAFD), which combined approximate, conditional, and approximate/conditional characteristics into traditional functional dependencies to define corresponding FDs for probabilistic relational databases. Reference [19] proposed horizontal FDs and vertical FDs for uncertain relational databases, which extended the traditional relational FDs into the uncertain relational databases. Although these work of uncertain relational databases are meaningful and significant, they cannot directly be applied in uncertain XML datasets, as XML are more complicated in structure then in relational databases.

For uncertain XML datasets. As far as we know, there is little work on FDs for uncertain XML datasets, much less on lossless decompositions of uncertain XML datasets based on uncertain XML FDs. The most related work on FDs on uncertain XML datasets is Reference. [20], in which three kinds of FDs based on tree-tuple model are given. In this paper, we extended Reference. [20] and applied them to decompose uncertain XML dataset losslessly.

1.2 Contributions

The main contributions of the paper are detailed as follows: (1) We give three types of FDs of uncertain XML dataset based on XML tree-tuple model (Sect. 3): Full FDs, Tuple-level FDs, and In-tuple-level FDs. We also analyze the relationships among these three types of FDs. (2) We give three lossless decompositions of uncertain XML datasets based on Full FDs, Tuple-level FDs, and In-tuple-level FDs, respectively (Sect. 4).

2 Preliminaries

An uncertain XML tree $T_{\text{uncertain}}$ is an XML tree T with some distributional node types [21], such as IND and MUX types. A node v of type IND specifies for each child w, the probability of choosing w. This probability is independent of the other choices of children. If the type of v is MUX, then choices of different children are mutually exclusive. That is, v chooses at most one of its children, and it specifies the probability of choosing each child (so the sum of these probabilities is at most 1). We use IND_SET and MUX_SET to denote the set of IND types and MUX types, respectively. So, an uncertain XML tree $T_{\text{uncertain}}$ is defined as $T_{\text{uncertain}}$ = (V, lab, ele, att, val, root, IND_SET, MUX_SET). An uncertain XML tree conforming to a DTD D is similar to Definition 2.2 just omitting the distributional node types (IND and MUX) in the definition. In this paper, we only consider the IND type, so $T_{\text{uncertain}}$ = (V, lab, ele, att, val, root, IND_SET). For MUX type, the definitions and methods are similar to that of the paper. We omit the probability of each value in the paper for clarity in the paper.

Given a DTD D = (E, A, P, R, r) and an uncertain XML tree $T_{\text{uncertain}}$ = (V, lab, ele, att, val, root, *IND_SET*) conforming to D, a tree tuple t for a node v in $T_{\text{uncertain}}$ is a tree rooted on node v with all of its decedent nodes. For a path set $S \subseteq \text{paths}(D)$ such that S is not a part of any path in D, $t(S)$ denotes a value set of S of all possible values of S in the tree tuple t. We use $t_1(S) = t_2(S)$ to denote the case that if two tree tuples t_1 and t_2 of a given node have the same values of each path in S.

3 FDs of Uncertain XML Dataset

Based on Reference. [20], we give three modified definitions of uncertain XML FDs based on tree-tuple model. We modify the expression of the path set to be reasonably used in lossless decomposing uncertain XML datasets.

Definition 3.1 (*Full FD*) Given a DTD D, a Full FD (FFD) over D has the form $S_L \to S_R$, where $S_L = \{s_{L1}, s_{L2}, \ldots, s_{Lm}\}$ is the left path set (i.e., determinant path set) and $S_R = \{s_{R1}, s_{R2}, \ldots, s_{Rn}\}$ is the right path set (i.e., determined path set). For $\forall s_{Li} \in S_L, s_{\text{Rj}} \in S_R$, $s_{\text{Li}}, s_{\text{Rj}} \in \text{paths}(D)$, where $i = 1, \ldots, m$, $j = 1, \ldots, n$. For an uncertain XML tree $T| = D$, if T satisfies FFD $S_L \to S_R$ (denoted by $T| = S_L \to S_R$), then it implies that for any two tuples t_1 and t_2 in T, if for $i = 1, \ldots, m$, $t_1(S_{\text{Li}}) = t_2(S_{\text{Li}})$ then $t_1(S_{\text{Rj}}) = t_2(S_{\text{Rj}})$ for $j = 1, \ldots, n$.

Definition 3.2 (*Tuple-level FD*) Given a DTD D and an uncertain XML tree $T| = D$, a Tuple-level FD (TFD) has the form $\{s_h, [S_L \to S_R]\}$, where s_h is the head path, $S_L = \{s_{L1}, s_{L2}, \ldots, s_{Lm}\}$ is the left path set (i.e., determinant path set) and $S_R = \{s_{R1}, s_{R2}, \ldots, s_{Rn}\}$ is the right path set (i.e., determined path set). For $\forall s_{\text{Li}} \in S_L, s_{\text{Rj}} \in S_R$, $s_h, s_h.s_{\text{Li}}, s_h.s_{\text{Rj}} \in \text{paths}(D)$, where $i = 1, \ldots, m, j = 1, \ldots, n$. If T satisfies TFD $\{s_h, [S_L \to S_R]\}$ (denoted by $T| = \{s_h, [S_L \to S_R]\}$), then it implies that for each tree tuple t rooted on node last(s_h), $t(s_h.s_{L1}, s_h.s_{L2}, \ldots, s_h.s_{Lm})$ can uniquely determine $t(s_h.s_{R1}, s_h.s_{R2}, \ldots, s_h.s_{Rn})$, i.e., one $t(s_h.s_{L1}, s_h.s_{L2}, \ldots, s_h.s_{Lm})$ value set cannot correspond to two or more $t(s_h.s_{R1}, s_h.s_{R2}, \ldots, s_h.s_{Rn})$ value set in tree tuple t, where last$(s_h) = \omega_n$ for $s_h = \omega_1 \ldots .\omega_n$.

The difference between FFDs and TFDs are as follows: FFDs are satisfied by all tree tuples together, while TFDs are satisfied by each tree tuple separately.

Definition 3.3 (*In-tuple-level FD*) Given a DTD D and an uncertain XML tree $T| = D$, an In-tuple-level FD (IFD) has the form $\{s_h.ID = n, [S_L \to S_R]\}$, where s_h is the head path, $S_L = \{s_{L1}, s_{L2}, \ldots, s_{Lm}\}$ is the left path set (i.e., determinant path set) and $S_R = \{s_{R1}, s_{R2}, \ldots, s_{Rn}\}$is the right path set (i.e., determined path set). For $\forall s_{Li} \in S_L, s_{Rj} \in S_R,\ s_h, s_h.s_{Li}, s_h.s_{Rj} \in \text{paths}(D)$, where $i = 1, \ldots, m, j = 1, \ldots, n$. If T satisfies IFD $\{S_h.ID = n, [S_L \to S_R]\}$ (denoted by $T| = \{S_h.ID = n, [S_L \to S_R]\}$), then it implies that for the tuple rooted on node last(s_h) with ID $= n$, $t(s_h.s_{L1}, s_h.s_{L2}, \ldots, s_h.s_{Lm})$ can uniquely determine $t(s_h.s_{R1}, s_h.s_{R2}, \ldots, s_h.s_{Rn})$, i.e., one $t(s_h.s_{L1}, s_h.s_{L2}, \ldots, s_h.s_{Lm})$ value set cannot correspond to two or more $t(s_h.s_{R1}, s_h.s_{R2}, \ldots, s_h.s_{Rn})$ value sets in tree tuple t, where last$(s_h) = \omega_n$ for $s_h = \omega_1 \ldots . \ldots \omega_n$.

In the above definition, we assign each tree tuple with an ID number n to distinguish from each other. From Definitions 3.1–3.3, the relationships between FFDs, TFDs, and IFDs are given as the following theorem:

Theorem 1 *(1) Given a DTD D and an uncertain XML tree* $T| = D$, *if T satisfies FFDs* $S_1 \to S_2$, *then T also satisfies the corresponding TFDs* $\{s_h, [S_{10} \to S_{20}]\}$ *and IFDs* $\{s_h.ID = n, [S_{10} \to S_{20}]\}$, *where* $s_h.S_{10} = S_1$, $s_h.S_{20} = S_2$ *and* $s_h.ID = 1, 2, \ldots, n$*(n is the last tree tuple of* s_h*). (2) Given a DTD D and an uncertain XML tree* $T| = D$, *if T satisfies TFDs* $\{s_h, [S_1 \to S_2]\}$, *then T also satisfies the corresponding IFDs* $\{s_h.ID = n, [S_1 \to S_2]\}$, *where* $s_h.ID = 1, 2, \ldots, n$*(n is the last tree tuple of* s_h*).*

4 Lossless Decomposition of Uncertain XML Dataset

In this section, we give three lossless decomposition methods based on the previous proposed FDs: FFD, TFD, and IFD. For an uncertain XML tree T and its n decompositions $\{T_1, T_2, \ldots, T_n\}$, we define $\{T_1, T_2, \ldots, T_n\}$ as **a lossless decomposition** of T if for a query Q, the results returned from T and the results returned from $\{T_1, T_2, \ldots, T_n\}$ are equivalent. Here, **two query results are equivalent** means that the two query results represent the same set of possible world instances.

Decomposition 1 (**Lossless decomposition based on FFD**). For an uncertain XML tree T that satisfies FFD $S_L \to S_R$, it can be decomposed into two XML trees T_1 and T_2 as follows. T_1 includes all nodes of paths S_L and S_R, and T_2 includes all nodes of paths S_L and Paths(T)$-S_R$. Intuitively, T_1 stores all the mappings between S_L and S_R, and T_2 retains all the alternatives of T but projects them onto S_L and Paths(T)$-S_R$. The decomposition of T into T_1 and T_2 is lossless because we can combine T_1 and T_2 based on S_L to get the original XML tree T.

Decomposition 2 (**Lossless decomposition based on TFD**). For an uncertain XML tree T that satisfies TFD $\{s_h, [S_L \to S_R]\}$, it can be decomposed into n XML tree $T_1, T_2, \ldots, T_n$ as follows, where n stands for the number of last(s_h) tuples in T and T_i includes all nodes of paths of the ith last(s_h) tuple. Each T_i satisfies FFD $S_L \to S_R$. The decomposition of T into $\{T_1, T_2, \ldots, T_n\}$ is lossless because we can simply combine T_1, T_2, ..., T_n based on path S_h to get the original XML tree T. Of course, each T_i can again be decomposed losslessly using Decomposition 1.

Decomposition 3 (Lossless decomposition based on IFD). For an uncertain XML tree T that satisfies IFD $\{s_h.ID = n, [S_L \rightarrow S_R]\}$, it can be decomposed into two XML trees T_1 and T_2 as follows. T_1 includes all nodes of paths of the nth last(S_h) tree tuple and satisfies FFD $S_L \rightarrow S_R$ and T_2 includes the rest of the tree tuples. The decomposition of T into $\{T_1, T_2\}$ is lossless because we can simply combine T_1 and T_2 based on path s_h to get the original XML tree T. Of course, T_1 can again be decomposed losslessly using Decomposition 1.

5 Conclusions and Future Work

This paper studies the FDs of uncertain XML datasets, which extends the notions of FDs of uncertain relational databases and deterministic FDs of XML datasets based on tree-tuple model. Three kinds of FDs such as FFDs, TFDs, and IFDs are given to capture three kinds of data dependencies of uncertain XML datasets. Then, three kinds of lossless decompositions are given to decompose an uncertain XML dataset losslessly based on FFDs, TFDs, and IFDs, respectively. Finally, three kinds of keys such as FKeys, TKeys, and IKeys of uncertain XML datasets are also given according to FFDs, TFDs, and IFDs, respectively.

An interesting works in the future is to study the FDs and keys of uncertain XML datasets with probability values. In this paper, we do not consider the probability values of uncertain XML datasets. When probability values are added in the distributional nodes of uncertain XML datasets, how to compute the probability value of FDs and keys is an interesting work.

Acknowledgments The work is supported by Natural Science Foundation of Anhui Province (No. 1208085MF110), Natural Science Foundation of China (No. 11201002), Science Research Plan Project of Anhui Agricultural University (No. YJ2010-12), and Education Reform and Development Research Project of Anhui Agricultural University (No. Jf2012-27).

References

1. Abiteboul, S., Hull, R., Vianu, V.: Foundations of Databases. Addison-Wesley, Boston (1995)
2. Ullman, J.D.: Principles of Database and Knowledge-Base Systems, vol. 1. Computer Science Press, New York (1988)
3. Koudas, N., Saha, A., Srivastava, D., Venkatasubramanian, S.: Metric functional dependencies. In: Proceedings of the 25th International Conference on Data Engineering, pp. 1275–1278. IEEE Computer Society Press, New York (2009)
4. Bohanno, P., Fan, W., Geerts, F., Jia, X., Kementsietsidis, A.: Conditional functional dependencies for data cleaning. In: Proceedings of the 23rd International Conference on Data Engineering, pp. 746–755. IEEE Computer Society Press, New York (2007)

5. Janosi-Rancz, K.T., Varga, V., Nagy, T.: Detecting XML functional dependencies through formal concept analysis. In: Proceedings of the 14th East European Conference on Advances in Databases and Information Systems, pp. 595–598. Springer, Heidelberg (2010)
6. Lee, M.L., Ling, T.W., Low, W.L.: Designing functional dependencies for XML. In: Proceedings of the 8th International Conference on Extending Database Technology: Advances in Database Technology, pp. 124–141. Springer, London (2002)
7. Liu, J., Vincent, M., Liu C.: Functional dependencies, from relational to XML. In: Proceedings of 5th International Andrei Ershov Memorial Conference, pp. 531–538. Springer, Heidelberg (2003)
8. Liu, J., Vincent, M., Liu, C.: Local XML functional dependencies. In Proceedings of 5th ACM CIKM International Workshop on Web Information and Data Management, pp. 23–28. ACM, New York (2003)
9. Vincent, M., Liu, L.: Functional dependencies for XML. In: Proceedings of 5th Asian-Pacific Web Conference, pp. 22–34. Springer, Heidelberg (2003)
10. Vincent, M., Liu, J., Liu, C.: Strong functional dependencies and their application to normal forms in XML. TODS **29**, 445–462 (2004)
11. Yan, P., Lv, T.: Functional dependencies in XML documents. In: Proceedings of the 8th Asia Pacific Web Conference Workshop, pp. 29–37. Springer, Heidelberg (2006)
12. Hartmann, S., Link, S.: More functional dependencies for XML. In: Proceedings of the 7th East European Conference on Advances in Databases and Information Systems, pp. 355–369. Springer, Heidelberg (2003)
13. Hartmann, S., Link, S., Trinh, T.: Solving the implication problem for XML functional dependencies with properties. In: Proceedings of the 18th Workshop on Logic, Language, Information and Computation, pp. 161–175. Springer, Heidelberg (2010)
14. Lv, T., Yan, P.: Removing XML data redundancies by constraint-tree-based functional dependencies. In: Proceedings of ISECS International Colloquium on Computing, Communication, Control, and Management, pp. 595–599. IEEE Computer Society, Washington, DC (2008)
15. Lv, T., Yan, P.: XML normal forms based on constraint-tree-based functional dependencies. In: Proceedings of Joint 9th Asia-Pacific Web Conference and 8th International Conference on Web-Age Information Management Workshops, pp. 348–357. Springer, Heidelberg (2007)
16. Vo, L. T. H., Cao, J., Rahayu, W.: Discovering conditional functional dependencies in XML data. In: Proceedings of the 22nd Australasian Database Conference, pp. 143–152. Australian Computer Society, Sydney (2010)
17. Wang, D.Z., Dong, L., Sarma, A.D., Franklin, M.J., Halevy, A.Y.: Functional dependency generation and applications in pay-as-you–go data integration systems. Proceedings of the 12th International Workshop on the Web and Databases. http://www.cs.berkeley.edu/~daisyw/webdb09.pdf (2009)
18. De, S., Kambhampati, S.: Defining and mining functional dependencies in probabilistic databases. arXiv.org. http://arxiv.org/pdf/1005.4714v2 (2010)
19. Sarma, A.D., Ullman, J., Widom, J.: Schema design for uncertain databases. Proceedings of Alberto Mendelzon Workshop on Foundations of Data Management. http://ilpubs.stanford.edu:8090/820/ (2009). Accessed 20 Oct 2011
20. Lv, T., He, W., Yan, P.: Uncertain XML Functional Dependencies based on Tree Tuple Models. Proceedings of The 13th International Conference on Web-Age Information Management (WAIM2012) Workshop, LNCS, Springer, Heidelberg (2012)
21. Kimelfeld, B., Sagiv, Y.: Modelling and querying probabilistic XML data. ACM SIGMOD Rec. **37**, 69–77 (2008)

Space–Time Clustering Analysis of Emergency Incidents in Nanning, Guangxi

Peng Chen, Hongzhi Huang and Jinguang Sui

Abstract In practical emergency management, understanding the emergency incidents distribution in time and space is quite significant to emergency response and resource deployment. Thus, in this paper, the emergency incidents, which are categorized into police, first aid, traffic, and firefighting in Nanning, Guangxi, were studied from space and time perspectives. The results indicate that the incidents are non-equilibrium distributed by hour of day. The police, first aid, and firefighting incidents are clustered, but the first aid ones are dispersed across the day. In space, the police, first aid, and traffic incidents are proved to be clustered, but the firefighting incidents are distributed uniformly.

Keywords Emergency management · Space–time analysis · Clustering pattern

1 Introduction

Since Nanning City Emergency Response Center was constructed in 2003, million calls for emergency service were received every year, and this number keeps increasing year by year. This brought great challenge to the emergency agencies which own limited resources such as police officers, firefighters, vehicles [1]. Therefore, exploring the emergency incidents distribution pattern in time and

P. Chen (✉) · H. Huang · J. Sui
Institute for Police Information Engineering, Chinese People's Public Security University, Daxing district, Beijing, China
e-mail: uctzpch@gmail.com

H. Huang
e-mail: huanghongzhi@126.com

J. Sui
e-mail: sjinguang@sina.com

S. Patnaik and X. Li (eds.), *Proceedings of International Conference on Computer Science and Information Technology*, Advances in Intelligent Systems and Computing 255, DOI: 10.1007/978-81-322-1759-6_90,

space is significant to emergency response works such as forecasting, risk assessment, and resource deployment. In this paper, the space–time pattern of emergency incidents across Nanning was studied and the significance and inference of hotspot identification to emergency management were discussed.

2 Data Framework and Analysis Strategy

The emergency incidents data were provided by Nanning City Emergency Response Center. They are the emergency records received via calls for emergency service, which includes police help, first aid, traffic help, and firefighting in 2003. A total of 1,308 recordings were collected from totally 9,223 items because the left ones are in short of attributes as date, time, locations, and types, which are important to the data analysis. Among the processed recordings, police department has 451 ones, first aid agency has 408 recordings, traffic department has 384 recordings, and firefighting has the least 48 recordings.

Considering that the main task of Nanning City Emergency Response Center is receiving the calls for emergency service and deploying resources to cope with emergency problems, the temporal distribution of four category incidents by time of day was investigated and pattern was studied. Then, based on the locations existed in data attribute table, the incident data were projected in GIS software and its concentration pattern in space was analyzed using spatial data mining ways.

3 Temporal Analysis

The four types of emergency recordings were integrated by hour, and so namely the temporal distribution results of police, first aid, traffic, and firefighting by hour of day was created. As shown in Fig. 1, it can be seen that police, first aid, and traffic are similar in temporal pattern, but the firefighting data have its unique feature by hour of day for its fewer recordings. The lowest level of emergency incidents recorded across the day lies around 05:00 am, but the peak level lies at the time around midnight. It is not strange to observe this pattern because according to the human's routine activity, 05:00 am is the time when fewest people still keep awake. So the four emergency incidents that are closely related to human activities are at the lowest level of the day. On the other hand, in addition to the midnight, several top levels appear in the morning, noon, and afternoon separately, which are corresponding to the time when people on the way to work, noon break, and way to home.

In further, the pattern of temporal distribution is studied using quartile analysis. The quartile is a simple and cogent way to study hourly dispersion of incidents [2]. The most direct and clearest way to study the incident dispersion by day is to find quartile minutes. The first quartile minute is the time where 25 % of incidents

Fig. 1 Temporal distribution of four emergency incidents by hour of day

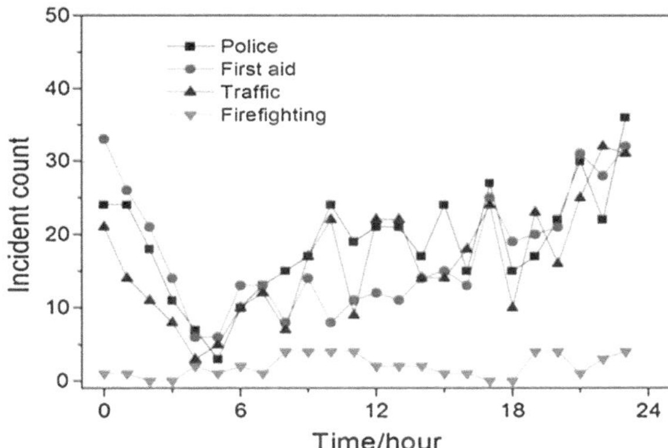

Table 1 Quartile minutes of the incidents data

Item	Police	First aid	Traffic	Firefighting
First quartile minute	08:22	05:12	09:18	08:57
Median minute	13:54	14:43	14:49	11:47
Third quartile minute	19:51	20:27	20:28	19:38
Daily time span (minutes)	689	915	670	641

occur between it and starting minute, 00:00. The second quartile minute is the time where 25 % of incidents occur between it and first quartile minute. The second quartile minute is called median minute as well. By recursive operation, the third quartile minute could be found. The daily time span is the minutes between first quartile minute and third quartile minute, which measures the dispersion extent of incidents. Using the quartile minutes, the dispersion of incidents can be known.

Table 1 displays the first, second, and third quartile minutes to the four emergency incidents. It could be seen that the median minute of firefighting incidents is 11:47 am, which is the only one which appears in the first half of the day. It indicates that more incidents of firefighting concentrate in the first half of the day, while the other three ones are in the second half of the day. The daily time spans of police, traffic, and firefighting are less than 720 min, but the first aid extends to as many as 915 min. From the results, it could be referred that first aid incidents are more dispersed, but police, traffic, and firefighting incidents are more clustered across the day.

4 Spatial Analysis

There are three patterns of point process exist in space, namely clustering, random, and uniform, but the clustering pattern is more significant because clustering pattern is helping a lot in hotspot identification analysis. One frequently used method in spatial point process clustering analysis is quadrat analysis (QA), which realizes the work by producing the lattices in space and calculating the frequency of incidents count in lattice and then compares the observed distribution to complete spatial random distribution generated from Poisson process [3]. The principle of Poisson process is as follows: Assuming the number of incidents in lattice is k, the expected number of lattices with k hits was

$$p(x = k) = \frac{e^{-\lambda}\lambda^k}{k!} \tag{1}$$

where $\lambda = n/m$, n is the total count of incidents in space, and m denotes the number of lattices. The cumulative frequency that corresponds to count levels in lattice is calculated, and then, the difference between observed and the expected cumulative levels $D_{\max} = |Q_i - E_i|$ is computed and compared against the Kolmogorov–Smirnov statistic $D_{\alpha=0.05}$ (see Eq. 2). If $D_{\max} > D_{\alpha=0.05}$, the null will be rejected, and observed pattern is proved to be in clustering pattern.

$$D_{\alpha=0.05} = \frac{1.36}{\sqrt{m}} \tag{2}$$

The studying area in Nanning is 515 km^2 and the generated lattice is 651 (31 × 21), so the size of the lattice is 0.89 km. After that, the count of incidents in each lattice was calculated and both the observed and Poisson cumulative frequency level against incident count levels were computed. The computed $|Q_i - E_i|$ between observed pattern and expected pattern is shown in Table 2. Based on Eq. (2), the $D_{\alpha=0.05}$ is computed to be 0.533. So according to Table 2, it can be inferred that the police, first aid, and traffic incidents distributed in space are significantly clustered, but the firefighting incidents are shown to be randomly distributed.

The kernel density estimation (KDE) to the four types of incidents in space is shown in Fig. 2. As demonstrated, more than one hot spot are identified in police, first aid, and traffic incidents distribution [4]. Among them, several hot spots of police, first aid, and traffic incidents are overlapped, which indicates that some correlations exist among three types of incidents in space. While the firefighting incidents hot spot being apart away means that the spatial features that influence the firefighting incidents distribution are different from the police, first aid, and traffic, so it is worthy further exploring in next work.

Table 2 Computed $|Q_i - E_i|$ between observed pattern and expected pattern

Incident count	Police	First aid	Traffic	Firefighting
0	**0.3404**	**0.3125**	**0.2999**	**0.0238**
1	0.0388	0.0199	0.0219	0.0158
2	0.0590	0.0673	0.0592	0.0045
3	0.0666	0.0696	0.0658	0.0031
4	0.0575	0.0516	0.0533	0
5+	0.0459	0.0398	0.0352	0

The figures in bold are $D_{\max}$

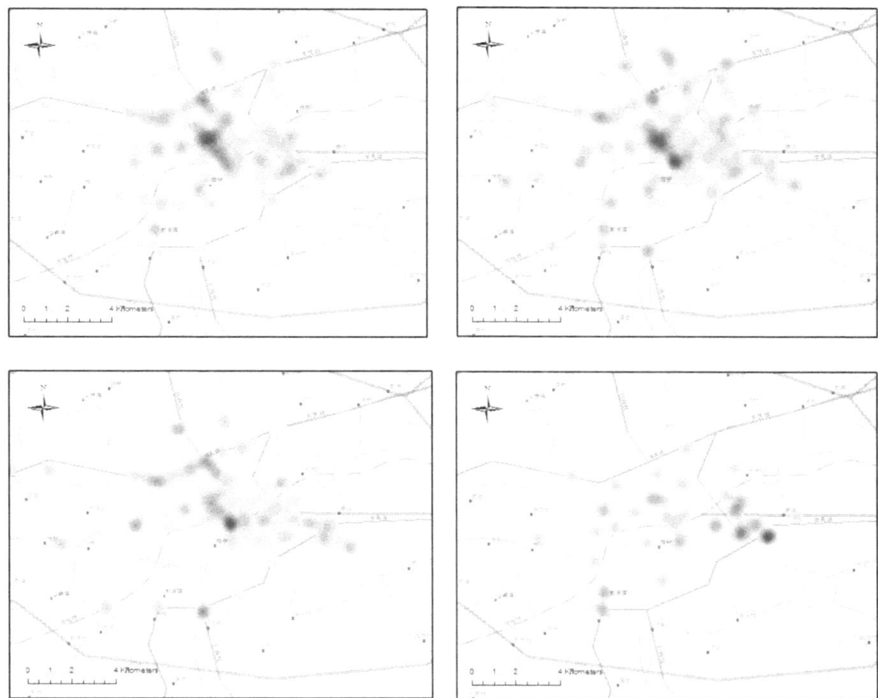

Fig. 2 The kernel density estimation of the incidents in space

5 Conclusion

In this paper, the space–time pattern of emergency incidents in Nanning, Guangxi, was studied using temporal and spatial clustering methods. The results demonstrate that the distribution patterns between agencies are different. The first aid incidents being more dispersed by hour of day indicate that the risk is uniformly distributed, so the resources such as ambulance should be deployed more intensely. In space, the police, first aid, and traffic incidents were shown to be clustered, which indicates that some significant hot spot exists. Using the kernel density estimation, the hot spots were detected and found to be overlapped in space, which infers that the incidents seem to be correlated. Therefore, the spatial features such as population, land use, and property pattern in hotspot areas should be further explored to find schemes that interact with crime, traffic accidents, or injuries. Based on this, the reasonable emergency plan and resource deployment could be made to improve the emergency response efficiency.

Acknowledgments This work was sponsored by Chinese People's Public Security University Natural Science project (2013LG03).

References

1. Wang, X., Ma, Y.: Spatial and temporal statistical character of gale in Xinjiang. Xinjiang Meteorol. **25**(1), 1–3 (2002)
2. Felson, M., Poulsen, E.: Simple indicators of crime by time of day. Int. J. Forecast. **19**(4), 595–601 (2003)
3. Miles, R.E.: on the homogenous planar Poisson point process. Math. Biosci. **6**, 85–127 (1970)
4. Ned, L.: Crime Stat II: A Spatial Statistics Program for the Analysis of Crime Incident Locations (version 2.0). Ned Levine & Associates/National Institute of Justice: Houston, TX/Washington, DC (2002)

The Study and Application of Protocol Conformance Testing Based on TTCN-3

Meijia Xu, Honghui Li and Jiwei Zheng

Abstract This paper is mainly focused on the TTCN-3-based protocol conformance test method and test system implementations. In this paper, based on the analysis of the SSL protocol software, and combined the architecture of TTCN-3 testing system, according to TTCN-3 Control Interface specification and SSL protocol software data structure characteristics, design and implement the SSL codec which used to implement the function of data conversion between abstract testing suit and actual network transmission in TTCN-3 testing system.

Keywords TTCN-3 · Abstract testing suit · Codec · SUT adapter

1 Introduction

With the popularization of information technology networks, security of data transmission becomes more and more important. Since the plaintext data transmitted over the network is easy to be criminals eavesdropping, interception or tampering, so a lot of computer systems and software are no longer transmitted some higher safe data in clear text, before turning data authentication is used both entities. During data transmission, using an encryption algorithm to encrypt the transmitted data means to ensure the safety and reliability of data transmission.

M. Xu (✉)
Beijing Jiaotong University School of Computer and Technology, Beijing, China
e-mail: 11120494@bjtu.edu.cn

H. Li
Beijing Jiaotong University School of Research Center of Network Management, Beijing, China
e-mail: hhli@bjtu.edu.cn

J. Zheng
Daqin Railway Co., Ltd., Taiyuan, China
e-mail: 510241874@qq.com

S. Patnaik and X. Li (eds.), *Proceedings of International Conference on Computer Science and Information Technology*, Advances in Intelligent Systems and Computing 255, DOI: 10.1007/978-81-322-1759-6_91,

In fact, this is the main operating way of the SSL protocol. In order to achieve transmission security of sensitive data, more and more computer systems and software begin to use the SSL protocol to protect data during transmission.

Conformance testing of SSL protocol layer mainly tests whether using SSL protocol, whether according the SSL protocol during software transmission. If the data transmission process of the software meets the requirement of SSL protocol specification, then you think data transmission process is safe, its security is mutual negotiation by entity communication with data encryption algorithm and key length, and so on.

In addition to the tests on the transport layer of the SSL protocol software, at the same time also need to test application layer. However, it is due to the diversification of application layer designment and implementation. According to own requirements, every application software will be for specific implementation. So the design and implementation of SSL protocol software has a bigger difference, and thus, testing the application layer needs to be more flexible.

2 Related Concepts

2.1 Protocol Conformance Testing

Now, there is a big gap in network test technology level and ability between our country and the world, most of the design of protocol conformance testing product stays on theoretical research of the testing methods. Protocol conformance testing in terms of theory is mainly defined by the International Standards Organization (ISO) ISO/IEC—9646 conformance testing methodology and framework (CTMF). Protocol conformance test model [1] is shown in Fig. 1.

Protocol conformance test model mainly consists of three parts: system under testing (SUT), testing system (TS), and underlying service provider. SUT mainly consists of two parts: upper tester (UT) and implementation under test (IUT). Point of control and observation (PCO) gets the output of the IUT in the process of the test, at the same time, according to the specific conditions enters some data and observe and control the input and output of implementation under testing.

According to the relationship of the PCO and the IUT, the tester can be divided into two types: Upper tester (UT) is a control and observation point tester in the upper of IUT. Its main function is to control the execution of the test case and use abstract service primitives (ASP) and protocol data unit (PDU) to observe and control the input and output of implementation under testing at the PCO; lower tester (LT) is a control and observation point tester in the lower of IUT. It mainly corresponds to the end point of an upper tester, receives data from IUT transmission, compares with expected data, and makes a decision.

Test coordination procedure (TCP) is mainly to coordinate communication between upper tester and lower tester, allowing them to keep pace. The implementation of TCP can adopt IUT or other protocol.

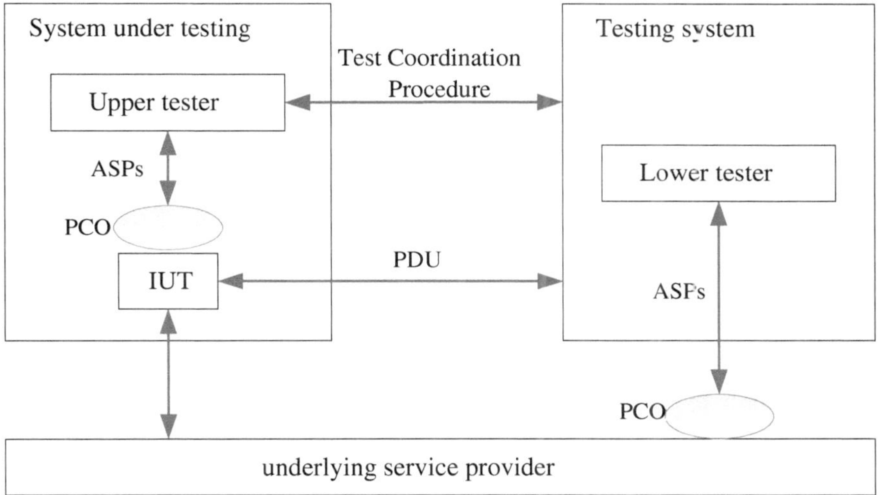

Fig. 1 Protocol conformance test model

Underlying service provider provides the underlying communication services for IUT and ensures accuracy of services they provide, so you can ensure that test results can reflect the IUT.

Protocol conformance testing process [2] can be divided into four stages (Fig. 2):

1. conformance testing generation phase: Text description in accordance with the protocol generates consistent abstract test suite;

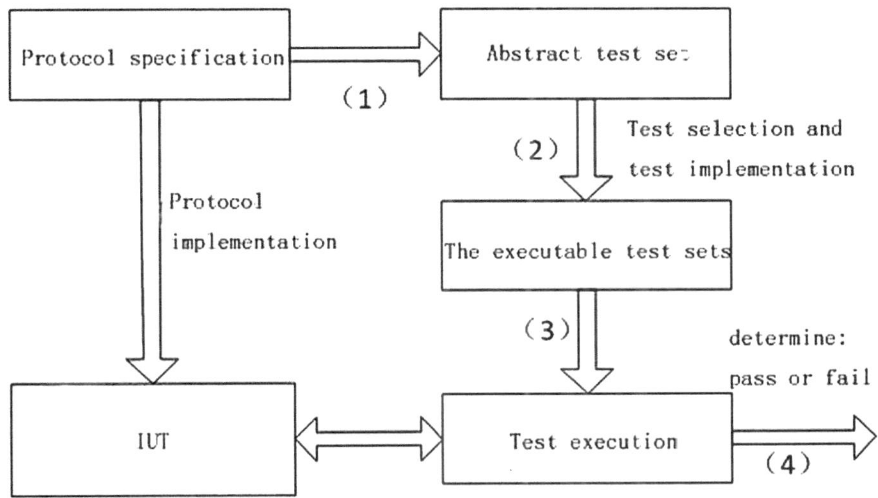

Fig. 2 Protocol conformance testing process

2. conformance testing implementation phase: According to abstract test suite, combined with the specific environment generates executable test suit (ETS);
3. conformance testing execution phase: Perform the ETS and at the same time observe and record the external behavior of the implementation under test protocol;
4. conformance testing determination phase: Judge the test result and record the results into test report.

2.2 Introduction to TTCN-3

TTCN-3 [3] is a language designed specifically for testing. Many constructs are similar to those in other programming languages but are extended with additional concepts not available elsewhere. These concepts include built-in data matching, distributed test system architecture, and concurrent execution of test components. TTCN-3 has a larger type system than normal programming languages and includes native types for lists, test verdicts, and test system components. In addition, TTCN-3 provides direct support for timers as well as for message-based and procedure-based communication mechanisms.

2.3 TTCN-3 System Architecture

A TTCN-3 system includes different test entities [4], and these entities interact with each other in execution of a test set. In Figure 3, TTCN-3 system architecture is mainly composed of three-layer structure [5]. TTCN-3 executable (TE) controls TTCN 3 execution of test cases, and its operation depends on the services provided by the other two layers. TTCN-3 management and control (TMC) is responsible for interactions between the test users, the data encoding and decoding, etc. The interaction between TE and SUT is done by platform adapter (PA) and SUT adaptor (SA). TE entities communicate with the upper and lower level through TCI and TRI interface. The two interfaces define a series of functions. They communicate by calling each other's implementations of the interface for communication.

3 The Design of SSL Protocol Conformance Testing Based on TTCN-3

The focus of the SSL protocol conformance testing is design and implementation of codec. This section mainly introduces the realization of codec.

In the process of designing SSL protocol test system based on TTCN-3, the design and development of SSL codec is one of the most important links. Codec is an entity

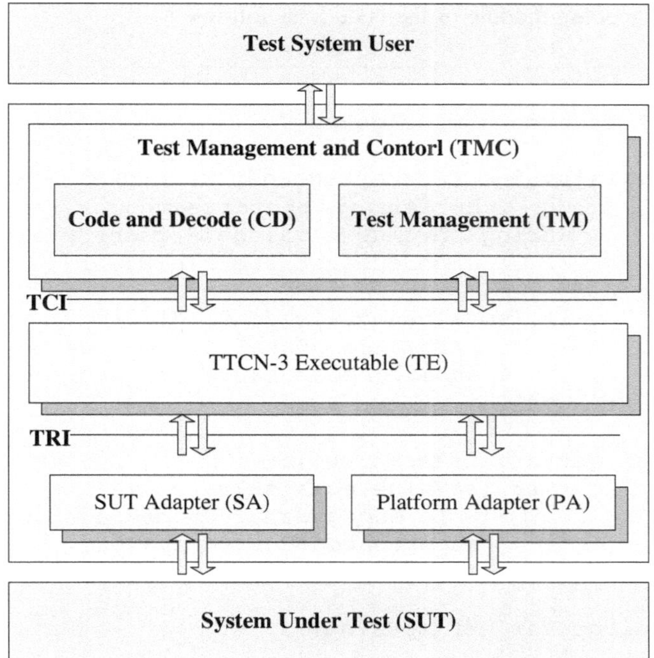

Fig. 3 TTCN-3 system architecture

with translation function. Encoding or decoding is to transform between TTCN 3 data types and data of the network protocol conversion, the purpose is to make the test system and the system under test identify each other. In the decoding and encoding process, test system determines the codec configuration information. When data need to be encoded, TTCN 3 actuators sent TTCN-3 data to the encoding module; similarly, when data need to be decoded, TTCN 3 actuators sent the system under test data to the decoding module. Encoding module put TTCN-3 specification data values into network transmission byte streams, and decoding module put the byte stream data returned by the system under test into TTCN-3 specification data, so you can complete proper communication between the test system and the system under test.

SSL is a security protocol on the transport layer, there is an application layer on top of it. The system under test contains the implementation of the SSL protocol and application layer above SSL. Codec module of SSL protocol software testing system contains implementation of the SSL layer codec and application layer codec. The SSL protocol encrypt network connection at the transport layer, the application layer achieve data transmission security through the SSL protocol. From the system under test to the test system the data is a cipher text, so it need the decoding through the SSL layer, then the data need the decoding through the application layer, forming the abstract application layer data so that the test system and the user understand [6].

Part of encoding module of the code is as follows:

```
    /**
     * To encode messageBody.
     */
    privatevoidencodeMessageBody(RecordValueval,
                ByteArrayOutputStream out) {
          String spValue = val.getPresentSpValue();
          // To encodecharstring fields of
messageBody
          if (spValue.equals("data1")) {
                //To encodedata1 of messageBody
                super.encodeCharstring(out,
(CharstringValue)

    val.getVariant(spValue));
          } if (spValue.equals("data2")){
                //To encodedata2 of messageBody
                super.encodeCharstring(out,
(CharstringValue)

    val.getVariant(spValue));
          }
    }
    /**
     * To encode charstring Basic data types fields.
     */
privatevoidencodeCharstring(RecordValueval, String
fieldName,
                ByteArrayOutputStream out) {
          CharstringValuecs = (CharstringValue)
val.getField(fieldName);

          int length = cs.getLength();
          IntegerValuelengthValue = (IntegerValue)
RB.getTciCDRequired()

    .getInteger().newInstance();
          lengthValue.setInt(length);
          ......
// Call base class method of TCI interface to encode
charstring data
          super.encodeCharstring(out, cs);
    }
```

Part of decoding module of the code is as follows:

```
/**
 *Data analysis according to the protocol format
 * @parammessage  decode data.
 * @paramresult  the data after decoding.
 */
privatevoiddecodeResponse(byte[] message,
RecordValue result) {
        int length;
        // To encodecodeVal
        IntegerValuecodeVal=
decodeInteger(message);
        result.setField("codeVal", codeVal);

        //To encodestatusContent
        UnionValuecont =
(UnionValue)result.getField("statusContent");
        decodeMessageBody(message, statusContent);
        result.setField("statusContent ",
statusContent);
                    ......
}
```

4 Conclusion

SSL protocol is a special kind of security protocol. The way it works is different from common communication protocols. So, firstly, start from the working characteristics of the SSL protocol software and combine the structure of TTCN 3 test system, work out the method of the SSL protocol conformance test software. This paper mainly introduces the function design and implementation of the codec. The others, Writing the test case and Configuring the adapter are successful implementation.

Because of my own knowledge level and other conditions limit, this paper that put forward the SSL protocol software testing methods also need some extension and improvement. Needing further in-depth study and improving the content is as follows:

1. Test coverage of the abstract test set based on SSL needs in-depth study;
2. General codec based on SSL needs further study.

Acknowledgments The research has been supported by Project No. 2012X010-C of Information system connectivity and security maintain technology application research- technology research of the railway information system evaluation.

References

1. Chen, J., Wang, L., Gong, Z.: Architecture research of protocol conformance testing execution system. Comput. Eng., 55–60 (2003)
2. ISO/IEC. Information system technology open system interconnection conformance testing methodology and framework part 3: The tree and tabular combined notation (TTCN), pp. 79–52
3. ETSI ES 201 873-1 V4.4.1: Methods for testing and specification (MTS); The testing and test control notation version 3, Part l: TTCN-3 Core Language, pp. 98–103 (2012)
4. Gao, X., Jiang, F.,Yang, J.: TTCN-3 research review. Comput. Eng. Sci. **26**(6), 17–20 (2004)
5. Jiang, F., Ji, X., Ceng, F.: Design and implementation of TTCN-3 test system. Comput. Eng. **31**(11), 80–81 (2005)
6. Glaeser, M., Mueller, S., Rennoch, A., Schmitting, P.: Standardized TTCN-3 specifications for Sip-isup/isdn interworking testing. Int. J. Softw. Tools Technol. Transf. **10**(4), 353–358 (2008)

Modeling and Analyses of Operational Software System with Rejuvenation and Reconfiguration

Xiaozhi Du, Huimin Lu and Yuan Rao

Abstract In this paper, a software rejuvenation model with reconfiguration is proposed to improve the software performance. Firstly, continuous-time Markov chain is adopted to describe the system model. Then, the formal definitions and analyses of system availability and throughput are given. Finally, some numeric experiments are done. The results show that the presented method is effective and adopting reconfiguration can improve the system throughput though the availability has a trivial reduction.

Keywords Software rejuvenation · Software performance · Reconfiguration

1 Introduction

The importance of software reliability and availability has been well recognized, and these performance measures are increasingly being demanded in present software systems [1]. While recent researches show that when software application is executed continuously for long intervals of time, some error conditions in them are accumulated to result in performance degradation or even a crash failure, which is called software aging [2]. The phenomenon has been observed in many software systems, such as Web server [3], SOAP server [4]. To counteract

X. Du (✉) · Y. Rao
School of Software Engineering, Xian Jiaotong University, Xian, China
e-mail: smart_zhi@163.com

Y. Rao
e-mail: yuanrao@163.com

H. Lu
School of Software Engineering, Changchun University of Technology, Changchun, China
e-mail: hmlu.cc@gmail.com

S. Patnaik and X. Li (eds.), *Proceedings of International Conference on Computer Science and Information Technology*, Advances in Intelligent Systems and Computing 255, DOI: 10.1007/978-81-322-1759-6_92,

software aging and its related transient software failures, a preventive and proactive technique, called software rejuvenation, has been proposed and is becoming popular [2].

Over the recent years, quantitative studies of software aging and rejuvenation have been taken, and many different approaches have been developed and the effects of software rejuvenation have been studied. These studies can be categorized into two kinds, time-based rejuvenation policy and measure-based rejuvenation policy. The time-based rejuvenation policy is characterized by the fact that the software is periodically rejuvenated every time a predefined time constant has elapsed. In these approaches, a certain failure time distribution is assumed and continuous-time Markov chain (CTMC) [2], semi-Markov process [5], Markov regenerative process (MRGP) [6], stochastic Petri net (SPN) model [7], etc. are developed to compute and optimize system availability or related measures. The measure-based rejuvenation policy applies statistical analysis to the measured data on resource availability to predict the expected time to resource exhaustion [8], and it provides a window of time during which a rejuvenation action is advised. The basic idea of the measure-based rejuvenation policy is to monitor and collect data on the attributes and parameters, which are responsible for determining the health of the running software system. Grottke et al. [3] used nonparametric statistical methods to detect and estimate trends of aging and adopted AR model to predict the aging of a Web server. Hoffmann et al. [9] gave a practice guide to resource forecasting, they adopted several methods to model and predict the software aging of a Web server, they found that probabilistic wrapper (PWA) was a better method for variable select, and support vector machine (SVM) was a better approach for resource forecasting.

In this paper, we present a software rejuvenation policy with reconfiguration. The aim of this study is to observe the effects of reconfiguration on systems performance, in order to distinguish an optimal rejuvenation policy. System performance is expressed through availability and throughput. Continuous-time Markov chain is adopted to describe the behavior model of the software system with multiple performance degradation levels.

The rest of this paper is organized as follows. In Sect. 2, the software system behavior with rejuvenation and reconfiguration is presented. In Sect. 3, the formal definitions and analyses of system availability and throughput are given. The experimental results are presented in Sect. 4. Section 5 provides the concluding remarks.

2 Software System Behavior Model

The behavior of software with software rejuvenation and reconfiguration is shown in Fig. 1. Circles represent states and directed arcs represent transitions. At the beginning, the software system is in the normal state U, in which the system is highly efficient and highly robust, and works perfectly without failure. A full

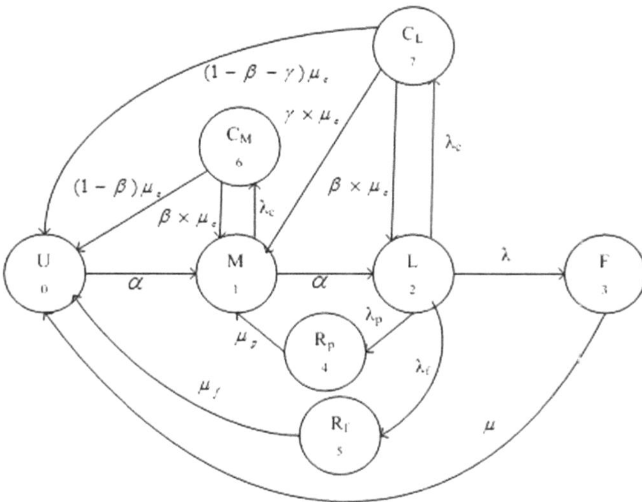

Fig. 1 Software system model with rejuvenation and reconfiguration

rejuvenation and a reboot after a crash bring the system back to this state. And the successful reconfiguration operation can also bring the system back into this normal state. After running for some time, the software system goes into the medium efficient state *M*, in which the service capacity of the software system decreases. In this state, the reconfiguration operation can be taken. If there is no reconfiguration operation, the system enters into the low efficient state *L* finally, in which the service capacity of the software system is lower than that in the state *M*. When the software is in the low efficient state *L*, in which the service capacity is very low, failure can occur at any time. In this state, the partial rejuvenation, the full rejuvenation, or the reconfiguration operation can be taken. If there is no rejuvenation operation and reconfiguration operation, the system enters into the failure state *F* finally, in which the system is down and does not provide any service. After the repair is finished, the system returns into the normal state and a new round begins.

Because the failure occurs stochastically, the loss caused by the failure is very high. In order to reduce the loss, the software rejuvenation operation is taken after the system has been running for a while, which makes the system go into the partial rejuvenation state R_p or the full rejuvenation state R_f from the low efficient state *L*. When the system is in the rejuvenation states, it stops work and does not provide service, but the loss caused by the rejuvenation is low and the recovery time is short. After the partial rejuvenation operation is finished, the software system returns into the medium efficient state *M*. After the full rejuvenation operation is finished, the system returns into the normal state *U* and a new round begins. The loss caused by the partial rejuvenation is lower than that by the full rejuvenation, and the recovery time of the partial rejuvenation is shorter than that of the full rejuvenation.

However, sometimes the system performance degradation is caused by the inefficient allocation of the system resources not by the software aging. Then, the system resources are reconfigured to make the system performance return to a high level. Therefore, when the system is in the degradation state (M or L), if the reconfiguration operation is taken, the system enters into the reconfiguration state (C_M or C_L). If the reconfiguration operation is successful, the system returns into the better state, but if it fails, the system continues running in the degradation state.

Before the system runs from the medium efficient state M into the low efficient state L, the reconfiguration operation can be taken, and then, the system enters into the first reconfiguration state C_M. If the reconfiguration is effective, the system returns into the normal state U and a new round begins. But, if there configuration operation fails, the system returns into the former state M. When the software system is in the low efficient state L, the reconfiguration operation can also be taken, and then, the system enters into the second reconfiguration state C_L. If the reconfiguration operation is very effective, the system returns into the normal state U and a new round begins. And if the reconfiguration operation is not very effective, the system returns into the medium efficient state M. But, if the reconfiguration operation fails, the system returns into the low efficient state L.

When the software system is in the reconfiguration state (C_M or C_L), the system is down and does not provide service for the user. The operation time of the reconfiguration is less than that of the rejuvenation.

In Fig. 1, $\alpha, \lambda, \mu, \lambda_p, \mu_p, \lambda_f, \mu_f, \lambda_c, \mu_c$ are the state transition rates, β is the unsuccessful probability of the reconfiguration operation, and γ is the probability from the second reconfiguration state C_L into the medium efficient state M. α corresponds to the resource degradation rate, thus, from the normal state to the medium efficient state, as well as from medium efficient state to low efficient state, transitions occur with rate α.

3 System Performance Analyses

Let $\{X(t), t \geq 0\}$ be the stochastic process describing the software system's evolution in time in one of the states included on the state space $\Omega = \{0, 1, 2, 3, 4, 5, 6, 7\}$. And we assume that the firing times of all the transitions are exponentially distributed. Therefore, the process $\{X(t), t \geq 0\}$ is continuous-time markov chain.

Usually, we are interested in the long-run behavior of a software system, and we need the steady-state probability distribution. Let Q be the generator matrix of the software system, where $Q = [q_{ij}]$, with q_{ij} denoting the transition rate from state i to state j. The asymptotic probability distribution of each state can be determined by solving the linear system of Eq. (1).

$$\pi \cdot Q = 0, \quad \sum_{i \in \Omega} \pi_i = 1 \tag{1}$$

where π is the steady-state probability vector, $\pi_0, \pi_1, \pi_2, \pi_3, \pi_4, \pi_5, \pi_6, \pi_7$ is the steady-state probability of the normal state U, the medium efficient state M, the low efficient state L, the failure state F, the partial rejuvenation state R_p, the full rejuvenation state R_f, the first reconfiguration state C_M, and the second reconfiguration state C_L, respectively.

Specially, according to the operational scenario of the software system shown in Fig. 1, the steady probability of each state follows the Eq. (2).

$$\begin{cases} \pi_0 + \pi_1 + \pi_2 + \pi_3 + \pi_4 + \pi_5 + \pi_6 + \pi_7 = 1 \\ \pi_1(\alpha + \lambda_c) = \pi_0 \cdot \alpha + \pi_4 \cdot \mu_p + \pi_6 \cdot \beta \cdot \mu_c + \pi_7 \cdot \gamma \cdot \mu_c \\ \pi_2(\lambda + \lambda_c + \lambda_p + \lambda_f) = \pi_1 \cdot \alpha + \pi_7 \cdot \beta \cdot \mu_c \\ \pi_3 \cdot \mu = \pi_2 \cdot \lambda \\ \pi_4 \cdot \mu_p = \pi_2 \cdot \lambda_p \\ \pi_5 \cdot \mu_f = \pi_2 \cdot \lambda_f \\ \pi_6 \cdot \mu_c = \pi_1 \cdot \lambda_c \\ \pi_7 \cdot \mu_c = \pi_2 \cdot \lambda_c \end{cases} \tag{2}$$

By resolving the Eq. (2), we get the steady probability of each state. The asymptotic probability of state i will be of the form $\pi_i(\lambda_C, \beta)$.

3.1 Availability Analysis

Based on the above description and analysis, the system is working and can provide service for users when it is in the normal state U, the medium efficient state M and the low efficient state L. And the system stops work and does not provide service when it stays in the failure state, the rejuvenation states (partial and full), and the reconfiguration states (the first and the second).

Therefore, the system availability is calculated by the Eq. (3).

$$P_{\text{avail}}(\lambda_C, \beta) = \pi_0(\lambda_C, \beta) + \pi_1(\lambda_C, \beta) + \pi_2(\lambda_C, \beta) \tag{3}$$

In order to estimate whether the software rejuvenation model with reconfiguration is effective, we compare it to the rejuvenation model without reconfiguration. Let λ_c equals to 0, that is, the reconfiguration interval is ∞, then the system model becomes to the software rejuvenation model without reconfiguration. Thus, the system availability of the software rejuvenation without reconfiguration is computed by the Eq. (4).

$$P_{\text{avail\_without\_reconfig}} = \pi_0(0, \beta) + \pi_1(0, \beta) + \pi_2(0, \beta) \tag{4}$$

Table 1 Default parameters

Parameter	Comment	Value
α	Degradation rate	1/240 (h^{-1})
λ	Failure rate	1/10,000 (h^{-1})
$1/\mu$	Recovery time	2 (h)
λ_p	Partial rejuvenation rate	1/24 (h^{-1})
$1/\mu_p$	Partial rejuvenation time	1/20 (h)
λ_f	Full rejuvenation rate	1/72 (h^{-1})
$1/\mu_f$	Full rejuvenation time	1/6 (h)
$1/\mu_c$	Reconfiguration time	1/20 (h)
λ_c	Reconfiguration rate	Variable
β	Unsuccessful probability of reconfiguration	Variable
γ	Successful probability from state C_L to state M	0.4
μ_U	Service rate of the normal state	100 (h^{-1})
μ_M	Service rate of the medium efficient state	60 (h^{-1})
μ_L	Service rate of the low efficient state	20 (h^{-1})
λ_R	Arrival rate	50 (h^{-1})

3.2 Throughput Analysis

The system service rate is different when it is in the normal state *U*, the medium efficient state *M* and the low efficient state *L*. Assume that the service rate of the normal state, the medium efficient state, and the low efficient state is μ_U, μ_M, and μ_L, respectively, the arrival rate of the clients is λ_R, then the average throughput of the software system is calculated by Eq. (5).

$$\begin{aligned} Th(\lambda_C, \beta) = \min(\lambda_R, \mu_U)\pi_0(\lambda_C, \beta) + \min(\lambda_R, \mu_M)\pi_1(\lambda_C, \beta) \\ + \min(\lambda_R, \mu_L)\pi_2(\lambda_C, \beta) \end{aligned} \quad (5)$$

Let λ_c equals to 0, then we get the system throughput of the software rejuvenation without reconfiguration by Eq. (6).

$$\begin{aligned} Th\_\text{without\_reconfig} = \min(\lambda_R, \mu_U)\pi_0(0, \beta) \\ + \min(\lambda_R, \mu_M)\pi_1(0, \beta) + \min(\lambda_R, \mu_L)\pi_2(0, \beta) \end{aligned} \quad (6)$$

4 Experiment Results

The default parameters along with the rejuvenation and reconfiguration are shown in Table 1. And the system service capacity of the normal state, medium efficient state, and low efficient state is also shown in Table 1. Notice that all the transition rates are measured in hours.

Fig. 2 System availability

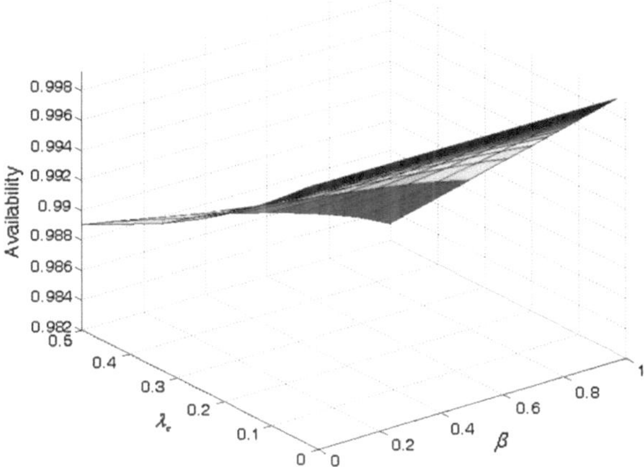

Because the aim of this paper is to study the effect of reconfiguration rate and the unsuccessful probability of reconfiguration operation on the system performance, we set λ_C fluctuate between once every 1 day (0.0417 h^{-1}) to once every 2 h (0.5 h^{-1}) and β from 0 to 1.

The relationship between the system availability and the reconfiguration rate λ_c and the unsuccessful probability of reconfiguration β is shown in Fig. 2. From Fig. 2, it is observed that with the increasing of the unsuccessful probability of reconfiguration β, the system availability decreases. This fact is obvious, because with the increasing of β, the system has decreasing chance of transition from reconfiguration states to better operational states. For certain β, it is also observed that with the decreasing of the reconfiguration rate λ_c, the system availability decreases. The reason is that the smaller the reconfiguration rate λ_c, the longer the reconfiguration interval. Because when the system is in the reconfiguration state, the system is unavailable, with the increasing of λ_c, the probability of the system entering into the reconfiguration states increases, which results in decreasing availability.

The relationship between the system throughput and the reconfiguration rate λ_c and the unsuccessful probability of reconfiguration β is shown in Fig. 3. From Fig. 3, it is observed that at first, the system throughput increases with the decreasing of the reconfiguration rate λ_c; when λ_c arrives to a certain value, the system has the maximum throughput; then, with the decreasing of λ_c, the system throughput decreases. The reason is that when the system is in the reconfiguration states, it does not provide service. Therefore, when λ_c is large, the system runs into the reconfiguration states very frequently, and then, the decreasing of the system throughput caused by reconfiguration is obvious. On the other hand, when λ_c is very small, the probability of system running into the failure and rejuvenation states is large, and then, the decreasing of the system throughput caused by failure and rejuvenation is tremendous, because when the system is in the failure and rejuvenation states, the system is also not provide service. For a fixed λ_c, with the

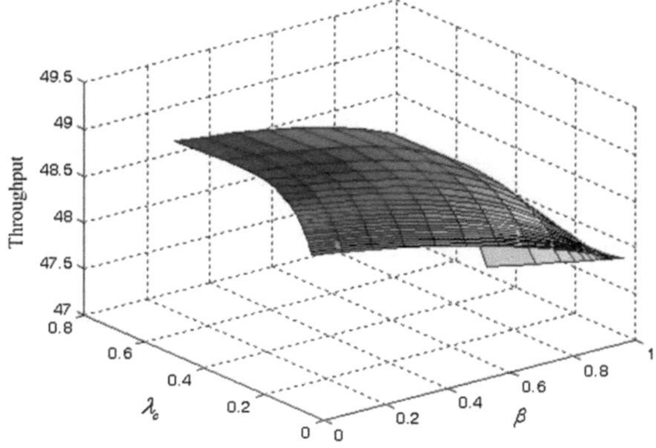

Fig. 3 System throughput

increasing of the unsuccessful probability of reconfiguration β, the system throughput decreases. The reason is that with the increasing of β, more and more invalid reconfiguration operations are taken, and thus, the throughput decreases.

5 Conclusions

In this paper, we apply reconfiguration into the research of software aging and rejuvenation. For a software system with multiple degradation levels, we present a new software rejuvenation model with reconfiguration. The experimental results show that when the unsuccessful probability of the reconfiguration operation is low, the rejuvenation model with reconfiguration is superior to the model without reconfiguration for throughput. When the unsuccessful probability of the reconfiguration operation is high, though the rejuvenation model with reconfiguration is inferior to the model without reconfiguration for the system availability, the rejuvenation model with reconfiguration is also superior to the model without reconfiguration for throughput.

Acknowledgments This work was supported in part by NSFC under Grant No. 60933003, the Fundamental Research Funds for the Central Universities, the National Torch Plan Project in China (2012GH571817), the Fundamental Research Funds for the Central Universities in China (08143003), the Shanxi Province Key Scientific and Technological Project (2012K11-08, 2012K06-18), and XI'AN Science and Technology Project (CX12178(3)).

References

1. Paulson, L.D.: Computer system, heal thyself. IEEE Comput. **35**, 20–22 (2002)
2. Huang,Y., Kintala,C., Kolettis, N., et al.: Software rejuvenation: Analysis, module and applications. Proceedings of 25th Symposium on Fault Tolerant Computing, pp. 381–390 (1995)

3. Grottke, M., Li, L., Vaidyanathan, K., Trivedi, K.S.: Analysis of software aging in a web server. IEEE Trans. Reliab. **55**, 411–420 (2006)
4. Silva, L., Madeira, H., Silva, J.G.: Software aging and rejuvenation in a soap-based server. IEEE-NCA:. Network Computer and Applications, pp. 56–65 (2006)
5. Bao, Y., Sun, X., Trivedi, K.S.: A workload-based analysis of software aging, and rejuvenation. IEEE Trans. Reliab. **54**, 541–548 (2005)
6. Garg, S., Puliafito, A., Telek, M., et al.: Analysis of preventive maintenance in transactions based software systems. IEEE Trans. Comput. **47**, 96–107 (1998)
7. Wang, D., Xie, W., Trived, K.S.: Performability analysis of clustered systems with rejuvenation under varying workload. Perform. Eval. **64**, 247–265 (2007)
8. Vaidyanathan, K., Trivedi, K.S.: A comprehensive model for software rejuvenation. IEEE Trans. Dependable Secure Comput. **2**, 124–137 (2005)
9. Hoffmann, G.A., Trivedi, K.S., Malek, M.: A best practice guide to resource forecasting for computing systems. IEEE Trans. Reliability **56**, 615–628 (2007)

Research and Design of Performance Monitoring Tool for Hadoop Clusters

Chongyang Xue, Feng Liu, Honghui Li, Jun Xiao and Zhen Liu

Abstract Hadoop is an open source platform. Because of its open source, high fault tolerance, and scalability, it has been widely used to deal with big data in many IT industries. It becomes particularly important and difficult to monitor and analyze the performance of a Hadoop cluster as the scale of the cluster grows. Based on Hadoop source code, this paper proposed a performance monitoring tool named Hadoop Monitor. Integrated with a job scheduling framework, such as Quartz, the Hadoop Monitor grasps the performance data from Hadoop cluster periodically and then generates a performance report. This tool can provide a great help to analyze the performance of Hadoop cluster.

Keywords Hadoop · Performance monitoring · MapReduce · HDFS

1 Introduction

Hadoop is an open source distributed computing platform in the Apache Software Foundation [1]. With the advantages such as high fault tolerance and high extensibility, Hadoop makes it easy to build and manage a distributed computing

C. Xue (✉) · F. Liu · H. Li · J. Xiao · Z. Liu
School of Computer and Information Technology, Beijing Jiaotong University, Beijing 100044, China
e-mail: xuechongyang@139.com

F. Liu
e-mail: fliu@bjtu.edu.cn

H. Li
e-mail: hhli@bjtu.edu.cn

J. Xiao
e-mail: 09112070@bjtu.edu.cn

Z. Liu
e-mail: zhliu@bjtu.edu.cn

S. Patnaik and X. Li (eds.), *Proceedings of International Conference on Computer Science and Information Technology*, Advances in Intelligent Systems and Computing 255, DOI: 10.1007/978-81-322-1759-6_93,

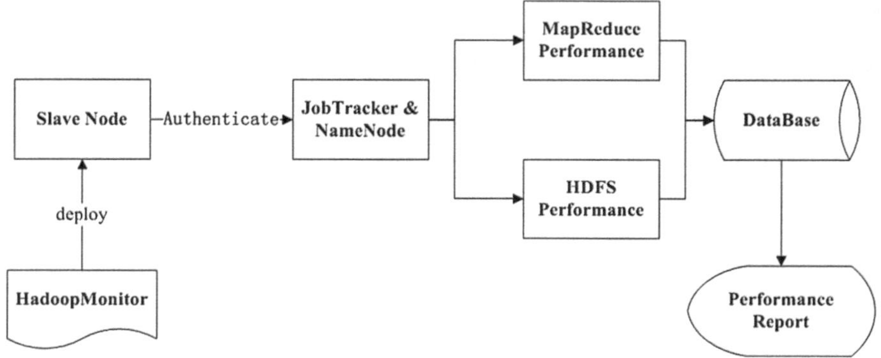

Fig. 1 Business flow chart of Hadoop Monitor

system with ordinary PCs. At the same time, Hadoop makes it easy and quick to develop parallel applications without caring about the implementation of the distributed computing system.

As the number of nodes and tasks in Hadoop cluster increasing, the monitoring and maintenance of Hadoop cluster are becoming more and more difficult. This paper puts forward a Hadoop performance monitoring tool named Hadoop Monitor, which is integrated with a job scheduling framework. This tool gets the performance data of Hadoop cluster periodically and renders the performance report of the cluster for analyze. To use this monitoring tool, first integrate it with a time scheduling framework and deploy them in a slave node of the Hadoop cluster. Through this slave node, the Hadoop Monitor can get permission to monitor the performance of the Hadoop cluster. Secondly, Hadoop Monitor periodically grasps the performance data of the cluster and deposited in the database. Finally, the tool would generate performance report from the performance data in the database. The business flow chart is shown in Fig. 1.

The two main functions of Hadoop are distributed parallel computing which is based on MapReduce and the Hadoop distributed file system (HDFS). The performance monitoring of the Hadoop cluster is mainly aimed at these two parts. Section 2 discusses the performance monitoring of MapReduce; Sect. 3 discusses the performance monitoring of HDFS; Sect. 4 presents an experiment using the performance monitoring tool “Hadoop Monitor.” Section 5 makes a summary of this paper.

2 Performance Monitoring of MapReduce

MapReduce is a parallel programming model with which developers can easily create distributed parallel programs [1]. In the Hadoop, MapReduce tasks are scheduled by JobTracker which is running in the master node and executed by many TaskTrackers that are running on slave nodes. The JobTracker provides many API to get information about the tasks running in the Hadoop cluster, so the Hadoop Monitor can use the JobTracker API to get some performance data of the Hadoop cluster. This paper discusses the performance monitoring of MapReduce in two directions as follows:

1. Performance monitoring of MapReduce overall. The performance data of the cluster should be monitored are running job number, running map task number, running reduce task number, TaskTracker nodes number, and blacklist nodes number.
2. Performance monitoring of detailed MapReduce jobs. For each finished job, the Hadoop Monitor should monitor the submit time, launch time, finish time, status, success map and reduce task number, failed map and reduce task number, and killed map and reduce task number.

2.1 Performance Monitoring of MapReduce Overall

To monitor the performance of Hadoop cluster, Hadoop Monitor must be deployed on a node which has access to the Hadoop cluster through Kerberos [2] network authentication protocol [3]. When the Hadoop Monitor gets the permission to access cluster, it can get a JobClient object which is defined in Hadoop API from the JobTracker. With the JobClient object, Hadoop Monitor can get many job information of Hadoop cluster. Figure 2 shows the class diagram, which shows how to get the JobClient object.

JobClient has a function to get a Cluster Status object, which is also defined in Hadoop API and contains the status information of Hadoop cluster. The object Cluster Status has methods to get running job number, running map task number,

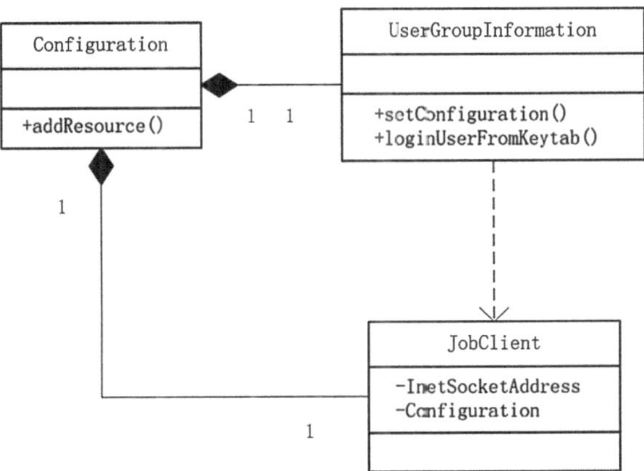

Fig. 2 Class diagram of obtaining JobClient

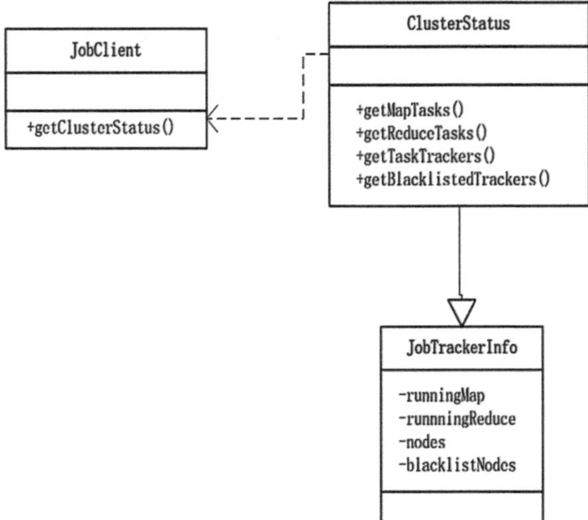

Fig. 3 Class diagram of JobTracker Info

running reduce task number, TaskTracker nodes number, and blacklist nodes number. With these data, we can build an object named JobTracker Info which is defined by Hadoop Monitor and contains the performance information of Hadoop cluster. Class diagram of JobTracker Info is shown in Fig. 3.

2.2 *Performance Monitoring of Detailed MapReduce Jobs*

There are two kinds of MapReduce jobs to monitor in Hadoop cluster. One is the currently running tasks, and the other is the tasks that have just finished.

For the currently running tasks, the JobClient object can provide an array of JobStatus objects, which contains the status information of running jobs. With this status information, a JobDetail Info object could be built. The most important property of JobDetail Info is the running Time, which describe the length of task execution time. For the jobs whose running Time is too long, the Hadoop Monitor would generate an alarm signal to the maintenance staff. Class diagram of JobDetail Info is shown in Fig. 4.

For the tasks that have just finished, Hadoop writes their information in corresponding logs. In order to get the performance information of jobs that just finished, Hadoop Monitor needs to parse the job log to a Job History. JobInfo object is defined by Hadoop API from the object of Job History. JobInfo, the tool needs to get some important data to analyze the performance of a job. For example, job's ID, name and user, and job's submit time, launch time, finish time, status, success map and reduce task number, failed map and reduce task number, and killed map and reduce task number. With these data, Hadoop Monitor can

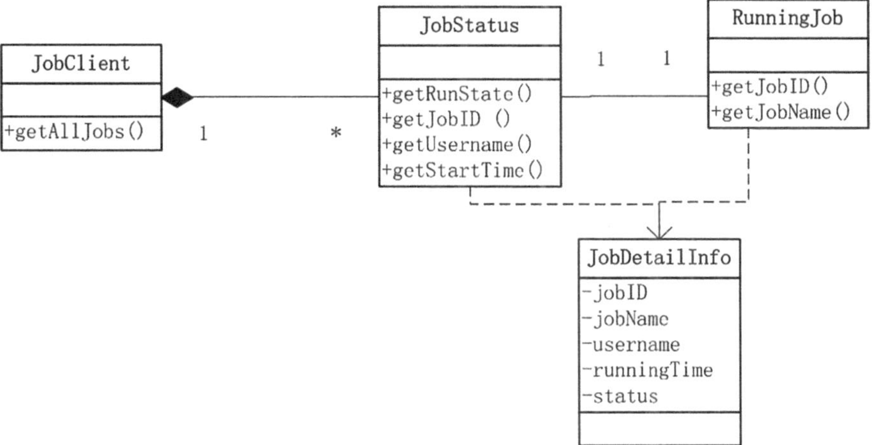

Fig. 4 Class diagram of JobDetail Info

analyze whether a job's duration is too long or has too many invalid subtasks. The result of the analysis can help developers to improve their MapReduce job's efficiency and help improve the utilization of Hadoop cluster. To get the submit time, launch time and finish time of a job from the Job History. JobInfo, Hadoop API provided a static method from class hadoop. util. String Utils.

To get the status, success map and reduce task number, failed map and reduce task number, and killed map and reduce task number of a job from the Job History. JobInfo, Hadoop Monitor needs to do some simple statistics work. Class Job History. JobInfo has a method named get All Tasks () to return all map and reduce task records of a specific job. With these map and reduce records, Hadoop Monitor needs to judge each task type and execution result and also need to count the number of tasks that was success, failed, or killed.

To get the number of failed and killed map and reduce tasks, first need to get a task list by the method get All Tasks (). When the task list was obtained, the tool needs to traverse the list to judge and count each task in the list.

3 Performance Monitoring of HDFS

Hadoop distributed file system (HDFS) is consisted of one Name Node and many Data Nodes, each node is a usual PC. Name Node is the core of HDFS, and it records the memory usage of the whole cluster and the status of every Data Nodes. The paper discusses the performance monitoring of HDFS in two directions as follows:

1. Performance monitor of the whole status of HDFS. In this section, the total capacity of HDFS, used capacity of HDFS, and remain capacity of HDFS should be monitored periodically to analyze the performance.
2. Performance monitor of all the Data Nodes in Hadoop cluster. Hadoop Monitor needs to check each Data Node periodically to see if the heartbeat of Data Node is normal. In addition, in consideration of load balance, the tool needs to check the used capacity of each live Data Node.

3.1 Performance Monitoring of the Whole Status of HDFS

To monitor the performance of the whole status of HDFS, Hadoop Monitor needs to get some data periodically: configured total capacity, used capacity, and remain capacity of the HDFS. The JobClient object has a function get Fs () to return a File System object, the function get Status () from the File System can return a Fs Status object. The Fs Status object contains the status of the whole HDFS, and it provides several methods to directly get the total capacity, used capacity, and remain capacity of HDFS.

3.2 Performance Monitoring of all the Data Nodes in Hadoop Cluster

To monitor all the Data Nodes of HDFS, Hadoop Monitor needs to monitor the heartbeat and capacity of each Data Node. The object File System which is got by the method get Fs () from JobClient can be converted to a Distributed File System object. This object has a method called get Data Node Stats () to return the status of every Data Nodes in the form of an Data node Info object array.

Hadoop Monitor check the heartbeat of a Data Node to determine whether the node is live or not. According to Hadoop source code, the default interval that Name Node checks a Data Node is 5 min, and the default interval that Data Node makes a heartbeat activity is 3 s, which is represented by "long heartbeat Interval." The source code of checking the heartbeat of Data Node is shown in Table 1.

To monitor the used capacity of each Data Node, the class Data node Info has simple interfaces to get the capacity information of the Data Node. In consideration of load balance, if the used capacity of Data Node is much higher or lower than the average level, the Data Node should be considered anomalous and the operations staff should be alarmed.

Table 1 Source code of checking the heartbeat

```
longheartbeatExpireInterval = 2 *
heartbeatRecheckInterval + 10 * heartbeatInterval;
for(DatanodeInfo dataNodeInfo :
dfs.getDataNodeStats()) {
if(dataNodeInfo.getLastUpdate() <
 (System.currentTimeMillis()-
 heartbeatExpireInterval))
deadNode ++;
else
liveNode ++;
}
```

4 The Experiment and Performance Analysis

This paper developed the monitoring tool called Hadoop Monitor in java language and integrated the tool with a job scheduling framework called Quartz [4]. Hadoop Monitor was used to do an experiment with a Hadoop cluster from a famous Internet company in China. Table 2 details information of the experiment environment.

Hadoop Monitor gets the performance data of the Hadoop cluster and stores the data to database. With performance data of Hadoop cluster in everyday, Hadoop Monitor can do many analytics and draw performance curves to generate performance report of the Hadoop cluster. For example, Figs.5 and 6 are from a performance report generated by Hadoop Monitor.

Figure 5 shows the changing trends of Job number, map task number, reduce task number, and nodes number of the Hadoop cluster in one day. In this chart, we

Table 2 Experiment environment

Hadoop edition	Hadoop 0.20.2-cdh3u1
Hadoop nodes number	77
Operation system	RedHat Linux 6
DataBase	MySQL 5.5

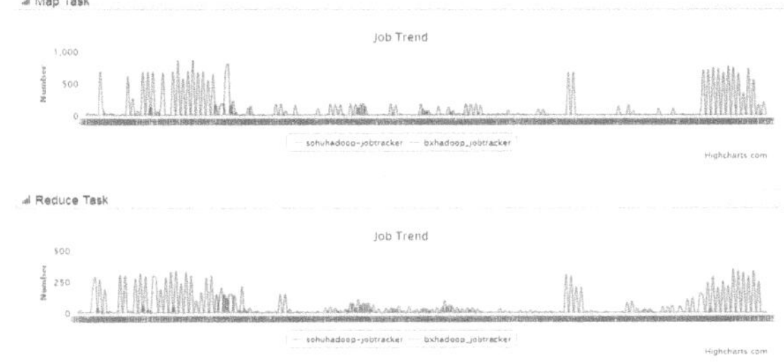

Fig. 5 Tendency charts of MapReduce

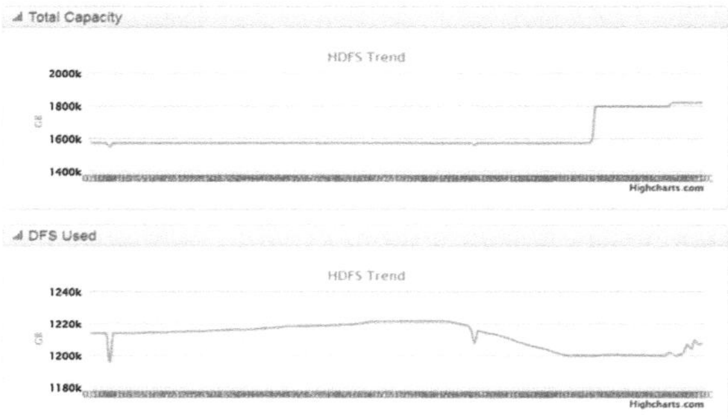

Fig. 6 Tendency charts of HDFS

can see that the peak time of tasks is the first half of the day, that is because the most jobs are executed during 1 a.m. to 8 a.m. Figure 6 shows the changing trends of the total capacity and used capacity of HDFS in one day.

5 Summary

The core content of this paper is the proposition of a method to develop a tool called Hadoop Monitor to monitor the performance of Hadoop cluster. The Hadoop Monitor referenced the structure of master and slave and some source code of Hadoop and is integrated with a job scheduling framework. This tool is very practical and original, because it gives the maintenance staff a great convenience to know the performance of Hadoop cluster. It is very useful for maintenance staff to maintain a large scale of Hadoop cluster.

Acknowledgments The research has been supported by Project No. 2012X010-C of Information system connectivity and security maintain technology application research–technology research of the railway information system evaluation.

References

1. Lu, J.: Hadoop in Action, 2nd edn. China Machine Press, Beijing (2011)
2. Kerberos: The network authentication protocol (EB/OL). http://web.mit.edu/kerberos/. Accessed on 5 May 2013
3. Cloudera (EB/OL): http://www.cloudera.com/. Accessed on 15 May 2013
4. QUARTZ (EB/OL): http://quartz-scheduler.org/. Accessed on 5 May 2013
5. Apache Hadoop (EB/OL): http://hadoop.apache.org. Accessed on 15 May 2013

The DHU-SAM: Modeling MIS Software Architecture

Mingyou Ying, Min Xie, Qiong Wu, Zehui Chen and Jingxian Chen

Abstract In this paper, focusing on MIS software, we proposed a complete software architecture model and studied the elements consisting of the architecture and the rationale that associates the elements, thus forming the structure pattern of the architecture, named as the DHU-SAM. We found that the frame of the architecture is obviously featured with a so-called entrance-Passageway-Interior-Exit (e-PIE) structure and that there are only three kinds of basic components to compose the architecture, namely user interfaces, data interfaces, and functional calculators. The architecture possesses a certain hierarchical pattern that is determined by the client's requirements only. With the help of the architecture model presented in this paper, many analyses that are too difficult to conduct currently in software engineering can be smoothly carried out. It can also highly assure quality of software requirement specifications for software engineers.

Keywords Software architecture · Software architecture model · Management information system · Structure pattern

M. Ying (✉) · J. Chen
Glorious school of Management, Donghua University, Shanghai, China
e-mail: 1_mail@dhu.edu.cn

J. Chen
e-mail: xchen33@dhu.edu.cn

M. Xie
APTEAN, 99 Tianzhou Road, Shanghai, China
e-mail: superfelix@gmail.com

Q. Wu · Z. Chen
IBM, 399 Keyuan Road, Shanghai, China
e-mail: cchurch005@gmail.com

Z. Chen
e-mail: kkpop.kkpop@gmail.com

S. Patnaik and X. Li (eds.), *Proceedings of International Conference on Computer Science and Information Technology*, Advances in Intelligent Systems and Computing 255, DOI: 10.1007/978-81-322-1759-6_94,

1 Introduction

Perry and Wolf [1] proposed a triplet model (elements, form, rationale) for software architectures and expected that the software architecture could be deployed as a critical toolkit that could play a role as the framework for satisfying requirements, the technical basis for design, the managerial basis for cost estimation and process management, an effective basis for reuse, and the basis for dependency and consistency analyses as well. Otherwise, software development might cause immense loss. Unfortunately, this is the reality in software engineering. Boehm's book "Software Engineering Economics" [2] and the CHAOS Reports issued by the Standish Group [3] offered evidence. Since the birth of the Perry and Wolf's architecture model, no one has suggested any revision to the model except Boehm [4] who suggested adding "constraints" into the model. The meaning of the terms "elements," "forms," and "rationale" popping out from the model remains almost unexpanded. The goals for studying software architecture, which are set up by Perry and Wolf, have not been approached even a step ahead. The reason why the study of software architecture in this direction is at a standstill lies in the extreme diversity of software but never the resources insufficiency.

In 2012, we addressed an issue on the structure pattern of a user requirement assembly (URA), a small piece of an MIS software system for realizing an item of users' requirements [5]. Later, we further addressed an issue on the structure pattern of a user module (UM) [6]. In this paper, we are going to study the structure pattern of a whole MIS software architecture, in terms of a hierarchical structure embedded in a gross frame. For convenience, the pattern is named as the DHU-SAM, where DHU is the abbreviation of Donghua University and SAM stands for software architecture model.

The main findings of this paper are as follows: (1) the gross frame of the DHU-SAM is featured with an entrance-Passageway-Interior-Exit (or the e-PIE) structure; (2) there are only three kinds of component elements for the SAM: the user interfaces, the data interfaces, and the functional calculators, with which any MIS software architecture can be described as a hierarchical structure embedded in a gross frame; (3) the SAM can facilitate the analyses as Perry and Wolf have expected; (4) the SAM can also facilitate to assure the software requirement specifications of high quality.

The paper is organized as follows. Section 2 deals with terms and icons. Section 3 briefs Ying et al. [5, 6]. Section 4 reveals the structure of the DHU-SAM. Section 5 explores some important issues related to the SAM. Finally, the last section concludes the paper.

2 Terms and Icons

In software engineering, there is no consensus on lots of terms, yet, say, software architecture is an example. Hence, it is necessary to clearly define some terms beforehand to avoid misunderstanding.

Software Architecture: A gross structure of software, which is depicted in terms of components and logical relationships among them.

Component: An essential constituent of software and an element of a software architecture as well. It has a relatively complete functionality rather than a single sentence in code or a basic operation. Only three kinds of components are available in this paper: the user interfaces, the data interfaces, and the functional calculators.

Assembly: A small piece of software. It logically groups several components to possess a single applied function. For example, a URA possesses a single function to meet an item of the users' requirements.

User Module: A piece of software. Each UM corresponds to a specific user. It logically groups several relevant assemblies and components so that it can meet all the requirements of the correspondent user. There are clear-cut boundaries between different modules except they are possibly associated with the relevant data and the logics of related business processes only.

User Interface: A component of software used for human–computer interaction. In this paper, we divide user interfaces into three categories: the simple user interfaces (SUIs), the switch user interfaces (SwUIs), and the creation user interfaces (CUIs), according to their uses and features of how they are adjacent to their neighbor components. An SUI is adjacent downward to one component only. Components functioning log-in, logout, or evaluation by a user are typical examples of the SUIs. An SwUI is deployed when a user needs to pick one from several independent functions at a time. It is adjacent downward to one of the several candidate components, depending on the user's will. A CUI is deployed when a user needs to initiate a new computation task without interrupting the computation tasks running currently. It can be adjacent downward to one or more components with the help of a "create" command.

Data Interface: A component of software used for data input or output. Data interfaces can be broken down into two types: data input interfaces and data output interfaces. We name the data interface after the source of the input data or destination of the output data. For example, a database data input interface is the one getting data from a database. And a Windows file data output interface handles data output to files in Windows.

Functional Calculator: A component of software that automatically carries out computation without the need of user intervention, and data input or output.

Adjacency of Components: Components in an assembly are placed in order based on the rationale of this assembly, without the need of any connector among the components. If A and B are two adjacent components, and A is placed before B according to the algorithm logic, then we say A is prior to B, or B is posterior to A.

If it does not matter, no matter A is either prior or posterior to B according to the algorithm logic.

User Requirement: A want or desire for a software system from a user. The users include the staffs of the client organization, who use the software system to process daily business information, and the IT engineers in the client organization, who monitor and support the software. User's wants are not limited in processing business information, and they also cover the desires for conveniently operating the software system. Therefore, a user's requirements include both the app ones and the facility ones. The former addresses the wants for business information processing, and the latter deals with the wants for facilities.

General View and Local View: A blueprint depicting the gross structure of a software architecture is called a general view. A blueprint depicting the structure of a part of a software architecture is called a local view. A MIS system is often used by hundreds of users. It means there will be hundreds of UMs in the software architecture and much more number of components. Hence, the general view cannot express all the details as fine as components are, in one blueprint, and it has to depict the structure of the software architecture in term of assemblies or UM groups. Then, local views are applied to express the structures of the assemblies or UM groups in the general view. Thus, a well-organized set of blueprints is indispensable to present the whole structure of the software architecture in every detail.

Business Process and Business Process View: A business process is often regarded as a sequence of activities to achieve a business goal [7]. Actually, a business process is also a standard or a rule of the client organization that the staffs of the organization must follow. In case a business process is executed with the aid of a computer system, the involved components are not located in a same UM. On the contrary, they are scattered in different UMs and are driven by different users. When being driven, they should be arranged in the same order as their counterpart activities that are arranged in the correspondent business process. However, they cannot be arranged in an adjacent way in the software since they belong to different UMs. Thus, we have to apply conditional logics to control them in the right order when running. In this case, it is necessary to introduce a business process view to illustrate all the relevant components and their logic relations for a business process.

Shared Task, Individual Task, and the States of a Task: Large amounts of daily work in a client organization are shared tasks. That is, they have to be split into subtasks that are assigned to different staffs, and then, the staffs will complete their own subtasks, or their individual tasks, following a certain business process. As above mentioned, when a business process is executed with the aid of a computer system, invocation of the relevant components must be controlled by conditional logics. Thus, the states of shared tasks are able to play an important part in the control mechanism. Specifically speaking, they can be used as the conditions in the conditional logics.

Shared tasks and business process are two closely related but totally different concepts. Let us take a manufacturing system as a metaphor. The business process

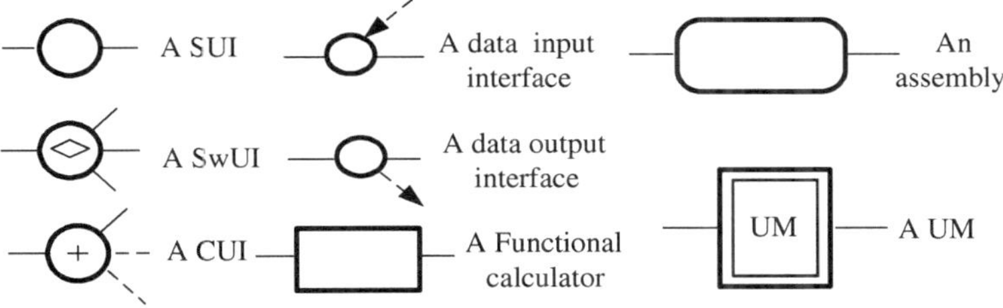

Fig. 1 The icons

looks like the production line, while the shared task plays a role of work pieces processed in the system.

In this paper, we use the following icons as shown in Fig. 1.

Sometimes, we may insert certain letters or symbols into icons to emphasize certain features of the components or the assemblies that the icons denote. For examples, an assembly icon with letters "UR" inside denotes a URA and that with letters "Fac" inside stands for a facility assembly. Similarly, a functional calculator icon with a diamond symbol inside represents a conditional logic calculator and that with a refresh symbol means that this is a data update or state update calculator. A constraint binding on a component or an assembly is usually marked with words near the icon.

3 The Structure Patterns for the Partial Software Architectures

Ying et al. [5, 6] addressed the issues how the structure patterns of a URA and a UM are obtained. The two papers suggest that they ought to be obtained through "inducing rationale from scenarios and deducing structure from rationale." Their achievements are quite ideal such that the resulted structure patterns can strictly conform themselves to the Perry and Wolf's triplet model and that they can facilitate the analyses as Perry and Wolf have expected.

3.1 The Structure Pattern of a URA

The rationale of a URA is illustrated in Fig. 2. If a user wants to use the computer system to complete an individual task, the user must clearly assign the task to the system at first. Then, the computer system fetches right data from right data sources. Next, the system executes computation automatically. As soon as the computation finishes, the system will send the data to right storages if needed. At

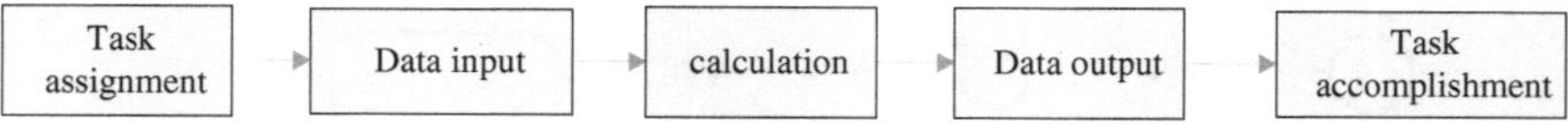

Fig. 2 The rationale for a URA

last, the system delivers the result to the user in a right form and reports that the task is done. So, the rationale can be easily turned into a five-block flowchart, as shown in Fig. 2.

Obviously, each block represents a subcalculation task, and it determines what kind of components should be used. Block 1 needs and only needs an SUI. Block 2 needs and only needs data input interfaces, and the number of the interfaces equals to the number of data sources. Block 3 needs and only needs a functional calculator. Block 4 needs and only needs data output interfaces, and the number of the interfaces equals to the number of the data storage devices. The last block needs and only needs an SUI to receive the result that the user concerns. Aligning these components in accordance with the rationale will result in a structure of the URA.

Usually, data input interfaces are presented in parallel way if there are several ones. Similarly, data output interfaces are presented in parallel way if there are several ones.

Thus, it can be seen that a URA always possesses a structure consisting of multiple interfaces and a single functional calculator and that the first and the last components must be SUIs, and a functional calculator is always in the middle, with data input interfaces ahead and data output interfaces behind if they are available, no matter whatever the URA is. This structure form is called "multiple interfaces plus a functional calculator structure" or "MI + C" structure in short. Figure 3 depicts an MI + C structure for a URA.

The only exception occurs when the functional calculator cannot automatically deliver a result as ideal as the user wants, such as the case of daily scheduling in production management and the case of stock depot allocation in warehouse

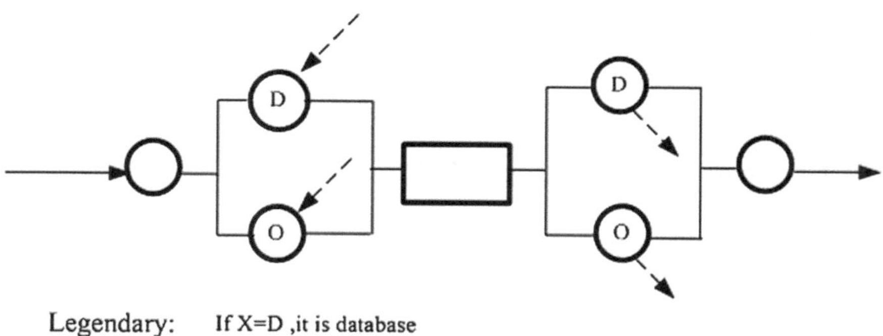

Fig. 3 The structure for a URA

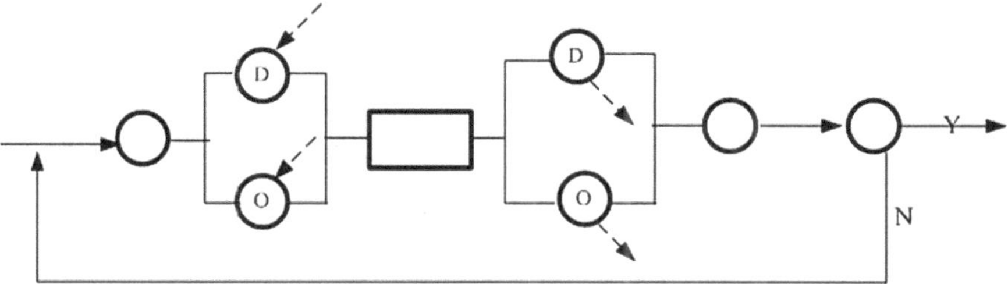

Fig. 4 The MI + CII structure

management. Under these circumstances, a user has to adopt a so-called trial-and-error method to get a feasible solution that makes the user happy, and the URA of the MI + C structure does not work. Hence, we need to revise the MI + C structure through adding an SwUI after the last SUI in the previous MI + C structure and forming a trial-and-error loop, as shown in Fig. 4. We name this new pattern as the "MI + CII" structure.

MIS software often grants the client's requests of management through setting up many management components to assure security or to monitor the system. If we insert some management components in a URA, will they interfere in the app function of the URA or vice versa? Here is an example to show that both of them would not interfere in each other's functions. If we insert five management components in the URA of Fig. 3, thus, we obtain a URA with management functions as shown in Fig. 5, where components M1 and M5 are used for recording starting time and ending time, respectively, and the records may be used for daily availability analyses. Components M2, M3, and M4 are settled down from security considerations. M2 is an event recorder and outputs relevant data to the log. M3 and M4 are security guards. The former prevents attacks from hackers or viruses, makes records, and sends records to the right storages. The latter prevents data leaks, makes records, and sends records to the right storages too. Obviously, they will not affect each other's stated functions during the runtime.

It must not be neglect that the MI + C or MI + CII structure is a minimal realization of an item of user requirement. Here, term "minimal realization" simply means a feasible software solution to an application problem with a minimal number of components.

3.2 The Structure Pattern of a UM and the Business Process View

The structure of a UM is much more complicated, because a user may have several or even couple of dozens of application requirements. Besides, the user may also have certain amount of facility requirements. To simplify our discussion, we assume that a user can only pick one app function and one individual task to run at

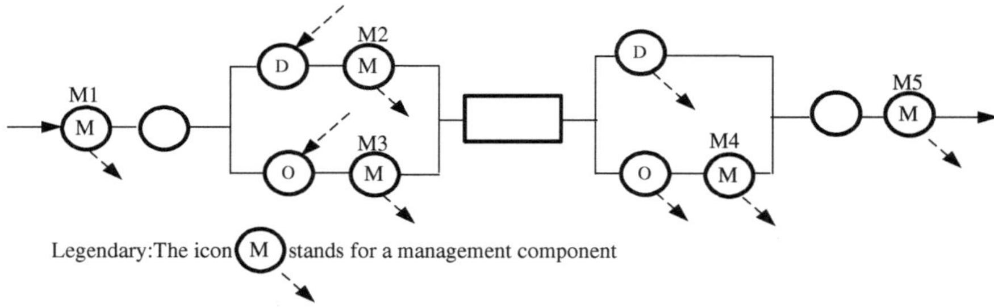

Fig. 5 The structure of a URA with management components

a time but is allowed to run one or more facility assemblies with the URA running concurrently.

Figure 6 shows the rationale of running a UM: Firstly, the system should inform the user that the module is ready for use. For example, it can prompt some information such as the app function menu or a list of shared tasks created that the user involved, involves, or is going to involve in, on the screen of a monitor. Then, the user can pick one of the available app functions and choose an individual task. The system will verify the state of the shared task. If the state is OK, the system invokes a correspondent URA to process the selected individual task; otherwise, the system will deny the user's claim. No matter while or not while processing the individual task, the user may have leisure to get task information, at her or his will, by calling a facility assembly via the "create" command. In Fig. 6, the icon—a circle with "+" inside—denotes the "create" command.

It is not difficult to deduce the structure of a UM from the rationale described above. Let us look at an illustrative example. Assume that a user specifies three app requirements and two facility requirements without the loss of generality.

So, there must be three URAs and two facility assemblies in the UM.

When the user starts to run the UM, firstly the UM requires a CUI to deal with the subcalculation task represented with Block 1 of Fig. 6 that can display the information on app functions and facilities on the monitor screen. The CUI is required to adjacent to a SwUI that allows the user to choose an app function and an individual task, which is what the Block 2 of the upper branch of Fig. 6 just wants. Also, the CUI can be selectively adjacent to a facility assembly via a built-in device

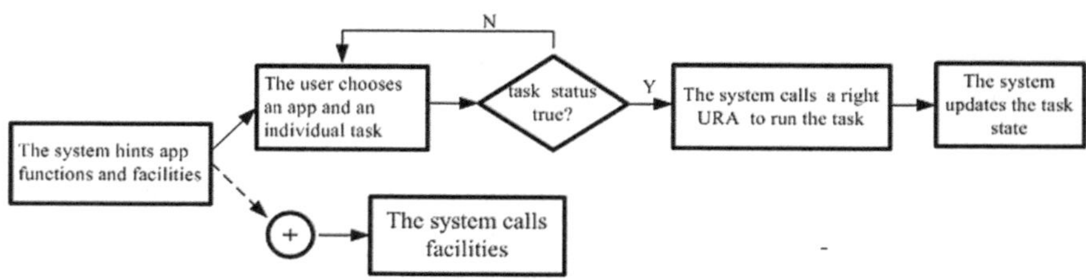

Fig. 6 The rationale for a UM

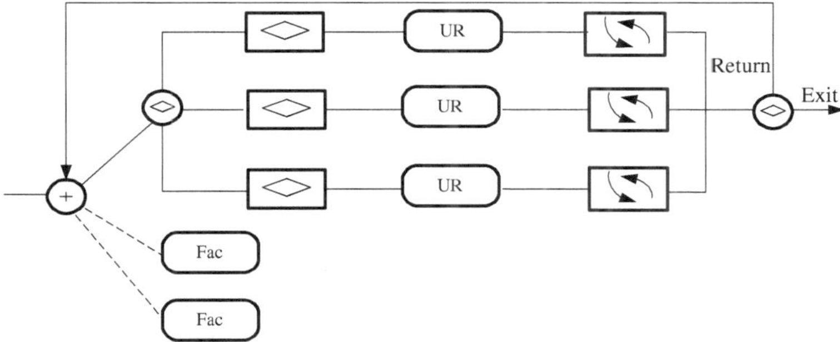

Fig. 7 The structure of a UM

of function of creating new task. This is just the subcalculation task of the lower branch of Fig. 6. There needs a conditional logic calculator between the SwUI and the URA chosen for doing the individual task because whether the chosen URA can be invoked or not depends on the state of the shared task. After a URA, there must be a state update calculator. Finally, the last component is an SwUI that allows the user to decide whether or not to quit the module. If not, the SwUI will be adjacent to the first SwUI. Figure 7 tells a should-be structure of a UM.

It is better to explain a business process view with a nice example. Consider a business process of three users, which processes shared tasks. Each user is in charge of the individual tasks that are different from other users. Now, suppose that a shared task is ready to be processed with the process. Let variable x describe the state of the shared task in such a way that $x = 0$, 0.5, 1, 1.5, 2, 2.5, 3 means, respectively, the state that user A has created the task but before starting to work on it, the state that user A is processing his or her individual task but does not finish it, the state that user A finishes his or her individual task but user B does not start to work on the task, the state that user B is processing his or her individual task but does not finish it, the state that user B finishes his or her individual task but user C does not start to work on the task, the state that user C is processing his or her individual task but does not finish it, and at last, the state that user C finishes his or her individual task, and thus, the shared task is finished.

Set the condition for the conditional logic calculator in the UM for user A as $0< =x < 1$, set the condition for the conditional logic calculator in the UM for user B as $1< =x < 2$, and set the condition for the conditional logic calculator in the UM for user C as $2< =x < 3$.

Now, we draw the components and relations relevant to the business process from the three UMs and link the three-state update calculator with a line of dashes (it means that the three calculators update a same state variable). Thus, the business process view for our example is formed, and it is shown in Fig. 8.

Next, we are going to explain how the control mechanism is established. Currently, if $x = 0.5$, it means that user A is processing and the shared task is not ready for user B or user C to work. If user B or user C wants to work on the shared

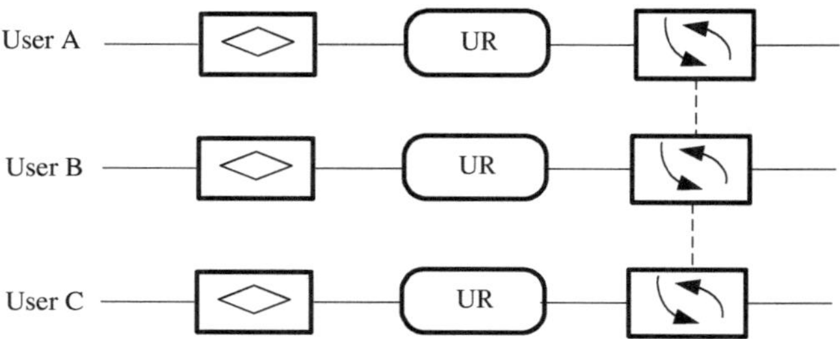

Fig. 8 A business process view

task, the conditional logic calculator before the URA of the UM for user B or user C will deny the claim from user B or user C. Similarly, we can verify the mechanism easily.

It is obvious that the control mechanism can be established, in a similar way, for any complicated business processes, except the state variable may be replaced with a proper state vector.

Obviously, the structure pattern of a UM discussed above is a minimal realization.

The authors of the two papers [5, 6] also found that the structure patterns for URAs, facility assemblies, and UMs as well are unique because of the intangibility of software. Thus, they concluded that design of software architecture does not make much sense.

4 The DHU-SAM

Our study about the structure of software architecture model can be dated back from 2007, when we proposed the "e-PIE model" for the gross structure of software on the basis of investigating several kinds of application software. Term "e-PIE" abbreviates **e**ntrance, **P**assageway, **I**nterior, and **E**xit [8]. It was inspired by term *architecture*. If the software is confidential or restricted for use, then a user must log on the system. It just looks like going through a "gate" with security guards. After checking ID, only legal one is allowed to enter, and the user is guided to a certain interface, by certain means, which tells the user what he or she can use. Then, the user can use the software as restricted. Here, the guiding means play a passageway role as in a building, and the part of the software that the user can use simulates the user's own office or cabinet. Thus, all these parts sound a business area inside an office building. Hence, the term interior is introduced. Of course, when the user wants to quit the system, he or she needs an exit. Metaphor is metaphor after all. The key problem in studying our software architecture is to find out its structure pattern in terms of software components and their relationships rather than an intuitive metaphor. However, due to lack of sharp ideas, this

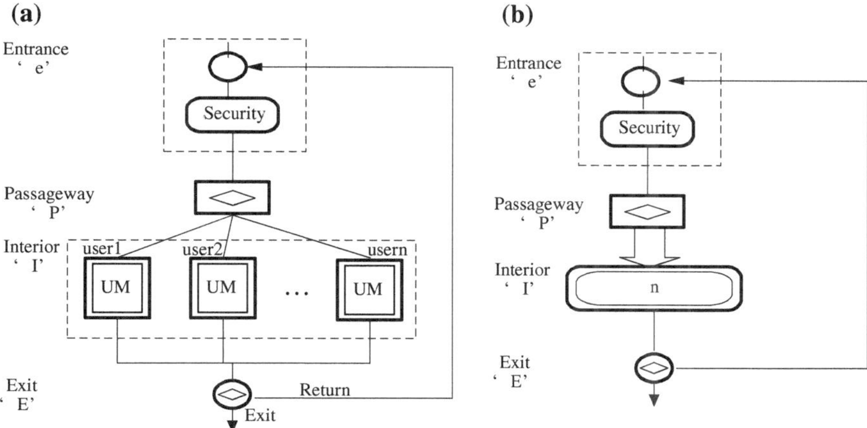

Fig. 9 The DHU-SAM

study has been suspended for several years. Until recently, we resume a new round of the research work after we carefully review literature again. We perceive that there must be some links among structure patterns of the software architecture, the "rationale" in Perry and Wolf's model, and the "scenario" in Kruchten's "4 + 1" view model [9]. Then, an idea comes upon us, that is, to induce the rationale from scenarios and to deduce the structure from the rationale. Consequently, it brings up the breakthroughs in Ying et al. [5, 6]. It also helps us to obtain: (1) the structure pattern of e-PIE for software architecture; and (2) a complete structure model for the software architecture or the DHU-SAM.

Obviously, it can be seen that (1) the entrance consists of a simple interface and a security assurance assembly; (2) the passageway is a large conditional logic calculator, (3) the interior aggregates all UMs, (4) the exit is a simple SwUI, (5) the UMs are structured as Sect. 3.2 describes, and (6) the DHU-SAM is a minimal realization. Figure 9(a) shows the structure pattern for the complete software architecture, where the interior is depicted with number of UMs. Since the MIS software usually has hundreds or thousands of UMs, it is impossible for a blueprint to show all the UMs. Therefore, we suggest using Fig. 9(b) to replace Fig. 9(a). In Fig. 9(b), the interior is simplified as a rounded rectangle, inside which the number "n" represents the numbers of UMs contained in the interior, and the links between the passageway and the interior are simplified as a hollow arrow.

5 Discussion

1. *Whether or not can the above software architecture model realize Perry and Wolf's expectations, such that software architecture can play a role as the framework for satisfying requirements, as the technical basis for design and the*

managerial basis for cost estimation and process management, as an effective basis for reuse; and as the basis for dependency and consistency analysis?

Obviously, the first expectations can easily be achieved because the structure follows the logic for operations and each UM covers all the requirements from the corresponding user. The second expectation can be reached, since the structure pattern is of the "work breakdown structure (WBS in short)" feature. The WBS feature makes the FMEA (a reliability analysis tool) feasible in software design phase and hence can be turned into the technical basis for design. For software develop projects, efficient and effective cost control, quality control, and time control require a nice WBS job as their prerequisites. Doubtlessly, the structure pattern can play a role as a managerial basis for cost estimation and process management. For the third expectation, as we discussed above, there are only three types of basic components in the architecture. This fact can lay down a solid foundation for standardization or serialization of software components so that the resulted reusability could easily exceed what Perry and Wolf expected. Finally, it is straightforward that the architecture can work as the basis for dependency and consistency analysis.

2. *Is it necessary for a software architect to design software architecture? Are there several different architecture styles in a software system?*

If you agree with the definition of software architecture in this paper, and if appreciate Perry and Wolf's thought on studying the software architecture model, you will find the issues on architecture design and architecture style do not make sense and they are pseudo ones since the DHU-SAM is the unique minimal realization. Its pattern depends only on the rationale, and its scale depends only on the number of users and the number of the business processes that the software serves. However, due to different interests of software architecture research people, now there are at least over one hundred definitions for software architecture. This phenomenon may make the research area appear prosperous, but it will not really help software quality improvement. So, it is the time to filter all noises from this research area and to move more resources on the direction that may have significant impact on improving productivity and quality for software industry.

3. *How good is the modifiability of this DHU-SAM?*

To some degree, software reliability problems and software quality loss problems as Boehm revealed in his famous book "Software Engineering Economics" [2], or as the Standish Group's CHAOS Reports stated [3], can refer to the poor modifiability of software. Regardless of how to efficiently find out software bugs, even if the bugs are found, people still cannot assure that their correction can be free from new bugs. On the other hand, any change in user's requirements or in human resource allocation or any shake-up from the client will request for modification. Besides, any change in technical environment will also call for modification of the software. To software engineers' regret, currently there is no guarantee granted for the modification that would not have negative impact on the software. Software modification work looks like hardware repair work to certain extent. It is broken down into two successive actions: finding out all the components that need modified (repaired) first and then fixing them. So, in the real world, why it is too difficult to modify software is rooted in the first action. Software engineers are not capable to find out exactly which components need modified without a tool to learn the structure of software and to analyze the performance of software before and after modification. Now, with the DHU-SAM, software modification will never ever become software engineers' nightmare because of the obvious reason. As for the modification of software architecture, life is even easier. According to the changes in the business environment, judge whether the e-PIE frame needs modified, if some user modules need modified and if some business processes need modified then all the relevant components the user modules and the related business process views will be fixed according to the structure patterns of the e-PIE frame fix.

4. *Is it difficult to completely describe the architecture of complex software in graphs?*

Yes, it is extremely difficult if without the help of the DHU-SAM. However, even if with the help of the DHU-SAM, it is yet a tedious and burdensome task despite its simplicity in nature. For instance, if an MIS software system consists of 3,000 users and 500 different business processes, a complete graphical description of its architecture would consist of at least a general view of the e-PIE frame, 3,000 different views of user modules, and 500 views of the business processes. Such sophistication is unavoidable. Nevertheless, the tool needed is not available in software engineering. It is necessary to set up a new graphical tool to solve the representation problem. Actually, the work for setting up the new tool may not be hard if we learn the relevant principles of engineering graphics.

5. *When should we model the architecture of an MIS software system with the DHU-SAM?*

Think about how software engineers start a project for developing an MIS software system. The requirements' engineers initiate the project with elicitation application requirements from potential users. Then, they document all the users' application requirements into a set of specifications of software requirements (SRS). Then, they deliver the SRS document to some software designers to analyze and to make design decisions. Anyone who is familiar with software quality assurance (SQA) would easily find out number of serious deficiencies in terms of SQA during the requirement specification stage. There is no way to turn the users' application requirements into an SRS of high quality. Obviously, the DHU-SAM can be a right bridge crossing over the gap because it can turn users' application requirements into component functions and logic relations of the components.

By the way, it is worth to mention Royce's waterfall model here [10]. Before the end of 1980s, it was required by the DOD of the United States that development of military software must follow the waterfall model to ensure quality of the software to be developed. Because of poor approaches to specifying users' requirements, projects for the development of military software often delayed delivery of ideal results seriously. It caused deployment of the waterfall model ceased. However, it is a law that quality is shaped gradually. The DOD's change in policy of management of development of military software does not help quality control because it violates the quality law. Therefore, if a right method for specifying users' requirements can be found, the quality problem for SRS may be solved with the help of the DHU-SAM. When the day comes, we believe that the waterfall model will be rehabilitated.

6. *Why does not the DHU-SAM address databases or likewise?*

We did think of databases as constituents of the DHU-SAM. However, the fact that repositories are indispensable to MIS software systems is too trivial to word. Any data interface has an end that links some repository. We do think that the DHU-SAM implies repository constituents, and we do not think that it is necessary to throw them into the DHU-SAM explicitly.

6 Conclusions and Prospects

The DHU-SAM works well for modeling the architecture of MIS software. Its gross frame is featured with an e-PIE structure. There are only three kinds of component elements for it: the user interfaces, the data interfaces, and the functional calculators, with which any MIS software architecture can be described as a hierarchical structure embedded in a gross frame. The DHU-SAM can facilitate the analyses as Perry and Wolf have expected. It can help assure the software requirement specifications of high quality.

Since whether the DHU-SAM stands or falls depends on the quality of client's requirement specifications, and the issue on well specifying client's requirements remains as one of the toughest unsolved problem in software engineering, our further study will be divided into two aspects. On the one hand, we are going on refining what we have achieved on software architecture. On the other hand, we should speed up our research work on user requirement specification, especially on summing up and improving systematically lots of our significant jobs that we did in the area.

References

1. Perry, D., Wolf, L.: Foundations for the study of software architecture. Softw. Eng. Notes, ACM SIGSOFT **17**(4), 40–52 (1992)
2. Boehm, B.: Software Engineering Economics. Prentice-Hall, New York (1981)

3. The Standish Group: CHAOS Report. http://www.standishgroup.com/chaos_2009.php accessed 10 November 2011 (2009)
4. Gacek, C., Abd-Allah, A., Clark, B., Boehm, B.: On the definition of software system architecture. In: Garlan D(ed) First International Workshop on Architectures for Software System, Technical report CMU-CS-95-151, Carnegie Mellon University, Pittsburgh (1995)
5. Ying, M., et al.: A pattern for partial software architecture. Appl. Mech. Mater. **263–266**, 1838–1843 (2012)
6. Ying, M., et al.: Structure patterns for user modules and business process views (in Chinese). Technical Report of the SQA Club of DHU, Serial No. TR 2013-01, the SQA Club of DHU (2013)
7. Hammer, M., Champy, J.: Reengineering the Corporation. Nicholas Brealey Publishing, Boston (2001)
8. Zhu, Yu.: Discussion on Automated Service System Development Method (in Chinese). Bachelor Thesis, Donghua University (2007)
9. Kruchten, P.B.: Architectural blueprints—the "4 + 1" view model of software architecture. IEEE Softw. **12**(6), 42–50 (1995)
10. Royce, W.: Managing the development of large software systems. In: Proceedings of IEEE WESCON. pp. 328–338 (1970)

Research on Applicability of SVM Kernel Functions Used in Binary Classification

Yi Bao, Tao Wang and Guoyong Qiu

Abstract Support vector machine (SVM) has been often used in binary classification. In order to seek the guidance principles of the kernel function selection, this paper analyzed a variety of kernel functions used to construct the SVM classifiers and carried out comparative studies on the 4 data sets for binary classification of UCI Machine Learning Repository. The experimental results show that, using the nu-SVC with radial basis kernel function (RBF) has the optimal classification accuracy, but using the C-SVC with RBF kernel function has the best generalization ability.

Keywords Binary classification · SVM · Kernel function

1 Introduction

The task of binary classification is to divide objects to two different parts according to some attributes of the objects. Binary classification has already been applied in many areas, such as medical test, credit assessment, and spam filtering, and even some kinds of multiple classification tasks can also be treated as an iteration of several binary classifications [1]. Lots of machine learning methods have been applied in binary classification, such as Naïve Bayes, Bayes network, artificial

Y. Bao (✉)
Postgraduate Education College, Northwest University of Politics and Law,
Xi'an 710063 Shaanxi, China
e-mail: wt423@sina.com

T. Wang · G. Qiu
School of Computer Science, Shaanxi Normal University, Xi'an 710062 Shaanxi, China
e-mail: water@snnu.edu.cn

G. Qiu
e-mail: qgyqgy@snnu.edu.cn

S. Patnaik and X. Li (eds.), *Proceedings of International Conference on Computer Science and Information Technology*, Advances in Intelligent Systems and Computing 255, DOI: 10.1007/978-81-322-1759-6_95,

neural networks, decision tree, and support vector machine (SVM). "One or the other" is the consequence of the binary classification. For example, to spot the intrusion behavior from Internet, the signature patterns of intrusion traffic are needed, but all the signature patterns of normal traffic are not. When new visit traffic is coming, we can only spot it as "Normal" if we are not sure to figure out it as "Intrusion." Therefore, here is another perspective on binary classification problem, which is to make decision for classify an object to a certain category.

SVM is a classic machine learning method which is proposed by Cortes and Vapnik in 1995 [2]. SVM is based on statistical learning theory and is applied successfully to solve nonlinear, finite samples and large dimensions pattern recognition. Although SVM can be used in multiple classification, it was designed for binary classification at the very beginning [3]. To the SVM applied in binary classification, the construction and selection of the kernel functions and their parameters have always been a research focus. Up to now, there is not a kind of principles, which can guide researchers to select the kernel function and related parameters more appropriate, especially in the binary classification problems. We believe that the applicability of the SVM kernel functions used in binary classification is needed to further study [4].

To this end, this paper analyzed the theoretical principles of the SVM and then discussed the kernel function selection methods of the SVM used in binary classification. Finally, we implemented a comparative experiment carried out on the four UCI data sets [5] to figure out the impact on the classification accuracy and generalization ability of the SVM classifiers, which have different kernel functions, as well as the different parameters.

2 The Support Vector Machine

The theoretical basis of SVM is the *Structural Risk Minimization* criterion of statistical learning theory, which makes the set of functions to be a sequence, and this sequence is consisted of a subset of the function set. In each subset, least empirical risk is taken. With considering the confidence interval and minimization of the empirical risk, the subset with the least actual risk is selected. And the implementation of this criterion is just the SVM method.

SVM is developed from *Linearly Separable* problems. The main idea of the linearly separable SVM can be described as follows. There are two categories of samples C_1 and C_2 just as shown in Fig. 1. S is a hyperplane used to classify samples, and S_1 and S_2 are parallel with S and individually pass through the samples, which are nearest to the S. We call the distance between S_1 and S_2 as *Classification Interval*, ***Margin***. The plane is the most optimal classification hyperplane, which can not only accurately identify the samples, which belong to C_1 and C_2, but also make the ***Margin*** as maximum. The most optimal classification hyperplane can be described as this equation: $\mu \times x + b = 0$, where μ is weight coefficient, and b is deviation.

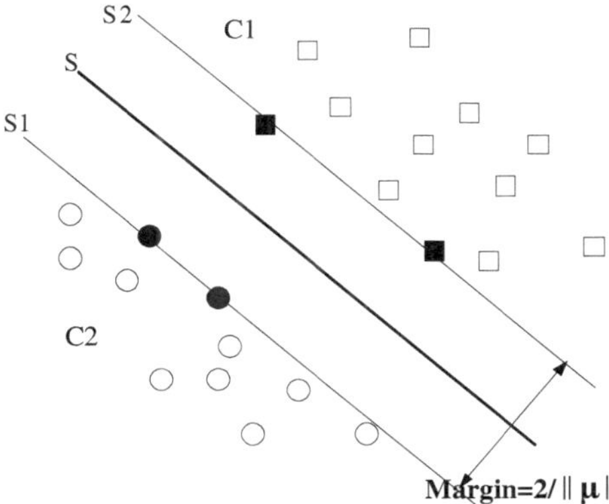

Fig. 1 Linearly separable samples in 2-dimensional plane

We let $\{(x_1, y_1), (x_2, y_2), \ldots, (x_i, y_i), \ldots\}$ be the set of samples, where x_i belongs to the set of real numbers $\mathbb{R}$, $y_i \in \{-1, 1\}$, $i = 1, 2, \ldots, n$. We have $\mu \times x + b \geq 0$ and x_i belongs to C_1, if $y_i = 1$, or we have $\mu \times x + b \leq 0$ and x_i belongs to C_2, if $y_i = -1$. That also means $y_i(\mu \times x_i + b) - 1 \geq 0$, $i = 1, 2, \ldots, n$. The classification interval is $2/\|\mu\|$, which satisfies requirements mentioned above. Solving the most optimal classification hyperplane is equivalent to solve the minimum $\|\mu\|^2/2$, that is, to solve min $\phi(\mu) = \|\mu\|^2/2$.

We can use the *Lagrange* function shown as follows to solve the best solution,

$$L(\mu, b, \alpha) = \frac{1}{2}\|\mu\|^2 - \sum_{i=1}^{n} \alpha_i [y_i(\mu \times x_i + b) - 1],$$

and then to differentiate μ and b separately,

$$\mu = \sum_{i=1}^{n} \alpha_i y_i x_i, \quad \sum_{i=1}^{n} y_i \alpha_i = 0, \quad \alpha \geq 0, i = 1, 2, \ldots, n.$$

where n is the number of the support vectors, x_i is the training vector, and the specific x_i is a support vector if $\alpha_i > 0$. For a new vector x_i which is still not classified, we can use the following function to identify its category,

$$f(x) = \text{sgn}\left\{ \mu \times x + b\} = \text{sgn}\{\sum_{i=1}^{n} \alpha_1 y_i (x_i \cdot x) + b_1 \right\},$$

where sgn is the symbolic function.

3 Construction of the Kernel Functions

If the training data sets are not completely linearly separable, any hyperplanes used to classification will make some mistakes for classification tasks. For this reason, we can introduce a slack variable σ_i and a punishment factor constant C to change the constraint prerequisite to $y_i(\mu \times x_i + b) - 1 + \sigma_i \geq 0, i = 1, 2, \ldots, n$. Then, the objective function will be

$$\phi(\mu, \sigma) = \frac{1}{2}\|\mu\|^2 + C\left[\sum_{i=1}^{n} \sigma_i\right], \quad 0 \leq \alpha_i \leq C, i = 1, 2, \ldots, n.$$

Consequently, the original problem will be transformed to a solving optimization problem by using quadratic programming,

$$\max_{\alpha} Q(\alpha) = \sum_{i=1}^{n} \alpha_i - \frac{1}{2}\sum_{i,j=1}^{n} \alpha_i \alpha_j y_i y_j (x_i \cdot x_j), \ 0 \leq \alpha_i \leq C, \text{ then } \quad \sum_{i=1}^{n} y_i \alpha_i = 0.$$

For nonlinear separable problems, we can also introduce the kernel functions to equivalently map the input space of nonlinear separable problems to the linear separable input space. Then, we can use $K(x, y) = (\phi(x) \cdot \phi(y))$ to replace (x, y) of the linear separable case, so the classification function will be transformed to

$$f(x) = \text{sgn}\left\{\sum_{i=1}^{n} \alpha_i y_i K(x_i, x) + b\right\}$$

Construction of the kernel functions in SVM is very important. There is not a good solution to judge the validity of the selected kernel function K so far [6]. According to the ***Mercer*** theorem [7], K is a valid kernel function, if and only if the matrix corresponding to K is symmetric positive semi-definite for the training data sets $\{(x_1, y_1), (x_2, y_2), \ldots, (x_i, y_i), \ldots\}$. Nevertheless, to judge a matrix corresponding to a kernel function symmetric positive semi-definite or not for a specific training data set is hard. Therefore, there are some kernel functions which were already validated the effectiveness in the application of the SVM to meet the ***Mercer*** theorem, which include the linear kernel function (Linear), radial basis kernel function (RBF), polynomial kernel function (Polynomial), and the sigmoid kernel function used neuron nonlinear function to be the inner product [8]. These kernel functions have the following formula to compute,

Linear kernel function:

$$K(x, x_i) = (x, x_i),$$

which will generate a linear classifier.

Radial basis kernel function:

$$K(x, x_i) = \exp\left\{-\frac{|x - x_i|^2}{2\sigma^2}\right\}, \quad \sigma > 0.$$

Polynomial kernel function:

$$K(x, x_i) = [(x \cdot x_i) + 1]^p,$$

where p is the order of the polynomial, and it will generate a p-order polynomial classifier, be different with RBF, which every basis function corresponds to a support vector determined by the algorithm.

Sigmoid kernel function:

$$K(x, x_i) = \tanh(\upsilon(x \cdot x_i) + c).$$

4 Experiments and Results

4.1 Experiments Setting

In this paper, all the experiments were performed in a PC, which has Inter Core Duo E7500 2.9G CPU, 2G RAM memory. We chose a classic SVM implementation package so-called *libSVM* [9] as the experimental tool. All experiments employed four data sets selected from the UCI Machine Learning Repository, and these four data sets are all public for researchers in binary classification. The specific information of these four data sets is shown in Table 1. Furthermore, all the experiments also employed 10-folder cross-validation method [10], which is popular and recognized in data mining research community. That is, the experimental data set will be divided into 10 parts, which are not overlapped each other. Each round of experiment will employ 9 of 10 as the training set, the remaining one as a test set, and every experiment will be performed 10 times, and the test result of classification is the average value.

4.2 Accuracy Indicators

With considering the confusion matrix as shown in Table 2, the accuracy indicators to evaluate the performance of classifier are used in this paper is the classification accuracy rate, *Accuracy* (AC) = (TP + TN)/(TP + FP + TN + FN). We also employed another two evaluation indicators, which are ROC curve and

Table 1 Specific information of data sets for experiments

Data sets	Abbreviation	Attributes type	Class label	Include missing value	Number of attributes	Number of samples
Breast-cancer	***bc***	Text	Text	Yes	10	286
Wisconsin-breast-cancer	***wbc***	Text	Text	Yes	10	699
Lung-cancer	***lc***	Text	Text	Yes	57	32
Tic-tac-toe	***ttt***	Numeric	Text	No	10	958

Table 2 Confusion matrix in binary classification

Classified as→	True	False
Is true	TP	FN
Is false	FP	TN

the area under curve (AUC) [10]. The *X*-axis of ROC curve represents the false alarm rate, the *Y*-axis represents the recall rate, and a set of points on the curve is obtained by adjusting the classifier decision threshold, which can avoid the inconvenience of results comparison caused by application of specific strategies between different researchers. ROC curve is closer to the upper left, and the stronger generalization ability of the corresponding classifier will be indicated. AUC refers to the area under the ROC curve, which is the integral curve of ROC. ROC curve and AUC can more accurately describe the classification performance of the model, especially in the case of imbalance density of class distribution and of the asymmetry costs between all kinds of misclassification. AUC can quantitatively describe the generalization ability of the classifier corresponding to a ROC curve.

4.3 Parameters Selection

The *libSVM* has implemented two kinds of binary classification models: *C-SVC* and *nu-SVC* (*ν-SVC* [9]). These two models have the same configurations except the punishment factor *C*. In the *C-SVC*, $C \in (0, \infty)$, and $C \in (0, 1]$ instead in *nu-SVC*. The *nu-SVC* also relates the ratio of the support vector and the ratio of the training error correlation. This paper used both models to compare the results. According to the choice of different kernel function, we got eight different combinations of classifiers, which are ***C_l*** (i.e., *C-SVC* with linear kernel function), ***C_p*** (i.e., *C-SVC* with polynomial kernel function), ***C_R*** (i.e., *C-SVC* with RBF), ***C_s*** (i.e., *C-SVC* with sigmoid kernel function), ***nu_l*** (i.e., *nu-SVC* with linear kernel function), ***nu_p*** (i.e., *nu-SVC* with polynomial kernel function), ***nu_R*** (i.e., *nu-SVC* with RBF), and ***nu_s*** (i.e., *nu-SVC* with sigmoid kernel function). For the selection of key parameters of models, we set the *degree* be 3 (only for polynomial kernel function), set *gamma* be 0.5 (for the polynomial, RBF, and Sigmoid kernel function), and set *coef0* be 0 (for the polynomial, RBF, and Sigmoid kernel function). For the *C-SVC* classifier, we set the *cost* parameter be 1, and for the *nu-SVC* classifier, set the *nu* parameter be 0.8. In addition, in order to analyze the generalization ability of classifiers, we also set all classifiers to work by using probability estimation generation approach, rather than a fixed threshold approach to make decisions.

4.4 Results and Analysis

a. *The Accuracy comparison between classifiers with various kernel functions*We first compared the classification accuracy on the four data sets of eight classifiers with various kernel functions and different parameters, and the results are shown in Table 3. Obviously, on the ***bc*** data set, ***nu_p*** got the best classification accuracy rate, that is 72.36 %; on the ***wbc*** data set, ***nu_s*** got the best classification accuracy rate, that is 96.38 %; on ***lc*** data set, ***C_p*** got the best classification accuracy rate, that is 76.42 %; on ***ttt*** data set, ***nu_R*** got the best classification accuracy rate, that is 93.17 %. Although different classifiers got the best AC on four data sets, from the perspective of average, ***nu_R*** got the best classification accuracy rate, which is 83.05 %, ***C_R*** got 82.67 % as follows.

From the perspective of different kernel functions, polynomial kernel function got the best AC twice on the four data sets. The RBF and sigmoid each got the best AC once, and linear kernel function showed the worst performance of the classification accuracy obviously. Furthermore, by examining the average classification accurate rate on the four data sets, SVM classifiers with the RBF kernel function have the better performance in *Accuracy*. By analyzing the classifier's applicability for different samples, we found on the small sample set, ***lc*** (only 32 samples), ***C_p*** which uses polynomial kernel function has significant superiority than other classifiers, whereas on the large sample set ***ttt*** (958 samples), ***nu_R*** which uses the RBF function has significant superiority than others.

b. *The generalization ability comparison between classifiers with various kernel functions*

We also compared the generalization ability on the four data sets of eight classifiers with various kernel functions and different parameters, and the results are shown in Table 4. On the ***bc*** data set, ***C_R*** achieved the best AUC value 0.70, and the ROC curve produced by classifiers on the ***bc*** as shown in Fig. 2. On the ***wbc*** data set, all classifiers except ***C_p*** achieved the best AUC value 0.99, and the ROC curve produced by classifiers on the ***wbc*** as shown in Fig. 3. On the ***lc*** data set, ***C_R*** achieved the best AUC value 0.82, and the ROC curve produced by classifiers on the ***lc*** as shown in Fig. 4. On the ***ttt*** data set, ***nu_R*** achieved the best AUC value 0.98, and the ROC curve produced by classifiers on the ***ttt*** as shown in Fig. 5.

From the perspective of average, ***C_R*** got the best average AUC value 0.86, which is obviously higher than others. ***C_p*** achieved the AUC value 0.83 as follows. From the perspective of different kernel functions, the RBF kernel function achieved the best AUC value on all the four data sets. The difference with *Accuracy* performance is that ***C_R*** with RBF kernel function has the best ability of generalization instead of ***nu_R***.

Table 3 AC results of eight classifiers on four data sets [% (standard deviation)]

	C_l	*C_p*	*C_R*	*C_s*	*nu_l*	*nu_p*	*nu_R*	*nu_s*
bc	69.03(4.53)	71.06(2.75)	71.91(5.83)	69.71(2.68)	69.70(4.32)	**72.36(3.96)**	72.25(5.39)	70.26(1.44)
wbc	96.28(2.39)	90.50(3.30)	96.31(2.22)	90.97(3.31)	96.32(2.11)	95.05(2.60)	95.35(2.36)	**96.38(2.24)**
lc	71.75(15.12)	**76.42(17.00)**	74.00(17.62)	72.92(16.26)	71.83(12.91)	71.50(10.87)	71.42(11.13)	71.25(11.32)
ttt	65.34(0.41)	90.50(2.95)	88.46(3.18)	63.44(3.98)	64.68(3.60)	87.32(3.55)	**93.17(2.67)**	65.34(0.41)
Mean	75.60	82.12	82.67	74.26	75.63	81.56	**83.05**	75.81

Table 4 AUC results of eitht classifiers on four data sets (standard deviation)

	C_l	*C_p*	*C_R*	*C_s*	*nu_l*	*nu_p*	*nu_R*	*nu_s*
bc	0.68(0.12)	0.66(0.11)	**0.70(0.12)**	0.52(0.13)	0.59(0.14)	0.67(0.11)	0.69(0.11)	0.48(0.11)
wbc	**0.99(0.01)**	0.93(0.05)	**0.99(0.01)**	**0.99(0.01)**	**0.99(0.01)**	**0.99(0.01)**	**0.99(0.01)**	**0.99(0.01)**
lc	0.74(0.36)	0.77(0.36)	**0.82(0.34)**	0.78(0.35)	0.55(0.19)	0.56(0.21)	0.56(0.20)	0.53(0.16)
ttt	0.52(0.08)	0.96(0.02)	0.96(0.02)	0.55(0.07)	0.51(0.08)	0.93(0.03)	**0.98(0.01)**	0.53(0.08)
Mean	0.73	0.83	**0.86**	0.71	0.66	0.79	0.80	0.63

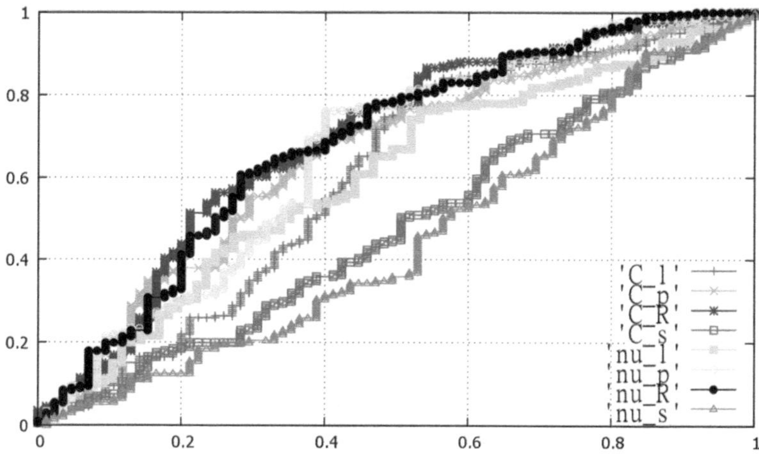

Fig. 2 ROC curve produced by eight classifiers on ***bc***

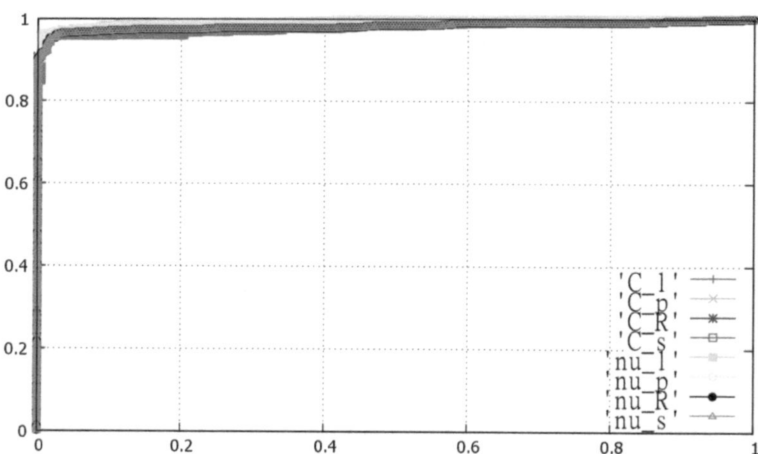

Fig. 3 ROC curve produced by eight classifiers on ***wbc***

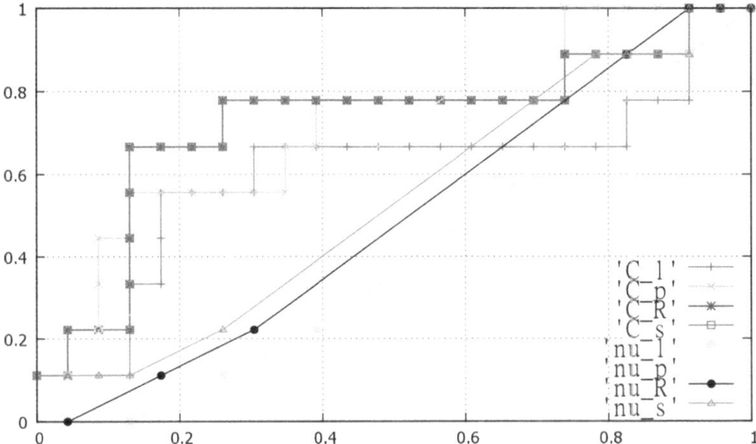

Fig. 4 ROC curve produced by eight classifiers on ***lc***

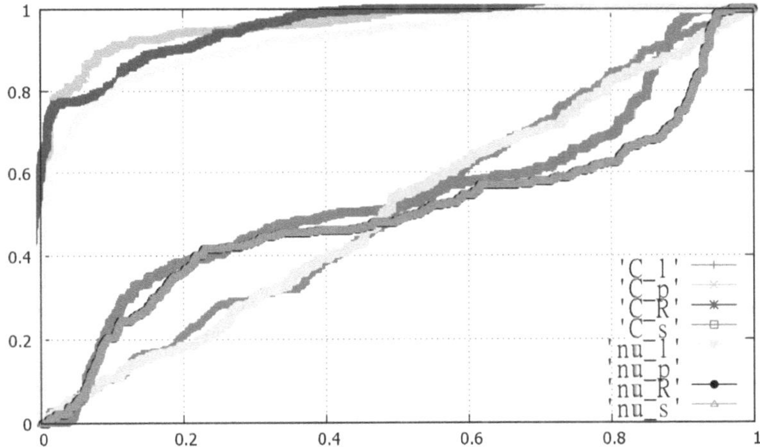

Fig. 5 ROC curve produced by eight classifiers on ***ttt***

5 Conclusions

Although some literatures pointed out that, when using the SVM to do classify, the RBF kernel function is a better choice [4]. Our research in this paper shows that, in the experimental methodology way the classification accuracy rate in some small data set, the SVM with RBF is not better than the one with polynomial kernel function. On the other hand, to get better generalization ability of the classifier, the SVM classifier with RBF kernel function should also need to select the key parameters carefully.

Therefore, we believe that employing the kernel functions to solve the SVM model in binary classification tasks will not necessarily construct a better classifier if we only rely on a certain type of kernel function. We need to adjust the construction strategies and the key parameters for different type of samples. In the future, we will extensively figure out the applicability of kernel functions on more various and larger data sets.

Acknowledgments This work was supported by the National Natural Science Foundation of China under Grant No. 61003129 and the Planned Science and Technology Project of Shanxi Province, China, under Grant No. 2010JM8039 and also supported by the Fundamental Research Funds for the Central Universities of China under Grant No. GK201302055.

References

1. Muhammad, F., Iram, F., Sungyoung, L., et al.: Activity recognition based on SVM kernel fusion in smart home. Computer Science and its Applications, Lecture Notes in Electrical Engineering. 203, 283–290 (2012)
2. Cortes, C., Vapnik, V.: Support-vector network. Mach. Learn. **20**(3), 273–297 (1995)
3. Hsu, C.-W., Lin, C.-J.: A comparison of methods for multiclass support vector machines. IEEE Trans. Neural Netw. **13**(2), 415–425 (2002)
4. Keerthi, S.S., Lin, C.-J.: Asymptotic behaviors of support vector machines with Gaussian kernel. Neural Comput. **15**(7), 1667–1689 (2003)
5. Frank, A., Asuncion, A.: UCI Machine Learning Repository. University of California, School of Information and Computer Science, Irvine, CA http://archive.ics.uci.edu/ml. (2010-7-8)
6. Bhavsar, H., Panchal, M.H.: A review on support vector machine for data classification. Int. J. Adv. Res. Comput. Eng. Technol. (IJARCET) **1**(10), 185 (2012)
7. Kurisu, M., Mera, K., Wada, R., Kurosawa, Y., et al.: A method using acoustic features to detect inadequate utterances in medical communication. In: IEEE International Conference on Systems, Man, and Cybernetics (SMC), pp. 116–119 (2012)
8. Scholkopf, B., Smola, J.: Learning with Kernels: Support Vector Machines, Regularization, Optimization, and Beyond. MIT Press, Cambridge (2002)
9. Chang, C.-C., Lin, C.-J,: LIBSVM:a library for support vector machines. ACM Trans. Intell. Syst. Technol. 2(**3**): Article No. 27, 27 pp. Software available at http://www.csie.ntu.edu.tw/~cjlin/libsvm
10. Witten, H., Frank, E.: Data Mining Practical Machine Learning Tools and Techniques, 2nd edn. Morgan Kaufmann Publisher, San Francisco (2005)

The Auxiliary Control System from Railway Operation Safety Based on Beidou Intelligent Navigation Services

Xianglei Zhao, Hongying Lu and Gang Dai

Abstract Currently, the safety of the railway operation mainly depends on ground devices; however, the security and reliability of railway operation face serious challenges. When the ground signals and wireless communication devices are severely damaged, the use of traditional GPS satellite positioning combined with terrestrial wireless communications technology is a solution to ensure safety. However, GPS has no communication function so that in the case of extreme natural disasters the information cannot be fed back to the control center, so it still cannot meet the requirements. Beidou satellite navigation system has a short message communication function that meets the demand to send messages. The auxiliary control system from railway operation safety based on Beidou intelligent navigation services is proposed. The system is consisted of vehicle intelligent terminal subsystem and train smart location-based services platform subsystem. The vehicle intelligent terminal subsystem achieves real-time positioning through Beidou, also obtains the train-running information that will be sent to the ground subsystem by short message, such as speed, kilometer mark, front and rear-train distance and so on. The train smart location-based services platform subsystem dynamically monitors the train's running status after receiving the running information and makes right dispatch so as to ensure the railway operation safe. Thus, the system plays an important role in securing the safety of railway operation and distress rescue aids decision-making.

X. Zhao (✉) · H. Lu · G. Dai
School of Computer and Information Technology, Beijing Jiaotong University, Beijing 100044, China
e-mail: 11120515@bjtu.edu.cn

H. Lu
e-mail: Hylu@bjtu.edu.cn

G. Dai
e-mail: gdai@bjtu.edu.cn

S. Patnaik and X. Li (eds.), *Proceedings of International Conference on Computer Science and Information Technology*, Advances in Intelligent Systems and Computing 255, DOI: 10.1007/978-81-322-1759-6_96,

Keywords Beidou satellite navigation system · Railway operation safety · Position · Short message

1 Introduction

Safety is the eternal theme of railway transport. Especially with the railway transport developing toward high speed, high density, technology-intensive, and complex technical system, the railway transport safety and security system are facing new demands and challenges.

At present, the railway operation safety is ensured by wireless train dispatching, cab signal, and automatic stopping devices. The basic idea is to avoid railway accidents through the mutual control of the ground devices and locomotive signal devices. However, in the extreme situations such as natural disasters, terrestrial signals, wireless communication devices are likely to be severely damaged, resulting in the train dispatching blind spots appearing, so that the train runs under the risk of lack of necessary protection. "7 • 23" Yongwen line particularly serious accident is a painful lesson. Therefore, it is necessary to develop auxiliary railway operation safety protection technologies independent on the ground devices and unaffected by the natural weather.

Satellite positioning is one of the options to solve these problems. However, the solution such as the use of traditional GPS satellite positioning technology combined with terrestrial wireless communications still cannot meet the requirements because in extreme situations such as natural disasters, message cannot be fed back to the control center.

Beidou Navigation Satellite System (BDS) is China's homegrown navigation system. In addition to basic navigation position function, Beidou has a short message communication function which other navigation systems do not have. With the unique function, ground command devices, satellites, and terminals can form a complete command system. Then to build auxiliary control system from railway operation safety based on Beidou intelligent navigation services based on the above is feasible. In this paper, the auxiliary control system from railway operation safety based on Beidou intelligent navigation services is designed and implemented under the research of railway operation safety auxiliary system and the Beidou satellite navigation technology. Meanwhile, the design and implementation of the two main subsystems that are vehicle intelligent terminal subsystem and train smart location-based services platform subsystem are introduced in detail. Then, the key technology in the implementation of the system is summarized. Finally, the system significance and prospects of the system are described simply.

2 Technical Overview

Navigation Satellite System has been used in railway operation very early. Such as, the Advanced Railroad Electronic System (ARES) developed by Burlington Northern Company and Rock Well in 1980s in America, the Japanese Computer And Radio Aided Train Control System (CARAT), the German Satellite-Handy (SANDY), and Passenger Information System (PIS), they are all based on GPS technology. Later in 1990s, Intelligent Railway System based on GPS and GPS railway (GSM-R) is researched in China. In the development of railway transportation, safety is always the most common concern. In the world, the railway operation safety systems, such as latest digital ATP of Japan, TVM300, and TVM430 from France are promoting high-speed railway at present.

Summarizing the current railway operation safety system, the railway operation safety system architecture is consisted of two parts: the vehicle system part and the ground system part. The vehicle system monitors the train real-time running status and early warns when necessary. Meanwhile, the vehicle system sends messages to the ground system for processing and analyzing. The ground system processes the obtained train dynamic data then displays the train's running status in real time and sends commands and messages to the vehicle system. However, the traditional vehicle-ground communication depends on ground devices so that in case of natural disasters, messages cannot be transmitted to each other. Fortunately, the Beidou satellite navigation system has a unique feature that meets the requirement.

BDS is China's homegrown navigation system. It is expected to achieve full-scale global coverage by around 2020 [1] within three steps [2].

The BDS can provide highly accurate and reliable positioning, navigation and timing service with the aid of a constellation of 35 satellites. So far, China has successfully launched 16 navigation satellites and four other experimental ones for BDS.

Beidou has since started providing licensed services for China's government and military users in transport, weather forecasts, fishing, forestry, telecommunications, hydrological monitoring, and mapping.

To compete with foreign rivals, the Beidou terminal can communicate with the ground station by sending and receiving short messages, 120 Chinese characters in each, in addition to the navigation and timing functions that the world's other major navigation systems can provide [3].

With the advantage of the wide application of Beidou in China and Beidou's unique function to send short messages, it has the technical conditions and application conditions to build auxiliary control system from railway operation safety based on Beidou intelligent navigation services.

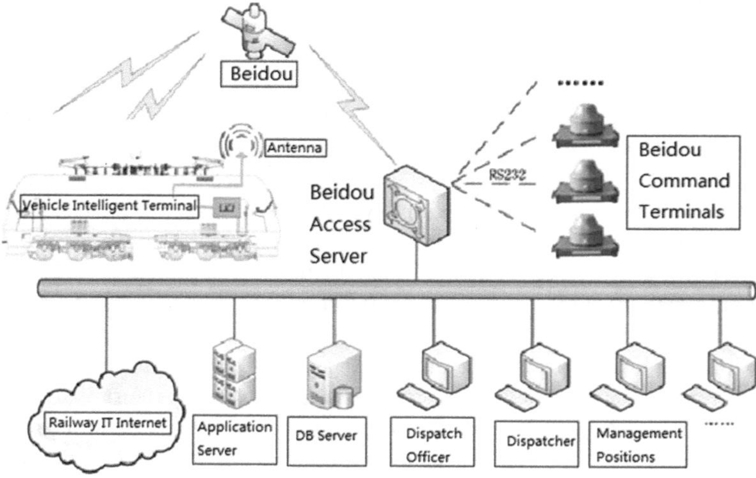

Fig. 1 The structure of the system

3 The Design of the System

The auxiliary control system from railway operation safety based on Beidou intelligent navigation services makes full use of the functions of navigation positioning and sending short messages of Beidou navigation system to connect the train's running information with the ground command system thus to form an intelligent auxiliary control system from railway operation safety independent on the ground signal system. According to the operation organization and management structure of railway departments, the system uses a structure that the Railway Bureau is the center of an operation control system and cooperated with three information management systems, Ministry of Railways, Railway Bureau, and the locomotive depot. Figure 1 reveals the structure of the system.

The ground system working on the existing information Internet system is consisted of database server, application server, Beidou command, and control access server, Beidou command terminal and user clients while the vehicle intelligent terminal on the train is consisted of safety host computer, DMI, and Beidou communications module.

4 The Implementation of the System

The auxiliary control system from railway operation safety based on Beidou intelligent navigation services is consisted of vehicle intelligent terminal subsystem and train smart location-based services platform.

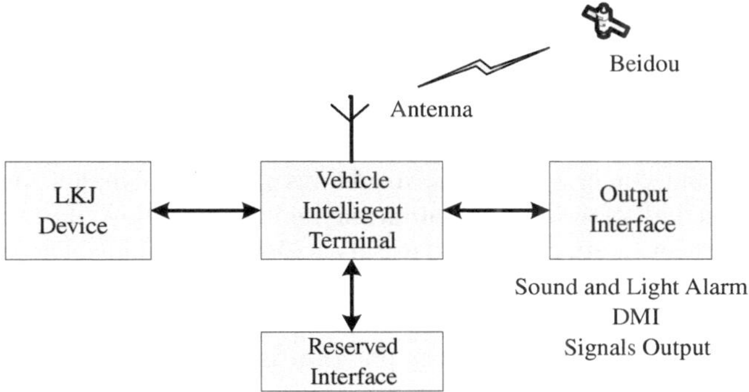

Fig. 2 The structure of vehicle intelligent terminal

4.1 The Vehicle Intelligent Terminal Subsystem

The vehicle intelligent terminal subsystem is consisted of security host computer, DMI (monitor), and Beidou communication module as shown in Fig. 2. The security computer that saves the basis data of the line including ramps, corners, bridges, and tunnels, and other data is used to collect and process the safety data. DMI is used for human–computer interaction between driver and the system. Beidou communication module is mainly used for positioning and communication between vehicle and ground.

On one hand, the vehicle intelligent terminal subsystem uses security computer to connect LKJ via the CAN industrial bus, and gets the current models, license plate number, train, speed, line number, kilometer mark, train length, and other information from LKJ device, and then sends the information to the ground system in real time through the Beidou system together with the positioning information obtained through the Beidou communication module. On the other hand, the vehicle intelligent terminal subsystem receives data including vehicle location and status information, alarm information, dispatch instructions, and other information sent by the ground system, and then analyzes and processes the data and instructions later used in human–computer interaction through DMI.

The vehicle intelligent terminal subsystem uses RNSS and RDSS dual-mode positioning system that combines the RDSS communication function with the RNSS fast location function so as to improve overall equipment reliability.

The vehicle intelligent terminal subsystem can operate in two modes one is single-position mode and the other one is combinations mode. The main work mode of the vehicle intelligent terminal subsystem is single-position mode that is RNSS positioning while RDSS is mainly used in short message communication between the vehicle and ground system.

4.2 Train Smart Location-Based Services Platform Subsystem

Train smart location-based services platform is a safety auxiliary system running at ground command center. The platform obtains the train dynamic position data through Beidou then shows it as operation diagram in the DMI. At the same time, the platform alarms for safety risk thus to avoid accidents happening. The subsystem is mainly used as an auxiliary safety control means while the CTC/TDCS and wireless train dispatching system are unavailable in the extreme natural disaster.

Train smart location-based services platform is consisted of Beidou command terminal access server and business-processing subsystem. The access server processes the standardized access and automated data exchange between Beidou terminal devices and railway-integrated IP network while the business-processing subsystem processes the position information and completes visual display. Business-processing subsystem includes three main functions that are operation diagram representation, conflict alarm, and sending emergency operation command. Besides, the subsystem also has safety analysis and dispatch monitoring function.

4.3 Main System Performance

1. System Capacity: Train smart location-based services platform can process simultaneously train dynamic position data not less than 2,000/s;
2. Number of users: Every server supports not less than 3,000 intelligent end-users simultaneously accessing;
3. Location processing time: less than 500 ms;
4. Positioning accuracy: less than 10 m;
5. Data storage requirements: train's dynamic data related to database stores not less than six months and static data can be permanently saved.

4.4 Technical Problems and Solutions

Beidou civilian satellite has an error of 10 m that is perfectly acceptable for the single-track railway traffic control, but for the double-track section and the station walk line, it is difficult to confirm which line the position is on. Meanwhile, the problem will also directly affect the front and rear-distance calculation. So there are several technical problems as follows:

1. Train's precise positioning meets the operating environment of railway;
2. Front and rear-train distance calculation of complex lines;
3. Solve another train approaching from the relationship between the degree and security alarms, handling manual intervention and direct intervention in the timing of the decision problem.

These above major technical problems are regarded as difficulties of the system implement. So how to solve these problems is the key to complete the system. Then, the following solutions are put forward:

1. According to Beidou's position information, combining the train's line number, running direction, the line's kilometer mark information and the accurate electronic map data, the corresponding positioning models and algorithms are studied to achieve accurate positioning of the train;
2. According to vehicle baseline databases and Beidou satellite positioning data, on the basis of solution (1), after precise positioning of the train, the accurate front and rear-train distance calculation of complex lines can be got;
3. By analyzing the characteristics of the tasks undertaken by motorcycle, speed, track density, block section, information transmission frequency, and other factors, studying the relationship between these factors and early warning, alarm, and emergency braking three levels of security system so as to set up security system model.

5 Conclusion

The auxiliary control system from railway operation safety based on Beidou intelligent navigation services that uses the Beidou satellite navigation system to achieve real-time positioning of the train and monitoring and does not rely on GPS, GPRS, GSM-R [4] and other systems, has a strong resilience and security. Meanwhile, the system using the short message function of Beidou satellite navigation system can achieve bidirectional message communication function that can effectively meet the small amount of information communication and real-time demanding communication needs. To meet the needs of railway operation, the system ensures a great extent on railway operation safety and plays an important role in distress rescue. The application of Beidou satellite navigation system to the railway sector has great significance because it can not only improve the efficiency of rail transport but also promote the development of Beidou satellite navigation system in China.

References

1. Introduction of the BeiDou Navigation Satellite System. BeiDou.gov.cn. 15 Jan 2010 (in Chinese)
2. The construction of BeiDou navigation system steps into important stage, "Three Steps" development guideline clear and certain. China National Space Administration. 19 May 2010 (in Chinese)
3. Satellite navigation system launched. China Daily. 28 Dec 2010. Retrieved 29 Dec 2011
4. Fan, Y., Zhang, B.: The Application Overview and Prospects of Chinese Beidou satellite navigation system in Mitigation. Aerosp. Chin. (2010) (in Chinese)

MapReduce Performance Optimization Based on Block Aggregation

Jun Li, Lihua Ai and Ding Ding

Abstract MapReduce is a distributed programming model for large-scale data processing. Hadoop as an open source implementation of the MapReduce programming model has been widely used due to its good scalability and fault tolerance. However, the default size of the split and Hadoop distributed file system (HDFS) block are the same, which makes the number of map tasks of the job increase linearly with the number of blocks. When input is large, the time for managing splits and initializing map tasks is considerable. In this paper, we propose a scheme, Block Aggregation MapReduce (BAMR), which automatically increases the split size appropriately according to input's size in order to reduce the number of map tasks. With this scheme, the time of managing splits and initializing map tasks will be shorten. Experiment shows that BAMR reduces the execution time significantly.

Keywords HDFS · MapReduce · Split · Block aggregation

1 Introduction

Internet services, such as search engines, portal sites, e-commerce Web sites and social networking sites, not only deal with enormous volumes of data, but also generate a large amount of data which needs to be processed everyday. MapReduce

J. Li (✉) · L. Ai · D. Ding
Department of Computer and Information Technology, Beijing Jiaotong University, Beijing 100044, China
e-mail: juenlee@outlook.com

L. Ai
e-mail: lhai@bjtu.edu.cn

D. Ding
e-mail: dding@bjtu.edu.cn

S. Patnaik and X. Li (eds.), *Proceedings of International Conference on Computer Science and Information Technology*, Advances in Intelligent Systems and Computing 255, DOI: 10.1007/978-81-322-1759-6_97,

is a programming model that supports distributed and parallel processing for large-scale data-intensive applications. MapReduce divides a computation into multiple small tasks and let them run on different machines in parallel [1].

Hadoop is an open source implementation of MapReduce programming model [2]. Many Internet companies, including Yahoo, Amazon, and Facebook, have already used it because of its excellent scalability and fault tolerant. Hadoop relies on its own distributed file system called Hadoop distributed file system (HDFS), which provides input data to MapReduce and store output of a job. MapReduce provides Map and Reduce function for distributed computation.

As a fundamental data processing platform, Hadoop has been accepted by more and more people because of its application value, but there's still plenty of room to be improved in its processing performance. There have been some researches on improving the processing performance of MapReduce. Literature [3] provides an index mechanism and an efficient join algorithm for Hadoop, and experiments show that it has more efficiency than Hadoop and HadoopDB. Literature [4] proposes an idea of auto-tuning parameters to optimize the performance of Hadoop. Literature [5] analyzes five factors that affect the performance of Hadoop and proposes methods of improving performance for every influencing factor; experiments show that these methods can increase the performance of MapReduce significantly. Literature [6] proposes using mechanisms of Prefetching and Pre-shuffling in sharing clusters, which can shorten execution time significantly by the way of data prefetching when job is running and moving map tasks to the node which is close to the reduce task.

But Prefetching proposed in the literature [6] is in the stage of running jobs, so it needs substantial network overhead. This paper proposes a scheme that is Block Aggregation MapReduce (BAMR). It makes data aggregated at the moment of write the input to HDFS and properly increases split size according to the aggregation degree during running MapReduce job, to reduce the cost time of managing splits and initializing map tasks, so it can shorten the whole execution time of MapReduce jobs and improve the performance of MapReduce further.

The typical workflow of MapReduce is illustrated in Fig. 1. Input files are saved in HDFS and a part of input file, which we call an input split, is disseminated to the corresponding map task. The size of the input split is limited to the block size, and each task is mapped to only one input split. The output of the map task, called intermediate output, is sent to the combiner. The combiner aggregates multiple intermediate outputs generated from the node to a list. Reduce combines those combined intermediate values into one or more final values for that same output key.

Hadoop divides the input to a MapReduce job into fixed-size pieces called input split and creates one map task for each split. The number of split of a job and the split size have a direct impact to the performance of MapReduce: Having many splits means the time taken to process each split is small compared to the time to process the whole input; but if splits are too small, then the overhead of managing the splits and of map creation begins to dominate the total job execution time [7]. The JobTracker takes the location information of the input into account and

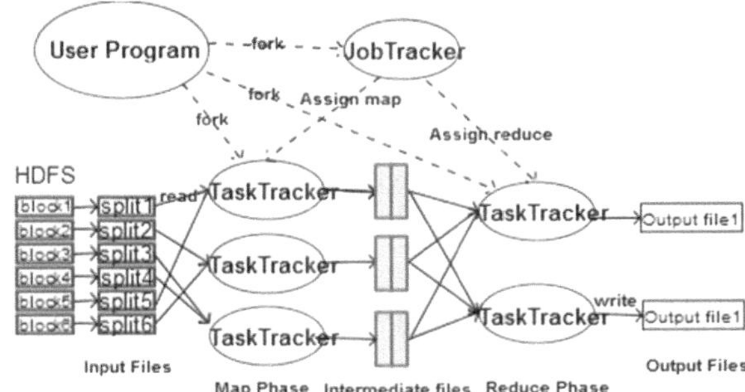

Fig. 1 MapReduce execution overview

attempts to schedule a map task on a machine that contains a replica of the corresponding input. This is called the data locality optimization.

Looked at the code for the details, the splits are handled by the client by *InputFormat.getSplits*, so a look at *FileInputFormat* gives the following info:

(a) For each input file, get the file length, the block size and calculate the split size as:

$$\max(minSize, \min(maxSize, blockSize)) \tag{1}$$

where *maxSize* corresponds to *mapred.max.split.size* and *minSize* is *mapred. min.split.size*.

(b) Divide the file into different FileSplits based on the split size calculated above. Each FileSplit is initialized with a start parameter corresponding to the offset in the input file.

Consider this scenario, we keep *blocksize* stays the same and increase the *minSize* to make the split larger enough to spanned two blocks. If the split spanned two blocks, it would be unlikely that any HDFS node stored both blocks, so some of the split would have to be transferred across the network to the node running the map task. Because of network bandwidth is a relatively scarce resource in a computing environment. So it is less efficient than running the whole map task using local data.

So the optimal split size in native Hadoop is the same as the block size. It is the largest size of input that can be guaranteed to be stored on a single node.

We should realize that HDFS block size has been configured before the start of the Hadoop cluster. And the value has been maintained when the cluster is running. It means that the number of map task linearly depends on input size of a job. This will leads to performance degradation when we run a job, which input size is remarkably large. The overhead of managing the splits and of map creation is considerable.

In this paper, we propose an optimization scheme, block aggregation, to overcome the aforementioned problem. This scheme is implemented in BAMR. In brief, under the premise of ensuring load balance and the data locality, BAMR makes a few (aggregation degree) adjacent blocks store in a node and accordingly increases the size of splits based on the degree of data aggregation, as the configuration of split size of a given MapReduce job. As a consequence, split size is not limited to the block size; it is set dynamically according to the input size of a job.

2 Block Aggregation

When client write data to HDFS, the data are partitioned into HDFS blocks firstly and saved independently in blocks. As saving each block, NameNode chooses some DataNode to save for blocks (Redundant Copy). If data are large, writing in a file needs to independently choose DataNode in several rounds. Considering the situation of unequal data size of different MapReduce jobs in the actual operation environment, the size of a HDFS block is set smaller in the strategy of BAMR. In this case, running MapReduce jobs with a small amount of data will not decrease degree of parallelism of tasks. When data are much larger, making data in each n ($n \in N^*$) contiguous blocks saved into the same node and increasing the size of a split to n^* *blocksize* at the moment of performing jobs. Because a split is increased to n times than ever, a map task will be decreased to $1/n$ than ever. Therefore, this scheme can reduce the total time of managing splits and initializing map tasks and the cost time of jobs.

The execution flow of MapReduce after applying the scheme of block aggregation to Hadoop is shown in Fig. 2. In the stage of running MapReduce, block size is invariable but split size increases. Because that blocks in a split are all located the same DataNode, when the size of a split is larger than a block, it will not lead to additional network overhead.

NameNode maintains the namespace of the whole file system in HDFS. When a client need to write in a file, NameNode creates metadata of the file in the namespace at first, including an object named INode file. This object maintains a list of blocks for each stored file in HDFS. After finishing writing a block, the corresponding list of blocks will be updated during writing in the file. With the help of this character of INode file, we can apply block aggregation to the

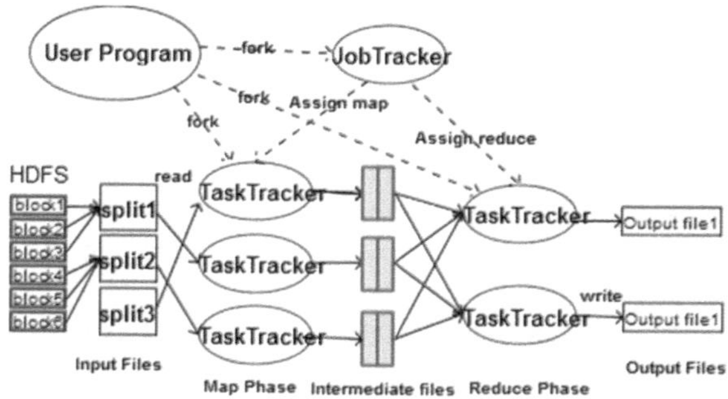

Fig. 2 BAMR execution overview

procedure of writing in file to HDFS. When client write the blocks to HDFS, NameNode chooses a DataNode for each block to store it. At this time, it needs to judge whether the current block and the last block should be store in a same DataNode. And according to the judgement to determine whether to choose another DataNode to store the current block or store it in last choose DataNode.

For explaining the scheme, we assume a file need to write to HDFS is composed of s blocks (b_1, b_2, b_3, …, b_s). Aggregation degree is n; the replication is 1 to facilitate illustration (actual implementation mechanism similar to this). So every n adjacent block need to be store in a same DataNode and composed of a split, e.g., b_1, b_2, …, b_n, composed split$_1$, is store in a same DataNode, b_{n+1}, b_{n+2}, …, b_{2n}, composed split$_2$, is store in a same DataNode, …. When client write b_i, if b_t is the first block of split$_j$, then we call a function to choose an appropriate DataNode to store b_t, and the follow-up of $n - 1$ blocks (b_{t+1}, b_{t+2}, …, b_{t+n-1}) are store in the same DataNode. In this way, the MapReduce job which input is this file can configure the split size is n* *blocksize*. And the map tasks are reduced to $1/n$ of the number of blocks.

Algorithm 1 shows the main implement steps of choosing the list of DataNode for blocks in BAMR.

Algorithm 1 the block-aggregation algorithm

```
Input: processFile, n     /* n aggregation degree*/
   Output: node     /* list is the list of DataNode which are used to store a block*/
1: numOfBlock←length[processFile.blocks] % n
2: if file != null and numOfBlock > 0
3:   then blocks←processFile.blocks
4:              index←length[blocks] – 1
5:              node←the DataNode list of blocks[index]
6: else        node←chooseRandom()
7: return node
```

Algorithm 1 is applied to each procedure of chose DataNode for store each current block (in function of choose Node of Replication Target Chooser.class). The first line of the algorithm above computes the number of blocks has been written in by DataNode(s) which was chosen last time, *numOfBlock*. The second line shows that if $0 < numOfBlock < n$ is true, the data written in by DataNode chosen last time is not enough to fill one split. Client should continue writing in the current block to this DataNode. The third to the fifth line return the choosing result of the DataNode; if the condition of the second line is false, $numOfBlock = 0$ shows that DataNode chosen last time have been filled in one split or it is the first time to choose a DataNode to store the first block of a file. In this case, the algorithm will jump to the sixth line, choose a DataNode randomly, and return it. We should realize that, if the *replica* is more than 1, each time NameNode should choose a few (replica) DataNode to store a block, the algorithm is also suitable for this condition. In this way, the node in lines 5, 6, and 7 should be replaced with a list of DataNode. The algorithm can be sure that the blocks in a split are stored in the same list DataNode.

The input of Algorithm 1, n is the degree of block aggregation which means the number of the blocks in the offset of one split. The value of n is related to n_m that is the number of configured map task slots during executing MapReduce job in clusters, n_d that is the number of nodes available for jobs, n_f that is the number of files which as the input of MapReduce job, and m_i that is the size of a file ($block_i = 1,2, \ldots, n_f$). Compute n at the moment that users issue the order of writing in files to HDFS. Through the path of data source allocated by users—src, the degree of block aggregation—n is determined by Formula (2):

$$n=\begin{cases} \frac{splitsize}{blocksize}, & \text{if user defined split size} \\ \left\lceil \frac{\sum_{i=1}^{n_f} m_i}{\max(n_m * n_{\partial},\, n_f)} \right\rceil, & \text{else} \end{cases} \tag{2}$$

Algorithm 2 shows the main steps of Formula (2)

Algorithm 2 the block-aggregation-level algorithm

```
Input:  src     /*input file path*/
Output:  n      /*aggregation degree*/
1:  if  getConfig("split.size") != null
2:    then return   getConfig("split.size") / blocksize
3:  SPLIT-TO-BLOCKS(src, list)
4:  sum← 0
5:  for i←0 to length[list] - 1
6:    do sum←sum + list[i]
7:  n←Math.ceil(sum / max(nm * nd , nf ))
8:  return  n
SPLIT-TO-BLOCKS(src, list)
1:  if src.isfile()
2:      then blocks←Math.ceil(file.length / blocksize)
3:               list.add(blocks)
4:  else if src.isdirectory()
5:    then for each file in src
6:             do SPLIT-TO-BLOCKS(file, list)
```

Algorithm 2 fully considers the number of files and configured map task slots, the number of nodes available for jobs and the size of a file. The input of Algorithm 2 is the file path (a catalog or a file) of written HDFS, and the output is the degree of block aggregation that is n. The first and second line judges whether there is a configure size of a split in the configuration file. If the size has been configured, the degree of block aggregation is determined by the configuration parameters; otherwise, execute the third to the seventh line. In the Algorithm 2, if the number of files is less than the sum of task slots, the process will consider the parallelism of tasks preferentially and distribute data equally to each slot in blocks. In this case, the size of a split is the data size processed by each slot. If the number of files is more than the sum of task slots, then the size of a split will be set as the average size of files. The sub-process, SPLIT-

TO-BLOCKS (src, list), represents the size of input data by blocks, so the input src can be a single file or a set of files (one or more catalogs).

3 Evaluation

The experiment adopts the environment, which consists of 12 node clusters, a master node and 11 nodes named DataNode. Each DataNode is configured with Intel i5 3.10GHZ quad-core processor, 2.0 GB RAM and 500 GB hard disk. The master node is configured with Intel Xeon E5606 2.13GHZ eight-core processor, 4.0 GB RAM and 1 TB hard disk. The operating system adopts Ubuntu 12.04 LTS. Each TaskTracker can execute at most 4 map tasks or 2 reduce tasks at the same time. The version of Hadoop system in our experiment is Apache hadoop-0.20.2.

The execution time of jobs is used to measure the performance. We can compare the change of MapReduce performance before and after adopting the scheme proposed in this paper through the experiment. Experimental test cases are exceptions of Web site traffic and analyses of relevant user behaviors. These cases are mainly used to analyze user behaviors, leading to exceptions of Web site traffic. The data come from CGBT (http://cgbt.cn/).

We conduct an experiment with one file, and the size of HDFS blocks is set at 64 MB. There are five groups of tests in the experiment, and each group of tests needs to execute the test case three times and get the average value of three times. The change of split size in the experiment is shown in the Fig. 3. As shown in Fig. 3, BAMR split size increases with input data of jobs, but split size of native Hadoop remains unchanged. When tasks are executed, the map tasks of BAMR will decrease relatively. The comparison for average runtime overhead of jobs is shown in Fig. 4. When the data size is 1.14 GB, the degree of block aggregation in BAMR is 1 and the size of split is the same as a HDFS block. When the size of input increases, the degree of block aggregation in BAMR algorithm also increases. For the same input, as split size increases, the number of map tasks decrease in BAMR and the total time of managing splits and initializing map tasks

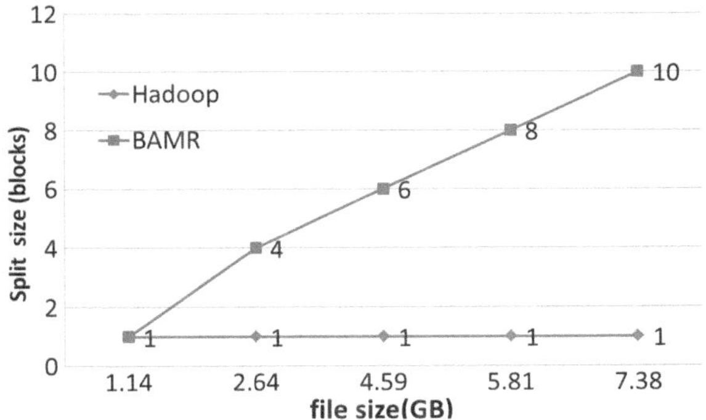

Fig. 3 Comparison of split size

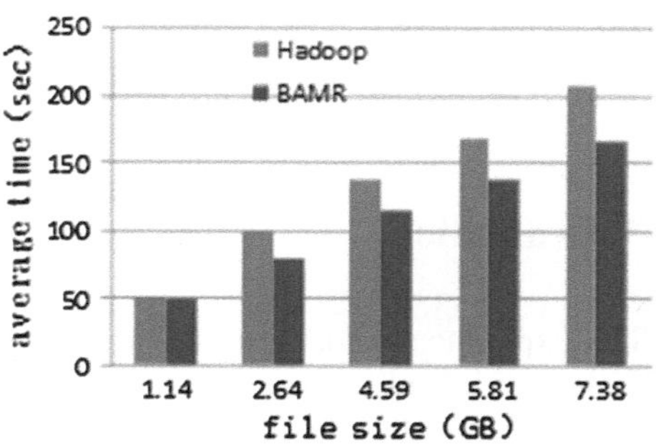

Fig. 4 Comparison of average time

and the cost time of jobs is decreased. The executing time decreases about 20 % relative to MapReduce implementation of native Hadoop.

4 Conclusion

In this paper, we analyze the implement mechanism of Hadoop and the practical application environment. We explain the dependency of split size to HDFS blocks in native Hadoop and propose an optimization scheme. We have evaluated the performance of BAMR with different workloads. The result shows that BAMR reduces the execution time obviously. Meanwhile, the scheme can ensure the load balancing to be broken and the parallelism of tasks. Because that the scheme proposed in this paper improves the implementation mechanism of Hadoop, its implementation has general applicability.

Acknowledgments This work was supported in part by the Ph.D. Programs Foundation of Ministry of Education of China (No. 20110009110032), the Fundamental Research Funds for the Central Universities (No. 2013JBM019), and the National Natural Science Foundation of China (No. 61300176).

References

1. Dean, J., Sanjay, G.: MapReduce: Simplified data processing on large clusters. Commun. ACM **51**(1), 107–113 (2008)
2. Apache. Hadoop[EB/OL]. http://hadoop.apache.org/,2013-4-2
3. Dittrich, J., Jorge-Arnulfo, Q., Alekh, J., et al.: Hadoop++: Making a yellow elephant run like a cheetah(without it even noticing). Proc. the VLDB Endowment **3**(1–2), 515–529 (2010)
4. Babu, S.: Towards automatic optimization of MapReduce programs. In: Proceedings of the 1st ACM symposium on cloud computing, pp. 137–142 (2010)
5. Dawei, J., Beng, C.Q., Lei, S., et al.: The performance of MapReduce: An in-depth study. Proc. VLDB Endowment **3**(1–2), 472–483 (2010)

6. Sangwon, S., Kyungchang, W., Inkyo, K., Seo, S., et al.: HPMR: prefetching and pre-shuffling in shared MapReduce computation environment. IEEE international conference, pp. 1–8 (2009)
7. White, T.: Hadoop: The Definitive Guide, 3rd edn. O'reilly, Yahoo! pp. 31–34 (2013)

An R-Calculus for the Logic Programming

Wei Li and Yuefei Sui

Abstract The AGM postulates are for the belief revision (revision by a single belief), and the DP postulates are for the iterated revision (revision by a finite sequence of beliefs). Li (The Computer Journal 50:378–390, 2007) gave an R-calculus for R-configurations $\Delta|\Gamma$, where Δ is a set of atomic formulas or the negations of atomic formulas, and Γ is a finite set of formulas in the first-order logic. This paper will give a set of axioms and deduction rules for the revision of logic programmings, based on the structure of formulas, so that the deduction system is sound and complete with respect to the maximal consistent subsets of the revised sets by the revising sets of formulas. Moreover, it will be showed that the deduction system satisfies the AGM postulates and the DP postulates.

Keywords Belief revision · R-calculus · AGM postulates · Maximal consistent subset

1 Introduction

The AGM postulates [1, 2, 3, 4] are for the revision of belief sets, where a belief set is a logically closed theory. Let K be a theory and φ a formula to revise K in the propositional logic. $K \circ \varphi$ is a theory which should be a selection function γ applied on the set of all the maximal consistent subsets of $K \cup \{\varphi\}$ containing φ. Somehow, the revision is reduced to get the maximal consistent subsets of $K \cup \{\varphi\}$.

W. Li
State Key Laboratory of Software Development Environment, Beijing University of Aeronautics and Astronautics, Beijing, China

Y. Sui (✉)
Key Laboratory of Intelligent Information Processing, Institute of Computing Technology, Chinese Academy of Sciences, Beijing, China
e-mail: yfsui@ict.ac.cn

S. Patnaik and X. Li (eds.), *Proceedings of International Conference on Computer Science and Information Technology*, Advances in Intelligent Systems and Computing 255, DOI: 10.1007/978-81-322-1759-6_98,

Given two theories Δ and Γ, to get a maximal consistent subset of $\Delta \cup \Gamma$ containing Δ, we can use the following procedure:

Assume that $\Gamma = \{\varphi_1, \ldots, \varphi_n\}$ is finite.

$$\Theta_0 = \Delta;$$
$$\Theta_i = \begin{cases} \Theta_{i-1} \cup \{\varphi_i\} & \text{if } \varphi_i \cup \Theta_{i-1} \text{ is consistent} \\ \Theta_i = \Theta_{i-1} & \text{otherwise} \end{cases}$$

Then, $\Gamma' = \Theta_n - \Delta \subseteq \Gamma$ is a maximal subset such that $\Gamma' \cup \Delta$ is consistent.

The procedure can be replaced by a deduction by a set of deduction rules such that $\Delta|\Gamma \Rightarrow \Theta$ is provable by the set of deduction rules if and only if Θ is a maximal consistent subset of $\Delta \cup \Gamma$ containing Δ.

The R-calculus [5] attempted to give a Gentzen-type deduction system to deduce a consistent one $\Gamma' \cup \Delta$ from an inconsistent theory $\Gamma \cup \Delta$, where each deduction rule is of the form

$$\frac{\Delta|\varphi_1, \Gamma \Rightarrow \Delta|\Gamma \quad \Delta|\varphi_2, \Gamma \Rightarrow \Delta|\Gamma}{\Delta|\varphi_1 \wedge \varphi_2, \Gamma \Rightarrow \Delta|\Gamma}$$

Unfortunately, it is not ensured that whether $\Delta|\Gamma \Rightarrow \Delta|\Gamma'$ is provable and there are no $\Gamma'' \subset \Gamma'$ such that $\Delta|\Gamma' \Rightarrow \Delta|\Gamma''$ is provable; then, $\Delta \cup \Gamma''$ is consistent and a maximal consistent subset of $\Delta \cup \Gamma$. The deduction system of the R-calculus is for the following contracting procedure:

Assume that $\Gamma = \{\varphi_1, \ldots, \varphi_n\}$ is finite.

$$\Theta_0 = \Delta \cup \Gamma;$$
$$\Theta_i = \begin{cases} \Theta_{i-1} - \{\varphi_i\} & \text{if } \Theta_{i-1} - \{\varphi_i\} \vdash \neg\varphi_i \\ \Theta_{i-1} & \text{otherwise} \end{cases}$$

Let $\Gamma' = \Theta_n - \Delta \subseteq \Gamma$. Then, $\Gamma' \cup \Delta$ is a contraction of $\Gamma \cup \Delta$.

In this paper, we shall give a deduction system for the theories in the logic programming, where a literal is an atomic formula or the negation of an atomic formula, a clause is the disjunction of literals, and a theory is the conjunction of clauses. We shall prove that the deduction system for the revision of logic programmings is sound and complete, that is, for any theories t and t', if $t|t' \Rightarrow t''$ is provable; then, t'' is a maximal consistent subtheory of t, t' including t, where t, t' is the theory $\{t, t'\}$, and conversely, $t|t' \Rightarrow t''$ is provable for any maximal consistent subtheory t'' of t, t' containing t.

The paper is organized as follows: Sect. 2 gives the basic definitions of the R-calculus and AGM postulates; Sect. 3 defines an R-calculus for the revision of theories, a set of deduction rules, and deductions; Sect. 4 proves that the deduction system is sound and complete with respect to the revision operator; Sect. 5 proves that if $\Delta|\Gamma \Rightarrow \Theta$ is provable, then Θ satisfies the AGM postulates and the DP postulates, and Sect. 6 concludes the whole paper. For the limit to the length of this paper, we omit the most proofs of theorems.

2 The AGM Postulates for the R-Calculus

The R-calculus is defined on a first-order logical language. Let L be a logical language of the first-order logic; φ, ψ be formulas and Γ, Δ be sets of formulas (theories), where Δ is a set of atomic formulas or the negations of atomic formulas.

Given two theories Γ and Δ, let $\Delta|\Gamma$ be an R-configuration.

The R-calculus consists of the following axiom and inference rules:

$$(\mathbf{A}^{\neg})\quad \Delta, \varphi|\neg\varphi, \Gamma \Rightarrow \varphi, \Delta|\Gamma$$

$$(R^{\mathrm{cut}})\quad \frac{\Gamma_1, \varphi \vdash \psi \quad \varphi \mapsto T\ \psi \quad \Gamma_2, \psi \vdash \chi \quad \Delta|\chi,\ \Gamma_2 \Rightarrow \Delta|\Gamma_2}{\Delta|\varphi,\ \Gamma_1, \Gamma_2 \Rightarrow \Delta|\Gamma_1, \Gamma_2}$$

$$(R^{\wedge})\quad \frac{\Delta|\varphi, \Gamma \Rightarrow \Delta|\Gamma}{\Delta|\varphi \wedge \psi, \Gamma \Rightarrow \Delta|\Gamma}$$

$$(R^{\vee})\frac{\Delta|\varphi, \Gamma \Rightarrow \Delta|\Gamma \quad \Delta|\psi,\ \Gamma \Rightarrow \Delta|\Gamma}{\Delta|\varphi \vee \psi,\ \Gamma \Rightarrow \Delta|\Gamma}$$

$$(R^{\rightarrow})\frac{\Delta|\neg\varphi,\ \Gamma \Rightarrow \Delta|\Gamma \quad \Delta|\psi,\ \Gamma \Rightarrow \Delta|\Gamma}{\Delta|\varphi \rightarrow \psi,\ \Gamma \Rightarrow \Delta|\Gamma}$$

$$(R^{\forall})\frac{\Delta|\varphi[t/x],\ \Gamma \Rightarrow \Delta|\Gamma}{\Delta|\forall x\varphi,\ \Gamma \Rightarrow \Delta|\Gamma}$$

where in R^{cut}, $\varphi \mapsto {}_{\mathrm{T}}\psi$ means that φ occurs in the proof tree T of ψ from $\Gamma 1$ and φ; and in $R^{\forall}$, t is a term and is free in φ for x.

Definition 2.1 $\Delta|\Gamma \Rightarrow \Delta'|\Gamma'$ is an R-theorem, denoted by $\vdash^{R}\Delta|\Gamma \Rightarrow \Delta'|\Gamma'$, if there is a sequence $\{(\Delta_i, \Gamma_i, \Delta'_i, \Gamma'_i) : i \leq n\}$ such that

1. $\Delta|\Gamma \Rightarrow \Delta'|\Gamma' = \Delta_n|\Gamma_n \Rightarrow \Delta'_n|\Gamma'_n$,
2. for each $1 \leq i \leq n$, either $\Delta_i|\Gamma_i \Rightarrow \Delta'_i|\Gamma'_i$ is an axiom, or $\Delta_i|\Gamma_i \Rightarrow \Delta'_i|\Gamma'_i$ is deduced by some R-rule of form $\frac{\Delta_{i-1}|\Gamma_{i-1} \Rightarrow \Delta'_{i-1}|\Gamma'_{i-1}}{\Delta_i|\Gamma_i \Rightarrow \Delta'_i|\Gamma'_i}$.

Definition 2.2 $\Delta|\Gamma \Rightarrow \Delta|\Gamma'$ is valid, denoted by $\models\Delta|\Gamma \Rightarrow \Delta|\Gamma'$, if for any contraction Θ of Γ' by Δ, Θ is a contraction of Γ by Δ

The AGM postulates for revision:

Closure: $K \circ p = \mathrm{Cn}\,(K \circ p)$
Success: $p \in \mathrm{Cn}(K \circ p)$
Inclusion: $K \circ p \subseteq K + p$
Vacuity: if $\neg p \notin K$ then $K \circ p = K + p$
Extensionality: $K \circ p$ is consistent if p is consistent
Extensionality: if $(p \leftrightarrow q) \in \mathrm{Cn}(\emptyset)$ then $K \circ p = K \circ q$
Superexpansion: $K \circ (p \wedge q) \subseteq (K \circ p) + q$
Subexpansion: if $\neg q \notin \mathrm{Cn}\,(K \circ p)$ then $(K \circ p) + q \subseteq K \circ (p \wedge q)$.

The AGM postulates are for the logically closed theory revision, and the R-calculus is for the set-theoretic theory revision. Therefore, the AGM postulates should be rewritten for the set-theoretic theory revision. The AGM postulates for the revision we shall give for the logic programming:

Success: $t \in t|t'$
Inclusion: $t,\ t' \vdash t|t'$
Vacuity: if $t \nvdash \neg t'$ then $t|t' = t, t'$
Extensionality: $t|t'$ is consistent if t is consistent
Extensionality: if $\Vdash t \leftrightarrow t''$ then $t|t' \equiv t''|t$
Superexpansion: $t, t''|t' \subseteq (t|t'), t''$
Subexpansion: if $t|t' \vdash \neg t''$ then $(t|t'), t'' \subseteq t, t''|t'$

3 An R-Calculus for the Revision

Literals, clauses, and theories are defined as follows:

Definition 3.1 A *literal* l is an atom p or the negation $\neg p$ of an atom. A *clause* $c = l_1 \vee \cdots \vee l_n$ (denoted by $c = l_1;\ldots;\ l_n$) is a disjunction of literals, and a *theory* $t = c_1 \wedge \cdots \wedge c_m$ is a conjunction of clauses. That is

$$\begin{aligned} l &::= p|\neg p \\ c &::= l|l \vee e = l|l; e \\ t &::= c|c \wedge t = c|c, t. \end{aligned}$$

Given two theories t and t', we define the revision $t \parallel t'$ as follows:

$$t \parallel t' = \begin{cases} t, l & \text{if } t' = l \text{ is consistent with } t \\ t & \text{if } t' = l \text{ is inconsistent with } t \\ t, c & \text{if } t' = c \text{ is consistent with } t \\ t & \text{if } t' = c \text{ is inconsistent with } t \\ (\cdots((t \parallel c'_1) \parallel c'_2)\cdots) \parallel c'_m & \text{if } t' = \{c'_1, \ldots, c'_m\} \end{cases}$$

Axioms and deduction rules for the R-calculus:

$$(triv)\, l_1, \neg l_1, l_2, \ldots, l_n | l \Rightarrow l_1, \neg l_1, l_2, \ldots, l_n.$$

$$(c^{\wedge}|l) \begin{cases} l_1, \ldots, l_n | l \Rightarrow l_1, \ldots, l_n & \textit{if } l = \neg l_i \text{ for some } i \leq n \\ l_1, \ldots, l_n | l \Rightarrow l_1, \ldots, l_n, l & \text{otherwise} \end{cases}$$

Hence, we denote it by $l_1, \ldots, l_n | l \Rightarrow l_1, \ldots, l_n, l^i$, where $i = 0$, 1 and $l^0 = \lambda$, $l^1 = l$.

$$(c^{\vee}|l) \quad l_1;\ldots;l_n|l \Rightarrow l_1;\ldots;l_n, l;$$
$$(c|c) \quad l_1;\ldots;l_n|l'_1;\ldots;l'_m \Rightarrow l_1;\ldots;l_n,\ l'_1;\ldots;l'_m$$

where $n > 1$.

$$(t|l)\begin{cases} \dfrac{l_{i_1},\ldots,l_{i_n}|l \Rightarrow l_{i_1},\ldots,l_{i_n}, \text{for any } 1 \le i_1 \le n_1,\ldots,1 \le i_n \le n_n}{c_1,\ldots,c_n|l \Rightarrow c_1,\ldots,c_n} \\ \dfrac{l_{i_1},\ldots,l_{i_n}|l \Rightarrow l_{i_1},\ldots,l_{i_n}, \text{for some } 1 \le i_1 \le n_1,\ldots,1 \le i_n \le n_n}{c_1,\ldots,c_n|l \Rightarrow c_1,\ldots,c_n,l} \end{cases}$$

where $c_1 = l_{11};\ldots l_{1_{n1}}, c_2 = l_{21};\ldots;l_{2_{n2}},\ldots,c_n = l_{n1};\ldots;l_{nn_n}$

$$(t|c)\frac{c_1|,\ldots,c_n|l_1 \Rightarrow c_1,\ldots,c_n, l_1^{i_1}\cdots c_1,\ldots,c_n|l_m \Rightarrow c_1,\ldots,c_n, l_m^{i_m}}{c_1|,\ldots,c_n|l_1;\ldots;l_m \Rightarrow c_1,\ldots,c_n, l_1^{i_1};\ldots;l_m^{i_m}}$$

where

$$i_j = \begin{cases} 0 & \text{if } c_1,\ldots,c_n|l_j \Rightarrow c_1,\ldots,c_n \\ 1 & \text{otherwise} \end{cases}$$

$$(t|t) \quad c_1,\ldots,c_n|c'_1,\ldots,c'_m = (\cdots((c_1,\ldots,c_n|c'_1)|c'_2\cdots)|c'_m.$$

We have the following set of rules:

|	l	c	t
l			
c	$(c^{\wedge}\|l)$, $(c^{\vee}\|l)$	$(c\|c)$	
t	$(t\|l)$	$(t\|c)$	$(t\|t)$

where $(l|l)$ is a special case of $(c|l)$; $(l|c)$ is dual to $(c|l)$; $l|t = (\cdots(l|c_1)\cdots)|c_m$; and $c|t = (\cdots((c|c_1)|c_2)\cdots)|c_n$ where $t = c_1,\ldots, c_n$.

Definition 3.2 Let $t = c1, \ldots, c_n$ and $t' = c'_1,\ldots,c'_m$. $c_1,\ldots,c_n|c'_1,\ldots,c'_m \Rightarrow c_1,\ldots,c_n, c_1'^{,i_1},\ldots,c_m'^{,i_m}$ is deducible, denoted by $\vdash^R c_1,\ldots,c_n|c \Rightarrow c_1,\ldots,c_n, c_1'^{,i_1},\ldots,c_m'^{,i_m}$ if there is a sequence $\{\varphi_1, \ldots, \varphi_k\}$ of statements such that $\varphi_1 = t_1|t^1,\ldots,\varphi_k = t_m, t^m$; $t_1 = c_1,\ldots,c_n$; t_m, $t^m = c_1,\ldots,c_n, c_1'^{,i_1},\ldots,c_m'^{,i_m}$ and for each $i<m$, $\varphi_i \Rightarrow \varphi_{i+1}$ is either an axiom or a deduction rule.

Example 3.3 Revision p, $\neg r$, $p \to q|q \to r$ has the following subrevisions:

$$p,\ \neg r,\ \neg p|\neg q \Rightarrow p,\ \neg r,\ \neg p$$
$$p,\ \neg r,\ q|\neg q \Rightarrow p,\ \neg r,\ q$$
$$p,\ \neg r,\ \neg p|r \Rightarrow p,\ \neg r,\ \neg p$$
$$p,\ \neg r,\ q|r \Rightarrow p,\ \neg r,\ q$$

Therefore, we have

$$p,\ \neg r,\ \neg p;\ q|\neg q;\ r \Rightarrow p,\ \neg r,\ \neg p;\ q.$$

The examples show that (*triv*) is necessary.

Example 3.4 Revision $p,\ \neg p;\ q,\ \neg q;\ r,\ \neg s|\neg r;\ s$ has the following subrevisions:

$$p,\ \neg p,\ \neg q,\ \neg s|\neg r \Rightarrow p,\ \neg p,\ \neg q,\ \neg s$$
$$p,\ \neg p,\ \neg q,\ \neg s|s \Rightarrow p,\ \neg p,\ \neg q,\ \neg s$$
$$p,\ \neg p,\ r,\ \neg s|\neg r \Rightarrow p,\ \neg p,\ r,\ \neg s$$
$$p,\ \neg p,\ r,\ \neg s|s \Rightarrow p,\ \neg p,\ r,\ \neg s$$
$$p,\ q,\ \neg q,\ \neg s|\neg r \Rightarrow p,\ q,\ \neg q,\ \neg s$$
$$p,\ q,\ \neg q,\ \neg s|s \Rightarrow p,\ q,\ \neg q,\ \neg s$$
$$p,\ q,\ r,\ \neg s|\neg r \Rightarrow p,\ q,\ r,\ \neg s$$
$$p,\ q,\ r,\ \neg s|s \Rightarrow p,\ q,\ r,\ \neg s$$

Therefore, $p,\ \neg p;\ q,\ \neg q;\ r,\ \neg s|\neg r; s \Rightarrow p,\ \neg p;\ q,\ \neg q;\ r,\ \neg s$.

Lemma 3.5 *If* $t_2 \models t_1$ *then* $t_2,\ t_1 \| t \Rightarrow t_2 \| t$.

Lemma 3.6 $t_1,\ t_2 \| t\ \equiv\ t_2,\ t_1 \| t$.

$\|$ satisfies the principle of the minimal change with respect to inference, not to set.

Theorem 3.7 *Assume that* $\vdash^R t|t' \Rightarrow t,\ t''$ and $t,\ t'$ *is inconsistent. Then,*

1. $t \vdash t''$; *and*
2. *for any theory* t_1, *if* $t' \vdash t_1$, $t_1 \vdash t''$ *and* $t'' \nvdash t_1$ *then* $t,\ t_1$ *is inconsistent.*

4 The Soundness and Completeness of the *R*-Calculus

Theorem 4.1 (The termination theorem) *For any theories t and* t', *there is a theory* t'' *such that* $\vdash^R t|t' \Rightarrow t, t''$.

Proposition 4.2 *If* $c_1,\ \ldots,\ c_n,\ l_1;\ldots;\ l_m$ *are consistent then*

$$c_1,\ldots,c_n,l_1;\ldots;l_m \equiv c_1,\ldots,c_n,l_1^{i_1};\ldots;l_m^{i_m}.$$

We cannot have the following form of the soundness theorem:

For any theories t,t',t'', *if* $\vdash^R t|t' \Rightarrow t,t''$ *then* $t \parallel t' \Rightarrow t,\ t''$.

For example, let $t = p$, $t' = \neg p \vee q$. Then, $t \parallel t' = p,\ \neg p \vee q$; and $\vdash^R t|t' \Rightarrow p,\ q$, where $q \vdash t'$, $t' \nvdash t$. But as theories, $\{p,\ \neg p \vee q\}$ is logically equivalent to $\{p,\ q\}$.

Theorem 4.3 (The soundness theorem) *For any theories* $t,\ t',\ t''$, *if* $\vdash^R t|t' \Rightarrow t,\ t''$ *then there is a theory* t_0 *such that* $t'' \equiv t_0$ *and* $t \parallel t' \Rightarrow t,\ t_0$.

Theorem 4.4 (The completeness theorem) *For any theories* $t,\ t',t''$, *if* $t \parallel t' \Rightarrow t,\ t''$ *then there is a theory* t_0 *such that* $t_0 \equiv t''$ *and* $\vdash^R t|t' \Rightarrow t,\ t_0$.

5 The Basic Logical Properties of |

Theorem 5.1 | *satisfies the AGM postulates.*

The DP postulates:

$$\begin{aligned}&(\mathrm{C1})\ \text{If}\ t_2 \vDash t_1\ \text{then}\ t_2|(t_1|t) \equiv t_2|t;\\&(\mathrm{C2})\ \text{If}\ t_2 \vDash \neg t_1\ \text{then}\ t_2|(t_1|t) \equiv t_2|t;\\&(\mathrm{C3})\ \text{If}\ (t_2|t) \vDash t_1\ \text{then}\ t_2|(t_1|t) \vDash t_1;\\&(\mathrm{C4})\ \text{If}\ (t_2|t) \nvDash \neg t_1\ \text{then}\ t_2|(t_1|t) \nvDash \neg t_1.\end{aligned}$$

In order to prove that | satisfies the DP postulates, we assume the following

Assumption 5.2 $t_2, t_1, \neg t_1|t \Rightarrow t_2|t.$

Theorem 5.3 | *satisfies the DP postulates.*

Proof (C1) Assume that $t_2 \vDash t_1$. Then, by Lemma 3.2, $t_2|(t_1|t) \equiv t_2, t_1|t \equiv t_2|t$;

(C2) Assume that $t_2 \vDash \neg t_1$. Then, by Assumption 3.4, $t_2|(t_1|t) \equiv t_2, t_1|t \equiv t_2, \neg t_1, t_1|t \equiv t_2|t$;

(C3) Assume that $t_2|t \vDash t_1$. Then, by Lemma 3.3, $t_2|(t_1|t) = t_2, t_1|t \equiv t_1, t_2|t \equiv t_1|(t_2|t) \vDash t_1$;

(C4) Assume that $t_2|t \vDash \neg t_1$. Then, $t_2|(t_1|t) \equiv t_1|(t_2|t) \equiv t_1, t_2|t \vDash \neg t_1$. □

6 Conclusions

This paper gave an axiomatic system for the revision in the logic programming, which is sound and complete with respect to the maximal consistent subtheories of the revised theories, that is, if $t|t' \Rightarrow t, t''$ is provable, then t'' is a theory such that $t' \vdash t''$ and for any t_0 such that $t' \vdash t_0 \vdash t''$ and $t'' \nvdash t_0$, t, t_0 is inconsistent, and $t|t' \Rightarrow t, t''$ is provable for any t'' such that $t' \vdash t''$ and for any t_0 such that $t' \vdash t_0 \vdash t''$ and $t'' \nvdash t_0$, t, t_0 is inconsistent. Moreover, we proved that if $t|t' \Rightarrow t, t''$ is provable, then t, t'' satisfies the AGM postulates and the DP postulates.

Acknowledgments This work was supported by the Open Fund of the State Key Laboratory of Software Development Environment under Grant No. SKLSDE-2010KF-06, Beijing University of Aeronautics and Astronautics, and by the National Basic Research Program of China (973 Program) under Grant No. 2005CB321901.

References

1. Alchourrón, C.E., Gärdenfors, P., Makinson, D.: On the logic of theory change: partial meet contraction and revision functions. J. Symbolic Logic **50**, 510–530 (1985)
2. Fermé, E., Hansson, S.O.: AGM 25 years, twenty-five years of research in belief change. J. Philos. Logic **40**, 295–331 (2011)
3. Friedman, N., Halpern, J.Y.: Belief revision: a critique, to appear in Journal of Logic, Language and Information. In: Aiello, L.C., Doyle, J:, Shapiro, S.C. (eds.) Principles of Knowledge Representation and Reasoning: Proceedings of 5th Conference, pp. 421–431 (1996)
4. Gärdenfors, P., Rott, H.: Belief revision. In: Gabbay, D.M., Hogger, C.J., Robinson, J.A. (eds.) Handbook of Logic in Artificial Intelligence and Logic Programming: Epistemic and Temporal Reasoning, vol. 4, pp. 35–132. Oxford Science Publishing, Oxford (1995)
5. Li, W.: R-calculus: an inference system for belief revision. Comput. J. **50**, 378–390 (2007)

Incremental Composition of Dynamic Programs

Minghui Wu and Jia Lv

Abstract Owing to the introduction of dynamic language features, modular analysis and composition of dynamic programs are very difficult. To separate dynamic features from dynamic programs, we divide dynamic programs into two parts: static modules and dynamic modules. Static modules are specified with modular contracts, which can be verified before runtime, and dynamic modules are composed into static modules. To ensure the safe composition of dynamic modules and static modules, some runtime checks are inserted into dynamic programs, which will be executed at runtime.

Keywords Dynamic programs · Moudular analysis · Modular composition

1 Introduction

The advent of Web 2.0 has led to the proliferation of dynamic languages. The introduction of dynamic features into the programs is often desired. However, using dynamic features raises many questions of reliability and correctness. Most dynamic languages are dynamically typed or weakly typed, which will cause runtime errors such as access to nonexistent members. Some dynamic languages support runtime evaluation and reflection, which is difficult to static type analysis. Most dynamic languages do not support encapsulation enough, which make modular reasoning about the behavior of dynamic programs difficult.

M. Wu (✉)
Department of Computer Science and Engineering, Zhejiang University City College, Hangzhou 310015, China
e-mail: mhwu@zucc.edu.cn

J. Lv
College of Internet of Things Engineering, Hohai University, Changzhou 213022, China
e-mail: samlv2000@163.com

S. Patnaik and X. Li (eds.), *Proceedings of International Conference on Computer Science and Information Technology*, Advances in Intelligent Systems and Computing 255, DOI: 10.1007/978-81-322-1759-6_99,

In order to modular compose dynamic programs, we separate dynamic language features from static language features and divide the dynamic program modules into two kinds: static modules and dynamic modules. The static modules do not include dynamic features, which can be statically analyzed. The dynamic modules include dynamic features, which are composed into static modules and must be checked at runtime.

To specify the interference of the composition between dynamic modules and static modules, we use assume–guarantee paradigm [1, 2] to specify the behavior of static modules and verify whether the static modules satisfy their modular contracts before runtime. If there are dynamic modules which are composed into the one static module, some runtime checks will be inserted into dynamic modules. These runtime checks will be executed at runtime to ensure the correct composition of dynamic modules and static modules.

This paper consists of seven sections. The next section models dynamic programs as open systems. Section 3 defines modular contracts for dynamic programs. Section 4 introduces our method of incrementally composing dynamic programs based on their modular contracts. Section 5 discusses the related work. The Sect. 5 is the conclusion and future work.

2 Model Dynamic Programs as Open Systems

Local reasoning allows correctness to be dealt with one module at a time. Each module has a specification that describes its expected behavior. The goal is to prove that each module satisfies its specification, using only the specifications but not code of other modules. This way the complexity of the proof effort (formal or informal) can be kept under control. This local reasoning approach is sound if separate verification of individual modules suffices to ensure the correctness of the composite program [3].

The method of sound local reasoning in process-oriented languages is process encapsulation. The method of sound reasoning in object-oriented languages is object encapsulation. And the method of sound reasoning in aspect-oriented language is aspect encapsulation. In traditional program paradigm, owning to dynamic features in dynamic languages, it is difficult to use traditional encapsulation mechanism to support modular encapsulation and local reasoning.

To support modular encapsulation and local reasoning of dynamic programs, we model dynamic programs as open systems, which distinguishes two kinds of transitions: internal transitions and external transitions. Internal actions specify the static part of execution process, while external transitions specify the dynamic part of execution process. In order to specify some invariables of external transitions, we introduce **stable** operator to constrain external transition.

Definition 1 Internal transition: A state change is caused by internal actions of the open system.

Definition 2 External transition: A state change is caused by external actions of the open system.

Definition 3 Stable (denoted by the letter *S*): The system should ensure some properties when interfered by external actions.

The semantics of **stable** operator is interpreted as follows:

$\varphi_1 \ S \ \varphi_2$ is true iff (1) no new external parts are added to any execution processes and (2) some new external parts which satisfy formula φ_2 are added into or removed from one execution process, and the return state σ' satisfies formula φ_1.

$\varphi_1 \ S \ \varphi_2$ is false iff (1) some new external parts which do not satisfy formula φ_2 are added to the execution process and (2) some new external parts which satisfy formula φ_2 are added into or removed from the execution process, and the return state σ' does not satisfy formula φ_I any longer.

An open system $M = (S, \rightarrow, L)$ includes a set of states *S* endowed with internal transitions $\longrightarrow I$ or external transitions $\stackrel{O}{\longrightarrow}$ (binary relations on *S*), such that every state $s' \in S$ has some state $s' \in S$ with $s \stackrel{I}{\longrightarrow} s'$ or $s \stackrel{O}{\longrightarrow} s'$ and a labeling function $L : S \rightarrow P$ (Atoms). The distinction of internal transitions and external transitions leads to a compositional semantics.

Dynamic programs can be interpreted over open system, which are divided into static modules and dynamic modules. Static modules are interpreted as internal transitions, and dynamic modules of dynamic programs are interpreted as external transitions. **Stable** operators are interpreted as some behavior properties of the composition of static modules and dynamic modules.

3 Modular Contracts for Dynamic Programs

To support incremental composing of dynamic programs, a module of dynamic programs should satisfy two conditions: First, the module should have a well-formed interface, which specifies how other modules interact with themselves; second, modular specification or modular contract should be a part of modular interface. There should be a mechanism to ensure that modular behavior satisfies its interface and interfaces of other modules.

Similar to assume–guarantee paradigm [4, 5], we use assume condition to specify the precondition of a module execution and use guarantee condition to specify the post-condition of a module execution. To specify interference of dynamic modules in the process of module execution, rely condition is introduced to specify execution process of dynamic modules.

Definition 4 Rely Condition: An assertion over two states of the module execution; one is the state before the composition of dynamic modules, and the other is the state getting control back from the composition of dynamic modules.

When modular reasoning about the behavior of dynamic programs, we assume that any interleaved actions that the dynamic module may change the state of the module should be within the constraints specified by rely condition. We define modular contract of dynamic programs as follows.

Definition 5 A modular contract of dynamic programs consists of three parts: assume condition, rely condition, and guarantee condition, which are denoted by the formula: $(assume, rely, guarantee)$.

Similar to the concept of behavioral subtyping [6–8] in object-oriented language, we introduce a modular constraint named correctness to support modular composition of dynamic programs.

Definition 6 A module *M* satisfies **correctness** constraint relative to its modular contract $(assume, rely, guarantee)$ iff the state before execution of module *M* satisfies assume condition *assume* and the state after execution of module *M* satisfies guarantee condition *guarantee*. The rely condition *rely* should ensure the assume condition *assume* and guarantee condition *guarantee* are stable. The *correctness* constraint of module *M* can be specified as follows:

$$\frac{\begin{array}{c} assume \quad S \quad rely \\ guarantee \quad S \quad rely \end{array}}{M\ \underline{\text{correct}}\ (assume, rely, guarantee)}$$

4 Statically Verifying Static Modules

In order to incremental compose modules, a static module should be verified whether it satisfies **correctness** constraint or not. We illustrate how to verify whether a static module relative to its modular contract satisfies **correctness** constraint with the following example.

Example 1 A static module $M = \{x = x + 3;\ y = x + 3\}$; the contract of module *M* is $(x > 0, x > x' \wedge y > y', y > 6)$.

Situation 1: If there does not exist any dynamic module inserted into *M* in the execution process of module *M*, module *M* satisfies **correctness** relative to its modular contract.

Situation 2: If there exists a dynamic module inserted into *M* before the execution of the statement $x = x + 3$, the state before the composition satisfies $x > 0$. Since the rely condition is $x > x' \wedge y > y'$, the state after composition satisfies $x > 0$. Then, the process of composition ensures the formula $M, \sigma_0, \sigma_0 \models x > 0\,S\,x > x' \wedge y > y'$ is correct, which means the assume condition of module *M* will still be stable after the composition of the dynamic module.

Situation 3: If there exists a dynamic module inserted into M after the execution of the statement $x = x + 3$, the state before composition satisfies $x > 3$ and the rely condition is $x > x' \wedge y > y'$; then, the state after composition satisfies $x > 3$. The state after the execution of module M will still satisfy $y > 6$.

Situation 4: If there exists a dynamic module inserted into M after the execution of the statement $y = x + 3$, the state after the execution of statement $y = x + 3$ satisfies $y > 6$. Since the rely condition satisfies $x > x' \wedge y > y'$, the state after composition satisfies $y > 6$. Then, the process of composition satisfies the formula $M, \sigma_0, \sigma_t \models A(y > 6\, S\, x > x' \wedge y > y')$, which means the guarantee condition of module M will still be stable after composition.

To sum up, module M satisfies **correctness** constraint related to its modular contract $(x > 0, x > x' \wedge y > y', y > 6)$.

5 Incremental Composition of Dynamic Modules

5.1 Modular Composition Based on Correctness Constraint

Correctness constraint supports modular composition of dynamic modules with total obliviousness, which means that if the dynamic module satisfies **correctness** constraint, then the argument system will still satisfy **correctness** constraint.

Theorem 1 *Some dynamic modules are inserted into a module M which satisfies* **correctness** *constraint related to its contract. If these dynamic modules also satisfy* **correctness** *constraint related to their contracts, then the argument system will still satisfy* **correctness** *constraint.*

Proof A module M satisfies **correctness** constraint related to its contract:

$$\frac{\begin{array}{c} assume \quad S \quad rely \\ guarantee \quad S \quad rely \end{array}}{M\ \underline{\text{correct}}\,(assume,\ rely,\ guarantee)} \tag{1}$$

Following proof is by induction on n.

Basis: $n = 1$. A dynamic module D_1 is inserted into module M, and the argument module is $M \prec D_1$. Since module M satisfies **correctness** constraint relative to its modular contract, the argument module $M \prec D_1$ will ensure the formulae $M \prec D_1, \sigma_0 \models assume$ and $M \prec D_1, \sigma_t \models guarantee$ are correct according to formula (1).

Another dynamic module D_2 is inserted into the argument system module $M \prec D_1$, and the dynamic module D_2 also satisfies **correctness** constraint. The insert points of D_1 and D_2 are different, and the insert point of D_2 is in the execution process of module M. According to formula (1), $M, \sigma_0, \sigma_0 \models$

assume S rely and $M, \sigma_0, \sigma_t \models guarantee\, S\, rely$ will still be satisfied, which means the assume condition *assume* and the guarantee condition *guarantee* of module *M* will still be stable under the composition of the dynamic module D_2.

Thus, $n = 1$, the argument module $M \prec D_1$ still satisfies **correctness** constraint.

Induction step: $n = k$. We assume that the argument system module $M \prec A_1 \prec A_2, \ldots, A_k$ satisfies **correctness** constraint, and an dynamic module D_{k+1} which satisfies **correctness** constraint is inserted into *M*. According to formula (1), the state before the execution of argument system module $M \prec D_1 \prec D_2, \ldots, D_k \prec D_{k+1}$ will still satisfy the assume condition *assume* of module *M*, and the state after the execution of module *M* will still satisfy the guarantee condition *guarantee* of module *M*.

Another dynamic module D_{k+2} which satisfies *correctness* constraint is inserted into the argument module $M \prec D_1 \prec D_2, \ldots, D_k \prec D_{k+1}$. Since the argument module $M \prec D_1 \prec D_2, \ldots, D_k \prec D_{k+1}$ satisfies *correctness* constrain, formulae $M, \sigma_0, \sigma_0 \models assume\, S\, rely$ and $M, \sigma_0, \sigma_t \models guarantee\, S\, rely$ will still be true according to formula (1), which means the assume condition *assume* and guarantee condition *guarantee* of *M* will still be stable after the composition of the dynamic module D_{k+2}.

Then, when $n = k+1$, **theorem 1 is** still true.

To sum up, **theorem 1** is true.

5.2 Runtime Checking Dynamic Modules

According to **correctness** constraint, if a dynamic module *D* satisfies rely condition *rely* inserted into the module *M*, the execution process of dynamic module ensures the assume condition *assume* and guarantee condition *guarantee* are stable. To ensure that the assume condition *assume* and guarantee condition *guarantee* are stable, some runtime checks are inserted into dynamic modules, which ensure safe composition of dynamic modules and static modules.

6 Related Work

Researches [9, 10], in order to facilitate the development and verification of dynamically adaptive systems, separate functional concerns from adaptive concerns. Specifically, they model a dynamically adaptive program as a collection of (non-adaptive) steady-state programs and a set of adaptations that realize transitions among steady-state programs in response to environmental changes. They use linear temporal logic to specify properties of the non-adaptive portions of the system, and we use A-LTL (an adapt-operator extension to LTL) to concisely specify properties that hold during the adaptation process. They propose a modular model checking approach to verify that a formal model of an adaptive program satisfies its requirements specified in LTL and A-LTL, respectively.

Researches [4, 5, 11, 12] present assume–guarantee model checking, which is a novel technique for verifying correctness properties of loosely coupled multi-threaded software systems. Assume–guarantee model checking verifies each thread of a multithreaded system separately by constraining the actions of other threads with an automatically inferred environment assumption. Separate verification of each thread allows the enumeration of the local state of only one thread at a time, thereby yielding significant savings in the time and space needed for model checking.

Articles [2, 13] use rely–guarantee approach to support verifying and modular reasoning about aspect-oriented programs. Article [1] uses assume–guarantee method to settle the problem of behavioral problems caused by some aspects woven into a same point cut. They have defined semantic interference among aspects relative to their specifications and shown an effective way to detect interference or prove interference freedom of multiple aspects in a library.

7 Conclusion and Future Work

Owing to the introduction of dynamic language features, modular composition of dynamic programs is very difficult. To separate dynamic features from dynamic programs, we model dynamic programs as open systems, which distinguishes two kinds of modules: static modules and dynamic modules. To specify the behavior of static modules, we use rely–guarantee paradigm to define modular contracts for dynamic programs. Based on these modular contracts, dynamic programs can be statically verified and incrementally composed. To ensure the correct composition of dynamic modules at runtime, some runtime checks will be inserted into dynamic modules, which will be executed at runtime.

We propose a method of incremental composition of dynamic programs, but we do not introduce any tools to support our method. We will implement some tools to support our method in future. We also plan to use this method to incremental compose other program paradigms.

Acknowledgments Our research work is partly supported by the Science Foundation of Zhejiang Province under Grant No. 2010R50009.

References

1. Katz, E., Katz, S.: Incremental analysis of interference among aspects. In: Proceeding of foundations of aspect languages workshop, pp. 29–38 (2008)
2. Khatchadourian, R., Dovland, J., Soundarajan, N.: Enforcing behavioral constraints in evolving aspect-oriented programs. In: Proceedings of the 7th workshop on foundations of aspect-oriented languages, pp. 19–28 (2008)

3. Leino, K.R.M., Nelson, G.: Data abstraction and information hiding. ACM Trans. Program. Lang. Sys. **24**(5), 491–553 (2002)
4. Jonsson, B., Tsay, Y.-K.: Assumption/guarantee specifications in linear-time temporal logic. Theor. Comput. Sci. **167**(1), 47–72 (1996)
5. Flanagan, C., Qadeer, S.: Assume-guarantee model checking. SPIN, p. 11 (2003)
6. America, P.: Designing an object-oriented programming language with behavioural subtyping. In: Proceedings of the REX school/workshop on foundations of object-oriented languages, vol. 489, pp. 60–90 (1991)
7. Liskov, B.H., Wing, J.M.: A behavioral notion of subtyping. ACM Trans. Program. Lang. Syst. **16**(6), 1811–1841 (1994)
8. Liskov, B.H., Wing, J.M.: Behavioural subtyping using invariants and constraints. 254–280 (2001)
9. Zhang, J., Cheng, B.H.C.: Using temporal logic to specify adaptive program semantics. J. Syst. Softw. **79**(10), 1361–1369 (2006)
10. Zhang, J., Goldsby, H.J., Cheng, B.H.C.: Modular verification of dynamically adaptive systems. AOSD 161–172 (2009)
11. Pasareanu, C.S., Dwyer, M.B., Huth, M.: Assume-guarantee model checking of software: a comparative case study. In: Proceedings of the 5th and 6th international SPIN workshops on theoretical and practical aspects of SPIN model checking, vol. 1680, pp. 168–183 (1999)
12. Giannakopoulou, D., Pasareanu, C.S., Cobleigh, J.M.: Assume-guarantee verification of source code with design-level assumptions. In: Proceedings of the 26th international conference on software engineering, pp. 211–220 (2004)
13. Khatchadourian, R., Soundarajan, N.: Rely-guarantee approach to reasoning about aspect-oriented programs. In: Proceeding of the 5th workshop on software engineering properties of languages and aspect technologies, p. 217 (2007)

Dynamic Sleep State Schedule for Idle Nodes in Clusters

Yongpeng Liu, Yongyan Liu and Wanqing Chi

Abstract There are often many idle nodes in a computer cluster due to its workload varies with time. Dynamic sleep mechanisms are helpful to save energy wastage caused by such active idle nodes. By scheduling the sleep states of idle nodes, this paper proposes an approach to balance between the cluster's energy consumption and response times. Idle nodes are classified into different groups according to their sleep states. The numbers of nodes in each group are adjusted by tuning the sleep states of the nodes to match the need of computation resources. An effective metric for power management techniques on idle node is introduced. The significance of our approach on energy efficiency (EE) improvement in computer clusters is demonstrated by experiments.

Keywords Computer clusters · Energy management · Sleep state

1 Introduction

Large-scale computer clusters consume tremendous energy. In 2006, US data centers consumed about 1.5 % of the total US electricity consumption or the output of about 15 typical power plants [1]. In 2007, the electricity consumption of

Y. Liu (✉) · W. Chi
School of Computer, National University of Defense Technology, Changsha, People's Republic of China
e-mail: liuyp@nudt.edu.cn

W. Chi
e-mail: chiwq@nudt.edu.cn

Y. Liu
Information Center, Ministry of Science and Technology, Beijing, People's Republic of China
e-mail: liuyy@most.gov.cn

S. Patnaik and X. Li (eds.), *Proceedings of International Conference on Computer Science and Information Technology*, Advances in Intelligent Systems and Computing 255, DOI: 10.1007/978-81-322-1759-6_100,

global cloud computing was larger than India [2]. Many data center projects have been canceled or delayed because of an inability to meet such enormous power requirements.

However, data centers' workload varies with time and the average resource utilization of large-scale systems typically is low. Consequently, a quite number of nodes are idle in most time, and these idle nodes cause huge energy waste.

Dynamic sleep mechanism is proposed to reduce the power consumption of a node in its idle state [3]. The deeper the node sleeps, the less power it consumes, but the more energy and the more time delay are needed to wake it up. In this paper, we propose a schedule solution of multiple sleep states of idle nodes in computer clusters to make an effective tradeoff between energy and performance.

2 Related Works

For computer clusters, the dynamic sleeping of idle nodes, i.e., put idle nodes into sleep states and wake them on demand, is a typical power management technique.

The importance of supporting multiple sleep state for servers in data centers has been investigated by Gandhi et al. [4]. However, the sleep states of idle servers in their approach are not dynamically scheduled. Horvath et al. [5] exploit the multiple sleep states of idle servers and predicate the incoming workload based on history resource utilization change. In their solution, the optimal number of spare servers for each power states is selected in an ad hoc manner. Differently, our scheduling on the state transition of idle nodes works in an adaptive model. Xue et al. [6] dynamically regulate the capacity of the active resource pool in accordance with the time-varying workload demand. However, multiple sleep states mechanism is not exploited in their power management solution and spare nodes are simply turned off. In this paper, we explore the benefits of multiple sleep states mechanism to improve the energy efficiency (EE) of computer clusters.

3 Dynamic Sleep State Management

In a computer cluster, an idle node should be put into low-power sleep state to avoid the energy waste. To improve the EE of the whole system, an effective energy management solution is required to schedule the sleep or wake timing of idle nodes.

A sleep state schedule for idle nodes in computer clusters, called ASDMIN, is proposed in this paper. ASDMIN classifies the idle nodes into a number of node groups according to their sleep depths. Each group is therefore a *pool* of nodes of certain sleep level. The pool of sleep level i, denoted as B_i, is composed of all of the nodes with the same power consumption level P_i, and the number of nodes in B_i is N_i.

Dynamic sleep mechanism is exploited in ASDMIN to reduce the power consumption of idle nodes. However, it must be taken into account that deeper sleep state means longer wakeup latency. For the shortest wakeup latency, ASDMIN preferentially allocate nodes from the highest pool. When the nodes in the pool of the highest readiness level are not sufficient, the nodes in the pool(s) of next level(s) are allocated. In the result, there should be enough nodes in the pools to meet the constraint of response time. Therefore, we set a *reserve capacity threshold*, denoted as R_i, to control the minimum number of nodes in pool B_i. Whenever the node number of B_i is less than R_i, nodes in the lower pools will be upgraded to fill the reserve capacity.

3.1 Adaptive Adjustment of the Reserve Capacity Threshold

By adjusting the reserve capability thresholds dynamically and adaptively, ASDMIN make trade-off between energy conservation and system performance.

In general, if all the nodes in B_i are allocated but the resource is still insufficient to meet the amount of required nodes, its reserve capacity threshold R_i should be increased. Here, we propose the following formula (1a) to guide the adjustment of reserve capacity threshold increase.

$$R_i = \begin{cases} R_i + \alpha \times (C_i - N_i) & C_i > N_i \quad \text{(a)} \\ \max\{R_i - \beta \times (N_i - C_i), 0\} & C_i < N_i \quad \text{(b)} \end{cases} \tag{1}$$

where C_i is the number of nodes required to allocate from B_i, and α is a *performance weight factor* to reflect the user's preference for system performance. The bigger the α is, the faster the reserve capacity threshold increases. Consequently, the more idle nodes will stay in higher pool, and the more weight is given to system performance.

On the other hand, if there may be some residual nodes in a pool after it providing nodes to the application, the reserve capacity of the pool is larger than the requirement and its threshold should be decreased so that more nodes will be put into deeper sleep state to save energy. Formula (1b) is used to decrease the reserve capacity threshold, where β is an *energy weight factor*. The bigger the β is, the faster the reserve capacity threshold decreases, and the energy conservation is more preferable.

4 Metrics for Power Management on Idle Nodes

There are a number of metrics to evaluate EE [3]. However, these metrics do not cover the evaluation for idle nodes. Here, we conclude a new metric for power management of idle nodes based on MFLOPS/W that is used by the Green 500 [7].

Formally, let x be the number of instructions executed in a period of time t while consumes w energy. The EE of the computation is

$$\mathrm{EE} = \left(\frac{x}{t}\right) \Big/ w \tag{2}$$

When a power management solution S_i is applied to the same computation, it takes time t_i to complete the tasks while consumes w_i energy where S_0 means in particular that all idle nodes keep alive. Regarded to the S_0 scenario, the energy efficiency improvement ratio (EEIR) for solution S_i is defined as follows:

$$\mathrm{EEIR}(S_i) = \left(\frac{\mathrm{EE}_0}{\mathrm{EE}_i}\right) = \frac{\left(\frac{x}{t_0}\right) \Big/ w_0}{\left(\frac{x}{t_i}\right) \Big/ w_i} = \left(\frac{t_i}{t_0}\right) \times \left(\frac{w_i}{w_0}\right) \tag{3}$$

For a computer cluster, let C_i be the set of tasks executed in the period of time with power management solution i, and for each a in C_{i}, t_i^a and x_i^a be the time to complete the task and the number of instructions executed for task a, we have that

$$\mathrm{EEIR}(S_i) = \frac{\frac{\sum_{a\in C_o} x_0^a}{\sum_{a\in C_0} t_0^a} \Big/ w_0^{C_0}}{\frac{\sum_{a\in C_i} x_i^a}{\sum_{a\in C_i} t_i^a} \Big/ w_i^{C_i}} = \left(\frac{\sum_{a\in C_i} t_i^a}{\sum_{a\in C_0} t_0^a}\right) \times \left(\frac{w_i^{C_i}}{w_0^{C_0}}\right) \times \left(\frac{\sum_{a\in C_o} x_0^a}{\sum_{a\in C_i} x_i^a}\right) \tag{4}$$

When the period of time is significantly longer than the lengths of tasks, we have that $C_o \approx C_i$. Thus, we have that $\sum_{a\in C_o} x_0^a \approx \sum_{a\in C_o} x_0^a$. Therefore, the following can be used as a measure of EEIR.

$$\mathrm{EEIR}(S_i) \approx \left(\frac{\sum_{a\in C_i} t_i^a}{\sum_{a\in C_0} t_0^a}\right) \times \left(\frac{w_i^{C_i}}{w_0^{C_0}}\right) = \mathrm{SD}(S_i) \times \mathrm{ES}(S_i) \tag{5}$$

where $\mathrm{SD}(S_i) = \frac{\sum_{a\in C_i} t_i^a}{\sum_{a\in C_0} t_0^a}$ is the overall *slowdown rate* and $\mathrm{ES}(S_i) = \left(\frac{w_i^{C_i}}{w_0^{C_0}}\right)$ is the overall *energy saving rate* over a given period of time.

5 Implementation and Evaluation

Parallel Workload Archive [8] publishes some workload logs on clusters. Each log contains the following information on the jobs: submit time, wait time, run time, and number of allocated processors. In a system with power management on idle nodes, the time to complete a job, denoted as *execution time*, i.e., its end time minuses its submit time, consists of three parts: the *wait time*, the *run time*, and the *wakeup delay* that is the additional delay caused by awakening idle nodes allocated to the job. In this paper, we use these workload logs as the workloads of our simulation experiments.

There is no data about the power characters of idle nodes in the Parallel Workload Archive logs. We measured the power consumption and wakeup time of a real node and the results are shown in Table 1. These data are used in the simulations.

The computed nodes have four different idle states, *S*0, *S*1, *S*3, and *S*4. *S*0 is the active idle state, and *S*1, *S*3, and *S*4 are sleep states ranking on sleep depth.

The simulation experiments' results with all of the workload logs are shown in Fig. 1.

In Fig. 1, BUSY is the number of busy nodes in the system, *N*0, …, *N*3 are the number of nodes in pool *B*0, …, *B*3, respectively. On average, there are 94.98 % of idle nodes in the lowest pool (N_3). It means that most idle nodes are in the deepest sleep state in most of the time.

The effects of the four static solutions that put all idle nodes into corresponding sleep state and ASDMIN on the four systems are shown in Fig. 2. In comparison with scenario *S*0, ASDMIN reduces overall energy consumption by 56.32 % at the cost of increasing the average job execution time by 1.45 % on average of four workload logs. Thus, we improve the EE by 55.94 % while applying formula (5).

Table 1 Power characteristics of compute node

State	Power (W)	Wakeup latency (s)
Busy	350	
*S*0	207	0
*S*1	171	2
*S*3	32	10
*S*4	26	190

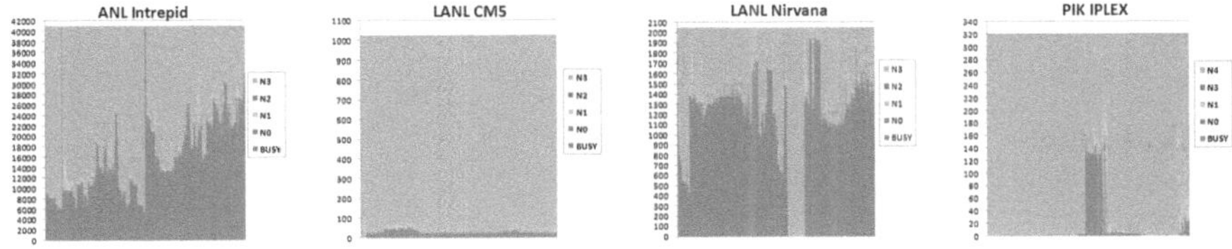

Fig. 1 The variation of the numbers of nodes in the pools in the ASDMIN scenario

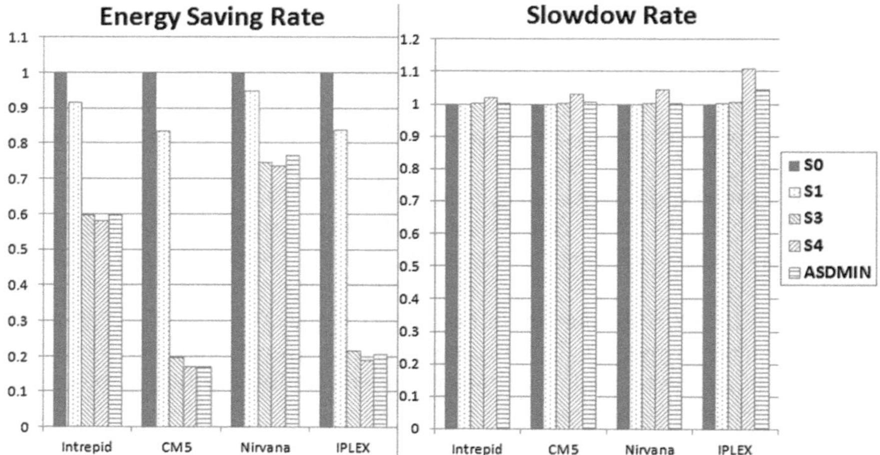

Fig. 2 The results of different management scenarios

6 Conclusion

To save the energy wasted by active idle nodes in computer clusters, we propose a schedule solution on the sleep states of idle nodes to balance between energy consumption and system response speed. Our experiments demonstrated that the EE is upgraded by 55.94 % on average.

For our future work, various policies will be explored in the selection of idle node for sleeping states transition. For example, temperature can be used as a guide to save cooling power consumption.

Acknowledgments This work is partly supported by the National High Technology Research and Development Program of China (863 Program) under Grant No. 2012AA01A301, the NSF of China under Grant No. 61272141, No. 60903059, and No. 61003075.

References

1. U.S. Environmental Protection Agency: Report to congress on server and data center energy efficiency. http://www.energystar.gov/ia/partners/prod_development/downloads/EPA_Datacenter_Report_Congress_Final1.pdf. 2 Aug 2007
2. Cook, G.: How Clean is Your Cloud?. Greenpeace International, Amsterdam (2012)
3. Liu, Y., Zhu, H.: A survey of the research on power management techniques for high performance systems. Softw. Pract. Experience **40**(1), 943–964 (2010)
4. Gandhi, A., Harchol-Balter, M., Kozuch, MA.: The case for sleep states in servers. In: Proceedings of the HotPower '11, Cascais, Portugal (2011)

5. Horvath, T., Skadron, K.: Multi-mode energy management for multi-tier server clusters. In: Proceedings of the 17th International Conference on Parallel Architecture and Compilation Techniques, Toronto, Canada, pp. 270–279 (2008)
6. Xue, Z., Dong, X., Ma, S., et al.: An energy-efficient management mechanism for large-scale server clusters. In: Proceedings of the 2007 IEEE Asia-Pacific Services Computing Conference, pp. 509–516 (2007)
7. Green 500. http://www.green500.org/. Jun 2013
8. Parallel Workloads Archive. http://www.cs.huji.ac.il/labs/parallel/workload/l_anl_int/ANL-Intrepid-2009-1.swf.gz. Apr 2011

Moving Human Detection Based on Depth Interframe Difference

Hongwei Xu, Jie Liu and Yue Ming

Abstract The objective of this paper is to propose a method of moving human detection based on depth video. The method used the interframe difference algorithm extract moving human contour from depth video. Due to the depth data provided by depth image, the image noise in the detection result is significantly reduced and the problem caused by human shadow in the detection based on ordinary video is solved. Experiments show that the method can improve the accuracy of the detection result and enhance robustness of moving human detection system.

Keywords Depth video · Interframe difference · Moving human detection

1 Introduction

Moving human detection and tracking is an important research content in visual analysis of human motion and has been an active research area in computer vision. It is widely used in many applications, such as protecting pedestrian from traffic accident in intelligent vehicles and automatic monitoring throughout the day in monitoring system.

Moving human detection is the basic part of moving human detection and tracking system. Many video human detection algorithms have been proposed so far which can be classified into two kinds: detection based on motion information and detection based on feature. Background difference and interframe difference belong to the first kind and have salient advantages in terms of detection speed and feasibility [1, 2]. In [3], the author introduced the background subtraction

H. Xu (✉) · J. Liu · Y. Ming
Beijing Key Laboratory of Work Safety Intelligent Monitoring, School of Electronic Engineering, Beijing University of Posts and Telecommunications, Beijing 100876, People's Republic of China
e-mail: buptxhw@gmail.com

S. Patnaik and X. Li (eds.), *Proceedings of International Conference on Computer Science and Information Technology*, Advances in Intelligent Systems and Computing 255, DOI: 10.1007/978-81-322-1759-6_101,

algorithms and gauged performances of a few frequently used background models. An improved algorithm based on frame difference and edge detection was presented in [4]. However, both of these methods will easily fail, while people overlap with others. Markus Enzweiler [5] introduced the main components of moving human detection system based on machine learning and performed an experiment to get performances of different machine learning methods. Although detection based on feature can obtain results of high accuracy, but it takes a very long time to train the classifier and has bad real-time performance.

Data of each pixel in depth image record the distance between target and stereo camera, meaning that the influence of light on depth image is slimmer than that on ordinary image. The movement of target will always keep out part of the background image, leading to changes in depth data in corresponding region. Thus, it is easy to get information of target's motion according to the changes in the depth data. Our paper puts forward a method that uses depth data to detect moving human. The proposed method can reduce the image noise greatly and overcome the human shadow problem.

The rest of this paper is organized as follows: After introducing the flowchart of our system in Sects. 2, 3 gives the comparative result and our discussion of the result. We conclude the paper in Sect. 4.

2 The System of Moving Human Detection Based on Depth Video and Interframe Difference

Human detection is to segment humans from the background in the video sequence image. The flowchart of the system is shown in Fig. 1.

In image preprocessing, we used reduced Gaussian filter to reduce the image noise and performed color space conversion. Moving edge image was obtained by depth interframe difference, the core part of the system. The purpose of result post-processing was to get rid of the interference area in the moving edge image through opening operation and Canny edge detection.

2.1 Image Preprocessing

It is very necessary to smooth the image to reduce the image noise. In this paper, we used a discrete sliding window convolution to reduce Gaussian noise in the image.

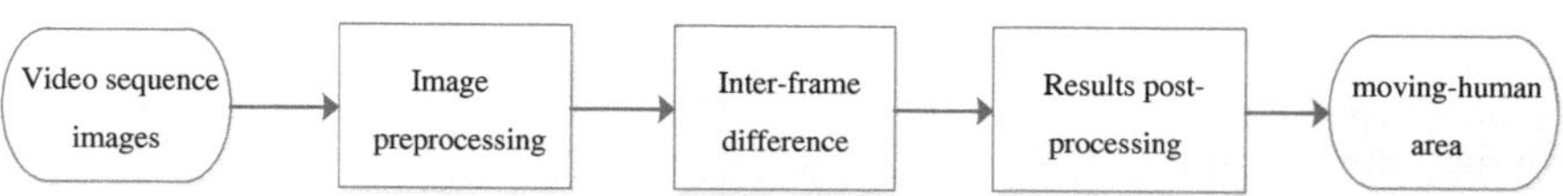

Fig. 1 The flowchart of the system

Another important operation is taking proper transformation of video image to meet the needs of the detection method. We converted color images to grayscale images after the operation of noise reduction according to the formula 1.

$$Y(x,y) = 0.299R(x,y) + 0.587G(x,y) + 0.114B(x,y) \tag{1}$$

where $R(x, y)$, $G(x, y)$, and $B(x, y)$ are RGB color components of Point (x, y).

2.2 Interframe Difference on Depth Video

The interframe difference algorithm has high detection speed and can be implemented on hardware easily. This sentence introduces the principle of the interframe difference. It has three steps:

1. Get two chronological frames
2. Subtract the two frames. The value of pixel in one frame minus the value of the corresponding pixel in the other frame
3. Take the absolute value of the results though step2

Through the three steps above, we can get the moving areas.

Lipton puts forward a strategy of two-frame-difference method [6]. The algorithm can be represented by

$$I_k(x,y) = |P_k(x,y) - P_{k-1}(x,y)|. \tag{2}$$

$$D_k(x,y) = \begin{cases} 255, & I_k(x,y) \geq T_h \\ 0, & \text{else} \end{cases}. \tag{3}$$

where $P_k(x,y)$ is the current frame image and $P_{k-1}(x,y)$ is the previous frame image.

The basis of interframe difference is the difference between consecutive frames, so there are a few insuperable defects: Firstly, it cannot detect stationary targets in the image. Secondly, it is more likely to detect two regions of interest (ROI) of the same fast-moving target and it is hard to determine the ROI's validity of the slow-moving target. Thirdly, interframe difference can only detect the moving target's contour [7, 8].

For interframe difference based on ordinary video, there are other problems: (1) The result is very sensitive to illumination intensity. If the environment light intensity changes in excess of the prescribed threshold, the result will contain a large amount of image noise. Although much of the noise could be removed by subsequent processing, it will cost a lot of computation resources and time, which affects the real-time performance of the system. (2) Shadows of person also appear in the results, and it is hard for computer to distinguish them from targets [9].

To solve the problems, this paper proposed to use depth video to perform human detection. In the research, we used two-frame-difference method to detect

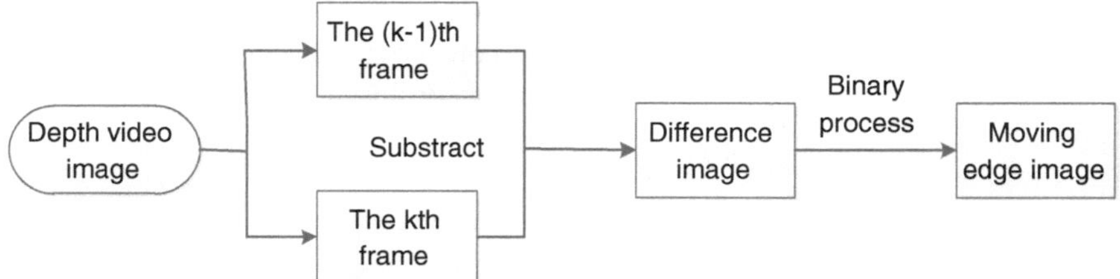

Fig. 2 The process of two-frame-difference

moving human in both depth video and ordinary video that are recorded by Kinect synchronously. The detection process of two-frame-difference is shown in Fig. 2.

2.3 Result of Post-processing

Due to the perturbations of objects in the background area and the slight shaking of the camera, the difference image contains residual noises, even error detection areas. It is very necessary to reduce noise and to remove error detection areas for the sake of recognition accuracy.

Opening operation is an image processing operation that uses the same structural elements to corrode image firstly and expand image then. It can eliminate small region and smooth large region's boundary with a precondition of barely changing the region's area. We used the opening operation to remove noises in the moving edge image [10].

The correct connected areas is the outline of the moving human. The error connected areas, such as outline of bookcase, should be removed. We used the Canny edge detection to further reduce the connected regions in the image. We calculated the area of each connected areas, and then, the areas that are too small or too large were ruled out. The effect of post-processing is shown in Fig. 3.

Image (a) is the result image without post-processing, and image (b) is the result image after post-processing. By comparing image (a) and image (b), we found that on the premise of not changing the content in the ROI, post-processing not only can eliminate much of the noise but also can remove some error detection areas.

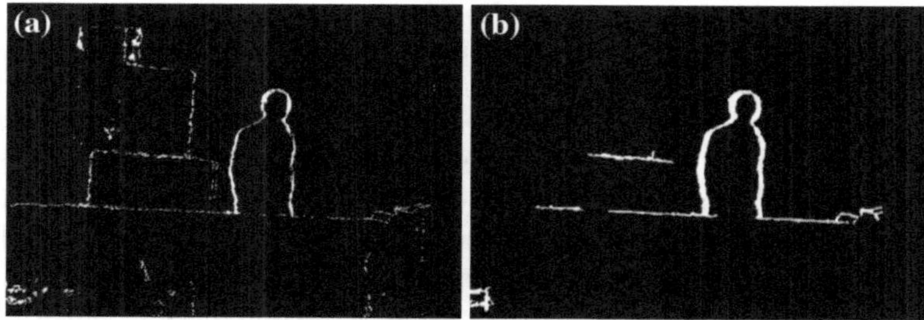

Fig. 3 The effect of post-processing

3 Experimental Results and Analysis

The experimental videos were synchronously acquired in indoor static background by Kinect with 15 FPS and 640 * 480 resolutions, which meet the requirements of real-time detection. Some screenshots are shown in Figs. 4, 5, 6, 7, 8, and 9.

Figure 4 shows three frames taken from the ordinary video randomly, and Fig. 5 shows the corresponding depth image. In Fig. 6, error detection areas caused by human shadows can be obviously seen around the moving human. These error detection areas are easy to cause identification errors and to reduce the accuracy of the system. However, detection results of two-frame-difference method on depth video images show complete outline of moving human without error detection areas around, which means that it overcomes the human shadow problem, as shown in Fig. 7. By comparing the Figs. 6 and 7, we found that the result of two-frame-difference method contains less image noise. The experimental results show that two-frame-difference based on depth video can effectively solve

Fig. 4 Ordinary video images

Fig. 5 Depth video images

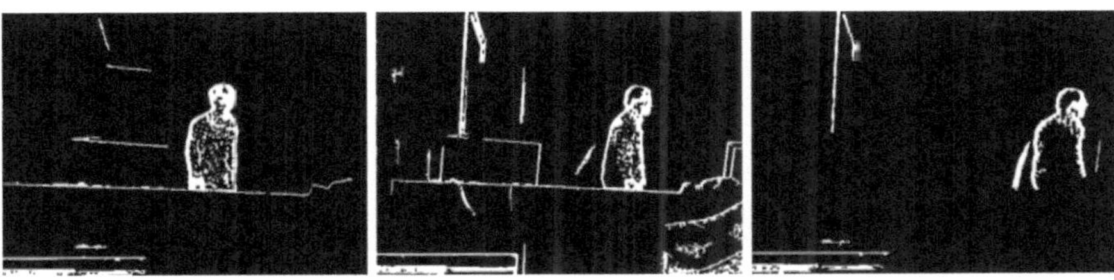

Fig. 6 Results of two-frame-difference method on ordinary video images

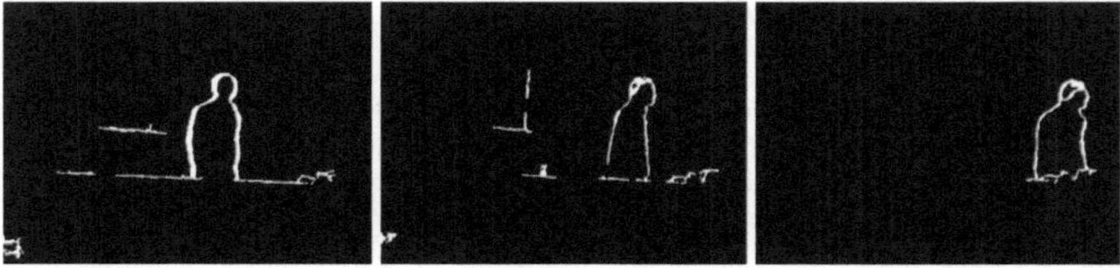

Fig. 7 Results of two-frame-difference method on depth video images

Fig. 8 Results of three-frame-difference method on ordinary video images

Fig. 9 Results of three-frame-difference method on depth video images

the problems in two-frame-difference based on ordinary video, such as image noise and human shadow. Two-frame-difference method based on depth video can detect moving human more accurately.

We can conclude by comparing Figs. 8 and 6 that the result of three-frame-difference based on ordinary video is better than the two-frame-difference's, but also suffers from the serious shadow problem. Figure 9 is the three-frame-difference method based on depth video's result which failed the detection and needed further research and improvement.

4 Conclusion

Aiming at solving problems of moving human detection based on ordinary video, we put forward to use the depth video to detect moving human and performed moving human detection based on depth video and interframe difference. The experimental results show that moving human detection based on depth video can improve system performance in two aspects: improve the accuracy owing to the solution of human shadow problem and enhance robustness owing to the reduction in image noise.

Further research will be focused on improvement in three-frame-difference method based on depth video and human segmentation.

Acknowledgments The work presented in this paper was supported by the National Natural Science Foundation of China (Grant No. NSFC-61170176), Fund for the Doctoral Program of Higher Education of China (Grant No. 20120005110002), National Great Science Specific Project (Grant Nos. 2011 ZX0300200301, 2012ZX03005008), and Beijing Municipal Commission of Education Build Together Project.

References

1. Enzweiler, M., Gavrila, D.M.: Monocular pedestrian detection: survey and experiments. Pattern Anal. Mach. Intell. **31**, 2175–2195 (2009)
2. Guo, L., Li, L., Zhao, Y., Zhang, M.: Study on pedestrian detection and tracking with monocular vision. In: Proceedings of 2nd International Conference on Computer Technology and Development, pp. 466–470 (2010)
3. Benezeth1, Y., Jodoin, P.M.: Review and evaluation of commonly-implemented background subtraction algorithms. In: Proceedings of 19th International Conference on Pattern Recognition, pp. 1–4 (2008)
4. Chaohui, Z.: An improved moving object detection algorithm based on frame difference and edge detection. In: Proceedings of 4th International Conference on Image and Graphics, 2007
5. Enzweiler, M.: Monocular pedestrian detection: survey and experiments. IEEE T. Pattern Anal. **31**(12), (2009)
6. Lipton, A.J., Fujiyoshi, H., Patil, R.S.: Moving target classification and tracking from real-time video. In: Proceedings of Applications of Computer Vision, pp. 8–14 (1988)
7. Xiaofeng, L.: Research on moving human detection based on streaming video. J. Beijing Univ. Ind. Commer. **27**(6), 40–44 (2009)
8. Chengru, W., Cuijun, L.: Research and implementation on moving human detection and tracking based on streaming video. TV technology (2012)
9. Tang, F., Harville, M., Tao, H., Robinson, I.N.: Fusion of local appearance with stereo depth for object tracking. In: Computer Vision and Pattern Recognition Workshops, pp. 1–8 (2008)
10. Rui, Z.: Design and implementation of moving human detection and tracking system based on openCV. Master Thesis of Wuhan University of Science and Technology (2011)

About the Editors

Prof. Srikanta Patnaik is Professor of Computer Science and Engineering at SOA University, Bhubaneswar, India, and also serves as adjunct/visiting professor to many universities inside India and abroad. He is the Chairman and Founder Director of Interscience Institute of Management and Technology, Bhubaneswar, India.

Presently, he is the Editor-in-Chief of the International Journal of Information and Communication Technology and International Journal of Computational Vision and Robotics, published by Inderscience Publishing House, England. He is also Editor-in-Chief of Book Series on "Modeling and Optimization in Science and Technology" (MOST), published by Springer, Germany, and other two series, namely "Advances in Computer and Electrical Engineering" (ACEE) and "Advances in Medical Technologies and Clinical Practice" (AMTCP) published by IGI-Global, USA. He has published more than 100 research papers in International Journals and Conference Proceedings. He is now actively engaged for the promotion of creativity and innovation among the engineering and management students in India.

Dr. Xiaolong Li received his bachelor's and master's degrees in Electrical and Information Engineering Department at the Huazhong University of Science and Technology, Wuhan, China, in 1999 and 2002, respectively. He received his Ph.D. degree in Electrical and Computer Engineering department at the University of Cincinnati in 2006. He joined the Morehead State University in 2006, where he was an Assistant Professor in the department of Industrial and Engineering Technology. In 2008, he joined the Department of Electronics and Computer Engineering Technology at the Indiana State University, where he is currently an assistant professor.

His current research interest are in the areas of wireless and mobile networking and microcontroller-based applications. Dr. Li has published more than 40 papers in International Journals and Proceedings. He is one of the editors of Springer book series on "Modeling and Optimization in Science and Technology" (MOST) and editor of book "Recent Development on Ad hoc and Sensor Networks" to be published by Springer in April 2014. He also serves as general chair and technical program chair of many international conferences. He is IEEE Communication Society Chapter chair at Central Indiana Section.

S. Patnaik and X. Li (eds.), *Proceedings of International Conference on Computer Science and Information Technology*, Advances in Intelligent Systems and Computing 255, DOI: 10.1007/978-81-322-1759-6,

Author Index

S. Patnaik and X. Li (eds.), *Proceedings of International Conference on Computer Science and Information Technology*, Advances in Intelligent Systems and Computing 255, DOI: 10.1007/978-81-322-1759-6,

Z